Structural and System Reliability

Based on material taught at the University of California, Berkeley, this textbook offers a modern, rigorous, and comprehensive treatment of the methods of structural and system reliability analysis. It covers the first- and second-order reliability methods for components and systems, simulation methods, time- and space-variant reliability, and Bayesian parameter estimation and reliability updating. It also presents more advanced, state-of-the-art topics such as finite-element reliability methods, stochastic structural dynamics, reliability-based optimal design, and Bayesian networks. A wealth of well-designed examples connect theory with practice, with simple examples demonstrating mathematical concepts and larger examples demonstrating their applications. End-of-chapter homework problems are included throughout.

Including all necessary background material from probability theory, and accompanied online by a solutions manual and PowerPoint slides for instructors, this is the ideal text for senior undergraduate and graduate students taking courses on structural and system reliability in departments of civil, environmental, and mechanical engineering.

Armen Der Kiureghian is the Taisei Professor of Civil Engineering Emeritus at the University of California, Berkeley, and a co-founder and President Emeritus of the American University of Armenia. He is also a Distinguished Alumnus of the Department of Civil and Environmental Engineering at the University of Illinois at Urbana-Champaign, and an elected member of the US National Academy of Engineering.

Structural and System Reliability

Armen Der Kiureghian
University of California, Berkeley

CAMBRIDGE
UNIVERSITY PRESS

University Printing House, Cambridge CB2 8BS, United Kingdom

One Liberty Plaza, 20th Floor, New York, NY 10006, USA

477 Williamstown Road, Port Melbourne, VIC 3207, Australia

314–321, 3rd Floor, Plot 3, Splendor Forum, Jasola District Centre, New Delhi – 110025, India

103 Penang Road, #05–06/07, Visioncrest Commercial, Singapore 238467

Cambridge University Press is part of the University of Cambridge.

It furthers the University's mission by disseminating knowledge in the pursuit of education, learning, and research at the highest international levels of excellence.

www.cambridge.org
Information on this title: www.cambridge.org/highereducation/isbn/9781108834148
DOI: 10.1017/9781108991889

© Armen Der Kiureghian 2022

This publication is in copyright. Subject to statutory exception and to the provisions of relevant collective licensing agreements, no reproduction of any part may take place without the written permission of Cambridge University Press.

First published 2022

Printed in the United Kingdom by TJ Books Limited, Padstow, Cornwall

A catalogue record for this publication is available from the British Library.

Library of Congress Cataloging-in-Publication Data
Names: Der Kiureghian, Armen, author.
Title: Structural and system reliability / Armen Der Kiureghian, University of California, Berkeley.
Description: New York, NY : Cambridge University Press, 2022. | Includes bibliographical references and index.
Identifiers: LCCN 2021027030 (print) | LCCN 2021027031 (ebook) | ISBN 9781108834148 (hardback) | ISBN 9781108991889 (epub)
Subjects: LCSH: Reliability (Engineering)–Mathematics. | Structural engineering–Mathematics. | System failures (Engineering) | BISAC: SCIENCE / Mechanics / General
Classification: LCC TA169 .D385 2022 (print) | LCC TA169 (ebook) | DDC 620/.00452–dc23
LC record available at https://lccn.loc.gov/2021027030
LC ebook record available at https://lccn.loc.gov/2021027031

ISBN 978-1-108-83414-8 Hardback

Additional resources for this publication at www.cambridge.org/DerKiureghian

Cambridge University Press has no responsibility for the persistence or accuracy of URLs for external or third-party internet websites referred to in this publication and does not guarantee that any content on such websites is, or will remain, accurate or appropriate.

To Nelly

Contents

Preface

This book is the outcome of teaching the subject of structural and system reliability at the University of California for more than 40 years. I was hired there as Assistant Professor in 1978 and tasked with the responsibility of developing the field of risk and reliability analysis in civil engineering. Over the subsequent years, I collected, produced/co-produced, and refined the material that forms the core of this book. Of course, I was influenced by the developments in the field as I encountered them in the literature or through interactions with colleagues both within and outside the university. I was also assisted by the many excellent students who did their research in this field under my supervision. Aside from a graduate course on structural and system reliability, I developed and taught undergraduate courses on engineering data analysis and engineering risk analysis, and a graduate course on stochastic structural dynamics, all of which employed probabilistic methods. Parts of this book are influenced by my teaching of these courses, also.

My motivation for writing this book lies in my strong belief that the topic of structural and system reliability is in need of a textbook that presents the material in a clear and well-organized way, without unnecessary mathematical formalism, and in a manner that facilitates self-learning through clear exposition of background theories followed by demonstrative examples. For this reason, the book contains many worked-out examples, each designed to demonstrate one or more fundamental concepts. The material seeks to provide a contemporary treatment of the subject.

The first 11 chapters of the book form the core of the teaching material. They include detailed numerical examples and end-of-chapter problems. The focus is on developing structural reliability methods, including the first- and second-order reliability methods (FORM and SORM), a host of sampling methods, models and methods for time- and space-variant reliability analyses, and Bayesian parameter estimation and reliability updating. Earlier methods, such as the second-moment reliability method (FOSM), are also presented for their historical interest. Methods for characterization and reliability assessment of systems are presented in a broad way, appropriate for both structural systems and infrastructure systems. Also included, in considerable detail, are methods for assessing reliability sensitivity and variable importance measures.

The last four chapters of the book, which are on the topics of finite-element reliability analysis, stochastic dynamics of nonlinear structures, reliability-based optimal design, and Bayesian networks for structural reliability assessment and updating, are presented more as state-of-the-art reviews than textbook material. These topics are still developing, and I hope that these chapters will motivate some readers to pursue further research and development in these areas.

My own entry into this field was somewhat accidental. As a new doctoral student at the University of Illinois at Urbana-Champaign, in the fall of 1972, I visited Professor Alfredo H.-S. Ang, who was looking for a research assistant. He asked whether I had a background in probability theory. I said that I did not. He told me that the project he had was in risk and reliability analysis and that it required a background in probability theory, going on to say that it was all right that I did not have such a background but that I would have to take his courses to learn it. He was so honest as to admit that some civil engineering faculty thought this field "useless," but he disagreed and saw a bright future. He then gave me Bruce Ellingwood's newly completed doctoral thesis to read, to give me an idea what the topic was about. With my lack of a background in probability, naturally I did not understand much of Bruce's thesis. But I was sufficiently intrigued to give the subject a try. I took two courses with Professor Ang that semester, one on risk analysis and decision-making and the other on structural reliability. The probabilistic approach in both courses appeared to me to be logical and easy to grasp and devoid of the somewhat dogmatic and mysterious approaches I had seen in other courses. That set the direction of my entire academic career.

Looking back, I am amazed at how far the field has developed. The need for risk and reliability assessment in engineering has become paramount in order to assure the safety and operability of engineered facilities and for optimal use of resources. With the advent of sensor technologies and advances in data analytics, even broader opportunities for the use of risk and reliability methods for decision-making are available. It is clear that Professor Ang's prediction in 1972 was correct.

To master this subject, the student must have a strong understanding of applied probability and statistics. Although it is desirable that the student have such a background before reading this book, it is possible to gain the necessary proficiency by thoroughly absorbing the material in Chapters 2 and 3. Additional material on random processes and random fields is presented in Chapters 11, 12, and 13. While that material is sufficient for the topics covered in this book, these are rich topics that go far beyond what is presented here. References for further reading are listed throughout the book.

The solution of non-trivial structural reliability problems requires the use of specialized software. In Chapter 1, several software packages for this purpose are mentioned. Among them are CalREL, FERUM, and OpenSees, which have been developed at the University of California, Berkeley (with FERUM further developed at the Université Clermont Auvergne, France), and are freely available as described in Chapter 1. Additional computational tools for reliability analysis and uncertainty quantification are being made available by the new generation of academics and researchers in this field, as also described in Chapter 1. Indeed, the resources available for studying and applying the methods of structural and system reliability are infinitely larger than what was available when I started.

It is my hope that students and instructors will find this book worthy of their consideration and use. I also hope that researchers and practicing engineers will find the exposition sufficiently clear for self-learning and use in their research and practice.

Acknowledgments

I have benefited from the counsel and assistance of many individuals who have indirectly contributed to the creation of this book. First among them is Alfredo H.-S. Ang, who introduced me to this field. His courses at the University of Illinois were instrumental in inspiring me to continue working in this area. He also encouraged me to apply for the position at UC Berkeley and I am sure his reference played a significant role in my hiring.

UC Berkeley has offered an ideal environment for academic pursuits. I was fortunate to be in a department with top-notch faculty. Even though the vast majority of them were working far from the probabilistic area, interactions with them were always inspiring and helped me broaden my perspective on the use of probabilistic methods. I greatly enjoyed collaborating with my colleagues Vitelmo Bertero, Yousef Bozorgnia, Jonathan D. Bray, James M. Kelly, Steve Mahin, Jack Moehle, Carl Monismith, Paulo Monteiro, Khalid Mosalam, Jerome L. Sackman, Ray Seed, Nick Sitar, Bozidar Stojadinovic, Robert Taylor, and Edward Wilson on various papers. I am particularly grateful to Jerome L. Sackman, who was a mentor, a collaborator, and a very dear friend throughout my career at Berkeley. I also benefited enormously from collaborations with colleagues from outside our department, including Richard Barlow of the Industrial Engineering and Operations Research Department, Douglas Dreger of the Department of Earth and Planetary Science, and Elijah Polak of the Department of Electrical Engineering and Computer Science. Outside UC Berkeley, I have been influenced by many colleagues in the field of structural and system reliability, chief among whom is Ove Ditlevsen at the Technical University of Denmark; discussions with him always helped my thinking reach new depths. I am grateful to my colleagues Robert Taylor and Sanjay Govindjee for offering valuable comments on Chapter 12.

There is no question that the major influence on my academic work and the writing of this book has come from my students at UC Berkeley. I have been extremely fortunate to have an exceptionally intelligent and dedicated group of students. The learning experience has always been mutual. In fact, a good teacher must be a good learner and I have learned as much from my students as they have learned from me. Those of my students who have had important impact on this book include Wiggo Smeby, Masoud Zadeh, Takeru Igusa, Alejandro Asfura, Jorge Crempien, Rodrigo Araya, Chih-Dao Wung, Pei-Ling Liu, Jyh-Bin Ke, Hong-Zong Lin, Yan Zhang, Dariush Mirfendereski, Chin-Man Mok, Charles Menun, Petros Keshishian, Mehrdad Sasani, Paolo Gardoni, Johannes O. Royset, Heonsang Koo, Kee-Jeung Hong, Terje Haukaas, Junho Song, Kazuya Fujimura, Sanaz Rezaeian, Michelle Bensi, Katerina Konakli, Iris Tien, Marco Broccardo, Mayssa Dabaghi, and Binbin Li. Aside from these doctoral students, I have had the good fortune of hosting an excellent group of postdoctoral visitors and visiting doctoral students in my research group. Among them, the following have had influence on what is presented in this book: Philippe Geyskens, Bruno Sudret, Jean-Marc Bourinet, Anthony Hanel, Daniel Straub, Salvatore Sessa, Luca Garrè, Umberto Alibrandi, Iason Papaioannou, Matteo Pozzi, Nikolay Dimitrov, James-A. Goulet and Ziqi Wang. Many of these doctoral students and postdocs are now prominent academics in the field of risk and reliability analysis. Furthermore, two of Junho Song's students from Seoul National University, Ji-Eun Byun and Eui-Hyun Choi, provided valuable assistance in developing the example problems in Chapter 8. In addition, Michelle Bensi, Iason Papaioannou, Junho Song, and Daniel Straub reviewed parts of the manuscript and made valuable comments. I express my sincere thanks to them all.

Finally, I must express my deep gratitude to my own family: my parents, who instilled in me the love for education and learning; my children, who thought their father was "always working" and tolerated my frequent absences; and to my dear wife Nelly who has steadfastly supported me all along and to whom this book is dedicated.

1 Introduction

1.1 Introduction

Uncertainties permeate many aspects of engineering decision-making in the design, construction, operation, and maintenance of structures such as buildings and bridges, and infrastructure systems such as transportation networks or distribution systems for electric power, gas, and water. Uncertainties are present in the characteristics of such systems including their material properties and dimensions, in the demands placed on these systems, in the mathematical models that are used to analyze their behaviors, in the measurements made to assess their health conditions, and even in the probabilistic models that are used to describe the relevant uncertain quantities. Under such conditions, the safety and serviceability of structures and infrastructure systems cannot be assured with certainty. Probabilistic and statistical methods are needed to define and assess measures of safety and serviceability.

In the engineering of constructed facilities under uncertainty, three goals are paramount: (1) to assess the safety of the system against failure; (2) to assess the serviceability of the system, i.e., the ability of the system to perform its intended function; and (3) to optimize decisions so that there is an optimal trade-off between safety and serviceability, on the one hand, and the cost of used resources on the other. The assurance of absolute safety and serviceability may be neither possible nor desired, and an optimal decision may allow for finite risks of failure and unserviceability. Whereas addressing the first two of the above goals is rooted in probabilistic analysis, the third requires methods of optimal design under uncertainty. All three goals are addressed in this book.

This book presents the probabilistic approach to assessing the reliability of structures and infrastructure systems. The word *reliability* is used to denote the complement of the failure probability or of the probability of unserviceability. A variety of methods are presented, providing alternatives depending on the nature of the problem and the computational tools available. One chapter addresses the topic of reliability-based optimal design. While the bulk of the book is tutorial in nature, a few chapters on recent developments are included that are more exploratory in nature and present opportunities for further advancement by researchers.

The field of structural reliability emerged from the discipline of structural engineering. However, the concepts and methods described in this book have much broader applicability. As we will see, mathematical formulation of the problem envisions characterization of structural components in terms of limit-state functions that define the transition between the states of each component and logical expressions of component states that define the state of a system. Any problem that can be formulated in terms of this framework can be solved by the methods described in this book. Although the vast majority of illustrative examples are taken from the field of civil engineering, the mathematical formulation of problems and solution methods are presented in as general a way as possible so as to allow consideration of possible applications in other fields.

1.2 Brief History of the Field

Uncertainties in the design of structures have been of concern for a long time. Historically, the issue was addressed through the *factor of safety* – the ratio of a measure of capacity over a measure of demand. Initially, the factor of safety was set on a purely judgmental basis with no explicit regard to variabilities in capacity and demand. The idea of using probability theory and statistics to arrive at an appropriate factor of safety was developed in the early- to mid-20th century in the works of Max Mayer (Mayer, 1926), Alfred M. Freudenthal (Freudenthal, 1947, 1956), and Alfred G. Pugsley (Pugsley, 1951). Mayer pioneered the idea of using probabilistic methods in mechanics, Freudenthal argued for employing probability distributions to model loads and capacities, and using elastic analysis for serviceability and inelastic analysis for safety assessment, and Pugsley provided statistical data on aircraft loads and argued for a safety factor defined in terms of stresses rather than capacities and loads. Vladimir V. Bolotin (Bolotin, 1967, 1971) in Russia was another pioneer, particularly in the development and use of random process theory in reliability analysis.

The next generation of researchers and developers in the field of structural reliability included Masanobu Shinozuka, a student of Freudenthal (Konishi and Shinozuka, 1956; Shinozuka, 1964, 1972, 1983), Jack Benjamin (Benjamin, 1964, 1968a, 1968b; Benjamin and Cornell, 1970), C. Allin Cornell, a student of Benjamin (Cornell, 1967, 1969), Alfredo H.-S. Ang, who was my doctoral advisor (Ang and Amin, 1968, Ang and Cornell, 1974), Niels Lind (Lind, 1971; Hasofer and Lind, 1974), and Emilio Rosenblueth (Rosenblueth and Esteva, 1972; Rosenblueth, 1975, 1976), among others. Shinozuka played a key role in developing simulation methods for reliability analysis, particularly under stochastic loads and random fields. Benjamin and Cornell's 1970 book, which was the first of its kind in civil engineering, facilitated learning and research in probabilistic methods and decision theory applied to civil engineering problems. Cornell's 1969 paper laid the foundation for development of probabilistic structural design code formats. This paper influenced Rosenblueth and Esteva (1972) to develop a probabilistic basis for the Mexican design code. Rosenblueth was also a pioneer in advocating optimal design methods and his 1975 paper on point estimation of probability moments was later employed by other researchers for approximate probabilistic analysis. The paper by Ang and Cornell (1974) marks the full development of what we now call the mean-centered, first-order, second-moment (MCFOSM) method. As described in Section 5.2.1, this method lacks invariance with respect to the formulation of the problem. Ang and Cornell were aware of this shortcoming and acknowledged it in a discussion section at the end of the paper. Interestingly, a paper by Hasofer and Lind (1974) published four months earlier had resolved this problem. That paper led to the development of what we now call the first-order, second-moment (FOSM) method (Section 5.2.2) and paved the way for further development of reliability methods in the next decade.

The two decades starting in the late 1970s marked a period of intensive development of the field. The theoretical foundations of the field were formulated and specialized software to carry out needed computations was developed. Specialized journals, including *Structural Safety*, *Probabilistic Engineering Mechanics*, and *Reliability Engineering and System Safety*, were established and other journals, including those of the American Society of Civil Engineers, published increasing numbers of research papers on probabilistic methods and structural reliability. Additionally, a host of books appeared that increased the level of interest in the field. These included Ang and Tang (1975, 1984), Ditlevsen (1981), Thoft-Christensen and Baker (1982), Augusti et al. (1984), Melchers (1987, 1999), Madsen et al. (1986), Thoft-Christensen and Murotsu (1986), Ditlevsen and Madsen (1996), Haldar and Mahadevan (2000a, 2000b), Nikolaidis et al. (2005), Lamaire (2005), Nowak and Collins (2012), and Melchers and Beck (2018). Many civil engineering departments started requiring courses on probability and statistics in undergraduate programs and offered senior-level or graduate courses on structural reliability. Major influencers of this development included Ove Ditlevsen and his students H. Madsen and P. Bjerager at the Technical University of Denmark, Rudiger Rackwitz and his coworkers at the Technical University of Munich, including M. Hohenbichler and K. Brietung, Gerhart Schueller, Y.-K. Wen, Mircea Grigoriu, Bruce Ellingwood, Ross Corotis, Dan Frangopol, John Sørensen, James Beck, Christian Bucher, Mark Stewart, Arvid Naess, Michael Faber, Marc Maes, the author's own group at the University of California, Berkeley, and others. Throughout this book, references are made to the works of these authors as the topics are developed.

The new generation of researchers in structural and system reliability is focused on broadening the field and integrating it with other fields such as structural health monitoring, finite elements, uncertainty propagation, optimization, early warning systems, data analytics, and artificial intelligence. Because the solution of complex problems usually requires the use of computationally intensive software, a topic of increasing interest is in developing surrogate (or meta-) models that can be used in place of implicit and computationally expensive models to assess uncertainty propagation and reliability analysis. Although much progress has been made in this direction, the development of efficient and accurate surrogate models for real-world complex systems is a challenge that needs much further work.

The topic of reliability-based optimal design is another area of current interest. The key here is developing a bridge between the fields of structural reliability and optimal design that takes advantage of the theoretical and operational strengths of both fields instead of developing specialized optimization algorithms that solve a particular application of structural reliability. In the area of structural health monitoring, with sensors becoming ubiquitous, naturally the interest is in continuous updating of reliability measures as information is gathered and processed, particularly for near-real-time decision-making. Bayesian techniques are particularly well-suited for this purpose and this is another area of current research and development. In particular, the Bayesian network approach can be a highly fruitful framework to pursue in this direction. Finally, the finite-element method has become the quintessential computational framework for solving all kinds of field problems defined by differential equations, including structural, mechanical, geotechnical, and environmental problems. Uncertainty quantification and reliability methods need to be integrated with finite elements in order to broaden the scope of their applications to real-world problems. Several chapters in this book present state-of-the-art reviews of methods in the above areas.

1.3 The Nature of Uncertainties and Probability

The nature of uncertainties and probability has been the subject of debate for a long time. It is important to present the philosophical standpoint of this book on that subject.

As described in Chapter 10, there are several sources of uncertainty in engineering analysis and design. These include inherent variability or randomness, such as that arising in properties of materials and intensities of environmental loads; model uncertainty that results from the imperfection of mathematical models describing complex physical phenomena; statistical uncertainty in the estimation of model parameters, due to limited data; measurement error due to imperfection of measuring devices; and human error resulting from mistakes and omissions made by engineers in the processes of design, construction or operation of facilities. In practice, these uncertainties are often categorized as aleatory or epistemic in nature, the former characterized as irreducible, inherent variability and the latter as reducible uncertainty due to lack of knowledge. The two types are treated differently in classical statistics: aleatory uncertainties through probability distributions and epistemic uncertainties through confidence intervals.

The approach in this book is different from this in two ways. First, as described in Der Kiureghian and Ditlevsen (2009), we do not consider the distinction between aleatory and epistemic uncertainties to be crisp. What may appear as aleatory uncertainty may be epistemic if a higher-order model is used. For example, the wind speed on the surface of a building may be considered aleatory if no predictive wind model is used and data from measurements only are employed; however, it can be considered as at least partly epistemic if a physics-based predictive model is used that provides an estimate of the wind speed depending on various environmental and atmospheric factors. Hence, the distinction between aleatory and epistemic uncertainties makes sense only within the universe of models that we use to formulate a problem. Second, we use the rules of probability theory to model and analyze both types of uncertainty. Furthermore, with the eventual aim of using the results of probabilistic analysis in decision-making, the end result of our analysis incorporates all the relevant uncertainties into what we call a predictive estimate. Hence, the measure of reliability we compute incorporates the aleatory uncertainty arising from inherently random phenomena as well as the contribution of epistemic uncertainties arising from model error or limited data size.

The second issue concerns the interpretation of the probability of an event. Let $\Pr(E)$ denote the probability of event E. In classical statistics, $\Pr(E)$ is interpreted as a property of E. In fact, the frequentist notion of probability defines $\Pr(E) = \lim_{n\to\infty} n_E/n$, where n is the number of independent experiments where event E may or may not occur, and n_E is the count of the experiments where E occurs. In engineering applications, it is often not feasible to conduct a large number of experiments: there are many cases where even a single experiment is impossible. Consider, for example, determining the depth of the bedrock at the site of a future building. If a borehole could be dug, we could measure the depth exactly and there would be no need for probabilistic analysis. But suppose that resources are not available to dig a borehole and the information about the depth is needed right away. The engineer then has no choice but to make a probabilistic estimate of the depth with whatever information is at hand – e.g., by fitting a Gaussian surface to borehole data from nearby sites (so-called Kriging) or by transforming the measured travel time of shock waves that bounce back from the bedrock. In such an exercise, the probability of event E (say, the event that the depth is in the range 20–25m) is the property of the engineer who assigns it based on the information gathered. Of course, the accuracy of the estimate depends on the quality of the models used and analyses performed to arrive at the probability estimate. But this is typical of all engineering analysis, be it for predicting the stress at a point in a structure or determining the stability of an embankment. The probability estimate by the engineer must, therefore, be considered as the engineer's degree of belief based on the available information and the modeling and analysis performed. Naturally, transparency in the modeling and analysis is of paramount importance.

The above notions of uncertainty and probability are consistent with the Bayesian philosophy as originally conceived by Thomas Bayes with modern interpretations by de Finetti (1974) and Lindley (2014). Further treatment of this subject is presented in Chapters 10 and 15.

1.4 Objectives

The main objective of this book is to present a state-of-the-art treatment of the methods of structural and system reliability to serve as a textbook for upper-division or graduate courses in engineering, particularly in civil and environmental engineering programs. With the aim of making the book maximally accessible, the narrative avoids excessive mathematical formalism and uses as simple a notation as possible. To facilitate learning, numerous examples are presented, some with detailed step-by-step derivations and intermediate numerical results. It is assumed that the student possesses good command of the basic principles of probability theory. Nevertheless, Chapter 2 presents the minimum on this topic that is necessary in order to follow the remainder of the book. In addition, the student must have strong knowledge of multivariate distributions, a topic that is covered in Chapter 3.

Other subjects, such as Bayesian statistics and random processes, are introduced in later chapters, where they are employed.

This book is also intended for use by researchers and practicing engineers. The illustrative examples are intended to facilitate self-learning. Furthermore, while the material in Chapters 2–11 is well established, Chapters 12–15 present snapshots of current research in still-evolving topics of finite-element reliability methods, nonlinear stochastic dynamic analysis, reliability-based optimal design, and Bayesian networks for risk and reliability assessment. These chapters offer opportunities for exploration to those interested in pursuing research and further development in these fields.

There are many other topics that could be included in this book. Examples include probabilistic design codes and code calibration, modeling of specific types of loads (e.g., live, wind, and earthquake loads), reliability assessment of existing structures, and decision theory. However, a decision was made to focus on the mathematical foundations of the methods of structural and system reliability rather than aspects that would focus too narrowly on specific application areas or topics whose proper treatment would require significant deviation from the main focus of the book. No doubt the experience and expertise of the author also had a role in the choice of topics.

1.5 Software

The structural and system reliability methods described in this book require the use of computer software for all but the most trivial problems. During the past three decades, a number of commercial and free software packages have been developed. Some of these were described in two special issues of the journal *Structural Safety* (Volume 28, Issues 1 and 2, 2006) and a comparative analysis of a larger collection is provided by Chehade and Younes (2020). Notable commercial software packages for structural and system reliability analysis include:

- PROBAN (https://manualzz.com/doc/7264297/proban---dnv-gl reached on December 2, 2020), developed by DNV GL, Oslo, Norway, as a part of the SESAM system.
- STRUREL (www.strurel.de/index.html reached December 2, 2020), developed by the late Professor Rudiger Rackwitz's group at the Technical University of Munich, Germany.
- COSSAN (https://cossan.co.uk/ reached December 2, 2020), developed by the late Professor Gerhart Schueller's group at the University of Innsbruck, Austria; now maintained at the University of Liverpool, UK.
- NESSUS (www.swri.org/nessus reached December 2, 2020), developed at the Southwest Research Institute, San Antonio, TX.

Notable free software packages for structural and system reliability analysis include:

- CalREL (https://bitbucket.org/sanjayg0/calrel/src/master/ reached December 2, 2020), developed by the author's group at the University of California, Berkeley.

- FERUM (www.sigma-clermont.fr/en/ferum reached December 2, 2020), a MATLAB toolbox originally developed by the author's group at the University of California, Berkeley, and further developed and maintained by Jean-Marc Bourinet at the Université Clermont Auvergne, France.
- OpenSees (https://opensees.berkeley.edu/ reached December 2, 2020), a general-purpose structural analysis program developed at the Pacific Earthquake Engineering Research (PEER) Center of the University of California, Berkeley, that contains some capabilities for structural reliability analysis.
- DAKOTA (https://dakota.sandia.gov/ reached December 2, 2020), developed at the Sandia National Laboratories, Albuquerque, NM.
- Rt (http://terje.civil.ubc.ca/the-computer-program-rt/ reached December 2, 2020), developed by Professor Terje Haukaas' group at the University of British Columbia, Vancouver, Canada.

In addition to the above, useful tools for reliability analysis, uncertainty quantification, Bayesian updating, and other probabilistic tools are available at the following sites:

- https://systemreliability.wordpress.com/software/ (reached December 2, 2020) contains tools for surrogate modeling and reliability analysis by sampling techniques developed by Professor Junho Song's Structural & System Reliability Group at the Seoul National University, South Korea.
- www.bgu.tum.de/era/software/ (reached December 2, 2020) contains tools for Bayesian inference, surrogate modeling and reliability and risk analysis developed by Professor Daniel Straub's Engineering Risk Analysis Group at the Technical University of Munich, Germany.
- https://sudret.ibk.ethz.ch/software.html (reached December 2, 2020) contains tools for metamodeling and uncertainty quantification developed by Professor Bruno Sudret's group at ETH, Zurich, Switzerland.

A variety of software is available for Bayesian network analysis (the topic of Chapter 15) for inference and learning purposes. A comprehensive but somewhat outdated list prepared by Murphy (2014) is available at www.cs.ubc.ca/~murphyk/Software/bnsoft.html (reached December 2, 2020). Among these is the MATLAB toolbox described in Murphy (2001):

- BNT (https://github.com/bayesnet/bnt reached December 2, 2020), freely downloadable MATLAB toolbox for Bayesian network analysis.

Furthermore, Murphy (2007) presents a review of several Bayesian network software. Among these, the following two commercial codes are noteworthy:

- GeNie (www.bayesfusion.com/ reached December 2, 2020), developed by Professor Mark J. Druzdzel's group at the University of Pittsburg and available through BayesFusion, LLC. A free version for educational purposes is provided.

- Hugin (www.hugin.com/ reached December 2, 2020), developed at the University of Aalborg and available through HuginExpert A/S. A free version for educational purposes is provided.

Details about the specifications and capabilities of the above software can be obtained from the listed websites.

1.6 Organization of Chapters

Following this introductory chapter, Chapter 2 presents a review of the basic concepts and rules of probability theory. Mastery of this material is essential for a thorough understanding of the material in the subsequent chapters. This chapter can be skipped if the student has a fairly advanced understanding of probability theory. However, in my own teaching, I have found it necessary to review this material, even for students who have taken a prior course. I normally spend six lecture hours reviewing this material in a semester-long course of 45 lecture hours.

Chapter 3 describes multivariate distribution models and the transformation of variables to the standard normal space. The inverse transforms as well as the Jacobians of both transforms are presented. Several reliability methods make use of these transformations. I have found it useful to cover this material in two lecture hours. A focused presentation of this material rather than covering it as a part of a particular reliability method gives it the weight that it deserves.

Chapter 4 introduces the basic formulation of the structural reliability problem. Exact solutions are provided for the special cases of capacity and demand values having jointly normal or jointly lognormal distributions. Formulations are presented in terms of the safety margin and safety factor. The important issue of sensitivity to the tail of the assumed distributions is investigated. Finally, the structural reliability formulation is generalized and expressed in terms of limit-state functions of basic random variables, a formulation that is used throughout subsequent chapters. Several example formulations of the limit-state function provide the link between this theory and classical domains of civil engineering. In my teaching, I normally spend one lecture covering this topic.

Chapter 5 presents methods of structural reliability analysis under incomplete probability information. These include second-moment methods, i.e., the MCFOSM method, the FOSM method, and the generalized second-moment method, as well as methods that employ information beyond the second moments such as higher-order moments or marginal distributions. An algorithm for finding the "design point," which is a point of approximation in several reliability methods, is also introduced in this chapter. In my opinion, the methods described in this chapter now have only historical value. In the concluding section of the chapter, I argue that the notion of having perfect information about a limited number of moments and no information on higher moments or other probabilistic characteristics is counter to reality. In practice, one estimates the moments from available data with decreasing accuracy for higher moments. Furthermore, the moment-based formulations do not provide a logical framework for accounting for statistical uncertainties. In my teaching of the course, I make only brief introduction of these methods (no more than one lecture) but require that students read the chapter on their own.

Chapter 6 develops the first-order reliability method (FORM), which is one of the linchpins of this book. This is a full-distribution reliability method that employs a first-order approximation of the limit-state function in the standard normal space, thus requiring transformation of random variables into the standard normal space, as described in Chapter 3. It is important to recognize the approximate nature of FORM, and this is discussed in a separate section dealing with measures of its accuracy. The chapter then presents FORM measures of variable importance and parameter sensitivities. Other topics in this chapter include addressing the problem of multiple design points, the inverse reliability problem, and determining the distribution of a function of random variables by FORM. In my teaching, I devote eight lecture hours to the material in this chapter.

Chapter 7 develops the second-order reliability method (SORM). This is a refinement of FORM in which a second-order approximation of the limit-state function is used. Aside from the classical SORM that requires computation of second derivatives of the limit-state function, two additional SORM approximations are developed that require first-order derivatives only. Several examples compare results obtained from the three SORM approximations together with those of FORM analysis. I devote two lecture hours to this chapter.

Chapter 8, another linchpin of this book, introduces system reliability analysis. It starts with the definition of a general system in terms of the system function and its characterization in terms of cut and link sets. For systems with statistically independent component states, methods are described for computing the system reliability in terms of the component failure probabilities. For systems with dependent components, methods for computing system reliability in terms of minimum cut and link sets are presented and bounding formulas are derived. For cases with incomplete probability information, bounds on system reliability are obtained by the use of linear programming. For the case of complete probability information, the latter formulation leads to an efficient matrix-based reliability method for certain classes of problems. Next, attention is focused on structural systems, where approximations and bounding formulas based on FORM and an event-tree approach are presented. The chapter ends with the development of component importance and sensitivity measures. I devote around nine lecture hours to the material in this chapter.

Chapter 9 presents simulation methods for assessing the reliability of structural systems. The chapter starts with the presentation of methods for generating pseudorandom numbers for specified distributions. Then the basic Monte Carlo simulation method is described, showing its shortcoming for estimating small probabilities that are typical of structural reliability problems. Next, several methods are presented for enhancing the efficiency of the Monte Carlo method. These include use of antithetic variates, importance sampling, directional sampling, orthogonal-plane sampling, and subset simulation. An adaptive importance sampling method for numerical integration in high dimensions is also presented, which is useful for Bayesian posterior analysis. The chapter ends with the presentation of parameter sensitivity analysis by simulation. My recommended time for this topic is four lecture hours.

Chapter 10 deals with Bayesian parameter estimation and reliability updating. The chapter starts with a discussion of the sources and types of uncertainties then develops the basic notions of Bayesian parameter estimation and model assessment. Next, analysis of structural reliability under parameter and model uncertainties is described and various measures of reliability that account for these uncertainties are introduced. Methods for Bayesian updating of structural reliability and the distribution of basic variables in light of observational data concludes the chapter. I recommend devoting four or five hours to this chapter.

Chapter 11 starts by introducing three types of time- and space-variant reliability problems: the outcrossing problem, the encroaching problem, and the outcrossing-encroaching problem. This provides motivation for the remainder of the chapter. After a brief review of random process theory, approximate or bounding solution approaches are presented for the three classes of time- and space-variant reliability problems. Next, the Poisson process and related distributions of waiting and inter-arrival times are derived. These are used to develop stochastic load models. The chapter ends with a presentation of the load combination problem based on the load-coincidence approach. Given the time constraints of the semester, I devote only two lectures to this chapter, focusing more on the Poisson process and stochastic load models and load combination.

Chapters 12–15 present the more advanced topics of finite-element reliability methods, stochastic structural dynamics, reliability-based optimal design, and Bayesian networks for system reliability and risk analyses. These are topics of current research and are still evolving. Time permitting, I devote about one lecture to introduce the more salient aspects of each topic, emphasizing example applications. I encourage students to read these chapters on their own, particularly if they are interested in conducting research in probabilistic methods.

2 Review of Probability Theory

2.1 Introduction

The theory of probability is a vast subject, well beyond the needs and objectives of this book. This chapter provides a relatively brief review of those topics that are needed for the development of the material in subsequent chapters of this book. Where necessary, additional material on probability or stochastic process theory is presented in later chapters. As probability theory deals with multiple outcomes of a random phenomenon, the chapter begins with a brief review of the essential concepts from set theory, which is useful in describing collections of objects. It then presents the fundamental axioms and rules of probability theory, followed by the development of the calculus of probability. The concepts of random variable, probability distribution, expectation, and moments are developed. Special emphasis is placed on the distributions and moments of functions of random variables due to their relevance to the theory of structural reliability. The chapter ends with a description of extreme-value distributions that are useful in developing capacity and demand models in structural and system reliability.

2.2 Elements of Set Theory

A *set* is a collection of well-defined objects. In the theory of probability, we collect the possible outcomes of a random phenomenon into a set S, which we denote the *sample space*. Each element x of this set is a *sample point*. We write $x \in S$, which defines x as a member or element of S. A sample space is said to be *discrete* or *countable* when its elements x are discrete and countable. A discrete sample space can be *finite* or *infinite*, the latter when it contains infinite sample points. A sample space is said to be continuous or uncountable when it contains a continuum of sample points. In this case, the sample space naturally has infinite sample points.

Any collection of sample points is an *event*. Thus, an event is a subset of the sample space and we can write $E \subseteq S$, which defines event E as a subset of or equal to S. When an event

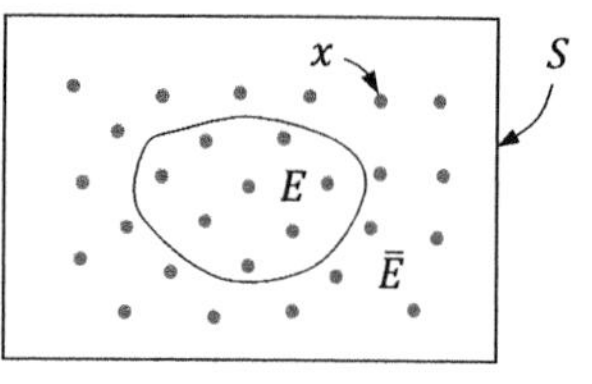

Figure 2.1 Sample space, sample point, and event in a Venn diagram

contains all the sample points, i.e., when $E = S$, it is said to be the *certain* event. When E contains no sample points, it is called the *null* or *impossible* event and is denoted $\emptyset$. The *complement* of event E, denoted $\bar{E}$, is an event that contains the sample points that are not in E. These definitions are conceptually illustrated for a discrete sample space in the Venn diagram shown in Figure 2.1.

Example 2.1 – Discrete Sample Space

Consider the state of a building after an earthquake. Theoretically speaking, the state of the building can be anywhere on a continuum from absolutely no damage to complete collapse, i.e., the sample space is continuous. However, as in many engineering applications, it is convenient to discretize the state of the building into a finite set of sample points. Suppose we consider the discretized states of no damage (ND), light damage (LD), heavy damage (HD), and complete collapse (CC). The sample space then is $S = \{ND, LD, HD, CC\}$. Various events can now be defined, such as $E_1 = \{ND, LD\}$, i.e., the occurrence of no or light damage, and $E_2 = \{LD, HD, CC\}$, i.e., the occurrence of light or heavy damage or complete collapse. It should be clear that $\bar{E}_1 = \{HD, CC\}$ and $\bar{E}_2 = \{ND\}$.

Example 2.2 – Continuous Sample Space

Consider a structural component having capacity R and subjected to demand S. The component fails if the demand exceeds the capacity, i.e., the failure event is $F = \{R < S\}$. Assuming R and S take on continuous non-negative values, Figure 2.2 shows the sample space in an (r, s) Cartesian coordinate system where r and s denote the outcomes of R and S, respectively. The failure event is marked as the gray area above the line $r - s = 0$. This is an example of a continuous sample space.

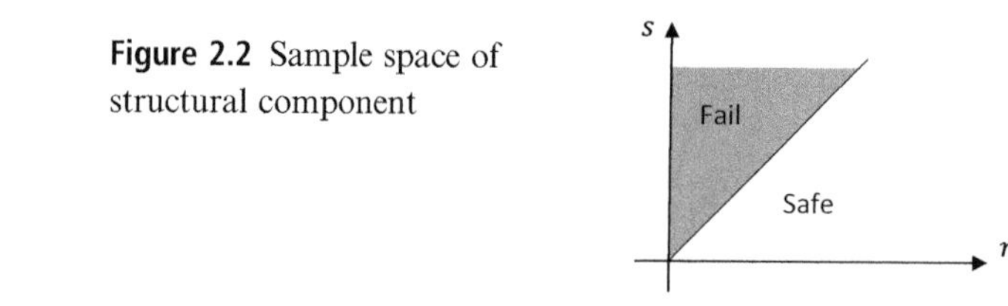

Figure 2.2 Sample space of structural component

We can operate on two or more events to create new events. There are two basic operations. One is the *union* of two events, which creates an event that contains all the sample points that are in either of the two events. For events E_1 and E_2, this operation is denoted

$$E_1 \cup E_2 \tag{2.1}$$

and the result is the shaded event shown in Figure 2.3(a). The second is the *intersection* operation, which results in an event that contains the sample points that are in both events. For events E_1 and E_2 the intersection operation is defined as

$$E_1 \cap E_2. \tag{2.2}$$

The result is the shaded event shown in Figure 2.3(b). In practice, we often remove the intersection operator so that E_1E_2 implies the same as the operation in (2.2). Because the result of the above basic operations is an event, it can be further operated with other events. For example, $(E_1 \cup E_2)E_3$ denotes the intersection of event E_3 with the event resulting from the union of events E_1 and E_2. Such operations are called *compound* operations.

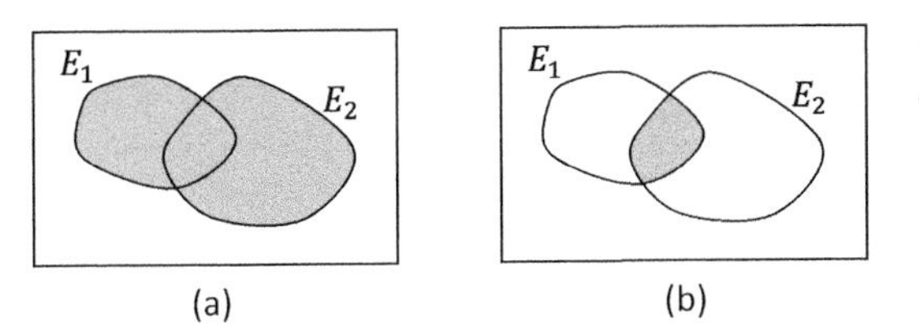

Figure 2.3 Shaded area indicates (a) $E_1 \cup E_2$, (b) $E_1 \cap E_2$

The above basic operations have the following properties:

- *Commutative*: Changing the order of events does not change the result of the union or intersection operations.
- *Associative*: The union or intersection of several events does not depend on the order of operation on the events.
- *Distributive*: An intersection (union) operation can be distributed into the union (intersection) of two events.

The commutative rule implies $E_1 \cup E_2 = E_2 \cup E_1$ and $E_1E_2 = E_2E_1$, so that the ordering of events within a basic operation is immaterial. The associative rule implies that for three events $(E_1 \cup E_2) \cup E_3 = E_1 \cup (E_2 \cup E_3)$ and $(E_1E_2)E_3 = E_1(E_2E_3)$. On the left-hand sides of these equations, we operate first on events E_1 and E_2, as indicated by the enclosing parentheses, and then on the resulting event and E_3. On the right-hand sides, we operate first on events E_2 and E_3 and then on the resulting event and E_1. These identities can be easily verified by examining the Venn diagrams of the three events. The order of these operations is immaterial, so there is no need to use the parentheses. Hence, the union and intersection of n events are written as

$$E_1 \cup E_2 \cup \cdots \cup E_n = \cup_{i=1}^{n} E_i \tag{2.3}$$

$$E_1E_2 \cdots E_n = \cap_{i=1}^{n} E_i, \tag{2.4}$$

where on the right-hand sides we have introduced compact notations for the union and intersection of events. The distributive law implies the following identities:

$$E_1(E_2 \cup E_3) = (E_1E_2) \cup (E_1E_3) \tag{2.5}$$

$$E_1 \cup (E_2E_3) = (E_1 \cup E_2)(E_1 \cup E_3). \tag{2.6}$$

The student is encouraged to verify these identities in a Venn diagram of the three events.

To simplify expressions involving mixed operations, we introduce a rule: The hierarchy of operations is intersection first, union second, unless indicated otherwise by parentheses. Following this rule, the right-hand side of (2.5) can be written as $E_1E_2 \cup E_1E_3$, and the left-hand side of (2.6) can be written as $E_1 \cup E_2E_3$. However, the parentheses on the left-hand side of (2.5) and the right-hand side of (2.6) cannot be removed.

A pair of events is said to be *mutually exclusive* if they share no sample points. For mutually exclusive events E_1 and E_2, $E_1E_2 = \varnothing$. A set of events $E_1, E_2, \ldots, E_n$ is said to be *collectively exhaustive* if $\cup_{i=1}^n E_i = S$, i.e., if the events span the entire sample space. It should be clear that an event E and its complement $\bar{E}$ are both mutually exclusive and collectively exhaustive.

We end this brief account of set theory by stating two identities known as De Morgan's rules (named after 19th century mathematician Augustus De Morgan). Given a set of events $E_1, E_2, \ldots, E_n$, these rules state:

$$\overline{\cup_{i=1}^n E_i} = \cap_{i=1}^n \bar{E}_i, \tag{2.7}$$

$$\overline{\cap_{i=1}^n E_i} = \cup_{i=1}^n \bar{E}_i. \tag{2.8}$$

According to the rule in (2.7), the complement of unions equals the intersection of complements, while according to (2.8), the complement of intersections equals the union of complements. These rules are useful when dealing with systems problems. Problem 2.3 at the end of this chapter asks for the proofs of these rules.

2.3 Basic Rules of Probability Theory

To each event E in the sample space S, we assign a numerical value called the probability of E and denote it $\Pr(E)$. The meaning of this assignment is the subject of much philosophical debate. Here, we will follow the so-called Bayesian notion of probability, wherein $\Pr(E)$ is a measure of our degree of belief that event E will occur, i.e., that the specific sample point to be realized in the sample space S will be contained within E. We will further discuss this and other notions of probability later in this section.

As with other branches of mathematics, probability theory is based on a set of axioms. Specifically, there are three axioms:

- Probability is bounded between 0 and 1, i.e., $0 \leq \Pr(E) \leq 1$.
- Probability of the certain event equals 1, i.e., $\Pr(S) = 1$.

- Probability of the union of two mutually exclusive events is the sum of their probabilities, i.e., $\Pr(E_1 \cup E_2) = \Pr(E_1) + \Pr(E_2)$, if E_1 and E_2 are mutually exclusive.

The first two axioms are obviously conventions. The third axiom can be verified by use of the frequency notion of probability, in which the probability of an event is seen as the limiting fraction of the number of occurrences of the event in an infinite series of experiments, each involving the random draw of a sample point from the sample space. However, for a more general notion of probability, such as the Bayesian notion mentioned above, there is no verification of the axiom. Indeed, this being an axiom, there is no need for verification. We accept it as a reasonable foundational rule. All other rules in probability theory are derived from the above axioms. Presented below are the most important rules of probability without providing detailed proofs.

The probability of the complement of an event is given by

$$\Pr(\bar{E}) = 1 - \Pr(E). \tag{2.9}$$

This is verified by noting that E and $\bar{E}$ are mutually exclusive and collectively exhaustive so that according to the third axiom $\Pr(E \cup \bar{E}) = \Pr(E) + \Pr(\bar{E}) = 1$. It follows from this rule that $\Pr(\varnothing) = 0$ as the null event is the complement of the certain event. We note that, whereas the probability of the null event is 0, an event with zero probability is not necessarily the null (impossible) event. This will become clear later in this chapter.

The *addition rule* concerns the probability of the union of events. For events E_1 and E_2, this rule states

$$\Pr(E_1 \cup E_2) = \Pr(E_1) + \Pr(E_2) - \Pr(E_1 E_2). \tag{2.10}$$

The validity of this rule can be easily verified by examining Figure 2.3. Specifically, the first two terms on the right-hand side of (2.10) double count the sample points in the intersection domain of the two events, hence the reason for subtracting the last term. When the events are mutually exclusive, the last term in (2.10) drops and the addition rule reduces to the third axiom stated above. For a set of n events, the addition rule generalizes to

$$\begin{aligned}\Pr(\cup_i E_i) = {} & \sum\nolimits_i \Pr(E_i) - \sum\nolimits_{i<j} \Pr(E_i E_j) + \sum\nolimits_{i<j<k} \Pr(E_i E_j E_k) - \cdots \\ & + (-1)^{n-1} \Pr(E_1 E_2 \cdots E_n),\end{aligned} \tag{2.11}$$

where all the indices go from 1 to n while satisfying the stated conditions. Essentially, the right-hand side consists of the sum of probabilities of all events, minus the sum of probabilities of all distinct pair-wise intersections, plus the sum of probabilities of all distinct triple intersections, and so on, all the way to the probability of intersection of all events with a sign that depends on the number of events (negative when the number of events is even and positive when it is odd). For this reason, this expression is commonly known as the *inclusion-exclusion formula*. Using the De Morgan rule in (2.7), the probability of the union of n events can also be written as

$$\Pr(\cup_i E_i) = 1 - \Pr(\bar{E}_1 \bar{E}_2 \cdots \bar{E}_n). \tag{2.12}$$

One can show that when the series in (2.11) is truncated after one of the sum terms, the result is a lower bound if the last included sum has a negative sign and an upper bound if the last included sum has a positive sign. Thus, when the sums in the series are sequentially calculated, the resulting partial summations zigzag around the true solution. Naturally, one can stop the calculations when the two bounds are sufficiently close. In many cases, narrow bounds can be obtained by including the first two or three sum terms of the series.

In order to go further, it is necessary that we introduce the important concept of *conditional probability*. We define the conditional probability of event E_1 given event E_2 as the ratio

$$\Pr(E_1|E_2) = \frac{\Pr(E_1E_2)}{\Pr(E_2)}. \tag{2.13}$$

Where $\Pr(E_1|E_2)$ is interpreted as the probability of event E_1 given that event E_2 has occurred, i.e., the probability that the realized sample point is contained within event E_1, given that it is contained within event E_2. Obviously, for E_1 to occur under this condition, the realized sample point must fall within the intersection event E_1E_2. This is the reason for the probability of the intersection event appearing in the numerator of the ratio. However, we need only to consider sample points that are within event E_2, and that explains the normalization by $\Pr(E_2)$. For now, the above definition is valid only when $\Pr(E_2) > 0$. Later, we will consider cases when the ratio is indefinite, with both the numerator and denominator being 0. The conditional probability then is obtained as the limit of the indefinite ratio.

The *multiplication rule* concerns the probability of the intersection of events. Using the definition in (2.13), the probability of intersection of two events E_1 and E_2 is written as

$$\begin{aligned} \Pr(E_1E_2) &= \Pr(E_1|E_2)\Pr(E_2) \\ &= \Pr(E_2|E_1)\Pr(E_1), \end{aligned} \tag{2.14}$$

where we have exchanged the indices to obtain the second line. It is evident that the probability of intersection of two events is the product of two probabilities, one conditional and one unconditional or *marginal*.

Two events are said to be *statistically independent* if the conditional probability of one event given the other is identical to its marginal probability, i.e., if $\Pr(E_1|E_2) = \Pr(E_1)$. In that case, the multiplication rule (2.14) reduces to

$$\Pr(E_1E_2) = \Pr(E_1)\Pr(E_2). \tag{2.15}$$

Thus, in the case of statistically independent events, the probability of the intersection is equal to the product of the marginal probabilities. For two events, the equality in (2.15) is also indicative of statistical independence.

For n events, repeatedly applying the multiplication rule, the probability of intersection can be written in the form

$$\Pr(E_1E_2\cdots E_n) = \Pr(E_1|E_2\cdots E_n)\Pr(E_2|E_3\cdots E_n)\cdots\Pr(E_{n-1}|E_n)\Pr(E_n). \tag{2.16}$$

Of course, permutation of the indices can be used in the conditioning, so that there are $n!$ ways of writing the intersection probability as the product of $n-1$ conditional and one marginal probabilities. The n events are statistically independent if for every subset $E_{1'}, \ldots, E_{r'}$, $2 \leq r' \leq n$ of the events the following equality holds:

$$\Pr(E_{1'}E_{2'}\cdots E_{r'}) = \Pr(E_{1'})\Pr(E_{2'})\cdots\Pr(E_{r'}). \tag{2.17}$$

Note that, for $2 < n$, the equality $\Pr(E_1E_2\cdots E_n) = \Pr(E_1)\Pr(E_2)\cdots\Pr(E_n)$ alone does not guarantee statistical independence of the n events.

Next, we introduce the *total probability* rule. Let the events $E_1, E_2, \ldots, E_n$ be mutually exclusive and collectively exhaustive so that $E_iE_j = \varnothing$ for $i \neq j$ and $\cup_{i=1}^{n} E_i = S$. Also consider an event A in the same sample space. One can write

$$\begin{aligned}\Pr(A) &= \textstyle\sum_{i=1}^{n}\Pr(AE_i)\\ &= \textstyle\sum_{i=1}^{n}\Pr(A|E_i)\Pr(E_i),\end{aligned} \tag{2.18}$$

where the first line is based on the third axiom of probability as the events AE_i, $i = 1, \ldots, n$, are mutually exclusive. Figure 2.4 demonstrates the idea behind this formula. Essentially, the problem of computing the probability of event A is broken down into computing the conditional probabilities $\Pr(A|E_i)$ with respect to the mutually exclusive and collectively exhaustive events E_i and multiplying them by the respective marginal probabilities $\Pr(E_i)$. Because the conditional probabilities are defined within smaller sample spaces, they are usually easier to compute than the unconditional probability $\Pr(A)$. Of course, one needs to be clever in selecting the events E_i so that the needed conditional and marginal probabilities are easy to compute. In essence, this formula offers a "divide-and-conquer" strategy.

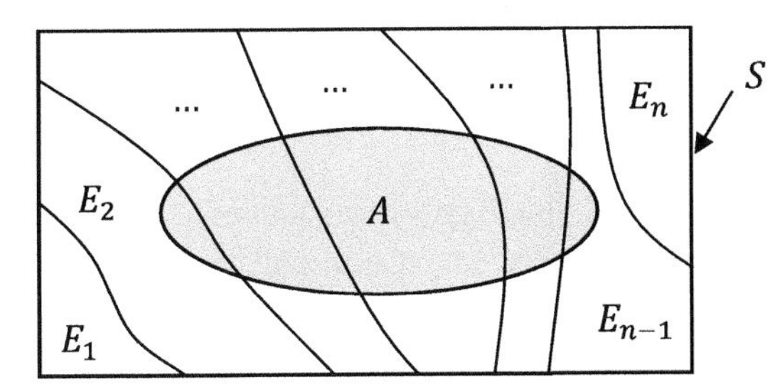

Figure 2.4 Event A and set of mutually exclusive, collectively exhaustive events $E_1, E_2, \ldots, E_n$

Example 2.3 – Application of the Total Probability Rule

The probability that a building will sustain damage due to an earthquake depends on the intensity of shaking. Suppose the intensity is discretized into the states of weak, moderate, and strong. Assume it is determined that the probability of damage to the building is 0.01, 0.10, and 0.60 for the three levels of shaking intensity, respectively. Furthermore, based on the history of past earthquakes in the region, it is estimated that, for a randomly occurring earthquake, the

Example 2.3 (cont.)

probabilities for the three levels of shaking intensity at the site of the building are 0.90, 0.08, and 0.02, respectively. We are interested in the probability that the building will sustain damage during the next earthquake in the region.

Let W, M, and S denote the events that the shaking intensity is weak, moderate, and strong, respectively. Observe that these events are necessarily mutually exclusive: A shaking intensity cannot be in more than one of the states W, M, or S. Furthermore, owing to our way of discretizing the intensity of shaking, the outcome of the intensity must be one of the three events. Thus, W, M, and S are also collectively exhaustive. Under these conditions, we must have $\Pr(W) + \Pr(M) + \Pr(S) = 1$. This requirement is satisfied by the numbers we have: $0.90 + 0.08 + 0.02 = 1$. Hence, we can use the total probability rule by conditioning on the three events W, M, and S. Denoting the event of damage D, we have

$$\begin{aligned} \Pr(D) &= \Pr(D|W)\Pr(W) + \Pr(D|M)\Pr(M) + \Pr(D|S)\Pr(S) \\ &= 0.01 \times 0.90 + 0.10 \times 0.08 + 0.60 \times 0.02 \\ &= 0.009 + 0.008 + 0.012 \\ &= 0.029. \end{aligned} \tag{E1}$$

Observe that the total probability rule allowed us to account not only for the uncertainty in the intensity of shaking but also for the uncertainty in the damage event for a given intensity. It is interesting to note that the contributions to the probability of damage from weak, moderate, and strong-intensity earthquakes (respectively 0.009, 0.008, and 0.012) are of similar magnitude, even though the corresponding conditional probabilities of damage are vastly different. Of course, this has to do with the higher frequency of weak- and moderate-intensity earthquakes relative to strong-intensity ones.

The final rule to introduce in this section is the Bayes rule, named after 18th century statistician Thomas Bayes. Consider event A and the set of mutually exclusive, collectively exhaustive events $E_1, \ldots, E_n$, all defined in the same sample space. Using the multiplication rule, we have $\Pr(AE_i) = \Pr(E_i|A)\Pr(A) = \Pr(A|E_i)\Pr(E_i)$. From the last equality, we obtain

$$\Pr(E_i|A) = \frac{\Pr(A|E_i)}{\Pr(A)}\Pr(E_i). \tag{2.19}$$

The significance of this expression lies in the appearance of the probability of E_i on both sides of the equation, conditional on the left and unconditional (marginal) on the right. Bayes interpreted this formula as a rule to update the probability of an event in the light of new information. Specifically, the formula shows how the probability of event E_i changes when we learn that event A has occurred. The term $\Pr(E_i)$ in this formula is known as the *prior* probability, being the probability of the event prior to knowing about the occurrence of event A; $\Pr(E_i|A)$ is known as the *posterior* probability; $\Pr(A|E_i)$ is known as the *likelihood*; and $\Pr(A)$ is essentially a normalizing factor. Note that $\Pr(A|E_i)$ should be interpreted as the probability of event A, if E_i were known to have occurred, and $\Pr(A)$ should be interpreted as the probability of occurrence of A prior to knowing that it has actually occurred.

Replacing the denominator in (2.19) by the total probability expression in (2.18), the Bayes rule is written in the form

$$\Pr(E_i|A) = \frac{\Pr(A|E_i)\Pr(E_i)}{\sum_{j=1}^{n}\Pr(A|E_j)\Pr(E_j)}. \tag{2.20}$$

It is now seen that the posterior probability of E_i among n mutually exclusive and collectively exhaustive events is proportional to the product of its likelihood and prior, with the denominator acting as a normalizing factor. In Chapter 10, we will see that the above formula is the basis for Bayesian statistics, which provide a powerful tool for estimating model parameters.

Example 2.4 – Updating Probability in the Light of New Information

Reconsider Example 2.3 and assume that we know that the building has sustained damage due to an earthquake. Using the Bayes rule, we can learn about the intensity of that earthquake. Specifically,

$$\Pr(W|D) = \frac{\Pr(D|W)\Pr(W)}{\Pr(D)} = \frac{0.01 \times 0.90}{0.029} = 0.310, \tag{E1}$$

$$\Pr(M|D) = \frac{\Pr(D|M)\Pr(M)}{\Pr(D)} = \frac{0.10 \times 0.08}{0.029} = 0.276, \tag{E2}$$

$$\Pr(S|D) = \frac{\Pr(D|S)\Pr(S)}{\Pr(D)} = \frac{0.60 \times 0.02}{0.029} = 0.414. \tag{E3}$$

Observe that the updated probabilities are significantly different from the prior probabilities ($\Pr(W) = 0.90, \Pr(M) = 0.08, \Pr(S) = 0.02$): Knowing that damage has occurred, the probabilities for the intensity to have been moderate or strong have sharply increased and the probability for the intensity to have been weak has sharply decreased. Note also that the three updated probabilities still sum to 1, as they should. This is a consequence of the normalization by $\Pr(D)$.

As the final item in this section, we note that all rules of probability apply to conditional probabilities as long as the proper sample space is maintained throughout the expression. For example, the addition rule for the union of two events conditioned on an event A is written as

$$\Pr(E_1 \cup E_2|A) = \Pr(E_1|A) + \Pr(E_2|A) - \Pr(E_1E_2|A) \tag{2.21}$$

and the multiplication rule is written as

$$\Pr(E_1E_2|A) = \Pr(E_1|E_2A)\Pr(E_2|A). \tag{2.22}$$

Likewise, the total probability of event A given event B can be written as

$$\Pr(A|B) = \sum_{i=1}^{n} \Pr(A|E_iB)\Pr(E_i|B), \tag{2.23}$$

where $E_1, \ldots, E_n$ are mutually exclusive and collectively exhaustive events within B. Indeed, all probabilities can be considered as conditioned on their respective sample spaces. Hence, by $\Pr(A)$ we really mean $\Pr(A|S)$, where S denotes the sample space. However, when the sample space is obvious and not changed, we normally do not explicitly specify it.

2.4 Random Variable

To facilitate the development of a probability calculus, we introduce the concept of a *random variable*. A random variable is defined by mapping the sample space onto a line (Figure 2.5). Each sample point is mapped onto a point on the line, so that the sample space and events represent intervals on the line. In many cases, the mapping is natural and obvious. For example, if the random phenomenon of interest is the number of earthquakes in a region over a given period of time, then the random variable takes on that number, which can be any non-negative integer. Likewise, if the phenomenon of interest is the magnitude of the next earthquake, then the numerical value of the magnitude represents the outcome of the random variable. However, there are cases where the mapping is not obvious. Consider Example 2.1, in which the possible states of a building after an earthquake were discretized into the four states of no damage (*ND*), light damage (*LD*), heavy damage (*HD*), and complete collapse (*CC*). These states do not have natural numerical values. Hence, for the purpose of the mapping we need to establish a rule. For example, the rule can be $ND \to \{X = 0\}$, $LD \to \{X = 1\}$, $HD \to \{X = 2\}$, and $CC \to \{X = 3\}$. The random variable X then has the possible outcomes $\{0, 1, 2, 3\}$.

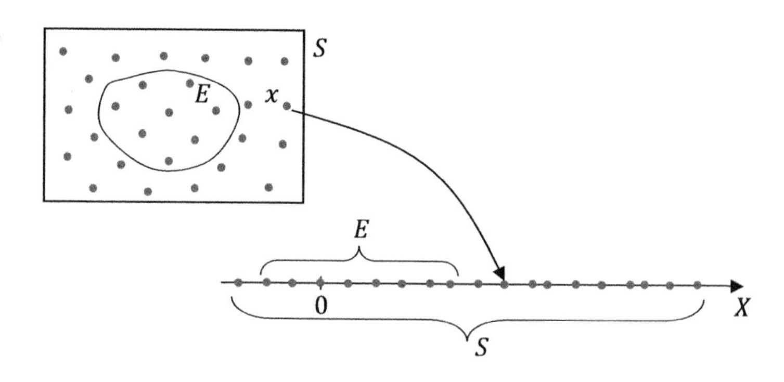

Figure 2.5 Random variable defined by mapping a sample space on a line

Depending on the nature of the sample space, the random variable can either be discrete with finite or infinite outcomes, or continuous. For example, the random variable describing the number of earthquakes is a discrete random variable with an infinite number of possible outcomes (one cannot put an upper bound on the possible number of earthquakes in the region during a given period of time), the random variable representing the earthquake magnitude is continuous, and the random variable representing the state of the building is discrete with a finite number of possible outcomes. Hereafter we denote the random variable by a capital letter and its specific outcome by the lower case of the same letter.

Different outcomes of a random variable usually have different probabilities of occurrence. We now define rules for assigning these probabilities. These rules are known as *probability distributions*.

For a discrete random variable, we define the *probability mass function* (PMF) as

$$p_X(x) = \Pr(X = x). \tag{2.24}$$

The function $p_X(x)$ takes on non-zero values only at the discrete outcome points of the random variable. Furthermore, it must satisfy the conditions $0 \leq p_X(x) \leq 1$ and $\sum_x p_X(x) = 1$. The latter condition, denoted *normalization*, is a consequence of the fact that the possible outcomes of the random variable constitute a set of mutually exclusive and collectively exhaustive events and, therefore, the sum of their probabilities should be 1. A bar diagram, as exemplified in Figure 2.6(a), is an appropriate form to illustrate the PMF.

For a continuous random variable, a definition as in (2.24) is meaningless because there are infinite possible outcomes and, regardless of their relative likelihoods, the probability that the random variable will assume any single outcome is 0. Hence, we define the *probability density function* (PDF) as a function $f_X(x)$ such that

$$f_X(x)\mathrm{d}x = \Pr(x < X \leq x + \mathrm{d}x). \tag{2.25}$$

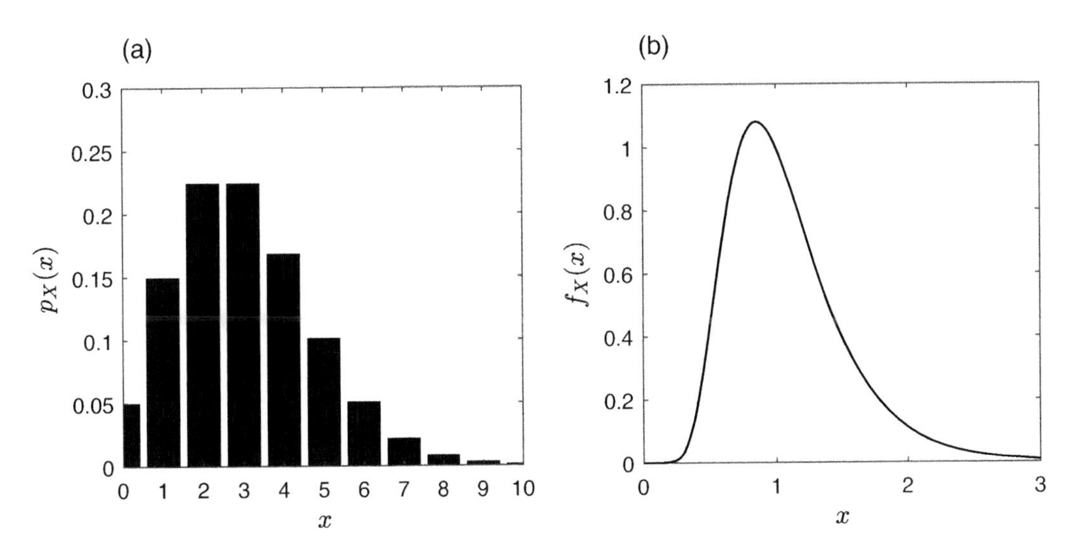

Figure 2.6 Examples for (a) PMF and (b) PDF

The PDF must satisfy the conditions $0 \leq f_X(x)$ and $\int_{-\infty}^{+\infty} f_X(x)\mathrm{d}x = 1$, the latter being the normalization condition. Figure 2.6(b) illustrates a PDF with the area underneath being identical to 1.

Another way to assign probabilities to different outcomes of a random variable is through the *cumulative distribution function* (CDF), which is defined as

$$F_X(x) = \Pr(X \leq x). \tag{2.26}$$

This definition applies to both discrete and continuous random variables. For a discrete random variable, the relation between the CDF and the PMF is

$$F_X(x) = \sum\nolimits_{x' \leq x} p_X(x'), \tag{2.27}$$

and for a continuous random variable the relation between the CDF and the PDF is

$$F_X(x) = \int_{-\infty}^{x} f_X(x)\mathrm{d}x. \tag{2.28}$$

Taking the derivative from both sides of (2.28), another useful relation for a continuous random variable is

$$f_X(x) = \frac{\mathrm{d}F_X(x)}{\mathrm{d}x}. \tag{2.29}$$

Thus, the PDF is the derivative of the CDF. It should be clear that the CDF is a non-decreasing function with the limiting values $F_X(-\infty) = 0$ and $F_X(+\infty) = 1$. Figure 2.7 shows the CDFs corresponding to the PMF and PDF in Figure 2.6.

Tables A2.1 and A2.2 in Appendix 2A list probability distribution models commonly used in engineering applications. Table A2.1 is for discrete random variables and Table A2.2 for

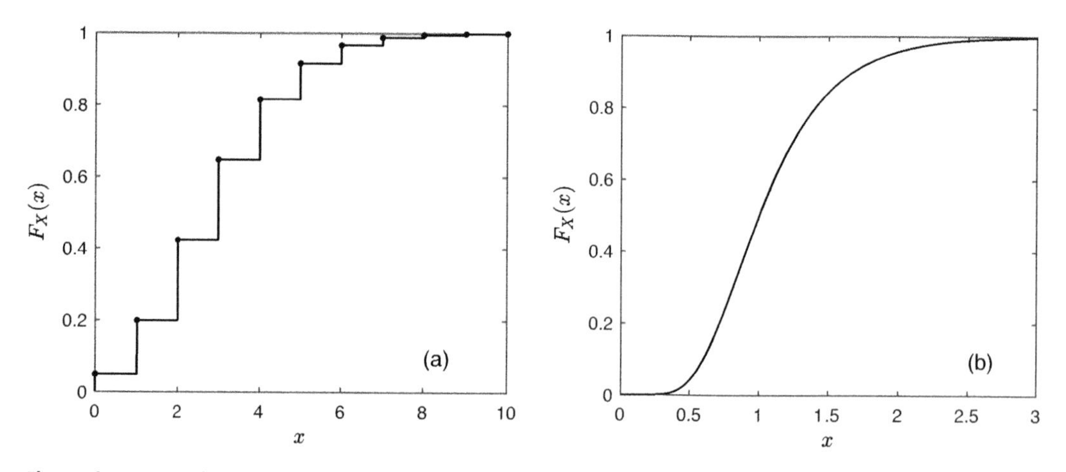

Figure 2.7 CDFs for (a) discrete variable, (b) continuous variable

continuous random variables. For each distribution model, the parameters, bounds, and other properties of the distribution are summarized. We will make frequent use of the information summarized in these tables throughout this book.

We now extend the concept of conditional probability to an event and a random variable. Recall that the conditional probability of an event E_1 given another event E_2 is defined as $\Pr(E_1|E_2) = \Pr(E_1E_2)/\Pr(E_2)$. Consider an event E and a random variable X. $\{X = x\}$ is an event, so one can write

$$\Pr(E|X = x) = \frac{\Pr(E \cap X = x)}{\Pr(X = x)}. \tag{2.30}$$

If X is a discrete random variable, the denominator is identical to the PMF and one has

$$\Pr(E|X = x) = \frac{\Pr(E \cap X = x)}{p_X(x)}. \tag{2.31}$$

On the other hand, if X is a continuous random variable, both the numerator and denominator in (2.30) are 0 and the ratio is undefined. In this case, we define the conditional probability as the limit

$$\Pr(E|X = x) = \lim_{\Delta x \to 0} \frac{\Pr(E \cap x < X \le x + \Delta x)}{f_X(x)\Delta x}. \tag{2.32}$$

Sometimes the above limit is computed approximately by using a sufficiently small Δx.

We can now define the total probability (see 2.18) of an event conditioned on a random variable. If the random variable is discrete, the rule is

$$\Pr(E) = \sum_x \Pr(E|X = x) p_X(x) \tag{2.33}$$

and if the random variable is continuous, the rule is

$$\Pr(E) = \int_{-\infty}^{+\infty} \Pr(E|X = x) f_X(x) \mathrm{d}x. \tag{2.34}$$

The rules in (2.33) and (2.34) are useful when the conditional probability $\Pr(E|X = x)$ is available as a function of x.

It is also possible to condition a random variable on an event. The result is a conditional distribution. Specifically, if X is a discrete random variable, we define its PMF conditioned on event E by

$$\begin{aligned} p_{X|E}(x) &= \Pr(X = x|E) \\ &= \frac{\Pr(X = x \cap E)}{\Pr(E)} \\ &= \frac{\Pr(E|X = x)}{\Pr(E)} p_X(x). \end{aligned} \tag{2.35}$$

If X is a continuous random variable, we define its PDF conditioned on event E as

$$\begin{aligned} f_{X|E}(x) &= \lim_{\Delta x \to 0} \frac{\Pr(x < X \le x + \Delta x \cap E)}{\Pr(E)\Delta x} \\ &= \frac{\Pr(E|X = x)}{\Pr(E)} f_X(x). \end{aligned} \tag{2.36}$$

The corresponding form of the conditional CDF is

$$\begin{aligned} F_{X|E}(x) &= \frac{\Pr(X \le x \cap E)}{\Pr(E)} \\ &= \frac{\Pr(E|X \le x)}{\Pr(E)} F_X(x). \end{aligned} \tag{2.37}$$

Later in this chapter we will also consider conditioning one random variable on another random variable.

2.5 Reliability and Hazard Functions

Let T denote the time to the failure of a system. In classical reliability theory (Barlow and Proschan, 1975), the function $R(t) = \Pr(t < T) = 1 - F_T(t)$, $0 \le t$, is known as the *reliability function*. For typical systems, this function decays with time (the longer the time, the higher the chance of failure). The solid line in Figure 2.8 depicts this function for a hypothetical system, where $R(0)$ is the reliability at the instant of putting the system into operation.

Suppose we observe no failures during the interval $[0, t_0]$. We wish to update the system reliability in light of this information. We achieve this by conditioning the distribution of T on the observed event $\{t_0 < T\}$. Using (2.37),

$$\begin{aligned} F_{T|t_0<T}(t) &= \frac{\Pr(T \le t \cap t_0 < T)}{\Pr(t_0 < T)} \\ &= \begin{cases} 0 & t \le t_0 \\ \dfrac{\Pr(t_0 < T \le t)}{\Pr(t_0 < T)} & t_0 < t \end{cases} \\ &= \begin{cases} 0 & t \le t_0, \\ \dfrac{F_T(t) - F_T(t_0)}{1 - F_T(t_0)} & t_0 < t. \end{cases} \end{aligned} \tag{2.38}$$

The corresponding conditional reliability function is

$$R(t|t_0 < T) = \begin{cases} 1 & t \le t_0, \\ 1 - \dfrac{F_T(t) - F_T(t_0)}{1 - F_T(t_0)} = \dfrac{1 - F_T(t)}{1 - F_T(t_0)} = \dfrac{R(t)}{R(t_0)}, & t_0 < t. \end{cases} \tag{2.39}$$

Thus, the observation of no failure during the interval $[0, t_0]$ results in upward scaling of the reliability function by the factor $1/R(t_0)$. The updated reliability function is shown as a dashed line in Figure 2.8.

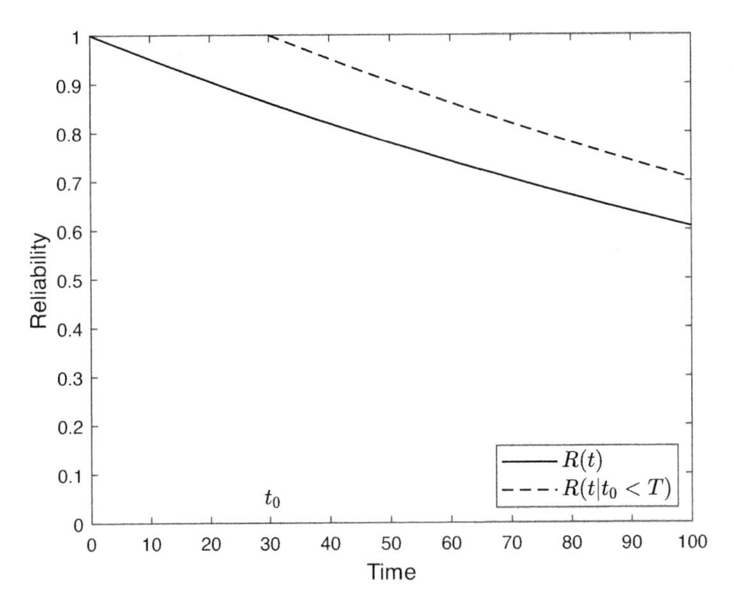

Figure 2.8 Reliability function before (solid line) and after (dashed line) observation of no failure up to time t_0

In the classical reliability theory, the *hazard function* is defined as the probability of failure in the next unit of time, given survival up to the present time. Denoting the hazard function by $h(t)$, we can write

$$\begin{aligned} h(t)\mathrm{d}t &= \Pr(t < T \leq t + \mathrm{d}t | t < T) \\ &= \frac{\Pr(t < T \leq t + \mathrm{d}t)}{\Pr(t < T)} \\ &= \frac{f_T(t)\mathrm{d}t}{1 - F_T(t)}. \end{aligned} \tag{2.40}$$

Thus, the hazard function is

$$h(t) = \frac{f_T(t)}{1 - F_T(t)}. \tag{2.41}$$

For most systems, the hazard function has the "bath-tub" shape depicted in Figure 2.9. It tends to be high initially, but rapidly decays over a short period. Failures during this phase are typically due to manufacturing ("birth") defects. The longer the system survives, the less is the chance of failure due to manufacturing defects. This is followed by a phase where the hazard function is essentially constant. Failures during this phase are primarily due to accidental effects, e.g., accidental overloads, which are independent of the system characteristics and entirely random in nature. In the last phase, the hazard function tends to grow with time. Failures during this phase are primarily due to system fatigue and wear and tear ("old age deaths"). It is interesting to note that this curve also adequately depicts the reliability of human life.

By integrating both sides of (2.41), one obtains

$$F_T(t) = 1 - \exp\left[-\int_0^t h(t)\mathrm{d}t\right]. \tag{2.42}$$

Using this result in (2.41) gives

$$f_T(t) = h(t)\exp\left[-\int_0^t h(t)\mathrm{d}t\right]. \tag{2.43}$$

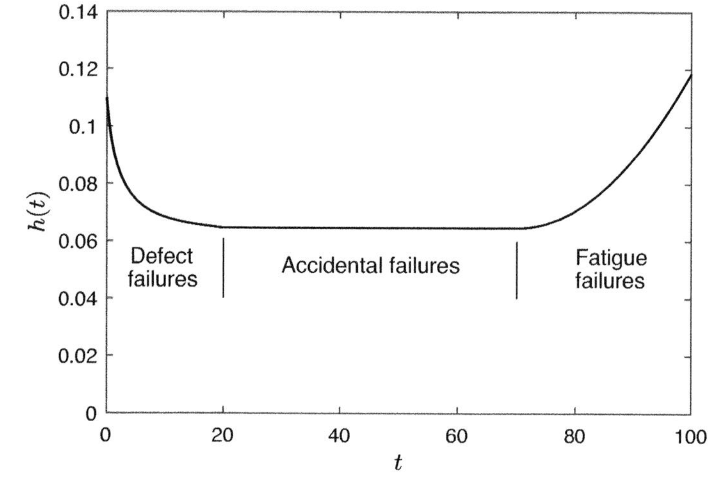

Figure 2.9 The hazard function

Thus, the hazard function offers an alternative way to specify the distribution of time to failure. For example, if the hazard function is taken to be a constant $h(t) = \nu$, the result is the exponential distribution (see Table A2.2). Of course, this model can only depict accidental failures. If the hazard function is taken to be of the form $h(t) = (k/u)(t/u)^{k-1}$ with k and u as positive parameters, then T has the Weibull distribution with CDF $F_T(t) = 1 - \exp\left[-(t/u)^k\right]$. Note that, depending on the value of parameter k, this model can depict all three phases of the bath-tub-shaped hazard curve. Specifically, for $0 < k < 1$ it depicts the early phase, for $k = 1$ it depicts the constant middle phase, and for $1 < k$ it depicts the "old-age" phase. More elaborate models for the hazard function can be found in the literature on classical reliability theory.

2.6 Multiple Random Variables

In most problems, one has to deal with more than one random variable. It is then necessary to define the random variables not only individually but also collectively in order to describe their possible inter-dependencies. In this section we discuss the rules that govern assignment of probability distributions to multiple random variables. As a compact notation, we define $\mathbf{X} = [X_1, \ldots, X_n]^{\mathrm{T}}$ as a vector of n random variables and $\mathbf{x} = [x_1, \ldots, x_n]^{\mathrm{T}}$ as its outcome.

For an n-vector of discrete random variables, we define the *joint PMF* as

$$p_{\mathbf{X}}(\mathbf{x}) = \Pr(X_1 = x_1 \cap \cdots \cap X_n = x_n). \tag{2.44}$$

This function has the following properties:

$$0 \leq p_{\mathbf{X}}(\mathbf{x}) \leq 1, \tag{2.45}$$

$$\sum\nolimits_{x_n} p_{X_1 \ldots X_n}(x_1, \ldots, x_n) = p_{X_1 \cdots X_{n-1}}(x_1, \ldots, x_{n-1}), \tag{2.46}$$

$$\sum\nolimits_{x_1} \cdots \sum\nolimits_{x_n} p_{\mathbf{X}}(\mathbf{x}) = 1. \tag{2.47}$$

The first and third properties are generalizations of the properties described for the PMF of a single random variable. The second condition is known as the *consistency* rule: by summing over all outcomes of one of the random variables, we obtain the joint PMF of the remaining random variables. One can repeatedly apply this rule to eliminate all but one random variable, thereby obtaining the marginal PMF $p_{X_i}(x_i)$ of each of the random variables. It follows that the joint PMF provides complete information about the random variables, including lower-order joint distributions and the marginal distributions.

For a vector of continuous random variables, the *joint PDF* is defined as a function $f_{\mathbf{X}}(\mathbf{x})$ such that

$$f_{\mathbf{X}}(\mathbf{x})\Delta\mathbf{x} = \Pr(x_1 < X_1 \leq x_1 + \Delta x_1 \cap \cdots \cap x_n < X_n \leq x_n + \Delta x_n), \tag{2.48}$$

where $\Delta\mathbf{x} = \Delta x_1 \Delta x_2 \cdots \Delta x_n$. This function has the following properties:

$$0 \leq f_{\mathbf{X}}(\mathbf{x}), \tag{2.49}$$

$$\int_{-\infty}^{+\infty} f_{X_1 \cdots X_n}(x_1, \ldots, x_n)\mathrm{d}x_n = f_{X_1 \cdots X_{n-1}}(x_1, \ldots, x_{n-1}), \tag{2.50}$$

$$\int_{-\infty}^{+\infty} \cdots \int_{-\infty}^{+\infty} f_{\mathbf{X}}(\mathbf{x})\mathrm{d}\mathbf{x} = 1, \tag{2.51}$$

where $\mathrm{d}\mathbf{x} = \mathrm{d}x_1 \cdots \mathrm{d}x_n$. Again, the first and third properties are generalizations of the properties described for the PDF of a single random variable. The second condition is the consistency rule: By integrating over the outcomes of one random variable, we obtain the joint PDF of the remaining random variables. One can repeatedly apply this rule to eliminate all but one random variable, thereby obtaining the marginal PDF $f_{X_i}(x_i)$ of each of the random variables. It follows that the joint PDF provides complete information about the random variables, including lower-order joint PDFs and the marginal distributions.

For the vector of random variables $\mathbf{X}$, the *joint CDF* is defined as

$$F_{\mathbf{X}}(\mathbf{x}) = \Pr(X_1 \leq x_1 \cap \cdots \cap X_n \leq x_n). \tag{2.52}$$

This function has the following properties:

$$0 \leq F_{\mathbf{X}}(\mathbf{x}) \leq 1, \tag{2.53}$$

$$F_{X_1 \cdots X_{n-1} X_n}(x_1, \ldots, x_{n-1}, -\infty) = 0, \tag{2.54}$$

$$F_{X_1 \cdots X_{n-1} X_n}(x_1, \ldots, x_{n-1}, +\infty) = F_{X_1 \cdots X_{n-1}}(x_1, \ldots, x_{n-1}), \tag{2.55}$$

$$F_{X_1 \cdots X_n}(+\infty, \ldots, +\infty) = 1. \tag{2.56}$$

In addition, $F_{\mathbf{X}}(\mathbf{x})$ is a non-decreasing function in each of its arguments. The third line above is a statement of the consistency rule.

For a vector of continuous random variables, the following relations hold between the joint PDF and the joint CDF:

$$F_{\mathbf{X}}(\mathbf{x}) = \int_{-\infty}^{x_1} \cdots \int_{-\infty}^{x_n} f_{\mathbf{X}}(\mathbf{x}) \mathrm{d}\mathbf{x}, \tag{2.57}$$

$$f_{\mathbf{X}}(\mathbf{x}) = \frac{\partial^n F_{\mathbf{X}}(\mathbf{x})}{\partial x_1 \cdots \partial x_n}. \tag{2.58}$$

We are now ready to introduce the concept of conditional probability distribution. For a pair of discrete random variables X_1 and X_2, we define the *conditional PMF* of X_1 given X_2 by

$$p_{X_1|X_2}(x_1|x_2) = \begin{cases} \dfrac{p_{X_1 X_2}(x_1, x_2)}{p_{X_2}(x_2)}, & \text{if } p_{X_2}(x_2) \neq 0 \\ 0, & \text{if } p_{X_2}(x_2) = 0. \end{cases} \tag{2.59}$$

This is a direct application of the conditional probability definition in (2.13). This becomes clear if we note that $p_{X_1 X_2}(x_1, x_2)/p_{X_2}(x_2) = \Pr(X_1 = x_1 \cap X_2 = x_2)/\Pr(X_2 = x_2)$, so that the left-hand side can be interpreted as $p_{X_1|X_2}(x_1|x_2) = \Pr(X_1 = x_1 | X_2 = x_2)$. Using (2.59), one can write the joint PMF of the two discrete random variables in the form

$$p_{X_1 X_2}(x_1, x_2) = p_{X_1|X_2}(x_1|x_2) p_{X_2}(x_2) = p_{X_2|X_1}(x_2|x_1) p_{X_1}(x_1), \tag{2.60}$$

where we have considered both possible orderings of the random variables. Thus, in general, the joint PMF is the product of one conditional and one marginal distribution. Two discrete random variables are said to be *statistically independent* if the conditional distribution of one random variable given the other is identical to its marginal distribution, i.e., if

$$p_{X_1|X_2}(x_1|x_2) = p_{X_1}(x_1) \tag{2.61}$$

for all values of x_1 and x_2. Using this relation in (2.60), we find that, for statistically independent random variables, the joint PMF equals the product of the marginal PMFs:

$$p_{X_1X_2}(x_1, x_2) = p_{X_1}(x_1)p_{X_2}(x_2). \tag{2.62}$$

In fact, for two random variables, (2.62) is sufficient to verify statistical independence.

The concept of conditional PMF can now be extended to a vector of random variables. Consider the n-vector of discrete random variables $\mathbf{X} = [X_1, \ldots, X_k, X_{k+1}, \ldots, X_n]^{\mathrm{T}}$. Conditioning the first k variables on the last $n - k$ variables, we can write

$$p_{X_1\cdots X_k|X_{k+1}\cdots X_n}(x_1, \ldots, x_k|x_{k+1}, \ldots, x_n) = \begin{cases} \dfrac{p_{X_1\cdots X_n}(x_1, \ldots, x_n)}{p_{X_{k+1}\cdots X_n}(x_{k+1}, \ldots, x_n)}, & \text{if } p_{X_{k+1}\cdots X_n}(x_{k+1}, \ldots, x_n) \neq 0, \\ 0, & \text{if } p_{X_{k+1}\cdots X_n}(x_{k+1}, \ldots, x_n) = 0. \end{cases} \tag{2.63}$$

This rule can be used to write the joint PMF as a product of conditional and marginal PMFs. For example, repeatedly applying the rule, one obtains

$$p_{X_1\cdots X_n}(x_1, \ldots, x_n) = p_{X_1|X_2\cdots X_n}(x_1|x_2, \ldots, x_n)p_{X_2|X_3\cdots X_n}(x_2|x_3, \ldots, x_n)\cdots p_{X_{n-1}|X_n}(x_{n-1}|x_n)p_{X_n}(x_n). \tag{2.64}$$

Of course, there are $n!$ ways of sequencing the random variables. The set of discrete random variables is said to be statistically independent if the equality

$$p_{X_1\cdots X_n}(x_1, \ldots, x_n) = p_{X_1}(x_1)\cdots p_{X_n}(x_n) \tag{2.65}$$

holds for all values of $x_1, \ldots, x_n$.

We now extend the above concepts to continuous random variables. For a pair of continuous random variables X_1 and X_2, we define the *conditional PDF* of X_1 given X_2 by

$$f_{X_1|X_2}(x_1|x_2) = \begin{cases} \dfrac{f_{X_1X_2}(x_1, x_2)}{f_{X_2}(x_2)}, & \text{if } f_{X_2}(x_2) \neq 0, \\ 0, & \text{if } f_{X_2}(x_2) = 0. \end{cases} \tag{2.66}$$

This definition is also a result of the fundamental definition in (2.13). To see this, multiply both sides of (2.66) by Δx_1 and, additionally, the numerator and denominator on the right side by Δx_2. The right-hand side then is equivalent to $f_{X_1X_2}(x_1, x_2)\Delta x_1 \Delta x_2 / f_{X_2}(x_2)\Delta x_2 = \Pr(x_1 < X_1 \leq x_1 + \Delta x_1 \cap x_2 < X_2 \leq x_2 + \Delta x_2)/\Pr(x_2 < X_2 \leq x_2 + \Delta x_2)$. Thus, the left-hand side can be interpreted as $f_{X_1|X_2}(x_1|x_2)\Delta x_1 = \Pr(x_1 < X_1 \leq x_1 + \Delta x_1 | x_2 < X_2 \leq x_2 + \Delta x_2)$. As Δx_2 can be arbitrarily small, we can further interpret the last expression as $f_{X_1|X_2}(x_1|x_2)\Delta x_1 = \Pr(x_1 < X_1 \leq x_1 + \Delta x_1 | X_2 = x_2)$. So, when multiplied by Δx_1, the conditional PDF represents the probability that X_1 lies in the interval $(x_1, x_1 + \Delta x_1]$ given that $X_2 = x_2$.

Using (2.66), the joint PDF of two random variables can be written as the product of a conditional and a marginal PDF:

$$f_{X_1X_2}(x_1, x_2) = f_{X_1|X_2}(x_1|x_2)f_{X_2}(x_2) = f_{X_2|X_1}(x_2|x_1)f_{X_1}(x_1). \tag{2.67}$$

Two continuous random variables are said to be statistically independent when the conditional PDF of one variable given the other is identical to its marginal PDF, i.e.,

$$f_{X_1|X_2}(x_1|x_2) = f_{X_1}(x_1). \tag{2.68}$$

This is equivalent to the condition

$$f_{X_1X_2}(x_1, x_2) = f_{X_1}(x_1)f_{X_2}(x_2). \tag{2.69}$$

It is only in the case of statistically independent random variables that the joint PDF is the product of the two marginals. When the variables are not statistically independent, one of the relations in (2.67) can be used to write the joint PDF as a product of two PDFs, one conditional and one marginal.

The above definitions are now extended to a vector of continuous random variables. Following the steps used for discrete random variables, the conditional joint PDF of random variables $X_1, \cdots, X_k$ given random variables $X_{k+1}, \ldots, X_n$ is defined by

$$f_{X_1\cdots X_k|X_{k+1}\cdots X_n}(x_1, \ldots, x_k|x_{k+1}, \ldots x_n)$$
$$= \begin{cases} \dfrac{f_{X_1\cdots X_n}(x_1, \ldots, x_n)}{f_{X_{k+1}\cdots X_n}(x_{k+1}, \ldots x_n)}, & \text{if } f_{X_{k+1}\cdots X_n}(x_{k+1}, \ldots x_n) \neq 0, \\ 0, & \text{if } f_{X_{k+1}\cdots X_n}(x_{k+1}, \ldots x_n) = 0. \end{cases} \tag{2.70}$$

This relation can be used to write the joint PDF of a set of continuous random variables as a product of conditional and marginal PDFs. For example, repeatedly applying the rule, one can write

$$f_{X_1\cdots X_n}(x_1, \ldots, x_n)$$
$$= f_{X_1|X_2\cdots X_n}(x_1|x_2, \ldots, x_n)f_{X_2|X_3\cdots X_n}(x_2|x_3, \ldots, x_n)\cdots f_{X_{n-1}|X_n}(x_{n-1}|x_n)f_{X_n}(x_n). \tag{2.71}$$

This is the counterpart of (2.64) for continuous random variables. The set of n continuous random variables is said to be statistically independent if the equality

$$f_{X_1\cdots X_n}(x_1, \ldots, x_n) = f_{X_1}(x_1)\cdots f_{X_n}(x_n) \tag{2.72}$$

holds for all values of $x_1, \ldots, x_n$.

In the remainder of this chapter, we will discuss continuous random variables only, because problems in structural reliability mostly involve such variables. Furthermore, in most cases the reader can easily extend any new concepts introduced for continuous random variables to the case of discrete random variables. In most cases, integrals will have to be replaced by summations.

2.7 Expectation and Moments

Consider the function $g(\mathbf{X})$ of the vector of random variables $\mathbf{X} = [X_1, \ldots, X_n]^{\mathrm{T}}$ with joint PDF $f_{\mathbf{X}}(\mathbf{x})$. We define the *expectation* of $g(\mathbf{X})$ by

$$\mathrm{E}[g(\mathbf{X})] = \int_{-\infty}^{+\infty} \cdots \int_{-\infty}^{+\infty} g(\mathbf{x}) f_{\mathbf{X}}(\mathbf{x}) \mathrm{d}\mathbf{x}. \tag{2.73}$$

Essentially, the expectation of the function is the weighted average of the function values over all outcomes of the random variables with the weights being the corresponding probabilities.

Expectation is a linear operation on the function $g(\mathbf{X})$. As such, its order can be exchanged with other linear operations, as long as individual limits exist. In particular, the order of the expectation operation can be exchanged with that of a summation, i.e., if the function of interest is a sum of several functions, then the expectation of the total function is the sum of expectations of the individual functions, thus:

$$\mathrm{E}\left[\sum_i g_i(\mathbf{X})\right] = \sum_i \mathrm{E}[g_i(\mathbf{X})]. \tag{2.74}$$

Computation of the expectation of a general multivariate function can be quite difficult because of the required multifold integration. Below, we focus on specific forms of the function, which carry useful information about the random variable(s).

Consider the function $g(\mathbf{X}) = X_i^m$. Owing to the consistency rule, the multifold integral in (2.73) reduces to

$$\mathrm{E}\left[X_i^m\right] = \int_{-\infty}^{+\infty} x_i^m f_{X_i}(x_i) \mathrm{d}x_i. \tag{2.75}$$

Where $\mathrm{E}\left[X_i^m\right]$ is known as the mth *moment* of X_i. In particular, $\mathrm{E}[X_i] = \mu_i$ is the *mean* and $\mathrm{E}\left[X_i^2\right]$ is the *mean-square* of X_i. It should be clear that these expectations represent the geometric moments of the PDF with respect to the zero axis. As the area underneath the PDF is unity, the mean represents the centroidal distance of the PDF – a measure of the center of the PDF relative to the zero axis – and the mean-square represents the moment of inertia of the PDF. Other central measures of the distribution are the *median*, which is the threshold at 50% cumulative probability and is written as $x_{0.5} = F_X^{-1}(0.5)$, where the superposed -1 indicates the inverse function, and the *mode*, which is the outcome at which the PDF is maximum, i.e., $\tilde{x} = \arg\max f_X(x)$.

Now consider the function $g(\mathbf{X}) = (X_i - \mu_i)^m$. The corresponding expectation

$$\mathrm{E}[(X_i - \mu_i)^m] = \int_{-\infty}^{+\infty} (x_i - \mu_i)^m f_{X_i}(x_i) \mathrm{d}x_i \tag{2.76}$$

is known as the mth *central moment of* X_i. One can easily verify that the first ($m = 1$) central moment is identical to zero. The second central moment is the *variance*, $\mathrm{E}\left[(X_i - \mu_i)^2\right] = \mathrm{Var}[X_i] = \sigma_i^2$, and its square root, σ_i, is the *standard deviation*. The latter, which is analogous to the radius of gyration of the PDF around its centroidal axis, is a measure of dispersion or spread of the PDF. As such, it is a measure of the extent of randomness or uncertainty in the random variable. A dimensionless measure of uncertainty is the *coefficient of variation* (c.o.v.), defined as $\delta_i = \sigma_i / |\mu_i|$. Using the property in (2.74), one can easily show that

$$\mathrm{Var}[X_i] = \mathrm{E}\left[X_i^2\right] - \mathrm{E}^2[X_i]. \tag{2.77}$$

That is, the variance of a random variable is equal to its mean-square minus the square of the mean. This expression is useful in computing the variance because often it is simpler to compute moments than central moments.

Next consider the function $g(\mathbf{X}) = X_i X_j$. The corresponding expectation, $\mathrm{E}\left[X_i X_j\right]$, is known as the *mean of product*. The expectation of the central product function, $g(\mathbf{X}) = (X_i - \mu_i)(X_j - \mu_j)$, defines the *covariance* of the two random variables,

$$\mathrm{E}\left[(X_i - \mu_i)(X_j - \mu_j)\right] = \mathrm{Cov}\left[X_i, X_j\right]. \tag{2.78}$$

Similarly to the relation in (2.77), one can easily show that

$$\mathrm{Cov}\left[X_i, X_j\right] = \mathrm{E}\left[X_i X_j\right] - \mathrm{E}[X_i]\mathrm{E}\left[X_j\right]. \tag{2.79}$$

That is, the covariance equals the mean of the product minus the product of the means. It should be clear that the covariance is a generalization of the concept of variance. If one replaces j with i, (2.79) reduces to (2.77).

Normalizing the covariance by the product of the two standard deviations, we obtain the dimensionless quantity known as the *correlation coefficient*

$$\rho_{ij} = \frac{\mathrm{Cov}\left[X_i, X_j\right]}{\sigma_i \sigma_j}. \tag{2.80}$$

This also yields the expression $\mathrm{Cov}\left[X_i, X_j\right] = \rho_{ij}\sigma_i\sigma_j$ for the covariance. One can show that $-1 \leq \rho_{ij} \leq 1$. The covariance and correlation coefficient provide a measure of linear dependence between two random variables. If X_i and X_j are linearly dependent, i.e., if one is a linear function of the other, then $\rho_{ij} = 1$ if the relation between the two random variables is increasing, i.e., $0 < \mathrm{d}X_i/\mathrm{d}X_j$, and $\rho_{ij} = -1$ if the relation is decreasing, i.e., $\mathrm{d}X_i/\mathrm{d}X_j < 0$. In all other cases, $-1 < \rho_{ij} < 1$. Figure 2.10 depicts scatter plots of data generated from several bivariate PDFs with the corresponding correlation coefficient values. Observe that when the data line up along a linear trend, the correlation coefficient is close to 1 if the trend has a positive slope and close to -1 if the trend has a negative slope.

There are two cases in Figure 2.10 in which the correlation coefficient is 0. In one, the data show no trend. In this case, the random variables happen to be statistically independent, in which case $\mathrm{E}\left[X_i X_j\right] = \mathrm{E}[X_i]\mathrm{E}\left[X_j\right]$ and $\mathrm{Cov}\left[X_i, X_j\right] = 0$. In the other, the data exhibit a strong nonlinear trend, but the trend has no linear component (for any given x_2 there is a pair of symmetric outcomes x_1) and again $\mathrm{Cov}\left[X_i, X_j\right] = 0$. It should be clear from the latter case that zero correlation does not necessarily imply statistical independence.

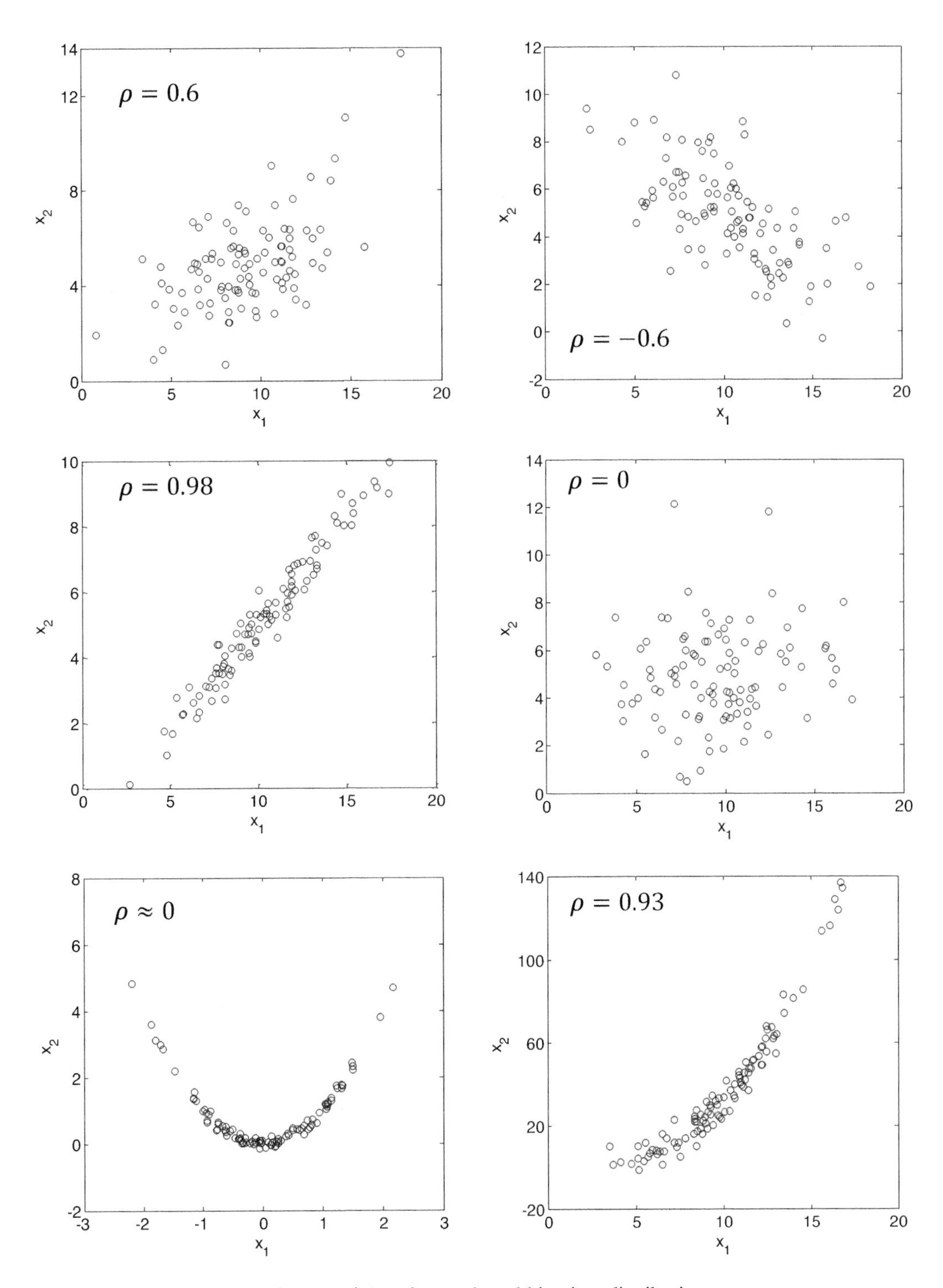

Figure 2.10 Correlation coefficients of data from selected bivariate distributions

The expectation quantities described above are collectively known as the second moments of the random variables. For the n-vector of random variables $\mathbf{X}$, these are collected into a *mean vector* $\mathbf{M_X}$ and *covariance matrix* $\mathbf{\Sigma_{XX}}$ as follows,

$$\mathbf{M_X} = \mathrm{E}[\mathbf{X}] = \begin{Bmatrix} \mu_1 \\ \vdots \\ \mu_n \end{Bmatrix}, \mathbf{\Sigma_{XX}} = \begin{bmatrix} \mathrm{Var}[X_1] & \cdots & \mathrm{Cov}[X_1, X_n] \\ \vdots & \ddots & \vdots \\ \mathrm{Cov}[X_n, X_1] & \cdots & \mathrm{Var}[X_n] \end{bmatrix}, \tag{2.81}$$

where the notation $\mathrm{E}[\mathbf{X}]$ implies taking expectation of each element of the vector. Furthermore, we define the *diagonal matrix of standard deviations* and the *correlation matrix* as follows:

$$\mathbf{D_X} = \begin{bmatrix} \sigma_1 & \cdots & 0 \\ \vdots & \ddots & \vdots \\ 0 & \cdots & \sigma_n \end{bmatrix}, \mathbf{R_{XX}} = \begin{bmatrix} \rho_{11} & \cdots & \rho_{1n} \\ \vdots & \ddots & \vdots \\ \rho_{n1} & \cdots & \rho_{nn} \end{bmatrix}. \tag{2.82}$$

It is easy to verify that the covariance matrix can be written in any of the following matrix forms

$$\begin{aligned} \mathbf{\Sigma_{XX}} &= \mathrm{E}\left[(\mathbf{X} - \mathbf{M_X})(\mathbf{X} - \mathbf{M_X})^{\mathrm{T}}\right] \\ &= \mathrm{E}\left[\mathbf{X}\mathbf{X}^{\mathrm{T}}\right] - \mathrm{E}[\mathbf{X}]\mathrm{E}\left[\mathbf{X}^{\mathrm{T}}\right] \\ &= \mathbf{D_X}\mathbf{R_{XX}}\mathbf{D_X}. \end{aligned} \tag{2.83}$$

Furthermore, one can show that $\mathbf{\Sigma_{XX}}$ and $\mathbf{R_{XX}}$ are not only symmetric but also positive definite provided there is no linear relation between the random variables. These matrices are singular when there is a linear relation between any subset of the random variables.

2.8 Distribution of Functions of Random Variables

In most engineering problems, system performance criteria are expressed as functions of random variables. For example, the displacement at a point in a structural system is a function of its material and geometric properties and the applied loads, which may all be uncertain. Naturally, a function of random variables is itself random. Thus, given the statistical properties of the random variables, we are interested in determining the corresponding properties of the function of random variables. In this section we derive formulas for the joint distribution of a set of functions of random variables. The following section describes formulas for computing the second moments of functions of random variables.

Consider $m \leq n$ functions of the vector of random variables $\mathbf{X} = [X_1, \ldots, X_n]^{\mathrm{T}}$,

$$Y_k = g_k(X_1, \ldots, X_n), k = 1, \ldots, m, \tag{2.84}$$

and assume the joint PDF $f_{\mathbf{X}}(\mathbf{x})$ is known. Our task is to derive the joint PDF $f_{\mathbf{Y}}(\mathbf{y})$ of the vector of transformed random variables $\mathbf{Y} = [Y_1, \ldots, Y_m]^{\mathrm{T}}$. We first consider the case $m = n$ and assume that there is one-to-one mapping between the outcome spaces of the two sets of random variables. That means for every $\mathbf{x}$ there is one and only one $\mathbf{y}$, and vice versa. In that case, it should be clear that the probability in a small neighborhood of an outcome $\mathbf{x}$ should map onto a small neighborhood of the corresponding point $\mathbf{y}$ in the transformed space. That is, we should have

$$f_{\mathbf{Y}}(\mathbf{y})\mathrm{d}\mathbf{y} = f_{\mathbf{X}}(\mathbf{x})\mathrm{d}\mathbf{x}. \tag{2.85}$$

What we need to determine is the size of the "neighborhood" $\mathrm{d}\mathbf{y}$ for a given $\mathrm{d}\mathbf{x}$. It is well known from the theory of transformations that the ratio $\mathrm{d}\mathbf{y}/\mathrm{d}\mathbf{x}$ is equal to the determinant of the Jacobian of the $\mathbf{x} \to \mathbf{y}$ transformation, i.e., the determinant of the matrix $\mathbf{J}_{\mathbf{y},\mathbf{x}}$ having the elements $J_{ki} = \partial y_k / \partial x_i, k, i = 1, \ldots, n$. Note that, in general, the Jacobian may depend on $\mathbf{x}$. Thus, the solution to our problem is

$$f_{\mathbf{Y}}(\mathbf{y}) = f_{\mathbf{X}}(\mathbf{x}) \left| \det \mathbf{J}_{\mathbf{y},\mathbf{x}} \right|^{-1}, \tag{2.86}$$

where we have used the absolute value of the determinant because the probability density must be non-negative. Any $\mathbf{x}$ appearing in the right-hand side must be expressed in terms of $\mathbf{y}$ using the inverse of the system of equations (2.84). In the special case when $m = n = 1$, the formula simplifies to

$$f_Y(y) = f_X(x) \left| \frac{\mathrm{d}x}{\mathrm{d}y} \right|. \tag{2.87}$$

Example 2.5 – Relation between Normal and Lognormal Distributions

Suppose X has the lognormal distribution with parameters λ and ζ so that $f_X(x) = \left(\sqrt{2\pi}\zeta x\right)^{-1} \exp\left[-(\ln x - \lambda)^2/2\zeta^2\right]$ (see Table A2.2). We wish to determine the distribution of $Y = \ln X$. First observe that $\mathrm{d}y/\mathrm{d}x = 1/x$ and $\mathrm{d}x/\mathrm{d}y = x$. Using (2.87), we have

$$\begin{aligned} f_Y(y) &= \frac{1}{\sqrt{2\pi}\zeta x} \exp\left[-\frac{(\ln x - \lambda)^2}{2\zeta^2}\right] |x| \\ &= \frac{1}{\sqrt{2\pi}\zeta} \exp\left[-\frac{(y - \lambda)^2}{2\zeta^2}\right]. \end{aligned} \tag{E1}$$

It is evident that the distribution of Y is normal with mean λ and standard deviation ζ. Thus, the logarithm of a lognormally distributed random variable is normal – a fact that explains the name of the distribution. It should also be clear that λ and ζ, respectively, represent the mean and standard deviation of $\ln X$, i.e., $\lambda = \mu_{\ln X}$ and $\zeta = \sigma_{\ln X}$.

Example 2.6 – Linear Functions of Jointly Normal Random Variables

Consider the linear function of random variables

$$\mathbf{Y} = \mathbf{AX} + \mathbf{B}, \tag{E1}$$

where $\mathbf{X}$ and $\mathbf{Y}$ are $n \times 1$ vectors of random variables, $\mathbf{A}$ an $n \times n$ nonsingular matrix of deterministic coefficients, and $\mathbf{B}$ an $n \times 1$ vector of deterministic constants. Let $\mathbf{X}$ have the joint normal distribution with mean vector $\mathbf{M_X}$ and covariance matrix $\mathbf{\Sigma_{XX}}$. As described in Chapter 3, the joint normal PDF is defined by

$$f_{\mathbf{X}}(\mathbf{x}) = \frac{1}{(2\pi)^{n/2}(\det \mathbf{\Sigma_{XX}})^{1/2}} \exp\left[-\frac{1}{2}(\mathbf{x} - \mathbf{M_X})^{\mathrm{T}}\mathbf{\Sigma_{XX}^{-1}}(\mathbf{x} - \mathbf{M_X})\right]. \tag{E2}$$

We wish to determine the distribution of $\mathbf{Y}$.

Because $\mathbf{A}$ is nonsingular, we have $\mathbf{X} = \mathbf{A}^{-1}(\mathbf{Y} - \mathbf{B})$ as the inverse of (E1). Furthermore, $\mathbf{J_{y,x}} = \mathbf{A}$ and $\mathbf{J_{x,y}} = \mathbf{A}^{-1}$. Thus,

$$\begin{aligned} f_{\mathbf{Y}}(\mathbf{y}) &= \frac{1}{(2\pi)^{n/2}(\det \mathbf{\Sigma_{XX}})^{1/2}} \\ &\quad \times \exp\left[-\frac{1}{2}(\mathbf{A}^{-1}(\mathbf{y} - \mathbf{B}) - \mathbf{M_X})^{\mathrm{T}}\mathbf{\Sigma_{XX}^{-1}}(\mathbf{A}^{-1}(\mathbf{y} - \mathbf{B}) - \mathbf{M_X})\right] \times |\det \mathbf{A}|^{-1} \\ &\propto \exp\left[-\frac{1}{2}(\mathbf{y} - \mathbf{B} - \mathbf{AM_X})^{\mathrm{T}}\mathbf{A}^{-\mathrm{T}}\mathbf{\Sigma_{XX}^{-1}}\mathbf{A}^{-1}(\mathbf{y} - \mathbf{B} - \mathbf{AM_X})\right], \end{aligned} \tag{E3}$$

where the superscript $-\mathrm{T}$ denotes the transpose of the inverse and we have dropped the normalizing factor of the distribution. Comparing (E2) and (E3), it is evident that $\mathbf{Y}$ has the joint normal distribution with mean vector $\mathbf{M_Y} = \mathbf{B} + \mathbf{AM_X}$ and covariance matrix $\mathbf{\Sigma_{YY}} = \mathbf{A\Sigma_{XX}A^T}$. Hence, linear functions of jointly normal random variables are also jointly normal. This is an important result that plays a crucial role in structural reliability methods.

When the mapping $\mathbf{x} \to \mathbf{y}$ is not one-to-one, then multiple values $\mathbf{x}_i$, $i = 1, \ldots, k$, are associated with the same $\mathbf{y}$. These are the roots of (2.84) for given $\mathbf{y}$. In that case, in the equality in (2.85), we must include the probabilities in the neighborhoods of all these roots and, hence, the right-hand side of (2.85) includes a summation over these roots. For the one-dimensional case in (2.87), the solution takes the form

$$f_Y(y) = \sum_{i=1}^{k} f_X(x_i)\left|\frac{\mathrm{d}x_i}{\mathrm{d}y}\right|, \tag{2.88}$$

where x_i are the roots of the equation $y = g(x)$, which must be expressed in terms of y.

Example 2.7 – Non One-to-One Mapping

Consider the function $Y = X^2$, where X is a normal random variable with zero mean and unit variance (commonly referred to as a *standard normal variable*). For a given outcome y, clearly there are two roots, $x_1 = -\sqrt{y}$ and $x_2 = +\sqrt{y}$. The corresponding derivatives are $\mathrm{d}x_1/\mathrm{d}y = -1/2\sqrt{y}$ and $\mathrm{d}x_2/\mathrm{d}y = +1/2\sqrt{y}$. Thus,

$$\begin{aligned} f_Y(y) &= \sum_{i=1}^{2} \frac{1}{\sqrt{2\pi}} \exp\left(-\frac{x_i^2}{2}\right)\left|\frac{\mathrm{d}x_i}{\mathrm{d}y}\right| \\ &= \frac{1}{\sqrt{2\pi}} \exp\left(-\frac{y}{2}\right)\left|-\frac{1}{2\sqrt{y}}\right| + \frac{1}{\sqrt{2\pi}} \exp\left(-\frac{y}{2}\right)\left|+\frac{1}{2\sqrt{y}}\right| \\ &= \frac{1}{\sqrt{2\pi y}} \exp\left(-\frac{y}{2}\right). \end{aligned} \tag{E1}$$

This is the chi-square distribution with one degree of freedom (see Table A2.2). It is instructive to note that, although Y is clearly dependent on X, the correlation coefficient between the two random variables is 0. This is because $\mathrm{Cov}[X, Y] = \mathrm{E}[XY] - \mathrm{E}[X]\mathrm{E}[Y] = \mathrm{E}[X^3] - \mathrm{E}[X]\mathrm{E}[X^2] = 0 - 0$. So, this is a case of uncorrelated but statistically dependent random variables. The reason for this is that the relation between X and Y lacks a linear component, because for every outcome x there is an outcome $-x$ of equal likelihood that produces the same y.

It is noted that, more generally, the sum of n squared standard normal random variables has the chi-square distribution with n degrees of freedom, which has the PDF $f_Y(y) = 0.5\Gamma^{-1}(n/2)(y/2)^{n/2-1}\exp(-y/2),\ 0 \le y$, where $\Gamma^{-1}(\cdot)$ is the inverse of the gamma function. This result, which is also noted in the footnotes of Table A2.2, plays an important role in several reliability analysis methods described in later chapters of this book.

Now consider the case $m < n$, i.e., we have fewer functions than the dimension of $\mathbf{X}$, but assume that the mapping is one-to-one, i.e., that each outcome $\mathbf{x}$ leads to a unique outcome $\mathbf{y}$. To obtain the solution, we introduce $n - m$ additional functions of the form $Y_k = X_k$, $k = m + 1, \ldots, n$, so that the number of functions equals the number of random variables and we can use (2.86). However, this produces an n-dimensional PDF, including the variables corresponding to the added variables $Y_{m+1}, \ldots, Y_n$. To obtain the PDF for the m variables of interest, we must integrate over the unwanted variables. This is an application of the consistency rule. The end result is

$$f_{\mathbf{Y}}(\mathbf{y}) = \int_{-\infty}^{+\infty} \cdots \int_{-\infty}^{+\infty} f_{\mathbf{X}}(x_1, \ldots, x_m, x_{m+1}, \ldots, x_n) \left|\det \mathbf{J}_{\mathbf{y},\mathbf{x}_m}\right|^{-1} \mathrm{d}x_{m+1} \cdots \mathrm{d}x_n, \tag{2.89}$$

where $\mathbf{J}_{\mathbf{y},\mathbf{x}_m}$ represents the Jacobian of $\mathbf{Y}$ with respect to the first m variables in $\mathbf{X}$. Note that the solution involves $(n - m)$-fold integration over the remaining variables in $\mathbf{X}$. Also note that variables $x_1, \ldots, x_m$ must be expressed in terms of $\mathbf{y}$ and the remaining variables $x_{m+1}, \ldots, x_n$ before integration is performed.

Example 2.8 – Distribution of Sum of Random Variables

Consider the function $Y = X_1 + \cdots + X_n$. This is a case with $m = 1$. The 1×1 Jacobian with respect to X_1 is 1. The corresponding inverse relation is $x_1 = y - x_2 - \cdots - x_n$. Thus,

$$\begin{aligned} f_Y(y) &= \int_{-\infty}^{+\infty} \cdots \int_{-\infty}^{+\infty} f_{\mathbf{X}}(x_1, x_2, \ldots, x_n)|1|^{-1}\mathrm{d}x_2 \cdots \mathrm{d}x_n \\ &= \int_{-\infty}^{+\infty} \cdots \int_{-\infty}^{+\infty} f_{\mathbf{X}}(y - x_2 - \cdots - x_n, x_2, \ldots, x_n)\mathrm{d}x_2 \cdots \mathrm{d}x_n. \end{aligned} \tag{E1}$$

It is clear that the solution requires $(n - 1)$-fold integration, which is not an easy task. Note that the solution remains equally difficult if the random variables are statistically independent, in which case $f_{\mathbf{X}}(y - x_2 - \cdots - x_n, x_2, \ldots, x_n)$ is replaced by $f_{X_1}(y - x_2 - \cdots - x_n)f_{X_2}(x_2) \cdots f_{X_n}(x_n)$. As the first term involves all the integration variables, the integrals cannot be decoupled. This example serves to demonstrate that deriving the PDF of the sum of a set of random variables is a difficult task. However, there are special cases where the distribution of the sum is readily known. One example is the sum of jointly normal random variables, as already demonstrated in Example 2.6. It is noted that there are more efficient methods for determining the distribution of the sum of random variables by use of characteristic functions.

2.9 Second Moments of Functions of Random Variables

We first consider a set of m linear functions of n random variables $\mathbf{X} = [X_1, \ldots, X_n]^{\mathrm{T}}$, defined in a matrix form

$$\mathbf{Y} = \mathbf{AX} + \mathbf{B}, \tag{2.90}$$

where $\mathbf{A}$ is an $m \times n$ matrix of deterministic coefficients and $\mathbf{B}$ an m-vector of deterministic constants. We derive the mean vector and covariance matrix of $\mathbf{Y}$ by applying the expectation operator. Following the rules described in Section 2.7, we have

$$\mathbf{M}_{\mathbf{Y}} = \mathbf{AM}_{\mathbf{X}} + \mathbf{B} \tag{2.91}$$

$$\begin{aligned} \boldsymbol{\Sigma}_{\mathbf{YY}} &= \mathrm{E}\left[(\mathbf{Y} - \mathbf{M}_{\mathbf{Y}})(\mathbf{Y} - \mathbf{M}_{\mathbf{Y}})^{\mathrm{T}}\right] \\ &= \mathrm{E}\left[(\mathbf{AX} - \mathbf{AM}_{\mathbf{X}})(\mathbf{AX} - \mathbf{AM}_{\mathbf{X}})^{\mathrm{T}}\right] \\ &= \mathrm{E}\left[\mathbf{A}(\mathbf{X} - \mathbf{M}_{\mathbf{X}})(\mathbf{X} - \mathbf{M}_{\mathbf{X}})^{\mathrm{T}}\mathbf{A}^{\mathrm{T}}\right] \\ &= \mathbf{A}\boldsymbol{\Sigma}_{\mathbf{XX}}\mathbf{A}^{\mathrm{T}}. \end{aligned} \tag{2.92}$$

In an expanded form, these equations read

$$\mu_{Y_k} = \sum_{i=1}^{n} \mu_{X_i} a_{ki} + b_k, k = 1, \ldots, m, \tag{2.93}$$

$$\begin{aligned} \sigma_{Y_k}^2 &= \sum_{i=1}^{n} \sum_{j=1}^{n} a_{ki} a_{kj} \rho_{X_i X_j} \sigma_{X_i} \sigma_{X_j} \\ &= \sum_{i=1}^{n} a_{ki}^2 \sigma_{X_i}^2 + 2 \sum_{i=1}^{n-1} \sum_{j=i+1}^{n} a_{ki} a_{kj} \rho_{X_i X_j} \sigma_{X_i} \sigma_{X_j}, k = 1, \ldots, m, \end{aligned} \tag{2.94}$$

$$\text{Cov}[Y_k Y_l] = \sum_{i=1}^{n} \sum_{j=1}^{n} a_{ki} a_{lj} \rho_{X_i X_j} \sigma_{X_i} \sigma_{X_j}, k, l = 1, \ldots, m, \tag{2.95}$$

where a_{ki} and b_k are the elements of **A** and **B**, respectively. Equation (2.94) represents the variance of Y_k, which is the kth diagonal element of the covariance matrix. In the second line of that equation, we have separated the contributions from the variances of the random variables X_i and those from the covariances of pairs of random variables X_i and X_j. If the random variables are uncorrelated, only the first, single-sum term of this expression needs to be considered. Note that in formulating the second term in (2.94), advantage is taken of symmetry to reduce the number of terms in the double sum. It is also useful to note that the expression in (2.95) for the covariance reduces to the expression in (2.94) for the variance when $k = l$.

In probabilistic analysis, it is often of interest to rank random variables according to their importance in contributing to the variability in a quantity of interest. We now introduce such importance measures for linear functions of random variables. Of interest is the importance ranking of random variables X_i, $i = 1, \ldots, n$, to the uncertainty in random variable Y_k. Focusing on the first term on the right-hand-side of the second line in (2.94), we see that the direct contribution of random variable X_i to the variance of Y_k is $(a_{ki}\sigma_i)^2$. Hence, we define

$$\text{Imp}(X_i, Y_k) \propto |a_{ki}| \sigma_{X_i} \tag{2.96}$$

as a measure of the importance of random variable X_i in contributing to the variance of Y_k. Observe that the importance of random variable X_i in contributing to the uncertainty in Y_k is not only related to its own uncertainty as represented by the standard deviation σ_{X_i}, but also in the sensitivity of Y_k relative to X_i as represented by the coefficient a_{ki}. More general sensitivity measures are defined by Sobol (2001) and include the contributions from interactions between the random variables.

Next, consider general, nonlinear functions of random variables. If we are able to determine the joint distribution of **Y** then that distribution can be used to compute the required moments. However, this is a difficult task, as it usually requires multifold integration. Instead, here we develop approximate results by use of first- and second-order Taylor series expansions. Expanding (2.84) in multivariate Taylor series around the mean point $\mathbf{M_X}$, we have

$$Y_k = g_k(\mathbf{M_X}) + \nabla_\mathbf{X} g_k(\mathbf{M_X})(\mathbf{X} - \mathbf{M_X}) + \frac{1}{2}(\mathbf{X} - \mathbf{M_X})^\text{T} \mathbf{H}(\mathbf{M_X})(\mathbf{X} - \mathbf{M_X}) + \cdots, \tag{2.97}$$

in which $\nabla_{\mathbf{X}} g_k(\mathbf{M}_{\mathbf{X}}) = [\partial g_k/\partial X_1 \cdots \partial g_k/\partial X_n]$ is the gradient row vector evaluated at the mean point and $\mathbf{H}(\mathbf{M}_{\mathbf{X}})$ is the Hessian matrix with elements $H_{ij} = \partial^2 g_k/\partial X_i \partial X_j$ and evaluated at the same point. Truncating the series after the second term, we obtain a first-order approximation. The resulting expression is linear in $\mathbf{X}$ so the solutions described earlier for linear functions apply. Hence, the first-order approximations of the mean and variance of Y_k are

$$\mu_{Y_k} \cong g_k(\mathbf{M}_{\mathbf{X}}), \tag{2.98}$$

$$\sigma^2_{Y_k} \cong \nabla_{\mathbf{X}} g_k(\mathbf{M}_{\mathbf{X}})\, \mathbf{\Sigma}_{\mathbf{XX}} \nabla_{\mathbf{X}} g_k(\mathbf{M}_{\mathbf{X}})^{\mathrm{T}}. \tag{2.99}$$

Furthermore, using a similar first-order expansion for Y_l, the corresponding approximation of the covariance between Y_k and Y_l is

$$\mathrm{Cov}[Y_k, Y_l] \cong \nabla_{\mathbf{X}} g_k(\mathbf{M}_{\mathbf{X}})\, \mathbf{\Sigma}_{\mathbf{XX}} \nabla_{\mathbf{X}} g_l(\mathbf{M}_{\mathbf{X}})^{\mathrm{T}}. \tag{2.100}$$

Collecting (2.99) and (2.100) for all indices k and l into a matrix, the covariance matrix of $\mathbf{Y}$ can be written in the compact form

$$\mathbf{\Sigma}_{\mathbf{YY}} \cong \mathbf{J}_{\mathbf{Y},\mathbf{X}}(\mathbf{M}_{\mathbf{X}})\, \mathbf{\Sigma}_{\mathbf{XX}}\, \mathbf{J}_{\mathbf{Y},\mathbf{X}}(\mathbf{M}_{\mathbf{X}})^{\mathrm{T}}, \tag{2.101}$$

where $\mathbf{J}_{\mathbf{Y},\mathbf{X}}(\mathbf{M}_{\mathbf{X}})$ is the Jacobian matrix with elements $J_{ki} = \partial g_k/\partial X_i$, $k = 1, \ldots, m$, $i = 1, \ldots, n$, evaluated at the mean point. Note that the kth row of $\mathbf{J}_{\mathbf{Y},\mathbf{X}}(\mathbf{M}_{\mathbf{X}})$ is the gradient vector $\nabla_{\mathbf{X}} g_k(\mathbf{M}_{\mathbf{X}})$. Furthermore, the expanded expressions for the variance and covariance of Y_k are similar to the expressions in (2.94) and (2.95) with a_{ki} replaced by $\partial g_k/\partial X_i$. Moreover, as in (2.96), the first-order importance measure of random variable X_i in contributing to the uncertainty in random variable Y_k is given by

$$\mathrm{Imp}(X_i, Y_k) \propto \left| \frac{\partial g_k}{\partial X_i} \right|_{\mathbf{X}=\mathbf{M}_{\mathbf{X}}} \sigma_{X_i}. \tag{2.102}$$

Again, it is worth noting that the importance of X_i is related not only to its variability but also to the sensitivity of Y_k with respect to X_i as represented by the partial derivative.

If we truncate the Taylor series in (2.97) after the third term, we obtain a second-order approximation of Y_k. Taking the expectation of this expression and noting that the second term is identical to zero, we obtain the following second-order approximation of the mean:

$$\mu_{Y_k} \cong g_k(\mathbf{M}_{\mathbf{X}}) + \frac{1}{2} \sum_{i=1}^{n} \sum_{j=1}^{n} \left(\frac{\partial^2 g_k}{\partial X_i \partial X_j} \right)_{\mathbf{X}=\mathbf{M}_{\mathbf{X}}} \rho_{X_i X_j} \sigma_{X_i} \sigma_{X_j}. \tag{2.103}$$

The corresponding approximations of the variance and covariances involve the third and fourth moments of $\mathbf{X}$, which are not usually available. Because the mean is more important than the variance/covariances, it is good practice to use the second-order approximation of the mean along with the first-order approximation of the variance/covariances. Note, however, that the use of (2.103) requires computing the Hessian.

The accuracy of the first- and second-order approximations of the mean and variance/covariance depends on the strength of nonlinearity of the functions $g_k(\mathbf{X})$ within the range of variability (say mean plus/minus one standard deviation) of the random variables $\mathbf{X}$.

Example 2.9 – First- and Second-Order Approximations of Mean and Standard Deviation

Consider the nonlinear function $Y = \ln X$, where X is a random variable with mean μ_X and standard deviation σ_X. The first-order approximations of the mean and standard deviation of Y are

$$\mu_Y \cong \ln \mu_X, \tag{E1}$$

$$\sigma_Y \cong \frac{1}{\mu_X}\sigma_X = \delta_X, \tag{E2}$$

where δ_X denotes the c.o.v. of X. The second-order approximation of the mean is

$$\mu_Y \cong \ln \mu_X + \frac{1}{2}\left(-\frac{1}{\mu_X^2}\sigma_X^2\right) = \ln \mu_X - \frac{\delta_X^2}{2}. \tag{E3}$$

Now suppose X has the lognormal distribution with parameters λ and ζ. As we have seen in Example 2.5, Y then has the normal distribution with mean λ and standard deviation ζ. Expressing these parameters in terms of the mean and standard deviation of X (see note 20 in Table A2.2), we have the exact results

$$\mu_Y = \lambda = \ln \mu_X - \frac{\zeta^2}{2}, \tag{E4}$$

$$\sigma_Y = \zeta = \sqrt{\ln\left(1 + \delta_X^2\right)}. \tag{E5}$$

Figure 2.11 shows plots of the errors (exact minus approximate values) in the estimates of μ_Y and σ_Y as a function of δ_X. It can be seen that the first-order approximation overestimates the mean, the second-order approximation underestimates the mean but offers a significant improvement over the first-order approximation, and the first-order approximation overestimates the standard deviation. All errors increase with increasing c.o.v. of X. This is because for a larger c.o.v. a larger portion of the nonlinear function contributes to the moments of Y.

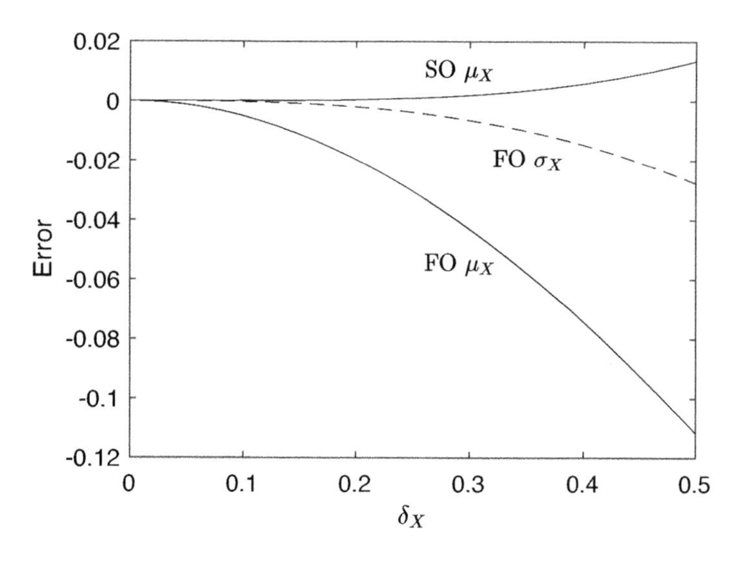

Figure 2.11 Errors in the first- and second-order approximations of the mean (solid lines) and in the first-order approximation of standard deviation (dashed line)

2.10 Extreme-Value Distributions

In some applications, we are interested in the smallest or the largest of a set of random variables. For example, consider a chain of links with random strengths. Under a tensile load, the chain will break at the weakest link. Hence, the strength of the chain is identical to the smallest of the link strengths. For another example, consider the reliability of a bridge against inundation by flooding. Over its lifespan, if it is inundated, the bridge will be inundated during the year with the highest flood level. So, in this case, we are interested in the largest of the annual flood levels during the lifespan of the bridge. In this section we derive exact and asymptotic results for the distribution of these extreme values for the case where the random variables are statistically independent and identically distributed (SIID). The pioneer of this area of study is Gumbel (1958).

Consider the set of n SIID random variables X_i, $i = 1, \ldots, n$, with the common CDF $F_X(x)$ and PDF $f_X(x)$. We define their extreme values as

$$Y_1 = \min(X_1, X_2, \ldots, X_n), \tag{2.104}$$

$$Y_n = \max(X_1, X_2, \ldots, X_n). \tag{2.105}$$

The CDF of Y_1 is derived as follows:

$$\begin{aligned} F_{Y_1}(y) &= 1 - \Pr(y < Y_1) \\ &= 1 - \prod_{i=1}^{n} \Pr(y < X_i) \\ &= 1 - [1 - F_X(y)]^n. \end{aligned} \tag{2.106}$$

In the second line, we have used the fact that if the smallest of n variables is to be greater than y, then each of the variables must be greater than y. The product of the probabilities is a consequence of the assumption of statistical independence of the random variables. The last line is obtained by using the common complementary CDF of variables X_i. In a similar way, the CDF of Y_n is derived as

$$\begin{aligned} F_{Y_n}(y) &= \Pr(Y_n < y) \\ &= \prod_{i=1}^{n} \Pr(X_i < y) \\ &= [F_X(y)]^n. \end{aligned} \tag{2.107}$$

The corresponding PDFs are obtained by differentiation:

$$f_{Y_1}(y) = nf_X(y)[1 - F_X(y)]^{n-1} \tag{2.108}$$

$$f_{Y_n}(y) = nf_X(y)[F_X(y)]^{n-1}. \tag{2.109}$$

The above results are exact, and they are useful when the number of random variables and their common distribution are known. However, in general, numerical integration is necessary to compute the means and standard deviations of these distributions. Later in this section we derive asymptotic distributions for the extreme values as $n \to \infty$.

Example 2.10 – Extreme-Value Distributions for Lognormal Random Variables

Consider the case where SIID random variables X_i, $i = 1, \ldots, n$, have the common lognormal distribution with parameters λ and ζ so that $F_X(x) = \Phi[(\ln x - \lambda)/\zeta]$. Figure 2.12 shows plots

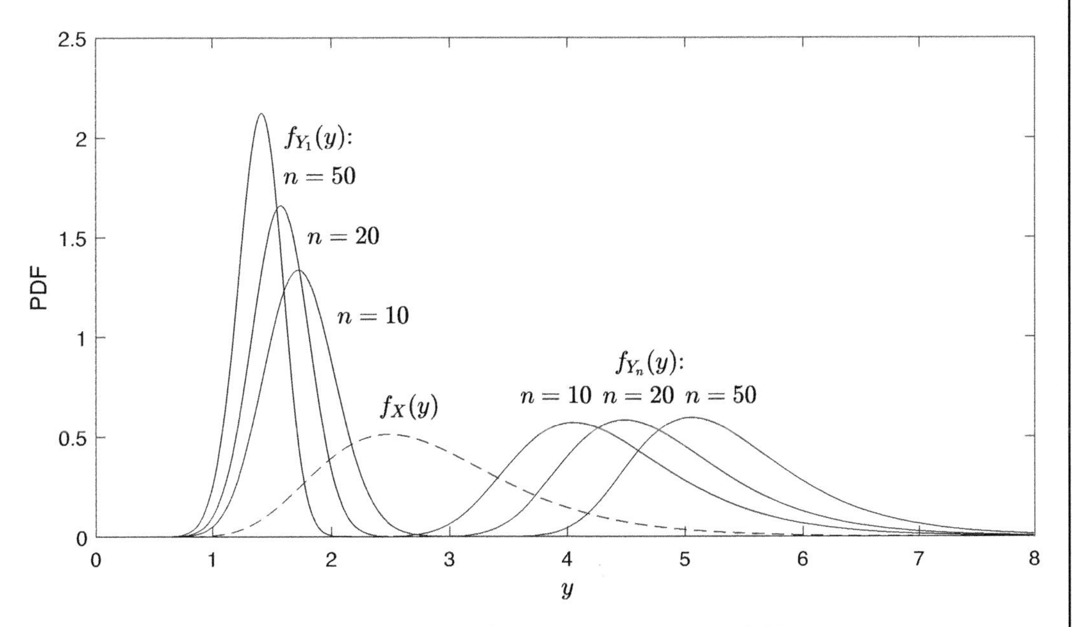

Figure 2.12 Extreme-value distributions for lognormal random variables

Example 2.10 (cont.)

of the PDFs of Y_1 and Y_n together with the common PDF of X_i for $\lambda = 1, \zeta = 0.3$, and $n = 10$, 20, and 50. Observe that the extreme-value distributions shift towards the relevant tail region (the lower tail for Y_1 and the upper tail for Y_n) with increasing n. It follows that, for a large n, the behavior of the relevant tail of the common distribution of X_i has a decisive bearing on the shape of the extreme-value distribution.

As we have seen in Example 2.10, for large n, the distribution of the extreme value lies in the relevant tail region of the underlying distribution $F_X(x)$. Gumbel (1958) showed that, as $n \to \infty$, the extreme-value distributions asymptotically approach one of three types: Type I distribution (also known as the Gumbel distribution), Type II distribution, and Type III distribution.

The Type I distribution is asymptotically approached if the tail of interest is unbounded and has an exponential decay. This behavior is verified by the conditions

$$\text{For } Y_1\text{:} \lim_{x \to -\infty} \frac{\mathrm{d}}{\mathrm{d}x} \left[\frac{F_X(x)}{f_X(x)} \right] = 0, \tag{2.110}$$

$$\text{For } Y_n\text{:} \lim_{x \to \infty} \frac{\mathrm{d}}{\mathrm{d}x} \left[\frac{1 - F_X(x)}{f_X(x)} \right] = 0. \tag{2.111}$$

Observe that the ratios inside the square brackets in both expressions become indefinite, i.e., $0/0$, as x approaches the corresponding extreme tail. The resulting distributions have the CDFs

$$F_{Y_1}(y) = 1 - \exp\{-\exp[\alpha_1(y - u_1)]\}, \tag{2.112}$$

$$F_{Y_n}(y) = \exp\{-\exp[-\alpha_n(y - u_n)]\}, \tag{2.113}$$

where α_1, u_1, and α_n, u_n are the distribution parameters. As shown in Table A2.2, their relations with the means and standard deviations of the two distributions are

$$\mu_{Y_1} = u_1 - \frac{\gamma}{\alpha_1}, \quad \sigma_{Y_1} = \frac{\pi}{\sqrt{6}\alpha_1}, \tag{2.114}$$

$$\mu_{Y_n} = u_n + \frac{\gamma}{\alpha_n}, \quad \sigma_{Y_n} = \frac{\pi}{\sqrt{6}\alpha_n}, \tag{2.115}$$

where $\gamma = 0.577$ is Euler's constant. Furthermore, one can show that u_1 and u_n are the modes of Y_1 and Y_n, respectively. (The derivations of the respective PDFs are zero at these points.) We now show that the asymptotic distribution of the largest extreme approaches (2.113), provided the underlying common distribution satisfies (2.111).

We first define the so-called *characteristic values*

$$u_n = F_X^{-1}\left(1 - \frac{1}{n}\right), \tag{2.116}$$

$$\alpha_n = nf_X(u_n), \tag{2.117}$$

where $F_X^{-1}(\cdot)$ denotes the inverse of the CDF. It should be clear that for large n, u_n lies in the upper tail of the distribution because $(1 - 1/n)$ is close to 1. Thus, u_n provides an anchor in the upper tail region. To understand the behavior of the distribution in the tail, we expand $F_X(x)$ in the Taylor series around u_n:

$$F_X(x) = F_X(u_n) + (x - u_n)f_X(u_n) + \frac{1}{2!}(x - u_n)^2 f'_X(u_n) + \frac{1}{3!}(x - u_n)^3 f''_X(u_n) + \cdots. \tag{2.118}$$

To obtain the derivatives of the PDF, we apply L'Hôpital's rule to the indefinite ratio in (2.111):

$$\lim_{x\to\infty} \frac{1 - F_X(x)}{f_X(x)} = -\frac{f_X(x)}{f'_X(x)}. \tag{2.119}$$

Solving for $f'_X(x)$,

$$\lim_{x\to\infty} f'_X(x) = -\frac{[f_X(x)]^2}{1 - F_X(x)}. \tag{2.120}$$

Taking the derivative of the above expression, we have

$$\begin{aligned} \lim_{x\to\infty} f''_X(x) &= -\frac{2f'_X(x)f_X(x)[1 - F_X(x)] + [f_X(x)]^3}{[1 - F_X(x)]^2} \\ &= \frac{[f_X(x)]^3}{[1 - F_X(x)]^2}, \end{aligned} \tag{2.121}$$

where we have used the relation in (2.120). At $x = u_n$ with large n (u_n in the tail region), using the definitions in (2.116) and (2.117), the above derivatives simplify to $f'_X(u_n) = -\alpha_n^2/n$ and $f''_X(u_n) = \alpha_n^3/n$. Substituting these expressions in (2.118), we have

$$\begin{aligned} F_X(x) &= 1 - \frac{1}{n} + (x - u_n)\frac{\alpha_n}{n} - \frac{1}{2!}(x - u_n)^2\frac{\alpha_n^2}{n} + \frac{1}{3!}(x - u_n)^3\frac{\alpha_n^3}{n} - \cdots, \\ &= 1 - \frac{1}{n}\left[1 - \frac{(x - u_n)\alpha_n}{1!} + \frac{(x - u_n)^2\alpha_n^2}{2!} - \frac{(x - u_n)^3\alpha_n^3}{3!} + \cdots\right] \\ &= 1 - \frac{1}{n}\exp\left[-\alpha_n(x - u_n)\right], \end{aligned} \tag{2.122}$$

where the last line is obtained by observing that the series inside the square brackets in the second line is identical to the Taylor series expansion of the exponential function $\exp\left[-\alpha_n(x - u_n)\right]$. Finally, using the above expression of the underlying CDF in (2.107), we have

$$\lim_{n\to\infty} F_{Y_n}(y) = \lim_{n\to\infty}\left\{1 - \frac{1}{n}\exp\left[-\alpha_n(x - u_n)\right]\right\}^n \\ = \exp\{-\exp\left[-\alpha_n(x - u_n)\right]\}, \tag{2.123}$$

where we have used $\lim_{n\to\infty}(1 - x/n)^n = \exp(-x)$. This proves convergence to the Type I largest-value distribution. We note that common distributions such as the exponential, gamma, normal, and lognormal all satisfy the condition in (2.111) and, therefore, their extreme values follow the Type I largest-value distribution.

A similar derivation leads to the Type I smallest-value distribution defined in (2.112). However, that distribution is seldom used in engineering practice because it assumes an unbounded lower tail, which is not realistic for physical quantities of interest.

Example 2.11 – Comparison of Exact and Asymptotic Distributions for Type I Largest Value

To get an understanding of the nature of the asymptotic distribution, Figure 2.13 compares the exact (solid lines) and asymptotic (dash-dotted lines) PDFs of $Y_n = \max(X_1, X_2, \ldots, X_n)$ for the case of a common underlying lognormal distribution with $\lambda = 1$ and $\zeta = 0.3$ for $n = 10$, 20, and 50, as in Example 2.10. The asymptotic PDF is obtained by differentiating the CDF in (2.123) and using u_n and α_n values computed from (2.116) and (2.117), respectively. Observe that the asymptotic distributions get closer to the exact distributions as n grows, with there being little difference between the two distributions for $n = 50$.

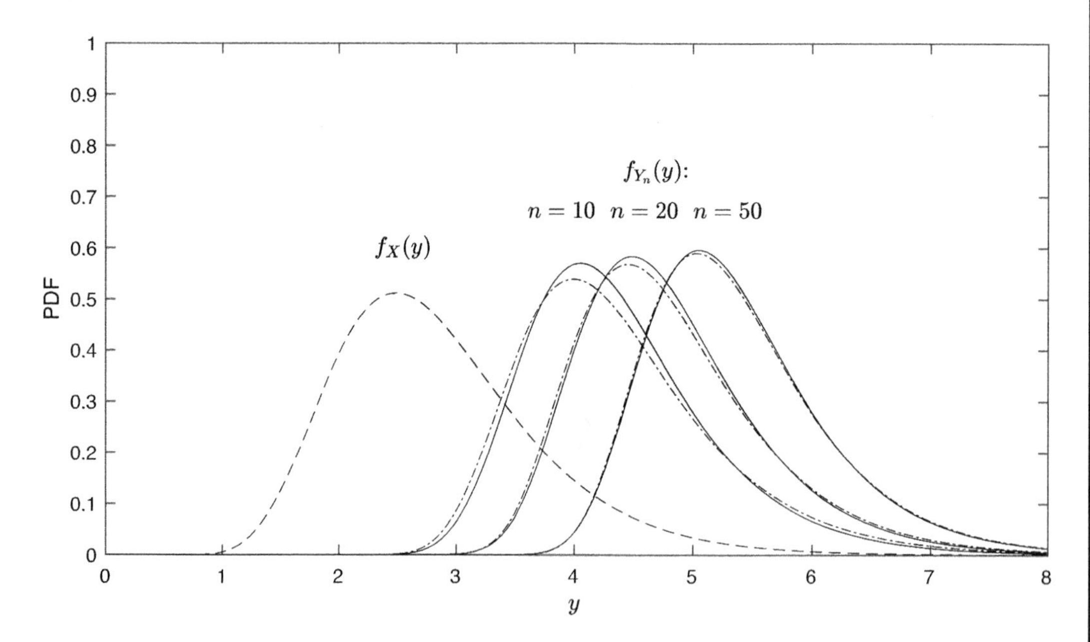

Figure 2.13 Exact (solid lines) and asymptotic (dash-dotted lines) PDFs of Y_n

The Type II asymptotic extreme-value distribution is approached if the tail of interest is unbounded and decays according to a power law. The distribution of the smallest value again is not of interest in engineering because of the requirement for an unbounded lower tail. Hence, we only consider the largest extreme, Y_n. The upper tail has a power decay if it satisfies the condition

$$\lim_{x\to\infty} x^k [1 - F_X(x)] = c, \tag{2.124}$$

where c and k are positive constants. This expression can be reformulated to read

$$\lim_{x\to\infty} F_X(x) = 1 - \frac{c}{x^k}. \tag{2.125}$$

For large n, u_n is in the tail region, so we can write

$$\begin{aligned} F_X(u_n) &= 1 - \frac{1}{n} \\ &\cong 1 - \frac{c}{u_n^k}, \end{aligned} \tag{2.126}$$

which gives $c \cong u_n^k/n$ for large n. Thus, using this relation in (2.125) and substituting the result in (2.107), we have

$$\begin{aligned} \lim_{n\to\infty} F_{Y_n}(y) &= \lim_{n\to\infty} \left[1 - \frac{1}{n}\left(\frac{u_n}{y}\right)^k\right]^n \\ &= \exp\left[-\left(\frac{u_n}{y}\right)^k\right], 0 < y, \end{aligned} \tag{2.127}$$

where we have again used $\lim_{n\to\infty}(1 - x/n)^n = \exp(-x)$. The above defines the Type II largest-value distribution. The mode of this distribution is $\tilde{y} = u_n[k/(k+1)]^{1/k}$, which is slightly different from u_n. See Table A2.2 for other properties of this distribution.

Finally, the Type III asymptotic extreme-value distribution occurs when the tail of interest of the underlying distribution is bounded. This distribution is of interest only for the smallest values. The criterion for the tail behavior is stated as

$$\lim_{x\to\epsilon} (x - \epsilon)^{-k} F_X(x) = c, \tag{2.128}$$

where ϵ is the lower bound and c and k are positive constants. The derivation of the asymptotic distribution is similar to that of the Type II distribution and is not presented here. The final result is

$$F_{Y_1}(y) = 1 - \exp\left[-\left(\frac{y-\epsilon}{u_1-\epsilon}\right)^k\right], \epsilon \le y, \tag{2.129}$$

where $u_1 = F_X^{-1}(1/n)$ is the location parameter and k is a shape parameter. When $\epsilon = 0$, the above distribution reduces to

$$F_{Y_1}(y) = 1 - \exp\left[-\left(\frac{y}{u_1}\right)^k\right], 0 \leq y, \tag{2.130}$$

which is known as the Weibull distribution and is widely used to model the strength of materials. Table A2.2 provides other properties of the Type III smallest-value and Weibull distributions.

Example 2.12 – Comparison of Exact and Asymptotic Distributions for Type III Smallest Value

Consider a chain with n links. Let the SIID variables $X_i, i = 1, \ldots, n$, denote the link strengths. The strength of the chain under a tension load is given by $Y_1 = \min(X_1, X_2, \ldots, X_n)$. Suppose the link strengths have a common gamma distribution with mean 100 and standard deviation 50. From the expressions for the mean and standard deviation of the gamma distribution given in Table A2.2, we find the parameter values $k = 4$ and $\lambda = 0.04$. We wish to compute and compare the exact and asymptotic distributions of Y_1 for $n = 100$ and 1000.

To determine the asymptotic distribution, we need the parameters ϵ, k, and u_1. Because the gamma distribution has a lower bound at zero, $\epsilon = 0$. To determine the shape parameter k, we examine the lower tail of the gamma distribution by expanding the PDF near $x = 0$ while disregarding the normalizing factor:

$$\begin{aligned}\lim_{x\to 0} f_X(x) &\propto x^{k-1}\exp(-\lambda x) \\ &= x^{k-1}\left[1 - \lambda x + \frac{(\lambda x)^2}{2} - \cdots\right].\end{aligned} \tag{E1}$$

In the second line we have expanded the exponential term in Taylor series. It is clear that, as x approaches 0, the leading term of the series is x^{k-1}. Integrating, we find that as $x \to 0$, $F_X(x)$ approaches 0 in proportion to x^k. Hence, the k parameter of the asymptotic distribution is identical to the k parameter of the gamma distribution, i.e., $k = 4$. Finally, we compute the parameter u_1 for each value of n by using the relation $u_1 = F_X^{-1}(1/n)$, where $F_X^{-1}(\cdot)$ is the inverse of the gamma CDF.

Figure 2.14 compares the exact and asymptotic PDFs of the strength of the chain, Y_1. Observe that the PDFs are located in the far lower tail of the gamma distribution of X_i. Note also that the exact and asymptotic distributions are in closer agreement for larger values of n.

Example 2.12 (cont.)

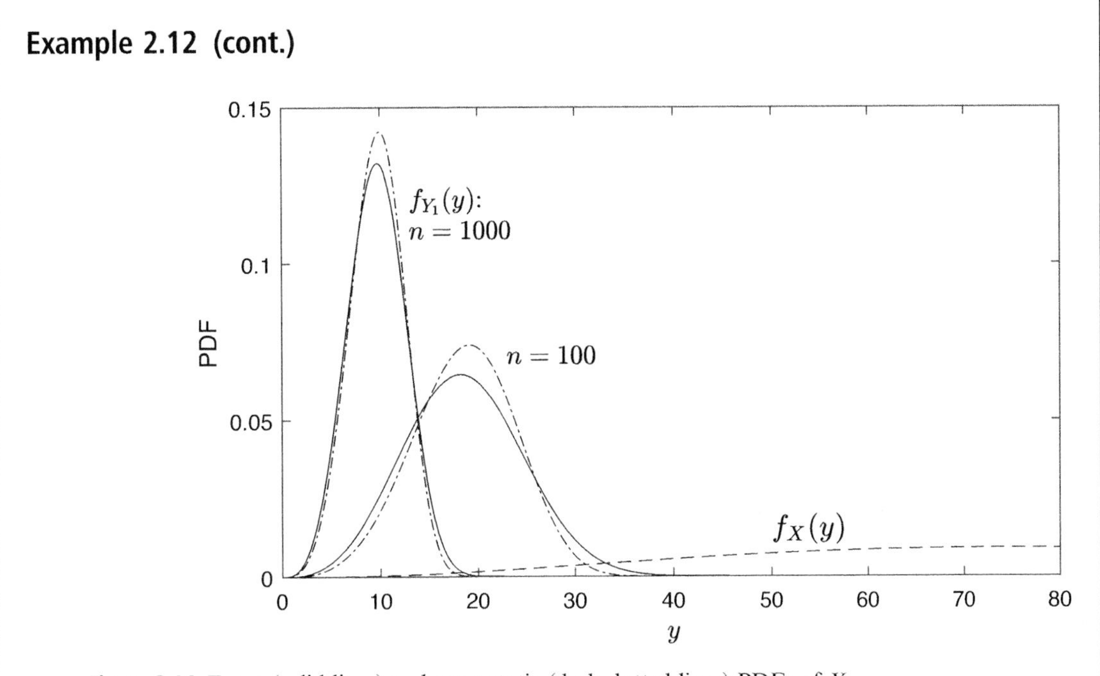

Figure 2.14 Exact (solid lines) and asymptotic (dash-dotted lines) PDFs of Y_1

2.11 Probability Distribution Models

There is a rich collection of models for univariate and multivariate probability distributions. Some of these models arise from mathematical considerations, such as the asymptotic extreme-value distributions described in the preceding section, some arise from statistical data analysis, and others are purely mathematical models that satisfy the conditions imposed on probability distributions. For the sake of brevity, a detailed description of these models is not presented here. Standard books on applied probability theory can be consulted to read these details. Instead, Tables A2.1 and A2.2 summarize the forms of selected distribution models and their essential properties for discrete and continuous random variables, respectively. These tables are useful when the reader needs to know the form of a distribution or the relations between its parameters and the mean and standard deviation. Throughout this book we will make use of these distributions in example problems and will refer to these tables for the necessary formulas. Nevertheless, details on some of the more important distributions, such as the normal and lognormal distributions and the Gaussian and Poisson processes, are presented in later chapters of this book.

APPENDIX 2A

Probability Distribution Models

Table A2.1 presents models for selected discrete random variables. The first column gives the name and short symbol of the distribution together with the list of distribution parameters in parentheses; the second column lists the PMF together with the limit(s) of the variable outcomes; column three specifies the applicable range of parameter values; column four lists the relations between the parameters and the mean and standard deviation; and the final column specifies the numbers for notes below the table that provide clarifying details.

Table A2.2 presents models for selected distributions of continuous random variables. The first column gives the name and short symbol of the distribution together with the list of

Table A2.1 Selected models for discrete random variables

Name and symbol	Probability mass function, $p_X(x)$	Parameter ranges	Mean and standard deviation	Notes
Bernoulli trial, $\mathrm{B}(p)$	$p^x(1-p)^{1-x}, x=0,1$	$0<p<1$	$p,\ \sqrt{p(1-p)}$	1
Binomial, $\mathrm{Bin}(n,p)$	$\binom{n}{x}p^x(1-p)^{n-x}$, $x=0,\ldots,n$	$0<p<1$, n a positive integer	$np,\ \sqrt{np(1-p)}$	2, 3
Geometric, $\mathrm{Geo}(p)$	$p(1-p)^{x-1}, x=1,2,\ldots$	$0<p<1$	$\frac{1}{p},\ \frac{\sqrt{1-p}}{p}$	4
Negative binomial, $\mathrm{NBin}(k,p)$	$\binom{x-1}{k-1}p^k(1-p)^{x-k}$, $x=k,k+1,\ldots$	$0<p<1$, k a positive integer	$\frac{k}{p},\ \frac{\sqrt{k(1-p)}}{p}$	3, 5
Hypergeometric, $\mathrm{HypGeo}(N,n,k)$	$\frac{\binom{k}{x}\binom{N-k}{n-x}}{\binom{N}{n}}$, x integer, $x\le k, n-x\le N-k$	$0<k,n\le N$, k,n,N positive integers	$\frac{nk}{N},\ \sqrt{\frac{nk(N-k)(N-n)}{N^2(N-1)}}$	3, 6
Poisson, $\mathrm{Psn}(v)$	$\frac{v^x}{x!}\exp(-v), x=0,1,\ldots$	$0<v$	$v,\ \sqrt{v}$	7

Notes

1. This is the distribution of a random trial with outcomes $X=1$ ("success") with probability p and $X=0$ ("failure") with probability $1-p$.
2. Binomial is the distribution of the number of "success" outcomes in a sequence of n independent Bernoulli trials.
3. $\binom{n}{x}=\frac{n!}{x!(n-1)!}$ is the binomial coefficient.
4. This is the distribution of the number of independent Bernoulli trials to the first "success" outcome.
5. This is the distribution of the number of independent Bernoulli trials to the kth "success" outcome.
6. Consider N independent trials with k "successes." Suppose we randomly select $n\le N$ of the trials. The hypergeometric distribution describes the number X of "successes" out of the n selections without replacement.
7. $\mathrm{Psn}(\lambda t)$ with $\lambda t=v$ is the homogeneous Poisson process that describes the number of random events happening in time or space interval $(0,t]$ with mean rate λ per unit time or length. The Poisson process is described in Chapter 11.

Table A2.2 Selected models for continuous random variables

Name and symbol	Probability density function, $f_X(x)$, and cumulative distribution function, $F_X(x)$	Parameter ranges	Mean and standard deviation	Note
Uniform, $\text{Uni}(a,b)$	$\frac{1}{b-a},\ \frac{x-a}{b-a},\ a \le x \le b$	$a < b$	$\frac{a+b}{2},\ \frac{b-a}{2\sqrt{3}}$	
Beta, $\text{Beta}(a,b,q,r)$	$\frac{(x-a)^{q-1}(b-x)^{r-1}}{B(q,r)(b-a)^{q+r-1}},\ a \le x \le b$ No closed-form CDF	$a < b$, $0 < q, r$	$\frac{ar+bq}{q+r},\ \frac{b-a}{q+r}\sqrt{\frac{qr}{q+r+1}}$	1, 2
Exponential, $\text{Exp}(\lambda)$	$\lambda \exp(-\lambda x),\ 1-\exp(-\lambda x),\ 0 < x$	$0 < \lambda$	$\frac{1}{\lambda},\ \frac{1}{\lambda}$	3
Gamma, $\text{Gam}(k,\lambda)$	$\frac{\lambda(\lambda x)^{k-1}}{\Gamma(k)}\exp(-\lambda x),\ \frac{\Gamma(k,\lambda x)}{\Gamma(k)},\ 0 < x$	$0 < k, \lambda$	$\frac{k}{\lambda},\ \frac{\sqrt{k}}{\lambda}$	3, 4, 5, 6
Chi-square, $\chi^2(v)$	$\frac{\left(\frac{x}{2}\right)^{\frac{v}{2}-1}}{2\Gamma\left(\frac{v}{2}\right)}\exp\left(-\frac{x}{2}\right),\ \frac{\Gamma(v/2,\,x/2)}{\Gamma(v/2)},\ 0 < x$	$0 < v$	$2v,\ \sqrt{2v}$	4, 6, 7
Rayleigh, $\text{Ray}(u)$	$\frac{2x}{u^2}\exp\left[-\left(\frac{x}{u}\right)^2\right], 1-\exp\left[-\left(\frac{x}{u}\right)^2\right], 0 < x$	$0 < u$	$\frac{\sqrt{\pi}u}{2},\ \frac{\sqrt{4-\pi}u}{2}$	8
Gumbel, $\text{Gmb}(v,\alpha)$	$\alpha \exp\{-\alpha(x-v) - \exp[-\alpha(x-v)]\}$, $\exp\{-\exp[-\alpha(x-v)]\}$,	$0 < \alpha$	$v+\frac{\gamma}{\alpha},\ \frac{\pi}{\sqrt{6}\alpha}$	9, 10
Weibull, $\text{Wbl}(u,k)$	$\frac{k}{u}\left(\frac{x}{u}\right)^{k-1}\exp\left[-\left(\frac{x}{u}\right)^k\right],\ 1-\exp\left[-\left(\frac{x}{u}\right)^k\right],\ 0 < x$	$0 < u, k$	$u\Gamma(1+\frac{1}{k}),\ u\sqrt{\Gamma.(1+\frac{2}{k}) - \Gamma^2(1+\frac{1}{k})}$	4, 11
Type I smallest value, $\text{TIS}(u,\alpha)$	$\alpha \exp\{\alpha(x-u) - \exp[\alpha(x-u)]\}$, $1-\exp[-\exp[\alpha(x-u)]\}$	$0 < \alpha$	$u-\frac{\gamma}{\alpha},\ \frac{\pi}{\sqrt{6}\alpha}$	10
Type II largest value, $\text{TIIL}(u,k)$	$\frac{k}{u}\left(\frac{u}{x}\right)^{k+1}\exp\left[-\left(\frac{u}{x}\right)^k\right],\ \exp\left[-\left(\frac{u}{x}\right)^k\right],\ 0 < x$	$0 < u, k$	$u\Gamma(1-\frac{1}{k}),\ u\sqrt{\Gamma(1-\frac{2}{k}) - \Gamma^2(1-\frac{1}{k})}$	4
Type III smallest value, $\text{TIIIS}(\epsilon,u,k)$	$\frac{k}{u-\epsilon}\left(\frac{x-\epsilon}{u-\epsilon}\right)^{k-1}\exp\left[\left(\frac{x-\epsilon}{u-\epsilon}\right)^k\right],\ 1-\exp\left[\left(\frac{x-\epsilon}{u-\epsilon}\right)^k\right]$, $\epsilon < x$	$0 < u, k$	$\epsilon + (u-\epsilon)\Gamma(1+\frac{1}{k})$, $(u-\epsilon)\sqrt{\Gamma(1+\frac{2}{k}) - \Gamma^2(1+\frac{1}{k})}$	4, 11
Normal, $\text{N}(\mu,\sigma)$	$\frac{1}{\sqrt{2\pi}\sigma}\exp\left[-\frac{1}{2}\left(\frac{x-\mu}{\sigma}\right)^2\right],\ \Phi\left(\frac{x-\mu}{\sigma}\right)$	$0 < \sigma$	$\mu,\ \sigma$	12

Table A2.2 (cont.)

Name and symbol	Probability density function, $f_X(x)$, and cumulative distribution function, $F_X(x)$	Parameter ranges	Mean and standard deviation	Note
Lognormal, $\mathrm{LN}(\lambda,\zeta)$	$\frac{1}{\sqrt{2\pi}\zeta x}\exp\left[-\frac{1}{2}\left(\frac{\ln x-\lambda}{\zeta}\right)^2\right]$, $\Phi\left(\frac{\ln x-\lambda}{\zeta}\right)$, $0<x$	$0<\zeta$	$\exp\left(\lambda+\frac{\zeta^2}{2}\right)$, $\exp\left(\lambda+\frac{\zeta^2}{2}\right)\sqrt{\exp\left(\zeta^2\right)-1}$	12, 13, 14
Student's t, $T(\nu)$	$\frac{\Gamma[(\nu+1)/2]}{\sqrt{\nu\pi}\Gamma(\nu/2)}\left(1+\frac{x^2}{\nu}\right)^{-(\nu+1)/2}$	$0<\nu$	0, $\sqrt{\frac{\nu}{\nu-2}}$ for $2<\nu$	15
Laplace, $\mathrm{Lap}(\alpha,\beta)$	$\frac{\beta}{2}\exp\left(-\beta\lvert x-\alpha\rvert\right)$, $\frac{1}{2}\exp\left(-\beta\lvert x-\alpha\rvert\right)$ for $x\le\alpha$, $1-\frac{1}{2}\exp\left(-\beta\lvert x-\alpha\rvert\right)$ for $\alpha\le x$	$0<\beta$	α, $\frac{\sqrt{2}}{\beta}$	
Pareto, $\mathrm{Par}(k,u)$	$\frac{k}{u}\left(\frac{u}{x}\right)^{k+1}$, $1-\left(\frac{u}{x}\right)^k$, $u<x$	$0<u,k$	$\frac{ku}{k-1}$ for $1<k$, $\sqrt{\frac{k}{k-2}}\frac{u}{k-1}$ for $2<k$, otherwise values are ∞	16

Notes

1. $B(q,r)=\Gamma(q)\Gamma(r)/\Gamma(q+r)$ is the beta function. See note 11 for the definition of $\Gamma(\cdot)$.
2. $\mathrm{Beta}(a,b,1,1)$ is identical to $\mathrm{Uni}(a,b)$.
3. $\mathrm{Exp}(\lambda)$ is identical to $\mathrm{Gam}(1,\lambda)$.
4. $\Gamma(k)=\int_0^\infty u^{k-1}\exp(-u)\mathrm{d}u$ is the gamma function.
5. $\Gamma(k,x)=\int_0^x u^{k-1}\exp(-u)\mathrm{d}u$ is the incomplete gamma function.
6. $\mathrm{Gam}(\nu/2,1/2)$ is identical to $\chi^2(\nu)$.
7. $\chi^2(\nu)$ is the distribution of the sum of ν squared standard normal random variables.
8. $\mathrm{Ray}(u)$ is identical to $\mathrm{Wbl}(u,2)$.
9. $\mathrm{Gmb}(\nu,\alpha)$ is the same as Type I largest-value distribution.
10. $\gamma=0.5772156649$ is Euler's constant.
11. $\mathrm{Wbl}(u,k)$ is identical to $\mathrm{TIIIS}(0,u,k)$.
12. $\Phi(x)=\frac{1}{\sqrt{2\pi}}\int_{-\infty}^x\exp\left(-\frac{x^2}{2}\right)\mathrm{d}x$ is the standard normal CDF.
13. Relations for parameters λ and ζ in terms of mean and standard deviation are $\zeta=\sqrt{\ln\left(1+\delta^2\right)}$ and $\lambda=\ln\mu-\zeta^2/2$, where $\delta=\sigma/\mu$ is the coefficient of variation.
14. If X is $\mathrm{LN}(\lambda,\zeta)$, then $Y=\ln X$ is $\mathrm{N}(\lambda,\zeta)$. Conversely, if X is $\mathrm{N}(\mu,\sigma)$, then $Y=\exp(X)$ is $\mathrm{LN}(\mu,\sigma)$.
15. The $T(\nu)$ distribution arises in statistics as the distribution of $\sqrt{n}(\bar{X}-\mu)/s$, where $\bar{X}$ and s are the sample mean and standard deviation of a sample of size n drawn from a normal population with unknown mean μ and standard deviation σ; $\nu=n-1$ is the *degrees of freedom*. The variance of the distribution is infinite when $\nu\le2$.
16. For $\mathrm{Par}(k,u)$, the mean exists when $0<k$ and the variance exists when $2<k$; otherwise, these values are infinite.

distribution parameters; the second column lists the PDF and the CDF, if available in closed form, together with the limit(s) of the variable outcomes; column three specifies the applicable range of parameter values; column four lists the relations between the parameters and the mean and standard deviation; and the final column specifies numbers for the notes below the table that provide clarifying details.

PROBLEMS

2.1 Let A, B, and C be three arbitrary events. Find symbolic expressions for the events that, of A, B, and C: (a) A occurs; (b) only A occurs; (c) A occurs and either B or C occur, but not both; (d) all three occur; (e) at least two of the three events occur; (f) only two of the three events occur; (g) one or two of the three events occur; (h) only one of the three events occurs; and (i) none of the three events occurs. For each case, show the event in a Venn diagram.

2.2 A weight W, somewhere between 1 and 10 tons, is randomly placed along a beam of 10 m span. Let X denote the distance from the point of application of the load to the *nearest* support (see figure below). One can show that the maximum shear force in the beam caused by this load is $V = W(1 - 0.1X)$ tons and the maximum bending moment is $M = W(1 - 0.1X)X$ tons-m.

1) Sketch the sample space of W and X in a two-dimensional coordinate system. Show the event that the maximum shear force in the beam exceeds 6 tons. Also show the event that the maximum bending moment in the beam exceeds 15 tons-m.
2) Suppose it is known that the shear force does not exceed 4 tons. How does this information alter the sample space? Make a new sketch of the conditioned sample space and show the events described above.

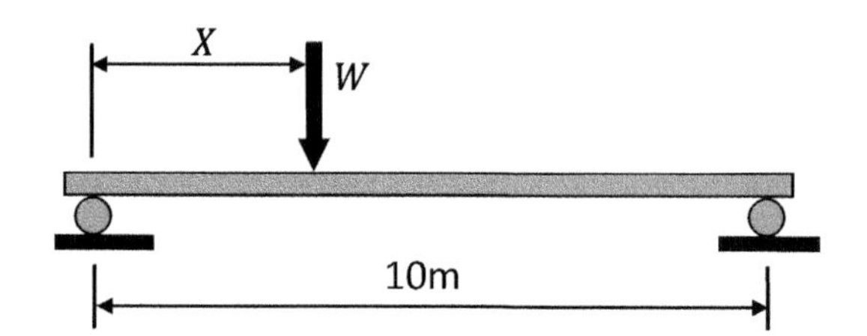

2.3 Prove the De Morgan's rules in (2.7) and (2.8). Hint: First verify the rules for $n = 2$ events by use of a Venn diagram. Then prove that if the rules are valid for k events, they must be valid for $k + 1$ events. Use induction to generalize the rule for $2 < n$.

2.4 A traveler going from Startville to Endville has to go over Road 1 and the Bridge, and either Road 2 or Road 3. Let R_i, $i = 1, 2, 3$, denote the event that Road i is open and $\bar{R}_i$ denote the event that Road i is closed. Likewise, let B denote the event that the Bridge is open and $\bar{B}$ denote the event that the Bridge is closed.

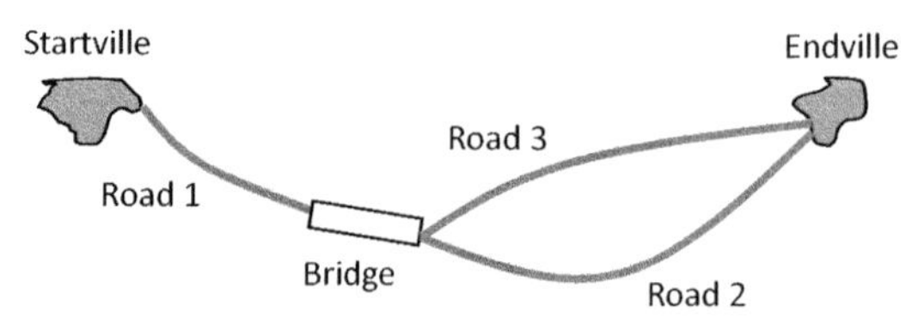

1) Write a symbolic expression in terms of R_i and B for the event that the traveler will be able to reach Endville.

2) Write a symbolic expression in terms of $\bar{R}_i$ and $\bar{B}$ for the event that the traveler will *not* be able to reach Endville.
3) Using De Morgan's rules, show that the above two events are complementary.

2.5 The following probabilities are known for events A, B, and C: $\Pr(A) = 0.30$, $\Pr(B) = 0.60$, $\Pr(C) = 0.10$, $\Pr(A\bar{C}) = 0.25$, $\Pr(B \cup C) = 0.64$, and $\Pr(A\bar{B}|C) = 0.20$. In addition, events A and B are known to be statistically independent.
1) Are events A and C statistically independent? How about events B and C?
2) Determine $\Pr(ABC)$.

2.6 Ice floes in the Alaskan Beaufort Sea are a source of hazard for an offshore platform. In the event of collision, the probability of failure depends on the size S (e.g., diameter) of the floe and is given by

$$\Pr(\text{failure}|S = s) = 0.0025s^2, 0 < s \leq 20 \text{ m}$$
$$= 1, 20 \text{ m} < s.$$

From measured data of many floes, the probability distribution of S is found to be exponential with mean $\mu_S = 2$ m.
1) Determine the probability of failure of the platform if an ice floe of unknown size hits it.
2) Suppose an ice floe of unknown size collides with the platform and the platform survives. What is the PDF of the size of the floe that hit the platform?
3) Suppose an ice floe of unknown size collides with the platform and the platform fails. What is the PDF of the size of the floe that hit the platform?
4) Plot and compare the PDFs of the sizes of a randomly selected floe, a floe that hit the platform but did not fail it, and a floe that hit the platform and failed it.

2.7 Compute and plot the hazard function for the time period 0–10 years for the following distributions of time to failure T:
1) Exponential distribution with mean = 10 years.
2) Gamma distribution with mean = 10 years and standard deviation = 4 years.
3) Lognormal distribution with mean = 10 years and standard deviation = 4 years.
4) Weibull distribution with scale parameters $u = 10$ years and shape parameters $k = 0.75$, 1, and 1.5. Note that the case with $k = 1$ is identical to the exponential distribution.
5) Comment on the appropriateness of each distribution model to describe failures due to manufacturing defects, accidental overloads, and wear and tear.

2.8 Random variables X and Y have the bivariate exponential CDF

$$F_{XY}(x, y) = 1 - \exp(-x) - \exp(-y) + \exp(-x - y - xy), 0 < x, y.$$

Determine:
1) The joint PDF of X and Y.
2) The marginal PDF of X.
3) The conditional PDF of X given $Y = y$.

4) The probability that $4 < X$ given that $Y = 2$.
5) The probability that $4 < X$ given that $Y \leq 2$.
6) The means, standard deviations, and correlation coefficient of X and Y. You may use numerical integration to compute the correlation coefficient.

2.9 Let U_1 and U_2 be statistically independent random variables having the standard uniform distribution, i.e., with PDFs $f_{U_i}(u) = 1, 0 \leq u \leq 1, i = 1, 2$. Show that random variables X_1 and X_2 defined by the relations

$$X_1 = \sqrt{-2 \ln U_1} \sin(2\pi U_2)$$

$$X_2 = \sqrt{-2 \ln U_1} \cos(2\pi U_2)$$

have the standard, uncorrelated normal distribution. (Note: These relations are used to simulate normally distributed random variables. One first generates pairs of uniformly distributed random numbers within the range [0,1] and then uses the above relations to compute the corresponding standard normal random variables. See Chapter 10.)

2.10 Consider the linear transform $\mathbf{Y} = \mathbf{AX} + \mathbf{B}$, where $\mathbf{X} = [X_1, X_2, X_3]^T$ is a vector of random variables with mean vector $\mathbf{M_X}$ and covariance matrix $\boldsymbol{\Sigma}_{\mathbf{XX}}$, $\mathbf{A}$ is a 3×3 coefficient matrix, and $\mathbf{B}$ is a 3×1 vector of constants. For

$$\mathbf{M_X} = \begin{Bmatrix} 10 \\ 50 \\ 30 \end{Bmatrix}, \boldsymbol{\Sigma}_{\mathbf{XX}} = \begin{bmatrix} 16 & 20 & 4 \\ 20 & 29 & 11 \\ 4 & 11 & 26 \end{bmatrix}, \mathbf{A} = \begin{bmatrix} 5 & -4 & 3 \\ 2 & 9 & 10 \\ 4 & -5 & 6 \end{bmatrix}, \mathbf{B} = \begin{Bmatrix} 3 \\ 5 \\ -6 \end{Bmatrix}.$$

1) Determine the mean vector, covariance matrix, diagonal matrix of standard deviations, and correlation matrix of $\mathbf{Y}$.
2) Determine the order of importance of random variables $\mathbf{X}$ in contributing to the variances of Y.

2.11 The bending strength of a reinforced concrete beam is given by

$$M_u = A_s f_y d \left(1 - 0.6 \frac{A_s f_y}{b d f'_c}\right)$$

where A_s is the cross-sectional area of the reinforcing steel, f_y the yield strength of steel, and f'_c the compressive strength of concrete, and $b = 0.25$ m and $d = 0.50$ m are the dimensions of the cross section. Suppose A_s, f_y, and f'_c are random variables with the means and c.o.v. listed below.

Variable	Mean	c.o.v
A_s m^2	0.0025	0.05
f_y kN/m^2	350,000	0.10
f'_c kN/m^2	25,000	0.20

1) Make first-order approximate estimates of the mean and standard deviation of M_u, assuming the three random variables are uncorrelated.
2) Make a second-order approximate estimate of the mean of M_u.
3) What are the results in (1) and (2) if A_s and f_y are correlated with $\rho_{Af_y} = 0.30$?
4) Determine the order of importance of the random variables in contributing to the variance of M_u.

2.12 The concentration of a toxic substance in the daily discharge from a chemical factory has the Rayleigh distribution with PDF

$$f_X(x) = \frac{2x}{a^2} \exp\left[-\left(\frac{x}{a}\right)^2\right], 0 \le x,$$

where $a = 3$ parts per billion (ppb) is the scale parameter. Assume there are 300 working days in a year and that the daily concentration values are statistically independent.

1) Derive an exact expression for the PDF of the annual maximum concentration level. What are the probabilities that the annual maximum concentration level will exceed the thresholds 8 ppb and 9 ppb?
2) Determine the characteristic values of the annual maximum, u_{300} and α_{300}. What is the form of the asymptotic distribution of the annual maximum concentration? What are its mean, standard deviation, and mode? What are the probabilities that the annual maximum concentration level will exceed the thresholds 8 ppb and 9 ppb? Compare these probabilities with those computed in part (1). Comment on the differences.
3) Plot and compare the exact and asymptotic PDFs of the annual maximum concentration together with the PDF of X.

2.13 A chain is made of 10 links, each link having a random tensile capacity. The link capacities are statistically independent and have the common beta PDF

$$f_X(x) = \frac{3}{4000}(x - 90)(110 - x), 90 \le x \le 110.$$

1) Determine the probability that the tensile capacity of the chain is less than 92.
2) Derive the asymptotic distribution of the capacity of the chain and determine its mean and standard deviation. What is the probability that the capacity of the chain is less than 92 according to this distribution?
3) Plot and compare the exact and asymptotic PDFs of the tensile capacity of the chain together with the PDF of X.

3 Multivariate Distributions

3.1 Introduction

Multivariate distribution models are essential ingredients of reliability and risk analysis. They are used to describe sets of dependent random variables that are present in models of engineering systems. In this chapter, we present selected joint distribution models and their properties. Included are joint normal, joint lognormal, joint distribution as product of conditionals, and Morgenstern, Nataf, and copula families of distributions with prescribed marginals. A more extensive review of multivariate distributions can be found in Kotz et al. (2004). Also presented are methods to transform random variables to the standard normal space. This kind of transformation is used in several computational reliability methods, as described in later chapters of this book.

3.2 The Multinormal Distribution

A random variable X has the *normal* or *Gaussian* distribution when its probability density function (PDF) is defined by

$$f_X(x) = \frac{1}{\sqrt{2\pi}\sigma} \exp\left[-\frac{1}{2}\left(\frac{x-\mu}{\sigma}\right)^2\right], \tag{3.1}$$

where μ and σ are, respectively, the mean and standard deviation. We use the notation $X = \mathrm{N}(\mu, \sigma)$ to indicate that X has the normal distribution with mean μ and standard deviation σ. A normal random variable with zero mean and unit standard deviation is called a *standard normal variable*. We denote such a variable as U and its PDF by

$$\phi(u) = \frac{1}{\sqrt{2\pi}} \exp\left(-\frac{u^2}{2}\right). \tag{3.2}$$

The corresponding cumulative distribution function (CDF) is denoted

$$\Phi(u) = \int_{-\infty}^{u} \frac{1}{\sqrt{2\pi}} \exp\left(-\frac{u^2}{2}\right) du. \tag{3.3}$$

The CDF for the general case in (3.1) can be expressed in terms of the standard normal CDF as

$$F_X(x) = \Phi\left(\frac{x-\mu}{\sigma}\right). \tag{3.4}$$

There is no closed-form solution for $\Phi(u)$. However, one can show that it is related to the error function through $\Phi(u) = 0.5 + 0.5\,\mathrm{erf}\big(u/\sqrt{2}\big)$. Note that the standard normal PDF is symmetric with respect to the zero axis, so that $\Phi(-u) = 1 - \Phi(u)$.

A vector of continuous random variables $\mathbf{X} = [X_1, \ldots, X_n]^{\mathrm{T}}$ is said to have the *multivariate normal* or *multinormal* distribution when its joint PDF is defined by

$$f_{\mathbf{X}}(\mathbf{x}) = \frac{1}{(2\pi)^{n/2}(\det \boldsymbol{\Sigma}_{\mathbf{XX}})^{1/2}} \exp\left[-\frac{1}{2}(\mathbf{x}-\mathbf{M_X})^{\mathrm{T}}\boldsymbol{\Sigma}_{\mathbf{XX}}^{-1}(\mathbf{x}-\mathbf{M_X})\right], \tag{3.5}$$

where $\mathbf{M_X}$ is the mean vector and $\boldsymbol{\Sigma}_{\mathbf{XX}}$ is the covariance matrix. As a brief notation, we use $\mathbf{X} = \mathrm{N}(\mathbf{M_X}, \boldsymbol{\Sigma}_{\mathbf{XX}})$ to indicate that the vector $\mathbf{X}$ has the multinormal distribution with mean vector $\mathbf{M_X}$ and covariance matrix $\boldsymbol{\Sigma}_{\mathbf{XX}}$.

It is useful to examine the joint PDF in (3.5) for $n = 2$. Expanding the matrix multiplication in the exponent and denoting the random variables as X and Y, the bivariate normal PDF is written as

$$\begin{aligned} f_{XY}(x,y) = \frac{1}{2\pi\sigma_X\sigma_Y\sqrt{1-\rho^2}} \exp\Bigg\{-\frac{1}{2(1-\rho^2)}\Bigg[\left(\frac{x-\mu_X}{\sigma_X}\right)^2 \\ -2\rho\left(\frac{x-\mu_X}{\sigma_X}\right)\left(\frac{y-\mu_Y}{\sigma_Y}\right) + \left(\frac{y-\mu_Y}{\sigma_Y}\right)^2\Bigg]\Bigg\}, \end{aligned} \tag{3.6}$$

where ρ is the correlation coefficient between the two random variables. It should be clear from the form of the exponent in (3.6) that the contours of constant probability density are concentric ellipses centered at (μ_X, μ_Y) in the (x, y) plane. Figure 3.1 shows the contours in the normalized space of $u_1 = (x - \mu_X)/\sigma_X$ and $u_2 = (y - \mu_Y)/\sigma_Y$ for $\rho = 0.5$. It should also be clear that when $\rho = 0$ the joint PDF becomes a product of the two normal marginal PDFs, which indicates that the two random variables are statistically independent. The contours in the normalized space then become circles. It follows that in the case of bivariate normal random variables, uncorrelatedness implies statistical independence. This result also applies in the multivariate case, where uncorrelatedness implies a diagonal covariance matrix.

The standard multinormal PDF is the special case of (3.5) for multinormal random variables $\mathbf{U} = [U_1, \ldots, U_n]^{\mathrm{T}}$ with zero means and unit standard deviations. We denote it by

$$\phi_n(\mathbf{u}, \mathbf{R}) = \frac{1}{(2\pi)^{n/2}(\det \mathbf{R})^{1/2}} \exp\left[-\frac{1}{2}\mathbf{u}^{\mathrm{T}}\mathbf{R}^{-1}\mathbf{u}\right], \tag{3.7}$$

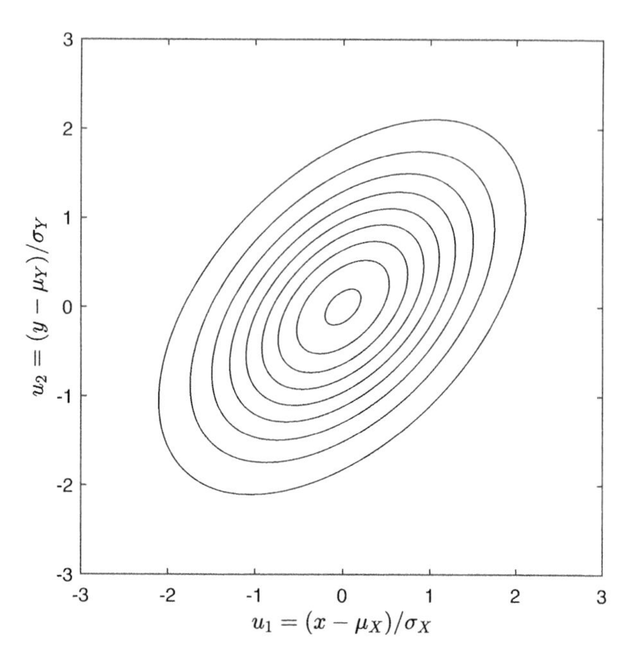

Figure 3.1 Contours of bivariate normal PDF with $\rho = 0.5$

where the subscript n denotes the number of random variables and $\mathbf{R}$ is the $n \times n$ correlation matrix. Figure 3.1 represents a contour plot of $\phi_2(u_1, u_2, \rho)$ for $\rho = 0.5$. When these standardized random variables are uncorrelated, i.e., $\mathbf{R}$ is an identity matrix, then we have the standard uncorrelated multinormal PDF

$$\begin{aligned}\phi_n(\mathbf{u}) &= \frac{1}{(2\pi)^{n/2}} \exp\left[-\frac{1}{2}\|\mathbf{u}\|^2\right] \\ &= \prod\nolimits_{i=1}^{n} \phi(u_i).\end{aligned} \tag{3.8}$$

Observe that in this case the random variables are statistically independent, and the joint PDF is a product of the marginal PDFs. The contours of the above PDF in any intersecting plane are concentric circles.

The joint CDF corresponding to the standard multinormal PDF in (3.7) is denoted

$$\Phi_n(\mathbf{u}, \mathbf{R}) = \int_{-\infty}^{u_n} \cdots \int_{-\infty}^{u_1} \phi_n(\mathbf{u}, \mathbf{R}) d\mathbf{u}. \tag{3.9}$$

No closed-form solution of this function is available. For $n = 1$, $\Phi(u)$, where we have dropped the subscript 1, is related to the error function, as mentioned following (3.4). For $n = 2$, one can show (see Section 8.11) that

$$\Phi_2(u_1, u_2, \rho) = \Phi(u_1)\Phi(u_2) + \int_0^{\rho} \phi_2(u_1, u_2, \rho) d\rho. \tag{3.10}$$

Thus, computation of the bivariate standard normal CDF requires numerical evaluation of a single-fold integral. In a similar way, one can show that computation of the n-variate standard normal CDF requires evaluation of an $(n-1)$-fold integral. Obviously, this is impractical for large n. For this reason, several algorithms have been developed for computing the multinormal CDF in high dimensions. The best known among these is the quasi–Monte Carlo method developed by Genz (1992) and Genz and Bretz (2002), which is also used in the statistical toolbox of MATLAB (`mvncdf` and related commands). Because of the importance of this function in system reliability analysis, a number of specialized algorithms have been developed by researchers in this field. These include Breitung (1984), Gollwitzer and Rackwitz (1988), Breitung and Hohenbichler (1989), Ambartzumian et al. (1998), Pandey (1998), and Kang and Song (2010). The method developed by Ambartzumian et al. is implemented in the CalREL software (Der Kiureghian et al., 2006).

The general multinormal CDF corresponding to the joint PDF in (3.5) is obtained in terms of $\Phi_n(\mathbf{u}, \mathbf{R})$ by setting $U_i = \left(X_i - \mu_{X_i}\right)/\sigma_{X_i}$. Note that this linear transformation does not affect the correlation coefficients, so that $\mathbf{R}$ is the correlation matrix of both $\mathbf{X}$ and $\mathbf{U}$.

The multinormal distribution has several important properties, which we summarize here without providing detailed explanations or proofs:

1) The multinormal distribution is completely defined by the mean vector and covariance matrix.
2) Subsets of jointly normal random variables are also jointly normal. That is, if $\mathbf{X} = [X_1, \ldots, X_n]^{\mathrm{T}}$ is jointly normal, then any subset of the variables $X_1, \ldots, X_n$ is also jointly normal. To specify the distribution, we only need to construct the mean vector and covariance matrix from the corresponding elements of $\mathbf{M}_{\mathbf{X}}$ and $\mathbf{\Sigma}_{\mathbf{XX}}$. In particular, each individual random variable X_i is normal with mean μ_i (the ith element of $\mathbf{M}_{\mathbf{X}}$) and standard deviation σ_i (the square root of the ith diagonal element of $\mathbf{\Sigma}_{\mathbf{XX}}$).
3) Consider the partition $\mathbf{X} = [X_1, \ldots, X_k | X_{k+1}, \ldots, X_n]^{\mathrm{T}} = \left[\mathbf{X}_1^{\mathrm{T}} | \mathbf{X}_2^{\mathrm{T}}\right]^{\mathrm{T}}$ of the vector of random variables and the corresponding partition of the mean vector and covariance matrix

$$\mathbf{M}_{\mathbf{X}} = \begin{Bmatrix} \mathbf{M}_1 \\ \mathbf{M}_2 \end{Bmatrix}, \mathbf{\Sigma}_{\mathbf{XX}} = \begin{bmatrix} \mathbf{\Sigma}_{11} & \mathbf{\Sigma}_{12} \\ \mathbf{\Sigma}_{21} & \mathbf{\Sigma}_{22} \end{bmatrix}. \tag{3.11}$$

An important property of the multinormal distribution is that the conditional distribution of $\mathbf{X}_1$ given $\mathbf{X}_2 = \mathbf{x}_2$ is also normal. The corresponding conditional mean vector and covariance matrix are

$$\mathbf{M}_{\mathbf{X}_1|\mathbf{X}_2=\mathbf{x}_2} = \mathbf{M}_1 + \mathbf{\Sigma}_{12}\mathbf{\Sigma}_{22}^{-1}(\mathbf{x}_2 - \mathbf{M}_2), \tag{3.12}$$

$$\mathbf{\Sigma}_{\mathbf{X}_1\mathbf{X}_1|\mathbf{X}_2=\mathbf{x}_2} = \mathbf{\Sigma}_{11} - \mathbf{\Sigma}_{12}\mathbf{\Sigma}_{22}^{-1}\mathbf{\Sigma}_{21}. \tag{3.13}$$

It is worth noting that, whereas the conditional mean vector depends on the observed values $\mathbf{x}_2$, the conditional covariance matrix is independent of the given outcomes. In the bivariate case of (3.6), the conditional distribution of X given $Y = y$ is normal with the mean and standard deviation

$$\mu_{X|Y=y} = \mu_X + \rho\sigma_X\left(\frac{y - \mu_Y}{\sigma_Y}\right) \tag{3.14}$$

$$\sigma_{X|Y=y} = \sigma_X\sqrt{1 - \rho^2}. \tag{3.15}$$

Note that observing Y always reduces the uncertainty in X (the standard deviation decreases) if the variables are correlated.

4) For jointly normal random variables, uncorrelatedness implies statistical independence. This is readily evident in (3.12) and (3.13), because for uncorrelated random variables $\boldsymbol{\Sigma}_{12} = \mathbf{0}$ and the conditional mean vector and covariance matrix are identical to the unconditional mean vector and covariance matrix.
5) Linear functions of jointly normal random variables are also jointly normal (see Example 2.6 in Chapter 2). Thus, if $\mathbf{X}$ is jointly normal and $\mathbf{Y} = \mathbf{AX} + \mathbf{B}$, where $\mathbf{A}$ and $\mathbf{B}$ are deterministic, then $\mathbf{Y}$ is jointly normal with mean vector $\mathbf{M_Y} = \mathbf{AM_X} + \mathbf{B}$ and covariance matrix $\boldsymbol{\Sigma}_{\mathbf{YY}} = \mathbf{A}\boldsymbol{\Sigma}_{\mathbf{XX}}\mathbf{A}^{\mathrm{T}}$.
6) *Central limit theorem*: Let $Y_i, i = 1, \ldots, n$, be a set of statistically independent and identically distributed random variables and $X = Y_1 + \cdots + Y_n$. According to this theorem, the distribution of X asymptotically approaches the normal distribution as $n \to \infty$, regardless of the form of the common distribution of Y_i. This result also holds when Y_i have arbitrary distributions, as long as their variances are of similar magnitude. Based on this theorem, random quantities that arise from the accumulation of a large number of independent random effects of equal variance are expected to have a normal distribution.

The multinormal distribution has an important role in computational reliability methods. Thorough understanding of the material in this section is essential for mastering the reliability methods described in later chapters of this text.

3.3 The Multivariate Lognormal Distribution

A positive-valued random variable Y has the lognormal distribution if its logarithm $X = \ln Y$ has the normal distribution – hence its name (see Example 2.5 in Chapter 2). Using the method described in Section 2.7.1 of Chapter 2 for determining the distribution of a function of random variables, one can easily show that the PDF of Y has the form

$$f_Y(y) = \frac{1}{\sqrt{2\pi}\zeta y}\exp\left[-\frac{1}{2}\left(\frac{\ln y - \lambda}{\zeta}\right)^2\right], 0 < y, \tag{3.16}$$

where $\lambda = \mu_X = \mu_{\ln Y}$ is the mean and $\zeta = \sigma_X = \sigma_{\ln Y}$ is the standard deviation of the logarithm of the random variable, often called the "logarithmic" mean and standard deviation. One can show that the mean and coefficient of variation (c.o.v.) of Y have the following relations with the two parameters (see Table A2.2):

$$\mu_Y = \exp\left(\lambda + \frac{1}{2}\zeta^2\right), \ \delta_Y = \sqrt{\exp\left(\zeta^2\right) - 1}. \tag{3.17}$$

The inverse relations are

$$\lambda = \ln \mu_Y - \frac{1}{2}\zeta^2, \ \zeta = \sqrt{\ln\left(1 + \delta_Y^2\right)}. \tag{3.18}$$

These inverse relations are useful when the mean and standard deviation (or c.o.v.) are given and one needs to determine the parameters of the distribution. Note that for small δ_Y (say $\delta_Y \leq 0.30$), $\zeta \cong \delta_Y$ in (3.18). Hence, ζ is approximately equal to the c.o.v. of the lognormal random variable. Owing to its relationship with the normal distribution, the CDF of the lognormal distribution can be computed in terms of the standard normal CDF,

$$F_Y(y) = \Phi\left(\frac{\ln y - \lambda}{\zeta}\right). \tag{3.19}$$

A vector of positive-valued, continuous random variables $\mathbf{Y} = [Y_1, \ldots, Y_n]^{\mathrm{T}}$ is said to have the *multivariate lognormal* distribution when its logarithm $\mathbf{X} = \ln \boldsymbol{Y}$ with elements $X_i = \ln Y_i$, $i = 1, \ldots, n$, has the joint normal distribution. To specify the distribution of $\mathbf{X}$, we need to specify its mean vector and covariance matrix. Let $\mu_{X_i} = \mu_{\ln Y_i} = \lambda_i$ and $\sigma_{X_i} = \sigma_{\ln Y_i} = \zeta_i$, $i = 1, \ldots, n$, where each of the pairs of parameters λ_i and ζ_i are related to the mean and c.o.v. of the corresponding Y_i according to the relations in (3.17) and (3.18). To complete the specification of the joint distribution, we also need to specify the correlation coefficient between pairs of the random variables $X_i = \ln Y_i$ and $X_j = \ln Y_j$. One can show that this correlation coefficient is given in terms of the correlation coefficient and c.o.v. of Y_i and Y_j through

$$\rho_{\ln Y_i \ln Y_j} = \frac{1}{\zeta_i \zeta_j} \ln\left(1 + \rho_{Y_i Y_j} \delta_{Y_i} \delta_{Y_j}\right). \tag{3.20}$$

Thus, given the mean vector and covariance matrix of $\mathbf{Y}$, one can determine the set of parameters λ_i, ζ_i, and $\rho_{\ln Y_i \ln Y_j}$, $i,j = 1, \ldots, n$, which collectively define the mean vector and covariance matrix of $\mathbf{X}$ and, hence, its distribution. Although it is possible to write the formula for the joint PDF of $\mathbf{Y}$, this is not necessary because any probability statement about the variables Y_i can be expressed in terms of the logarithms $\ln Y_i$ and therefore in terms of the multinormal distribution. For example, the joint CDF can be written as $F_{\mathbf{Y}}(\mathbf{y}) = \Pr\left(\bigcap_{i=1}^n \{Y_i \leq y_i\}\right) = \Pr\left(\bigcap_{i=1}^n \{\ln Y_i \leq \ln y_i\}\right)$. As $\ln Y_i$ are jointly normal, we have $\Pr\left(\bigcap_{i=1}^n \{\ln Y_i \leq \ln y_i\}\right) = \Phi_n(\mathbf{u}, \mathbf{R})$, where $u_i = (\ln y_i - \lambda_i)/\zeta_i$ and $\mathbf{R}$ denotes the correlation coefficient matrix of $\ln \boldsymbol{Y}$. Nevertheless, for the sake of greater insight, we present the bivariate lognormal PDF, which for variables X and Y is

$$f_{XY}(x,y) = \frac{1}{2\pi\zeta_X\zeta_Y xy\sqrt{1-\rho^2}} \exp\left\{-\frac{1}{2(1-\rho^2)}\left[\left(\frac{\ln x-\lambda_X}{\zeta_X}\right)^2 - 2\rho\left(\frac{\ln x-\lambda_X}{\zeta_X}\right)\left(\frac{\ln y-\lambda_Y}{\zeta_Y}\right) + \left(\frac{\ln y-\lambda_Y}{\zeta_Y}\right)^2\right]\right\}, \tag{3.21}$$

where $\rho = \rho_{\ln X \ln Y}$ is determined according to (3.20). Figure 3.2 shows contour plot of this PDF for $\lambda_X = \lambda_Y = 0$, $\zeta_X = \zeta_Y = 0.5$, and $\rho_{XY} = 0.5$. Note that the contours are not ellipses.

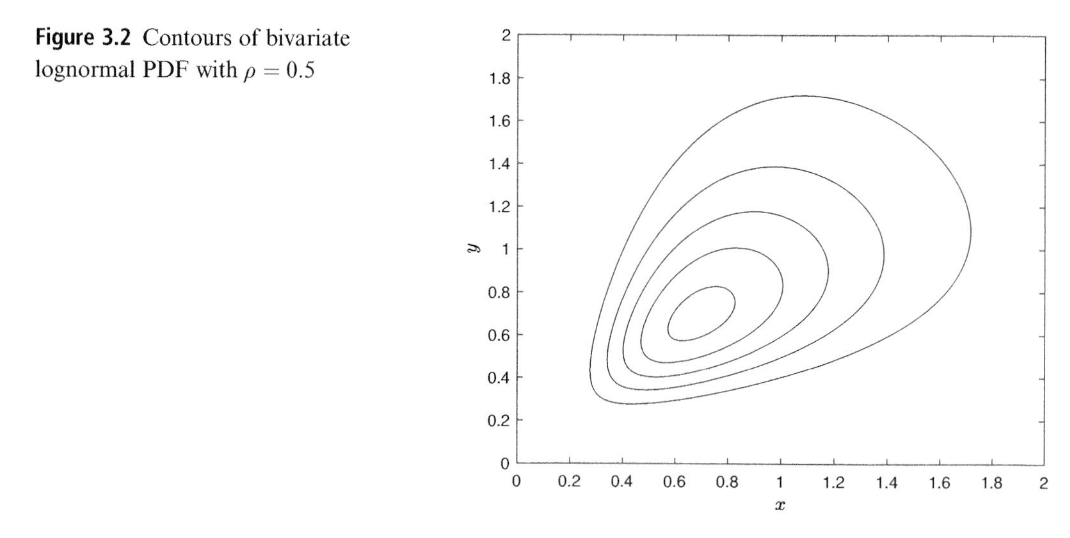

Figure 3.2 Contours of bivariate lognormal PDF with $\rho = 0.5$

The multivariate lognormal distribution has a number of important properties that we derive from the properties of the multinormal distribution described in the preceding section. They are as follows:

1) The multivariate lognormal distribution is completely defined by the mean vector and the covariance matrix of the random variables.
2) If $\mathbf{Y} = [Y_1, \ldots, Y_n]^{\mathrm{T}}$ is jointly lognormal then any subset of the variables $Y_1, \ldots, Y_n$ is also jointly lognormal. In particular, each individual variable Y_i has a lognormal marginal distribution.
3) Consider the partition $\mathbf{Y} = [Y_1, \ldots, Y_k | Y_{k+1}, \ldots, Y_n]^{\mathrm{T}} = \left[\mathbf{Y}_1^{\mathrm{T}} | \mathbf{Y}_2^{\mathrm{T}}\right]^{\mathrm{T}}$ of the vector of random variables. If $\mathbf{Y}$ is jointly lognormal then the conditional distribution of $\mathbf{Y}_1$ given $\mathbf{Y}_2 = \mathbf{y}_2$ is also lognormal. The set of distribution parameters is obtained from (3.12)–(3.13), where $\mathbf{X} = \ln \mathbf{Y}$, $\mu_{X_i} = \lambda_i$, $\sigma_{X_i} = \zeta_i$, and $\rho_{X_iX_j} = \rho_{\ln Y_i \ln Y_j}$.
4) For jointly lognormal random variables, uncorrelatedness implies statistical independence.
5) Product functions of jointly lognormal random variables are also jointly lognormal. Thus, if $\mathbf{Y}$ is jointly lognormal and $Z_k = a_k Y_1^{m_{k1}} Y_2^{m_{k2}} \cdots Y_n^{m_{kn}}$, $k = 1, \ldots, K$, in which $0 < a_k$ and m_{ki} are arbitrary powers, then the vector $\mathbf{Z} = [Z_1, \ldots, Z_K]^{\mathrm{T}}$ is also jointly lognormal. This result is derived from the fact that $\ln Z_k = \ln a_k + m_{k1} \ln Y_1 + \cdots + m_{kn} \ln Y_n$, $k = 1, \ldots, K$, are linear functions of the random variables $\ln Y_i$. With Y_i

jointly lognormal, $\ln Y_i$ are jointly normal and, given the linear relations, $\ln Z_k$ are also jointly normal. It follows that Z_k are jointly lognormal. The set of parameters for the distribution of $\mathbf{Z}$ is obtained by computing the means, standard deviations, and correlation coefficients of $\ln Z_k$. They are

$$\lambda_{Z_k} = \ln a_k + m_{k1}\lambda_{Y_i} + \cdots + m_{kn}\lambda_{Y_n}, \tag{3.22}$$

$$\zeta_{Z_k}^2 = \sum_{i=1}^{n}\sum_{j=1}^{n} m_{ki}m_{kj}\rho_{\ln Y_i \ln Y_j}\zeta_{Y_i}\zeta_{Y_j}, \tag{3.23}$$

$$\rho_{\ln Z_k \ln Z_l} = \frac{1}{\zeta_{Z_k}\zeta_{Z_l}}\sum_{i=1}^{n}\sum_{j=1}^{n} m_{ki}m_{lj}\rho_{\ln Y_i \ln Y_j}\zeta_{Y_i}\zeta_{Y_j}. \tag{3.24}$$

6) *Central limit theorem*: Let $Z_i, i = 1, \ldots, n$, be a set of positive-valued, statistically independent, and identically distributed random variables and Y be their product, i.e., $Y = Z_1Z_2\cdots Z_n$. By taking the logarithm of the product, we find $\ln Y$ to be the sum of n statistically independent and identically distributed random variables. According to the central limit theorem, the distribution of $\ln Y$ asymptotically approaches the normal distribution and, therefore, the distribution of Y approaches the lognormal distribution as $n \to \infty$ regardless of the form of the common distribution of Z_i. This result also holds when Z_i have arbitrary distributions, as long as their c.o.v. are of similar magnitude. Based on this theorem, random quantities that arise from the product of a large number of independent random effects tend to have the lognormal distribution.

3.4 Joint Distribution as Product of Conditionals

As we have seen in Section 2.6, the joint PDF of a set of random variables can be written as the product of conditional PDFs in the form

$$\begin{aligned} &f_{X_1\cdots X_n}(x_1, \ldots, x_n) \\ &= f_{X_1|X_2\cdots X_n}(x_1|x_2, \ldots, x_n) f_{X_2|X_3\cdots X_n}(x_2|x_3, \ldots, x_n)\cdots f_{X_{n-1}|X_n}(x_{n-1}|x_n) f_{X_n}(x_n). \end{aligned} \tag{3.25}$$

One way to define the joint PDF of a set of random variables is to define their conditional PDFs and obtain the joint PDF by their multiplication. In some applications, the distribution of a set of random variables $\mathbf{X}$ is available conditioned on another set of random variables $\mathbf{Y}$, i.e., we have the conditional PDF $f_{\mathbf{X}|\mathbf{Y}}(\mathbf{x}|\mathbf{y})$. Provided the joint distribution $f_{\mathbf{Y}}(\mathbf{y})$ of the given variables, the joint PDF of the two sets of random variables is the product $f_{\mathbf{XY}}(\mathbf{x}, \mathbf{y}) = f_{\mathbf{X}|\mathbf{Y}}(\mathbf{x}|\mathbf{y})f_{\mathbf{Y}}(\mathbf{y})$. Below, we present two areas where this formulation is useful.

Suppose $\mathbf{X}$ represents the vector of responses of a structural system subjected to stochastic excitation. Using the methods of random vibration theory (presented in Chapter 13), it is possible to determine the distribution of $\mathbf{X}$ for specified properties of the structure – e.g., mass, damping, stiffness. Now, if the structural properties are uncertain and described by random variables $\mathbf{Y}$, then the random vibration analysis yields the conditional distribution $f_{\mathbf{X}|\mathbf{Y}}(\mathbf{x}|\mathbf{y})$. The joint distribution of $\mathbf{X}$ and $\mathbf{Y}$ is then obtained as described in the preceding paragraph and the marginal distribution of $\mathbf{X}$ is obtained by use of the consistency rule (see (2.50)) $f_{\mathbf{X}}(\mathbf{x}) = \int_{\mathbf{y}} f_{\mathbf{XY}}(\mathbf{x}, \mathbf{y})d\mathbf{y}$. Written in the form $f_{\mathbf{X}}(\mathbf{x}) = \int_{\mathbf{y}} f_{\mathbf{X}|\mathbf{Y}}(\mathbf{x}|\mathbf{y}) f_{\mathbf{Y}}(\mathbf{y})d\mathbf{y}$, this expression can also be viewed as an application of the total probability rule. Note that the resulting distribution of the responses $\mathbf{X}$ accounts not only for the stochastic nature of the excitation but also for randomness in the structural properties.

The second application area concerns the treatment of uncertainties in model parameters. Suppose the distribution of random variables $\mathbf{X}$ involves a set of parameters $\boldsymbol{\Theta}$, which are unknown and must be estimated from observed data. Such estimation invariably involves uncertainty. Using a Bayesian formulation (to be presented in Chapter 10), it is possible to construct a "posterior" distribution $f_{\boldsymbol{\Theta}}(\boldsymbol{\theta})$ that represents the uncertainties in the estimation of the parameters. Now, if we interpret the parameterized distribution of $\mathbf{X}$ as a conditional distribution $f_{\mathbf{X}|\boldsymbol{\Theta}}(\mathbf{x}|\boldsymbol{\theta})$, the joint distribution takes the form $f_{\mathbf{X}\boldsymbol{\Theta}}(\mathbf{x}, \boldsymbol{\theta}) = f_{\mathbf{X}|\boldsymbol{\Theta}}(\mathbf{x}|\boldsymbol{\theta}) f_{\boldsymbol{\Theta}}(\boldsymbol{\theta})$ and the marginal distribution of $\mathbf{X}$ is obtained as $\tilde{f}_{\mathbf{X}}(\mathbf{x}) = \int_{\boldsymbol{\theta}} f_{\mathbf{X}|\boldsymbol{\Theta}}(\mathbf{x}|\boldsymbol{\theta}) f_{\boldsymbol{\Theta}}(\boldsymbol{\theta})d\boldsymbol{\theta}$. This distribution, denoted the *predictive* distribution of $\mathbf{X}$, incorporates not only the randomness in $\mathbf{X}$ but also the uncertainties in the estimation of the distribution parameters.

Example 3.1 – The Normal-Gamma Distribution

Suppose X has the normal distribution with mean μ and variance $1/Y$, where Y has the gamma distribution with parameters (k, λ), i.e., $f_Y(y) = \left[\lambda(\lambda y)^{k-1}/\Gamma(k)\right] \exp(-\lambda y)$. The joint distribution of X and Y is

$$\begin{aligned} f_{XY}(x, y) &= \frac{\sqrt{y}}{\sqrt{2\pi}} \exp\left[-\frac{1}{2}(x-\mu)^2 y\right] \frac{\lambda(\lambda y)^{k-1}}{\Gamma(k)} \exp(-\lambda y) \\ &= \frac{\lambda^k y^{k-1/2}}{\sqrt{2\pi}\Gamma(k)} \exp\left\{-\left[\frac{(x-\mu)^2}{2} + \lambda\right] y\right\}, 0 < y. \end{aligned} \tag{E1}$$

Figure 3.3 shows the contour plot of the above bivariate PDF for $\mu = 0$, $k = 3$, and $\lambda = 4$. Note that the dispersion in X strongly depends on the outcome of Y.

The marginal distribution of X is obtained by integration over y, i.e.,

$$\begin{aligned} f_X(x) &= \int_0^\infty \frac{\lambda^k y^{k-1/2}}{\sqrt{2\pi}\Gamma(k)} \exp\left\{-\left[\frac{(x-\mu)^2}{2} + \lambda\right] y\right\} dy, \\ &= \frac{\lambda^k}{\sqrt{2\pi}\Gamma(k)} \int_0^\infty y^{k-1/2} \exp\left\{-\left[\frac{(x-\mu)^2}{2} + \lambda\right] y\right\} dy, \\ &= \frac{\Gamma(k+1/2)}{\sqrt{2\pi\lambda}\Gamma(k)} \left[1 + \frac{(x-\mu)^2}{2\lambda}\right]^{-(k+1/2)}. \end{aligned} \tag{E2}$$

Example 3.1 (cont.)

The solution in the bottom line is obtained by observing that the integrand in the second line is proportional to the gamma CDF with the shape parameter $(k + 1/2)$ and scale parameter $\left[(x - \mu)^2/2 + \lambda\right]$. One can show that the final solution is equivalent to the T distribution with $2k$ degrees of freedom for the normalized random variable $(X - \mu)\sqrt{k/\lambda}$. Figure 3.4 shows a plot of the above PDF.

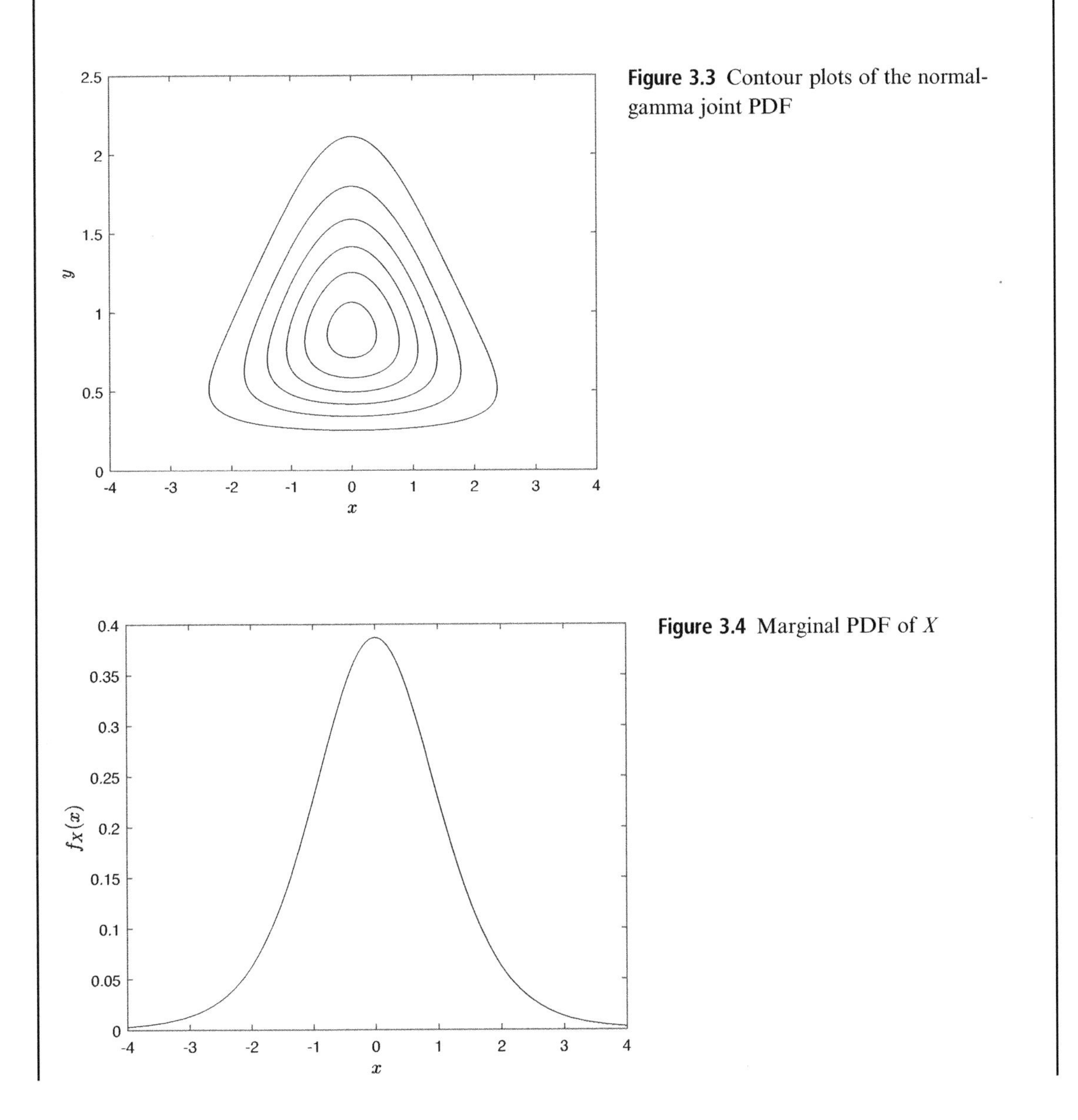

Figure 3.3 Contour plots of the normal-gamma joint PDF

Figure 3.4 Marginal PDF of X

3.5 Multivariate Distributions with Prescribed Marginals

Often, sufficient data are available to select appropriate marginal distributions but not enough to determine the dependence between the random variables beyond the correlation coefficients. In such cases, joint distribution models that are consistent with the prescribed marginal distributions and correlation coefficients are of interest. In this section we introduce three families of such distribution models; namely, the Morgenstern, Nataf, and copula families of joint distributions. For the sake of simplicity of notation, from here on we denote the marginal PDF and CDF of random variable X_i as $f_i(x_i)$ and $F_i(x_i)$, respectively; the mean, standard deviation, and c.o.v. values as μ_i, σ_i, and δ_i, respectively, and the correlation coefficient between variables X_i and X_j as ρ_{ij}.

Before proceeding with the three families of joint distributions, it is worth noting the following two special cases: (a) If all the marginals are known to be normal then the multinormal distribution is a viable choice because we know that that joint distribution has normal marginals; and (b) if all the marginals are known to be lognormal then the multivariate lognormal distribution is a viable choice because it produces lognormal marginals. However, we emphasize that normal marginals does not necessarily imply the multinormal joint distribution, nor does lognormal marginals imply the multivariate lognormal distribution. It is in fact possible to have a non-normal joint distribution that produces normal marginals, and a non-lognormal joint distribution that produces lognormal marginals. One such example is described in the following section.

3.5.1 Morgenstern Family of Multivariate Distributions

For two random variables X_1 and X_2, the Morgenstern bivariate CDF is defined as

$$F_{X_1X_2}(x_1, x_2) = F_1(x_1)F_2(x_2)\{1 + \alpha_{12}[1 - F_1(x_1)][1 - F_2(x_2)]\}, \tag{3.26}$$

where α_{12} is a parameter that controls the correlation between the two variables (Morgenstern 1956). It is easily verified that $F_{X_1X_2}(x_1, \infty) = F_1(x_1)$ and $F_{X_1X_2}(\infty, x_2) = F_2(x_2)$, so that the model is consistent with the specified marginal distributions. The joint PDF is obtained by differentiating the above expression with respect to x_1 and x_2, resulting in

$$f_{X_1X_2}(x_1, x_2) = f_1(x_1)f_2(x_2)\{1 + \alpha_{12}[1 - 2F_1(x_1)][1 - 2F_2(x_2)]\}. \tag{3.27}$$

As we must have $0 \leq f(x_1, x_2)$ and the product of the two square brackets in the above expression has a value within the interval $[-1, +1]$, we must have $-1 \leq \alpha_{12} \leq 1$. To determine α_{12}, we use the relation

$$\rho_{12} = \int_{-\infty}^{+\infty}\int_{-\infty}^{+\infty}\left(\frac{x_1 - \mu_1}{\sigma_1}\right)\left(\frac{x_2 - \mu_2}{\sigma_2}\right)f_{X_1X_2}(x_1, x_2)\mathrm{d}x_1\mathrm{d}x_2. \tag{3.28}$$

Substituting for $f_{X_1X_2}(x_1, x_2)$ from (3.27) and noting that integrals of the form $\int_{-\infty}^{+\infty}[(x_i - \mu_i)/\sigma_i]f_i(x_i)\mathrm{d}x_i$ are identical to zero, after some derivation one obtains (see Liu and Der Kiureghian 1986)

$$\rho_{12} = 4\alpha_{12} Q_1 Q_2, \tag{3.29}$$

where

$$Q_i = \int_{-\infty}^{+\infty} \left(\frac{x_i - \mu_i}{\sigma_i} \right) f_i(x_i) F_i(x_i) \mathrm{d}x_i. \tag{3.30}$$

Thus,

$$\alpha_{12} = \frac{\rho_{12}}{4 Q_1 Q_2}. \tag{3.31}$$

Table 3.1 lists the values of the integral in (3.30) for commonly used distribution models as reported in Liu and Der Kiureghian (1986). For distributions that can be linearly transformed to a standard form, Q_i has a constant value. For distributions that cannot be linearly transformed to a standard form, Q_i is a function of the shape parameter of the distribution. Liu and Der Kiureghian (1986) showed that in general $0 < Q_i < 0.5$. However, as can be seen in Table 3.1, Q_i is less than 0.29 for all the distributions considered. Because $|\alpha_{12}|$ is bounded by 1, it follows from (3.29) that the Morgenstern model can accommodate correlation coefficients no greater in magnitude than about $4 \times 1 \times 0.29 \times 0.29 = 0.336$. Thus, the Morgenstern model is valid for correlation coefficients in the interval $-0.336 < \rho_{ij} < +0.336$. This rather limited range is a shortcoming of this model.

Table 3.1 Statistic $Q = \int_{-\infty}^{+\infty} \left(\frac{x-\mu}{\sigma}\right) f_X(x) F_X(x) \mathrm{d}x$ for selected distributions (after Liu and Der Kiureghian, 1986)

Distribution	CDF	Standardization	Q
Normal	$\Phi\left(\frac{x-\mu}{\sigma}\right)$	$u = \frac{x-\mu}{\sigma}$	$\frac{1}{2\sqrt{\pi}} = 0.282$
Uniform	$\frac{x-a}{b-a}, a \le x \le b$	$u = \frac{x-a}{b-a}$	$\frac{\sqrt{3}}{6} = 0.289$
Shifted exponential	$1 - \exp[-\alpha(x - x_0)], x_0 \le x$	$u = \alpha(x - x_0)$	0.25
Shifted Rayleigh	$1 - \exp\left[-\frac{1}{2}\left(\frac{x-x_0}{\alpha}\right)^2\right], x_0 \le x$	$u = \frac{x-x_0}{\alpha}$	$\frac{\sqrt{2\pi} - \sqrt{\pi}}{\sqrt{32 - 8\pi}} = 0.280$
Type I largest vale	$\exp\{-\exp[-\alpha(x - v)]\}$	$u = \alpha(x - v)$	$\frac{\sqrt{6}\ln 2}{2\pi} = 0.270$
Type I smallest value	$1 - \exp\{-\exp[\alpha(x - v)]\}$	$u = \alpha(x - v)$	$\frac{\sqrt{6}\ln 2}{2\pi} = 0.270$
Lognormal	$\Phi\left(\frac{\ln x - \lambda}{\zeta}\right), 0 < x$	NA	$\frac{2\Phi(\zeta/\sqrt{2}) - 1}{2\sqrt{\exp(\zeta^2) - 1}} < 0.282$
Gamma	$\frac{\Gamma(k, \lambda x)}{\Gamma(k)}, 0 < x$	NA	$\frac{\Gamma(2k)}{4^k \sqrt{k}\Gamma^2(k)} < 0.282$
Type II largest value	$\exp\left[-\left(\frac{v}{x}\right)^k\right], 0 < x$	NA	$\frac{(2^{1/k} - 1)\Gamma(1 - 1/k)}{2\sqrt{\Gamma(1-2/k) - \Gamma^2(1-1/k)}} < 0.270$
Type III smallest value	$1 - \exp\left[-\left(\frac{x-e}{v-e}\right)^k\right], e < x$	NA	$\frac{(1 - 2^{1/k})\Gamma(1 + 1/k)}{2\sqrt{\Gamma(1+2/k) - \Gamma^2(1+1/k)}} < 0.270$

NA, not applicable

Generalization of the Morgenstern model to the multivariate case has the form (Kotz, 1975)

$$F_{\mathbf{X}}(\mathbf{x}) = \left[\prod_i F_i(x_i)\right]\left\{1 + \sum_{i<j}^{n} \alpha_{ij}[1 - F_i(x_i)][1 - F_j(x_j)] + \sum_{i<j<k}^{n} \alpha_{ijk}[1 - F_i(x_i)][1 - F_j(x_j)][1 - F_k(x_k)] + \cdots\right\}, \tag{3.32}$$

where α_{ij}, α_{ijk}, etc., are parameters. The corresponding joint PDF has the form

$$f_{\mathbf{X}}(\mathbf{x}) = \left[\prod_i f_i(x_i)\right]\left\{1 + \sum_{i<j}^{n} \alpha_{ij}[1 - 2F_i(x_i)][1 - 2F_j(x_j)] + \sum_{i<j<k}^{n} \alpha_{ijk}[1 - 2F_i(x_i)][1 - 2F_j(x_j)][1 - 2F_k(x_k)] + \cdots\right\}. \tag{3.33}$$

Conditions on the parameters α_{ij}, α_{ijk}, etc., are obtained by requiring that the joint PDF for all sets of the random variables be non-negative. Obviously, we must have $|\alpha_{ij}| \leq 1$ to assure non-negativity of the bivariate PDFs. Conditions on α_{ijk}, α_{ijkl}, etc., are obtained by requiring non-negativity of the corresponding trivariate, quatrovariate, etc., PDFs. These relations involve the higher order distribution parameters and can be quite complicated. One can show (see Liu and Der Kiureghian, 1986) that a generalization of (3.31) applies to the higher-order parameters, i.e.,

$$\alpha_{12\cdots k} = \frac{\rho_{12\cdots k}}{(-2)^k Q_1 Q_2 \cdots Q_k}, \tag{3.34}$$

where

$$\rho_{12\cdots k} = \mathrm{E}\left[\left(\frac{X_1 - \mu_1}{\sigma_1}\right)\left(\frac{X_2 - \mu_2}{\sigma_2}\right)\cdots\left(\frac{X_k - \mu_k}{\sigma_k}\right)\right] \tag{3.35}$$

is the k-dimensional normalized joint central moment of the random variables. In practice, these joint moments are seldom available. Hence, one usually sets all the parameters higher than bivariate to zero. It should be clear that in that case the model in (3.32)–(3.33) can handle correlation coefficients no greater than about 0.336 in magnitude.

Example 3.2 – Bivariate Morgenstern Distribution with Rayleigh and Lognormal Marginals

Suppose X has the shifted Rayleigh distribution with parameters α and x_0 (see Table 3.1) and Y has the lognormal distribution with parameters λ and ζ. The bivariate Morgenstern PDF with these marginals is defined by

$$f_{XY}(x,y) = \frac{x - x_0}{\alpha}\exp\left[-\frac{1}{2}\left(\frac{x - x_0}{\alpha}\right)^2\right] \times \frac{1}{\sqrt{2\pi}\zeta y}\exp\left[-\frac{1}{2}\left(\frac{\ln y - \lambda}{\zeta}\right)^2\right] \times \left\{1 + \alpha_{12}\left\{2\exp\left[-\frac{1}{2}\left(\frac{x - x_0}{\alpha}\right)^2\right] - 1\right\}\left[1 - 2\Phi\left(\frac{\ln y - \lambda}{\zeta}\right)\right]\right\}. \tag{E1}$$

Example 3.2 (cont.)

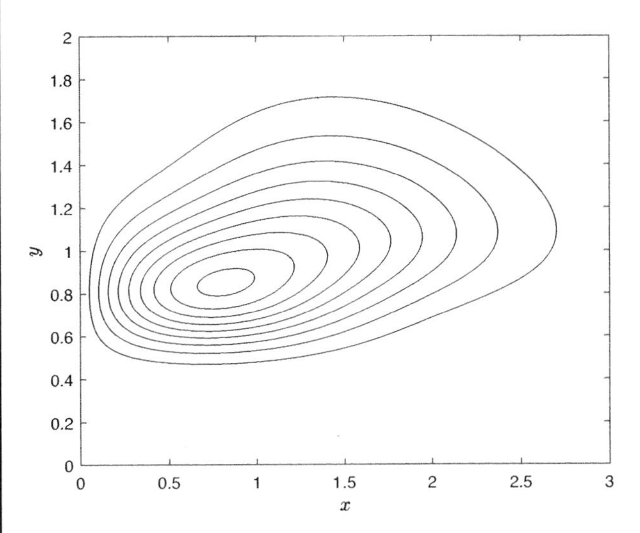

Figure 3.5 Contour plot of bivariate Morgenstern PDF with Rayleigh and lognormal marginals

For $\alpha = 1$, $x_0 = 0$, $\lambda = 0$, $\zeta = 0.3$, and $\alpha_{12} = 1$, Figure 3.5 shows contour plots of the above bivariate PDF. Using (3.29) and the Q_i values in Table 3.1, the correlation coefficient between the two random variables is $\rho_{12} = 4 \times 1 \times 0.280 \times \left[\Phi(0.3/\sqrt{2}) - 0.5\right]/\sqrt{\exp\left(0.3^2\right) - 1} = 0.307$. The moderate positive correlation between the two random variables is evident in the slightly right-tilted shape of the contours in Figure 3.5.

Example 3.3 – Bivariate Morgenstern Distribution with Normal Marginals

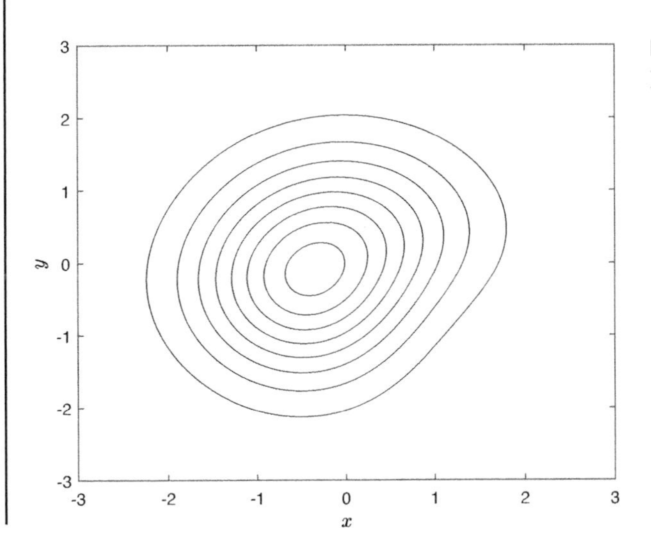

Figure 3.6 Contour plot of bivariate Morgenstern PDF with normal marginals

Example 3.3 (cont.)

Consider a bivariate Morgenstern distribution for random variables X and Y with standard normal marginals. For $\alpha_{12} = 1$, using (3.29) and the value of Q_i from Table 3.1, the correlation coefficient is $\rho_{12} = 4 \times 0.282 \times 0.282 = 0.318$. Figure 3.6 shows the contours of the bivariate Morgenstern PDF. It should be clear from the non-elliptical shapes of the contours that the joint distribution is not normal. This example demonstrates that having normal marginals does not necessarily imply a normal joint distribution.

3.5.2 Nataf Family of Multivariate Distributions

Consider the set of random variables $\mathbf{Z} = [Z_1, \ldots, Z_n]^{\mathrm{T}}$ obtained by marginally transforming random variables $\mathbf{X} = [X_1, \ldots, X_n]^{\mathrm{T}}$ according to the rule

$$Z_i = \Phi^{-1}[F_i(X_i)],\ i = 1, 2, \ldots, n, \tag{3.36}$$

where $\Phi^{-1}[\cdot]$ is the inverse of the standard normal CDF. We assume each CDF $F_i(x_i)$ is a continuous and strictly increasing function of its argument. Using rules for the distribution of functions of random variables described in Section 2.8, one can easily show that each Z_i has the standard normal distribution. The Nataf (1962) distribution for the random variables $\mathbf{X}$ is obtained by assuming that $\mathbf{Z}$ has the multinormal distribution with zero means, unit variances, and yet-to-be-determined correlation matrix $\mathbf{R}_0$. Using the rules in Section 2.8, one can easily show that the corresponding joint PDF of $\mathbf{X}$ is (Liu and Der Kiureghian 1986)

$$f_{\mathbf{X}}(\mathbf{x}) = \phi_n(\mathbf{z}, \mathbf{R}_0) \prod_{i=1}^{n} \frac{f_i(x_i)}{\phi(z_i)}, \tag{3.37}$$

where $z_i = \Phi^{-1}[F(x_i)]$, $\phi(\cdot)$ is the standard normal PDF, and $\phi_n(\cdot, \mathbf{R}_0)$ is the n-dimensional standard multinormal PDF with correlation matrix $\mathbf{R}_0$. The correlations between Z_i variables are obviously related to the correlations between the X_i variables. To determine the relationship between the two, we write the expression for the correlation coefficient ρ_{ij} between X_i and X_j in terms of the bivariate version of (3.37):

$$\begin{aligned}
\rho_{ij} &= \int_{-\infty}^{+\infty}\int_{-\infty}^{+\infty} \left(\frac{x_i - \mu_i}{\sigma_i}\right)\left(\frac{x_j - \mu_j}{\sigma_j}\right) \frac{f_i(x_i) f_j(x_j)}{\phi(z_i)\phi(z_j)} \phi_2\left(z_i, z_j, \rho_{0,ij}\right) \mathrm{d}x_i \mathrm{d}x_j \\
&= \int_{-\infty}^{+\infty}\int_{-\infty}^{+\infty} \left(\frac{x_i - \mu_i}{\sigma_i}\right)\left(\frac{x_j - \mu_j}{\sigma_j}\right) \phi_2\left(z_i, z_j, \rho_{0,ij}\right) \mathrm{d}z_i \mathrm{d}z_j,
\end{aligned} \tag{3.38}$$

where we have used the relation $\mathrm{d}z_i/\mathrm{d}x_i = f_i(x_i)/\phi(z_i)$ obtained from (3.36). Observe that ρ_{ij} is related to $\rho_{0,ij}$ through a double integral, the relation also possibly depending on the distribution parameters μ_i, μ_j, σ_i, and σ_j. The distribution model in (3.37) is valid provided the mappings in (3.36) are one-to-one and the correlation matrix $\mathbf{R}_0$ is positive definite. The first condition is satisfied when the CDFs $F_i(x_i)$ are continuous and strictly increasing. The second condition is discussed below.

Unfortunately, no closed-form solution of $\rho_{0,ij}$ in terms of ρ_{ij} and the distribution parameters is available in the general case. Because usually ρ_{ij} is known and $\rho_{0,ij}$ to be determined, a numerical trial-and-error type solution scheme seems to be the only choice. However, close examination of the relation in (3.38) reveals the following findings (see Liu and Der Kiureghian 1986):

1) In general, one can write $\rho_{0,ij} = C\rho_{ij}$, where $1 \leq C$ is a function of ρ_{ij} and possibly the distribution parameters. This finding implies that for uncorrelated variables, i.e., for $\rho_{ij} = 0$, we have $\rho_{0,ij} = 0$, which implies statistical independence of Z_i and Z_j and, therefore, of X_i and X_j. Therefore, for Nataf-distributed random variables, uncorrelatedness implies statistical independence. Furthermore, as C has a lower bound of 1, the correlation coefficients in the $\mathbf{Z}$ space are equal to or greater in magnitude than the corresponding correlation coefficients in the $\mathbf{X}$ space. In fact, the largest allowable value of $|\rho_{ij}|$ is that which yields $|\rho_{0,ij}| = 1$. For values of $|\rho_{ij}|$ larger than this limit, the Nataf distribution model is invalid.
2) C is independent of ρ_{ij} if either X_i or X_j is normal.
3) C is independent of the distribution parameters of X_i if X_i can be linearly transformed into a standard form, e.g., if X_i has one of the first six distributions listed in Table 3.1.
4) C is dependent on the c.o.v. of X_i if X_i cannot be linearly transformed into a standard form, e.g., if X_i has one of the last four distributions listed in Table 3.1.

Based on findings (2)–(4), we can state the following for the distributions listed in Table 3.1: (a) $C =$ constant if X_i is normal and X_j has a standardizable distribution; (b) $C = C(\delta_j)$ if X_i is normal and X_j has a non-standardizable distribution; (c) $C = C\left(\rho_{ij}\right)$ if both X_i and X_j have standardizable but non-normal distributions; (d) $C = C\left(\rho_{ij}, \delta_j\right)$ if X_i has a non-normal standardizable distribution and X_j has a non-standardizable distribution; and (e) $C = C\left(\rho_{ij}, \delta_i, \delta_j\right)$ if both X_i and X_j have non-standardizable distributions. Based on these findings, Liu and Der Kiureghian (1986) developed polynomial approximations of the function C by fitting to exact solutions obtained by numerical integration of (3.38). These approximate solutions for the above five cases are summarized in Tables A3.1 to A3.5 of Appendix 3A.

To provide further insight, Figure 3.7 shows plots of $\rho_{0,ij}$ versus ρ_{ij} for selected pairs of marginal distributions, including the case where both marginals are normal. For the lognormal marginals, $\delta_i = \delta_j = 0.5$ is assumed. For the case of normal marginals, obviously $\rho_{0,ij} = \rho_{ij}$ and the curve plots as a straight diagonal line. Observe that for all the other cases $|\rho_{ij}| \leq |\rho_{0,ij}|$, i.e., the correlation coefficients in the normal space are greater in magnitude than those in the original space. The difference in the case of the uniform marginals is extremely small so that the curve almost coincides with the diagonal line of the normal-normal case. Observe further that for the exponential-exponential and lognormal-lognormal cases, $\rho_{0,ij}$ reaches -1 for values of ρ_{ij} above -1, whereas for the uniform-uniform case no such limit exists. This has to do with the symmetric nature of the latter distribution and the skewed nature of the exponential and lognormal distributions. Naturally, for the exponential and lognormal distributions with correlation coefficient ρ_{ij} below these limits, the Nataf distribution model is not valid. Table 3.2 lists the upper and lower limits of admissible correlation coefficients ρ_{ij} for each pair of the standardizable distributions. For the non-standardizable distributions, the admissible interval depends on the c.o.v. but the limits are more or less similar to those listed in Table 3.2. In general, there is a limiting negative correlation coefficient when the two marginal distributions are similarly skewed (both to the right or both to the left) and a limiting positive correlation coefficient when the two

Table 3.2 Admissible interval of correlation coefficients for the Nataf model (after Liu and Der Kiureghian, 1986)

Marginal distribution	Normal	Uniform	Shifted exponential	Shifted Rayleigh	Type I largest	Type I smallest
Normal	$[-1, 1]$					
Uniform	$[-0.977, 0.977]$	$[-1, 1]$				
Shifted exponential	$[-0.903, 0.903]$	$[-0.866, 0.866]$	$[-0.645, 1]$			
Shifted Rayleigh	$[-0.986, 0.986]$	$[-0.970, 0.970]$	$[-0.819, 0.957]$	$[-0.947, 1]$		
Type I largest	$[-0.969, 0.969]$	$[-0.936, 0.936]$	$[-0.760, 0.981]$	$[-0.915, 0.993]$	$[-0.886, 1]$	
Type I smallest	$[-0.969, 0.969]$	$[-0.936, 0.936]$	$[-0.981, 0.780]$	$[-0.983, 0.915]$	$[-1, 0.886]$	$[-0.886, 1]$

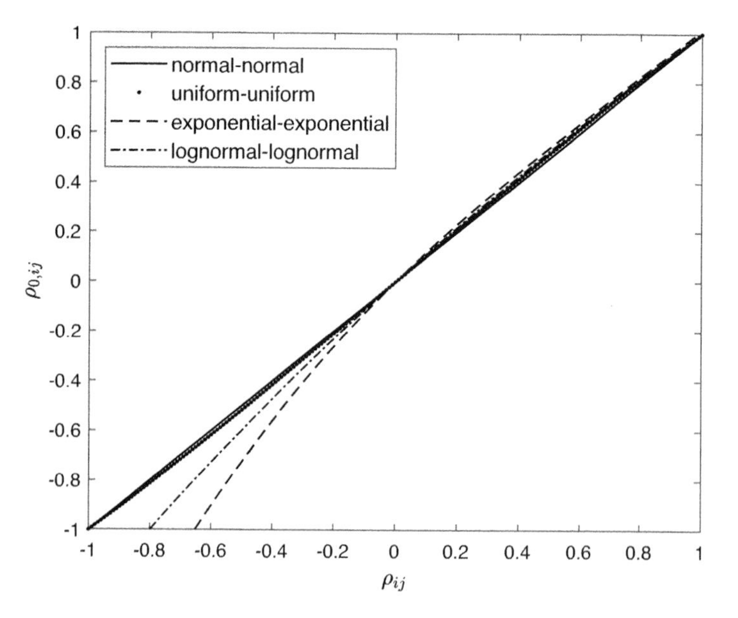

Figure 3.7 Relations between correlation coefficients for Nataf model with selected marginal distributions

distributions are skewed in opposite directions. It is important to note that the admissible interval of correlation coefficients is much wider for the Nataf model than for the Morgenstern model. This is a major advantage of the Nataf model over the Morgenstern model.

Example 3.4 – Bivariate Nataf Distribution with Rayleigh and Lognormal Marginals

Reconsider Example 3.2 with X having the shifted Rayleigh distribution with parameters α and x_0 and Y having the lognormal distribution with parameters λ and ζ. Let ρ denote the correlation coefficient between the two variables. The bivariate Nataf PDF with these marginals and correlation coefficient is defined as

$$f_{X_1X_2}(x_1, x_2) = \frac{x_1 - x_0}{\alpha} \exp\left[-\frac{1}{2}\left(\frac{x_1 - x_0}{\alpha}\right)^2\right] \times \frac{1}{\sqrt{2\pi}\zeta x_2} \exp\left[-\frac{1}{2}\left(\frac{\ln x_2 - \lambda}{\zeta}\right)^2\right] \times \frac{\phi_2(z_1, z_2, \rho_0)}{\phi(z_1)\phi(z_2)}. \tag{E1}$$

The correlation coefficient ρ_0 in the normal space is computed using the formula $\rho_0 = C\rho$. Because the shifted Rayleigh distribution is standardizable and the lognormal distribution is not, C must be a function of ρ and the c.o.v. of the lognormal distribution, δ. The latter is given by $\delta = \sqrt{\exp(\zeta^2) - 1}$. The fitted polynomial approximation for the case of shifted Rayleigh-lognormal marginals is given in Table A3.4 as $C(\rho, \delta) = 1.011 + 0.001\rho + 0.014\delta + 0.004\rho^2 + 0.231\delta - 0.130\rho\delta$. For $\alpha = 1$, $x_0 = 0$, $\lambda = 0$, $\zeta = 0.3$, and $\rho = 0.307$ (the same as the correlation coefficient for the Morgenstern model in Example 3.2), we obtain $\delta = 0.307$, $C = 1.026$, and $\rho_0 = 0.315$. Note that the correlation coefficient in the normal space is slightly larger than that in the original space. Figure 3.8 shows a contour plot of the above bivariate PDF for these

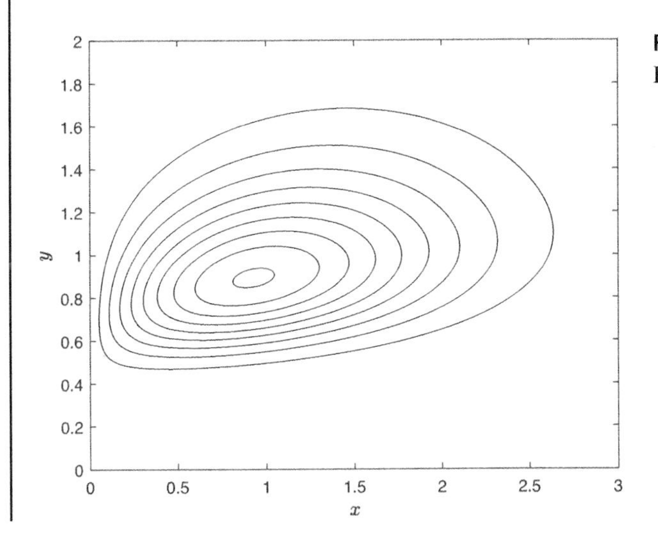

Figure 3.8 Contour plot of bivariate Nataf PDF with Rayleigh and lognormal marginal

Example 3.4 (cont.)

numerical values. Observe that these contours are slightly different from those of the bivariate Morgenstern model shown in Figure 3.5, indicating the difference between the two joint distributions.

Remark
It is easy to show that when all the marginal distributions $f_i(x_i)$ are normal, the Nataf model yields the multinormal distribution. This result is obtained by observing that when X_i are normal, the marginal transforms in (3.36) are linear. In that case, assuming Z_i are jointly normal implies the multinormal distribution for X_i. Furthermore, it is easily verified that when all the marginals are lognormal then the Nataf model yields the multivariate lognormal distribution.

3.5.3 Copula Distributions

Consider the marginal transformation of random variables X_i, $i = 1, \ldots, n$, in the form

$$Z_i = F_i(X_i). \tag{3.39}$$

One can easily verify that, provided the CDF $F(x_i)$ is continuous, each Z_i has the uniform distribution within the interval $[0, 1]$. Thus, the domain of the transformed random variables $\mathbf{Z} = [Z_1, \ldots, Z_n]^{\mathrm{T}}$ is the n-dimensional unit hyper-cube $[0, 1]^n$. To make the transformation in (3.39) reversible, we make the additional assumption that each marginal CDF $F_i(x_i)$ is a strictly increasing function of its argument.

A copula distribution is assigned to random variables $\mathbf{X} = [X_1, \ldots, X_n]^{\mathrm{T}}$ by selecting an appropriate joint CDF for $\mathbf{Z}$ within its unit hyper-cube domain. Denoted $C(z_1, \ldots, z_n)$, this is usually called a copula function. Owing to the one-to-one mapping in (3.39), we have $\Pr(X_1 \leq x_1 \cap \cdots \cap X_n \leq x_n) = \Pr(Z_1 \leq z_1 \cap \cdots \cap Z_n \leq z_n)$. It follows that

$$F_{\mathbf{X}}(\mathbf{x}) = C[F_1(x_1), \ldots, F_n(x_n)]. \tag{3.40}$$

The word *copula* means coupling or connecting things, here connecting the joint CDF to the marginal CDFs. To qualify as the joint CDF of $\mathbf{Z}$, the copula function must satisfy the following requirements:

1) $C(z_1, \ldots, z_n)$ must be non-decreasing with respect to each of its arguments;
2) $0 \leq C(z_1, \ldots, z_n) \leq 1$ for $\mathbf{z} \in [0, 1]^n$;
3) $C(z_1, \ldots z_{i-1}, 0, z_{i+1} \ldots, z_n) = 0$ for any i;
4) $C(1, \ldots, 1, z_i, 1, \ldots, 1) = z_i$ for any i.

Any function that satisfies the above requirements is a copula function. Figure 3.9 shows a typical copula function for two variables.

One particular class of copulas is the Archimedean class, which is defined by

$$C(z_1, \ldots, z_n) = \psi^{-1}(\psi(z_1) + \cdots + \psi(z_n)), \tag{3.41}$$

where the function ψ, known as the *generator*, must satisfy the following requirements: (a) $\lim_{z \to 0}\psi(z) = \infty$, (b) $\psi(1) = 0$, (c) $\psi'(z) < 0$, and (d) $0 < \psi''(z)$, where the prime indicates the derivative. Several well-known generator functions and their inverses are listed in Table 3.3. Observe that a single parameter controls these copula functions. In the bivariate case, this parameter can be related to the correlation coefficient between the two random variables. In higher dimensions, the single parameter obviously cannot describe the full correlation matrix of $\mathbf{X}$, if it were known. Hence, the copulas are more useful for modeling bivariate distributions. A particular feature of these copula distributions is that they can model tail dependence. This is described through an example below.

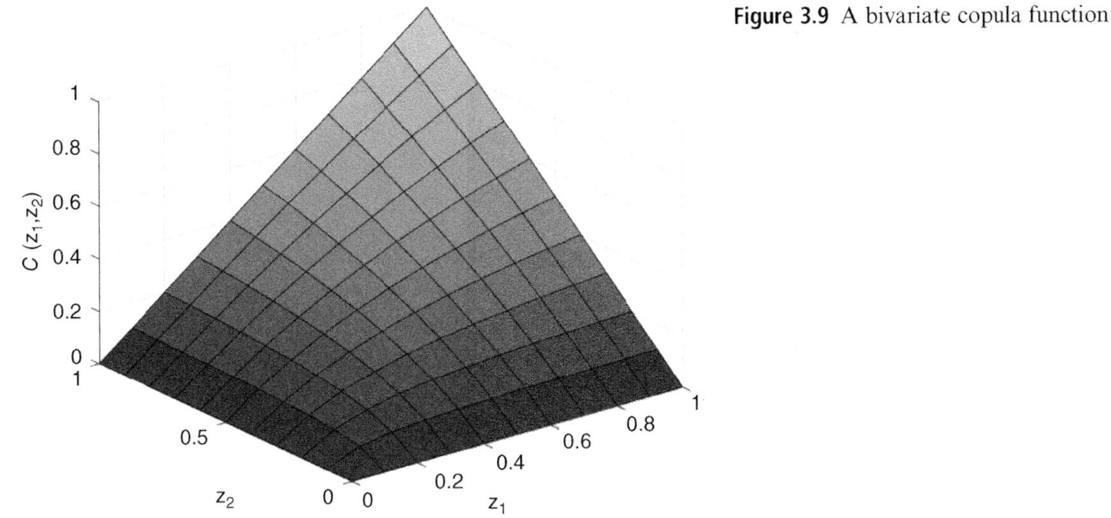

Figure 3.9 A bivariate copula function

Table 3.3 Selected generators for Archimedean copulas

Name	Generator $\psi(z)$	Generator inverse $\psi^{-1}(z)$	Parameter
Clayton	$\frac{1}{\theta}\left(z^{-\theta} - 1\right)$	$(1 + \theta z)^{-1/\theta}$	$0 < \theta$
Frank	$-\ln\left[\frac{\exp(-\theta z) - 1}{\exp(-\theta) - 1}\right]$	$\frac{1}{\theta}\ln\{1 - [1 - \exp(-\theta)\exp(-z)]\}$	$0 < \theta$
Gumbel	$(-\ln z)^{\theta}$	$\exp\left(-z^{1/\theta}\right)$	$0 < \theta$
Joe	$-\ln\left[1 - (1 - z)^{\theta}\right]$	$1 - [1 - \exp(-z)]^{1/\theta}$	$0 < \theta$

Another class of copulas is the elliptical class. Included in this class is the Gaussian copula, which is identical to the Nataf distribution, and the Student's T copula. These copulas have elliptical contours in a transformed space and, therefore, cannot describe tail dependence. When available data on a set of random variables suggests tail dependence, copulas other than elliptical ones should be selected. For a more in-depth coverage of copula distributions see Nelson (2006).

Example 3.5 – Bivariate Clayton Copula Distribution

Consider the Clayton copula with the generator defined in Table 3.3. For two random variables X and Y, the joint CDF is defined by

$$\begin{aligned} F_{XY}(x,y) &= \left\{1+\theta\left[\frac{1}{\theta}\left(F_X(x)^{-\theta}-1\right)+\frac{1}{\theta}\left(F_Y(y)^{-\theta}-1\right)\right]\right\}^{-1/\theta} \\ &= \left\{1+F_X(x)^{-\theta}-1+F_Y(y)^{-\theta}-1\right\}^{-1/\theta} \\ &= \left[F_X(x)^{-\theta}+F_Y(y)^{-\theta}-1\right]^{-1/\theta}. \end{aligned} \tag{E1}$$

The corresponding joint PDF is obtained by differentiation of the joint CDF with respect to x and y, which yields

$$f_{XY}(x,y) = (\theta+1)f_X(x)f_Y(y)[F_X(x)F_Y(y)]^{-\theta-1}\left[F_X(x)^{-\theta}+F_Y(y)^{-\theta}-1\right]^{-1/\theta-2}. \tag{E2}$$

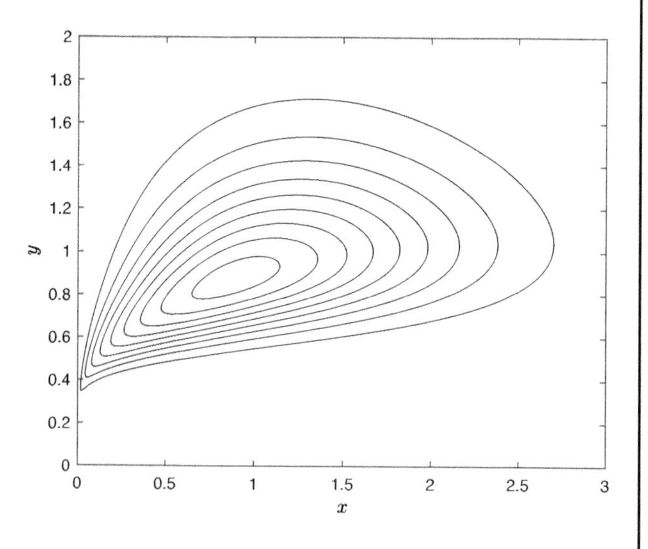

Figure 3.10 Contour plot of bivariate Clayton copula PDF with Rayleigh and lognormal marginals

As in Examples 3.2 and 3.4, we assume X has a shifted Rayleigh distribution with parameters $\alpha = 1$ and $x_0 = 0$, and Y has a lognormal distribution with parameters $\lambda = 0$ and $\zeta = 0.3$. Given these marginals and the above joint PDF, we can compute the correlation coefficient for any given value of θ. Through trial and error, we find that $\theta = 0.638$ yields the correlation

Example 3.5 (cont.)

coefficient $\rho = 0.307$, which is the value used in the previous examples. Figure 3.10 shows the contours of the above joint PDF for this value of θ. Observe that this joint PDF is significantly different from those in Figures 3.5 and 3.8 for the Morgenstern and Nataf models, respectively. In particular, the domain of the joint PDF is much narrower for the lower tail outcomes of x and y than for the central or large outcomes. This indicates strong dependence in the lower tail of the distribution. This kind of tail dependence (either lower or upper) is a special feature of copula distributions. When observed data indicate this kind of dependence, copula distributions are good candidates for fitting.

Remark

Example 3.5 shows that the correlation coefficient we have been using, known as the Pearson product-moment correlation coefficient, is not a good measure to describe tail dependence, which is essentially nonlinear in nature. There are other measures of dependence, such as Spearman's rank correlation coefficient and Kendall's rank correlation coefficient, that better describe nonlinear dependence (Corder and Foreman, 2014). We will not expand on this topic in this text, but the reader should be aware that there are measures of dependence between two random variables other than the Pearson correlation coefficient.

3.6 Transformation to the Standard Normal Space

In several reliability methods, advantage is gained by transforming the random variables $\mathbf{X} = [X_1, \ldots, X_n]^{\mathrm{T}}$ into standard normal random variables $\mathbf{U} = [U_1, \ldots, U_n]^{\mathrm{T}}$. In the most general case, this transformation can be written in the vector form $\mathbf{U} = \mathbf{T}(\mathbf{X})$, where the ith element of the vector is $U_i = T_i(X_1, \ldots, X_n)$. As shown below, in all cases the transformation can be formulated in a triangular form so that $U_i = T_i(X_1, \ldots, X_i)$. This section describes this transformation for different types of distributions, starting with a univariate distribution. In all cases, a one-to-one transformation is sought. That is, we would like to be able to go from either space to the other and have a unique solution. Conditions on the probability distributions to satisfy this requirement are described.

In addition to the transformation itself, we are interested in its inverse, $\mathbf{X} = \mathbf{T}^{-1}(\mathbf{U})$, and in its Jacobian matrix, $\mathbf{J}_{\mathbf{u},\mathbf{x}}$. The latter is the matrix of partial derivatives with elements $J_{ij} = \partial u_i / \partial x_j$, $i, j = 1, \ldots, n$. For each type of transformation, the inverse and the Jacobian matrix are derived.

The student may skip this section in first reading because this material will not be used until we introduce advanced reliability methods starting in Chapter 6. Detailed examples of transformations to the standard normal space and their Jacobians are given in Chapter 6 in the context of the first-order reliability method.

3.6.1 Single Random Variable

Consider a random variable X with CDF $F_X(x)$. We seek a one-to-one transformation $U = T(X)$ such that $U = \text{N}(0,1)$. Provided $T(\cdot)$ is a continuous and strictly increasing function, we can write $\Phi(u) = \Pr(U \le u) = \Pr(T(X) \le u) = \Pr\left(X \le T^{-1}(u)\right) = F_X(x)$, where $x = T^{-1}(u)$ is the inverse transform. It follows that the desired transformation is

$$U = \Phi^{-1}[F_X(X)], \tag{3.42}$$

where $\Phi^{-1}[\cdot]$ denotes the inverse of the standard normal CDF. We have in fact seen this transformation earlier, when defining the Nataf distribution (see Section 3.5.2). Two conditions on $F_X(x)$ are necessary to satisfy the requirement of one-to-one mapping of the above transformation: (a) $F_X(x)$ must not have discontinuities (jumps). Note that at such points a delta function appears in the PDF of X. If a discontinuity in the CDF of X occurs, then a range of outcomes of U map onto a single outcome of X, thus invalidating the one-to-one mapping requirement. (b) $F_X(x)$ must be strictly increasing, i.e., when plotted as a function of x, $F_X(x)$ should not be flat over any interval. If the curve is flat over an interval, then all outcomes of X over that interval map onto a single outcome of U, again invalidating the one-to-one mapping requirement.

It is important to note that (3.42) is not the only solution for a mapping of the random variable X onto the standard normal space. For example, $U = -\Phi^{-1}[F_X(X)]$ is an alternative solution. The difference between the two is that, whereas (3.42) is a strictly increasing function, $U = \Phi^{-1}[F_X(X)]$ is a strictly decreasing function. For our purposes, either will do. However, for consistency, we will use transformations that are strictly increasing.

The transformation in (3.42) is essentially defined by equating the cumulative probabilities of U and X. For this reason, it is called a *probability-preserving* transformation. Figure 3.11 shows this transformation when X is gamma-distributed with parameters $k = 3$ and $\lambda = 3$. The PDFs of X and U are also shown in the figure, attached to their respective axes. As can

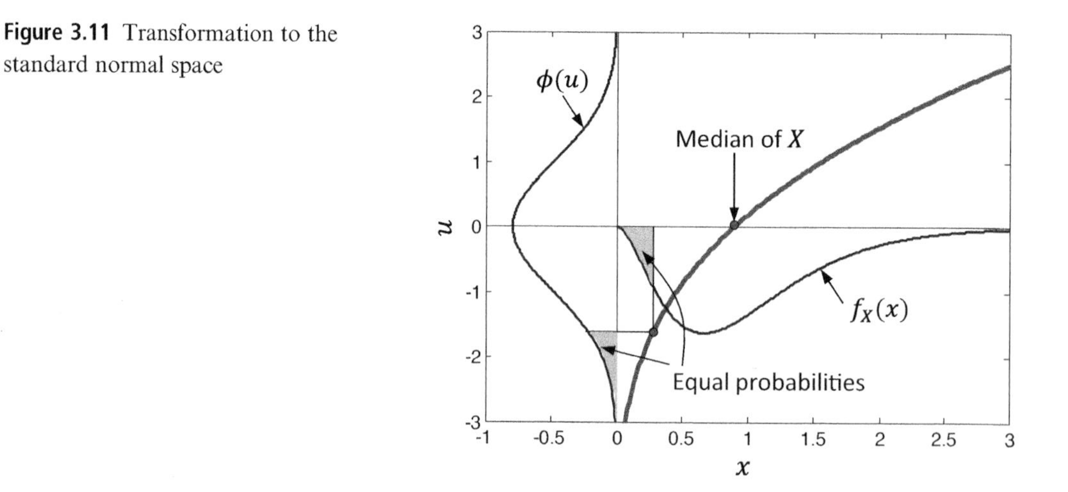

Figure 3.11 Transformation to the standard normal space

be seen, each pair of transformed values (x, u) corresponds to equal cumulative probabilities. The following points on the transformation curve are noteworthy:

- As x approaches its lower bound, u asymptotically approaches $-\infty$.
- As x approaches $+\infty$ in the extreme right tail of its distribution, u also approaches $+\infty$. Note that if the distribution of x were bounded on the right, u would asymptotically approach $+\infty$ as x approached its upper bound.
- The intercept of the transformation curve with the x axis marks the median of X.

The inverse of the transformation in (3.42) is

$$X = F_X^{-1}[\Phi(U)], \tag{3.43}$$

where $F_X^{-1}[\cdot]$ is the inverse of the CDF of X. The 1×1 Jacobian matrix is obtained by taking derivative of both sides of (3.42), yielding

$$\frac{\mathrm{d}u}{\mathrm{d}x} = \frac{f_X(x)}{\phi(u)}, \tag{3.44}$$

where $\phi(\cdot)$ is the standard normal PDF.

Two special cases are worthy of consideration. The first is the case of a normal random variable with mean μ_X and standard deviation σ_X. In this case, the transformation takes the linear form $U = (X - \mu_X)/\sigma_X$. The corresponding inverse transform is $X = \mu_X + \sigma_X U$ and the Jacobian is the scalar quantity $\mathrm{d}u/\mathrm{d}x = 1/\sigma_X$. The second is the case of a lognormal random variable with parameters λ and ζ. The transformation takes the form $U = (\ln X - \lambda)/\zeta$. Its inverse is $X = \exp(\lambda + \zeta U)$ and the scalar Jacobian is $\mathrm{d}u/\mathrm{d}x = 1/(\zeta x)$. In this case the transformation is nonlinear. Indeed, for all non-normal distributions, the transformation to the standard normal space is nonlinear. In some cases, closed-form solutions usually involving the functions $\Phi(\cdot)$ or $\phi(\cdot)$ can be obtained for the transform, its inverse, and the Jacobian. In other cases, these quantities must be computed numerically. Many examples are presented in later chapters of this book.

3.6.2 Statistically Independent Random Variables

Consider a vector of statistically independent random variables $\mathbf{X} = [X_1, \ldots, X_n]^{\mathrm{T}}$ with marginal CDFs $F_i(x_i)$, $i = 1, \ldots, n$. As the random variables are statistically independent, these marginal distributions completely define the joint distribution. To construct a transformation to the standard normal space, we use (3.42) for each member of the vector. Thus, the transformation takes the form

$$U_i = \Phi^{-1}[F_i(X_i)], i = 1, \ldots, n \tag{3.45}$$

and the inverse takes the form

$$X_i = F_i^{-1}[\Phi(U_i)], i = 1, \ldots, n. \tag{3.46}$$

The above transformations are valid provided each marginal CDF is continuous and strictly increasing. Furthermore, the $n \times n$ Jacobian matrix $\mathbf{J}_{\mathbf{u},\mathbf{x}}$ in this case is diagonal and, using (3.44), has the elements

$$J_{ij} = \frac{\partial u_i}{\partial x_i} = \begin{cases} \dfrac{f_i(x_i)}{\phi(u_i)}, & i = j \\ 0, & i \neq j, \end{cases} \tag{3.47}$$

where $f_i(x_i)$ is the PDF of X_i.

3.6.3 Multinormal Random Variables

Consider a vector of jointly normal random variables $\mathbf{X} = [X_1, \ldots, X_n]^{\mathrm{T}}$ with mean vector $\mathbf{M}_{\mathbf{X}}$ and covariance matrix $\mathbf{\Sigma}_{\mathbf{XX}}$. Because linear functions of jointly normal random variables are jointly normal (see Example 2.6), the transformation to the standard normal space is linear in this case. Let the transformation be of the form

$$\mathbf{U} = \mathbf{AX} + \mathbf{B}, \tag{3.48}$$

where $\mathbf{A}$ is an $n \times n$ deterministic coefficient matrix and $\mathbf{B}$ is an $n \times 1$ vector of deterministic constants. From (2.91) we have $\mathbf{M}_{\mathbf{U}} = \mathbf{AM}_{\mathbf{X}} + \mathbf{B}$ and from (2.92) we have $\mathbf{\Sigma}_{\mathbf{UU}} = \mathbf{A\Sigma}_{\mathbf{XX}}\mathbf{A}^{\mathrm{T}}$. $\mathbf{U}$ has zero means, so we must have $\mathbf{B} = -\mathbf{AM}_{\mathbf{X}}$. Furthermore, because U_i are uncorrelated and have unit variances, the covariance matrix of $\mathbf{U}$ must be an $n \times n$ identity matrix. Thus, we must have $\mathbf{A\Sigma}_{\mathbf{XX}}\mathbf{A}^{\mathrm{T}} = \mathbf{I}$. It follows that $\mathbf{A}$ should be a matrix that diagonalizes the covariance matrix of $\mathbf{X}$. One option for $\mathbf{A}$ is the eigenmatrix of $\mathbf{\Sigma}_{\mathbf{XX}}$. However, a more efficient solution is the Cholesky decomposition of the covariance matrix. Provided $\mathbf{\Sigma}_{\mathbf{XX}}$ is positive definite, one can write $\mathbf{\Sigma}_{\mathbf{XX}} = \mathbf{LL}^{\mathrm{T}}$, where $\mathbf{L}$ is the lower triangular Cholesky decomposition matrix of $\mathbf{\Sigma}_{\mathbf{XX}}$. Substituting for $\mathbf{\Sigma}_{\mathbf{XX}}$, we have $\mathbf{ALL}^{\mathrm{T}}\mathbf{A}^{\mathrm{T}} = \mathbf{I}$. It follows that selecting $\mathbf{A} = \mathbf{L}^{-1}$ accomplishes our aim. Hence, the desired transformation has the form

$$\mathbf{U} = \mathbf{L}^{-1}(\mathbf{X} - \mathbf{M}_{\mathbf{X}}). \tag{3.49}$$

Appendix 3B presents the steps for computing the lower-triangular matrix $\mathbf{L}$ and its inverse.

Although the above formulation is valid, from the viewpoint of numerical analysis it is better to perform the Cholesky decomposition on the correlation matrix. Recall that the covariance matrix can be written as $\mathbf{\Sigma}_{\mathbf{XX}} = \mathbf{D}_{\mathbf{X}}\mathbf{R}_{\mathbf{XX}}\mathbf{D}_{\mathbf{X}}$, where $\mathbf{D}_{\mathbf{X}}$ is the diagonal matrix of standard deviations and $\mathbf{R}_{\mathbf{XX}}$ the correlation matrix of $\mathbf{X}$. Applying the Cholesky decomposition to the correlation matrix, $\mathbf{R}_{\mathbf{XX}} = \mathbf{LL}^{\mathrm{T}}$, and substituting for the covariance matrix in the earlier expression, we obtain $\mathbf{AD}_{\mathbf{X}}\mathbf{LL}^{\mathrm{T}}\mathbf{D}_{\mathbf{X}}\mathbf{A}^{\mathrm{T}} = \mathbf{I}$. It follows that $\mathbf{A} = \mathbf{L}^{-}\mathbf{D}_{\mathbf{X}}^{-1}$ should be selected. Hence, the desired transformation now takes the form

$$\mathbf{U} = \mathbf{L}^{-1}\mathbf{D}_{\mathbf{X}}^{-1}(\mathbf{X} - \mathbf{M}_{\mathbf{X}}), \tag{3.50}$$

where $\mathbf{L}$ now denotes the Cholesky decomposition of the correlation matrix. The advantage of (3.50) over (3.49) is in the more favorable behavior of the correlation matrix for performing the decomposition. Whereas the elements of $\boldsymbol{\Sigma}_{\mathbf{XX}}$ can have vastly different magnitudes depending on the selected dimensions of the elements $\mathbf{X}$, the correlation matrix is dimensionless and its elements are bounded within the interval $[-1, +1]$. Because of this, the decomposition of $\mathbf{R}_{\mathbf{XX}}$ is numerically more stable than that of $\boldsymbol{\Sigma}_{\mathbf{XX}}$. For this reason, the transformation in (3.50) is preferred.

The inverse of the transformation in (3.50) is

$$\mathbf{X} = \mathbf{M}_{\mathbf{X}} + \mathbf{D}_{\mathbf{X}}\mathbf{L}\mathbf{U}. \tag{3.51}$$

Furthermore, the Jacobian of the transformation is given by

$$\mathbf{J}_{\mathbf{u},\mathbf{x}} = \mathbf{L}^{-1}\mathbf{D}_{\mathbf{X}}^{-1}. \tag{3.52}$$

Because $\mathbf{L}$ is triangular, its inverse is easy to compute. Furthermore, $\mathbf{D}_{\mathbf{X}}$ is diagonal and its inverse is readily available. The inverse of the above Jacobian, which is also used in reliability analysis, is given by

$$\mathbf{J}_{\mathbf{u},\mathbf{x}}^{-1} = \mathbf{J}_{\mathbf{x},\mathbf{u}} = \mathbf{D}_{\mathbf{X}}\mathbf{L}. \tag{3.53}$$

3.6.4 Nataf-Distributed Random Variables

Consider a vector of Nataf-distributed random variables $\mathbf{X} = [X_1, \ldots, X_n]^{\mathrm{T}}$ defined by the marginal CDFs $F_i(x_i)$, $i = 1, \ldots, n$, and correlation matrix $\mathbf{R}_{\mathbf{XX}}$ having the elements ρ_{ij}, $i,j = 1, \ldots, n$. Recall that the Nataf distribution is defined by assigning the multinormal distribution to the standard normal random variables $\mathbf{Z} = [Z_1, \ldots, Z_n]^{\mathrm{T}}$ obtained by the marginal transformations in (3.36). Here $\mathbf{Z}$ has zero means and unit variances so its distribution is completely defined by its correlation matrix $\mathbf{R}_0$. As seen in Section 3.5.2, the elements $\rho_{0,ij}$ of $\mathbf{R}_0$ are given in terms of ρ_{ij} and possibly the parameters of the marginal distributions, such as the c.o.v. δ_i and δ_j, when these distributions are not standardizable.

The transformation to the standard normal space for Nataf-distributed random variables is achieved by simply transforming the vector of multinormal random variables $\mathbf{Z}$ to the standard normal space. Using (3.50), and noting that the mean vector of $\mathbf{Z}$ is zero and its diagonal matrix of standard deviations is an identity matrix, we have

$$\begin{aligned}\mathbf{U} &= \mathbf{L}_0^{-1}\mathbf{Z} \\ &= \mathbf{L}_0^{-1}\left\{\begin{matrix}\Phi^{-1}[F_1(\mathbf{X}_1)] \\ \vdots \\ \Phi^{-1}[F_n(\mathbf{X}_n)]\end{matrix}\right\},\end{aligned} \tag{3.54}$$

where $\mathbf{L}_0$ is the Cholesky decomposition of $\mathbf{R}_0$, i.e., $\mathbf{R}_0 = \mathbf{L}_0\mathbf{L}_0^{\mathrm{T}}$, and use has been made of (3.36) in the second line. It is seen that this transformation is a combination of the transformation in (3.44) for statistically independent random variables and the transformation in (3.50) for correlated normal random variables.

The inverse transform for Nataf-distributed random variables is given by

$$X_i = F_i^{-1}[\Phi(Z_i)], i = 1, \ldots, n, \tag{3.55}$$

$$\mathbf{Z} = \mathbf{L}_0\mathbf{U}. \tag{3.56}$$

Furthermore, the Jacobian of the transformation is

$$\begin{aligned} \mathbf{J}_{\mathbf{u},\mathbf{x}} &= \mathbf{J}_{\mathbf{u},\mathbf{z}}\mathbf{J}_{\mathbf{z},\mathbf{x}} \\ &= \mathbf{L}_0^{-1}\mathrm{diag}\left[\frac{f_i(x_i)}{\phi(z_i)}\right], \end{aligned} \tag{3.57}$$

where $\mathrm{diag}[a_{ii}]$ denotes a diagonal matrix with a_{ii} as its ith diagonal element and $z_i = \Phi^{-1}[F_i(x_i)]$. Observe that the above Jacobian has a triangular form. The Jacobian of the inverse transform is given by

$$\mathbf{J}_{\mathbf{x},\mathbf{u}} = \mathrm{diag}\left[\frac{\phi(z_i)}{f(x_i)}\right]\mathbf{L}_0. \tag{3.58}$$

3.6.5 General Non-Normal Random Variables

Consider the general case of a vector of random variables $\mathbf{X} = [X_1, \ldots, X_n]^{\mathrm{T}}$ defined by a joint CDF $F_{\mathbf{X}}(\mathbf{x})$ and the corresponding joint PDF $f_{\mathbf{X}}(\mathbf{x})$. The transformation described below was first proposed for structural reliability analysis by Hohenbichler and Rackwitz (1981) based on the earlier work of Rosenblatt (1952). Although this transformation was formulated much earlier by Segal (1938), in the structural reliability community it is known as the "Rosenblatt transformation." We note that this transformation is the only choice available for Morgenstern and copula distributions.

We define the conditional CDFs

$$F_{i|1,\ldots,i-1}(x_i|x_1, \ldots, x_{i-1}) = \Pr(X_i \leq x_i|X_1 = x_1, \ldots, X_{i-1} = x_{i-1}), i = 1, \ldots, n. \tag{3.59}$$

As we have seen in (2.71) and Section 3.4, the joint distribution of random variables is sometimes specified in terms of these conditional distributions. There are two ways of deriving the above conditional CDFs, if they are not readily available. First, using the joint CDF, we can write

$$F_{i|1,\ldots,i-1}(x_i|x_1, \ldots, x_{i-1}) = \frac{\partial^{i-1}F_{1,\ldots,i}(x_1, \ldots, x_i)}{\partial x_1 \cdots \partial x_{i-1}} \frac{1}{f_{1,\ldots,n-1}(x_1, \ldots, x_{i-1})}, \tag{3.60}$$

where $F_{1,\ldots,i}(x_1, \ldots, x_i)$ is the joint CDF of the first i random variables. Note that this function can be obtained from $F_{\mathbf{X}}(\mathbf{x})$ by setting $x_{i+1} = \cdots = x_n = \infty$. The second formulation uses the conditional PDF,

$$F_{i|1,\ldots,i-1}(x_i|x_1,\ldots,x_{i-1}) = \int_{-\infty}^{x_i} f_{i|1,\ldots,i-1}(x_i|x_1,\ldots,x_{i-1})\mathrm{d}x_i, \tag{3.61}$$

in which $f_{i|1,\ldots,i-1}(x_i|x_1,\ldots,x_{i-1})$ is the conditional PDF of X_i given $X_1 = x_1,\ldots,$ $X_{i-1} = x_{i-1}$. This function is obtained by using the relationship

$$f_{i|1,\ldots,i-1}(x_i|x_1,\ldots,x_{i-1}) = \frac{f_{1,\ldots,i}(x_1,\ldots,x_{i-1},x_i)}{f_{1,\ldots,i-1}(x_1,\ldots,x_{i-1})}. \tag{3.62}$$

As can be seen, the formulation in (3.60) requires differentiation, whereas that in (3.61) requires integration not only of the conditional PDF but also for determining the lower-order joint PDFs needed in (3.62), which must be obtained from $f_{\mathbf{X}}(\mathbf{x})$ by integration over the unlisted variables. Clearly, the first option is more convenient if $F_{\mathbf{X}}(\mathbf{x})$ is given in an analytical form. We note, however, that in many practical cases the joint distribution of dependent random variables is available in the form of conditional PDFs, in which case the formulation in (3.61) is convenient.

Given the conditional CDFs defined in (3.59), the Rosenblatt transformation is formulated as

$$\begin{aligned} U_1 &= \Phi^{-1}[F_1(X_1)] \\ U_2 &= \Phi^{-1}\left[F_{2|1}(X_2|X_1)\right] \\ &\cdots \\ U_n &= \Phi^{-1}\left[F_{n|1,\ldots,n-1}(X_n|X_1,\ldots,X_{n-1})\right]. \end{aligned} \tag{3.63}$$

This transformation is valid and invertible provided each conditional CDF is a continuous and strictly increasing function of its arguments. Observe that the transformation has a triangular form so that U_i is a function of only X_1 to X_i. The inverse transform takes the form

$$\begin{aligned} X_1 &= F_1^{-1}[\Phi(U_1)] \\ X_2 &= F_{2|1}^{-1}[\Phi(U_2)|X_1] \\ &\cdots \\ X_n &= F_{n|1,\ldots,n-1}^{-1}[\Phi(U_n)|X_1,\ldots,X_{n-1}]. \end{aligned} \tag{3.64}$$

It is important to note that the inverse transform represents a set of equations in a triangular form. Given $\mathbf{U} = (U_1,\ldots,U_n)$, from the first equation one determines X_1 in terms of U_1. Then, given U_2 and X_1, one determines X_2 from the second equation, and so on. In this manner, each equation involves a single unknown. Therefore, when analytical solutions are not available, which most often is the case, a numerical root-finding scheme is used to determine the single unknown in each equation from top to bottom. A Newton–Raphson scheme is most effective for this purpose, particularly when the derivatives of the functions $F_{i|1,\ldots,i-1}(x_i|x_1,\ldots,x_{i-1})$, i.e., the conditional PDFs $f_{i|1,\ldots,i-1}(x_i|x_1,\ldots,x_{i-1})$, are available.

The Jacobian matrix, $\mathbf{J}_{\mathbf{u},\mathbf{x}}$, associated with the Rosenblatt transform is obtained by differentiating (3.63). Its elements are defined by

$$J_{ij} = \begin{cases} 0, & i < j, \\ \dfrac{f_{i|1,\ldots,i-1}(x_i|x_1,\ldots,x_{i-1})}{\phi(u_i)}, & i = j, \\ \dfrac{1}{\phi(u_i)} \dfrac{\partial F_{i|1,\ldots,i-1}(x_i|x_1,\ldots,x_{i-1})}{\partial x_j}, & j < i. \end{cases} \tag{3.65}$$

Observe that the above Jacobian has a triangular form. As a result, its inverse is easily obtained, as shown in (B3.2) of Appendix 3B.

It is important to note that the Rosenblatt transformation depends on the ordering of random variables. As we will see later, the results obtained with certain approximate reliability methods using this transformation may slightly change depending on the ordering of the random variables. On the other hand, reliability results obtained with the transformations described in the preceding sections, including that of Nataf-distributed random variables, are independent of the ordering of the random variables.

APPENDIX 3A

Correlation Coefficients for Nataf Distribution

The following five tables, adopted from Liu and Der Kiureghian (1986), present formulas for the factor C in the identity $\rho_{0,ij} = C\rho_{ij}$, where ρ_{ij} is the correlation coefficient between two random variables X_i and X_j and $\rho_{0,ij}$ is the correlation coefficient between their marginally transformed standard normal variates, $Z_i = \Phi^{-1}[F_i(X_i)], i = 1, 2$. As described in Section 3.5.2, C is independent of ρ_{ij} if either of the two random variables is normal and it is dependent on the c.o.v. δ_i of X_i if the distribution cannot be linearly transformed into a standard form. The five tables consider all pairs of the random variables listed in Table 3.1. Where exact solutions are not available, approximate formulas are developed by fitting second- or third-order polynomials to numerically obtained exact results. The underlying error in the value of C in all cases is smaller than 5% and in most cases it is much smaller than 1%. In Tables A3.3–A3.5, ρ stands for ρ_{ij}.

Table A3.1 Factor C for X_i normal and X_j a standardizable variable (after Liu and Der Kiureghian, 1986)

Distribution of X_j	C
Uniform	1.023
Shifted exponential	1.107
Shifted Rayleigh	1.014
Type I largest value	1.031
Type I smallest value	1.031

Table A3.2 Factor C for X_i normal and X_j a non-standardizable variable (after Liu and Der Kiureghian, 1986)

Distribution of X_j	C
Lognormal	$\frac{\delta_j}{\sqrt{\ln\left(1+\delta_j^2\right)}}$
Gamma	$1.001 - 0.007\delta_j + 0.118\delta_j^2$
Type II largest value	$1.030 + 0.238\delta_j + 0.364\delta_j^2$
Type III smallest value	$1.031 - 0.195\delta_j + 0.328\delta_j^2$

Table A3.3 Factor C for X_i and X_j both standardizable variables (after Liu and Der Kiureghian, 1986)

Distribution of X_i	Distribution of X_j				
	Uniform	Shifted exponential	Shifted Rayleigh	Type I largest value	Type I smallest value
Uniform	$1.047 - 0.047\rho^2$				
Shifted exponential	$1.133 + 0.029\rho^2$	$1.229 - 0.367\rho + 0.153\rho^2$			
Shifted Rayleigh	$1.038 - 0.008\rho^2$	$1.123 - 0.100\rho + 0.021\rho^2$	$1.028 - 0.029\rho$		
Type I largest value	$1.055 + 0.015\rho^2$	$1.142 - 0.154\rho + 0.031\rho^2$	$1.046 - 0.045\rho + 0.006\rho^2$	$1.064 - 0.069\rho + 0.005\rho^2$	
Type I smallest value	$1.055 + 0.015\rho^2$	$1.142 + 0.154\rho + 0.031\rho^2$	$1.046 + 0.045\rho + 0.006\rho^2$	$1.064 + 0.069\rho + 0.005\rho^2$	$1.064 - 0.069\rho + 0.005\rho^2$

Table A3.4 Factor C for X_i standardizable and X_j non-standardizable variables (after Liu and Der Kiureghian, 1986)

Distribution of X_i	Distribution X_j			
	Lognormal	Gamma	Type II largest value	Type II smallest value
Uniform	$1.019 + 0.014\delta_j + 0.010\rho^2 + 0.249\delta_j^2$	$1.023 - 0.007\delta_j + 0.002\rho^2 + 0.127\delta_j^2$	$1.033 + 0.305\delta_j + 0.074\rho^2 + 0.405\delta_j^2$	$1.061 - 0.237\delta_j - 0.005\rho^2 + 0.379\delta_j^2$
Shifted exponential	$1.098 + 0.003\rho + 0.019\delta_j + 0.025\rho^2 + 0.303\delta_j^2 - 0.437\rho\delta_j$	$1.104 + 0.003\rho - 0.008\delta_j + 0.014\rho^2 + 0.173\delta_j^2 - 0.296\rho\delta_j$	$1.109 - 0.152\rho + 0.361\delta_j + 0.130\rho^2 + 0.455\delta_j^2 - 0.728\rho\delta_j$	$1.147 + 0.145\rho - 0.271\delta_j + 0.010\rho^2 + 0.459\delta_j^2 - 0.467\rho\delta_j$
Shifted Rayleigh	$1.011 + 0.001\rho + 0.014\delta_j + 0.004\rho^2 + 0.231\delta_j^2 - 0.130\rho\delta_j$	$1.104 + 0.001\rho - 0.007\delta_j + 0.002\rho^2 + 0.126\delta_j^2 - 0.090\rho\delta_j$	$1.036 - 0.038\rho + 0.266\delta_j + 0.028\rho^2 + 0.383\delta_j^2 - 0.229\rho\delta_j$	$1.047 + 0.042\rho - 0.212\delta_j + 0.353\delta_j^2 - 0.136\rho\delta_j$
Type I largest value	$1.029 + 0.001\rho + 0.014\delta_j + 0.004\rho^2 + 0.233\delta_j^2 - 0.197\rho\delta_j$	$1.031 + 0.001\rho - 0.007\delta_j + 0.003\rho^2 + 0.131\delta_j^2 - 0.132\rho\delta_j$	$1.056 - 0.060\rho + 0.263\delta_j + 0.020\rho^2 + 0.383\delta_j^2 - 0.332\rho\delta_j$	$1.064 + 0.065\rho - 0.210\delta_j + 0.003\rho^2 + 0.356\delta_j^2 - 0.211\rho\delta_j$
Type I smallest value	$1.029 - 0.001\rho + 0.014\delta_j + 0.004\rho^2 + 0.233\delta_j^2 + 0.197\rho\delta_j$	$1.031 - 0.001\rho - 0.007\delta_j + 0.003\rho^2 + 0.131\delta_j^2 + 0.132\rho\delta_j$	$1.056 + 0.060\rho + 0.263\delta_j + 0.020\rho^2 + 0.383\delta_j^2 + 0.332\rho\delta_j$	$1.064 - 0.065\rho - 0.210\delta_j + 0.003\rho^2 + 0.356\delta_j^2 + 0.211\rho\delta_j$

Table A3.5 Factor C for X_i and X_j both non-standardizable variables (after Liu and Der Kiureghian, 1986)

Distribution of X_i	Distribution X_j			
	Lognormal	Gamma	Type II largest value	Type II smallest value
Lognormal	$\frac{\ln(1+\rho\delta_i\delta_j)}{\rho\sqrt{\ln(1+\delta_i^2)\ln(1+\delta_j^2)}}$			
Gamma	$1.001+0.033\rho+0.004\delta_i-0.016\delta_j+0.002\rho^2+0.223\delta_i^2+0.130\delta_j^2+0.029\delta_i\delta_j-0.104\rho\delta_i-0.119\rho\delta_j$	$1.002+0.022\rho-0.012(\delta_i+\delta_j)+0.001\rho^2+0.125(\delta_i^2+\delta_j^2)+0.014\delta_i\delta_j-0.077\rho(\delta_i+\delta_j)$		
Type II largest value	$1.026+0.082\rho-0.019\delta_i+0.222\delta_j+0.018\rho^2+0.288\delta_i^2+0.379\delta_j^2+0.126\delta_i\delta_j-0.441\rho\delta_i-0.277\rho\delta_j$	$1.029+0.056\rho-0.030\delta_i+0.225\delta_j+0.012\rho^2+0.174\delta_i^2+0.379\delta_j^2+0.075\delta_i\delta_j-0.313\rho\delta_i-0.182\rho\delta_j$	$1.086+0.054\rho+0.104(\delta_i+\delta_j)-0.055\rho^2+0.662(\delta_i^2+\delta_j^2)+0.203\delta_i\delta_j-0.570\rho(\delta_i+\delta_j)-0.020\rho^3-0.218(\delta_i^3+\delta_j^3)-0.371\rho(\delta_i^2+\delta_j^2)+0.257\rho^2(\delta_i+\delta_j)+0.141\delta_i\delta_j(\delta_i+\delta_j)$	
Type III smallest value	$1.031+0.052\rho+0.011\delta_i-0.210\delta_j+0.002\rho^2+0.220\delta_i^2+0.350\delta_j^2+0.009\delta_i\delta_j+0.005\rho\delta_i-0.174\rho\delta_j$	$1.032+0.034\rho-0.007\delta_i-0.202\delta_j+0.121\delta_i^2+0.339\delta_j^2+0.003\delta_i\delta_j-0.006\rho\delta_i-0.111\rho\delta_j$	$1.065+0.146\rho+0.241\delta_i-0.259\delta_j+0.013\rho^2+0.372\delta_i^2+0.435\delta_j^2+0.034\delta_i\delta_j+0.005\rho\delta_i-0.481\rho\delta_j$	$1.063-0.004\rho-0.200(\delta_i+\delta_j)-0.001\rho^2+0.337(\delta_i^2+\delta_j^2)-0.007\delta_i\delta_j+0.007\rho(\delta_i+\delta_j)$

APPENDIX 3B

Brief on Cholesky Decomposition

Let $\mathbf{A}$ be a symmetric, positive-definite matrix having the elements a_{ij}, $i,j = 1, \ldots, n$. We wish to construct a lower-triangular matrix $\mathbf{L}$ and its inverse $\mathbf{L}^{-1}$, which is also lower triangular, such that $\mathbf{LL}^{\mathrm{T}} = \mathbf{A}$. $\mathbf{L}$ is known as the Cholesky decomposition of $\mathbf{A}$. It should be clear that $\mathbf{L}^{-1}\mathbf{AL}^{-\mathrm{T}} = \mathbf{I}$, where the superposed $-\mathrm{T}$ denotes the transposed inverse and $\mathbf{I}$ denotes the identity matrix. Furthermore, the inverse of $\mathbf{A}$ is given by $\mathbf{A}^{-1} = \mathbf{L}^{-\mathrm{T}}\mathbf{L}^{-1}$. The following material can be found in many numerical analysis texts, e.g., Golub and van Loan (1996).

1) Determination of $\mathbf{L}$: Let λ_{ij}, $i,j = 1, \ldots, n$, denote the elements of $\mathbf{L}$. Compute

$$\begin{aligned} \lambda_{ij} &= \sqrt{a_{ij} - \sum_{k=1}^{i-1} \lambda_{ik}^2}, \quad i = j, \\ &= \frac{1}{\lambda_{jj}} \left(a_{ij} - \sum_{k=1}^{j-1} \lambda_{ik}\lambda_{jk} \right), \quad j < i, \\ &= 0, \quad i < j. \end{aligned} \tag{B3.1}$$

The elements are computed row-wise from left to right, i.e., in the order λ_{11}, λ_{21}, λ_{22}, λ_{31}, λ_{32}, λ_{33}, etc.

2) Determination of $\mathbf{L}^{-1}$: Let γ_{ij}, $i,j = 1, \ldots, n$, denote the elements of $\mathbf{L}^{-1}$. Compute

$$\begin{aligned} \gamma_{ij} &= \frac{1}{\lambda_{ii}}, \quad i = j, \\ &= -\frac{1}{\lambda_{ii}} \sum_{k=j}^{i-1} \lambda_{ik}\gamma_{kj}, \quad j < i, \\ &= 0, \quad i < j. \end{aligned} \tag{B3.2}$$

The elements are computed row-wise from the diagonal to the left, i.e., in the order γ_{11}, γ_{22}, γ_{21}, γ_{33}, γ_{32}, γ_{31}, etc.

PROBLEMS

3.1 Random variables X and Y are jointly lognormal with means $\mu_X = 100$ and $\mu_Y = 200$, standard deviations $\sigma_X = 40$ and $\sigma_Y = 60$, and correlation coefficient $\rho_{XY} = 0.5$.

1) Write an expression for the joint PDF $f_{XY}(x, y)$ and plot its contours.
2) Write an expression for the conditional PDF $f_{X|Y}(x|y)$ and plot it for $y = 100$, 200, and 300. On the same graph, plot the marginal PDF $f_X(x)$. Observe the change in the conditional PDF of X depending on the given value of Y.

3.2 Let X have the gamma distribution with shape parameter k and an uncertain scale parameter Λ. Suppose Λ has the exponential distribution $f_\Lambda(\lambda) = \alpha \exp(-\alpha\lambda)$, $0 < \lambda$. Note that $1/\alpha$ is the mean of Λ.

1) Derive the joint PDF of X and Λ and plot its contours for $k = 6$ and $\alpha = 0.8$.
2) Derive the marginal distribution of X.
3) Plot the marginal distribution of X for the above parameter values. On the same graph, plot a gamma PDF with parameters $k = 5$ and $\lambda = 1/0.8$. This is the PDF of X that uses the mean value of Λ and disregards the uncertainty in the scale parameter.

3.3 Random variable X has a gamma marginal distribution with mean $\mu_X = 10$ and standard deviation $\sigma_X = 3$, and random variable Y has the Gumbel distribution with mean $\mu_Y = 100$ and standard deviation $\sigma_Y = 40$. The correlation coefficient between the two variables is $\rho_{XY} = 0.3$. Construct the following joint PDF models of the two random variables and plot their contours: (a) Nataf distribution; (b) Morgenstern distribution; and (c) Clayton copula distribution.

3.4 Consider the transformation

$$U = \Phi^{-1}[F_X(X)],$$

where $F_X(x)$ is the CDF of random variable X and $\Phi^{-1}[\cdot]$ denotes the inverse of the standard normal CDF.

1) Show that, provided $F_X(x)$ is a continuous and strictly increasing function, U has the standard normal distribution.
2) Describe what happens if $F_X(x)$ is discontinuous at x_0, say if $F_X(x_0^-) = 0.3$ and $F_X(x_0^+) = 0.4$, where x_0^- and x_0^+ are infinitesimally close to x_0 from below and above.
3) Describe what happens if $F_X(x)$ is constant (non-increasing) over an interval, say $a \le x \le b$.
4) Let X have the mean $\mu_X = 100$ and standard deviation $\sigma_X = 30$. Plot the above transformations for the following distributions of X: (a) normal; (b) lognormal; (c) Gumbel; and (d) uniform.

3.5 The vector of random variables $\mathbf{X} = [X_1, X_2, X_3]^{\mathrm{T}}$ has the joint normal distribution with the mean vector and covariance matrix:

$$\mathbf{M_X} = \begin{Bmatrix} 10 \\ 50 \\ 30 \end{Bmatrix}, \mathbf{\Sigma_{XX}} = \begin{bmatrix} 16 & 20 & 4 \\ 20 & 29 & 11 \\ 4 & 11 & 26 \end{bmatrix}$$

Determine a transformation $\mathbf{U} = \mathbf{T}(\mathbf{X})$ so that $\mathbf{U}$ is a 3-vector of standard normal random variables. Also determine the inverse transform $\mathbf{X} = \mathbf{T}^{-1}(\mathbf{U})$ and the Jacobians $\mathbf{J_{u,x}}$ and $\mathbf{J_{x,u}}$.

3.6 For each of the bivariate distribution models developed in Problem 3.3, construct a transformation to the standard normal space. Also describe the inverse transform and the corresponding Jacobian. For each case, determine the mapping of the outcome $(x, y) = (15, 60)$ in the standard normal space.

4 Formulation of Structural Reliability

4.1 Introduction

Many structural or system reliability problems involve two opposing quantities: a capacity (or resistance, supply, strength, etc.) and a demand (or load, stress, load effect, etc.). The structure or system is said to have failed when the demand exceeds the capacity. The word "fail" is used here in a general sense. It does not necessarily imply fracture or collapse of the structure or system. Rather, it is the failure of the structure or system to perform according to a specified criterion – here, the criterion that the structure or system must meet the demand placed on it. This chapter closely examines this important problem and presents several exact and approximate solutions. The opportunity is taken also to introduce a number of useful definitions and concepts and highlight several important observations. At the end of the chapter, the capacity-demand formulation is generalized to a *structural component reliability* formulation. Methods for solution of this more general problem as well as the extension to systems problems are described in later chapters of this book.

4.2 The *R*-*S* Reliability Problem

Let R and S respectively denote the capacity of and the demand on a structure or system. We are interested in determining the *probability of failure* defined by

$$p_f = \Pr(R \leq S). \tag{4.1}$$

The complement of the failure probability, $1 - p_f$, is called the *reliability*. In problems of interest in civil engineering, the failure probability is usually small and the reliability close to 1, i.e., 0 with one or more 9s after the decimal point. As it is easier to record a number close to 0 than a number just below 1, the failure probability is usually reported even though the word "reliability" characterizes the field. Later on, other measures of reliability that have a more convenient numerical range will be introduced.

Let $f_{RS}(r, s)$ denote the joint probability density function (PDF) of R and S. Although in practice R and S are often statistically independent, there are cases where they are dependent, e.g., the supply and demand of water of a dam, which both depend on the rainfall during the year. Thus, for the sake of generality, we consider the dependent case. Assuming R and S take on only non-negative values, Figure 4.1 shows the failure domain in the outcome space of the two random variables together with the illustrative contour plots of the joint PDF. The failure probability is the integral of the joint PDF over the failure domain. Thus, we can write

$$p_f = \iint_{r \le s} f_{RS}(r, s) \mathrm{d}r \mathrm{d}s. \tag{4.2}$$

It appears that evaluating the probability of failure requires the computation of a double integral of the joint PDF over the failure domain. In general, this will have to be done numerically. Next, consider alternative formulations of the problem that lead to simpler results.

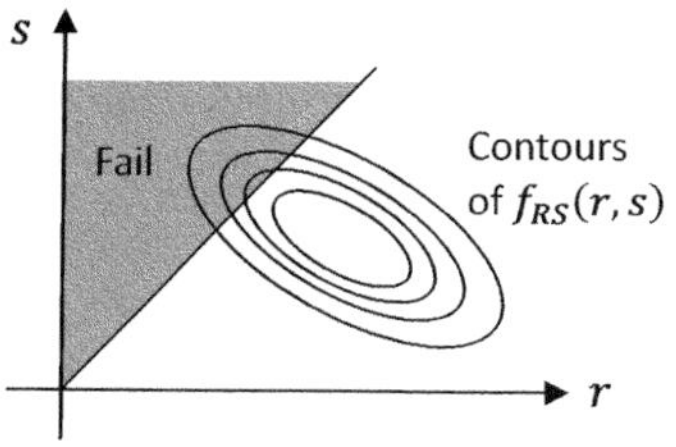

Figure 4.1 Outcome space for the R-S reliability problem

4.2.1 Solution by Conditioning on S

Let us replace the joint PDF in (4.2) by its equivalent $f_{R|S}(r|s) f_S(s)$, where $f_{R|S}(r|s)$ is the conditional PDF of R given $S = s$ and $f_S(s)$ is the marginal PDF of S. The integral in (4.2) then takes the form

$$\begin{aligned} p_f &= \int_0^{\infty} \int_0^{s} f_{R|S}(r|s) f_S(s) \mathrm{d}r \mathrm{d}s \\ &= \int_0^{\infty} F_{R|S}(s|s) f_S(s) \mathrm{d}s. \end{aligned} \tag{4.3}$$

It is apparent that, provided the conditional cumulative distribution function (CDF) $F_{R|S}(r|s)$ is available, the probability calculation reduces to a single-fold integral. In the special case when R and S are statistically independent, (4.3) takes the form

$$p_f = \int_0^{\infty} F_R(s) f_S(s) \mathrm{d}s. \tag{4.4}$$

Figure 4.2 shows a plot of the PDFs $f_R(r)$ and $f_S(s)$ on a single axis when R is lognormal with mean $\mu_R = 25$ and coefficient of variation (c.o.v.) $\delta_R = 0.3$, and S is gamma distributed with mean $\mu_S = 10$ and c.o.v. $\delta_S = 0.4$. It is apparent that the integral in (4.4) is the accumulation of the area underneath $f_R(r)$ to the left of each s times the differential probability $f_S(s)\mathrm{d}s = \Pr(s < S \leq s + \mathrm{d}s)$. For a given s, the area marks the conditional failure probability $\Pr(R \leq S|S = s) = F_R(s)$. Hence, (4.4) can be seen as a statement of total probability, where the failure event is conditioned on the outcome of the demand S. It is important to note that, contrary to what is often reported in the literature, the failure probability is not equal to the overlap area between the two PDFs, although the larger that area, the larger the failure probability is likely to be.

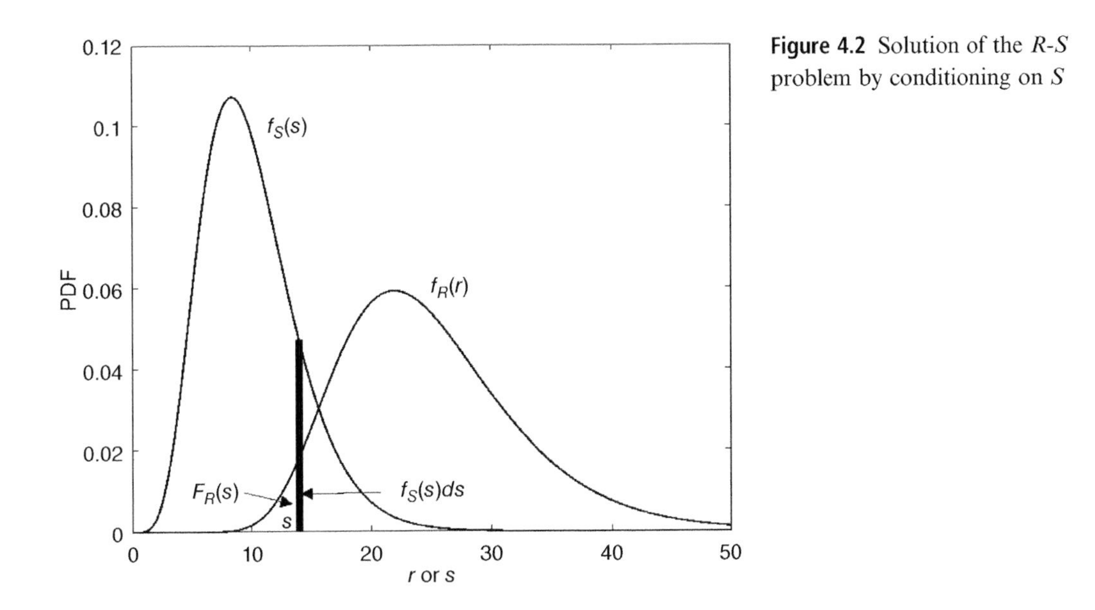

Figure 4.2 Solution of the *R-S* problem by conditioning on *S*

4.2.2 Solution by Conditioning on *R*

Now we replace the joint PDF in (4.2) by its equivalent $f_{S|R}(s|r)f_R(r)$, where $f_{S|R}(s|r)$ is the conditional PDF of S given $R = r$ and $f_R(r)$ is the marginal PDF of R. The integral in (4.2) then takes the form

$$\begin{aligned} p_f &= \int_0^\infty \int_r^\infty f_{S|R}(s|r)f_R(r)\mathrm{d}s\mathrm{d}r \\ &= \int_0^\infty [1 - F_{S|R}(r|r)]f_R(r)\mathrm{d}r. \end{aligned} \tag{4.5}$$

It is again apparent that, provided the conditional CDF $F_{S|R}(s|r)$ is available, the probability calculation reduces to a single-fold integral. In the special case when R and S are statistically independent, (4.5) takes the form

$$p_f = \int_0^\infty [1 - F_S(r)]f_R(r)\mathrm{d}r. \tag{4.6}$$

Figure 4.3 shows an interpretation of the above integral for the same example as shown in Figure 4.2. It is apparent that the integral in (4.6) is the accumulation of the area underneath $f_S(s)$ to the right of each r times the differential probability $f_R(r)\mathrm{d}r = \Pr(r < R \leq r + \mathrm{d}r)$. For a given r, the area marks the conditional failure probability $\Pr(R \leq S|R = r) = 1 - F_S(r)$. Hence, (4.6) can also be seen as a statement of total probability, where the failure event is conditioned on the outcome of the capacity R.

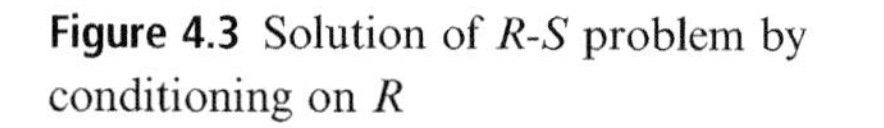
Figure 4.3 Solution of R-S problem by conditioning on R

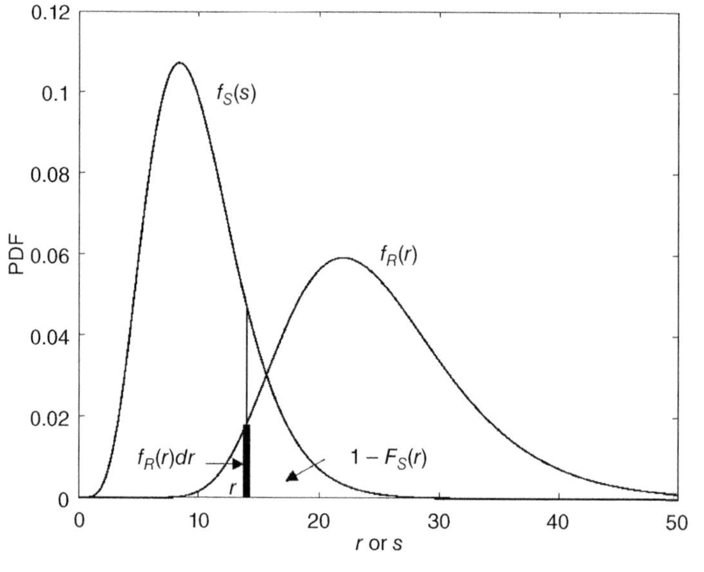

4.2.3 Formulation in Terms of Safety Margin

An alternative formulation of the R-S reliability problem is in terms of the *safety margin*

$$Z = R - S. \tag{4.7}$$

As $R \leq S$ implies $Z \leq 0$, the failure probability now is given by

$$\begin{aligned} p_f &= \Pr(Z \leq 0) \\ &= F_Z(0), \end{aligned} \tag{4.8}$$

where $F_Z(z)$ is the CDF of Z. This expression appears to be much simpler than the one in (4.2). However, one must realize that determining the CDF of a function of two random variables involves a twofold integral, one for determining the PDF and one for determining the CDF (see Example 2.8). Thus, we have not overcome the need for the twofold integration.

Owing to the linear dependence of Z on R and S, we can easily determine its moments. In particular, the mean of Z is given by

$$\mu_Z = \mu_R - \mu_S \tag{4.9}$$

and its variance is given by

$$\sigma_Z^2 = \sigma_R^2 - 2\rho_{RS}\sigma_R\sigma_S + \sigma_S^2, \tag{4.10}$$

where μ_R and μ_S are the means, σ_R and σ_S are the standard deviations, and ρ_{RS} is the correlation coefficient of R and S. Figure 4.4(a) is a schematic diagram of the PDF of Z, showing the area that constitutes the failure probability. Note that the mean safety margin is the distance from the mean of the distribution to the failure (zero) threshold, while the standard deviation indicates the dispersion of the distribution.

Consider the standard variate $U = (Z - \mu_Z)/\sigma_Z$, where U has a zero mean and a unit standard deviation. The failure probability can be written as

$$\begin{aligned} p_f &= \Pr(Z \le 0) \\ &= \Pr\left(U \le -\frac{\mu_Z}{\sigma_Z}\right) \\ &= F_U(-\beta) \end{aligned} \tag{4.11}$$

where

$$\beta = \frac{\mu_Z}{\sigma_Z}. \tag{4.12}$$

Figure 4.4(b) is a schematic representation of the PDF of U showing the area that constitutes the failure probability. We see that this area is located at a distance β from the origin. Owing to (4.12), this distance can be interpreted as the mean safety margin measured in units of the corresponding standard deviation. Thus, β, at once, accounts for the location parameter (the mean) as well as the dispersion (the standard deviation) of the distribution of the safety margin. Obviously a larger β corresponds to a smaller failure probability. For these reasons, β is called the *reliability index*. Of course, β alone does not provide sufficient information to determine the failure probability, because (4.11) also requires knowledge of the CDF of U.

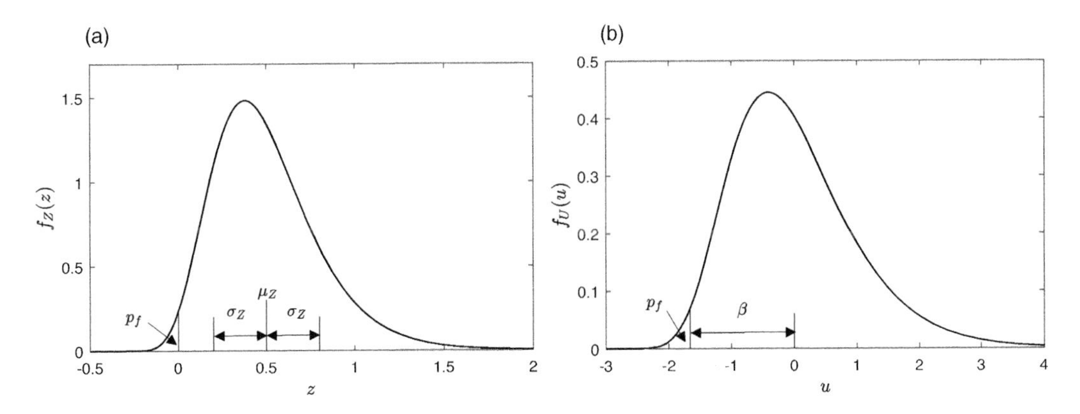

Figure 4.4 Safety margin formulations of the failure probability and reliability index: (a) in the outcome space of Z, (b) in the outcome space of U

Determining the CDF of U requires the same effort as determining that of Z. However, because U is a standard variate, approximate estimates of the failure probability can be obtained by making an appropriate distribution assumption. In particular, if we assume U has the standard normal distribution, the corresponding approximation of the failure probability is

$$p_f \cong \Phi(-\beta). \tag{4.13}$$

The accuracy of this approximation depends on how sensitive the failure probability is to the choice of the distribution. This issue will be discussed later in this chapter. For now, a special case for which an exact solution with the safety margin formulation is available is presented.

Example 4.1 – Special Case of Jointly Normal *R* and *S*

Earlier we assumed that R and S are non-negative. This, of course, is reasonable because capacity and demand values are typically non-negative. Hence, assigning the joint normal distribution to these variables does not seem to be reasonable. However, if the c.o.v. are small (say not exceeding 30%), the probability masses in the negative domains are negligible. With this caveat, it can be assumed that R and S are jointly normal with the means, standard deviations, and correlation coefficient defined earlier. As Z is a linear function of R and S, it follows that Z has the normal distribution and, therefore, U has the standard normal distribution. Thus, for this special case (4.13) provides the exact solution of the failure probability. Furthermore, using (4.9) and (4.10) in (4.12), the expression for the reliability index is

$$\begin{aligned}\beta &= \frac{\mu_R - \mu_S}{\sqrt{\sigma_R^2 - 2\rho_{RS}\sigma_R\sigma_S + \sigma_S^2}} \\ &= \frac{r-1}{\sqrt{r^2\delta_R^2 - 2r\rho_{RS}\delta_R\delta_S + \delta_S^2}},\end{aligned} \tag{4.14}$$

where $\delta_R = \sigma_R/\mu_R$ and $\delta_S = \sigma_S/\mu_S$ are the c.o.v. and $r = \mu_R/\mu_S$ is the safety factor relative to the mean capacity and demand values. The second line in (4.14) shows that the reliability index can be expressed in terms of the three dimensionless quantities r, δ_R, and δ_S instead of the means and variances, and the correlation coefficient ρ_{RS}, which is also dimensionless.

4.2.4 Formulation in Terms of Safety Factor

An alternative formulation of the R-S reliability problem is in terms of the *safety factor*, R/S. However, it is advantageous to consider the logarithm of this quantity, which yields the logarithmic safety margin

$$Z = \ln R - \ln S. \tag{4.15}$$

As $R \le S$ implies $\ln R \le \ln S$, the probability of failure is still given as $p_f = F_Z(0)$. Furthermore, if we introduce the standard variate $U = (Z - \mu_Z)/\sigma_Z$, we again have

$p_f = F_U(-\beta)$, where $\beta = \mu_Z/\sigma_Z$. However, (4.15) is a nonlinear function of R and S, so closed-form expressions for the mean and standard deviation of Z are not readily available. One option is to use first- and second-order approximations (see Section 2.9) to obtain approximate expressions for these values and, hence, for β. Here, we choose to use a second-order approximation for μ_Z and a first-order approximation for σ_Z. Noting that $\partial Z/\partial R = 1/R$, $\partial^2 Z/\partial^2 R = -1/R^2$, $\partial Z/\partial S = -1/S$, $\partial^2 Z/\partial^2 S = 1/S^2$, and $\partial^2 Z/\partial R\partial S = 0$, using the relation in (2.103), we obtain

$$\mu_Z \cong \ln\mu_R - \ln\mu_S - \frac{1}{2}\left(\delta_R^2 - \delta_S^2\right). \tag{4.16}$$

Furthermore, using (2.99), we obtain for the variance of Z

$$\sigma_Z^2 \cong \delta_R^2 - 2\rho_{RS}\delta_R\delta_S + \delta_S^2. \tag{4.17}$$

Hence, an approximation of the reliability index is

$$\beta \cong \frac{\ln r - \frac{1}{2}\left(\delta_R^2 - \delta_S^2\right)}{\sqrt{\delta_R^2 - 2\rho_{RS}\delta_R\delta_S + \delta_S^2}}, \tag{4.18}$$

where we have used the relation $r = \mu_R/\mu_S$. Note that this expression is defined in terms of the same four dimensionless quantities as the expression in the second line of (4.14). With this β, one can use (4.13) to obtain an approximate estimate of the failure probability.

We now consider a special case for which the formulation in (4.15) leads to an exact solution.

Example 4.2 – Special Case of Jointly Lognormal *R* and *S*

Suppose R and S are jointly lognormal with the means, standard deviations, and correlation coefficient defined earlier. The mean of Z is

$$\begin{aligned}\mu_Z &= \mu_{\ln R} - \mu_{\ln S} \\ &= \ln\mu_R - \frac{1}{2}\ln\left(1+\delta_R^2\right) - \ln\mu_S + \frac{1}{2}\ln\left(1+\delta_S^2\right) \\ &= \ln\left(r\sqrt{\frac{1+\delta_S^2}{1+\delta_R^2}}\right),\end{aligned} \tag{4.19}$$

where $r = \mu_R/\mu_S$. Using (3.20), the variance of Z is

$$\begin{aligned}\sigma_Z^2 &= \sigma_{\ln R}^2 - 2\rho_{\ln R \ln S}\sigma_{\ln R}\sigma_{\ln S} + \sigma_{\ln S}^2 \\ &= \ln\left(1+\delta_R^2\right) - 2\ln\left(1+\rho_{RS}\delta_R\delta_S\right) + \ln\left(1+\delta_S^2\right) \\ &= \ln\left[\frac{\left(1+\delta_R^2\right)\left(1+\delta_S^2\right)}{\left(1+\rho_{RS}\delta_R\delta_S\right)^2}\right].\end{aligned} \tag{4.20}$$

Example 4.2 (cont.)

Thus, the solution for the reliability index in this special case is

$$\beta = \frac{\ln\left(r\sqrt{\frac{1+\delta_S^2}{1+\delta_R^2}}\right)}{\sqrt{\ln\left[\frac{\left(1+\delta_R^2\right)\left(1+\delta_S^2\right)}{\left(1+\rho_{RS}\delta_R\delta_S\right)^2}\right]}}. \tag{4.21}$$

Observe that the solution is again given in terms of the four, dimensionless quantities r, δ_R, δ_S, and ρ_{RS}.

The exact solution in (4.21) for the special case of jointly lognormal R and S offers an opportunity to examine the accuracy of the approximation in (4.18). We know that for $x \ll 1$, $\ln(1+x) \cong x$ in first-order approximation. If we assume $\delta_R^2 \ll 1$, $\delta_S^2 \ll 1$, and $\rho_{RS}\delta_R\delta_S \ll 1$, and replace the corresponding logarithmic terms in (4.21) with their first-order approximations, we obtain (4.18). Hence, the approximation in (4.18) is good for small c.o.v. values. Shown in Figure 4.5 is the error $\left(\beta_{\text{exact}} - \beta_{\text{appr}}\right)/\beta_{\text{exact}}$ of the approximation in (4.18) for jointly lognormal random variables with $\rho_{RS} = 0$, $r = 2$ and 3, and a wide range of c.o.v. values. It can be seen that for c.o.v. values smaller than about 40%, the error in the approximate expression in (4.18) is smaller than 5%.

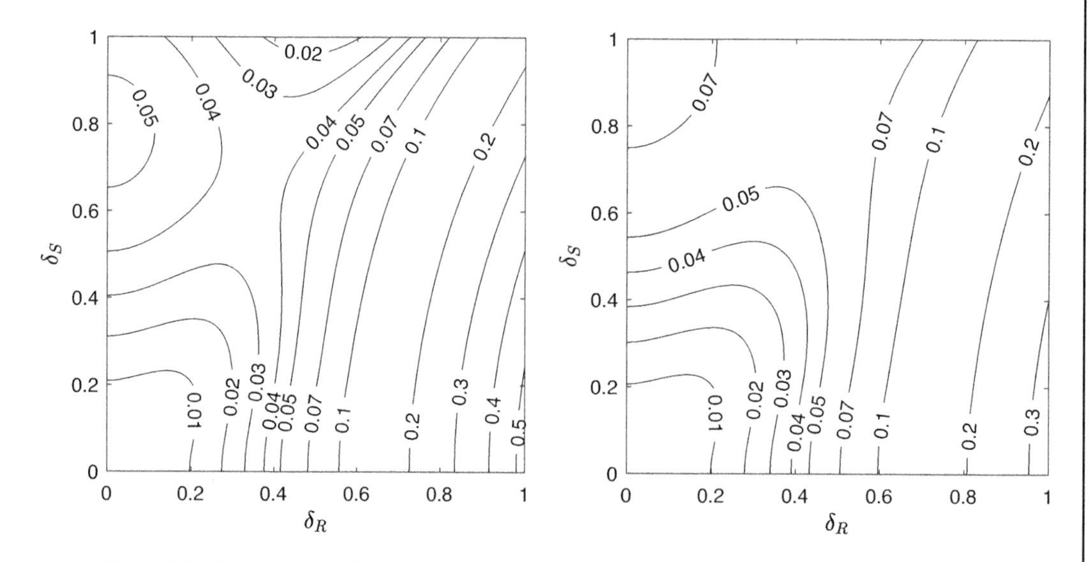

Figure 4.5 Contour plot of percent error in the approximate reliability index relative to the exact result for the case of lognormal R and S

4.3 The Tail-Sensitivity Problem

Equation (4.13) presents an approximate estimate of the failure probability by assigning a normal distribution to the standard variate U. What is the nature of this approximation? Under what conditions is it a good approximation, and when is it a poor approximation? Because this expression uses only the first- and second-moments of Z, the question to ask is how sensitive the failure probability is to the assumed distribution of U, or more generally to those of R and S. The exact solutions in (4.14) and (4.21) offer an opportunity to examine this issue.

Figure 4.6 shows plots of the reliability index and failure probability according to the solutions in (4.14) and (4.21) for the special cases of statistically independent normal and lognormal distributions of R and S, respectively. The reliability index and failure probability are shown as functions of $r = \mu_R/\mu_S$ for the case of $\delta_R = \delta_S = 0.3$. It is observed that small values of the reliability index and large values of the failure probability are insensitive to the choice of the distributions, whereas large values of the reliability index and small values of the failure probability are strongly sensitive to the choice of the distributions. This has been termed the *tail-sensitivity problem* by Ditlevsen (1979a). We know that the contributions to the failure probability come from the lower tail of the capacity distribution and the upper tail of the demand distribution (see Figures 4.2 and 4.3). When the probability of failure is small, these contributions come from the far tail regions of the respective distributions. The mean and variance provide measures of location and dispersion only, so the choice of the distribution can have a profound influence in the far tail and, therefore, on a small failure probability and the corresponding large reliability index. This observation suggests that caution must be exercised in the selection of distributions when dealing with

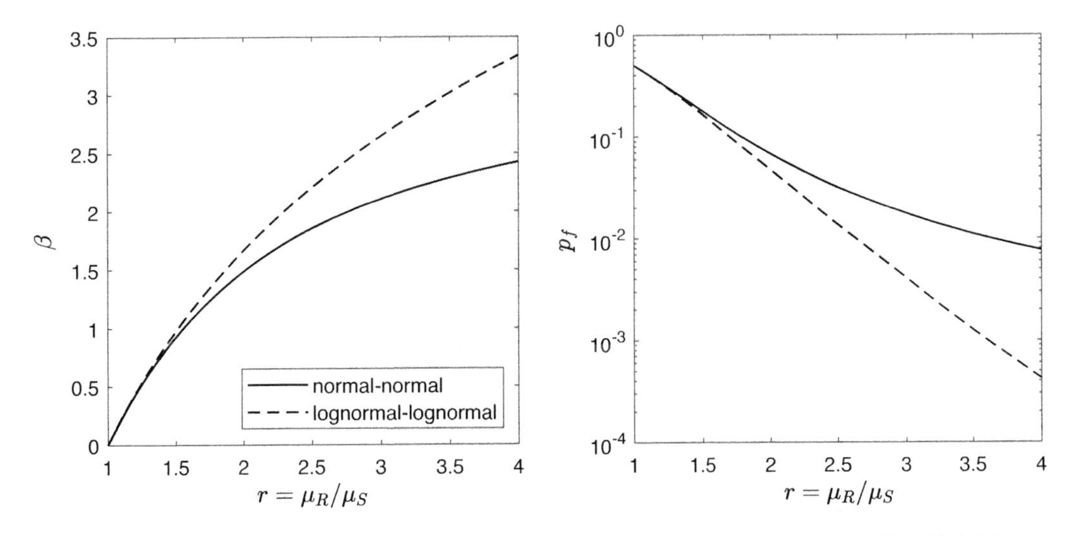

Figure 4.6 Influence of distribution choice on reliability index (left) and failure probability (right) for $\delta_R = \delta_S = 0.3$

highly reliable systems. If the distribution is not known, a precise measure of reliability for such a system cannot be ascertained. For this reason, in the context of probabilistic codified design, Ditlevsen (1979a) recommends standardizing the choice of distributions by a code committee to avoid arbitrariness in their selection and consequent results.

4.4 The Generalized Structural Reliability Problem

Reconsider the formulation in (4.7), where the reliability problem is defined in terms of the safety margin $Z = R - S$. Underlying this formulation is the assumption that statistical information (observed data, distributions, or second-moment information) are directly available for the capacity and demand values. In practice, this is often not the case, because the capacity and demand values are usually not directly observable. For example, the capacity of a structure against excessive deformation is a function of its material properties, geometry, and configuration. Similarly, the demand placed on a structure is a function of the magnitudes and spatial distribution of loads acting on the structure as well as the way the structure responds to these loads. It is often these basic quantities, i.e., material properties, geometric variables, load magnitudes, spatial distribution of loads, that are observable. Hence, more generally, R and S should be viewed as functions of more basic random variables. Let $\mathbf{X} = (X_1, \ldots, X_n)$ denote the set of basic random variables of a structural reliability problem. The word *basic* is used here to denote variables for which observational data are available or can be obtained. As just mentioned, this set may include such variables as material properties, geometric quantities, load magnitudes, environmental effects, etc., or any other relevant quantities for which statistical data can be gathered. We now see the capacity and demand as functions of these variables. Hence, in a more general description, we rewrite (4.7) as

$$Z = R(\mathbf{X}) - S(\mathbf{X}). \tag{4.22}$$

However, this formulation is still not the most general, as it assumes that there are distinct capacity and demand values. In many structural problems such a distinction is not possible. Consider, for example, the capacity of a reinforced concrete column under axial force and bending moment. As is well known, the capacity of the column against the bending moment depends on the axial force, which itself is a load (demand) value. For this reason, in designing such a column, structural engineers use interaction diagrams, which describe the safe domain of the column in the space of axial force and bending moment. To allow for such cases, we define the general structural reliability problem in terms of a *limit-state function*

$$g = g(\mathbf{X}), \tag{4.23}$$

which is formulated in such a way that $\{g(X) \leq 0\}$ defines the failure event and the probability of failure is given by

$$p_f = \Pr[g(\mathbf{X}) \leq 0]. \tag{4.24}$$

When the capacity and demand are separable, one can write $g = R(\mathbf{X}) - S(\mathbf{X})$. However, $g = g(\mathbf{X})$ is more general.

In the outcome space of random variables $\mathbf{X}$, the hyper-surface $g(\mathbf{x}) = 0$ is known as the *limit-state surface*. As schematically shown in the two-dimensional diagram in Figure 4.7, this surface (line in two dimensions) delineates the boundary between the safe and fail states of the structure. Also schematically shown in Figure 4.7 are the contours of the joint PDF of $\mathbf{X}$, $f_{\mathbf{X}}(\mathbf{x})$. The probability of failure is the integral of the PDF over the failure domain, $\Omega \equiv \{\mathbf{x}: g(\mathbf{x}) \leq 0\}$,

$$p_f = \int_{\Omega} f_{\mathbf{X}}(\mathbf{x})d\mathbf{x}. \tag{4.25}$$

It is apparent that determination of the failure probability in the general case involves an n-fold integral of the joint PDF of the basic variables $\mathbf{X}$ over a sub-domain of the outcome space $\mathbf{x}$, where n denotes the number of random variables. For a large n, this is obviously not an easy task, and this poses one of the most challenging problems in structural reliability analysis. Later chapters introduce several computational methods for efficiently but approximately evaluating the above integral.

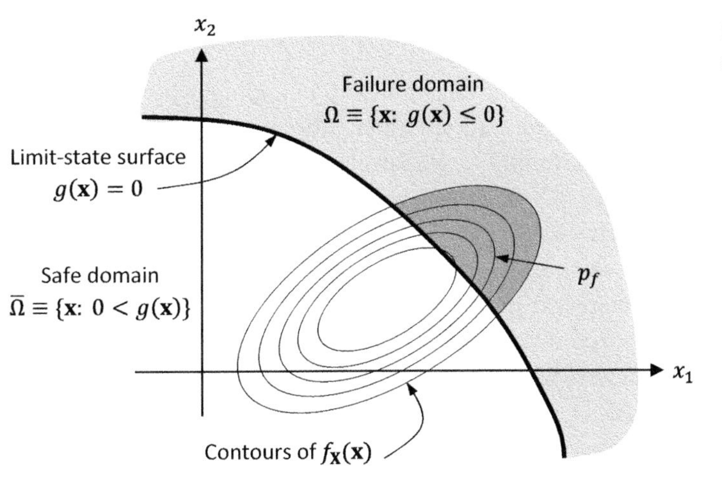

Figure 4.7 Illustration of structural reliability problem

The function $g(\mathbf{x})$ must satisfy certain conditions to qualify as a limit-state function. Furthermore, depending on its behavior, two classes of structural reliability problems are defined. First, the function must be continuous in its arguments; otherwise, there are outcomes of $\mathbf{X}$ for which the state of the structure remains undefined. Furthermore, the function must normally take both positive and negative values: if $g(\mathbf{x})$ is positive for all $\mathbf{x}$, then the structure has zero possibility of failing, and if it is negative for all $\mathbf{x}$, then the structure is certain to fail. These are degenerate cases and not of real interest. If $g(\mathbf{x})$ is a continuous and continuously differentiable function, a structural *component* reliability problem is defined. If the definition of $g(\mathbf{x})$ involves or, max, or min operations, then a structural *system* reliability problem is defined. For example, the function $g(\mathbf{x}) = g_1(\mathbf{x})$ if $x_1 \leq a$ and $g(\mathbf{x}) = g_2(\mathbf{x})$ if $a < x_1$, provided it is continuous, defines a system problem if the derivatives as x_1 approaches a from below and from above are not identical. The function $g(\mathbf{x}) = \min[g_1(\mathbf{x}), \ldots, g_m(\mathbf{x})]$, where each $g_i(\mathbf{x})$ is continuous and continuously differentiable, defines a *series system* reliability problem with m components. Similarly, the function $g(\mathbf{x}) = \max[g_1(\mathbf{x}), \ldots, g_m(\mathbf{x})]$ defines a *parallel system* reliability problem with m components. Finally, the function $g(\mathbf{x}) = \min\{\max_{i \in c_1}[g_i(\mathbf{x})], \ldots, \max_{i \in c_k}[g_i(\mathbf{x})]\}$, where c_j, $j = 1, \ldots, k$, denote subsets of component indices and each $g_i(\mathbf{x})$ is continuous and continuously differentiable, defines a general system reliability problem with k *cut sets*. System reliability problems are addressed in Chapter 8.

Before closing this section, several example limit-state functions are presented, to allow the student to gain better understanding of the relevance of the above formulations to practical problems:

- Consider the reliability of an elastic column against buckling under axial load. The limit-state function can be formulated as $g = \pi^2 EI/(kL)^2 - P$, where E is the elastic modulus, I the moment of inertia of the cross section, L the column length, k the effective length factor depending on the end conditions of the column, and P the axial load. All of these quantities can be random variables. This is an example of a component reliability problem, where capacity and demand are distinct.
- Let $\Delta_i(\mathbf{X})$ denote the ith interstory drift in a multistory building, where $\mathbf{X}$ denotes the set of uncertain material, geometry, and load variables. To determine the probability that the interstory drift will exceed a threshold δ_i, we solve a reliability problem with the limit-state function $g(\mathbf{x}, \delta_i) = \delta_i - \Delta_i(\mathbf{x})$, where δ_i can be viewed as a parameter in the limit-state function. Noting that $\Pr[\Delta_i(\mathbf{X}) \leq \delta_i] = 1 - \Pr[g(\mathbf{X}, \delta_i) \leq 0]$, the CDF of Δ_i can be obtained by repeatedly solving the reliability problem for a range of δ_i values. Furthermore, if we are interested in the probability that the interstory drift limit in *any* story of the building will exceed its respective safe threshold, then the reliability problem becomes that of a series system with $g_i(\mathbf{x}, \delta_i) = \delta_i - \Delta_i(\mathbf{x})$ being the limit-state function for the ith component and $g(\mathbf{x}) = \min[g_1(\mathbf{x}, \delta_1), \ldots, g_m(\mathbf{x}, \delta_m)]$ being the limit-state function for a structure with m stories. The solution of a problem like this would benefit from coupling a structural analysis code together with a reliability analysis code.
- Consider the event of yielding according to the von Mises criterion at a specified point in a continuum. The limit-state function can be expressed as $g(\mathbf{x}) = \sigma_y^2 - [(\sigma_{11} - \sigma_{22})^2 + (\sigma_{22} - \sigma_{33})^2 + (\sigma_{11} - \sigma_{33})^2 + 6(\sigma_{12}^2 + \sigma_{23}^2 + \sigma_{31}^2)]/2$, where $\mathbf{x}$ represents the set of uncertain material properties, including the yield stress σ_y and geometry and load variables, and where the stress components σ_{ij} are implicit functions of $\mathbf{x}$. If there is likelihood of yielding at more than one point, then a system reliability problem must be solved. Note that the solution of a problem like this would benefit from coupling a stress analysis code, e.g., a finite-element software, together with a reliability analysis code. Chapter 12 addresses this topic.

- Suppose $X(t)$ denotes the response of a structure to stochastic dynamic loading. Where $X(t)$ is a random process, which is essentially a parameterized (by t) sequence of random variables. The probability that $X(t)$ will up-cross a threshold ζ during a small time interval $(t, t + \mathrm{d}t]$ can be solved as a parallel system reliability problem with component limit-state functions $g_1 = X(t) - \zeta$ and $g_2 = \zeta - X(t + \mathrm{d}t) \cong \zeta - X(t) - \dot{X}(t)\mathrm{d}t$, where we have introduced the approximation $X(t + \mathrm{d}t) \cong X(t) + \dot{X}(t)\mathrm{d}t$ with $\dot{X}(t)$ denoting the derivative process. $g_1 \leq 0$ states that the process is at or below the threshold ζ at time t, and $g_2 \leq 0$ states that the process is at or above the threshold ζ at time $t + \mathrm{d}t$ so that a crossing must have occurred during the small time interval $(t, t + \mathrm{d}t]$. This problem can be solved by reliability methods, provided the joint distribution of $X(t)$ and $\dot{X}(t)$ is available. We note that when the computed probability is divided by $\mathrm{d}t$, the result approximates the mean up-crossing rate at time t, a quantity that is of great interest in stochastic reliability analysis. This example demonstrates a case where benefit will be gained from coupling a dynamic analysis code together with a reliability analysis code. Chapters 11 and 13 describe this type of reliability problem.
- The reliability methods developed in this book are not limited to structural engineering and mechanics problems. Any two-state (e.g., fail and safe) reliability problem that can be formulated in terms of one or more continuously differentiable functions of random variables potentially can be solved by these methods. This may include problems in engineering, sciences, economics, and other fields. As an example, consider the problem of contaminant transport due to a toxic spill, where the concern is the exceedance above a safe threshold of the concentration of the toxic material in the ground water at a location of interest. The limit-state function for this problem may be formulated as $g = C_0 - C(\mathbf{X}, \mathbf{s}, t)$, where C_0 is the safe threshold, $\mathbf{s}$ the spatial coordinates of the location of interest, t time, and $\mathbf{X}$ the vector of random variables defining uncertain quantities such as the amount of toxic material spilled and the properties of ground, e.g., dispersivity, hydraulic conductivity, and porosity. The concentration value at location $\mathbf{s}$ and time t, $C(\mathbf{X}, \mathbf{s}, t)$, is obtained from ground water contaminant transport analysis, typically by numerical solution of a set of differential equations. Applications of reliability methods to this type of problem can be found in Sitar et al. (1987) and Jang et al. (1994).

As the above examples demonstrate, structural reliability analysis is intimately related to the main domains of structural engineering and mechanics, i.e., structural analysis and finite elements, which require specialized software for numerical solutions. The limit-state functions in these problems are typically implicit in nature and require numerical solutions. Integration of reliability methods with such software is essential for broad application of structural reliability methods. For other domains of application, reliability methods may need to be combined or integrated with the relevant numerical analysis algorithms. It is important to keep this broad view of the reliability methods described in this book rather than limit one's horizon to problems defined by explicit limit-state functions. Many examples of both kinds are presented in later chapters of this book.

4.5 Concluding Remarks

This chapter introduced the notion of structural reliability starting from its most elementary formulation in terms of the so-called R-S (capacity-demand) problem. Determination of the failure probability requires a single-fold integral, provided the conditional distribution of one variable given the other is available. The chapter also introduced the concept of the reliability index as a measure of safety, considering the safety margin and safety factor formulations. Exact and approximation expressions of the reliability index were derived for special cases.

An important concept demonstrated in this chapter is the sensitivity of the failure probability to the choice of distribution. For small failure probabilities, the region of relevance is the far-right tail of the demand distribution and the far-left tail of the capacity distribution. As a result, small failure probability values are sensitive to the choice of these distributions. This has two important consequences: (a) Reliability methods that do not make use of distribution information can only provide nominal measures of reliability; deriving estimates of the failure probability in such cases is unwarranted. (b) Reliability methods should make use of probability distributions, when they are known. Furthermore, any uncertainty in the selection of probability distributions must be reflected in the estimate of the failure probability. This topic is addressed in Chapter 10.

Going beyond the limited scope of the R-S problem, this chapter introduced the general formulation of a structural component reliability problem in terms of a limit-state function of a vector of random variables. In this general formulation, uncertain quantities affecting the state of a component are represented as random variables without characterizing them as capacity or demand. When the limit-state function is set equal to zero, the resulting surface divides the outcome space of the random variables into safe and fail domains. Computing the probability of failure then requires a multifold integration over the failure domain in the outcome space of the random variables. Solution methods for this formulation are pursued in later chapters.

PROBLEMS

4.1 The capacity, R, of a structural component has the Raleigh distribution with the PDF $f_R(r) = (2r/u^2)\exp\left[-(r/u)^2\right]$, $0 \le r$, and the demand, S, has the exponential distribution with the PDF $f_S(s) = \lambda\exp(-\lambda s)$, $0 \le s$, where u and λ are the distribution parameters. Assume R and S are statistically independent. Derive an expression for the failure probability $p_f = \Pr(R \le S)$ and compute its value for $u = 100$ and $\lambda = 0.05$.

4.2 The capacity, R, of a structural component has the Gamma distribution with mean 60 and standard deviation 10, and the corresponding demand, S, has the Gumbel distribution with mean 30 and standard deviation 10. Assume R and S are statistically independent. Using (4.4) and (4.6), determine the failure probability by numerical integration. Your results from the two formulas should be identical.

4.3 The annual capacity, R, and demand, S, of a water reservoir are jointly lognormal with means $\mu_R = 1{,}500{,}000\ \text{m}^3$ and $\mu_S = 750{,}000\ \text{m}^3$, standard deviations $\sigma_R = 250{,}000\ \text{m}^3$ and $\sigma_S = 200{,}000\ \text{m}^3$, and correlation coefficient $\rho_{RS} = -0.3$, the latter due to the dependence of both capacity and demand on the amount of annual rainfall.

1) Determine the probability that the reservoir will not meet the demand placed on it.
2) How will the result in (1) change if we disregard the correlation between the capacity and demand variables?
3) How will the results in (1) and (2) change if we assume the capacity and demand are jointly normal?

4.4 The capacity, R, of a structural component has the gamma distribution with the PDF $f_R(r) = v(vr)^{k-1}\exp(-vr)/\Gamma(k)$, $0 \le r$, and the demand, S, has the exponential distribution with the PDF $f_S(s) = \lambda\exp(-\lambda s)$, $0 \le s$, where k, v, and λ are the distribution parameters.

1) Derive the PDF of the safety margin $Z = R - S$ and plot it for the parameter values $k = 3$, $v = 0.2$, and $\lambda = 1$.
2) Derive an expression for the failure probability and compute it for the above parameter values.

4.5 For the distributions R and S described in Problem 4.4,

1) Derive the PDF of the safety factor $Z = R/S$ and plot it for the parameter values $k = 3$, $v = 0.2$, and $\lambda = 1$.
2) Compute the failure probability $p_f = \Pr(Z \le 1)$ and compare it with the result obtained in Problem 4.4. You may use numerical integration for this part.

5 Analysis of Structural Reliability Under Incomplete Probability Information

5.1 Introduction

Early developers of structural reliability theory were cognizant of the fact that complete distributional information on basic random variables of a reliability problem may not exist. This understanding and the need for computationally simple solutions led to the development of second-moment reliability methods, which assume knowledge of the first two moments of the random variables (means, variances, and correlation coefficients) and nothing else. Later, methods that make use of higher moments, bounds, marginal distributions, or other distributional information were developed. It was only in the late 1970s that full distributional information was included in the analysis of general structural reliability problems.

This chapter introduces second-moment reliability methods as well as methods that make use of partial information beyond the second moments. The main purpose of this introduction is to provide a historical review of the subject and illuminate the shortcomings of these methods. The author believes that second-moment reliability methods and their more advanced counterparts have exhausted their usefulness in advancing the theory and application of structural reliability. Therefore, they should not be used in practice. The argument for this opinion is presented at the end of this chapter. In essence, the argument is based on the understanding that, in the real world, probabilistic information is available in the form of observed samples of random quantities. From such samples, one can estimate means, variances, correlation coefficients, third moments, etc. However, these estimates are all associated with errors, depending on the sample size, which tend to increase with the order of the moment. The idea that one has perfect knowledge of the first two moments and nothing else is unrealistic. In addition, the framework for second-moment and similar methods does not allow proper modeling and incorporation of statistical and model uncertainties in reliability analysis. Modern methods of structural reliability, which are based on the Bayesian notion of probability and uncertainty analysis, provide a framework that allows consistent treatment of all types of uncertainties. It is believed that an understanding of the shortcomings of second-moment and similar methods will offer the student an opportunity to better appreciate the power and rationale of the Bayesian approach to the analysis of structural reliability. Distributional reliability methods are presented in Chapters 6–9 and the Bayesian approach is presented in Chapter 10.

5.2 Second-Moment Reliability Methods

Consider the structural component reliability problem defined by the n-vector of basic random variables $\mathbf{X} = [X_1, \ldots, X_n]^\mathrm{T}$, characterized by its mean vector $\mathbf{M}_\mathbf{X}$ and covariance matrix $\boldsymbol{\Sigma}_{\mathbf{XX}}$, and the limit-state function $g = g(\mathbf{X})$. Failure occurs when $\{g(\mathbf{X}) \leq 0\}$, so we can view the limit-state function as the safety margin. With partial probability information, one cannot expect exact estimation of failure probability. Instead, the goal is to compute the reliability index.

Based on the analysis of the safety margin in Section 4.2.3, the reliability index is defined as

$$\beta = \frac{\mu_g}{\sigma_g}, \tag{5.1}$$

where μ_g and σ_g, respectively, denote the mean and standard deviation of the limit-state function. As described in Section 4.2.3, this result is obtained by defining the standard variate $U = \left[g(\mathbf{X}) - \mu_g\right]/\sigma_g$ and setting the failure probability as $p_f = \Pr[g(\mathbf{X}) \leq 0] = F_U\left(-\mu_g/\sigma_g\right) = F_U(-\beta)$. As shown in Figure 4.4(a), β denotes the distance from the mean of the limit-state function to its zero value, as measured in units of the corresponding standard deviation. As also described in Section 4.2.3, if an estimate of the failure probability is desired, the approximation

$$p_f \cong \Phi(-\beta) \tag{5.2}$$

may be used. However, this estimate should be regarded as a notional value of the failure probability because, with the distribution of random variables unavailable, there is no way to determine its accuracy.

It is clear from the above analysis that determining the mean and standard deviation of the limit-state function is a key element of second-moment reliability methods. As the limit-state function in the general case is nonlinear, the analysis must necessarily involve approximations. The type of approximation employed turns out to have an important bearing on the reliability measure. The following subsections describe the second-moment reliability methods that have been developed, starting in the late 1960s.

5.2.1 The Mean-Centered, First-Order, Second-Moment Reliability Method

As its name implies, this method uses a mean-centered, first-order approximation to compute the mean and variance of the limit-state function. Using (2.98) and (2.99), these approximations are

$$\mu_g \cong g(\mathbf{M_X}) \tag{5.3}$$

and

$$\sigma_g^2 \cong \nabla g(\mathbf{M_X})\boldsymbol{\Sigma}_{\mathbf{XX}}\nabla g(\mathbf{M_X})^{\mathrm{T}}, \tag{5.4}$$

respectively, where $\nabla g(\mathbf{M_X}) = [\partial g/\partial x_1, \ldots, \partial g/\partial x_n]_{\mathbf{x}=\mathbf{M_X}}$ is the gradient row-vector of $g(\mathbf{X})$ evaluated at the mean point. Thus, the mean-centered, first-order, second-moment (MCFOSM) reliability index is

$$\beta_{\mathrm{MCFOSM}} = \frac{g(\mathbf{M_X})}{\sqrt{\nabla g(\mathbf{M_X})\boldsymbol{\Sigma}_{\mathbf{XX}}\nabla g(\mathbf{M_X})^T}}. \tag{5.5}$$

Much of the structural reliability literature in the 1960s and 1970s employed the above measure of the reliability index. The most significant of these publications is a paper by Ang and Cornell (1974), which represents the report of the American Society of Civil Engineer's Technical Committee on Structural Safety. At the time, the investigators were aware of a major shortcoming of this measure, namely that this solution lacked invariance with respect to the formulation of the limit-state function (Ditlevsen, 1973). That is, different but equivalent formulations of the limit-state function could lead to different results for β_{MCFOSM}. Indeed, in an appendix of their paper, Ang and Cornell (1974) acknowledged this problem and argued that codes of practice should mandate how engineers formulate the limit-state function so there is no ambiguity in the solution. The following simple example highlights this problem. This "toy" example is used throughout this and the next chapter in order to illustrate and compare different reliability methods. The example includes only two variables so that the solution can be graphically illustrated. More realistic and elaborate examples are presented in subsequent chapters.

Example 5.1 – Lack of Invariance of MCFOSM

Consider the structural reliability problem defined by

$$g(X_1, X_2) = X_1^2 - 2X_2, \mathbf{M_X} = \begin{Bmatrix} 10 \\ 20 \end{Bmatrix}, \boldsymbol{\Sigma}_{\mathbf{XX}} = \begin{bmatrix} 4 & 5 \\ 5 & 25 \end{bmatrix}. \tag{E1}$$

The gradient vector evaluated at the mean point is $\nabla g = [2x_1 \;\; -2]_{\mathbf{x}=\mathbf{M_X}} = [20 \;\; -2]$. Thus, the reliability index is

$$\beta_{\mathrm{MCFOSM}} = \frac{10^2 - 2 \times 20}{\sqrt{[20 \;\; -2]\begin{bmatrix} 4 & 5 \\ 5 & 25 \end{bmatrix}\begin{Bmatrix} 20 \\ -2 \end{Bmatrix}}} = \frac{60}{36} = 1.66. \tag{E2}$$

Now, suppose we consider the equivalent limit-state function

$$g(X_1, X_2) = 1 - \frac{2X_2}{X_1^2}. \tag{E3}$$

Example 5.1 (cont.)

The limit-state functions in (E1) and (E3) are equivalent in the sense that they describe the same fail and safe domains in the outcome space of the random variables X_1 and X_2. The gradient vector evaluated at the mean point for the alternative limit-state function is $\nabla g = \left[4x_2/x_1^3 - 2/x_1^2\right]_{\mathbf{x}=\mathbf{M}_\mathbf{X}} = [0.08 \; -0.02]$. Hence, the corresponding reliability index is

$$\beta_{\text{MCFOSM}} = \frac{1 - 2 \times 20/10^2}{\sqrt{[0.08 \; -0.02]\begin{bmatrix} 4 & 5 \\ 5 & 25 \end{bmatrix}\begin{Bmatrix} 0.08 \\ -0.02 \end{Bmatrix}}} = \frac{0.6}{0.14} = 4.29. \tag{E4}$$

The difference between the two reliability indices is enormous. Clearly, this kind of variability relative to the formulation of the reliability problem is unacceptable under all circumstances. This example provides sufficient evidence to discourage use of this method in research and practice.

5.2.2 The First-Order, Second-Moment Reliability Method

In a paper published the same year as that of Ang and Cornell, Hasofer and Lind (1974) developed a second-moment reliability method possessing invariance relative to the formulation of the limit-state function. To understand the idea behind their method, first consider the linear limit-state function

$$g(\mathbf{X}) = a_0 + \sum\nolimits_{i=1}^{n} a_i X_i = a_0 + \mathbf{a}^\mathrm{T}\mathbf{X}, \tag{5.6}$$

where $\mathbf{a} = [a_1 \ldots a_n]^\mathrm{T}$ is a vector of deterministic coefficients. Now consider transforming the random variables $\mathbf{X}$ into corresponding uncorrelated standard variables $\mathbf{U}$. The needed transformation is the same as that developed for correlated normal random variables in (3.47), i.e.,

$$\mathbf{U} = \mathbf{L}^{-1}\mathbf{D}_\mathbf{X}^{-1}(\mathbf{X} - \mathbf{M}_\mathbf{X}), \tag{5.7}$$

where $\mathbf{D}_\mathbf{X}$ is the diagonal matrix of standard deviations and $\mathbf{L}$ is the lower-triangular Cholesky decomposition of the correlation matrix so that $\mathbf{R}_{\mathbf{XX}} = \mathbf{L}\mathbf{L}^\mathrm{T}$. The inverse of this transform is

$$\mathbf{X} = \mathbf{M}_\mathbf{X} + \mathbf{D}_\mathbf{X}\mathbf{L}\mathbf{U}. \tag{5.8}$$

Substituting this relation for $\mathbf{X}$ in (5.6), we have

$$\begin{aligned} G(\mathbf{U}) &= a_0 + \mathbf{a}^\mathrm{T}(\mathbf{M}_\mathbf{X} + \mathbf{D}_\mathbf{X}\mathbf{L}\mathbf{U}) \\ &= b_0 + \mathbf{b}^\mathrm{T}\mathbf{U}, \end{aligned} \tag{5.9}$$

where $G(\mathbf{U})$ denotes the limit-state function in the $\mathbf{U}$ space, $b_0 = a_0 + \mathbf{a}^T\mathbf{M}_X$, and $\mathbf{b}^T = \mathbf{a}^T\mathbf{D}_X\mathbf{L}$. For the limit-state function $G(\mathbf{U})$, because $\mathbf{U}$ are uncorrelated standard variables, we have $\mu_G = b_0$ and $\sigma_G^2 = \mathbf{b}^T\mathbf{I}\mathbf{b} = \|\mathbf{b}\|^2$ so that

$$\beta = \frac{b_0}{\|\mathbf{b}\|}. \tag{5.10}$$

This result has an important geometric interpretation. Consider the limit-state hyperplane $G(\mathbf{u}) = b_0 + \mathbf{b}^T\mathbf{u} = 0$ in the outcome space of $\mathbf{U}$, as shown in two dimensions in Figure 5.1(a). As is well known from the theory of analytic geometry, the algebraic value of the perpendicular distance from the origin to the hyperplane is given by $-b_0/\|\mathbf{b}\|$. Aside from the sign, this expression is identical to the expression for the reliability index in (5.10). Thus, β can be interpreted as the perpendicular distance from the origin of the standard space to the limit-state surface. The reason for the opposing signs is that the algebraic distance is measured in the positive direction of the gradient. In our case, the value of the limit-state function decreases as we move from the origin towards the failure domain – thus, a negative distance if $0 < b_0$. On the other hand, the reliability index increases with increasing distance of the limit-state surface from the origin, thus the reason for the opposite sign.

Based on the above observation, Hasofer and Lind (1974) argued that the reliability index for a nonlinear limit-state function should be taken as the nearest distance from the origin to the limit-state surface in the transformed space of the standard variables. This idea is illustrated in Figure 5.1(b). As can be seen, this is tantamount to linearizing the nonlinear limit-state function at point $\mathbf{u}^*$, which is located at the foot of the normal from the origin. The linearized limit-state function can be written as

$$\begin{aligned} G(\mathbf{U}) &\cong G(\mathbf{u}^*) + \nabla G(\mathbf{u}^*)(\mathbf{U} - \mathbf{u}^*) \\ &= \nabla G(\mathbf{u}^*)(\mathbf{U} - \mathbf{u}^*), \end{aligned} \tag{5.11}$$

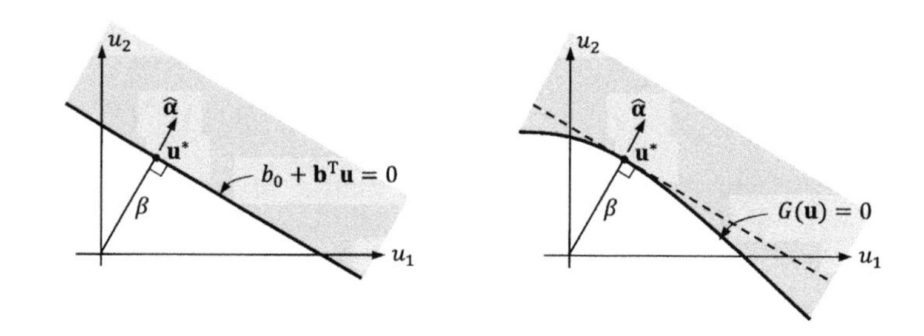

Figure 5.1 Hasofer–Lind reliability index: (a) linear limit-state function, (b) nonlinear limit-state function

where $G(\mathbf{u}^*) = 0$ because $\mathbf{u}^*$ is located on the limit-state surface and $\nabla G(\mathbf{u}^*) = [\partial G/\partial u_1 \cdots \partial G/\partial u_n]_{\mathbf{u}=\mathbf{u}^*}$ is the gradient vector at $\mathbf{u}^*$. Comparing this linear equation with (5.9), it is clear that we have $b_0 = -\nabla G(\mathbf{u}^*)\mathbf{u}^*$ and $\mathbf{b}^T = \nabla G(\mathbf{u}^*)$. Thus, using (5.10), the reliability index according to the Hasofer–Lind definition is

$$\beta = \frac{-\nabla G(\mathbf{u}^*)\mathbf{u}^*}{\|\nabla G(\mathbf{u}^*)\|} = \widehat{\boldsymbol{\alpha}}\mathbf{u}^*, \tag{5.12}$$

where

$$\widehat{\boldsymbol{\alpha}} = -\frac{\nabla G(\mathbf{u}^*)}{\|\nabla G(\mathbf{u}^*)\|} \tag{5.13}$$

is the normalized negative gradient (row) vector at the linearization point. Observe that $\widehat{\boldsymbol{\alpha}}$ is a unit vector normal to the limit-state surface at $\mathbf{u}^*$ and directed towards the failure domain (see Figure 5.1b). Furthermore, the point of linearization is given by

$$\mathbf{u}^* = \beta\widehat{\boldsymbol{\alpha}}^{\mathrm{T}}. \tag{5.14}$$

The linearization point, $\mathbf{u}^*$, plays an important role in several computational reliability methods. Most commonly, it is known as the *design point*; however, other terms such as *most probable point* (MPP), *most likely failure point*, or *locally most dangerous point* have also been used. In this book the term *design point* is used. This name comes from the idea that a structure can be designed by making sure that this point is sufficiently far from the origin in the standard space. The mapping of this point in the original space is denoted as $\mathbf{x}^*$. In the present case, using (5.8), we have $\mathbf{x}^* = \mathbf{M}_{\mathrm{X}} + \mathbf{D}_{\mathrm{X}}\mathbf{L}\mathbf{u}^*$. Note also that $g(\mathbf{x}^*) = G(\mathbf{u}^*) = 0$. Other properties of this point and the reason to avoid calling it MPP or the most likely failure point are discussed in Chapter 6.

It is worth noting that the Hasofer–Lind reliability index can be computed in the original space if $\mathbf{x}^*$ is known. Specifically, the limit-state function linearized at $\mathbf{x}^*$ is $g(\mathbf{X}) \cong \nabla g(\mathbf{x}^*)(\mathbf{X} - \mathbf{x}^*)$, where $\nabla g(\mathbf{x}^*) = [\partial g/\partial x_1 \cdots \partial g/\partial x_n]_{\mathbf{x}=\mathbf{x}^*}$ is the gradient row-vector at $\mathbf{x}^*$, and approximations of the mean and variance are $\mu_g \cong \nabla g(\mathbf{x}^*)(\mathbf{M}_{\mathrm{X}} - \mathbf{x}^*)$ and $\sigma_g^2 \cong \nabla g(\mathbf{x}^*)\boldsymbol{\Sigma}_{\mathrm{XX}}\nabla g(\mathbf{x}^*)^{\mathrm{T}}$, respectively, so that the corresponding reliability index is

$$\beta = \frac{\nabla g(\mathbf{x}^*)(\mathbf{M}_{\mathrm{X}} - \mathbf{x}^*)}{\sqrt{\nabla g(\mathbf{x}^*)\boldsymbol{\Sigma}_{\mathrm{XX}}\nabla g(\mathbf{x}^*)^{\mathrm{T}}}}. \tag{5.15}$$

This expression is equivalent to that in (5.12). This can be verified by using the identity

$$\nabla G(\mathbf{u}) = \nabla g(\mathbf{x})\mathbf{J}_{\mathbf{x},\mathbf{u}}, \tag{5.16}$$

where $\mathbf{J}_{\mathbf{x},\mathbf{u}}$ denotes the Jacobian of the $\mathbf{U}$ to $\mathbf{X}$ transformation. Here, $\mathbf{J}_{\mathbf{x},\mathbf{u}} = \mathbf{D}_{\mathrm{X}}\mathbf{L}$, as can be seen in (5.8). Substituting for $\mathbf{u}^*$ and $\nabla G(\mathbf{u}^*)$ in (5.12), one obtains (5.15).

How does the Hasofer–Lind reliability index assure invariance of β relative to the formulation of the limit-state function? To answer this question, consider two equivalent limit-state functions $g_1(\mathbf{X})$ and $g_2(\mathbf{X})$. These limit-state functions are said to be equivalent if they represent identical safe and failure domains. For example, $g_1 = R - S$ and $g_2 = \ln R - \ln S$ (assuming R and S are non-negative) are equivalent limit-state functions. By this definition, obviously, the limit-state surfaces defined by two equivalent limit-state functions must be identical in both the original and standard spaces. It follows that the design points $\mathbf{u}^*$ for the two formulations must be identical as well and, therefore, the reliability indices must be equal. It is evident that the lack of invariance of the MCFOSM method has to do with the selection of the linearization point. Linearization at the mean point leads to a result that lacks invariance, whereas linearization at the design point leads to a result that is invariant with respect to the formulation of the limit-state function.

The reliability index in (5.12) is often called the Hasofer–Lind reliability index and denoted β_{HL}. Here, it is called the *first-order, second-moment* (FOSM) reliability index and denoted β_{FOSM}. This designation highlights the fact that this index is based on a first-order approximation of the limit-state function and employs second-moment information. It does not refer to the point of linearization (as in MCFOSM), as the design point $\mathbf{u}^*$ is considered the only logical choice.

A question that remains to be answered is how one obtains the design point. The following subsection describes an algorithm for finding $\mathbf{u}^*$.

5.2.3 Algorithm for Finding the Design Point

The design point is the solution to the constrained optimization problem

$$\mathbf{u}^* = \arg\min\{\|\mathbf{u}\| \;|G(\mathbf{u}) = 0\}, \tag{5.17}$$

in which "arg min" denotes the argument that minimizes the objective function, here $\|\mathbf{u}\|$. The constraint $G(\mathbf{u}) = 0$ requires that the solution point be on the limit-state surface. Hence, the solution finds the point on the limit-state surface in the standard space, which has minimum distance from the origin.

Hasofer and Lind (1974) proposed a simple algorithm to solve the above problem. The algorithm was later generalized by Rackwitz and Fiessler (1978) for application to problems with distribution information. For this reason, the method has been called the *HL-RF algorithm* (Liu and Der Kiureghian, 1990). The algorithm constructs a sequence of trial points by solving the reliability problem for the linearized limit-state function at each iteration point. Performing calculations in the standard space, the limit-state function linearized at the ith trial point, $\mathbf{u}_i$, is $G(\mathbf{U}) \cong G(\mathbf{u}_i) + \nabla G(\mathbf{u}_i)(\mathbf{U} - \mathbf{u}_i)$. The corresponding reliability index is the distance from the origin to the hyperplane $G(\mathbf{u}_i) + \nabla G(\mathbf{u}_i)(\mathbf{u} - \mathbf{u}_i) = 0$, i.e.,

$$\begin{aligned}\beta_i &= \frac{G(\mathbf{u}_i) - \nabla G(\mathbf{u}_i)\mathbf{u}_i}{\|\nabla G(\mathbf{u}^*)\|} \\ &= \frac{G(\mathbf{u}_i)}{\|\nabla G(\mathbf{u}^*)\|} + \widehat{\boldsymbol{\alpha}}_i\mathbf{u}_i,\end{aligned} \tag{5.18}$$

where $\widehat{\boldsymbol{\alpha}}_i = -\nabla G(\mathbf{u}_i)/\|\nabla G(\mathbf{u}_i)\|$ is a unit row vector. The solution of this problem is taken as the next trial point $\mathbf{u}_{i+1}$. Using (5.14), one has

$$\mathbf{u}_{i+1} = \left[\frac{G(\mathbf{u}_i)}{\|\nabla G(\mathbf{u}^*)\|} + \widehat{\boldsymbol{\alpha}}_i\mathbf{u}_i\right]\widehat{\boldsymbol{\alpha}}_i^{\mathrm{T}}. \tag{5.19}$$

This formulation is essentially a variant of the Newton–Raphson method for finding the root of a set of nonlinear equations. If the sequence of points converges, then a solution of (5.17) is obtained. However, as is well known from the classical Newton-Raphson method, the sequence may not converge if the step size $\|\mathbf{u}_{i+1} - \mathbf{u}_i\|$ is too large. To improve the stability of the algorithm, Liu and Der Kiureghian (1990) and Zhang and Der Kiureghian (1995) introduced a step size control scheme, as described below.

A general scheme for solving constrained optimization problems is to construct a sequence of trial points according to the rule (Polak, 1997)

$$\mathbf{u}_{i+1} = \mathbf{u}_i + \lambda_i \mathbf{d}_i, i = 0, 1, \ldots, \tag{5.20}$$

in which i is the iteration step with $\mathbf{u}_0$ denoting the initial trial point, $\mathbf{d}_i$ is a search direction vector, and λ_i the step size. Algorithms differ in their choices of the search direction vector and step size. From (5.19), it is clear that the HL-RF algorithm uses $\lambda_i = 1$ and

$$\mathbf{d}_i = \left[\frac{G(\mathbf{u}_i)}{\|\nabla G(\mathbf{u}_i)\|} + \widehat{\boldsymbol{\alpha}}_i \mathbf{u}_i\right] \widehat{\boldsymbol{\alpha}}_i^{\mathrm{T}} - \mathbf{u}_i. \tag{5.21}$$

Liu and Der Kiureghian (1990) and Zhang and Der Kiureghian (1995) used the above search direction but employed a merit function $m(\mathbf{u})$ to monitor the adequacy of the step size. Provided the merit function assumes its minimum at the solution of (5.17) and provided $\mathbf{d}_i$ is a descent direction of $m(\mathbf{u}_i)$, then step size λ_i is adequate if $m(\mathbf{u}_i + \lambda_i \mathbf{d}_i) < m(\mathbf{u}_i)$. The best step size is of course that which minimizes the merit function along the search direction. But finding this minimum itself is an optimization problem. Therefore, in practice, heuristic rules are used to select the "best" step size.

Zhang and Der Kiureghian (1995) showed that the merit function

$$m(\mathbf{u}) = \frac{1}{2}\|\mathbf{u}\|^2 + c|G(\mathbf{u})| \tag{5.22}$$

satisfies the requirements mentioned above, provided that at each trial point $\mathbf{u}_i$ the constant c is assigned a value that satisfies the condition

$$c > \frac{\|\mathbf{u}_i\|}{\|\nabla G(\mathbf{u}_i)\|}. \tag{5.23}$$

For the step size, Zhang and Der Kiureghian recommended using the Armijo rule (Luenberger, 1986), which selects λ_i according to the rule

$$\lambda_i = \max_k \left\{ b^k \middle| m(\mathbf{u}_i + b^k \mathbf{d}_i) - m(\mathbf{u}_i) \leq -ab^k \nabla m(\mathbf{u}_i)\mathbf{d}_i \right\}, \tag{5.24}$$

in which $a, b \in (0, 1)$ are selected constants and k is a non-negative integer. It is important to note that this rule does not require the gradient of the limit-state function at the new trial point $\mathbf{u}_i + b^k \mathbf{d}_i$ to be computed. This rule is implemented in the structural reliability software CalREL (Der Kiureghian et al., 2006).

In manual calculations, the above rule might be too cumbersome. Instead, it is recommended to reduce the step size by half if the value of the merit function does not go down, i.e., to successively use the values $\lambda_i = 1$, 0.5, 0.25, etc., until the criterion $m(\mathbf{u}_i) > m(\mathbf{u}_i + \lambda_i \mathbf{d}_i)$ is satisfied.

The convergence of the above algorithm is checked by making sure that the trial point $\mathbf{u}_i$ is on the limit-state surface and that it is an origin-project point, i.e., that the vector from the origin is normal to the tangent plane of the limit-state surface at the trial point. This is verified by satisfying the conditions

$$\left| \frac{G(\mathbf{u}_i)}{G_0} \right| < \epsilon_1, \tag{5.25}$$

$$\left\| \mathbf{u}_i - \widehat{\boldsymbol{\alpha}}_i \mathbf{u}_i \widehat{\boldsymbol{\alpha}}_i^{\mathrm{T}} \right\| < \epsilon_2, \tag{5.26}$$

where G_0 is a benchmark non-zero value of the limit-state function, e.g., $G_0 = G(\mathbf{u}_0)$, and ϵ_1 and ϵ_2 are selected tolerances, e.g., $\epsilon_1 = \epsilon_2 = 10^{-3}$. The above algorithm with step size control is known as the *improved HL-RF algorithm*, or iHL-RF in short (Zhang and Der Kiureghian, 1995).

As can be seen, the algorithm requires the limit-state function $G(\mathbf{u}_i)$ and its gradient $\nabla G(\mathbf{u}_i)$ to be computed at a sequence of trial points. The computation is performed in the original space by mapping the trial points $\mathbf{u}_i$ onto the corresponding trial points $\mathbf{x}_i$ and using the relations $G(\mathbf{u}_i) = g(\mathbf{x}_i)$ and $\nabla G(\mathbf{u}_i) = \nabla g(\mathbf{x}_i)\mathbf{J}_{\mathbf{x},\mathbf{u}}$. Thus, although the trial points are selected in the standard space, they are mapped back into the original space where the limit-state function is defined and can be computed. The gradient vector $\nabla g(\mathbf{x}_i)$ is computed either analytically or numerically depending on the form of the limit-state function, and the Jacobian $\mathbf{J}_{\mathbf{x},\mathbf{u}}$ is computed by inverting the Jacobian $\mathbf{J}_{\mathbf{u},\mathbf{x}}$ of the $\mathbf{X}$ to $\mathbf{U}$ transformation. As will be seen later, this scheme is also used in cases where the transformation from $\mathbf{X}$ to $\mathbf{U}$ space is nonlinear, which is the case when the basic variables $\mathbf{X}$ have non-normal distributions. Furthermore, the limit-state function need not have an explicit analytical form. It can be an implicit function, defined algorithmically through a computer program, such as a finite-element or finite-difference code. For the sake of convenience and easy reference, Figure 5.2 shows a flow-chart of the iHL-RF algorithm for the general case of nonlinear transformation from the $\mathbf{X}$ to $\mathbf{U}$ space. This algorithm is used in several reliability methods described in later chapters of this book.

Provided $G(\mathbf{u})$ is continuously differentiable, the above algorithm is guaranteed to converge to a solution of (5.17) (see the proof in Zhang and Der Kiureghian, 1995). However, it is not possible to guarantee convergence to the global solution if there are multiple points that satisfy the conditions of being on the limit-state surface and having a gradient vector that passes through the origin. Indeed, short of making an exhaustive search of the entire space, there is no optimization algorithm that can guarantee convergence to the global solution for a general, non-convex limit-state function.

To make sure that the global solution has been obtained for a reliability problem, the following practical steps are suggested: (a) Make sure that the solution makes sense from a physical standpoint by examining the design point $\mathbf{x}^*$ in the original space. For a typical structural reliability problem with high reliability, one would normally expect above-mean

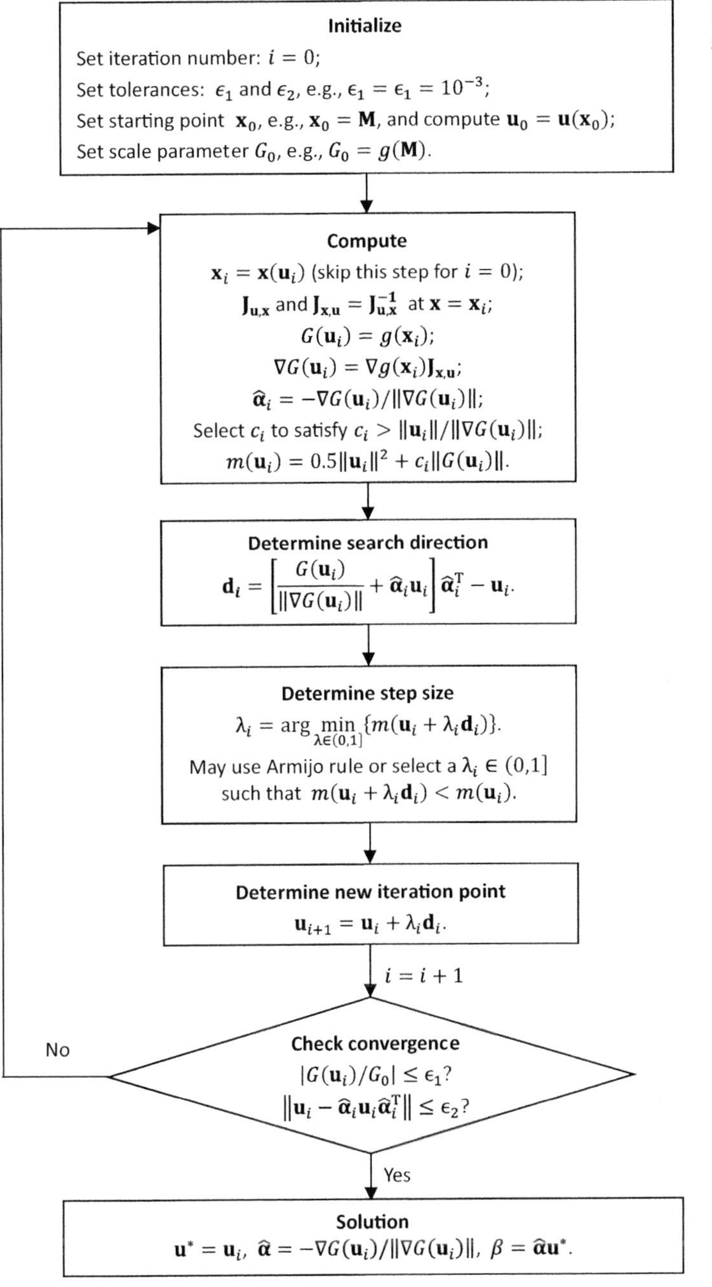

Figure 5.2 The iHL-RF algorithm

demand values and below-mean capacity values at the design point. This, however, is not always the case, as exemplified below. (b) Repeatedly solve the problem starting from different initial trial points $\mathbf{u}_0$ and take the solution that corresponds to the smallest distance from the origin.

In practice, it is seldom that component limit-states have multiple design points, and even rarer that convergence to a local design point is achieved. When multiple design points do occur, their occurrence can often be anticipated from the nature of the problem. There are ways of finding the multiple design points, when they do occur. Section 6.9 in the next chapter describes one such method.

In certain problems, it is necessary to obtain a sequence of design points for a parametrized limit-state function. In such cases, it is effective to use the design point for a previous value of the parameter as a warm starting point in the search for the design point for the new parameter value, provided the parameter values are selected as a sequence. Wang et al. (2015) describe an efficient algorithm for this purpose. Several example applications in later chapters employ this approach.

Example 5.2 – Solution by FOSM

Reconsider the toy structural reliability problem in Example 5.1 defined by the limit-state function $g(X_1, X_2) = X_1^2 - 2X_2$. For the given covariance matrix, the diagonal matrix of standard deviations and the correlation matrix are

$$\mathbf{D_X} = \begin{bmatrix} 2 & 0 \\ 0 & 5 \end{bmatrix}, \mathbf{R_{XX}} = \mathbf{D_X}^{-1}\Sigma_{\mathbf{XX}}\mathbf{D_X}^{-1} = \begin{bmatrix} 1 & 0.5 \\ 0.5 & 1 \end{bmatrix}. \tag{E1}$$

Using the formulas in Appendix 3B, the lower-triangular Cholesky decomposition of the correlation matrix and its inverse are obtained as

$$\mathbf{L} = \begin{bmatrix} 1 & 0 \\ 0.500 & 0.866 \end{bmatrix}, \mathbf{L}^{-1} = \begin{bmatrix} 1 & 0 \\ -0.577 & 1.155 \end{bmatrix}. \tag{E2}$$

Thus, the transformation to the standard space is defined by

$$\mathbf{U} = \mathbf{L}^{-1}\mathbf{D_X}^{-1}(\mathbf{X} - \mathbf{M_X}) = \begin{bmatrix} 0.500 & 0 \\ -0.289 & 0.231 \end{bmatrix}\left(\mathbf{X} - \begin{Bmatrix} 10 \\ 20 \end{Bmatrix}\right) \tag{E3}$$

and its inverse is

$$\mathbf{X} = \mathbf{M_X} + \mathbf{DLU} = \begin{Bmatrix} 10 \\ 20 \end{Bmatrix} + \begin{bmatrix} 2 & 0 \\ 2.500 & 4.330 \end{bmatrix}\mathbf{U}. \tag{E4}$$

It should be clear that the Jacobian matrix of the transformation and its inverse are the coefficient matrices of the above linear transforms, i.e.,

$$\mathbf{J_{u,x}} = \begin{bmatrix} 0.500 & 0 \\ -0.289 & 0.231 \end{bmatrix}, \mathbf{J_{x,u}} = \begin{bmatrix} 2 & 0 \\ 2.500 & 4.330 \end{bmatrix}. \tag{E5}$$

Example 5.2 (cont.)

Furthermore, the gradient vector of the limit-state function is $\nabla g = [2x_1 - 2]$. We are now ready to use the iHL-RF algorithm to find the design point.

Let us start the iterations at the mean point, i.e., $\mathbf{x}_0 = [10\ 20]^{\mathrm{T}}$. Using the $\mathbf{X}$ to $\mathbf{U}$ transform, this gives the starting point in the standard space as $\mathbf{u}_0 = [0\ 0]^{\mathrm{T}}$. We need to compute the limit-state function and its gradient at this point. The results are: $G(\mathbf{u}_0) = g(\mathbf{x}_0) = 10^2 - 2 \times 20 = 60$ and

$$\nabla G(\mathbf{u}_0) = \nabla g(\mathbf{x}_0)\mathbf{J}_{\mathbf{x},\mathbf{u}} = [20\ -2]\begin{bmatrix} 2 & 0 \\ 2.500 & 4.330 \end{bmatrix} = [35.00\ -8.66]. \tag{E6}$$

The norm of the gradient vector is $\|\nabla G(\mathbf{u}_0)\| = \sqrt{35.00^2 + 8.66^2} = 36.1$. Normalizing the negative gradient vector, we have $\widehat{\boldsymbol{\alpha}}_0 = -\nabla G(\mathbf{u}_0)/\|\nabla G(\mathbf{u}_0)\| = [-0.971\ 0.240]$. Now, we compute the search direction vector as

$$\mathbf{d}_0 = \left[\frac{60}{36.056} + [-0.971\ 0.240]\begin{Bmatrix} 0 \\ 0 \end{Bmatrix}\right]\begin{Bmatrix} -0.971 \\ 0.240 \end{Bmatrix} - \begin{Bmatrix} 0 \\ 0 \end{Bmatrix} = \begin{Bmatrix} -1.615 \\ 0.400 \end{Bmatrix}. \tag{E7}$$

To construct the merit function, we must select a constant c such that $c > \|\mathbf{u}_0\|/\|\nabla G(\mathbf{u}_0)\|$. Because $\mathbf{u}_0 = \mathbf{0}$, any positive value for c is appropriate. Let us select $c = 10$. Then the value of the merit function (see 5.22) at $\mathbf{u}_0$ is $m(\mathbf{u}_0) = (0)^2/2 + 10 \times 60 = 600$. Now let us try $\lambda_0 = 1$. We have $\mathbf{u}_1 = \mathbf{u}_0 + \lambda_0 \mathbf{d}_0 = [-1.615\ 0.400]^{\mathrm{T}}$. We need to map this point back to the original space, so we can compute the value of the limit-state function. Using the inverse transform in (E4), we obtain $\mathbf{x}_1 = [6.769\ 17.693]^{\mathrm{T}}$. The value of the limit-state function at the new trial point is $G(\mathbf{u}_1) = g(\mathbf{x}_1) = 6.769^2 - 2 \times 17.693 = 10.439$. Before going further, we examine the merit function value at the new trial point. We have $m(\mathbf{u}_1) = \left(1.615^2 + 0.400^2\right)/2 + 10 \times 10.439 = 114$. (Note that we use the value of c selected at point i to check the adequacy of point $i + 1$; however, a new value of c must be selected for each new trial point once it has been determined to be viable.) The condition $m(\mathbf{u}_1) < m(\mathbf{u}_0)$ is satisfied and $\mathbf{u}_1$ is a viable trial point. We now proceed to compute the gradient vector at the new trial point as $\nabla G(\mathbf{u}_1) = \nabla g(\mathbf{x}_1)\mathbf{J}_{\mathbf{x},\mathbf{u}} = [22.08\ -8.66]$, which has the norm $\|\nabla G(\mathbf{u}_1)\| = 23.7$. The normalized negative gradient vector is $\widehat{\boldsymbol{\alpha}}_1 = [-0.931\ 0.365]$. We are now ready to compute a new search direction vector starting at $\mathbf{u}_1$. Table 5.1 summarizes the iteration steps for this problem. Iterations converge to three significant figures in six steps. The last row shows the design point as $\mathbf{u}^* = [-1.928\ 0.853]^{\mathrm{T}}$, its mapping in the original space as $\mathbf{x}^* = [6.14\ 18.9]^{\mathrm{T}}$, the normalized negative gradient vector at the design point as $\widehat{\boldsymbol{\alpha}} = [-0.971\ 0.240]$, and the FOSM reliability index as $\beta_{\mathrm{FOSM}} = 2.11$.

It is useful to examine the above solution by graphical means. Figure 5.3 shows the limit-state surface for this two-dimensional problem in the original and standard spaces together

Example 5.2 (cont.)

with the design point and the reliability index. Also shown are the steps taken by the iHL-RF algorithm to converge to the design point. Note that the axes in the standard space must have identical scales.

Table 5.1 Summary of iterations in the iHL-RF algorithm to find the design point

Iteration i	$\mathbf{x}_i$	$\mathbf{u}_i$	$G(\mathbf{u}_i)$	$\nabla G(\mathbf{u}_i)^{\mathrm{T}}$	Selected c	$m(\mathbf{u}_i)$	$\widehat{\boldsymbol{\alpha}}_i^{\mathrm{T}}$	$\widehat{\boldsymbol{\alpha}}_i\mathbf{u}_i$
1	$\begin{Bmatrix} 10 \\ 20 \end{Bmatrix}$	$\begin{Bmatrix} 0 \\ 0 \end{Bmatrix}$	60	$\begin{Bmatrix} 35.0 \\ -8.66 \end{Bmatrix}$	10	600	$\begin{Bmatrix} -0.971 \\ 0.240 \end{Bmatrix}$	0
2	$\begin{Bmatrix} 6.77 \\ 17.7 \end{Bmatrix}$	$\begin{Bmatrix} -1.615 \\ 0.400 \end{Bmatrix}$	10.4	$\begin{Bmatrix} 22.1 \\ -8.66 \end{Bmatrix}$	10	114	$\begin{Bmatrix} -0.931 \\ 0.365 \end{Bmatrix}$	1.65
3	$\begin{Bmatrix} 6.11 \\ 18.4 \end{Bmatrix}$	$\begin{Bmatrix} -1.946 \\ 0.763 \end{Bmatrix}$	0.436	$\begin{Bmatrix} 19.4 \\ -8.66 \end{Bmatrix}$	10	6.54	$\begin{Bmatrix} -0.913 \\ 0.407 \end{Bmatrix}$	2.09
4	$\begin{Bmatrix} 6.15 \\ 18.9 \end{Bmatrix}$	$\begin{Bmatrix} -1.926 \\ 0.858 \end{Bmatrix}$	1.6×10^{-3}	$\begin{Bmatrix} 19.6 \\ -8.66 \end{Bmatrix}$	10	2.24	$\begin{Bmatrix} -0.915 \\ 0.404 \end{Bmatrix}$	2.11
5	$\begin{Bmatrix} 6.14 \\ 18.9 \end{Bmatrix}$	$\begin{Bmatrix} -1.928 \\ 0.852 \end{Bmatrix}$	2.7×10^{-5}	$\begin{Bmatrix} 19.6 \\ -8.66 \end{Bmatrix}$	10	2.22	$\begin{Bmatrix} -0.915 \\ 0.405 \end{Bmatrix}$	2.11
6	$\begin{Bmatrix} 6.14 \\ 18.9 \end{Bmatrix}$	$\begin{Bmatrix} -1.928 \\ 0.853 \end{Bmatrix}$	4.6×10^{-7}	$\begin{Bmatrix} 19.6 \\ -8.66 \end{Bmatrix}$	10	2.22	$\begin{Bmatrix} -0.915 \\ 0.405 \end{Bmatrix}$	2.11

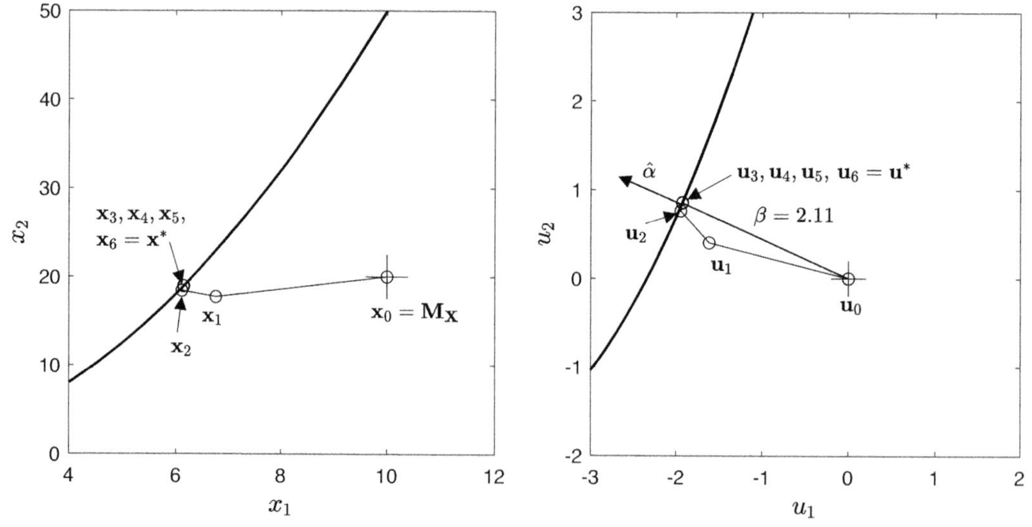

Figure 5.3 Convergence to the design point in original (left) and standard (right) spaces

Example 5.2 (cont.)

A rather unexpected result in the above solution is the following: In safety analysis, the most likely failure scenario is usually thought of as one in which the capacity is below its mean value and the demand is above its mean value. In the present case, X_1 is a capacity variable (the limit-state function increases with increasing X_1) and X_2 is a demand variable (the limit-state function decreases with increasing X_2). As the design point is the closest point to the origin in the standard space, it represents the most likely failure point within a second-moment context. For the above problem, we have the design point in the original space as $\mathbf{x}^* = [6.14\ 18.9]^{\mathrm{T}}$. Comparing with the mean point $\mathbf{M_X} = [10\ 20]^{\mathrm{T}}$, we find that both the capacity and demand variables are below their respective mean values at the design point. At first glance, this outcome appears to be unreasonable – we expect a higher-than-mean demand value at the design point. This result, however, is correct and reasonable in view of the assumed positive correlation between the demand and capacity values. Positive correlation between X_1 and X_2 implies that small (large) values of X_1 tend to occur concurrently with small (large) values of X_2. Hence, among all possible outcomes, the pair of outcomes (6.14, 18.9) is most likely to give rise to the failure event, even though the outcome for the demand is below its mean value. Indeed, if the variables in this problem are assumed to be uncorrelated, the solution is $\mathbf{u}^* = [-1.611\ 0.594]^{\mathrm{T}}$, $\mathbf{x}^* = [6.78\ 23.0]^{\mathrm{T}}$, and $\beta_{\mathrm{FOSM}} = 1.72$. Observe that the design-point value for the capacity variable is now below its mean and that for the demand variable is above its mean. Also observe that the correlation between the variables has a significant influence on the reliability index: $\beta_{\mathrm{FOSM}} = 2.11$ for correlated variables and $\beta_{\mathrm{FOSM}} = 1.72$ for uncorrelated variables.

It is important to point out that, had we defined the limit-state function as $g(X_1, X_2) = 1 - 2X_2/X_1^2$, the reliability solution would have been the same as obtained above. This is because the two surfaces $x_1^2 - 2x_2 = 0$ and $1 - 2x_2/x_1^2 = 0$ are identical in the two-dimensional outcome space of X_1 and X_2. This demonstrates the invariance property of the FOSM method.

5.2.4 The Generalized (Second-Moment) Reliability Index

As first pointed out by Ditlevsen (1979a), a fundamental shortcoming of β_{FOSM} is that it lacks *orderability*. That is, a greater value of this index does not necessarily imply greater reliability. This is so because all limit-state surfaces having equal minimum distance from the origin in the standard space have identical β_{FOSM} values, regardless of the shapes of the respective safe domains. Whether the limit-state surface is curved toward or away from the origin obviously must have an influence on the reliability measure. Thus, to have an orderable measure of reliability, one needs to consider the shape of the safe domain, as well as the relative weights of the points within that domain. Because the available probability information is restricted to the first and second moments and the probability distribution is unavailable, Ditlevsen (1979a) suggested using a formal weight function to characterize the likelihoods within the safe domain.

An important property of the standard space is its *rotational symmetry*. Within a second-moment context, this property can be explained as follows: Consider the probability p on the side of the hyperplane $\beta - \widehat{\boldsymbol{\alpha}}\mathbf{u} = 0$ away from the origin, where β denotes the distance of the hyperplane from the origin and $\widehat{\boldsymbol{\alpha}}$ denotes its unit normal row-vector and determines its orientation. Define the random variable $Z = \beta - \widehat{\boldsymbol{\alpha}}\mathbf{U}$ and observe that Z has the mean $\mu_Z = \beta$ and standard deviation $\sigma_Z = 1$, and that $p = \Pr(Z \leq 0)$. With only the mean and variance of Z known, it is not possible to determine p precisely. However, using the well-known Chebyshev bound (Ross, 2021), it is possible to derive the upper bound $p \leq 1/\beta^2$. Note that this bound does not involve the direction vector $\widehat{\boldsymbol{\alpha}}$. Thus, with only second-moment information available, an estimate of p must be independent of the orientation of the hyperplane. Only the distance from the origin matters. This defines the rotational symmetry property of the standard space in a second-moment context.

Based on the above rotational symmetry property, Ditlevsen (1979a) argued that the weight function over the safe domain must be rotationally symmetric. Considering additional requirements that the weight function be non-negative, have unit volume over the entire domain, have a value that decreases with increasing distance from the origin, and be regenerative with increasing dimension, he proposed the function $w_n(\mathbf{u}) = (2\pi)^{-n/2} \exp\left(-\|\mathbf{u}\|^2/2\right)$, which is identical to the n-variate standard normal probability density function (PDF). He then introduced the generalized second-moment reliability index as

$$\beta_{\text{GSM}} = \Phi^{-1}\left[\int_{0 < G(\mathbf{u})} w_n(\mathbf{u})d\mathbf{u}\right], \tag{5.27}$$

where the integral is over the safe domain and $\Phi^{-1}[\cdot]$ denotes the inverse of the standard normal cumulative distribution function (CDF). The above definition is motivated by the fact that, when $G(\mathbf{u})$ is a linear function of $\mathbf{u}$, (5.27) yields a result identical to that of FOSM.

It is important to understand that β_{GSM} is strictly a second-moment measure of reliability because it employs only second-moment information about the random variables – hence the subscript SM. This index is orderable within a second-moment context as it accounts for the shape of the safe domain with an appropriate weight function. However, as will shortly be discussed, the index loses this property in the light of any information beyond the second moments.

An obvious question to ask is how one computes the n-fold integral in (5.27). As seen in Chapter 4, this type of integral is common in reliability problems. Later chapters will present several computational methods for evaluating this type of integral.

Example 5.3 – Solution by GSM

Reconsider the structural reliability problem in Examples 5.1 and 5.2. Using simulation methods described in Chapter 9, the integral in (5.27) over the safe domain is computed to be 0.983. Hence, the generalized reliability index is

$$\beta_{\text{GSM}} = \Phi^{-1}(0.983) = 2.13. \tag{E1}$$

This result is slightly greater than the result obtained in Example 5.2. This is due to the fact that the limit-state surface is curved away from the origin, as can be seen in the right-side graph in Figure 5.3.

5.3 Reliability Methods with Beyond Second-Moment Information

A shortcoming of the FOSM method is that it cannot accommodate information beyond the second moments, such as bounds, higher than second moments (e.g., skewness and kurtosis coefficients), marginal distributions, etc. More critically, in the light of such information, the standard space defined by the transformation in (5.7) loses its rotational symmetry property and, as a result, the foundational assumptions behind β_{FOSM} and β_{GSM} become invalid. As a demonstration of this issue, consider a reliability problem with two uncorrelated standard random variables U_1 and U_2 (i.e., having zero means, unit variances, and zero correlation) and assume that we additionally know that $U_1 \leq 3$. If we consider the probability contents of several half-spaces defined by planes with equal distances b from the origin, as shown in Figure 5.4, clearly the probability contents of those planes that have normal vectors directed towards the positive direction of u_1 would be expected to be smaller than those with normal vectors in the opposite direction. Thus, we cannot claim that the probability contents defined by these planes of equal distance from the origin are independent of their orientations. It follows that the standard space defined by (5.7) is no longer rotationally symmetric.

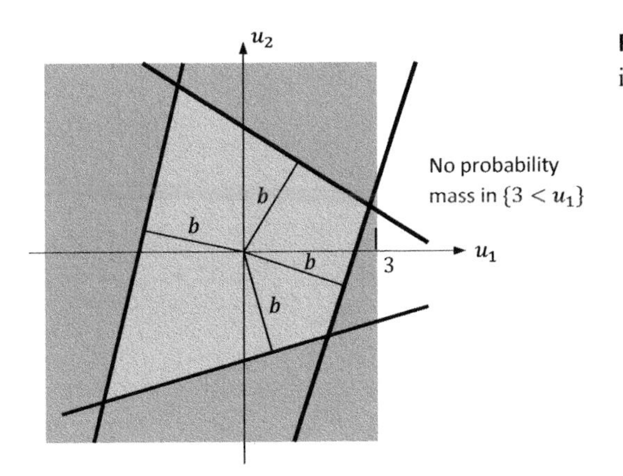

Figure 5.4 Standard space with additional information $U_1 \leq 3$

To overcome this problem, Ditlevsen (1979a) suggested using the beyond-the-second-moment information to construct a nonlinear transformation to a standard space that possesses the rotational symmetry property. For example, when a quantity is known to be non-negative (i.e., it has a known lower bound at 0), Ditlevsen suggests using the logarithm of the quantity as the random variable of interest in the reliability formulation. This idea was more formally applied by Winterstein and Bjerager (1987) and Zhao and Ono (2001) in the case where marginal third and fourth moments are available and by Der Kiureghian and Liu (1986) when the marginal distributions are known. An entirely different approach was proposed by Veneziano (1979) based on the concept of Chebyshev bounds. The following subsections briefly describe these three methods. Similarly to the second-moment methods, these methods assume that the available partial probability information, e.g., higher moments or marginal distributions, are precisely known. However, as mentioned earlier, this assumption is not realistic. Hence, the value of these methods lies in their historical role in the development of the structural reliability theory. For this reason, only brief reviews of these methods are presented.

Aside from the above methods, Wu et al. (1990) proposed the so-called advanced first-order, mean-value method to account for distribution information within the context of the MCFOSM method. However, their method cannot account for correlations between the random variables and, similar to MCFOSM, suffers from the lack of invariance problem.

5.3.1 Knowledge of Third and Fourth Moments

Winterstein and Bjerager (1987) used Hermite polynomials to model random variables for which the first to fourth marginal moments and the correlation coefficients are known. Hermite polynomials are defined by

$$He_k(z) = (-1)^k \exp\left(\frac{z^2}{2}\right)\frac{\mathrm{d}^k}{\mathrm{d}z^k}\exp\left(-\frac{z^2}{2}\right), k = 0, 1, 2, \ldots, \tag{5.28}$$

where k denotes the order of the polynomial. It is easy to verify that

$$He_0(z) = 1, \tag{5.29a}$$

$$He_1(z) = z, \tag{5.29b}$$

$$He_2(z) = z^2 - 1, \tag{5.29c}$$

$$He_3(z) = z^3 - 3z, \tag{5.29d}$$

$$He_4(z) = z^4 - 6z^2 + 3. \tag{5.29e}$$

The Hermite polynomials have the following properties:

- $\frac{d}{dz} He_k(z) = k\, He_{k-1}(z)$,
- If $Z_1, Z_2, \ldots$, are correlated standard normal random variables with $\rho_{0,ij}$ as the correlation coefficient between Z_i and Z_j, then

$$\begin{aligned} \mathrm{E}[He_k(Z_i)] &= 0, && \text{for } k = 1, 2, \ldots, \\ \mathrm{Cov}\left[He_k(Z_i)He_l(Z_j)\right] &= 0, && \text{for } k \neq l, \\ &= k!\rho_{0,ij}^k, && \text{for } k = l, \end{aligned}$$

Now suppose we model the random variables of the reliability problem in terms of Hermite polynomials of standard but correlated normal random variables Z_i, $i = 1, \ldots, n$, up to order 3 in the form

$$\frac{X_i - \mu_i}{\sigma_i} = a_i He_1(Z_i) + b_i He_2(Z_i) + c_i He_3(Z_i), i = 1, 2, \ldots, n, \tag{5.30}$$

where μ_i and σ_i are the mean and standard deviation of X_i, respectively, and a_i, b_i, and c_i are parameters to be determined. Using the properties described above, Winterstein and Bjerager (1987) showed that

$$\mathrm{E}\left[\frac{X_i - \mu_i}{\sigma_i}\right] = 0, \tag{5.31a}$$

$$\mathrm{E}\left[\left(\frac{X_i - \mu_i}{\sigma_i}\right)^2\right] = a_i^2 + 2b_i^2 + 6c_i^2 = 1, \tag{5.31b}$$

$$\mathrm{E}\left[\left(\frac{X_i - \mu_i}{\sigma_i}\right)^3\right] = 2b_i(2 + a_i^2 + 18a_ic_i + 42c_i^2) = \gamma_i, \tag{5.31c}$$

$$\begin{aligned} \mathrm{E}\left[\left(\frac{X_i - \mu_i}{\sigma_i}\right)^4\right] &= 15 + 288a_ic_i + 936c_i^2 - 12a_i^4 - 264a_i^3c_i \\ &\quad - 864a_i^2c_i^2 - 432a_ic_i^3 - 2808c_i^4 = \kappa_i, \end{aligned} \tag{5.31d}$$

where γ_i and κ_i are the skewness and kurtosis coefficients of X_i, respectively. For known values of these coefficients, the last three equations in (5.31) are solved to obtain the unknown parameters a_i, b_i, and c_i for each variable X_i. In this way, (5.30) defines a set of n random variables that have the specified first to fourth moments. To match the specified correlation coefficients ρ_{ij} between pairs of random variables X_i and X_j, we write

$$\mathrm{E}\left[\left(\frac{X_i - \mu_i}{\sigma_i}\right)\left(\frac{X_j - \mu_j}{\sigma_j}\right)\right] = a_ia_j\rho_{0,ij} + 2b_ib_j\rho_{0,ij}^2 + 6c_ic_j\rho_{0,ij}^3 = \rho_{ij}. \tag{5.32}$$

Once parameters a_i, b_i, and c_i, $i = 1, \ldots, n$, are determined, this equation is solved for the correlation coefficient $\rho_{0,ij}$ between each pair of standard normal random variables. In this manner, the set of standard but correlated normal variables $\mathbf{Z} = (Z_1, \ldots, Z_n)$ defines the set of basic random variables $\mathbf{X} = (X_1, \ldots, X_n)$ of the reliability problem through the relations in (5.30) so that the latter possess the specified first to fourth marginal moments and pairwise correlation coefficients.

Three conditions must be satisfied for the above solution to be viable: (a) The set of equations in (5.31) must yield real-valued solutions for a_i, b_i, and c_i, $i = 1, \ldots, n$; (b) the solution for each $\rho_{0,ij}$ must lie inside the interval $(-1, 1)$; and (c) the correlation matrix $\mathbf{R}_0$ of $\mathbf{Z}$, having the elements $\rho_{0,ij}$, must be positive definite. If these conditions are satisfied, then a transformation to a standard uncorrelated normal space is achieved by using $\mathbf{U} = \mathbf{L}_0^{-1}\mathbf{Z}$, where $\mathbf{L}_0$ is the Cholesky decomposition of $\mathbf{R}_0$. Note that the transformation from the $\mathbf{X}$ space to the $\mathbf{U}$ space is now nonlinear due to the form of (5.30). Also note that the space of $\mathbf{U}$ has the desired rotational symmetry property described earlier. The reliability solution in the standard normal space may be obtained in terms of the minimum distance from the origin to the limit-state surface, as described earlier. Although such a solution is a first-order approximation, it is no longer second moment, because it incorporates information on the third and fourth marginal moments. For that reason, the reliability index is denoted $\beta_{\text{FOSM+}}$, where the subscript + indicates that information beyond the second moments has been used. Note that this solution is also based on incomplete probability information. As a result, it is not possible to provide a precise statement about the failure probability.

Example 5.4 – Incorporating Higher Moment Information

Reconsider the structural reliability problem in Examples 5.1–5.3. Suppose, in addition to the second-moment information given previously, we have the skewness coefficients $\gamma_1 = 0.608$ and $\gamma_2 = 1.140$ and the kurtosis coefficients $\kappa_1 = 3.664$ and $\kappa_2 = 5.400$ of the two random variables. (These values correspond to a lognormal distribution for X_1 and a Type I largest-value distribution for X_2, though we assume we do not know these distributions.) Using the procedure described above, we solve for the parameters a_i, b_i, and c_i, $i = 1, 2$, in the set of equations (5.31) and use them in (5.30) to express X_1 and X_2 as follows:

$$\frac{X_1 - 10}{2} = 0.9902Z_1 + 0.09792\left(Z_1^2 - 1\right) + 0.006846\left(Z_1^3 - 3Z_1\right), \tag{E1}$$

$$\frac{X_2 - 20}{5} = 0.9694Z_2 + 0.1684\left(Z_2^2 - 1\right) + 0.02414\left(Z_2^3 - 3Z_2\right). \tag{E2}$$

Furthermore, solving (5.32) with $\rho_{ij} = 0.5$, we obtain $\rho_{0,12} = 0.512$. Figure 5.5 shows the solution of this problem in the standard space $\mathbf{U} = (U_1, U_2)$ obtained by linearly transforming random variables $\mathbf{Z} = (Z_1, Z_2)$. The design point is at $\mathbf{u}^* = [-2.140\ 1.625]^{\mathrm{T}}$ with the corresponding point in the original space at $\mathbf{x}^* = [6.417\ 20.59]^{\mathrm{T}}$; the unit normal vector is $\widehat{\boldsymbol{\alpha}} = [-0.796\ 0.605]$; and the reliability index is $\beta_{\text{FOSM+}} = 2.69$. The substantial difference from $\beta_{\text{FOSM}} = 2.11$ reflects the influence of the third- and fourth-moment information that we have incorporated into the analysis. It is noteworthy that the limit-state surface now is curved

Example 5.4 (cont.)

toward the origin, unlike the case in Example 5.3 where it was curved away from the origin. Furthermore, it is no longer a parabola because of the nonlinearity introduced in the transformation from the **X** to **Z** variables.

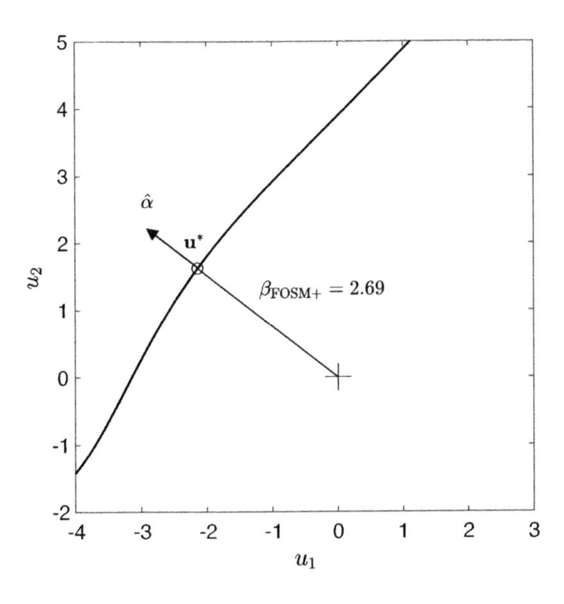

Figure 5.5 Solution of Example 5.4

5.3.2 Knowledge of Marginal Distributions

In many problems the available data are sufficient to determine second moments (including correlation coefficients) and marginal distributions, but not the joint distribution of the basic random variables. For such a case, to construct a transformation to a rotationally symmetric standard space, Der Kiureghian and Liu (1986) proposed formally employing the Nataf joint distribution model. As described in Chapter 3, this joint distribution model is constructed by assigning the multinormal distribution to the standard normal variables Z_i, $i = 1, \ldots, n$, obtained by marginally transforming the basic random variables $\mathbf{X} = [X_1, \ldots, X_n]^{\mathrm{T}}$ according to the rule

$$Z_i = \Phi^{-1}[F_i(X_i)], i = 1, \ldots, n. \tag{5.33}$$

It was shown in Chapter 3 that the correlation coefficient $\rho_{0,ij}$ between the pairs of variables Z_i and Z_j is related to the correlation coefficient ρ_{ij} between the pairs of variables X_i and X_j through the double integral

$$\rho_{ij} = \int_{-\infty}^{+\infty}\int_{-\infty}^{+\infty}\left(\frac{x_i - \mu_i}{\sigma_i}\right)\left(\frac{x_j - \mu_j}{\sigma_j}\right)\phi_2\left(z_i, z_j, \rho_{0,ij}\right)\mathrm{d}z_i\mathrm{d}z_j. \tag{5.34}$$

The unknown in this equation is $\rho_{0,ij}$, which is inside the integrand. Hence, a numerical approach must be used to solve for it. Alternatively, approximate formulas listed in Appendix 3A for common marginal distributions can be used. Provided $\rho_{0,ij}$ obtained from (5.34) satisfies the condition $-1 < \rho_{0,ij} < 1$ for all pairs of random variables and the correlation matrix $\mathbf{R}_0 = \left[\rho_{0,ij}\right]$ is positive definite, the Nataf distribution is applicable and the transformation $\mathbf{U} = \mathbf{L}_0^{-1}\mathbf{Z}$, where $\mathbf{L}_0$ denotes the Cholesky decomposition of $\mathbf{R}_0$, leads to a standard normal space that is rotationally symmetric. The solution for the reliability index, $\beta_{\text{FOSM+}}$, is then sought in the outcome space of $\mathbf{U}$.

Example 5.5 – Incorporating Marginal Distributions

Reconsider the structural reliability problem in Examples 5.1–5.4. Suppose, in addition to the second-moment information given previously, we know that X_1 has the lognormal distribution and X_2 has the Type I largest value (Gumbel) distribution. Note that the two variables are dependent with the correlation coefficient $\rho_{12} = 0.5$, but we do not have information about their joint distribution.

To construct the transformations in (5.33), we first determine the marginal CDFs. The parameters of the lognormal distribution of X_1 are $\zeta = \sqrt{\ln\left(1 + \delta_1^2\right)} = 0.198$ and $\lambda = \ln\mu_1 - \zeta^2/2 = 2.28$. The parameters of the Type I largest value distribution (see Table A2.2) are $\alpha = \pi/\left(\sqrt{6}\sigma_2\right) = 0.257$ and $v = \mu_2 - 0.577/\alpha = 17.7$. Thus,

$$Z_1 = \Phi^{-1}\left[\Phi\left(\frac{\ln X_1 - \lambda}{\zeta}\right)\right] = \frac{\ln X_1 - \lambda}{\zeta}, \tag{E1}$$

$$Z_2 = \Phi^{-1}[\exp\{-\exp[-\alpha(X_2 - v)]\}]. \tag{E2}$$

The correlation coefficient between Z_1 and Z_2 is computed as $\rho_{0,ij} = C\rho_{ij}$, where the factor C is determined from Table A3.4 as $C = 1.023$; thus, $\rho_{0,12} = 1.023 \times 0.5 = 0.511$. The correlation matrix of $\mathbf{Z}$ and its Cholesky decomposition are

$$\mathbf{R}_0 = \begin{bmatrix} 1 & 0.511 \\ 0.511 & 1 \end{bmatrix}, \mathbf{L}_0 = \begin{bmatrix} 1 & 0 \\ 0.511 & 0.859 \end{bmatrix}. \tag{E3}$$

The transformation to the standard space can now be written as

$$\mathbf{U} = \mathbf{L}_0^{-1}\mathbf{Z} = \mathbf{L}_0^{-1}\begin{Bmatrix} \dfrac{\ln X_1 - \lambda}{\zeta} \\ \Phi^{-1}[\exp\{-\exp[-\alpha(X_2 - v)]\}] \end{Bmatrix}. \tag{E4}$$

Observe that the transformation is nonlinear. For determining the design point with the iHL-RF algorithm, we also need to determine the Jacobian of the transformation. Taking partial derivatives in (E4), we find (see 3.57)

Example 5.5 (cont.)

$$\mathbf{J}_{\mathbf{u},\mathbf{x}} = \mathbf{L}_0^{-1}\begin{bmatrix} \dfrac{f_1(x_1)}{\phi(z_1)} & 0 \\ 0 & \dfrac{f_2(x_2)}{\phi(z_2)} \end{bmatrix}$$
$$= \mathbf{L}_0^{-1}\begin{bmatrix} \dfrac{1}{\zeta x_1} & 0 \\ 0 & \dfrac{\alpha \exp\{-\alpha(x_2 - v) - \exp[-\alpha(x_2 - v)]\}}{\phi(z_2)} \end{bmatrix}. \tag{E5}$$

We are now ready to proceed with the iHL-RF algorithm to find the design point. Starting at $\mathbf{x}_0 = \mathbf{M_X}$, from (E1)–(E2) we obtain $\mathbf{z}_0 = [0.0990\ 0.1773]^{\mathrm{T}}$ and from (E3) we obtain $\mathbf{u}_0 = [0.0990\ 0.1474]^{\mathrm{T}}$. The value of the limit-state function at this point is $G(\mathbf{u}_0) = g(\mathbf{x}_0) = 60$. The gradient vector is obtained according to the formula $\nabla G(\mathbf{u}_0) = \nabla g(\mathbf{x}_0)\mathbf{J}_{\mathbf{x},\mathbf{u}} = \nabla g(\mathbf{x}_0)\mathbf{J}_{\mathbf{u},\mathbf{x}}^{-1}$. Using numerical values in the above formulas, we obtain $\nabla G(\mathbf{u}_0) = [34.7181\ -8.2163]$ and $\widehat{\boldsymbol{\alpha}}_0 = -\nabla G(\mathbf{u}_0)/\|\nabla G(\mathbf{u}_0)\| = [-0.973\ 0.230]$. We are now ready to determine the next iteration point according to the iHL-RF rule (5.20). Using $\lambda_0 = 1$, the result is $\mathbf{u}_1 = [-1.576\ 0.373]^{\mathrm{T}}$. This point needs to be transformed back into the original space so that the limit-state function and its gradient can be computed. In the present case, the inverse transform is available in closed form as

$$\mathbf{Z} = \mathbf{L}_0\mathbf{U}, \tag{E6}$$

$$X_1 = \exp(\lambda + \zeta Z_1), \tag{E7}$$

$$X_2 = v - \frac{1}{\alpha}\ln[\ln\Phi(Z_2)]. \tag{E8}$$

The result is $\mathbf{z}_1 = [-1.576 - 0.486]^{\mathrm{T}}$ and $\mathbf{x}_1 = [7.177\ 17.173]^{\mathrm{T}}$. Calculations show that the value of the merit function at the new point is smaller than the value at $\mathbf{u}_0$ so that the new trial point is good. These steps are repeated until convergence is achieved at $\mathbf{u}^* = [-2.13\ 1.64]^{\mathrm{T}}$ with the normal vector $\widehat{\boldsymbol{\alpha}} = [-0.793\ 0.610]$ and reliability index $\beta_{\mathrm{FOSM+}} = 2.69$. The point corresponding to $\mathbf{u}^*$ in the original space is $\mathbf{x}^* = [6.43\ 20.7]^{\mathrm{T}}$. Figure 5.6 shows the solution in the standard space. Observe that the limit-state surface is again curved toward the origin. Also note that this solution is only slightly different from the solution of Example 5.4. As the skewness and kurtosis coefficients assumed there were those of the marginal distributions assumed in the present example, it follows that the small difference between the two sets of solutions is on account of the moments higher than the fourth order that are inherent in the marginal distributions.

Example 5.5 (cont.)

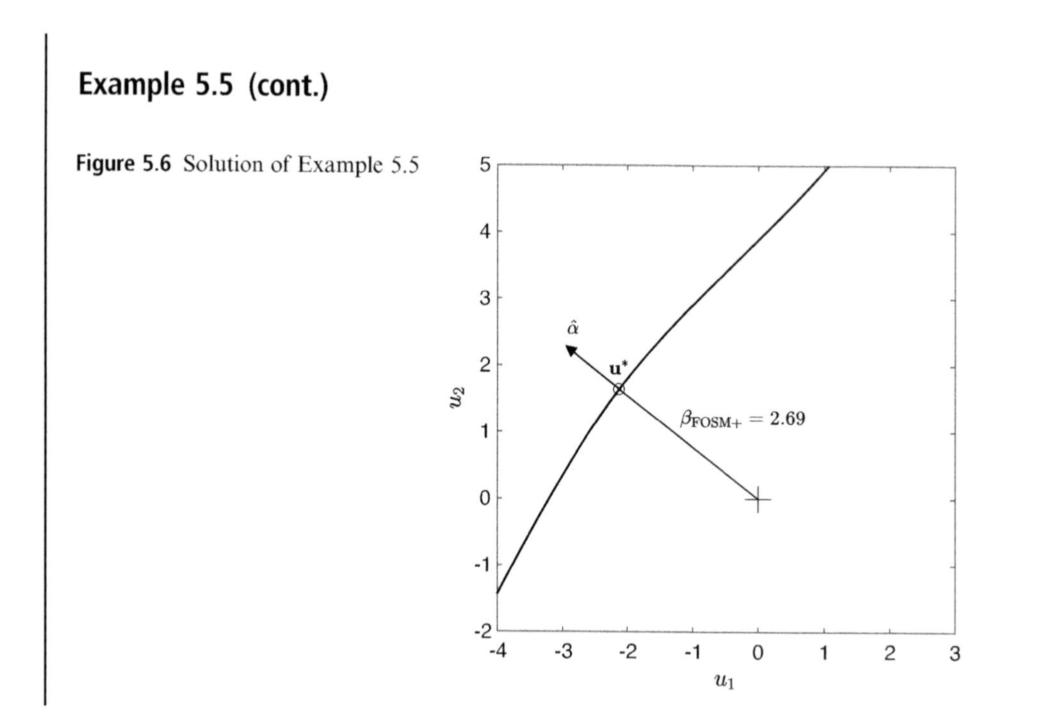

Figure 5.6 Solution of Example 5.5

5.3.3 Reliability Index Based on Upper Chebyshev Bound

Given partial probability information about a set of random variables, in some cases it is possible to formulate bounds on the probability contents of domains in the outcome space. The most famous of such bounds is the Chebyshev bound for the tail probability of a random variable for which only the mean and variance are known: $\Pr(k\sigma \le |(X-\mu)|) \le \min\left(1, 1/k^2\right)$, where μ and σ are the mean and standard deviation of X and k is a positive number. Generalized Chebyshev bounds exist for asymmetric domains in one dimension and for certain simple domains in multi-dimensional cases (Ross, 2021).

Veneziano (1979) was one of the first to identify some of the shortcomings of the FOSM method. He argued that β_{FOSM} did not properly account for the shape and dimension of the safe domain. For example, when $\mathbf{X} = \mathrm{N}(\mathbf{0}, \mathbf{I})$ and the safe domain is defined by the hypercube $\bar{\Omega} = [-\beta, \beta]^n$, then FOSM produces the reliability index β regardless of the dimension n. On the other hand, the upper Chebyshev bound of the failure probability in this case is $p_f^u = \min\left(1, n/\beta^2\right)$, which depends on the dimension n. He argued, furthermore, that FOSM did not have means to assess a measure of reliability when less or more than second-moment information about the random variables were available. He then proposed an alternative measure of reliability defined as

$$\gamma = \left(p_f^u\right)^{-1/2}, \tag{5.35}$$

where p_f^u is the upper Chebyshev bound on the failure probability for the particular state of information on **X** and the specific shape of the failure domain Ω. This form of the measure was motivated by the fact that it produces the FOSM reliability index in the special case of one random variable, with known mean and variance, and a symmetric failure domain defined by $\Omega = \{x: k\sigma \leq |x - \mu|\}$. Veneziano (1979) went on to derive the solutions of γ for several specific cases of one- and multi-dimensional problems for which Chebyshev upper bounds are available.

The above reliability measure by definition accounts for all the available probabilistic information, as well as the shape of the failure domain. However, although conceptually sound and satisfying, Veneziano's index lacks operability because there are very few cases for which Chebyshev bounds can be derived. Aside from a few highly idealized examples in his paper, no other solutions are known to exist. Furthermore, as with the earlier methods described in this chapter, Veneziano's index suffers from the myth that we know partial information about the random variables, e.g., a finite set of moments, exactly, and nothing else.

5.4 Concluding Remarks

This chapter introduced a number of reliability methods that employ partial probabilistic information about the random variables. It started with second-moment methods, which assume knowledge only of the first and second moments (means, standard deviations, correlation coefficients) of the random variables. It then discussed several methods that can incorporate information beyond the second moments, such as higher moments, bounds, and marginal distributions. These methods all provide measures of structural reliability given incomplete probability information (Der Kiureghian, 1989). In the introduction to this chapter, it was stated that the value of these methods lies in their historical role in the development of the field of structural reliability and that they should no longer be used in research or practice. The following paragraph provides the reasons for this opinion.

As we have seen, a structural reliability problem is defined in terms of a set of basic random variables **X**, which represent quantities for which observational data are available or can be gathered. An important step in reliability analysis is the formulation and assessment of a probabilistic model for **X**. This essentially amounts to selecting and validating a parameterized distribution model for **X**, say a joint PDF $f_{\mathbf{X}}(\mathbf{x}, \boldsymbol{\theta})$, and estimating its parameters $\boldsymbol{\theta}$ by statistical inference using the available data on **X**. The distribution parameters may include means, variances, correlation coefficients, or other parameters that are related to these or to higher moments of the random variables or to their bounds. Using statistical inference, one can estimate these parameters, but always with a degree of uncertainty. The larger the size of the available data, the smaller the errors in the parameter estimates. Furthermore, for given data, usually the lower moments are estimated more accurately than the higher ones. For example, with an observed sample of size n of a normal random variable, the variance in the estimate of the sample mean is σ^2/n, whereas the standard deviation in the estimate of the sample variance is approximately equal to $\sigma^2\sqrt{2/n}$ for large n, where σ is the standard deviation of the normal distribution. It can be seen that the error in the estimate of the variance is larger than that in the estimate of the mean, and that both errors diminish with increasing sample size. Hence, the idea that one knows a partial set of the moments, e.g., the means and variance/covariances, exactly and nothing else, which is the basic premise of the methods described in this chapter, is a myth. It was created with the good intention of developing simple methods for reliability analysis. However, as this chapter showed, this idealization creates its own problems. For example, the presence of any information beyond the second moments destroys the rotational symmetry property of the standard space, making the second-moment methods inoperable. Furthermore, these methods severely limit our ability to account for statistical and model uncertainties that are invariably present in reliability problems. Chapter 10 introduces the Bayesian framework for structural reliability analysis, which allows consistent and rigorous treatment of all uncertainties present in probabilistic models and estimates of their parameters. This approach requires consideration of full distributional information in reliability analysis. The balance of this book focuses on reliability methods that employ full distributional information.

PROBLEMS

5.1 A short column of rectangular cross section is subjected to an axial force P and a bending moment M. The column is made of an elastic-perfectly plastic material with yield stress Y in both tension and compression. The column fails in a fully plastic mode if

$$\frac{M}{M_u} + \left(\frac{P}{P_u}\right)^2 \geq 1,$$

where $P_u = bhY$ and $M_u = 0.25bh^2Y$ are the column plastic capacities under an axial load and a bending moment alone, respectively, and where $b = 0.25$ m is the width and $h = 0.50$ m is the depth of the cross section. P, M, and Y are random variables with the second moments listed below:

			Correlation coefficients		
Variable	Mean	Coefficient of variation	P	M	Y
P	2,200 kN	0.20	1	0.5	0
M	220 kNm	0.25	0.5	1	0
Y	35,000 kPa	0.10	0	0	1

Determine the MCFOSM reliability index, β_{MCFOSM}, using the following two equivalent formulations of the limit-state function:

$$g_1(P, M, Y) = 1 - \frac{M}{M_u} - \left(\frac{P}{P_u}\right)^2$$

$$g_2(P, M, Y) = M_u P_u^2 - M P_u^2 - M_u P^2.$$

Comment on the difference between the two results.

5.2 Solve Problem 5.1 using the FOSM reliability method. Use the iHL-RF algorithm to find the design point and list the quantities in the iteration steps in a table. Determine the design point in the original space and comment on the outcome values relative to the mean values of the random variables.

5.3 Solve the problem in Example 5.2 using the alternative formulation of the limit-state function $g(P, M, Y) = M_u P_u^2 - M P_u^2 - M_u P^2$. Show the steps of the iHL-RF algorithm in a table form. Your final result should be identical to that obtained in Example 5.2.

5.4 Consider the parabolic limit-state function

$$G(\mathbf{U}) = \beta + \frac{1}{2}\sum\nolimits_{i=1}^{n-1} \kappa_i U_i^2 - U_n,$$

where U_i, $i = 1, \ldots, n$, are standard variables (zero means, unit covariance matrix) with $0 < \beta$ and $0 < \kappa_i$ for all i.

1) Make a plot of the intersection of the limit-state surface in the outcome space of $\mathbf{U}$ with the (u_i, u_n) plane. Observe that the design point is at $\mathbf{u}^* = [0\,0\cdots 0\,\beta]^{\mathrm{T}}$.
2) Show that if $1 < \beta\kappa_i$ for any i, then the standard HL-RF algorithm with a unit step size will *not* converge to $\mathbf{u}^*$. (Hint: Start from the trial point $\mathbf{u}_0 = [\epsilon\,\epsilon\cdots\epsilon\,\beta]^{\mathrm{T}}$, where ϵ is arbitrarily small in relation to β, so that the trial point is close to $\mathbf{u}^*$. Then show that at least one coordinate of the next iteration point, $\mathbf{u}_1$, obtained by a unit step size falls farther away from the solution point than the same coordinate of $\mathbf{u}_0$. You may simplify the expression for $\mathbf{u}_1$ by dropping high-order terms in ϵ.)
3) For the above condition, suggest an appropriate step size λ_0 that will assure convergence of the algorithm. (Hint: To converge, all coordinates of $\mathbf{u}_1$ must be closer to the design point than the corresponding coordinates of $\mathbf{u}_0$.)

6 The First-Order Reliability Method

6.1 Introduction

This chapter considers structural component reliability problems defined by a continuous and continuously differentiable limit-state function $g = g(\mathbf{X})$ and assumes that the joint distribution of the basic random variables, $f_{\mathbf{X}}(\mathbf{x})$, is given and is transformable to the standard normal space. The focus is on developing the first-order reliability method (FORM). This method develops an approximation to the failure probability by replacing the limit-state surface by its tangent plane at the design point in the transformed standard normal space. The approximation takes advantage of several important properties of the standard normal space. Hence, the chapter starts by summarizing these properties. This is followed by the development of the FORM, including a series of examples with our "toy" problem to demonstrate different cases of the transformation. This is followed by a discussion of the accuracy of the FORM and possible measures of the underlying error. Next, measures of importance of random variables and reliability sensitivities with respect to parameters appearing in the limit-state function or the probability distribution of $\mathbf{X}$ are introduced. Other topics include the case of multiple design points, inverse reliability analysis, and approximate computation of the cumulative distribution function (CDF) and probability density function (PDF) of a function of random variables by use of the FORM. The next chapter develops the second-order reliability method (SORM), which also takes advantage of the properties of the standard normal space as described in this chapter.

FORM and SORM are the most widely used methods for structural reliability analysis. They are popular because they provide good approximations for many structural engineering problems with relatively simple calculations, and also because they provide useful insight into the nature of the reliability problem. In particular, the FORM provides measures of importance of the random variables in terms of their relative contributions to the variance of the limit-state function, and the SORM provides an understanding of the shape of the limit-state surface. The main weakness of FORM and SORM lies in the fact that one cannot determine the magnitude of the underlying error or put a practical bound on it. Therefore, it is important that the user of these methods be fully aware of the underlying assumptions and approximations and develop an insight as to the types of problems for which these methods may or may not work well. It is the author's opinion that such insight is gained through experience by applying these methods to a variety of problems.

6.2 Properties of the Standard Normal Space

The n-dimensaional standard normal space is defined by the set of statistically independent random variables $\mathbf{U} = [U_1, U_2, \ldots, U_n]^{\mathrm{T}}$ having the joint PDF

$$\begin{aligned}\phi_n(\mathbf{u}) &= \prod\nolimits_{i=1}^{n} \frac{1}{\sqrt{2\pi}} \exp\left(-\frac{u_i^2}{2}\right) \\ &= \frac{1}{(2\pi)^{n/2}} \exp\left(-\frac{1}{2}\sum\nolimits_{i=1}^{n} u_i^2\right) \\ &= \frac{1}{(2\pi)^{n/2}} \exp\left(-\frac{\|\mathbf{u}\|^2}{2}\right).\end{aligned} \tag{6.1}$$

The following properties of the above space are important in developing the FORM and SORM approximations:

1) The standard normal space is rotationally symmetric with respect to the origin, as all points of equal distance from the origin have the same probability density. It follows that the contours of the probability density are concentric spheres centered at the origin (see Figure 6.1(a)).
2) The probability density decreases exponentially with the square of the distance in the radial direction from the origin. It follows that for any domain $\mathcal{F}$ shown in Figure 6.1(a), the point nearest the origin, denoted $\mathbf{u}^*$, has the highest probability density among all outcomes within $\mathcal{F}$. Furthermore, if $\mathcal{F}$ is a failure domain, $\mathbf{u}^*$ can be considered as the most likely outcome, if failure were to occur. For this reason, this point, which is denoted the *design point*, can be considered to be the *most likely failure point in the standard normal space*. As we will see later, the most likely failure point is specific to the space of random variables considered and can be different in different spaces.
3) Consider the hyperplane

$$\beta - \widehat{\boldsymbol{\alpha}}\mathbf{u} = 0, \tag{6.2}$$

where β is the distance from the origin and $\widehat{\boldsymbol{\alpha}}$ the unit normal row vector directed away from the origin (see Figure 6.1(b) for a two-dimensional case). The probability density of points located on this plane decays exponentially with the square of the distance as we move away from the foot of the normal from the origin, $\mathbf{u}^*$. This has to do with the fact that the conditional distribution of n jointly normal random variables given a linear function of any subset of the variables is jointly normal in the $(n-1)$-dimensional space.

4) The probability content of the half space $\beta - \widehat{\boldsymbol{\alpha}}\mathbf{U} \leq 0$ is given by

$$\begin{aligned} p_1 &= \Pr(\beta - \widehat{\boldsymbol{\alpha}}\mathbf{U} \leq 0) \\ &= \Phi(-\beta), \end{aligned} \tag{6.3}$$

where $\Phi(\cdot)$ is the standard normal CDF. This is evident from the fact that $\mathrm{E}[\beta - \widehat{\boldsymbol{\alpha}}\mathbf{U}] = \beta$ and $\mathrm{Var}[\beta - \widehat{\boldsymbol{\alpha}}\mathbf{U}] = 1$ and that $\beta - \widehat{\boldsymbol{\alpha}}\mathbf{U}$, being a linear function of normal random variables, is normally distributed. It follows that $p_1 = \Phi[(0-\beta)/1] = \Phi(-\beta)$.

5) Consider the parabolic hyper-surface

$$\beta - u_n + \frac{1}{2}\sum_{i=1}^{n-1} \kappa_i u_i^2 = 0, \tag{6.4}$$

where β is the distance from the origin to the vertex $\mathbf{u}^*$ and κ_i are the principal curvatures of the hyper-paraboloid (see Figure 6.1(c) for a two-dimensional case). Tvedt (1990) showed that the probability content of the parabolic domain $\beta - u_n + \sum_{i=1}^{n-1} \kappa_i u_i^2/2 \leq 0$ is given by

$$\begin{aligned} p_2 &= \Pr\left(\beta - u_n + \frac{1}{2}\sum_{i=1}^{n-1} \kappa_i u_i^2 \leq 0\right) \\ &= \phi(\beta)\mathrm{Re}\left\{ \mathrm{i}\sqrt{\frac{2}{\pi}} \int_0^{\mathrm{i}\infty} \frac{1}{s} \exp\left[\frac{(s+\beta)^2}{2}\right] \prod_{i=1}^{n-1} \frac{1}{\sqrt{1+\kappa_i s}} \mathrm{d}s \right\}, \end{aligned} \tag{6.5}$$

where $\phi(.)$ denotes the standard normal PDF, $\mathrm{i} = \sqrt{-1}$ denotes the imaginary unit, and the integral inside the brackets is computed along the imaginary axis. Based on an asymptotic analysis with $\beta \to \infty$, Breitung (1984) derived the following asymptotic approximation of the above probability

$$p_2 \cong \Phi(-\beta) \prod_{i=1}^{n-1} \frac{1}{\sqrt{1+\beta\kappa_i}}. \tag{6.6}$$

It is noted that Breitung's formula has a minus sign before β in the denominator. Here, a plus sign is used in conformity with the definition of the parabolic function in (6.4). In our definition, κ_i is positive (negative) when the parabola in the (u_i', u_n) plane curves away from (toward) the origin. The formula is applicable for $-1 < \beta\kappa_i$. Hohenbichler and Rackwitz (1988) introduced an improvement in the above formula for finite β values in the form

$$p_2 \cong \Phi(-\beta) \prod_{i=1}^{n-1} \frac{1}{\sqrt{1+\psi(\beta)\kappa_i}}, \tag{6.7}$$

where $\psi(\beta) = \phi(\beta)/\Phi(-\beta)$. This formula is valid for $-1 < \psi(\beta)\kappa_i$.

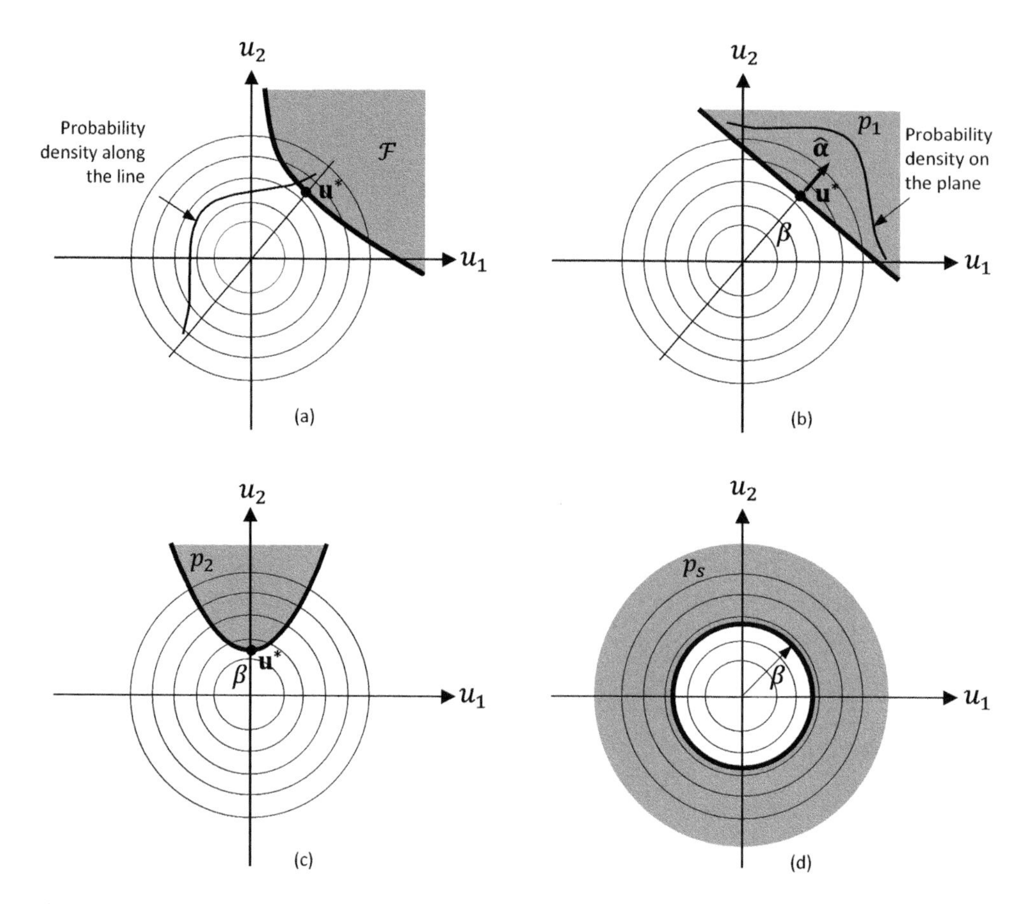

Figure 6.1 Properties of standard normal space

6) Consider the hyper-sphere

$$\beta^2 - \sum_{i=1}^{n} u_i^2 = 0 \tag{6.8}$$

of radius β. Owing to the fact that the sum of n squared standard normal random variables has the chi-square distribution with n degrees of freedom, the probability content outside the hyper-sphere (see Figure 6.1(d) for a two-dimensional case) is given by

$$\begin{aligned} p_s &= \Pr\left(\beta^2 - \sum_{i=1}^{n} u_i^2 \leq 0\right) \\ &= 1 - \chi_n^2(\beta^2) \end{aligned} \tag{6.9}$$

where $\chi_n^2(.)$ is the CDF of the chi-square distribution with n degrees of freedom.

The above properties of the standard normal space are used throughout this book to develop methods for structural reliability analysis. Other properties are described when needed for specific applications.

6.3 The First-Order Reliability Method

Consider the structural component reliability problem defined by the vector of n basic random variables $\mathbf{X} = [X_1, X_2, \ldots, X_n]^{\mathrm{T}}$ having the joint PDF $f_{\mathbf{X}}(\mathbf{x})$ and the limit-state function $g = g(\mathbf{X})$. We assume that the limit-state function is continuous and continuously differentiable. Our interest is in the probability of failure defined by

$$p_F = \int_{g(\mathbf{x}) \le 0} f_{\mathbf{X}}(\mathbf{x}) \mathrm{d}\mathbf{x}, \tag{6.10}$$

where the integration is over the outcomes in the failure domain, defined as $\mathcal{F} = \{g(\mathbf{X}) \le 0\}$. This is an n-fold integral and for the vast majority of problems requires numerical evaluation. (Some special cases where closed-form solutions are available were presented in Chapter 4; see Examples 4.1 and 4.2.)

To construct the FORM approximation, we transform the set of random variables into the standard normal space. This is accomplished through a transformation of the form

$$\mathbf{U} = \mathbf{T}(\mathbf{X}), \tag{6.11}$$

where $\mathbf{U} = [U_1, U_2, \ldots, U_n]^{\mathrm{T}}$ is the n-vector of standard normal variates. Various forms of this transformation were developed in Section 3.6. Recall that for the transformation to be one-to-one, i.e., for there to be one and only one $\mathbf{U}$ for each $\mathbf{X}$ and vice versa, it is necessary that all marginal and conditional CDFs of $\mathbf{X}$ be continuous and strictly increasing functions. In other words, the CDFs should not contain jumps (probability masses in the corresponding PDFs) or a constant value over a range of outcomes (gaps in the PDFs). Under these conditions, the transform is invertible and we write its inverse as

$$\mathbf{X} = \mathbf{T}^{-1}(\mathbf{U}). \tag{6.12}$$

To define the reliability problem in the standard normal space, we need to transform the limit-state function. We write

$$\begin{aligned} g(\mathbf{X}) &= g\big(\mathbf{T}^{-1}(\mathbf{U})\big) \\ &= G(\mathbf{U}), \end{aligned} \tag{6.13}$$

where $G = G(\mathbf{U})$ is the limit-state function expressed in terms of $\mathbf{U}$. The failure probability integral in the standard normal space now can be written as

$$p_F = \int_{\mathcal{F}} \phi_n(\mathbf{u}) \mathrm{d}\mathbf{u}, \tag{6.14}$$

where $\mathcal{F} = \{G(\mathbf{U}) \le 0\}$ is the failure domain.

Figure 6.2 illustrates the equivalent reliability problems in (6.10) and (6.14) in the original (left) and standard normal (right) spaces, respectively. Observe that, whereas the contours of the PDF in the original space can have any form, these contours are spherical in the standard normal space, centered at the origin.

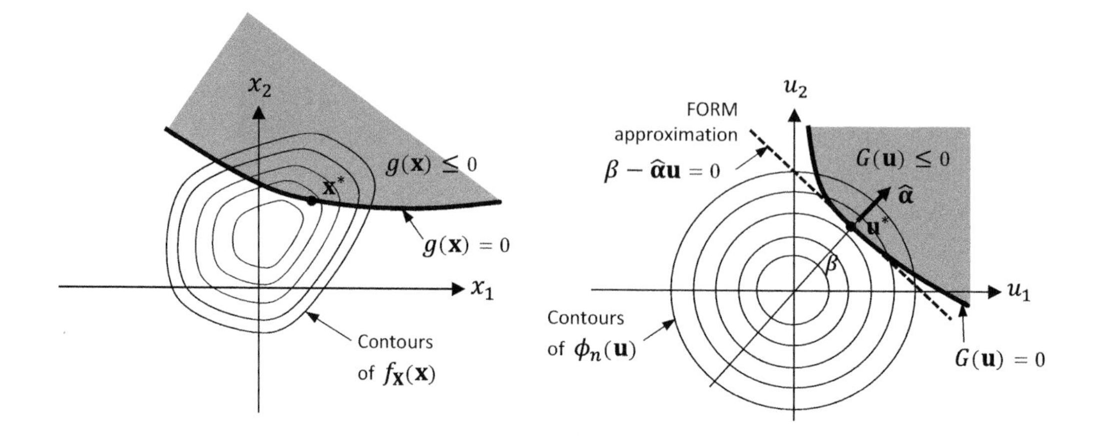

Figure 6.2 Structural component reliability problem in original (left) and standard normal (right) spaces and the FORM approximation

Let $\mathbf{u}^*$ denote the design point on the limit-state surface $G(\mathbf{u}) = 0$ in the standard normal space, i.e., the point with minimum distance from the origin. As described in Section 5.2.3, this point is the solution to the constrained optimization problem

$$\mathbf{u}^* = \arg\min\{\|\mathbf{u}\| \,|G(\mathbf{u}) = 0\}. \tag{6.15}$$

Among other options, the iHL-RF algorithm described in Section 5.2.3 solves this problem. As $\mathbf{u}^*$ is the nearest point in the failure domain from the origin in the standard normal space, it follows from property 2 in the previous section that this point has the highest likelihood of occurrence among all outcomes in the failure domain in the standard normal space. Furthermore, based on property 3, the probability density decays exponentially with the square of the distance from $\mathbf{u}^*$ in the tangential direction. It follows that the neighborhood of the design point is the major contributor to the failure probability integral in the standard normal space. Therefore, to approximate the limit-state function, it makes sense to approximate it at $\mathbf{u}^*$ in the standard normal space.

In the FORM, we use a first-order Taylor series approximation of the limit-state function at $\mathbf{u}^*$, i.e.,

$$\begin{aligned} G(\mathbf{U}) &\cong G(\mathbf{u}^*) + \nabla G(\mathbf{u}^*)(\mathbf{U} - \mathbf{u}^*) \\ &= \|\nabla G(\mathbf{u}^*)\|(\beta - \widehat{\boldsymbol{\alpha}}\mathbf{U}) \\ &\propto \beta - \widehat{\boldsymbol{\alpha}}\mathbf{U}, \end{aligned} \tag{6.16}$$

where $\nabla G(\mathbf{u})$ is the gradient row-vector at the outcome $\mathbf{u}$, $\widehat{\boldsymbol{\alpha}} = -\nabla G(\mathbf{u}^*)/\|\nabla G(\mathbf{u}^*)\|$ is the negative normalized gradient vector at the design point, and $\beta = \widehat{\boldsymbol{\alpha}}\mathbf{u}^*$ is the algebraic distance from the origin to the design point known as the *reliability index*. In the last line of the equation we have dropped the positive-valued scalar $\|\nabla G(\mathbf{u}^*)\|$ beccause the failure domain is not affected by any positive scaling of the limit-state function.

The hyperplane $\beta - \hat{\boldsymbol{\alpha}}\mathbf{u} = 0$ is tangent to the limit-state surface $G(\mathbf{u}) = 0$ at the design point. Hence, the FORM approximation amounts to replacing the true limit-state surface with its tangent plane at the design point. Using property 4 described in the preceding section, the FORM approximation is given by

$$\begin{aligned} p_F &\cong \int_{\beta - \hat{\boldsymbol{\alpha}}\mathbf{u} \le 0} \phi_n(\mathbf{u}) d\mathbf{u} \\ &= \Phi(-\beta). \end{aligned} \tag{6.17}$$

Given that the neighborhood of the design point makes a dominant contribution to the failure probability, this approximation is expected to provide good accuracy. The accuracy of the FORM approximation is discussed in more detail in the following section.

It is important to understand that the above is a strict approximation of the failure probability, in the present case a first-order approximation based on replacing the limit-state surface by its tangent hyperplane at the design point. This is in contrast with second-moment methods, where the probability distribution is not available, yet many reliability analysts use the approximation $p_F \cong \Phi(-\beta_{\mathrm{FOSM}})$ or even $p_F \cong \Phi(-\beta_{\mathrm{MVFOSM}})$. These approximations have no mathematical bases and are questionable. As argued in the concluding remarks of Chapter 5, they should not be used in practice.

The main computational effort in the FORM lies in finding the design point. This involves repeated computations of the limit-state function and its gradient at a series of trial points $\mathbf{u}_i$, $i = 1, 2, \ldots$, selected by an optimization scheme such as the iHL-RF algorithm (Section 5.2.3), until convergence is achieved. Specifically, as shown in the flow-chart in Figure 5.2, given a trial point $\mathbf{u}_i$, we successively compute

$$\mathbf{x}_i = \mathbf{T}^{-1}(\mathbf{u}_i), \tag{6.18a}$$

$$G(\mathbf{u}_i) = g(\mathbf{x}_i), \tag{6.18b}$$

$$\mathbf{J}_{\mathbf{u},\mathbf{x}} \text{ with elements } J_{jk} = \frac{\partial u_j}{\partial x_k}, \tag{6.18c}$$

$$\mathbf{J}_{\mathbf{x},\mathbf{u}} = \mathbf{J}_{\mathbf{u},\mathbf{x}}^{-1}, \tag{6.18d}$$

$$\nabla G(\mathbf{u}_i) = \nabla g(\mathbf{x}_i) \mathbf{J}_{\mathbf{x},\mathbf{u}}. \tag{6.18e}$$

The algorithm then gives a new trial point $\mathbf{u}_{i+1}$. The number of steps needed to converge depends on the nonlinearity of the limit-state function and the level of accuracy desired. In our experience with the iHL-RF algorithm, for most ordinary problems the sequence of trial points converges in 10 to 20 steps.

The results from a FORM solution consist of the design point $\mathbf{u}^*$, the reliability index β, the first-order approximation of the failure probability $p_{F1} = \Phi(-\beta)$, and the unit normal vector $\hat{\boldsymbol{\alpha}}$. These quantities provide a wealth of information about the reliability problem, which is commonly not available from other methods such as simulation. First, as described earlier, $\mathbf{u}^*$ denotes the most likely failure point in the standard normal space. Transforming this point back into the original space, we obtain $\mathbf{x}^* = \mathbf{T}^{-1}(\mathbf{u}^*)$. When the transformation $\mathbf{T}$ is nonlinear, as happens when the random variables $\mathbf{X}$ are non-normal, $\mathbf{x}^*$ does not necessarily coincide with the most likely failure point in the $\mathbf{X}$ space. Nevertheless, it is usually close to it and, hence, provides insight as to what combination of the random variable outcomes is most likely to lead to failure. This information often provides physical insight into the nature of the failure event. The reliability index β provides a measure of reliability of the structural component. It describes the distance from the origin to the most likely failure point in the standard normal space. Naturally, a larger β reflects a higher reliability. For most structural component problems, β falls in the range 2–5. As mentioned earlier, $p_{F1} = \Phi(-\beta)$ provides a first-order approximation of the failure probability in the sense of replacing the limit-state surface with its tangent hyperplane in the standard normal space. The accuracy of this approximation depends on the degree of nonlinearity of the limit-state function and also other factors as described in Section 6.4. Finally, the unit normal vector $\widehat{\boldsymbol{\alpha}}$ describes the relative importance of the standard normal random variables. This is shown in a subsequent section in this chapter. But first we will examine the FORM approximation for several examples.

Example 6.1 – Jointly Normal Random Variables

Consider our toy problem defined by the limit-state function $g(X_1, X_2) = X_1^2 - 2X_2$. Assume the second moments are the same as those given in Example 5.1, i.e., $\mu_1 = 10$, $\sigma_1 = 2$, $\mu_2 = 20$, $\sigma_2 = 5$, and $\rho_{12} = 0.5$. Here, additionally, we assume that the variables are jointly normal. In this case, the transformation to the standard normal space is identical to the linear transformation used in the first-order, second-moment (FOSM) solution of Example 5.2. As a result, the solutions for $\mathbf{u}^*$, β, and $\widehat{\boldsymbol{\alpha}}$ are identical to those obtained in that example. They are $\mathbf{u}^* = [-1.928\ 0.852]^{\mathrm{T}}$, $\beta = 2.11$, and $\widehat{\boldsymbol{\alpha}} = [-0.915\ 0.405]$. Furthermore, the mapping of $\mathbf{u}^*$ onto the original space yields $\mathbf{x}^* = [6.14\ 18.9]^{\mathrm{T}}$, which, due to the linearity of the $\mathbf{x}$ to $\mathbf{u}$ transformation, is indeed the most likely failure point in the original space. The first-order approximation of the failure probability is $p_{F1} = \Phi(-2.11) = 0.0175$. Figure 5.3 shows the solution of this problem in the original (left) and standard normal (right) spaces. A Monte Carlo solution of this problem yields $p_F \cong 0.0166$ (with a coefficient of variation [c.o.v.] of 2.5%). It appears that the FORM solution slightly overestimates the failure probability. This is on account of the slight outward curvature of the limit-state surface in the standard normal space (see Figure 5.3 [right]).

It is noteworthy that at the most likely failure point, $\mathbf{x}^*$, the outcomes of both capacity and demand variables are below their respective mean values. As explained in Example 5.2, this is due to the positive correlation between the two variables. If the variables are assumed to be statistically independent, the solution is $\mathbf{u}^* = [-1.61\ 0.595]^{\mathrm{T}}$, $\beta = 1.72$, $\widehat{\boldsymbol{\alpha}} = [-0.938\ 0.346]$, $\mathbf{x}^* = [6.78\ 23.0]^{\mathrm{T}}$, and $p_{F1} = 0.0430$. Note that, at the most likely failure point, the outcome of the capacity variable is below its mean value and that of the demand variable is above its mean value. A Monte Carlo solution of this problem yields $p_F \cong 0.0418$ (with a c.o.v. of 2.1%).

Example 6.2 – Nataf-Distributed Random Variables

Now assume that the random variables in the previous problem have the Nataf distribution with lognormal and Type I largest value (Gumbel) marginal distributions with the same second moments as in Example 6.1. In this case, the transformation to the standard normal space and, therefore, the solution of the design point is identical to those described in Example 5.5. Specifically, $\mathbf{u}^* = [-2.13\ 1.64]^{\mathrm{T}}, \beta = 2.69$, $\hat{\boldsymbol{\alpha}} = [-0.793\ 0.610]$, and $\mathbf{x}^* = [6.43\ 20.7]^{\mathrm{T}}$. Furthermore, we have the first-order approximation of the failure probability as $p_{F1} = \Phi(-2.69) = 0.00362$. The difference between the two examples is that the probability estimate here is a strict first-order approximation, whereas for Example 5.5 a strict probability approximation cannot be made. Figure 5.5 shows the solution of this problem in the standard normal space. A Monte Carlo solution of this problem yields $p_F \cong 0.00427$ (with a c.o.v. of 6.8%). The underestimation of the failure probability by the FORM is on account of the inward curvature of the limit-state surface, as can be seen in Figure 5.5.

Suppose the random variables of this problem are assumed to be statistically independent. The transformation to the standard normal space then is identical to that described in Example 5.5, except that $\mathbf{L}_0$ is replaced with an identity matrix. The solution for that case is $\mathbf{u}^* = [-1.63\ 1.10]^{\mathrm{T}}, \beta = 1.96$, $\hat{\boldsymbol{\alpha}} = [-0.829\ 0.559]$, $\mathbf{x}^* = [7.10\ 25.2]^{\mathrm{T}}$, and $p_{F1} = 0.00249$. A Monte Carlo solution of this problem yields $p_F \cong 0.0262$ (with a c.o.v. of 2.7%).

Example 6.3 – Morgenstern-Distributed Random Variables

Now assume that the random variables in our toy problem have the joint Morgenstern distribution with lognormal and Type I largest value marginals having the same means and standard deviations as described in the above examples. Specifically, the joint CDF and PDF are defined by

$$F_{X_1X_2}(x_1, x_2) = F_1(x_1)F_2(x_2)\{1 + \alpha_{12}[1 - F_1(x_1)][1 - F_2(x_2)]\} \tag{E1}$$

and

$$f_{X_1X_2}(x_1, x_2) = f_1(x_1)f_2(x_2)\{1 + \alpha_{12}[1 - 2F_1(x_1)][1 - 2F_2(x_2)]\}, \tag{E2}$$

where $F_1(x_1)$ and $f_1(x_1)$ are, respectively, the CDF and PDF of the lognormal distribution with parameters $\lambda = 2.28$ and $\zeta = 0.198$ and $F_2(x_2)$ and $f_2(x_2)$ are, respectively, the CDF and PDF of the Type I largest value distribution with parameters $\alpha = 0.257$ and $v = 17.7$ (see Example 5.5). We set $\alpha_{12} = 1$, the largest possible value this parameter can have. According to Table 3.1, $Q_1 = 0.289$ and $Q_2 = 0.270$ so that the correlation coefficient between the two random variables according to (3.29) is $\rho = 4 \times 1 \times 0.289 \times 0.270 = 0.312$. This is the largest correlation coefficient between the two random variables with the specified marginals that the Morgenstern distribution can accommodate.

Example 6.3 (cont.)

For this choice of the joint distribution, the transformation to the standard normal space is achieved by the Rosenblatt transformation described in Section 3.6.5. Using the order of random variables as X_1 and X_2, the transformation is defined by

$$U_1 = \Phi^{-1}[F_1(X_1)], \tag{E3a}$$

$$U_2 = \Phi^{-1}\left[F_{2|1}(X_2|X_1)\right], \tag{E3b}$$

where

$$\begin{aligned} F_{2|1}(x_2|x_1) &= \frac{\partial F(x_1, x_2)}{\partial x_1}\frac{1}{f(x_1)} \\ &= F_2(x_2)\{1 + [1 - 2F_1(x_1)][1 - F_2(x_2)]\} \end{aligned} \tag{E4}$$

is the conditional CDF of X_2 given $X_1 = x_1$. Furthermore, the Jacobian of the inverse transform is given by (see (3.65))

$$\mathbf{J}_{\mathbf{u},\mathbf{x}} = \begin{bmatrix} \dfrac{f_1(x_1)}{\phi(u_1)} & 0 \\ \dfrac{\partial F_{2|1}(x_2|x_1)}{\partial x_1}\dfrac{1}{\phi(u_2)} & \dfrac{f_{2|1}(x_2|x_1)}{\phi(u_2)} \end{bmatrix}, \tag{E5}$$

where $f_{2|1}(x_2|x_1) = f_2(x_2)\{1 + [1 - 2F_1(x_1)][1 - 2F_2(x_2)]\}$ is the conditional PDF of X_2 given $X_1 = x_1$. Using the iHL-RF algorithm starting from the mean point with $\epsilon_1 = \epsilon_2 = 10^{-3}$, the solution is obtained in seven iterations: $\mathbf{u}^* = [-2.10\ 1.10]^{\mathrm{T}}$, $\widehat{\boldsymbol{\alpha}} = [-0.887\ 0.463]$, $\beta = 2.37$, and $p_{F1} = 0.00890$. Furthermore, $\mathbf{x}^* = [6.47\ 20.9]^{\mathrm{T}}$. Figure 6.3 (left) shows this solution in the standard normal space.

As mentioned at the end of Section 3.6.5, the Rosenblatt transformation may depend on the selected order of the random variables. Suppose in this case we were to select the order as X_2 and then X_1. In that case the transformation has the form

$$U_1 = \Phi^{-1}[F_2(X_2)], \tag{E6a}$$

$$U_2 = \Phi^{-1}\left[F_{1|2}(X_1|X_2)\right], \tag{E6b}$$

where $F_{1|2}(x_1|x_2)$ is the conditional CDF of X_1 given $X_2 = x_2$. Solving with this transformation, we obtain the solution $\mathbf{u}^* = [0.413\ -2.22]^{\mathrm{T}}$, $\widehat{\boldsymbol{\alpha}} = [0.183\ -0.983]$, $\beta = 2.26$, and $p_{F1} = 0.0119$. Furthermore, $\mathbf{x}^* = [6.51\ 21.2]^{\mathrm{T}}$. Figure 6.3 (right) shows this solution in the

Example 6.3 (cont.)

standard normal space. It is seen that the FORM solutions obtained from the two transformations are different. These are both first-order approximations but carried out in different spaces. This dependence on the ordering of the random variables is specific to the Rosenblatt transformation and is not present when the random variables have the Nataf distribution and the Nataf transformation is used. It is also worth noting that the mapping of the design point onto the original space yields different results for $\mathbf{x}^*$ for the two transformations. This shows that $\mathbf{x}^*$ is not necessarily the most likely failure point in the original space.

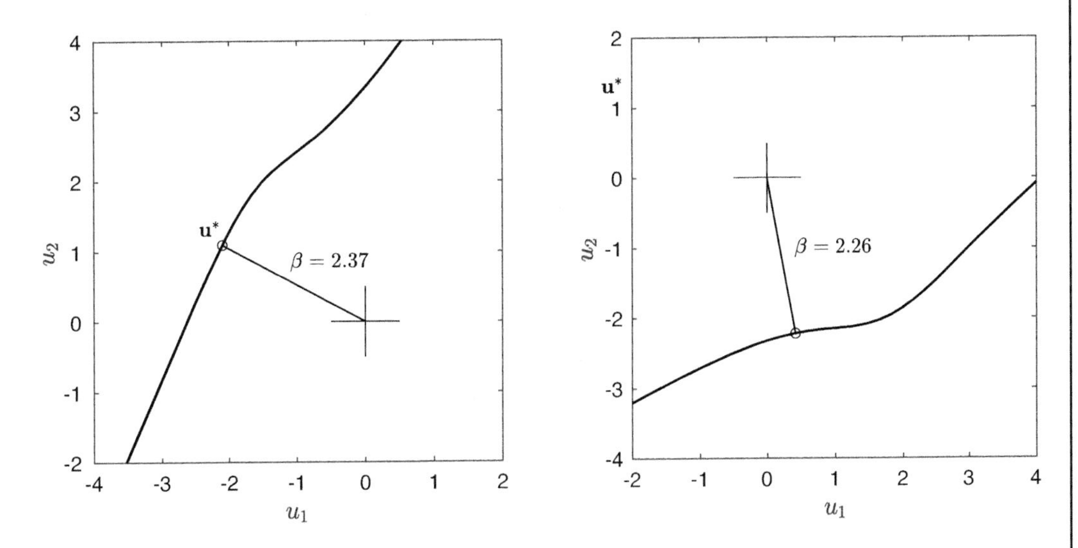

Figure 6.3 FORM solutions with different orderings of random variables: X_2 conditioned on X_1 (left); X_1 conditioned on X_2 (right)

A Monte Carlo solution of this problem produces $p_F \cong 0.0110$ (with a c.o.v. of 2.0%). It can be seen that the "exact" solution is somewhere between the two FORM solutions obtained by different orderings of the random variables.

Example 6.4 – Non-Normal Random Variables with Uncertain Mean

Now assume that the random variables X_1 and X_2 in our toy problem have independent lognormal and Type I largest value distributions, respectively, as before, but assume the mean of X_2 is unknown and is considered as a random variable X_3, having a lognormal distribution with mean $\mu_3 = 20$ and standard deviation $\sigma_3 = 2$. The mean vector of the random variables is

Example 6.4 (cont.)

$\mathbf{M_X} = [10\ X_3\ 20]^T$ and the diagonal matrix of standard deviations is $\mathbf{D_X} = \text{diag}[2\ 5\ 2]$. The Rosenblatt transformation for this case takes the form

$$U_1 = \Phi^{-1}[F_3(X_3)], \tag{E1a}$$

$$U_2 = \Phi^{-1}[F_1(X_1)], \tag{E1b}$$

$$U_3 = \Phi^{-1}\left[F_{2|3}(X_2|X_3)\right], \tag{E1c}$$

where we have selected the order of random variables as X_3, X_1, and X_2. Note that the order of X_1 and X_3 is immaterial, as they are statistically independent, but that X_3 should be selected before X_2 so that we can use the conditional distribution of X_2 given X_3. The conditional CDF is given by

$$F_{2|3}(x_2|x_3) = \exp\{-\exp[-\alpha(x_2 - v(x_3))]\} \tag{E2}$$

where $\alpha = \pi/5\sqrt{6} = 0.257$ and $v(x_3) = x_3 - 0.577/\alpha$. The gradient of the limit-state function is $\nabla g = [2x_1\ -2\ 0]$. Note that the last element of the gradient vector is 0 because X_3 does not appear in the limit-state function. The Jacobian of the inverse transform is given by

$$\mathbf{J_{u,x}} = \begin{bmatrix} \frac{f_3(x_3)}{\phi(u_1)} & 0 & 0 \\ 0 & \frac{f_1(x_1)}{\phi(u_2)} & 0 \\ \frac{\partial F_{2|3}(x_2|x_3)}{\partial x_3}\frac{1}{\phi(u_3)} & 0 & \frac{f_{2|3}(x_2|x_3)}{\phi(u_3)} \end{bmatrix}. \tag{E3}$$

For this problem, the IHL-RF algorithm converges after eight iterations, producing the results $\mathbf{u}^* = [0.326\ -1.60\ 1.05]^T$, $\widehat{\boldsymbol{\alpha}} = [0.167\ -0.825\ 0.540]$, $\beta = 1.94$, and $p_{F1} = 0.0260$. Furthermore, $\mathbf{x}^* = [20.6\ 7.14\ 25.5]^T$. A Monte Carlo solution produces $p_F \cong 0.0277$ (with a c.o.v. of 2.6%).

It is instructive to compare the above results with the results when the mean of X_2 is precisely known to be $\mu_2 = 20$. The results for that case were given in the last part of Example 6.2. Specifically, $\beta = 1.96$ and $p_{F1} = 0.00249$ were obtained. We see that the uncertainty in the mean of X_2 has resulted in a slight decrease in the reliability index (from 1.96 to 1.94) and a slight increase in the first-order approximation of the failure probability (from 0.00249 to 0.0260).

Example 6.5 – Reliability of a Column

Consider a short column subjected to biaxial bending moments M_1 and M_2 and axial force P. Assuming an elastic perfectly plastic material with yield strength Y, the failure of the column is defined by the limit-state function

$$g(M_1, M_2, P, Y) = 1 - \frac{M_1}{s_1 Y} - \frac{M_2}{s_2 Y} - \left(\frac{P}{aY}\right)^2 \tag{E1}$$

where $a = 0.190\ \text{m}^2$ is the cross-sectional area of the column and $s_1 = 0.030\ \text{m}^3$ and $s_2 = 0.015\ \text{m}^3$ are the flexural moduli of the fully plastic cross section of the column. Assume M_1, M_2, P, and Y are random variables having a Nataf joint distribution with the marginal distributions and second moments listed in Table 6.1.

Table 6.1 Description of random variables for Example 6.5

				Correlation coefficient			
Variable	Marginal distribution	Mean	c.o.v.	M_1	M_2	P	Y
M_1, kNm	Normal	250	0.3	1.0	1.0	1.0	1.0
M_2, kNm	Normal	125	0.3	0.5	0.3	0.0	
P, kN	Gumbel	2,500	0.2	0.3	0.0		
Y, kPa	Weibull	40,000	0.1	0.0			

We start by formulating the probability transformation. Using the tables in Appendix 3A, we compute the correlation coefficients $\rho_{0,ij}$ between the Nataf-distributed random variables in the normal space. As M_1 and M_2 have the normal distribution, $\rho_{0,12} = \rho_{12} = 0.5$. For M_1 and P, using Table A3.1 we have $\rho_{0,13} = 1.031 \times \rho_{13} = 0.309$. In a similar way, we obtain $\rho_{0,23} = 0.309$. Thus, the correlation matrix of the intermediate random variables $\mathbf{Z}$ is

$$\mathbf{R}_0 = \begin{bmatrix} 1 & 0.5 & 0.309 & 0 \\ 0.5 & 1 & 0.309 & 0 \\ 0.309 & 0.309 & 1 & 0 \\ 0 & 0 & 0 & 1 \end{bmatrix}. \tag{E2}$$

Using the formulas in Appendix 3B, the lower-triangular Cholesky decomposition of this matrix is computed as

$$\mathbf{L}_0 = \begin{bmatrix} 1.000 & 0.000 & 0.000 & 0.000 \\ 0.500 & 0.866 & 0.000 & 0.000 \\ 0.309 & 0.178 & 0.934 & 0.000 \\ 0.000 & 0.000 & 0.000 & 1.000 \end{bmatrix}. \tag{E3}$$

Example 6.5 (cont.)

The transformation to the standard normal space is defined by

$$\mathbf{U} = \mathbf{L}_0^{-1} \begin{Bmatrix} \dfrac{M_1 - \mu_1}{\sigma_1} \\ \dfrac{M_2 - \mu_2}{\sigma_2} \\ \Phi^{-1}[F_P(p)] \\ \Phi^{-1}[F_Y(y)] \end{Bmatrix}, \tag{E4}$$

where $F_P(p) = \exp\{-\exp[-\alpha(p - v)]\}$ is the Gumbel CDF of P and $F_Y(y) = 1 - \exp\left[-(y/u)^k\right]$ is the Weibull CDF of Y. From the moments given in Table 6.1 and the relations in Table A2.2, we compute $\alpha = 0.00257 \text{ kN}^{-1}$ and $v = 2{,}275$ kN for the Gumbel distribution and $k = 12.2$ and $u = 41{,}700$ kPa for the Weibull distribution. The Jacobian of the transformation is given by (see (3.52))

$$\mathbf{J}_{\mathbf{u},\mathbf{x}} = \mathbf{L}_0^{-1} \begin{bmatrix} \dfrac{1}{\sigma_1} & 0 & 0 & 0 \\ 0 & \dfrac{1}{\sigma_2} & 0 & 0 \\ 0 & 0 & \dfrac{f_P(p)}{\phi(z_3)} & 0 \\ 0 & 0 & 0 & \dfrac{f_Y(y)}{\phi(z_4)} \end{bmatrix}, \tag{E5}$$

where $f_P(p)$ is the Gumbel PDF of P and $f_Y(y)$ is the Weibull PDF of Y. We are now ready to use the iHL-RF algorithm to find the design point.

Starting from the mean point, the iHL-RF algorithm converges in nine iterations producing the FORM solution $\mathbf{u}^* = [1.21\ 0.699\ 0.941\ -0.180]^{\mathrm{T}}$, $\hat{\boldsymbol{\alpha}} = [0.491\ 0.283\ 0.381\ -0.731]$, $\beta = 2.47$, $p_{F1} = 0.00682$, and $\mathbf{x}^* = [341\ 170\ 3{,}223\ 31{,}770]^{\mathrm{T}}$. The "exact" estimate of the failure probability obtained by Monte Carlo simulations is $p_F \cong 0.00931$ (with a c.o.v. of 3%) with the generalized reliability index $\beta = \Phi^{-1}(1 - 0.00931) = 2.35$. Because the FORM approximation underestimates the failure probability, we anticipate that the limit-state surface bends toward the origin.

Example 6.5 (cont.)

Examining the mapping of the design point onto the original space $\mathbf{x}^*$, we see that the demand variables M_1, M_2, and P have values above their respective means whereas the capacity variable Y has a value below its mean. This is as expected. We note that, due to the nonlinear nature of the transformation from the $\mathbf{X}$ to $\mathbf{U}$ space, $\mathbf{x}^*$ may not represent the most likely failure point.

6.4 Accuracy of the FORM Approximation

It has been shown by Breitung (1984), Hohenbichler et al. (1987), and Breitung and Hohenbichler (1989) that the FORM approximation is asymptotically exact if the limit-state surface is "blown out" so that $\beta \to \infty$ while the curvatures at the design point diminish. Although this result is not relevant to practical problems that typically have β values in the range 2–5, it suggests that the FORM approximation would tend to be more accurate for larger β values than for smaller ones. For finite β values of practical interest, the question then arises as to what the sources of approximation in the FORM are and what can be said about the magnitude of the error inherent in the first-order approximation. These questions are addressed in this section. In the literature, the accuracy of the FORM approximation has been discussed in a number of papers, including Schuëller et al. (2004), Katafygiotis and Zuev (2008), and Valdebenito et al. (2010).

It is clear from property 4 of the standard normal space that the FORM solution is exact when the limit-state surface is a hyperplane in that space. This happens when $G(\mathbf{U})$ is a linear function of $\mathbf{U}$. Thus, the source of approximation in the FORM is the nonlinearity of the limit-state surface in the standard normal space. This nonlinearity arises from two sources: nonlinearity in the limit-state function $g(\mathbf{X})$ in the original space, and nonlinearity in the probability transformation $\mathbf{U} = \mathbf{T}(\mathbf{X})$. The nonlinearity in the limit-state function depends on the problem of interest and the way the basic random variables define the failure event. As the probability transformation is linear when the random variables are jointly normal, the nonlinearity in the probability transformation depends on the extent of non-normality of the random variables. Bounds or strong skewness in the distributions of the basic random variables result in stronger nonlinearity of the transformation and, hence, of the limit-state surface in the standard normal space. The FORM analyst must carefully examine these two sources of nonlinearity and be more cautious when they are strongly present.

Is it possible to determine an error bound on the FORM approximation of the failure probability? The answer is yes, but the bound is too wide to be useful in most applications.

Because $\mathbf{u}^*$ is the nearest point to the origin in the failure domain, the domain outside a hyper-sphere of radius β contains the entire failure domain. Thus, based on property 6 of the standard normal space described in Section 6.2, we can write

$$p_F \leq p_s = 1 - \chi_n^2(\beta), \tag{6.19}$$

where $\chi_n^2(.)$ is the CDF of the chi-square distribution with n degrees of freedom with n denoting the number of random variables. Figure 6.4 shows a plot of the above upper bound, together with the FORM approximation as a function of β for selected n. It can be seen that the upper bound quickly approaches 1 with increasing n and can be orders of magnitude greater than the FORM approximation. For this reason, the above upper bound is not useful in practice.

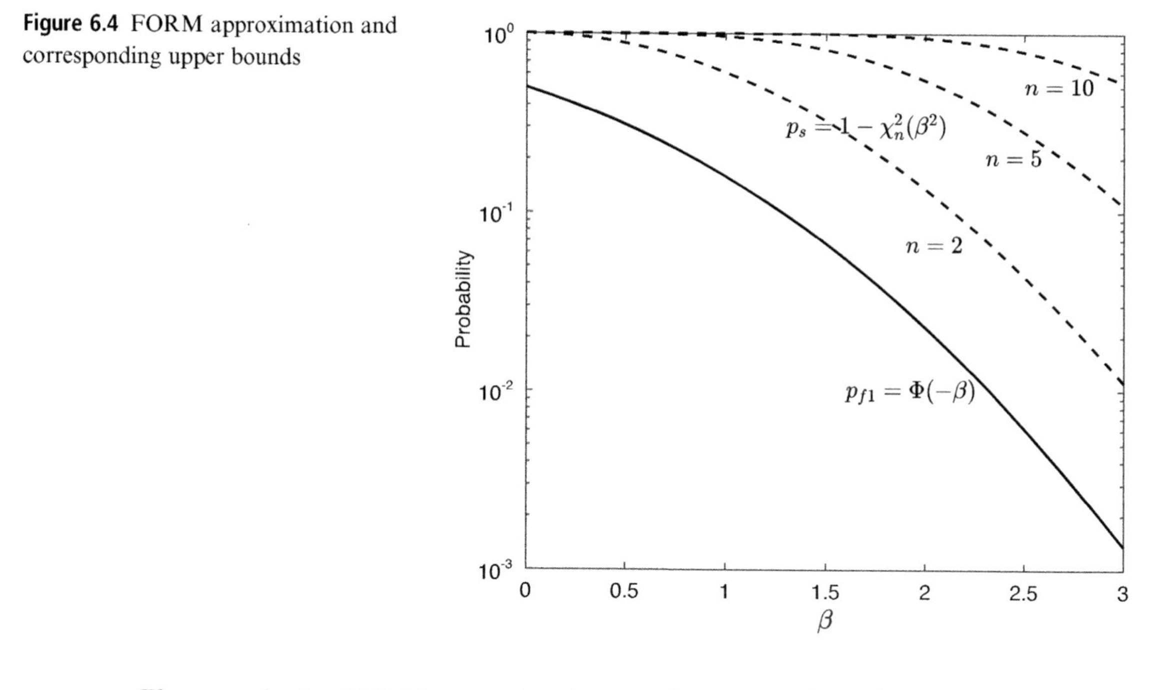

Figure 6.4 FORM approximation and corresponding upper bounds

The error in the FORM approximation can be measured relative to a higher order approximation. Der Kiureghian et al. (1987) derived the following error measure relative to a SORM approximation:

$$\frac{p_{F1} - p_{F2}}{p_{F2}} \cong (1 + \beta\bar{\kappa})^{\frac{n-1}{2}} \exp\left[-\frac{\delta_\kappa(n-1)}{4}\right] - 1. \tag{6.20}$$

In the above, p_{F2} is the approximation obtained by the SORM, $\bar{\kappa}$ is the average of the principal curvatures at the design point, and δ_κ is the "c.o.v." of the principal curvatures (computed as if the principal curvatures are observations of a random variable) providing a measure of variability in these quantities. It can be seen that the error measure consists of two terms. For a positive value of $\bar{\kappa}$ (on the average, the limit-state surface bends away from the origin), the term $(1+\beta\bar{\kappa})^{(n-1)/2}$ measures the overestimation of the failure probability by the FORM, whereas for a negative value of $\bar{\kappa}$ (on the average, the limit-state surface bends toward the origin), the same term predicts the underestimation by the FORM. The second term $\exp[-\delta_\kappa(n-1)/4]$ measures the effect of variability in the principal curvatures. The probability density in the standard normal space decays with distance from the origin, which means that positive and negative principal curvatures of equal magnitude do not have the same effect on the failure probability and their contributions are not canceled out. Hence, even when $\bar{\kappa}=0$, the variability in the principal curvatures has an effect and it is represented by the second term. Figure 6.5 shows the error measure as a function of $\bar{\kappa}$ for selected values of δ_κ for $\beta=3$ and $n=5$. Observe that the error is strongly dependent on the average curvature $\bar{\kappa}$. Of course, the above expression only measures the error in locally fitting the limit-state surface near the design point.

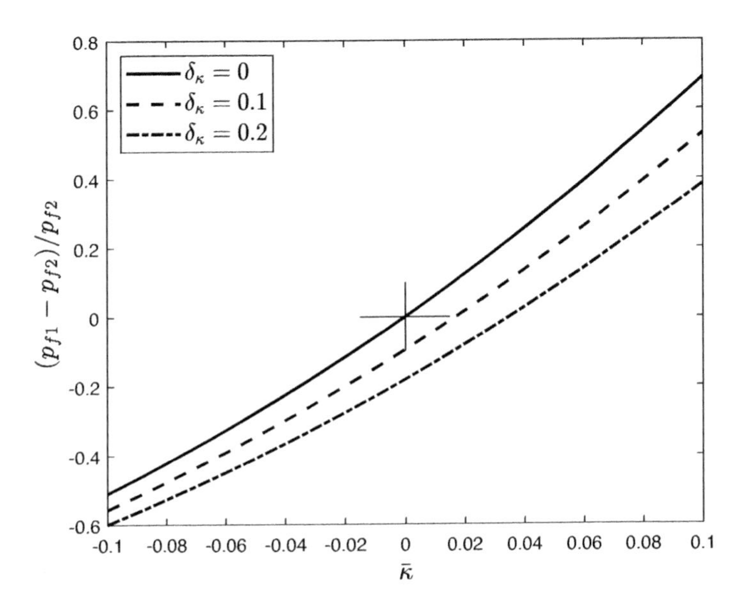

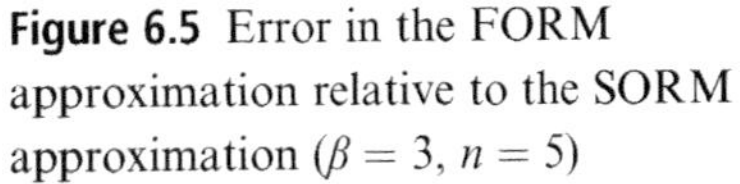
Figure 6.5 Error in the FORM approximation relative to the SORM approximation ($\beta=3$, $n=5$)

As described earlier, the power of the FORM approximation lies in the fact that the design point in the standard normal space has the highest probability density in the failure domain, and that the probability density decays exponentially with the square of the distance from the origin as you move in the radial direction and with the square of the distance from the design point as you move in the tangential direction (see Figures 6.1(a) and 6.1(b)). Based on these properties, it has been argued that the neighborhood of the design point is the major contributor to the failure probability integral and, therefore, a locally good approximation of the limit-state surface at that point would provide a good approximation of the failure probability. It turns out that this last argument may not hold in higher dimensions, i.e., for a large number of random variables n. To see this, consider a reformulation of the failure probability integral in (6.14) in the polar coordinates. We set $\mathbf{u}=r\boldsymbol{\theta}$, where r is the radial distance from the origin and $\boldsymbol{\theta}$ is a vector of $n-1$ rotation angles. The failure probability integral can now be written as (see Figure 6.6)

$$p_F = \int_{G(r\boldsymbol{\theta})\le 0} \phi_n(r\boldsymbol{\theta})\mathrm{d}r\mathrm{d}\boldsymbol{\theta}. \tag{6.21}$$

Owing to the rotational symmetry of the standard normal space, $\phi_n(r\boldsymbol{\theta})$ is independent of $\boldsymbol{\theta}$. Furthermore, the differential probability $\phi_n(r\boldsymbol{\theta})\mathrm{d}r\mathrm{d}\boldsymbol{\theta}$ is proportional to the probability of an annular domain of inner radius r and outer radius $r+\mathrm{d}r$. The latter is proportional to the PDF of the chi-square distribution with n degrees of freedom given by

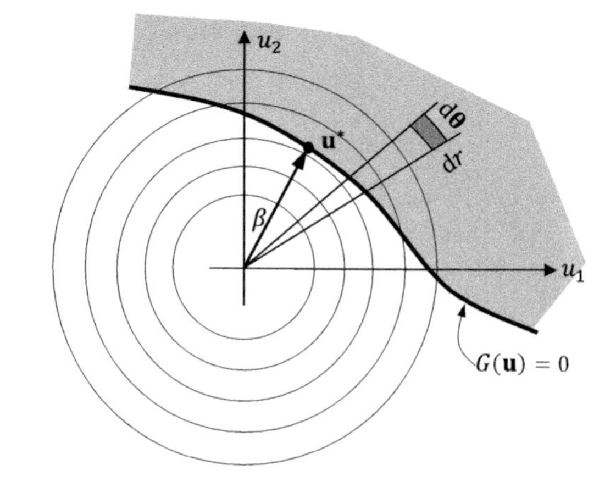

Figure 6.6 Failure probability integral in polar coordinates

$$f_X(x) = \frac{1}{2^{n/2}\Gamma(n/2)} x^{n/2-1} \exp\left(-\frac{x}{2}\right), 0 \le x, \tag{6.22}$$

where $x = r^2$ denotes the square of the distance from the origin. It is easy to verify that the mode of this distribution occurs at $x = n-2$. It follows that the differential probability $\phi_n(r\boldsymbol{\theta})\mathrm{d}r\mathrm{d}\boldsymbol{\theta}$ takes its maximum value at $r = \sqrt{n-2}$. Figure 6.7 shows plots of the chi-square PDF for selected n values as a function of $r = \sqrt{x}$. One can see that the mode of the distribution rapidly moves to the right with increasing n. It follows that the annular volume within the failure domain that contributes most to the failure probability is likely to be the one around $r = \sqrt{n-2}$. For large n, this value of r is likely to be greater than β. Hence, for large n, it is likely that the major contribution to the failure probability integral comes from regions further away from the origin than the design point. Although at such distances the probability density in the standard normal space is smaller than that at the design point, the volume of the corresponding annular domain can be much larger so that the contributed probability mass is larger than that coming from the annular domain near the design point. In such cases, clearly a good fit to the limit-state surface near the design point is not sufficient to obtain a good approximation of the failure probability.

The above analysis shows that the FORM approximation may experience deteriorating accuracy in higher dimensions, i.e., with increasing number of random variables. However, there is a caveat to this observation: In many structural reliability problems involving large numbers of random variables, the limit-state function tends to be linear or nearly linear in many of the random variables. If the distributions of these random variables are not strongly non-normal, the limit-state surface in the standard normal space tends to be flat in the corresponding directions. It is then only the random variables that produce strong non-linearity in the limit-state surface in the standard normal space that give rise to the error in the FORM approximation. Therefore, it is the number of these random variables that determines the critical dimension n in the above analysis. Later in this book, this problem will be demonstrated with an example involving nonlinear stochastic dynamic analysis that involves a large number of random variables.

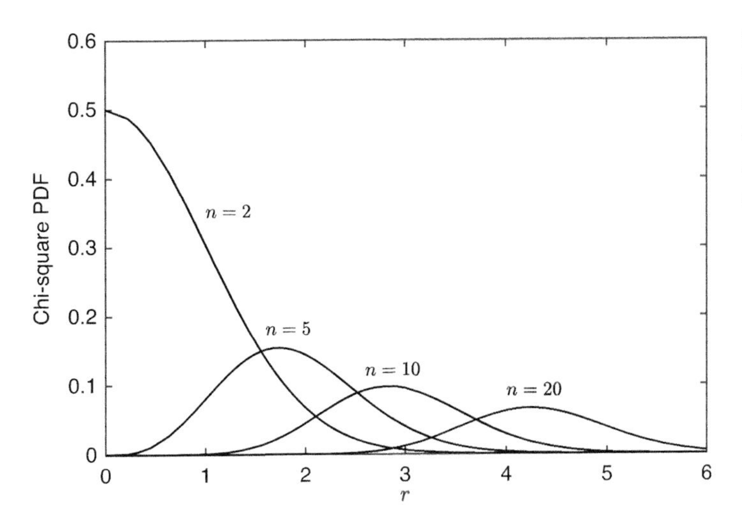

Figure 6.7 PDF of chi-square distribution for varying n; the mode of the distribution corresponds to the radius $r = \sqrt{x}$ of the most likely annular domain

6.5 FORM Measures of Importance of Random Variables

As was mentioned earlier, the unit normal vector $\widehat{\boldsymbol{\alpha}}$ provides measures of importance of the random variables $\mathbf{U}$. This property is examined in this section, which also introduces another unit vector that measures the relative importance of the basic random variables $\mathbf{X}$.

Using (6.16), we define the limit-state function linearized at the design point in the standard normal space as

$$\begin{aligned} G_1(\mathbf{U}) &= \beta - \widehat{\boldsymbol{\alpha}}\mathbf{U} \\ &= \beta - \widehat{\alpha}_1 U_1 - \widehat{\alpha}_2 U_2 - \cdots - \widehat{\alpha}_n U_n, \end{aligned} \tag{6.23}$$

where the positive factor $\|\nabla G(\mathbf{u}^*)\|$ has been dropped. For a linear limit-state function of normal random variables, the reliability index is the ratio of the mean to the standard deviation of the limit-state function. As U_i are statistically independent with zero means and unit variances, the mean and variance of the linearized limit-state function are

$$\mu_{G_1} = \beta, \tag{6.24}$$

$$\sigma^2_{G_1} = \widehat{\alpha}_1^2 + \widehat{\alpha}_2^2 + \cdots + \widehat{\alpha}_n^2 = 1, \tag{6.25}$$

where we have used the fact that $\widehat{\boldsymbol{\alpha}}$ is a unit vector. We see that the reliability index is equal to $\mu_{G_1}/\sigma_{G_1} = \beta/1 = \beta$, which is as expected. However, the last expression above presents an important finding: $\widehat{\alpha}_i^2$ is the contribution of the ith standard normal random variable U_i to the variance (uncertainty) of the linearized limit-state function. As such, it represents an importance measure of this random variable relative to the other random variables in the standard normal space. Furthermore, from (6.23), it can be seen that, for $\widehat{\alpha}_i < 0$, a larger outcome for U_i increases the value of the linearized limit-state function so that U_i acts as a capacity variable, whereas for $\widehat{\alpha}_i > 0$, a larger outcome for U_i decreases the value of the limit-state function so that U_i acts as a demand variable. It follows that the algebraic sign of α_i indicates the nature of the standard normal random variable U_i: if $\alpha_i < 0$, U_i is a capacity variable, and if $\alpha_i > 0$, U_i is a demand variable.

When the basic random variables $\mathbf{X}$ of the reliability problem are statistically independent, there is a one-to-one correspondence between the original random variables X_i and the standard normal random variables U_i. The nature (capacity or demand) and importance of the random variables X_i then are the same as those of U_i and can be determined in terms of the elements of $\widehat{\boldsymbol{\alpha}}$. However, when the basic random variables are statistically dependent, there is no such one-to-one correspondence. More specifically, for the triangular-type transformations used in structural reliability analysis (see Section 3.6), U_i is generally defined as a function of $X_1, X_2, \ldots, X_i$. Hence, the nature and importance order conveyed by the vector $\widehat{\boldsymbol{\alpha}}$ does not provide similar information about the random variables $\mathbf{X}$ in the original space.

To derive an importance vector for the basic random variables $\mathbf{X}$, we linearize the probability transformation at the design point (Der Kiureghian, 2005):

$$\begin{aligned}\mathbf{U} &= \mathbf{T}(\mathbf{X}) \\ &\cong \mathbf{T}(\mathbf{x}^*) + \mathbf{J}_{\mathbf{u},\mathbf{x}}(\mathbf{X} - \mathbf{x}^*) \\ &= \mathbf{u}^* + \mathbf{J}_{\mathbf{u},\mathbf{x}}(\mathbf{X} - \mathbf{x}^*),\end{aligned} \tag{6.26}$$

where the Jacobian $\mathbf{J}_{\mathbf{u},\mathbf{x}}$ must be computed at the design point. Of course, this relation is not exact when the transformation is nonlinear. If we set the two sides equal, i.e.,

$$\mathbf{U} = \mathbf{u}^* + \mathbf{J}_{\mathbf{u},\mathbf{x}}\left(\widetilde{\mathbf{X}} - \mathbf{x}^*\right), \tag{6.27}$$

then the random variables $\widetilde{\mathbf{X}}$ are somewhat different from $\mathbf{X}$. Because $\widetilde{\mathbf{X}}$ is a linear function of $\mathbf{U}$, it must have the joint normal distribution. Hence, $\widetilde{\mathbf{X}}$ is a normal approximation of $\mathbf{X}$ – a surrogate of $\mathbf{X}$. Solving for $\widetilde{\mathbf{X}}$ in (6.27), we obtain

$$\widetilde{\mathbf{X}} = \mathbf{x}^* + \mathbf{J}_{\mathbf{x},\mathbf{u}}(\mathbf{U} - \mathbf{u}^*), \tag{6.28}$$

where $\mathbf{J}_{\mathbf{x},\mathbf{u}} = \mathbf{J}_{\mathbf{u},\mathbf{x}}^{-1}$. The mean vector and covariance matrix of $\widetilde{\mathbf{X}}$ are

$$\widetilde{\mathbf{M}} = \mathbf{x}^* - \mathbf{J}_{\mathbf{x},\mathbf{u}}\mathbf{u}^*, \tag{6.29}$$

$$\widetilde{\boldsymbol{\Sigma}} = \mathbf{J}_{\mathbf{x},\mathbf{u}}\mathbf{J}_{\mathbf{x},\mathbf{u}}^{\mathrm{T}}. \tag{6.30}$$

These values depend on the design point and in general are somewhat different from the actual mean vector $\mathbf{M}_{\mathbf{X}}$ and covariance matrix $\boldsymbol{\Sigma}_{\mathbf{XX}}$ of $\mathbf{X}$. The magnitude of the difference depends on the degree of non-normality of $\mathbf{X}$.

Now substituting (6.27) in (6.23), we obtain

$$G_1(\mathbf{U}) = -\widehat{\boldsymbol{\alpha}}\, \mathbf{J}_{\mathbf{u},\mathbf{x}}\left(\widetilde{\mathbf{X}} - \mathbf{x}^*\right), \tag{6.31}$$

where $\widehat{\boldsymbol{\alpha}}\mathbf{u}^* = \beta$ has been used. The mean and variance of $G_1(\mathbf{U})$ are

$$\mu_{G_1} = -\widehat{\boldsymbol{\alpha}}\, \mathbf{J}_{\mathbf{u},\mathbf{x}}\left(\widetilde{\mathbf{M}} - \mathbf{x}^*\right) = \beta, \tag{6.32}$$

$$\sigma_{G_1}^2 = \widehat{\boldsymbol{\alpha}}\, \mathbf{J}_{\mathbf{u},\mathbf{x}}\widetilde{\boldsymbol{\Sigma}}\mathbf{J}_{\mathbf{u},\mathbf{x}}^{\mathrm{T}}\widehat{\boldsymbol{\alpha}}^{\mathrm{T}} = 1, \tag{6.33}$$

which are obtained by substituting for $\widetilde{\mathbf{M}}$ and $\widetilde{\boldsymbol{\Sigma}}$ from (6.29) and (6.30), respectively. It is seen that $\mu_{G_1}/\sigma_{G_1} = \beta$, as expected. The expansion of (6.33) involves both variances and covariances of $\widetilde{\mathbf{X}}$. To isolate the contributions coming from the individual variances, we write the covariance matrix in the form $\widetilde{\boldsymbol{\Sigma}} = \widetilde{\mathbf{D}}\widetilde{\mathbf{D}} + \left(\widetilde{\boldsymbol{\Sigma}} - \widetilde{\mathbf{D}}\widetilde{\mathbf{D}}\right)$, where $\widetilde{\mathbf{D}}$ is the diagonal matrix of standard deviations of $\widetilde{\mathbf{X}}$. Using this expression in (6.33), we have

$$\sigma_{G_1}^2 = \left\|\widehat{\boldsymbol{\alpha}}\, \mathbf{J}_{\mathbf{u},\mathbf{x}}\widetilde{\mathbf{D}}\right\|^2 + \widehat{\boldsymbol{\alpha}}\, \mathbf{J}_{\mathbf{u},\mathbf{x}}\left(\widetilde{\boldsymbol{\Sigma}} - \widetilde{\mathbf{D}}\widetilde{\mathbf{D}}\right)\mathbf{J}_{\mathbf{u},\mathbf{x}}^{\mathrm{T}}\widehat{\boldsymbol{\alpha}}^{\mathrm{T}} = 1. \tag{6.34}$$

It is clear that the squared elements of the row vector $\widehat{\boldsymbol{\alpha}}\, \mathbf{J}_{\mathbf{u},\mathbf{x}}\widetilde{\mathbf{D}}$ in the first term are the contributions to the variance of $G_1(\mathbf{U})$ arising from the individual elements of $\widetilde{\mathbf{X}}$, while the second term represents contributions from pairs of the random variables. Hence, the elements of the row vector can be considered as relative importance measures of the elements of $\widetilde{\mathbf{X}}$, which stands as a surrogate of the original random variables $\mathbf{X}$. Because of the second term in (6.34), the norm of this vector generally is not equal to unity. Based on these considerations, we introduce the importance vector of $\mathbf{X}$ as the unit row vector

$$\widehat{\boldsymbol{\gamma}} = \frac{\widehat{\boldsymbol{\alpha}}\, \mathbf{J}_{\mathbf{u},\mathbf{x}}\widetilde{\mathbf{D}}}{\left\|\widehat{\boldsymbol{\alpha}}\, \mathbf{J}_{\mathbf{u},\mathbf{x}}\widetilde{\mathbf{D}}\right\|}. \tag{6.35}$$

As $\widetilde{\mathbf{D}}$ is diagonal with positive elements, the algebraic signs of the elements of $\widehat{\boldsymbol{\alpha}}\, \mathbf{J}_{\mathbf{u},\mathbf{x}}\widetilde{\mathbf{D}}$ are the same as those of $\widehat{\boldsymbol{\alpha}}\, \mathbf{J}_{\mathbf{u},\mathbf{x}}$. Hence, from the expansion of (6.31), it is clear that a positive (negative) value of the element $\widehat{\gamma}_i$ of $\widehat{\boldsymbol{\gamma}}$ is an indication that the random variable $\widetilde{X}_i$ or X_i is of demand (capacity) type.

Observe that in the case where the basic random variables **X** are statistically independent, $\mathbf{J}_{\mathbf{u,x}}$ is diagonal and $\widetilde{\mathbf{D}} = \mathbf{J}_{\mathbf{x,u}}$ (see 6.30) so that (6.35) reduces to $\widehat{\gamma} = \widehat{\boldsymbol{\alpha}}$, as expected. When the basic random variables have the Nataf distribution, $\mathbf{J}_{\mathbf{u,x}} = \mathbf{L}_0^{-1}\text{diag}[f(x_i)/\phi(z_i)]$, where z_i are the intermediate variables (see 3.57). As shown by Kim and Song (2018), in that case $\widetilde{\mathbf{D}} = \text{diag}[\phi(z_i)f(x_i)]$ so that for Nataf-distributed random variables, (6.35) reduces to

$$\widehat{\gamma} = \frac{\widehat{\boldsymbol{\alpha}}\,\mathbf{L}_0^{-1}}{\left\|\widehat{\boldsymbol{\alpha}}\,\mathbf{L}_0^{-1}\right\|}. \tag{6.36}$$

Finally, when the random variables are jointly normal, $\mathbf{J}_{\mathbf{u,x}} = \mathbf{L}^{-1}\mathbf{D}^{-1}$ and $\widetilde{\mathbf{D}} = \mathbf{D}$ so that (6.35) reduces to

$$\widehat{\gamma} = \frac{\widehat{\boldsymbol{\alpha}}\,\mathbf{L}^{-1}}{\left\|\widehat{\boldsymbol{\alpha}}\,\mathbf{L}^{-1}\right\|}. \tag{6.37}$$

In this case, $\widetilde{\mathbf{X}} = \mathbf{X}$ and the elements of $\widehat{\gamma}$ precisely provide the relative individual contributions of the elements of **X** to the variance of the linearized limit-state function in the standard normal space.

Example 6.6 – Reliability of a Column: Variable Importance Measures

Consider the column reliability problem in Example 6.5 with the basic random variables being (M_1, M_2, P, Y), where M_1 and M_2 are the bending moments, P is the axial load, and Y is the yield stress of the material. The four variables have a Nataf joint distribution with the marginals and second moments listed in Table 6.1. Note that M_1 and M_2 have normal marginals whereas P has a Gumbel marginal distribution and Y has a Weibull marginal distribution. Using the FORM, we obtained $\mathbf{u}^* = [1.21\ 0.699\ 0.941\ -0.180]^{\mathrm{T}}$, $\mathbf{x}^* = [341\ 170\ 3{,}223\ 31{,}770]^{\mathrm{T}}$, and $\widehat{\boldsymbol{\alpha}} = [0.491\ 0.283\ 0.381\ -0.731]$. The $\widehat{\boldsymbol{\alpha}}$ vector indicates that the first three standard normal variables U_1, U_2, and U_3 are demand variables while U_4 is a capacity variable. Furthermore, of the four random variables in the standard normal space, U_4 is the most important in terms of contributing to the variance of the linearized limit-state function, followed by U_1, U_3, and then U_2.

To determine the natures and relative importances of the basic random variables **X**, we first compute the Jacobian $\mathbf{J}_{\mathbf{u,x}}$ and its inverse at the design point. Using (E5) in Example 6.5, these are obtained as

$$\mathbf{J}_{\mathbf{u,x}} = 10^{-3} \times \begin{bmatrix} 13.3 & 0 & 0 & 0 \\ -7.70 & 30.8 & 0 & 0 \\ -2.94 & -5.88 & 1.43 & 0 \\ 0 & 0 & 0 & 0.170 \end{bmatrix}, \tag{E1}$$

Example 6.6 (cont.)

$$\mathbf{J}_{\mathbf{x,u}} = \mathbf{J}_{\mathbf{u,x}}^{-1} = \begin{bmatrix} 75 & 0 & 0 & 0 \\ 18.7 & 32.5 & 0 & 0 \\ 231 & 133 & 670 & 0 \\ 0 & 0 & 0 & 5{,}889 \end{bmatrix}. \tag{E2}$$

Using the latter in (6.29), we have $\widetilde{\mathbf{M}} = [250\ 125\ 2{,}192\ 42{,}382]^{\mathrm{T}}$ as the mean vector of the surrogate variables $\widetilde{\mathbf{X}}$. Observe that the last two values are somewhat different from the means listed in Table 6.1, indicating the non-normal marginal distributions of variables P and Y. The corresponding covariance matrix is computed from (6.30), resulting in

$$\widetilde{\boldsymbol{\Sigma}} = 10^4 \times \begin{bmatrix} 0.562 & 0.141 & 1.74 & 0 \\ 0.141 & 0.141 & 0.867 & 0 \\ 1.74 & 0.867 & 56.0 & 0 \\ 0 & 0 & 0 & 3{,}468 \end{bmatrix}. \tag{E3}$$

The diagonal matrix of the standard deviations is obtained by taking the square roots of the diagonal elements of $\widetilde{\boldsymbol{\Sigma}}$, resulting in

$$\widetilde{\mathbf{D}} = \begin{bmatrix} 75.0 & 0 & 0 & 0 \\ 0 & 37.5 & 0 & 0 \\ 0 & 0 & 748 & 0 \\ 0 & 0 & 0 & 5{,}889 \end{bmatrix}. \tag{E4}$$

Observe again that the standard deviations of the last two variables are different from those listed in Table 6.1 on account of the non-normal nature of their marginal distributions.

The importance vector $\widehat{\gamma}$ is computed according to (6.35). The result is

$$\widehat{\gamma} = [\ 0.269 \quad 0.269 \quad 0.451 \quad -0.808\ ]. \tag{E5}$$

As the variables here are Nataf-distributed, the simpler expression in (6.36) can be used, and it gives an identical result. It is evident that the basic random variables M_1, M_2, and P are demand variables whereas Y is a capacity variable. This finding is intuitively obvious for this problem. However, in a more complex reliability problem, the nature of the random variables may not be so obvious and the information provided by the importance vector can be valuable. Furthermore, from the magnitudes of the elements of $\widehat{\gamma}$, it is clear that Y is the most important variable contributing to the variance of the linearized limit-state function in the standard

Example 6.6 (cont.)

normal space, P is the second most important variable, and M_1 and M_2 are equally ranked third in importance. It is remarkable that this order is counter to the order of the corresponding coefficients of variation of these variables as reported in Table 6.1. In particular, Y contributes most to the variance of the linearized limit-state function in the standard normal space even though its c.o.v. is only 0.1, while that of P is 0.2 and those of M_1 and M_2 are 0.3. Of course, the importance of Y has to do with the fact that it appears in the denominator of all three ratios in the limit-state function, as can be seen in (E1) of Example 6.5. Furthermore, P is more important than M_1 and M_2 because it is squared in the limit-state function.

6.6 FORM Parameter Sensitivities

In formulating structural reliability problems, often the limit-state function and the probability distribution include parameters whose values can be varied. Let $g = g(\mathbf{X}, \boldsymbol{\theta}_g)$ be the limit-state function with $\boldsymbol{\theta}_g$ as the *limit-state function parameters*, and $f_{\mathbf{X}}(\mathbf{x}, \boldsymbol{\theta}_f)$ be the joint PDF of the basic random variables with $\boldsymbol{\theta}_f$ denoting the *probability distribution parameters*. Also let $\boldsymbol{\theta} = (\boldsymbol{\theta}_g, \boldsymbol{\theta}_f)$ denote the set of all *parameters of the reliability problem*. An important advantage of the FORM is that, as a by-product of the analysis, it provides the sensitivities of the first-order approximation of the failure probability with respect to these parameters. More specifically, if θ is one of these parameters, we can easily compute $\partial\beta/\partial\theta$ and $\partial p_{F1}/\partial\theta$ as a by-product of the FORM analysis. These sensitivity measures are valuable in many different ways, as will be described in this and subsequent chapters of this text. Early papers on this subject include Hohenbichler and Rackwitz (1986a), Madsen (1988), Karamchandani et al. (1988), Bjerager and Krenk (1989), and Madsen and Tvedt (1990).

First consider the case of a single parameter θ, which is an element of either $\boldsymbol{\theta}_g$ or $\boldsymbol{\theta}_f$. Let $\mathbf{u}^*(\theta) = \beta(\theta)\hat{\boldsymbol{\alpha}}(\theta)^{\mathrm{T}}$, where the dependence on θ is explicitly shown. Because $\hat{\boldsymbol{\alpha}}(\theta)$ is a unit row vector, one has $\hat{\boldsymbol{\alpha}}(\theta)\hat{\boldsymbol{\alpha}}(\theta)^{\mathrm{T}} = 1$. Taking derivatives with respect to θ from both sides of this equality yields

$$\frac{\mathrm{d}\hat{\boldsymbol{\alpha}}(\theta)}{\mathrm{d}\theta}\hat{\boldsymbol{\alpha}}(\theta)^{\mathrm{T}} + \hat{\boldsymbol{\alpha}}(\theta)\frac{\mathrm{d}\hat{\boldsymbol{\alpha}}(\theta)^{\mathrm{T}}}{\mathrm{d}\theta} = 2\frac{\mathrm{d}\hat{\boldsymbol{\alpha}}(\theta)}{\mathrm{d}\theta}\hat{\boldsymbol{\alpha}}(\theta)^{\mathrm{T}} = 0. \tag{6.38}$$

It follows that $\hat{\boldsymbol{\alpha}}(\theta)$ and $\mathrm{d}\hat{\boldsymbol{\alpha}}(\theta)/\mathrm{d}\theta$ are orthogonal vectors. Now, taking the derivative of $\mathbf{u}^*(\theta) = \beta(\theta)\hat{\boldsymbol{\alpha}}(\theta)^{\mathrm{T}}$ with respect to θ, yields

$$\frac{\mathrm{d}\mathbf{u}^*(\theta)}{\mathrm{d}\theta} = \frac{\mathrm{d}\beta(\theta)}{\mathrm{d}\theta}\hat{\boldsymbol{\alpha}}(\theta)^{\mathrm{T}} + \beta(\theta)\frac{\mathrm{d}\hat{\boldsymbol{\alpha}}(\theta)^{\mathrm{T}}}{\mathrm{d}\theta}. \tag{6.39}$$

Multiplying the sides by $\hat{\boldsymbol{\alpha}}(\theta)$ and using (6.38), one obtains

$$\frac{d\beta(\theta)}{d\theta} = \widehat{\boldsymbol{\alpha}}(\theta)^{\mathrm{T}} \frac{d\mathbf{u}^*(\theta)}{d\theta}. \tag{6.40}$$

Furthermore, the design point is on the limit-state surface, so that

$$G(\mathbf{u}^*(\theta), \theta) = 0. \tag{6.41}$$

Taking the derivative with respect to θ yields

$$\nabla_{\mathbf{u}} G(\mathbf{u}^*, \theta) \frac{d\mathbf{u}^*(\theta)}{d\theta} + \frac{\partial G(\mathbf{u}^*, \theta)}{\partial \theta} = 0. \tag{6.42}$$

The remaining derivations depend on whether θ is a member of $\boldsymbol{\theta}_g$ or $\boldsymbol{\theta}_f$. The two cases are considered separately.

Case $\theta \in \boldsymbol{\theta}_g$:
Let $g = g(\mathbf{X}, \theta)$ be the limit-state function involving the parameter θ. Also, let $\mathbf{U} = \mathbf{T}(\mathbf{X})$ denote the probability transformation to the standard normal space and $\mathbf{X} = \mathbf{T}^{-1}(\mathbf{U})$ denote its inverse. Note that with $\theta \in \boldsymbol{\theta}_g$, both these transforms are independent of θ. One can write

$$\begin{aligned} g(\mathbf{X}, \theta) &= g[\mathbf{X}(\mathbf{U}), \theta] \\ &= G[\mathbf{U}(\mathbf{X}), \theta]. \end{aligned} \tag{6.43}$$

Taking a partial derivative with respect to θ,

$$\frac{\partial g(\mathbf{X}, \theta)}{\partial \theta} = \frac{\partial G(\mathbf{U}(\mathbf{X}), \theta)}{\partial \theta}. \tag{6.44}$$

The above expression is valid for any outcome of $\mathbf{U}$, including $\mathbf{u}^* = \mathbf{T}(\mathbf{x}^*)$. Thus,

$$\frac{\partial g(\mathbf{x}^*, \theta)}{\partial \theta} = \frac{\partial G(\mathbf{u}^*, \theta)}{\partial \theta}. \tag{6.45}$$

Using this in (6.42), one can write

$$\nabla_{\mathbf{u}} G(\mathbf{u}^*, \theta) \frac{d\mathbf{u}^*(\theta)}{d\theta} = -\frac{\partial g(\mathbf{x}^*, \theta)}{\partial \theta}. \tag{6.46}$$

As $\widehat{\boldsymbol{\alpha}}(\theta) = -\nabla_{\mathbf{u}} G(\mathbf{u}^*, \theta)/\|\nabla_{\mathbf{u}} G(\mathbf{u}^*, \theta)\|$, dividing (6.42) by $-\|\nabla_{\mathbf{u}} G(\mathbf{u}^*, \theta)\|$ and using the result together with (6.45) in (6.40), one finally obtains

$$\frac{d\beta}{d\theta} = \frac{1}{\|\nabla_{\mathbf{u}} G(\mathbf{u}^*, \theta)\|} \frac{\partial g(\mathbf{x}^*, \theta)}{\partial \theta}. \tag{6.47}$$

Extending this result to all members of the vector $\boldsymbol{\theta}_g$, we have

$$\nabla_{\boldsymbol{\theta}_g} \beta = \frac{1}{\left\|\nabla_{\mathbf{u}} G\left(\mathbf{u}^*, \boldsymbol{\theta}_g\right)\right\|} \nabla_{\boldsymbol{\theta}_g} g\left(\mathbf{x}^*, \boldsymbol{\theta}_g\right). \tag{6.48}$$

The above expression provides the gradient row vector of the reliability index with respect to the limit-state function parameters $\boldsymbol{\theta}_g$. The gradient vector $\nabla_{\mathbf{u}} G\left(\mathbf{u}^*, \boldsymbol{\theta}_g\right)$ is available from the search algorithm for finding the design point, so to compute the above sensitivity vector one only needs to compute the gradient of the limit-state function at the design point in the original space with respect to the parameters $\boldsymbol{\theta}_g$.

Case $\theta \in \boldsymbol{\theta}_f$:
Let $\mathbf{U} = \mathbf{T}(\mathbf{X}, \theta)$ denote the transformation to the standard normal space and $\mathbf{X} = \mathbf{T}^{-1}(\mathbf{U}, \theta)$ denote the inverse transform, where θ is a distribution parameter. One can write

$$\begin{aligned} g(\mathbf{X}) &= g\left[\mathbf{T}^{-1}(\mathbf{U}, \theta)\right] \\ &= G[\mathbf{T}(\mathbf{X}, \theta), \theta]. \end{aligned} \tag{6.49}$$

For any outcome $\mathbf{x}$, taking the derivative of both sides of the above expression with respect to θ and noting that $g(\mathbf{x})$ is independent of θ, one obtains

$$0 = \nabla_{\mathbf{u}} G(\mathbf{u}, \theta) \frac{\partial \mathbf{T}(\mathbf{x}, \theta)}{\partial \theta} + \frac{\partial G(\mathbf{u}, \theta)}{\partial \theta}. \tag{6.50}$$

The above expression is valid for any outcome $\mathbf{u}$, including $\mathbf{u}^* = \mathbf{T}(\mathbf{x}^*, \theta)$. Thus,

$$0 = \nabla_{\mathbf{u}} G(\mathbf{u}^*, \theta) \frac{\partial \mathbf{T}(\mathbf{x}^*, \theta)}{\partial \theta} + \frac{\partial G(\mathbf{u}^*, \theta)}{\partial \theta}. \tag{6.51}$$

Comparing (6.42) and (6.50), it is clear that

$$\frac{\mathrm{d}\mathbf{u}^*(\theta)}{\mathrm{d}\theta} = \frac{\partial \mathbf{T}(\mathbf{x}^*, \theta)}{\partial \theta}. \tag{6.52}$$

Substituting this relation in (6.40), one finally obtains

$$\frac{\mathrm{d}\beta}{\mathrm{d}\theta} = \widehat{\boldsymbol{\alpha}} \frac{\partial \mathbf{T}(\mathbf{x}^*, \theta)}{\partial \theta}. \tag{6.53}$$

Extending this result to all members of the vector of probability distribution parameters $\boldsymbol{\theta}_f$, yields

$$\nabla_{\boldsymbol{\theta}_f} \beta = \widehat{\boldsymbol{\alpha}}\, \mathbf{J}_{\mathbf{T},\boldsymbol{\theta}_f}\left(\mathbf{x}^*, \boldsymbol{\theta}_f\right), \tag{6.54}$$

where $\mathbf{J}_{\mathbf{T},\boldsymbol{\theta}_f}\left(\mathbf{x}^*, \boldsymbol{\theta}_f\right)$ is the Jacobian of the probability transformation with respect to the distribution parameters, evaluated at the design point $\mathbf{x}^*$. The above expression provides the gradient row vector of the reliability index with respect to the probability distribution parameters $\boldsymbol{\theta}_f$. Because $\widehat{\boldsymbol{\alpha}}$ is available from the FORM solution, only the gradient of the probability transformation at the design point with respect to the parameters $\boldsymbol{\theta}_f$ needs to be additionally computed.

It is important to note that, when computing the sensitivity with respect to a distribution parameter, it is necessary that we select other distribution parameters that are to be fixed. For example, in the case of a two-parameter distribution, the sensitivity with respect to the mean while the standard deviation is kept fixed is different from the sensitivity with respect to the mean when the coefficient of variation is kept fixed.

For any set of parameters $\boldsymbol{\theta} = (\boldsymbol{\theta}_g, \boldsymbol{\theta}_f)$, the FORM approximation of the failure probability is given by $p_{F1}(\boldsymbol{\theta}) = \Phi[-\beta(\boldsymbol{\theta})]$. Using the chain rule of differentiation, we have

$$\nabla_{\boldsymbol{\theta}} p_{F1} = -\phi(\beta) \nabla_{\boldsymbol{\theta}} \beta, \tag{6.55}$$

where $\phi(\beta) = \mathrm{d}\Phi(-\beta)/\mathrm{d}\beta$ is the standard normal PDF. (Note that $\phi(\beta) = \phi(-\beta)$ due to symmetry.) It is seen that the sensitivities of the FORM approximation of the failure probability are proportional to the sensitivities of the reliability index, with the proportionality factor being $-\phi(\beta)$. Observe that the sensitivities of the reliability index and the failure probability have opposite signs, so that a parameter that tends to increase the reliability index tends to decrease the failure probability and vice versa.

Example 6.7 – Reliability of a Column: Parameter Sensitivities

Consider the column reliability problem introduced in Example 6.5. We are interested in the sensitivities of the reliability index and the first-order approximation of the failure probability with respect to the limit-state and probability distribution parameters. Let $\boldsymbol{\theta}_g = (s_1, s_2, a)$ be the set of limit-state function parameters, where $s_1 = 0.030$ m^3, $s_2 = 0.015$ m^3, and $a = 0.190$ m^2, and $\boldsymbol{\theta}_f = (\mu_1, \sigma_1, \mu_2, \sigma_2, v, \alpha, u, k)$ be the set of probability distribution parameters, where $\mu_1 = 250$ kNm and $\sigma_1 = 75$ kNm are the mean and standard deviation of M_1, $\mu_2 = 125$ kNm and $\sigma_2 = 37.5$ kNm are the mean and standard deviation of M_2, $v = 2{,}275$ kN and $\alpha = 0.00257$ kN^{-1} are the parameters of the Gumbel distribution of P, and $u = 41.7$ MPa and $k = 12.2$ are the parameters of the Weibull distribution of Y (see Table A2.2 for definitions of the latter four parameters).

First consider the sensitivities with respect to $\boldsymbol{\theta}_g$. From the iHL-RF algorithm one obtains $\|\nabla_{\mathbf{u}} G(\mathbf{u}^*, \boldsymbol{\theta}_g)\| = 0.323$ at the convergence to the design point. The gradient of the limit-state function in (E1) of Example 6.5 with respect to the parameter set $\boldsymbol{\theta}_g$ is

$$\nabla_{\boldsymbol{\theta}_g} g(\mathbf{x}, \boldsymbol{\theta}_g) = \left[\frac{m_1}{s_1^2 y} \quad \frac{m_2}{s_2^2 y} \quad \frac{2}{a}\left(\frac{p}{ay}\right)^2 \right]. \tag{E1}$$

Evaluating at the design point $\mathbf{x}^* = [341\ 170\ 3{,}223\ 31{,}770]^{\mathrm{T}}$, one obtains $\nabla_{\boldsymbol{\theta}_g} g(\mathbf{x}^*, \boldsymbol{\theta}_g) = [11.9\ 23.8\ 3.00]$. Using (6.48), one has

$$\nabla_{\boldsymbol{\theta}_g} \beta = [\ 36.8 \quad 73.6 \quad 9.26]. \tag{E2}$$

The above sensitivities can be used, for example, to estimate the change in the reliability index for small changes in the parameters. For example, an increase of 0.01 m^2 in the cross-sectional area (to $a = 0.200$ m^2) would increase the reliability index by approximately $9.26 \times 0.01 = 0.0926$, i.e., to $\beta \cong 2.47 + 0.0926 \cong 2.56$. Furthermore, it is apparent that if the same amount of percent increase in the flexural moduli s_1 and s_2 can be implemented, it is better that the increase be in the latter than the former. This kind of analysis is useful in design situations, where adjustments in structural properties need to be made to achieve target reliability levels.

Example 6.7 (cont.)

More generally, these sensitivities are essential in reliability-based optimal design, which is the topic of Chapter 14.

To determine the sensitivities with respect to the distribution parameters, we need to compute the Jacobian of the probability transformation with respect to the parameters $\boldsymbol{\theta}_f$ at the design point. The applicable transformation is given in (E4) of Example 6.5. As pointed out by Bourinet and Lemaire (2008) and Bourinet (2017), matrix $\mathbf{L}_0$ in this transformation depends on the correlation coefficients of the Nataf distribution, which for non-normal marginal distributions depend on the coefficients of variation and, therefore, on the distribution parameters (see Appendix 3A). However, as evident from numerical examples examined by Bourinet (2017), this dependence is weak and for all practical purposes can be neglected. Therefore, we only differentiate the second term in the right-hand side of (E4) of Example 6.5. The result is

$$\mathbf{J}_{\mathbf{T},\boldsymbol{\theta}_f}(\mathbf{x}^*,\boldsymbol{\theta}_f) = \mathbf{L}_0^{-1}\begin{bmatrix} -\frac{1}{\sigma_1} & -\frac{m_1^*-\mu_1}{\sigma_1^2} & 0 & 0 & 0 & 0 & 0 & 0 \\ 0 & 0 & -\frac{1}{\sigma_2} & -\frac{m_2^*-\mu_2}{\sigma_2^2} & 0 & 0 & 0 & 0 \\ 0 & 0 & 0 & 0 & \frac{\mathrm{d}F_P(p^*)/\mathrm{d}v}{\phi(z_3^*)} & \frac{\mathrm{d}F_P(p^*)/\mathrm{d}\alpha}{\phi(z_3^*)} & 0 & 0 \\ 0 & 0 & 0 & 0 & 0 & 0 & \frac{\mathrm{d}F_Y(y^*)/\mathrm{d}u}{\phi(z_4^*)} & \frac{\mathrm{d}F_Y(y^*)/\mathrm{d}k}{\phi(z_4^*)} \end{bmatrix}. \tag{E3}$$

Using the expressions in Table A2.2, we (as above) can show that $\mathrm{d}F_P(p^*)/\mathrm{d}v = -f_P(p^*)$, $\mathrm{d}F_P(p^*)/\mathrm{d}\alpha = [(p^*-v)/\alpha]f_P(p^*)$, $\mathrm{d}F_Y(y^*)/\mathrm{d}u = -(y^*/u)f_Y(y^*)$, and $\mathrm{d}F_Y(y^*)/\mathrm{d}k = (y^*/k)\ln(y^*/u)f_Y(y^*)$. Using these expressions, the numerical result of (E3) is

$$\mathbf{J}_{\mathbf{T},\boldsymbol{\theta}_f}(\mathbf{x}^*,\boldsymbol{\theta}_f) = \mathbf{L}_0^{-1}\begin{bmatrix} -0.0133 & -0.0161 & 0 & 0 & 0 & 0 & 0 & 0 \\ 0 & 0 & -0.0267 & -0.0323 & 0 & 0 & 0 & 0 \\ 0 & 0 & 0 & 0 & -0.00134 & 494 & 0 & 0 \\ 0 & 0 & 0 & 0 & 0 & 0 & -0.000129 & -0.120 \end{bmatrix}. \tag{E4}$$

Now, using (6.54), we obtain

$$\nabla_{\boldsymbol{\theta}_f}\beta = 10^{-3}[-3.24 \quad -3.92 \quad -6.48 \quad -7.84 \quad -0.546 \quad 202{,}000 \quad 0.0951 \quad 88.8]. \tag{E5}$$

Example 6.7 (cont.)

Observe that the sensitivities with respect to the means of M_1 and M_2 and the scale parameter of P (elements 1, 3, and 5 of the above row vector) are negative, indicating that an increase in these values will decrease the reliability index. This is expected for these load variables. On the other hand, the sensitivity with respect to the scale parameter of Y (element 7) is positive, indicating that an increase in its value will increase the reliability index. This is again expected because Y is a capacity variable. Furthermore, the sensitivities with respect to the standard deviations of M_1 and M_2 (elements 2 and 4) are negative, indicating that an increase in the variances of these variables decreases the reliability index. The sensitivity with respect to the α parameter of the Gumbel distribution of P (element 6) is positive. This is because an increase in this parameter decreases the standard deviation of P; it also decreases the mean but by a lesser extent (see Table A2.2 for the relation of this parameter to the mean and standard deviation). Finally, the sensitivity with respect to shape parameter k of the Webull distribution of Y (element 8) is positive. This is because an increase in this parameter increases the mean and decreases the variance of the capacity variable Y.

To obtain the sensitivities of the first-order failure probability with respect to parameters $\boldsymbol{\theta}_g$ and $\boldsymbol{\theta}_f$, we need to scale the gradient vectors in (E2) and (E5) by $-\phi(\beta) = -0.0190$. The result is:

$$\nabla_{\boldsymbol{\theta}_g} p_{F1} = [-0.700 \quad -1.40 \quad -0.176], \tag{E6}$$

$$\nabla_{\boldsymbol{\theta}_f} p_{F1} = 10^{-4}[0.617 \quad 0.746 \quad 1.23 \quad 1.49 \quad 0.104 \quad -38{,}430 \quad -0.0181 \quad -16.9]. \tag{E7}$$

Observations similar to those discussed for the sensitivities of β can be made for the above sensitivities of p_{F1}.

6.7 Sensitivities with Respect to Alternative Set of Parameters

In formulating structural reliability problems, often there are alternative choices for the set of parameters. For example, the lognormal distribution for a basic random variable can be defined in terms of parameters λ and ζ (see Example 2.6), or the mean μ and standard deviation σ. Having the sensitivities of β with respect to one set of parameters, we wish to obtain the sensitivities with respect to an alternative set of parameters. Let $\boldsymbol{\theta}$ denote the set of parameters for which the gradient vector $\nabla_{\boldsymbol{\theta}}\beta$ is available and $\boldsymbol{\theta}'$ denote an alternative set of parameters with respect to which the sensitivities are sought. Let the two sets of parameters be related through the one-to-one transformation

$$\boldsymbol{\theta}' = \mathbf{F}(\boldsymbol{\theta}). \tag{6.56}$$

Using the chain rule of differentiation, we have

$$\nabla_{\theta'}\beta = \nabla_{\theta}\beta\, \mathbf{J}_{\theta,\theta'}, \tag{6.57}$$

where $\mathbf{J}_{\theta,\theta'} = \mathbf{J}_{\theta',\theta}^{-1}$ and $\mathbf{J}_{\theta',\theta}^{-1}$ is the inverse of the Jacobian of the transformation in (6.56). Note that the number of parameters in the two sets must be equal, so that $\mathbf{J}_{\theta,\theta'}$ is a square matrix.

Example 6.8 – Reliability of a Column: Sensitivities with Respect to Means and Standard Deviations

Suppose for the column problem in Example 6.7 we wish to compute the sensitivities with respect to the means and standard deviations of variables P and Y. For P, the relations between the parameters (v, α) and (μ_3, σ_3) (the inverse of the transform according to (6.56)) are given by (see Table A2.2)

$$v = \mu_3 + \frac{\gamma\sqrt{6}\sigma_3}{\pi}, \tag{E1}$$

$$\alpha = \frac{\pi}{\sqrt{6}\sigma_3}, \tag{E2}$$

where $\gamma = 0.577$ is Euler's constant. The needed Jacobian for variable P, therefore, is

$$\mathbf{J}_{(\alpha,v),(\mu_3,\sigma_3)} = \begin{bmatrix} 1 & \dfrac{\gamma\sqrt{6}}{\pi} \\ 0 & -\dfrac{\pi}{\sqrt{6}\sigma_3^2} \end{bmatrix}. \tag{E3}$$

Using (6.57) together with $\nabla_{(v,\alpha)}\beta = 10^{-3}[-0.546\ 202{,}000]$, one obtains

$$\nabla_{(\mu_3,\sigma_3)}\beta = 10^{-4}[-5.46\ -7.90]. \tag{E4}$$

For Y, the relations between the sets of distribution parameters (u, k) and (μ_4, σ_4) are

$$\mu_4 = u\Gamma(1 + 1/k), \tag{E5}$$

$$\sigma_4 = u\sqrt{\Gamma(1 + 2/k) - \Gamma^2(1 + 1/k)}. \tag{E6}$$

There is no closed-form solution for the inverse of this transformation. Hence, the Jacobian $\mathbf{J}_{(\mu_4,\sigma_4),(u,k)}$ is computed numerically by using finite differences and then inverted to obtain $\mathbf{J}_{(u,k),(\mu_4,\sigma_4)}$. Using the resulting matrix in (6.57) together with $\nabla_{(k,u)}\beta = 10^{-3}[0.0951\quad 88.8]$, one finally obtains

$$\nabla_{(\mu_4,\sigma_4)}\beta = 10^{-4} \times [1.24\ -2.45]. \tag{E7}$$

Example 6.8 (cont.)

Observe in (E4) that the sensitivity with respect to the mean of P is negative, whereas in (E7) the sensitivity with respect to the mean of Y is positive. These are as expected because P is a load variable and Y a capacity variable. Furthermore, the sensitivities with respect to both standard deviations are negative, indicating that an increase in the variance (uncertainty) of either variable decreases the reliability index.

As mentioned earlier, when computing the sensitivity with respect to a distribution parameter, it is necessary to decide what other parameters are being fixed. In the above analysis, the mean and standard deviation were the selected independent parameters so that when the mean was varied the standard deviation was kept fixed, and vice versa. Now, suppose we want to compute the sensitivity with respect to the mean of P while fixing its coefficient of variation, δ_3. The relations between the two sets of parameters are

$$\mu_3 = \mu_3, \tag{E8}$$

$$\sigma_3 = \delta_3 \mu_3. \tag{E9}$$

Thus, the Jacobian of interest is

$$\mathbf{J}_{(\mu_3,\sigma_3),(\mu_3,\delta_3)} = \begin{bmatrix} 1 & 0 \\ \delta_3 & \mu_3 \end{bmatrix} = \begin{bmatrix} 1 & 0 \\ 0.2 & 2{,}500 \end{bmatrix}. \tag{E10}$$

Now using (6.57) together with $\nabla_{(\mu_3,\sigma_3)}\beta = 10^{-4}[-5.46 \; -7.90]$, one obtains

$$\nabla_{(\mu_3,\delta_3)}\beta = 10^{-4}[-7.04 \; -19{,}700]. \tag{E11}$$

Observe that the sensitivities with respect to μ_3 are different when the fixed parameter is the standard deviation, shown in (E4), versus when the fixed parameter is the c.o.v., shown in (E11). As an example, an increase of 0.01 in the c.o.v. (from 0.20 to 0.21), while keeping the mean fixed, will result in a decrease of $1.97 \times 0.01 \cong 0.0197$ in the reliability index.

6.8 Importance Vectors with Respect to Means and Standard Deviations

The mean of a random variable represents a central measure of its outcomes and the standard deviation represents a measure of the variability in the outcomes. Naturally, sensitivities with respect to these parameters are of particular interest. Let

$$\nabla_{\mathbf{M}}\beta = \left[\frac{\partial\beta}{\partial\mu_1}\ \frac{\partial\beta}{\partial\mu_2}\cdots\frac{\partial\beta}{\partial\mu_n}\right] \tag{6.58}$$

denote the sensitivity vector with respect to the means of the basic random variables with the standard deviations fixed, and

$$\nabla_{\mathbf{D}}\beta = \left[\frac{\partial\beta}{\partial\sigma_1}\ \frac{\partial\beta}{\partial\sigma_2}\cdots\frac{\partial\beta}{\partial\sigma_n}\right] \tag{6.59}$$

denote the sensitivity vector with respect to the standard deviations with the means fixed. If the distribution has more than two parameters, we assume that those parameters are also fixed. It is desirable to compare the elements within these vectors in order to assess the relative importance of the random variables with respect to variations in their means and standard deviations. However, the basic random variables and, therefore, their means and standard deviations may have different dimensions and variations in their means and standard deviations need to be scaled in order to make them of similar influence. Furthermore, since the numerical values of the means and standard deviations for different random variables can be very different, a unit variation in one mean or standard deviation may not be statistically comparable to a unit variation in another mean or standard deviation. A useful way to scale the elements of the above vectors is to multiply each term by the corresponding standard deviation. This makes the elements of each vector dimensionless. Furthermore, because the error (variability) in the estimate of the mean of a random variable from observed data is proportional to its standard deviation, this scaling also makes the variations statistically equivalent.

Based on the above considerations, we define the following variable importance vectors

$$\boldsymbol{\delta} = \nabla_{\mathbf{M}}\beta\mathbf{D} = \left[\frac{\partial\beta}{\partial\mu_1}\sigma_1\ \frac{\partial\beta}{\partial\mu_2}\sigma_2\cdots\frac{\partial\beta}{\partial\mu_n}\sigma_n\right], \tag{6.60}$$

$$\boldsymbol{\eta} = \nabla_{\mathbf{D}}\beta\mathbf{D} = \left[\frac{\partial\beta}{\partial\sigma_1}\sigma_1\ \frac{\partial\beta}{\partial\sigma_2}\sigma_2\cdots\frac{\partial\beta}{\partial\sigma_n}\sigma_n\right], \tag{6.61}$$

where $\mathbf{D}$ denotes the diagonal matrix of standard deviations. The first vector, $\boldsymbol{\delta}$, describes the sensitivities with respect to scaled variations in the means, while the second vector, $\boldsymbol{\eta}$, describes the sensitivities with respect to scaled variations in the standard deviations. All elements are dimensionless and can be compared within each vector to determine the relative importance of the random variables with respect to the scaled variations in the means and in the standard deviations, respectively. The four sensitivity vectors $\widehat{\boldsymbol{\alpha}}$, $\widehat{\boldsymbol{\gamma}}$, $\boldsymbol{\delta}$, and $\boldsymbol{\eta}$ are standard outputs in FORM analysis by CalREL, FERUM, and OpenSees software (Der Kiureghian et al., 2006). Note that the first two are unit vectors while the latter two are not. Also note that by dividing an element within $\boldsymbol{\delta}$ or $\boldsymbol{\eta}$ by the corresponding standard deviation, one obtains the unscaled sensitivity with respect to the mean or standard deviation of the random variable, respectively.

Example 6.9 – Reliability of a Column: Importance Vectors with Respect to Means and Standard Deviations

From the results in Examples 6.7 and 6.8, we have the gradient vector with respect to the mean values with the standard deviations fixed as $\nabla_{\mathbf{M}}\beta = 10^{-3} \times [-3.24 - 6.48 - 0.546 - 0.124]$ and the gradient vector with respect to the standard deviations with the mean values fixed as $\nabla_{\mathbf{M}}\beta = 10^{-3} \times [-3.92 \quad -7.84 \quad -0.790 \quad -0.245]$. The diagonal matrix of standard deviations is $\mathbf{D} = \text{diag}[75 \quad 37.5 \quad 500 \quad 4]$. Using (6.60) and (6.61), one obtains

$$\boldsymbol{\delta} = [-0.243 \quad -0.243 \quad -0.273 \quad 0.495], \tag{E1}$$

$$\boldsymbol{\eta} = [-0.294 \quad -0.294 \quad -0.395 \quad -0.981]. \tag{E2}$$

Observe that in terms of variations in the means with the standard deviations fixed, Y is the most important random variable followed by P, with M_1 and M_2 being equally important and in the third place. In terms of variations in the standard deviations with the means fixed, Y is by far the most important random variable, distantly followed by P and then M_1 and M_2.

6.9 Multiple Design Points

It is possible for the limit-state surface in the standard normal space to have more than one minimum-distance point; i.e., for the constrained optimization problem in (6.15) to have more than one solution. The nearest of these points to the origin is the global solution and in many cases linearization of the limit-state surface at this point provides a good first-order approximation of the failure probability. However, when any of the remaining solution points is not too much farther from the origin than the global solution point, then its neighborhood may also make a significant contribution to the failure probability integral and, hence, the FORM solution obtained at the global solution point may not be sufficiently accurate. In addressing this problem, two issues arise: (a) How does one find the subsequent design points? (b) How does one use the multiple design points to obtain an improved approximation? Der Kiureghian and Dakessian (1998) proposed a method to find the multiple design points by adding a "bulge" at previously obtained design points so that the algorithm is forced to find a subsequent solution. They then used a system reliability approach (Chapter 8) to obtain an improved estimate of the failure probability. Here, a summary of the approach is provided. The interested reader should consult the original paper for further details.

Suppose in a reliability problem with the limit-state function $G = G(\mathbf{u})$ in the standard normal space, the iHL-RF algorithm has converged to a first design point denoted $\mathbf{u}_1^*$. We define the revised limit-state function

$$G_1(\mathbf{u}) = G(\mathbf{u}) + B_1(\mathbf{u}), \tag{6.62}$$

where $B_1(\mathbf{u})$ is a bulge function centered at $\mathbf{u}_1^*$ and defined as

$$B_1(\mathbf{u}) = s_1\left(r_1^2 - \|\mathbf{u} - \mathbf{u}_1^*\|^2\right)^2, \quad \text{for } \|\mathbf{u} - \mathbf{u}_1^*\| \le r_1, \\ = 0, \quad \text{elsewhere,} \tag{6.63}$$

where r_1 and s_1 are positive-valued parameters. Note that this function is continuously differentiable with respect to $\mathbf{u}$ so that $G_1(\mathbf{u})$ also has that property, provided $G(\mathbf{u})$ has it. Because of the addition of the positive-valued function $B_1(\mathbf{u})$, the revised limit-state surface moves away from the origin in the neighborhood of the design point $\mathbf{u}_1^*$ within a sphere of radius r_1. The search algorithm is then restarted with the revised limit-state function, leading to the second design point, $\mathbf{u}_2^*$, if one exists. A bulge function centered at $\mathbf{u}_2^*$ and having parameters r_2 and s_2 is then added to obtain a second revised limit-state function. Thus, to obtain the mth design point, the revised limit-state function is

$$G_m(\mathbf{u}) = G(\mathbf{u}) + \sum_{i=1}^{m-1} B_i(\mathbf{u}). \tag{6.64}$$

Based on some detailed considerations, Der Kiureghian and Dakessian (1998) recommended using $r_i = \gamma\beta_i$ and $s_i = \gamma\|\nabla G(\mathbf{u}_i^*)\|/(\gamma^2 - \delta^2)/\beta_i^3$, where β_i is the reliability index associated with the ith design point, $\gamma = 1.1$, and $\delta = 0.75$. They also discuss ways to avoid converging to spurious design points created by adding the bulges.

Next are two examples to demonstrate the above approach. The first is a classic problem in which the nonlinearity in the limit-state function is entirely due to the nonlinearity in the probability transformation. The second is a structural dynamics problem with uncertain mass, stiffness, and damping values and an implicit limit-state function.

Example 6.10 – Example with Multiple Design Points

Consider a reliability problem defined by the linear limit-state function

$$g(X_1, X_2) = 18 - 3X_1 - 2X_2, \tag{E1}$$

where the random variables have the joint exponential CDF

$$F_{X_2X_1}(x_1, x_2) = 1 - \exp(-x_1) - \exp(-x_2) + \exp[-(x_1 + x_2 + x_1x_2)], \quad 0 < x_1, x_2 \\ = 0, \quad \text{elsewhere.} \tag{E2}$$

Owing to its highly nonlinear limit-state surface in the standard normal space, this classic reliability problem has been studied by Hohenbichler and Rackwitz (1981), Madsen et al. (1986), Ditlevsen and Madsen (1996), and Der Kiureghian and Dakessian (1998), among others.

The Rosenblatt transformation to the standard normal space with the order of random variables selected as $(X_1, X_2) \to (U_1, U_2)$ is

$$U_1 = \Phi^{-1}[F_{X_1}(X_1)], \tag{E3a}$$

Example 6.10 (cont.)

$$U_1 = \Phi^{-1}\left[F_{X_2|X_1}(X_2|X_1)\right], \tag{E3b}$$

where $F_{X_1}(x_1) = 1 - \exp(-x_1)$ is the marginal CDF of X_1 and $F_{X_2|X_1}(x_2|x_1) = 1 - (1 - x_2)\exp[-(1 + x_1)x_2]$ is the conditional CDF of X_2 given X_1. Figure 6.8 (left) shows the limit-state surface in the standard normal space. Note that the nonlinearity in the limit-state surface is entirely due to the strong non-normality of the joint probability distribution of X_1 and X_2. Starting the iHL-RF algorithm at the origin, convergence is achieved in six steps to $\mathbf{u}_1^* = [2.78\ 0.0865]^{\mathrm{T}}$ with $\beta_1 = 2.78$. The iteration steps are shown in Figure 6.8 (left) with a sequence of line segments. Now we add a bulge function at $\mathbf{u}_1^*$ to obtain the revised limit-state surface shown in Figure 6.8 (right). Starting from the origin, the iHL-RF algorithm converges in 13 steps to $\mathbf{u}_2^* = [-1.30\ 3.25]^{\mathrm{T}}$ with $\beta_2 = 3.50$. The steps are shown in Figure 6.8 (right). Further analysis leads to convergence to spurious design points at the feet of the bulge, which is an indication that there are no more valid design points.

If we apply the FORM approximation at $\mathbf{u}_1^*$ the solution is $p_{F1} = 0.00269$, and if we apply it at $\mathbf{u}_2^*$ the solution is $p_{F1} = 0.00023$. With the two design points, a series-system reliability formulation can be used to obtain an improved FORM estimate of the failure probability. As shown in Chapter 8, this requires the reliability indices at the two design points as well as the corresponding unit normal vectors $\widehat{\boldsymbol{\alpha}}_1$ and $\widehat{\boldsymbol{\alpha}}_2$. The result is $p_{F1} = 0.00292$. A Monte Carlo solution of this problem yields the "exact" result as $p_F = 0.00294$. Observe that accounting for both design points significantly improves the failure probability estimate.

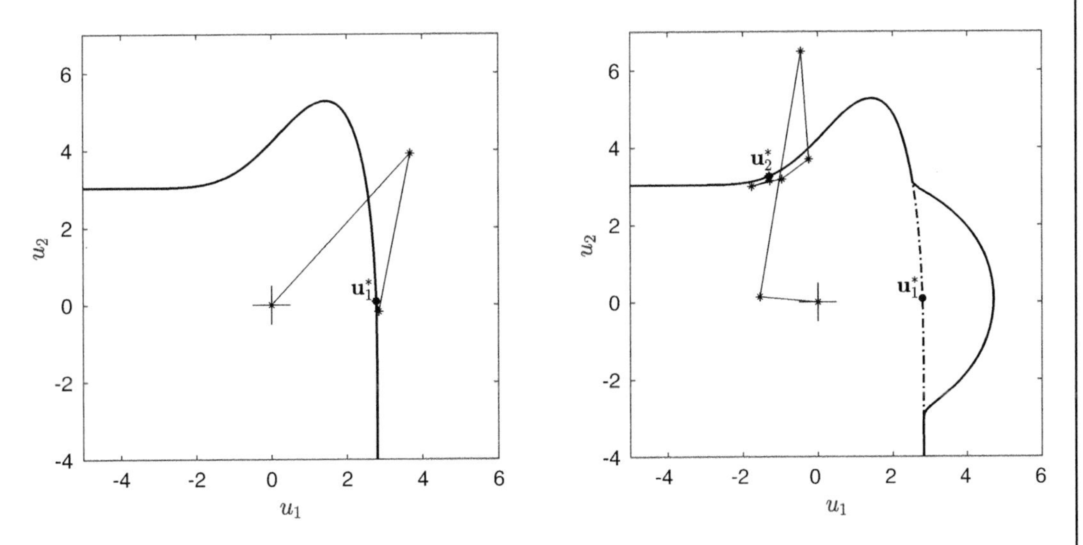

Figure 6.8 Finding multiple design points by use of a bulge function (after Der Kiureghian and Dakessian, 1988)

Example 6.11 – Seismic Reliability of a Building with a Tuned-Mass Damper

It is fortuitous that real structural reliability problems with multiple design points are uncommon. Furthermore, when they exist, it is often possible to anticipate them. The present example, adopted from Der Kiureghian and Dakessian (1998), is designed with anticipation of multiple design points.

One way to reduce the seismic demand on a tall structure is to install a pendulum at a high location, e.g., the roof or the top floor, and tune it to the fundamental period of the structure. This is known as a tuned-mass damper (Taniguchi et al., 2008). In the case of a dynamic excitation, the tuned-mass damper takes energy away from the structure and, hence, reduces the demand on the structure. However, the tuned-mass damper is effective when it is exactly tuned to the fundamental period. Any detuning reduces the effectiveness of the damper. In practice, the stiffness and mass properties of the structure and the damper are often uncertain, and this may lead to detuning.

Consider the 10-story building in Figure 6.9 with a tuned-mass damper, modeled as a single-degree-of-freedom oscillator, installed on its roof. Let M_i and K_i, $i = 1, \ldots, 10$, denote the floor masses and story stiffnesses of the structure and M_0 and K_0 denote the mass and stiffness of the tuned-mass damper, respectively. The combined system has the modal damping ratios Z_i, $i = 0, 1, \ldots, 10$. The structure is subjected to an earthquake ground motion characterized by a pseudo-acceleration response spectrum $A(T, Z) = 0.5H(Z)a(T)$, where $H(Z) = 0.5 + 1.5/(40Z + 1)$ is a damping-dependent amplification factor suggested by Kawashima and

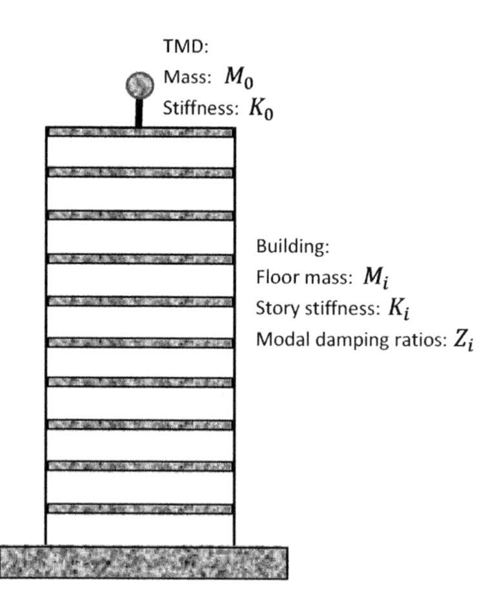

Figure 6.9 Ten story building with tuned-mass damper (after Der Kiureghian and Dakessian, 1998)

Example 6.11 (cont.)

Aizawa (1986) and $a(T)$ is a normalized pseudo-acceleration response spectrum as a function of period T for 5% damping ratio given in Der Kiureghian and Dakessian (1998). We assume the set of random variables $\mathbf{X} = (M_0, M_1, \ldots, M_{10}, K_0, K_1, \ldots, K_{10}, Z_0, Z_1, \ldots, Z_{10})$ with independent lognormal distributions with means $\mu_{M_0} = 158$ kips/g, $\mu_{M_1} = \cdots = \mu_{M_{10}} = 193$ kips/g, $\mu_{K_0} = 22$ kips/in, $\mu_{K_1} = \cdots = \mu_{K_{10}} = 1{,}200$ kips/in, and $\mu_{Z_0} = \mu_{Z_1} = \cdots = \mu_{Z_{10}} = 0.05$, and c.o.v. $\delta_{M_0} = \delta_{M_1} = \cdots = \delta_{M_{10}} = 0.2$, $\delta_{K_0} = \delta_{K_1} = \cdots = \delta_{K_{10}} = 0.2$, and $\delta_{Z_0} = \delta_{Z_1} = \cdots = \delta_{Z_{10}} = 0.3$. The limit-state function is defined as

$$g(\mathbf{X}) = V_0 - V_{\text{base}}(\mathbf{X}), \tag{E1}$$

where $V_{\text{base}}(\mathbf{X})$ is the base shear of the structure (equal to the drift in the first story times K_1) and $V_0 = 1{,}500$ kips is the safe threshold. The first story drift is computed by response spectrum analysis using the CQC (complete quadratic combination) modal combination rule (Der Kiureghian, 1981). For any set of random variables, this computation involves eigenvalue analysis of the 11-degree-of-freedom system to determine the modal frequencies and mode shapes, the base shear for each modal response, and their combination using the CQC rule. This example represents a case where the limit-state function is an implicit function of the random variables.

Analysis with the mean values of the mass and stiffness shows that the frequency of the tuned-mass damper is identical to the first-mode frequency of the structure and equal to 1.17 Hz. In this case the tuned-mass damper is effective, and the response of the structure is reduced (compared with the case without the tuned-mass damper). However, because of the uncertainty in the mass and stiffness properties, it is possible that the mass damper is "under-tuned" (the frequency of the oscillator is below the first-mode frequency of the building) or "over-tuned" (the frequency of the oscillator is above the first-mode frequency of the building), in which cases the structural response will be larger than it would be for the case of perfect tuning. It is possible that both these conditions will give rise to a design point – hence, the possibility of multiple design points.

Starting from the mean point, an initial run of the iHL-RF algorithm converges in six steps to a design point $\mathbf{u}_1^*$ with reliability index $\beta_1 = 1.06$. Eigenvalue analysis with the corresponding point $\mathbf{x}_1^*$ in the original space reveals that the first-mode frequency of the structure is 1.15 Hz and the frequency of the oscillator is 1.33 Hz. This is clearly a case where the mass damper is "over-tuned." Next, a bulge centered at $\mathbf{u}_1^*$ is added to the limit-state function in the standard normal space and the search with the iHL-RF algorithm is repeated. This leads to the second design point $\mathbf{u}_2^*$ with $\beta_2 = 1.71$. Eigenvalue analysis with the corresponding values $\mathbf{x}_2^*$ reveals a first-mode frequency of 1.20 Hz for the structure and frequency 1.03 Hz of the oscillator. This is clearly a case where the mass damper is "under-tuned." Further investigation with a bulge centered at $\mathbf{u}_2^*$ reveals no more valid design points.

Example 6.11 (cont.)

A FORM approximation at the design point $\mathbf{u}_1^*$ alone yields the failure probability $p_{F1} = 0.145$ and at the design point $\mathbf{u}_2^*$ alone yields $p_{F1} = 0.044$. Performing a FORM system reliability analysis (see Chapter 8) that accounts for both design points yields $p_{F1} = 0.188$. The "exact" solution of this problem by Monte Carlo simulation yields $p_F = 0.213$. We see that accounting for the second design point improves the FORM approximation. However, there is still a relatively large error. This is partly due to the asymptotic nature of the FORM approximation (see Section 6.4). Here, the reliability index is small (the design points are close to the origin) and the probability densities in the neighborhoods of the design points are decaying slowly. As a result, any discrepancy between the tangent planes at the design points and the actual limit-state surface has a sizable probability content, which leads to the error in the FORM approximation.

6.10 The Inverse Reliability Problem

Consider a structural reliability problem defined by the limit-state function

$$g = g(\mathbf{X}, \theta) \tag{6.65}$$

where θ is a parameter. The inverse reliability problem is defined as follows (Der Kiureghian et al., 1994): Determine θ such that the reliability index equals a target value β? This problem arises in a number of applications. For example, θ may represent a structural design parameter that must be set so as to achieve a specified reliability index. This is a special case of a more general reliability-based optimal design problem, where usually one has multiple design parameters, a cost function, and often one or more constraints, one of which could be the specified reliability index (see Chapter 14). Another example is a reliability problem of the form $g(\mathbf{X}, \theta) = \theta - R(\mathbf{X})$, where $R(\mathbf{X})$ is a critical structural response quantity and θ the corresponding safe threshold. The inverse reliability problem arises when the threshold θ needs to be determined for a specified reliability index β.

The inverse reliability problem can be solved by repeatedly solving the reliability problem in (6.65) for a range of selected θ values and finding the value of θ for which the reliability index matches the specified β. However, this is a brute force and tedious approach. A more elegant solution developed in Der Kiureghian et al. (1994) and Zhang and Der Kiureghian (1997) is presented below. The solution is an extension of the iHL-RF algorithm presented in Section 5.2.3.

The inverse reliability problem is defined by the following set of equations in the standard normal space

$$\widehat{\boldsymbol{\alpha}}\mathbf{u} - \beta = 0, \tag{6.66}$$

$$G(\mathbf{u},\theta)=0, \tag{6.67}$$

$$\mathbf{u}-\widehat{\boldsymbol{\alpha}}\mathbf{u}\widehat{\boldsymbol{\alpha}}^{\mathrm{T}}=0, \tag{6.68}$$

where $\widehat{\boldsymbol{\alpha}}=-\nabla_{\mathbf{u}}G(\mathbf{u},\theta)/\|\nabla_{\mathbf{u}}G(\mathbf{u},\theta)\|$ is the normalized negative gradient vector at the trial point $\mathbf{u}$. The first equation ensures that, upon convergence, the target reliability index is achieved; the second equation ensures that the solution point is on the limit-state surface; and the third equation states that the solution must be an origin-project point, i.e., the normal at the point must pass through the origin, which is a necessary condition for the design point. Using (6.66), the last equation is simplified to

$$\mathbf{u}-\beta\widehat{\boldsymbol{\alpha}}^{\mathrm{T}}=0. \tag{6.69}$$

We follow an approach similar to that of the iHL-RF algorithm developed in Section 5.2.3 but in the expanded space $(\mathbf{u},\theta)$. We construct a sequence of trial points according to the rule

$$\left\{\begin{matrix}\mathbf{u}_{i+1}\\ \theta_{i+1}\end{matrix}\right\}=\left\{\begin{matrix}\mathbf{u}_{i}\\ \theta_{i}\end{matrix}\right\}+\lambda_i\left\{\begin{matrix}\mathbf{d}_{i}\\ d_{\theta_i}\end{matrix}\right\},\ i=1,2,\ldots, \tag{6.70}$$

with $0<\lambda_i\le 1$ as the step size and $\left[\mathbf{d}_i^{\mathrm{T}}\ d_{\theta i}\right]^{\mathrm{T}}$ as the search direction vector. The search direction is obtained by finding the solution for the limit-state function linearized at the trial point,

$$G(\mathbf{u},\theta)\cong G(\mathbf{u}_i,\theta_i)+\nabla_{\mathbf{u}}G(\mathbf{u}_i,\theta_i)(\mathbf{u}-\mathbf{u}_i)+\frac{\partial G(\mathbf{u}_i,\theta_i)}{\partial\theta}(\theta-\theta_i). \tag{6.71}$$

The solution is

$$\mathbf{u}=\beta\widehat{\boldsymbol{\alpha}}_i^{\mathrm{T}}, \tag{6.72}$$

$$\theta=\theta_i+\left[\beta-\widehat{\boldsymbol{\alpha}}_i\mathbf{u}_i-\frac{G(\mathbf{u}_i,\theta_i)}{\|\nabla_{\mathbf{u}}G(\mathbf{u}_i,\theta_i)\|}\right]\frac{\|\nabla_{\mathbf{u}}G(\mathbf{u}_i,\theta_i)\|}{\partial G(\mathbf{u}_i,\theta_i)/\partial\theta}, \tag{6.73}$$

where $\widehat{\boldsymbol{\alpha}}_i=-\nabla_{\mathbf{u}}G(\mathbf{u}_i,\theta_i)/\|\nabla_{\mathbf{u}}G(\mathbf{u}_i,\theta_i)\|$. Thus, the search direction vector is

$$\left\{\begin{matrix}\mathbf{d}_i\\ d_{\theta_i}\end{matrix}\right\}=\left\{\begin{matrix}\beta\widehat{\boldsymbol{\alpha}}_i^{\mathrm{T}}-\mathbf{u}_i\\ \left[\beta-\widehat{\boldsymbol{\alpha}}_i\mathbf{u}_i-\dfrac{G(\mathbf{u}_i,\theta_i)}{\|\nabla_{\mathbf{u}}G(\mathbf{u}_i,\theta_i)\|}\right]\dfrac{\|\nabla_{\mathbf{u}}G(\mathbf{u}_i,\theta_i)\|}{\partial G(\mathbf{u}_i,\theta_i)/\partial\theta}\end{matrix}\right\}. \tag{6.74}$$

The step size is selected to assure reduction in a merit function. We use a merit function of the form

$$m(\mathbf{u}_i,\theta_i)=m_1(\mathbf{u}_i,\theta_i)+m_2(\mathbf{u}_i,\theta_i), \tag{6.75}$$

where $m_1(\mathbf{u}_i,\theta_i)=\|\mathbf{u}_i\|^2/2+c_1|G(\mathbf{u}_i,\theta_i)|$ is the merit function of the iHL-RF algorithm with $c_1>\|\mathbf{u}_i\|/\|\nabla_{\mathbf{u}}G(\mathbf{u}_i,\theta_i)\|$ and $m_2(\mathbf{u}_i,\theta_i)=c_2(\|\mathbf{u}_i\|-\beta)^2/2$ is motivated by (6.66), where assuming $0 < \beta$ the norm of $\mathbf{u}$ is used. Owing to the similar scales of the first term in $m_1(\mathbf{u}_i, \theta_i)$ and the second merit function, $c_2 = 1$ is selected. Note that the merit function takes its minimum value at the solution point. Furthermore, the merit function does not require computation of derivatives.

Experience with the above algorithm shows that the search direction vector in (6.74) tends to more strongly favor satisfying (6.66) than (6.67) and (6.68). This tends to slow down the convergence to the design point. To remedy this, the search direction vector in (6.74) is replaced with a weighted combination of that same vector and the search direction vector of the original iHL-RF algorithm in (5.21) (with the component d_{θ_i} set to 0) so that the revised search direction vector is

$$\begin{Bmatrix} \mathbf{d}_i \\ d_{\theta_i} \end{Bmatrix} = \begin{Bmatrix} a_1\left\{\left[\dfrac{G(\mathbf{u}_i, \theta_i)}{\|\nabla_{\mathbf{u}} G(\mathbf{u}_i, \theta_i)\|} + \widehat{\boldsymbol{\alpha}}_i \mathbf{u}_i\right]\widehat{\boldsymbol{\alpha}}_i^{\mathrm{T}} - \mathbf{u}_i\right\} + a_2\left\{\beta\widehat{\boldsymbol{\alpha}}_i^{\mathrm{T}} - \mathbf{u}_i\right\} \\ a_2\left[\beta - \widehat{\boldsymbol{\alpha}}_i \mathbf{u}_i - \dfrac{G(\mathbf{u}_i, \theta_i)}{\|\nabla_{\mathbf{u}} G(\mathbf{u}_i, \theta_i)\|}\right]\dfrac{\|\nabla_{\mathbf{u}} G(\mathbf{u}_i, \theta_i)\|}{\partial G(\mathbf{u}_i, \theta_i)/\partial\theta} \end{Bmatrix}. \tag{6.76}$$

In the above, a_1 and a_2 are positive weight factors with $a_1 + a_2 = 1$. A good choice for the weight factors is the ratio of the merit functions,

$$a_1 = \frac{m_1(\mathbf{u}_i, \theta_i)}{m(\mathbf{u}_i, \theta_i)} \text{ and } a_2 = \frac{m_2(\mathbf{u}_i, \theta_i)}{m(\mathbf{u}_i, \theta_i)}. \tag{6.77}$$

This choice produces a sequence that reduces the two parts of the merit function in a balanced manner. Finally, to check the convergence of the algorithm, in addition to the convergence criteria for the iHL-RF algorithm, we use the criterion

$$\frac{\|\mathbf{u}_{i+1} - \mathbf{u}_i\|}{\|\mathbf{u}_{i+1}\|} + \frac{|\theta_{i+1} - \theta_i|}{|\theta_{i+1}|} + \frac{|\widehat{\boldsymbol{\alpha}}_{i+1}\mathbf{u}_{i+1} - \beta|}{\beta} \le \epsilon, \tag{6.78}$$

where ϵ is a small tolerance value, e.g., $\epsilon = 10^{-3}$.

Example 6.12 – Inverse Reliability Analysis with an Explicit Limit-State Function

This example is adopted from Der Kiureghian et al. (1994). Consider the limit-state function in the standard normal space

$$G(\mathbf{U}, \theta) = \exp[-\theta(U_1 + 2U_2 + 3U_3)] - U_4 + 1.5, \tag{E1}$$

where θ is a parameter. We wish to determine θ such that $\beta = 2$. Starting from the initial trial point $(\mathbf{u}_1, \theta_1) = (0.2, 0.2, 0.2, 0.2, 0.1)$ and using $c_1 = 10$ at all steps, the above algorithm converges in 9 steps with the solution $\theta = 0.367$. Table 6.2 lists the results for the trial points $(\mathbf{u}_i, \theta_i)$, the norm $\|\mathbf{u}_i\|$, and the values of the constituents $m_1(\mathbf{u}_i, \theta_i)$ and $m_2(\mathbf{u}_i, \theta_i)$ of the merit function. It can be seen that the sequence converges in a balanced manner.

Example 6.12 (cont.)

Table 6.2 Summary of results for inverse reliability analysis (after Der Kiureghian et al., 1994, with permission from ASCE)

i	$\mathbf{u}_i$	$\|\mathbf{u}_i\|$	θ_i	$m_1(\mathbf{u}_i, \theta_i)$	$m_2(\mathbf{u}_i, \theta_i)$
1	(0.200,0.200,0.200,0.200)	0.400	0.100	21.949	1.28
2	(0.170,0.340,0.510,1.918)	2.021	0.443	2.737	2.15×10^{-2}
3	(0.256,0.513,0.769,1.622)	1.919	0.432	2.351	4.02×10^{-3}
4	(0.211,0.421,0.632,1.742)	1.912	0.441	2.132	3.87×10^{-3}
5	(0.220,0.440,0.658,1.810)	1.988	0.368	2.102	7.53×10^{-5}
6	(0.219,0.438,0.657,1.829)	2.004	0.359	2.045	8.69×10^{-6}
7	(0.218,0.437,0.655,1.828)	2.002	0.365	2.008	2.59×10^{-6}
8	(0.218,0.437,0.655,1.826)	2.000	0.367	2.004	1.01×10^{-7}
9	(0.218,0.436,0.655,1.826)	2.000	0.367	2.002	2.65×10^{-8}

6.11 FORM Approximation of the CDF and PDF of a Function of Random Variables

We saw in Section 2.8 that determining the PDF of a function of n random variables requires the solution of an $(n-1)$-fold integral. (See Example 2.8 for the case of a linear function.) Determining the CDF requires one additional integral, i.e., n-fold integration. Here, it is shown that structural reliability methods, including FORM and SORM, can be used to approximately obtain these distributions numerically.

Suppose we are interested in computing the PDF and CDF of the function $Z = h(\mathbf{X})$ of random variables $\mathbf{X}$. Let z denote the threshold of interest. Then, the CDF is defined as $F_Z(z) = \Pr(Z \leq z)$ and the PDF is $f_Z(z) = \mathrm{d}F_Z(z)/\mathrm{d}z$. We can solve this problem as a structural reliability problem defined by the limit-state function

$$g(\mathbf{X}, z) = h(\mathbf{X}) - z. \tag{6.79}$$

It should be clear that $F_Z(z) = p_F$ and $f_Z(z) = \mathrm{d}p_F/\mathrm{d}z$, where p_F is the failure probability of the structural reliability problem and $\mathrm{d}p_F/\mathrm{d}z$ is the sensitivity of the failure probability with respect to the limit-state function parameter z. Hence, a FORM solution together with sensitivity analysis with respect to the limit-state parameter z provides both the CDF and the PDF for the selected threshold. Often one is interested in the CDF and PDF for a range of threshold values. In that case, a series of FORM solutions together with sensitivity analysis should be performed for a sequence of incrementally increasing z values. Such an analysis is performed efficiently if the starting iteration point for each threshold z is selected as the design point for the preceding z value. Because the design points must be located close together, the search algorithm quickly converges to the design point for the new z value.

Example 6.13 – CDF and PDF of a Function of Random Variables

Consider the function $Z = \sum_{i=1}^{n} X_i$, where X_i are independent random variables having identical gamma distributions with parameters k and λ. We assume the parameter values $\lambda = 1$ and $k = 2$ and 5. (See Table A2.2 for the definition of these parameters.) The case with $k = 2$ corresponds to a stronger non-normal distribution than the case with $k = 5$. We also consider the cases with $n = 2$ and 5. We know that the distribution of Z is gamma with parameters (nk, λ). Hence, this example provides a good opportunity to examine the accuracy of the FORM in computing the distribution of a function of random variables.

To obtain the CDF of Z, we formulate the limit-state function as

$$g(\mathbf{X}, z) = \sum_{i=1}^{n} X_i - z. \tag{E1}$$

As mentioned earlier, the CDF is identical to the probability of failure of the above reliability problem and the PDF is equal to the sensitivity of the failure probability with respect to the limit-state function parameter, z.

For this problem, the distributions of the random variables and their roles in the limit-state function are identical, which means that the coordinates of the design point must also be identical. It follows that the mapping of the design point in the original space must have the coordinates $x_i^* = z/n$. Transforming to the standard normal space, we have $u_i^* = \Phi^{-1}[F(z/n)]$, where $F(x)$ is the gamma CDF with parameters (k, λ). Furthermore, the elements of the gradient vector are given by $\partial G(\mathbf{u}^*)/\partial u_i = \phi(u_i^*)/f(x_i^*)$, where $f(x)$ denotes the gamma PDF, and the elements of the unit normal vector are given by $\widehat{\alpha}_i = 1/\sqrt{n}$. It follows that the reliability index is $\beta = \Phi^{-1}[F(z/n)]/\sqrt{n}$. Furthermore, using (6.47), the sensitivity of the reliability index with respect to the parameter z is given by $\mathrm{d}\beta/\mathrm{d}z = -1/\|\nabla G(\mathbf{u}^*)\|$. Thus, the FORM approximations of the CDF and PDF of Z are given by $F_Z(z) \cong \Phi(-\beta)$ and $f_Z(z) \cong \phi(\beta)/\|\nabla G(\mathbf{u}^*)\|$, respectively. These results are easily computed without performing a search for the design point.

Figure 6.10 compares the exact CDF and PDF (solid lines) with the corresponding FORM approximations (dashed lines) for the four sets of parameter values. The top figures are for $n = 2$ and the bottom ones are for $n = 5$. The following observations are noteworthy: The FORM approximations are better in the tail regions than in the middle of the distribution. This is because the FORM is more accurate for computing small probabilities. The approximations are better for $k = 5$ than for $k = 2$. This is because in the latter case the gamma distribution is more strongly non-normal. Finally, the accuracy deteriorates with increasing n. This has to do with the effect of increasing dimension, but also because the limit-state surface in the standard normal space tends to curve toward the origin with respect to each random variable. This example will be reconsidered in Chapter 7 using the SORM to account for this curvature effect.

Example 6.13 (cont.)

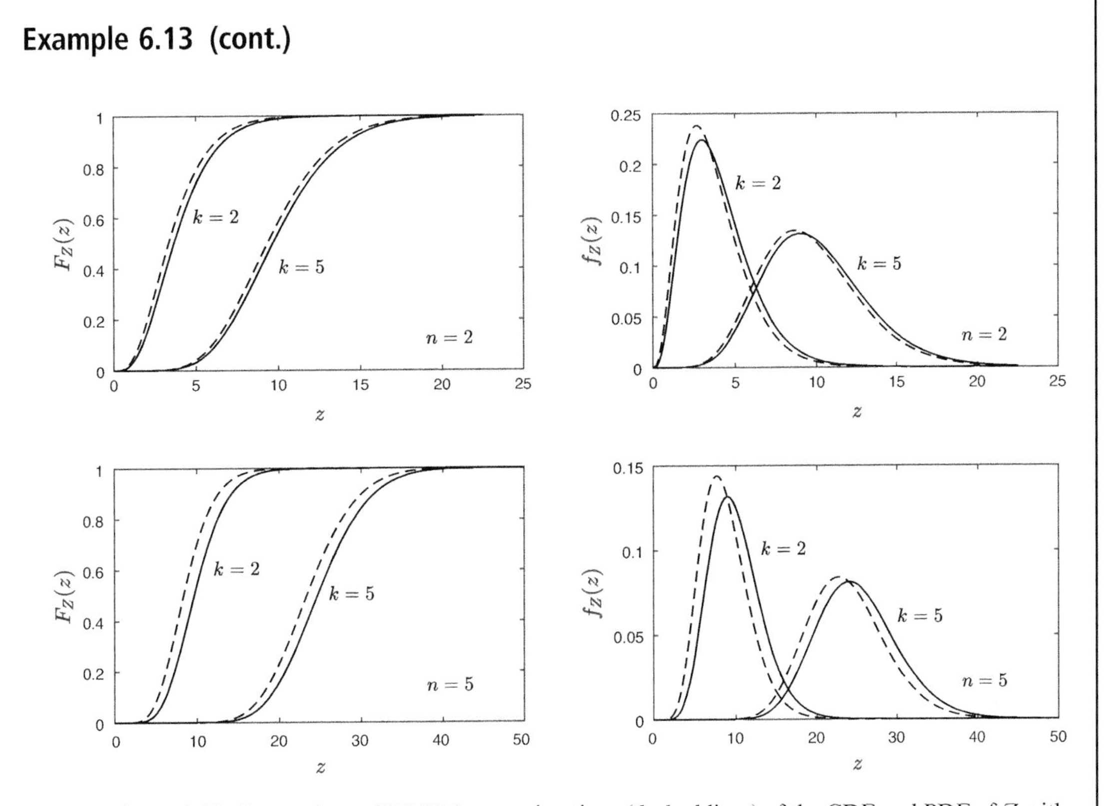

Figure 6.10 Comparison of FORM approximations (dashed lines) of the CDF and PDF of Z with exact results (solid lines)

6.12 Concluding Remarks

The FORM is perhaps the most widely used structural reliability method for component reliability analysis, i.e., problems that are defined by a single, continuous, and continuously differentiable limit-state function. The limit-state function can be implicit, i.e., require an algorithmic numerical evaluation, as long as the gradients can also be computed. Problems with multiple limit-state functions or functions that involve and, or, max, or min operations are system reliability problems and require additional steps, as described in Chapter 8.

The FORM provides a first-order approximation of the failure probability by replacing the limit-state surface with its tangent hyperplane at the point with minimum distance from the origin in the standard normal space. The FORM is asymptotically exact as the reliability index approaches infinity while the curvatures at the design point diminish. For practical problems with finite reliability index, the FORM approximation entails an error that depends on the nonlinearity of the limit-state surface in the standard normal space, and this error may grow with the dimension of the problem, i.e., the number of random variables. The nonlinearity in the standard normal space arises from two sources: nonlinearity in the limit-state function in the original space, and non-normality of the basic random variables. Owing to the asymptotic nature of the approximation, the FORM tends to be more accurate for larger reliability indices, i.e., smaller failure probabilities. Unfortunately, it is not possible to put a useful bound on the error inherent in the FORM approximation. However, experience has shown that, for the vast majority of structural engineering problems, the FORM provides an adequately accurate approximation.

The computational demand in the FORM is mainly in finding the design point, which requires repeated calculations of the limit-state function and its gradient with respect to the basic random variables. In most problems, no more than 10 to 20 iterations are needed to achieve convergence. Importantly, the number of iterations to converge is not affected by the dimension of the problem or the magnitude of the failure probability.

An important advantage of the FORM is that it provides measures of the relative importance of random variables as well as sensitivities with respect to parameters in the limit-state function or in the probability distribution of the basic random variables. Furthermore, the FORM provides the outcome of the random variables that are most likely to give rise to the failure event. These results provide rich insight into the nature of the reliability problem and offer opportunities to adjust the parameters or gather more data in order to enhance the reliability measure.

PROBLEMS

6.1 Consider the limit-state function

$$g(X_1, X_2) = X_1^{\theta} - X_2,$$

where $\theta = 0.5$. The marginal distribution of X_1 is lognormal with mean $\mu_1 = 1,200$ and standard deviation $\sigma_1 = 300$ and that of X_2 is gamma with mean $\mu_2 = 12$ and standard deviation $\sigma_2 = 4.8$.

1) Carry out FORM analysis and compute $\mathbf{u}^*$, $\mathbf{x}^*$, $\widehat{\boldsymbol{\alpha}}$, β, and p_{F1} for the following cases:
   (a) X_1 and X_2 are statistically independent.
   (b) X_1 and X_2 have the Nataf distribution with correlation coefficient $\rho_{12} = 0.5$.
   (c) X_1 and X_2 have the Morgenstern distribution with the above marginals and $\alpha_{12} = 1$.
2) For case (1b), plot the limit-state surfaces in the neighborhood of the design point in the original and standard normal spaces and the contours of the probability densities that pass through the design point in each space. Is $\mathbf{x}^*$ the most likely failure point in the original space?

6.2 The FORM solution of Problem 6.1 case (1b) yields $\beta = 3.81$, $\widehat{\boldsymbol{\alpha}} = [0.070\ 0.998]$, $\mathbf{x}^* = [1{,}24335.263]^{\mathrm{T}}$, and $\|\nabla G(\mathbf{u}^*)\| = 8.247$. Carry out the following analysis:

1) Determine the importance vector $\widehat{\boldsymbol{\gamma}}$ of the basic random variables. What are the orders of importance of the random variables in the standard normal space and in the original space? What is the nature (demand or capacity) of each random variable in the two spaces?
2) Compute the sensitivities of β and p_{F1} with respect to the limit-state parameter θ and distribution parameters λ and ζ of the lognormal random variable X_1 and parameters k and v of the gamma distribution of X_2. In computing the derivatives, neglect the dependence of the Nataf correlation coefficient $\rho_{0,12}$ on the parameters of the two random variables.
3) Determine the sensitivities of β with respect to the mean and standard deviation of each random variable. Also determine the sensitivities of β with respect to the mean and coefficient of variation of each random variable. Are the pairs of sensitivities with respect to each mean value different? Why?
4) Compute the importance vectors $\boldsymbol{\delta}$ and $\boldsymbol{\eta}$ (scaled sensitivities with respect to the means and standard deviations) of the two random variables. What is the order of importance of the two random variables in terms of scaled variations in their means and standard deviations?
5) Suppose the standard deviation of X_1 is increased to 310 and the mean of X_2 is increased to 12.5. Make a quick estimate of the reliability index for the revised parameter values.

6.3 Consider Problem 5.2 and assume P and M are jointly lognormal and Y has the gamma distribution. Using the iHL-RF algorithm, determine the design point, the reliability index, the unit vector $\widehat{\boldsymbol{\alpha}}$, the first-order approximation of the failure probability, and the mapping of the design point into the original space. If you have solved Problem 5.2, you may wish to start your iterations from the solution point of the FOSM analysis to speed up the convergence. It would be instructive to compare your solution with that of Problem 5.2. Note that the FOSM solution is numerically identical to the FORM solution for the case where the random variables are normally distributed. So, any difference in the results is due to the non-normal distribution of the random variables.

6.4 The buckling strength of a rectangular plate is given by

$$\sigma_c = \frac{\pi^2 EK}{12(1-\nu^2)}\left(\frac{t}{L}\right)^2,$$

where σ_c is the stress at which the plate buckles, $t = 1$ in is the plate thickness, $L = 100$ in is the span length, E is the Young's modulus, ν is the Poisson's ratio, and K is a dimensionless factor depending on the end conditions of the plate. E, ν, and K are uncertain and are modeled as random variables with the following distributions: E and ν have a joint Nataf distribution where the marginal distribution of E is normal with mean 30,000 ksi and standard deviation 3,000 ksi, the marginal distribution of ν is uniform between 0.25 and 0.35, and the correlation coefficient between E and ν is -0.30; K, independent of E and ν, has the lognormal distribution with mean 5 and a 15% c.o.v.

1) Formulate an appropriate limit-state function to determine the probability that the plate will buckle under an applied stress $s = 8$ ksi.
2) Formulate a transformation of the random variables into the standard normal space.
3) Starting from point $(E,\nu,K)_1 = (26{,}000, 0.290, 4.0)$, perform a full cycle of the iHL-RF algorithm for finding the design point. Your result should be the coordinates of a new iteration point $(E,\nu,K)_2$ that is closer to the solution point than the first trial point (the merit function value must be smaller at the new point).
4) The design point for the above problem is found to be $(E,\nu,K)^* = (24{,}930, 0.3062, 3.536)$. Determine the reliability index, β, the first-order approximation of the failure probability, p_{F1}, and the unit normal vector $\widehat{\boldsymbol{\alpha}}$.
5) Determine the importance vector $\widehat{\boldsymbol{\gamma}}$. What is the order of importance of the random variables?
6) Determine the sensitivity of p_{F1} with respect to the thickness, t, of the plate.
7) Determine the sensitivities of p_{F1} with respect to the mean and standard deviation of E, the lower and upper bounds of ν, and the mean and c.o.v. of K.
8) Without carrying out another reliability analysis, approximately estimate the reliability index if the applied stress is increased to $s = 8.5$ ksi.

6.5 An oscillator having frequency Ω_0 and damping ratio Z is subjected to a harmonic base excitation of unit amplitude and frequency Ω. All three quantities Ω_0, Z, and Ω are uncertain and are modeled as statistically independent random variables with the distributions, means and coefficients of variation listed below:

Variable	Mean	c.o.v.	Distribution
Ω_0 rad/s	6	15%	Normal
Z	0.2	50%	Lognormal
Ω rad/s	3	20%	Normal

It is well known that the amplitude of the steady-state harmonic response of the oscillator is given by $\left[\left(\Omega_0^2 - \Omega^2\right)^2 + 4Z^2\Omega_0^2\Omega^2\right]^{-1/2}$. The oscillator will fail if the response amplitude exceeds the threshold $\theta = 0.3$. The limit-state function for this problem is formulated as

$$g(\mathbf{X}, \theta) = \left(\Omega_0^2 - \Omega^2\right)^2 + 4Z^2\Omega_0^2\Omega^2 - \frac{1}{\theta^2},$$

where $\mathbf{X} = (\Omega_0, Z, \Omega)$ is the set of basic random variables.

1) Use the iHL-RF algorithm to find the design point. Determine $\mathbf{u}^*$, $\mathbf{x}^*$, $\widehat{\boldsymbol{\alpha}}$, β, and p_{F1}.
2) Determine the order of importance of the random variables and their nature, i.e., capacity or demand.
3) Determine the order of importance of the random variables with respect to statistically equivalent variations in their means and standard deviations.
4) Estimate the change in the failure probability if the safe threshold is reduced to $\theta = 0.28$.

6.6 Tension force T is applied to the end of a string to hold weight W in a pulley system with two cylinders. One can show that the load W slides down if $g(W, M_1, M_2, T) \leq 0$, where

$$g(W, M_1, M_2, T) = T - W \exp\left[-\frac{\pi}{4}(M_1 + M_2)\right],$$

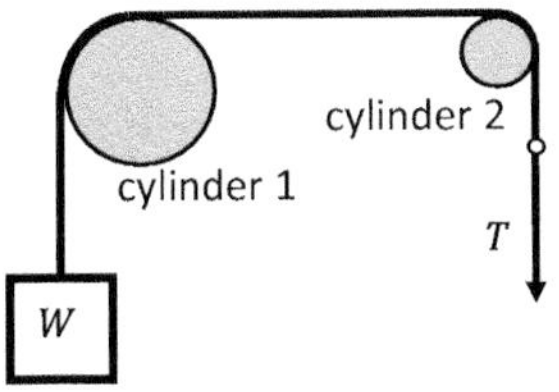

in which M_1 and M_2 are the friction coefficients between the string and the cylinders. Suppose W is an exponentially distributed random variable with mean $\mu_W = 5$ kN, and M_1 and M_2 are jointly lognormal random variables with common means $\mu_1 = \mu_2 = 0.4$, common standard deviations $\sigma_1 = \sigma_2 = 0.2$, and the correlation coefficient $\rho = 0.4$. $T = 10$ kN is deterministic. Carry out the following analysis:

1) Starting from the initial point $(w, m_1, m_2)_1 = (10, 0.2, 0.2)$, perform a full cycle of the iHL-RF algorithm for finding the design point. Your result should be the coordinates of a new iteration point $(w, m_1, m_2)_2$, which is closer to the design point than the first trial point (the merit function value must be smaller at the new point).

2) The design point for this problem is found to be at $(w, m_1, m_2)^* = (15.9, 0.296, 0.296)$. Determine $\mathbf{u}^*$, $\widehat{\boldsymbol{\alpha}}$, β, and p_{F1}.
3) Determine the order of importance of the random variables.
4) Determine the sensitivities of β with respect to the mean of W and the mean and c.o.v. of M_1.
5) Suppose we wish to reduce the probability of failure by half. By approximately how much should T be increased? Do not carry out additional reliability analysis.

6.7 The limit-state function describing the loss of stability of a soil slope is given by

$$g(\gamma_s, \tan\phi, C, \gamma_w, d, h, \theta) = \left[(\gamma_s d - \gamma_w h)\cos\theta \tan\phi + \frac{C}{\cos\theta}\right] - \gamma_s d \sin\theta$$

where

γ_s	weight density of soil, a random variable,
$\tan\phi$	tangent of soil friction angle, a random variable,
C	soil cohesion, a random variable,
$\gamma_w = 62$	lbs/ft^3 weight density of water,
$d = 30$ ft	depth
$h = 10$ ft	height of water table,
$\theta = 0.3$ rad	angle of inclination of the slope.

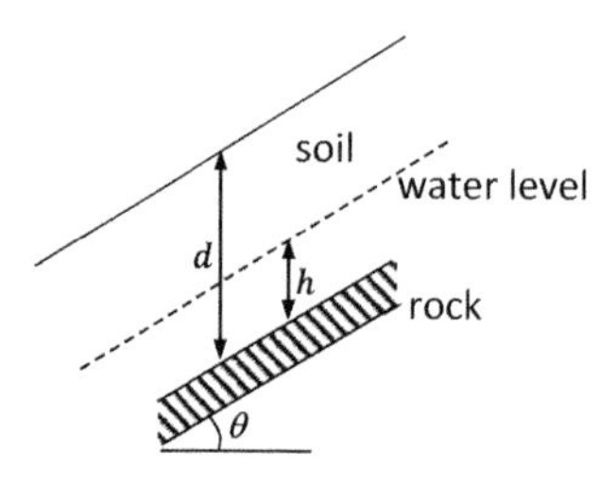

The three random variables have the following distributions: γ_s is normal with mean $\mu_{\gamma_s} = 120$ lbs/ft^3 and c.o.v. $\delta_{\gamma_s} = 0.15$; $\tan\phi$ and C are jointly lognormal with means $\mu_{\tan\phi} = 0.4$ and $\mu_C = 300$ lbs/ft^2, c.o.v. $\delta_{\tan\phi} = 0.15$ and $\delta_C = 0.50$, and correlation coefficient $\rho_{\tan\phi, C} = -0.50$. γ_s is independent of $\tan\phi$ and C.

1) Construct a transformation of the random variables into the standard normal space.
2) Starting from the point $(\gamma_s, \tan\phi, C)_1 = (130, 0.3, 250)$, perform a full cycle of the iHL-RF algorithm for finding the design point. Your result should be the coordinates of a new iteration point $(\gamma_s, \tan\phi, C)_2$, which is closer to the solution of the design point than the first trial point (the merit function value must be smaller at the new point).
3) The design point for the above problem is found to be $(\gamma_s, \tan\phi, C)^* = (121.8, 0.3011, 197.8)$. Determine $\mathbf{u}^*$, $\widehat{\boldsymbol{\alpha}}$, β, and p_{F1}.

4) Determine the importance vector $\widehat{\boldsymbol{\gamma}}$. What are the order of importance and the nature (capacity or demand) of the random variables?
5) Determine the sensitivities of p_{F1} with respect to d and h and the mean and standard deviation of $\tan\phi$.
6) Suppose it is desired to reduce the probability of failure by 20%. What course(s) of action would you recommend? Be specific and give quantitative measures.

6.8 A spacecraft is subject to hypervelocity impact of debris objects in the space. Based on high-velocity ballistic impact tests in the laboratory, a model has been developed to predict the depth of penetration of a projectile into a target. According to this model, the depth of penetration, D, is given in units of mm by

$$D = \theta_1 R \left(\frac{\gamma_P^2 V^2}{\gamma_T y} \right)^{\theta_2}$$

where $\theta_1 = 1.16$ and $\theta_2 = 0.33$ are model parameters, R the radius of the projectile in mm, V the velocity of the projectile in km/s, γ_P the mass density of the projectile in g/cm^3, γ_T the mass density of the target in g/cm^3, and y the yield strength of the target in MPa. The outer shield of the spacecraft has the thickness $d_0 = 20$ mm and is made of a high-strength alloy having mass density $\gamma_T = 3$ g/cm^3 and yield strength $y = 200$ MPa. The projectile characteristics $\mathbf{X} = (R, V, \gamma_P)$ are random variables having a Nataf joint distribution with the properties listed in the following table. We are interested in the probability that the projectile will puncture a hole in the outer shield of the spacecraft.

				Correlation coefficient		
Variable, units	Marginal distribution	Mean	c.o.v.	R	V	γ_P
R, mm	Exponential	2.5	1.00	1	0	−0.3
V, km/s	Lognormal	6	0.40	0	1	0
γ_P, g/cm^3	Lognormal	4	0.50	−0.3	0	1

1) Starting from the point $\mathbf{x}_1 = (5, 10, 6)$, perform a full cycle of the iHL-RF algorithm for finding the design point. The result of your analysis should be the new trial point $\mathbf{x}_2$. Verify that the new point is closer to the design point.
2) Suppose the algorithm converges to $\mathbf{x}^* = (11.9, 9.34, 4.57)$. Obtain the reliability index, the first-order approximation of the failure probability, and the unit vector $\widehat{\boldsymbol{\alpha}}$.
3) Determine the order of importance of the three random variables.
4) Determine the sensitivity of the reliability index and the first-order failure probability with respect to the mean of R and the mean and c.o.v. of V.
5) Without performing a complete reliability analysis, estimate the reliability index if the thickness of the shield, d_0, is increased by 1 mm.

6.9 In Problem 6.8, determine the required thickness d_0 of the outer shield of the spacecraft for a reliability index $\beta = 3.5$. Use the algorithm described in Section 6.10.

6.10 A structural reliability problem is defined by the limit-state function in the standard normal space

$$G(U_1, U_2) = 20 + 7U_1 - 2U_2 - U_1^2 - 3U_2^2 - 3.5U_1U_2.$$

Determine whether this problem has multiple design points and, if so, find them and determine the corresponding $\widehat{\boldsymbol{\alpha}}$ vectors and reliability indices.

7 The Second-Order Reliability Method

7.1 Introduction

This chapter develops the second-order reliability method (SORM). The method essentially fits a second-order approximation to the limit-state surface at the design point in the standard normal space, thus providing an improvement over the first-order reliability method (FORM) approximation. The classical SORM formulation requires twice-differentiability of the limit-state function $g = g(\mathbf{x})$ at the design point. However, an alternative SORM method is available that avoids computing the second derivatives. The method assumes that the joint distribution of the basic random variables, $f_X(\mathbf{x})$, is known and transformable to the standard normal space. It also assumes the design point is available, which implies the requirement that the limit-state function be continuous and continuously differentiable everywhere. In developing the SORM approximation, we will make use of the properties of the standard normal space described in Section 6.2.

7.2 Classical Formulation of the Second-Order Reliability Method

Following the pioneering work of Fiessler et al. (1979), we start by developing a second-order Taylor series approximation of the limit-state function at the design point $\mathbf{u}^*$ in the standard normal space:

$$\begin{aligned} G(\mathbf{u}) &\cong \nabla G(\mathbf{u}^*)(\mathbf{u}-\mathbf{u}^*) + \frac{1}{2}(\mathbf{u}-\mathbf{u}^*)^{\mathrm{T}}\mathbf{H}(\mathbf{u}-\mathbf{u}^*) \\ &= \|\nabla G(\mathbf{u}^*)\| \left[\beta - \widehat{\boldsymbol{\alpha}}\mathbf{u} + \frac{1}{2\|\nabla G(\mathbf{u}^*)\|}(\mathbf{u}-\mathbf{u}^*)^{\mathrm{T}}\mathbf{H}(\mathbf{u}-\mathbf{u}^*)\right]. \end{aligned} \tag{7.1}$$

In the above, $\mathbf{H}$ is the Hessian (second-derivative matrix) of $G(\mathbf{u})$ at the design point, having the elements $H_{ij} = \partial^2 G(\mathbf{u}^*)/(\partial u_i \partial u_j)$, $i,j = 1, 2, \ldots, n$, and we have used the relations $\widehat{\boldsymbol{\alpha}} = -\nabla G(\mathbf{u}^*)/\|\nabla G(\mathbf{u}^*)\|$ and $\beta = \widehat{\boldsymbol{\alpha}}\mathbf{u}^*$. Now consider a rotation of the axes $\mathbf{u}' = \mathbf{P}\mathbf{u}$, where $\mathbf{P}$ is an orthonormal matrix (rows are unit vectors orthogonal to each other) with $\widehat{\boldsymbol{\alpha}}$ as its last row. Such a matrix can be constructed by, for example, the well-known Gram–Schmidt algorithm (Cheney and Kincaid, 2009). This rotation positions the design point on the u'_n axis such that $\mathbf{u}'^* = [0 \ldots 0\,\beta]^{\mathrm{T}}$. Because $\mathbf{u} = \mathbf{P}^{\mathrm{T}}\mathbf{u}'$, defining $G'(\mathbf{u}') = G(\mathbf{P}^{\mathrm{T}}\mathbf{u}')/\|\nabla G(\mathbf{u}^*)\|$, one can write

$$G'(\mathbf{u}') \cong \beta - \widehat{\boldsymbol{\alpha}}\mathbf{P}^{\mathrm{T}}\mathbf{u}' + \frac{1}{2\|\nabla G(\mathbf{u}^*)\|}(\mathbf{u}' - \mathbf{u}'^*)^{\mathrm{T}}\mathbf{PHP}^{\mathrm{T}}(\mathbf{u}' - \mathbf{u}'^*). \tag{7.2}$$

Noting that $\widehat{\boldsymbol{\alpha}}\mathbf{P}^{\mathrm{T}}\mathbf{u}' = u'_n$ and defining the symmetric matrix $\mathbf{A} = \mathbf{PHP}^{\mathrm{T}}/\|\nabla G(\mathbf{u}^*)\|$, (7.2) takes the form

$$G'(\mathbf{u}') \cong \beta - u'_n + \frac{1}{2}(\mathbf{u}' - \mathbf{u}'^*)^{\mathrm{T}}\mathbf{A}(\mathbf{u}' - \mathbf{u}'^*). \tag{7.3}$$

Now, consider the partitionings

$$\mathbf{u}' = \left[\mathbf{u}_1'^{\mathrm{T}}\ u'_n\right]^{\mathrm{T}} \text{ and } \mathbf{A} = \begin{bmatrix} \mathbf{A}_{11} & \mathbf{A}_{12} \\ \mathbf{A}_{12}^{\mathrm{T}} & a_{nn} \end{bmatrix}, \tag{7.4}$$

where $\mathbf{u}'_1 = \left[u'_1\ u'_2 \ldots u'_{n-1}\right]^{\mathrm{T}}$, $\mathbf{A}_{11}$ is an $(n-1) \times (n-1)$ matrix formed by the first $n-1$ rows and columns of $\mathbf{A}$, and $\mathbf{A}_{12}$ is a column vector formed by the first $n-1$ elements of the nth column of $\mathbf{A}$. Expanding the matrix product in (7.3), one obtains

$$G'(\mathbf{u}') \cong \beta - u'_n + \frac{1}{2}\left[\mathbf{u}_1'^{\mathrm{T}}\mathbf{A}_{11}\mathbf{u}'_1 + 2(u'_n - \beta)\mathbf{A}_{12}^{\mathrm{T}}\mathbf{u}'_1 + a_{nn}(u'_n - \beta)^2\right]. \tag{7.5}$$

Because $u'_n - \beta = 0$ is the tangent plane at the design point in the $\mathbf{u}'$ space, for points on the limit-state surface in the neighborhood of the design point, the last two terms involving $(u'_n - \beta)$ are of smaller order than the first term inside the square brackets. Dropping these terms, we obtain the approximation

$$G'(\mathbf{u}') \cong \beta - u'_n + \frac{1}{2}\mathbf{u}_1'^{\mathrm{T}}\mathbf{A}_{11}\mathbf{u}'_1. \tag{7.6}$$

This is the equation of a paraboloid with its vertex at the design point, $\mathbf{u}'^* = [0 \ldots 0\,\beta]^{\mathrm{T}}$.

Now consider a rotation of the axes around u'_n achieved by the transformation $\mathbf{u}''_1 = \mathbf{Q}\mathbf{u}'_1$, where $\mathbf{Q}$ is an $(n-1) \times (n-1)$ orthonormal matrix. As $\mathbf{u}'_1 = \mathbf{Q}^{\mathrm{T}}\mathbf{u}''_1$, we can write $\mathbf{u}_1'^{\mathrm{T}}\mathbf{A}_{11}\mathbf{u}'_1 = \mathbf{u}_1''^{\mathrm{T}}\mathbf{Q}\mathbf{A}_{11}\mathbf{Q}^{\mathrm{T}}\mathbf{u}''_1$. It follows that by selecting $\mathbf{Q}^{\mathrm{T}}$ as the eigenmatrix of $\mathbf{A}_{11}$, the matrix product $\mathbf{Q}\mathbf{A}_{11}\mathbf{Q}^{\mathrm{T}}$ becomes diagonal so that (7.6) reduces to

$$G'(\mathbf{Q}^{\mathrm{T}}\mathbf{u}''_1, u_n) = G''(\mathbf{u}''_1, u_n) \cong \beta - u'_n + \frac{1}{2}\sum_{i=1}^{n-1}\kappa_i u_i''2, \tag{7.7}$$

where κ_i are the eigenvalues of $\mathbf{A}_{11}$. When set equal to 0, the above expression defines a paraboloid through its principal axes $(u''_1, \ldots, u''_{n-1}, u'_n)$ with κ_i, $i = 1, \ldots, n-1$, denoting its principal curvatures at the vertex. The paraboloid is tangent to the limit-state surface at the design point and its principal curvatures match those of the limit-state surface at the design point. Note that for $0 < \beta$, a positive curvature signifies a surface that curves away from the origin and a negative curvature signifies one that curves toward the origin. The vertex is at the design point, which is the point on the limit-state surface nearest the origin, which means that the inequality $-1 < \beta\kappa_i$ must hold for each principal curvature.

Observe that the failure domain defined by the limit-state function in (7.7) is identical to the parabolic domain defined in (6.5) so that the SORM approximation of the failure probability is given by the integral along the imaginary axis (Tvedt, 1990)

$$p_{F2} = \phi(\beta)\mathrm{Re}\left\{ \mathrm{i}\sqrt{\frac{2}{\pi}} \int_0^{\mathrm{i}\infty} \frac{1}{s} \exp\left[\frac{(s+\beta)^2}{2}\right] \prod_{i=1}^{n-1} \frac{1}{\sqrt{1+\kappa_i s}} \mathrm{d}s \right\}. \tag{7.8}$$

As described in Section 6.2, a simpler approximation is (Hohenbichler and Rackwitz, 1988)

$$p_{F2} \cong \Phi(-\beta) \prod_{i=1}^{n-1} \frac{1}{\sqrt{1+\psi(\beta)\kappa_i}}, \tag{7.9}$$

where $\psi(\beta) = \phi(\beta)/\Phi(-\beta)$. An even simpler approximation is (Breitung, 1984)

$$p_{F2} \cong \Phi(-\beta) \prod_{i=1}^{n-1} \frac{1}{\sqrt{1+\beta\kappa_i}}. \tag{7.10}$$

The above approximations are valid as long as $-1 < \beta\kappa_i$, which, as mentioned earlier, must be the case for a paraboloid fitted to the principal curvatures of the limit-state surface at the design point. It can be seen that the SORM approximation is given in terms of the reliability index β and the principal curvatures κ_i, $i = 1, \ldots, n-1$, of the limit-state surface at the design point in the standard normal space. In contrast to an alternative SORM approximation described later in this chapter, this classical SORM approximation is called the *curvature-fitting* SORM.

It is worth mentioning that pairs of curvatures of opposite sign do not cancel out their corrective effects on the FORM approximation. This is due to the fact that the probability content between the paraboloid and the tangent plane is bigger when the surface curves toward the origin than when it curves with the same curvature away from the origin. This is reflected in the above approximations. For example, let $\kappa_1 = -\kappa_2 = \kappa$. The product of the two terms in (7.10) then gives $1/(\sqrt{1+\beta\kappa}\sqrt{1-\beta\kappa}) = 1/\sqrt{1-(\beta\kappa)^2}$, which is greater than unity for any $\beta\kappa \neq 0$.

In summary, the curvature-fitting SORM requires the following calculations:

- Compute the Hessian $\mathbf{H}$ of the limit-state function at the design point. This is done by finite-difference calculations in the standard normal space, which requires computing $G(\mathbf{u})$ for a set of closely spaced points $\mathbf{u}_i$ around the design point. As the limit-state function can only be computed in the original space, this requires transforming the selected points $\mathbf{u}_i$ into the original space and computing the corresponding values of the limit-state function as $G(\mathbf{u}_i) = g(\mathbf{x}_i)$.
- Construct the orthonormal matrix $\mathbf{P}$ with $\widehat{\boldsymbol{\alpha}}$ as its last row. As mentioned earlier, the Gram–Schmidt algorithm (Cheney and Kincaid, 2009) can be used for this purpose.

- Compute the matrix $\mathbf{A} = \mathbf{PHP}^{\mathrm{T}}/\|\nabla G(\mathbf{u}^*)\|$ and discard its nth row and column to obtain $\mathbf{A}_{11}$.
- Compute the eigenvalues of $\mathbf{A}_{11}$. These are the principal curvatures κ_i, $i = 1, \ldots, n-1$, of the limit-state surface at the design point in the standard normal space.
- Use one of the expressions (7.8)–(7.10) to compute the SORM approximation of the failure probability.
- If a reliability index is required, compute the generalized reliability index $\beta_g = -\Phi^{-1}(p_{F2})$.

The most difficult part of the above analysis is computing the Hessian. This is particularly difficult when the limit-state function is implicit and involves numerical algorithms such as numerical integration or finite-element analysis. One faces two difficulties in such a problem: the cost of computing the full Hessian when the number of random variables is large, and the noise in the limit-state function arising from truncation errors inherent in numerical algorithms, such as in finite-element calculations. Later in this chapter, alternative SORM formulations that avoid computing the second derivatives are introduced.

Example 7.1 – Curvature-Fitting SORM Approximation with Non-Normal Dependent Random Variables

Consider our toy problem with the limit-state function $g(X_1, X_2) = X_1^2 - 2X_2$. Assume the distribution of the random variables is as described in Example 6.4, where X_1 and X_2 are statistically independent with lognormal and Type I largest value distributions, respectively, with the mean of X_2 being uncertain and represented by a lognormal random variable X_3. Analysis with CalREL using the above approach leads to $\kappa_1 = -0.0719$ and $\kappa_2 = 0.0160$. With $\beta = 1.94$ obtained in Example 6.4, (7.8) and (7.9) give $p_{F2} = 0.0279$ and (7.10) gives $p_{F2} \cong 0.0284$, while the FORM approximation obtained in Example 6.4 is $p_{F1} = 0.0260$. The SORM approximations favorably compare with the "exact" solution $p_F \cong 0.0277$ obtained by Monte Carlo simulations (with a coefficient of variation [c.o.v.] of 2.6%). The generalized reliability index based on the SORM approximation in (7.8) is $\beta_g = -\Phi^{-1}(0.0279) = 1.91$.

Example 7.2 – Reliability of a Column: Curvature-Fitting SORM Approximation

For the column reliability problem in Example 6.5, SORM analysis leads to $\kappa_1 = -0.155$, $\kappa_2 = -0.0399$, and $\kappa_3 \cong 0$. With these values and $\beta = 2.47$ from Example 6.5, (7.8) gives $p_{F2} = 0.00936$, (7.9) gives $p_{F2} \cong 0.00960$, and (7.10) gives $p_{F2} \cong 0.00914$. All three are close to the "exact" solution $p_F \cong 0.00931$ obtained from Monte Carlo simulations (with a c.o.v. of 3%). Recall that the FORM solution of this problem yielded $p_{F1} = 0.00682$. It is evident that the limit-state surface is curved toward the origin (the principal curvatures are negative) and the SORM provides a significant improvement. The generalized reliability index based on the

Example 7.2 (cont.)

SORM approximation for this problem is $\beta_g = -\Phi^{-1}(0.00936) = 2.35$. We see that the generalized reliability index is smaller than the FORM approximation on account of the inward curvature of the limit-state surface.

Example 7.3 – Reliability of a 10-Story Building with a Tuned-Mass Damper

For the problem described in Example 6.11, SORM analysis at the two design points yields $p_{F2} = 0.161$ and $p_{F2} = 0.058$. Systems analysis (see Chapter 8) yields $p_{F2} = 0.218$. This result is much closer to the "exact" solution $p_F \cong 0.213$ obtained by Monte Carlo simulations than the FORM solution $p_{F1} = 0.188$.

7.3 Gradient-Based SORM

Der Kiureghian and De Stefano (1991) showed that certain algorithms for finding the design point, including the iHL-RF algorithm, have the property that the trajectory of trial points $\mathbf{u}_i$, $i = 1, 2, \ldots$, asymptotically converges on the major principal axes of the limit-state surface at the design point, i.e., the axis u_i'', having the maximum $|\kappa_i|$ value, as the sequence converges to the design point. Using this property, the major principal curvature, $\max_{i=1,\ldots,n-1}|\kappa_i|$, can be approximately computed in terms of the quantities that are available in the last two iteration points of the algorithm as the sequence of trial points converges. Suppose the convergence to the design point is achieved after r iterations, with $\mathbf{u}_{r-1}$ and $\mathbf{u}_r$ denoting the last two trial points and $\widehat{\boldsymbol{\alpha}}_{r-1}$ and $\widehat{\boldsymbol{\alpha}}_r$ denoting the corresponding unit normal vectors (see Figure 7.1). In the plane of the two unit vectors, the curvature of the limit-state surface is given by the inverse of the radius of curvature measured from

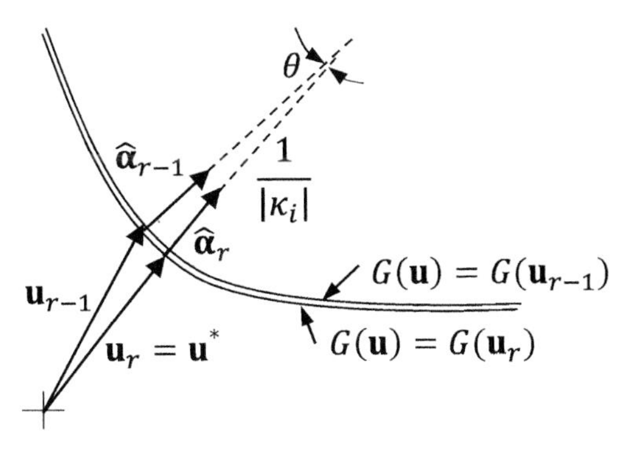

Figure 7.1 Determination of curvature using the last two iteration points

the intersection point of the two normal vectors. The small angle between the two vectors is $\theta = \cos^{-1}\left(\widehat{\boldsymbol{\alpha}}_{r-1}\widehat{\boldsymbol{\alpha}}_r^{\mathrm{T}}\right)$. The small arc between the two trial points is defined by the vector $\mathbf{u}_r - \mathbf{u}_{r-1}$. Its projection in the direction perpendicular to $\widehat{\boldsymbol{\alpha}}_r$ has the length $\sqrt{\|\mathbf{u}_r - \mathbf{u}_{r-1}\|^2 - [\widehat{\boldsymbol{\alpha}}_r(\mathbf{u}_r - \mathbf{u}_{r-1})]^2}$. As the tangent of the small angle θ is equal to the angle measured in radians, we have

$$|\kappa_i| \cong \frac{\cos^{-1}\left(\widehat{\boldsymbol{\alpha}}_{r-1}\widehat{\boldsymbol{\alpha}}_r^{\mathrm{T}}\right)}{\sqrt{\|\mathbf{u}_r - \mathbf{u}_{r-1}\|^2 - [\widehat{\boldsymbol{\alpha}}_r(\mathbf{u}_r - \mathbf{u}_{r-1})]^2}}. \tag{7.11}$$

Adopting the convention that κ_i is positive when the surface curves away from the origin and negative when it curves toward the origin, the geometry in Figure 7.1 shows that the sign of the curvature can be determined by the sign of the scalar $-\widehat{\boldsymbol{\alpha}}_r(\mathbf{u}_r - \mathbf{u}_{r-1})$. Hence,

$$\kappa_i \cong -\frac{\mathrm{sgn}[\widehat{\boldsymbol{\alpha}}_r(\mathbf{u}_r - \mathbf{u}_{r-1})]\cos^{-1}\left(\widehat{\boldsymbol{\alpha}}_{r-1}\widehat{\boldsymbol{\alpha}}_r^{\mathrm{T}}\right)}{\sqrt{\|\mathbf{u}_r - \mathbf{u}_{r-1}\|^2 - [\widehat{\boldsymbol{\alpha}}_r(\mathbf{u}_r - \mathbf{u}_{r-1})]^2}}, \tag{7.12}$$

where $\mathrm{sgn}[\cdot]$ denotes the algebraic sign of the quantity inside the square brackets.

To obtain the second major principal curvature, one repeats the search process from a randomly selected initial point in a subspace orthogonal to the identified major principal axis that lies in the plane of $\mathbf{u}_{r-1}$ and $\mathbf{u}_r$. The result, using (7.12), approximates the principal curvature having the second largest absolute value. Next, the search is repeated in a subspace orthogonal to the first two identified principal axes. The result is the third major principal curvature. This process is continued until all principal curvatures of significant magnitude have been obtained. Owing to the asymptotic nature of the approximation, the method produces more accurate results when curvatures are large. This is fortuitous because large curvatures are the important ones in the SORM approximation.

The advantages of the *gradient-based* SORM approach are twofold: One does not need to compute second derivatives, and one computes the principal curvatures in order of their importance and can stop the process when the curvature magnitudes are sufficiently small. The latter property is particularly important in problems with a large number of random variables, where often only a few principal curvatures are significant, i.e., the limit-state surface is significantly curved in only a few directions. The reader should consult Der Kiureghian and De Stefano (1991) for further details on the implementation of this method. This method is available in the CalREL software.

Example 7.4 – Gradient-Based SORM Approximation with Non-Normal Dependent Random Variables

The toy problem in Example 7.1 is solved using the gradient-based SORM approach. The principal curvatures found are $\kappa_1 = -0.0746$ and $\kappa_2 = 0.0287$. Comparing with the exact

Example 7.4 (cont.)

values $\kappa_1 = 0.0719$ and $\kappa_2 = 0.0159$ obtained by the classical approach, i.e., using the Hessian, it is evident that the gradient-based method is more accurate for the larger principal curvature. With $\beta = 1.94$ obtained in Example 6.4, (7.8) gives $p_{F2} = 0.0277$ compared with $p_{F2} = 0.0279$ of the classical approach. These estimates of the failure probability favorably compare with the "exact" solution $p_F \cong 0.0277$ obtained by Monte Carlo simulations (with a c.o.v. of 2.6%).

Example 7.5 – Reliability of a Structure with P-Δ Effect

Consider the six-story, two-bay frame building in Figure 7.2, which is subject to lateral and vertical loads $P_1, \ldots, P_9$ (Der Kiureghian and De Stefano, 1991). The structure's material behaves linearly, but we count for the geometric nonlinearity arising from the P-Δ effect, i.e., the effect of additional moments caused by the vertical loads and lateral displacements of the structure. All nodal loads and member properties, i.e., Young's modulus E, cross-sectional areas A, and moments of I, are considered as independent random variables with the distributions, means, and c.o.v. listed in Table 7.1. The problem, thus, is defined by 99 random variables. We wish to compute the probability that the structure drift exceeds the threshold $D_0 = 5$ in. The limit-state function is formulated as

$$g(\mathbf{X}) = D_0 - D(\mathbf{X}), \tag{E1}$$

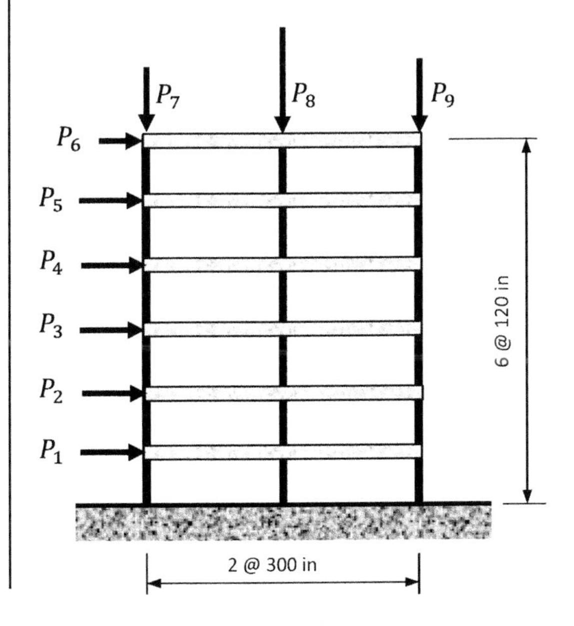

Figure 7.2 Six-story, two-bay example frame structure (after Der Kiureghian and De Stefano, 1991, with permission from ASCE)

Example 7.5 (cont.)

Table 7.1 Description of random variables (after Der Kiureghian and De Stefano, 1991, with permission from ASCE)

Variable	Distribution	Mean	c.o.v.
$P_1 - P_5$, kips	Type II largest	60	0.1
P_6, kips	Type II largest	30	0.1
P_7, P_9, kips	Type II largest	100	0.1
P_8, kips	Type II largest	200	0.1
E, side columns, ksi	Lognormal	28,000	0.1
E, center columns, ksi	Lognormal	30,000	0.1
E, beams, ksi	Lognormal	26,000	0.1
A, side columns, in^2	Lognormal	21.0	0.1
A, center columns, in^2	Lognormal	26.9	0.1
A, beams, in^2	Lognormal	16.0	0.1
I, side columns, in^4	Lognormal	2,100	0.1
I, center columns, in^4	Lognormal	2,690	0.1
I, beams, in^4	Lognormal	1,600	0.1

where $\mathbf{X}$ is the set of 99 random variables and $D(\mathbf{X})$ denotes the drift as an implicit function of $\mathbf{X}$. For each selected outcome $\mathbf{X}=\mathbf{x}$, a finite-element program is used to compute the drift. Furthermore, the direct-differentiation method (DDM) (Zhang and Der Kiureghian, 1993) is used to compute the gradient vector of $D(\mathbf{x})$ with respect to $\mathbf{x}$. This approach is described in Chapter 12.

Table 7.2 summarizes the results of the SORM analyses, considering separately the cases with and without the P-Δ effect . Listed are the principal curvatures obtained by the classical SORM and the gradient-based SORM, the number of iterations of the iHL-RF algorithm to converge to each principal curvature, and the corresponding reliability indices and failure probability estimates. In the gradient-based SORM, iterations are stopped when $|\beta\kappa_i| < 0.05$.

The following observations from Table 7.2 are noteworthy: (a) The gradient-based method accurately predicts all the significant principal curvatures; (b) Even though there are 99 random variables, hence 98 principal curvatures, only 6 principal curvatures are significant; (c) The first 5 principal curvatures are negative, indicating that the limit-state surface in the standard normal space mostly curves toward the origin; (d) Because of the inward curvature of the limit-state surface, the FORM approximation underestimates the failure probability; (e) Accounting for the P-Δ effect increases the failure probability. It is also noted that, for this problem, the computational time of the gradient-based SORM is a small fraction of that of the classical SORM (see Der Kiureghian and De Stefano, 1991). This is because the classical SORM needs to compute the entire 99×99 Hessian, even though only 6 principal curvatures

Example 7.5 (cont.)

Table 7.2 Results of SORM analysis for example structure with P-Δ effect (after Der Kiureghian and De Stefano, 1991, with permission from ASCE)

	With P-Δ effect			Without P-Δ effect		
Principal curvature	Classical SORM	Gradient-based SORM	No. iterations	Classical SORM	Gradient-based SORM	No. iterations
κ_1	−0.3038	−0.3038	16	−0.2944	−0.2944	18
κ_2	−0.1640	−0.1640	13	−0.1513	−0.1513	13
κ_3	−0.0977	−0.0978	13	−0.0899	−0.0899	13
κ_4	−0.0521	−0.0521	13	−0.0476	−0.0476	13
κ_5	−0.0317	−0.0318	13	−0.0285	−0.0285	13
κ_6	0.0190	0.0187	11	0.0174	0.0171	11
β	1.97		–	2.16		–
p_{F1}	0.0244			0.0155		
p_{F2}	0.0542	0.0551	–	0.0353	0.0361	–
β_g	1.61	1.60	–	1.81	1.80	–

are of significant value. In contrast, the gradient-based SORM stops the computations when the last calculated principal curvature, the sixth in this example, is sufficiently small.

7.4 Point-Fitting SORM

As mentioned earlier, computing the curvatures of the limit-state surface when there is computational noise is problematic. Such noise may arise from truncation errors in numerical algorithms used to compute the limit-state function. A good example is nonlinear finite-element analysis, where usually an iterative solution algorithm is used with a pre-set convergence tolerance. Using a finite-difference method to compute second derivatives in such a setting can lead to unstable or erroneous results. The *point-fitting* SORM described in this section overcomes this problem. The following is a slightly improved version of a method originally described in Der Kiureghian et al. (1987).

Consider a rotation of the axes according to the transformation $\mathbf{u}' = \mathbf{Pu}$, where $\mathbf{P}$ is the orthonormal matrix defined in the paragraph following (7.1) so that the design point falls on the u'_n axis. Consider the plane formed by the axes u'_i and u'_n. The intersection of the limit-state surface with this plane is a line, which may appear as in Figure 7.3. For each axis u'_i,

$i = 1, \ldots, n - 1$, we select two points on this curve to fit an approximating limit-state surface. The points for axis u'_i are selected by searching along a path consisting of the lines $u'_i = \mp b$ and a semicircle of radius b centered at the design point, shown as a dashed line in Figure 7.3. The parameter b is selected according to the rule

$$\begin{aligned} b &= 1, & &\text{if } |\beta| < 1, \\ &= |\beta|, & &\text{if } 1 \le |\beta| \le 3, \\ &= 3, & &\text{if } 3 < |\beta|. \end{aligned} \tag{7.13}$$

Figure 7.3 Point-fitting SORM

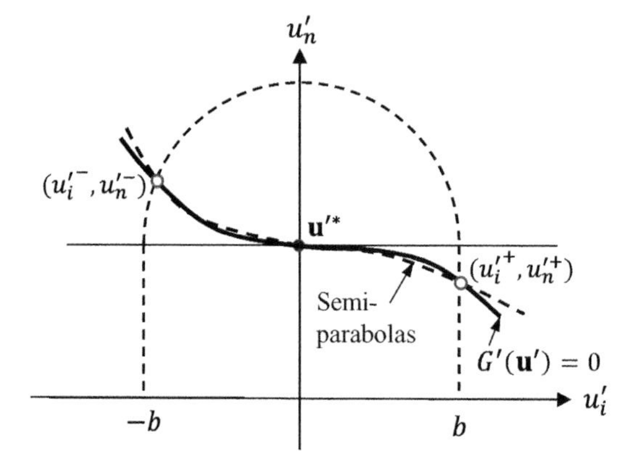

This rule assures that the fitting points are neither too close to nor too far from the design point. (It is noted that the original method developed by Der Kiureghian et al. (1987) did not include the semicircular portion of the path. This part was added later to assure that intersection points are always found, regardless of the extent of nonlinearity of the intersecting curve.) Let (u'^-_i, u'^-_n) denote the coordinates of the fitting point on the negative side of axis u'_i, and (u'^+_i, u'^+_n) denote the coordinates of the fitting point on the positive side of the same axis. Through each fitting point, a semi-parabola is defined that is tangent at the design point (see Figure 7.3). The curvatures of the two fitting semi-parabolas at the design point are given by

$$\kappa_i^{\operatorname{sgn}(u'_i)} = \frac{2\left(u_n'^{\operatorname{sgn}(u'_i)} - \beta\right)}{\left(u_i'^{\operatorname{sgn}(u'_i)}\right)^2}, \tag{7.14}$$

where $\operatorname{sgn}(u'_i)$ denotes the sign of the coordinate on the u'_i axis. The approximating limit-state function is now defined as

$$G'(\mathbf{u}') \cong \beta - u'_n + \frac{1}{2}\sum_{i=1}^{n-1} \kappa_i^{\operatorname{sgn}(u'_i)} u_i'^2. \tag{7.15}$$

When set equal to zero, the above expression defines a piecewise paraboloid surface that is tangent to the limit-state surface at the design point and is coincident with each of the fitting points. Interestingly, this function is continuous and twice differentiable in spite of the fact that the coefficients $\kappa_i^{\text{sgn}(u'_i)}$ are discontinuous. This is because this discontinuity occurs at points where u'_i takes on a zero value. The advantages of the above approximating limit-state surface are that it is insensitive to noise in the calculation of the limit-state function and it does not require calculation of derivatives. On the other hand, it requires finding $2(n-1)$ fitting points on the surface. Furthermore, as the fitting points are obtained in the $\mathbf{u}'$ space, the solution may depend on the selected rotation matrix **P**, which is not unique. Of course, one can do the fitting in the further rotated space $\mathbf{u}''$ as defined in Section 7.2, where the axes are along the principal axes of the surface at the design point. However, this would require computation of the Hessian, which defeats the purpose because we want to avoid computing the second derivatives. Der Kiureghian et al. (1987) have shown that the maximum error resulting from the worst choice of **P** is much smaller than the error in the FORM approximation. Thus, even for the worst choice of **P**, this method improves the FORM approximation.

The SORM approximation of the probability of failure in this approach is computed by adding the probability contents in each of the hyper-quadrants. Because the surface in each hyper-quadrant is a semi-paraboloid, any of (7.8)–(7.10) can be used to compute these probabilities.

The above approach accounts for higher-order nonlinearities in the limit-state surface. For example, when the design point is an inflection point of the limit-state surface, as exemplified in Figure 7.3, the classical SORM makes no improvement to the FORM approximation because the curvature at such a point is zero. On the other hand, the point-fitting SORM provides an improvement by accounting for the difference in the probability contents of the two semi-parabolas.

Example 7.6 – Point-Fitting SORM Approximation with Non-Normal Dependent Random Variables

The toy problem in Example 7.1 is solved using the point-fitting SORM approach. The fitting points along axis u'_1 are $(-1.94, 1.89)$ and $(1.94, 1.88)$ and those along axis u'_2 are $(-1.94, 1.90)$ and $(1.94, 1.91)$. Recall that the reliability index for this problem is $\beta = 1.94$. It follows that all four points are below the tangent plane, but more so along the u'_1 axis than along u'_2. The corresponding curvatures of the semi-parabolas are $\kappa_1^- = -0.0314$ and $\kappa_1^+ = -0.0343$ for u'_1 and $\kappa_2^- = -0.0213$ and $\kappa_1^+ = -0.0170$ for u'_2. Note that these curvatures are between the principal curvatures found in Example 7.1. This is because the axes $\mathbf{u}'$ used in the point-fitting method are not aligned with the principal directions of the limit-state surface. Adding the probability contents in the four

Example 7.6 (cont.)

semi-paraboloids, we obtain $p_{F2} \cong 0.0276$. This result is in close agreement with the classical SORM solution, $p_{F2} = 0.0279$.

Example 7.7 – Reliability of a Column: Point-Fitting SORM Approximation

Consider the column reliability problem in Examples 6.5 and 7.2 with four random variables, where the FORM reliability index is $\beta = 2.47$. Using the point-fitting SORM, the fitting points along axis u'_1 are $(-2.47, 2.27)$ and $(2.47, 2.34)$, those along axis u'_2 are $(-2.47, 2.43)$ and $(2.47, 2.44)$, and those along axis u'_3 are $(-2.47, 2.05)$ and $(2.47, 2.13)$. The corresponding curvatures of the semi-parabolas are $\kappa_1^- = -0.0630$ and $\kappa_1^+ = -0.0405$ for u'_1, $\kappa_2^- = -0.0120$ and $\kappa_1^+ = -0.00974$ for u'_2, and $\kappa_3^- = -0.138$ and $\kappa_3^+ = -0.111$ for u'_3. Note that all fitting points are below the tangent plane and the curvatures are all negative, indicating that the surface bends toward the origin. Adding the probability contents in the 12 semi-paraboloids, we obtain $p_{F2} \cong 0.00913$ and the generalized reliability index $\beta_g = 2.36$. This result is in close agreement with the classical SORM solution, $p_{F2} = 0.00936$ and $\beta_g = 2.35$, obtained in Example 7.2.

Example 7.8 – CDF and PDF of a Function of Random Variables by SORM

Consider Example 6.12, where we computed by the FORM the cumulative distribution function (CDF) and probability density function (PDF) of the function $Z = \sum_{i=1}^{n} X_i$, where

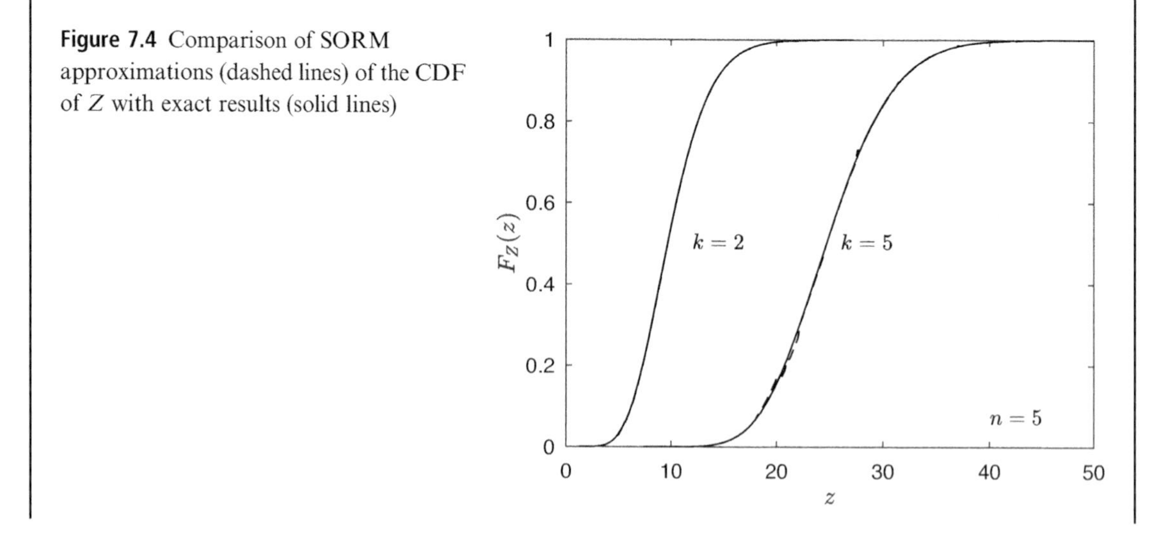

Figure 7.4 Comparison of SORM approximations (dashed lines) of the CDF of Z with exact results (solid lines)

Example 7.8 (cont.)

X_i are statistically independent, identically gamma-distributed random variables with parameters (k, λ). Here, we compute the CDFs for the cases $n = 5$ with $\lambda = 1$ and $k = 2$ and 5 by the point-fitting SORM. As can be seen in the bottom left chart in Figure 6.10, the FORM approximation tends to overestimate the CDFs in the central region of the distribution. We compare the SORM approximations with the exact results in Figure 7.4. As can be seen, the SORM approximations (dashed lines) are virtually indistinguishable from the exact solutions (solid lines), demonstrating the superior accuracy of the SORM over the FORM.

7.5 Concluding Remarks

The SORM provides an improvement over the FORM approximation by accounting for the curved nature of the limit-state surface around the design point. The classical SORM fits a hyper-paraboloid surface with the vertex at the design point and the principal curvatures matching the principal curvatures of the limit-state surface at that point. The main computational effort is in determining the principal curvatures, which requires computing the Hessian (the matrix of second derivatives) of the limit-state function at the design point. This can be computationally demanding, particularly when the limit-state function requires costly evaluation of implicit functions. Additionally, the second-derivative calculations can be inaccurate when the limit-state function is implicit and involves approximate numerical algorithms, such as nonlinear finite-element analysis.

The gradient-based SORM is an iterative method that finds the principal curvatures asymptotically through sequences of trial points that converge on the design point. Because the principal curvatures are found in the order of the absolute magnitude, the calculations can be stopped when the last principal curvature found is sufficiently small. This approach avoids computing second derivatives and is efficient when the number of significant principal curvatures is small in relation to the number of random variables.

The point-fitting SORM fits a collection of semi-parabolas at the design point, which pass through selected points on the limit-state surface along the coordinate axes. This method does not require computation of second derivatives and is insensitive to noise in the computation of the limit-state function, which may arise from inexact numerical computations such as those employed in nonlinear finite-element analysis.

SORM analysis in general is computationally intensive and requires the use of specialized software. To the best of the author's knowledge, CalREL is the only reliability software that has all three described methods implemented. FERUM has the curvature-fitting and point-fitting SORMs. See Section 1.5 for more information about these software packages.

PROBLEMS

7.1 Consider the limit-state function (as in Problem 6.1)

$$g(X_1, X_2) = X_1^{1/2} - X_2.$$

Where X_1 and X_2 have the Nataf distribution with the correlation coefficient $\rho_{12} = 0.5$; X_1 has a lognormal marginal distribution with mean $\mu_1 = 1,200$ and standard deviation $\sigma_1 = 300$, and X_2 has a gamma marginal distribution with mean $\mu_2 = 12$ and standard deviation $\sigma_2 = 4.8$ (as in case (1b) in Problem 6.1). The design point for this problem is at $\mathbf{u}^* = [0.266\ 3.801]^{\mathrm{T}}$.

1) Implement a transformation $\mathbf{u}' = \mathbf{Pu}$ of the axes in the standard normal space so that the design point falls on the u_2' axis.
2) Compute the curvature, κ, of the limit-state surface at the design point. For the two-dimensional case here, κ equals the second derivative of the limit-state function at the design point in the rotated space scaled by the norm of the gradient vector (see Section 7.2). Is the limit-state surface bending toward or away from the origin in the standard normal space?
3) Use (7.9) to compute the SORM approximation of the failure probability. Compare this result with the FORM approximation.

7.2 Employ the point-fitting SORM to solve Problem 7.1. For this purpose, first find fitting points on the limit-state surface at distances $\pm b = \min(3, \beta)$ from the u_2' axis. Then fit two semi-parabolas that are tangent to the surface at the design point and pass through the fitting points. Use (7.14) to compute the curvatures of the two semi-parabolas and compare their average with the curvature computed in Problem 7.1. Use (7.9) to compute the failure probability defined by the two semi-parabolas.

7.3 For the column reliability problem in Example 6.5, setting the tolerances in the iHL-RF algorithm at $\epsilon_1 = \epsilon_2 = 10^{-4}$, as convergence to the design point is achieved the last two trial points are $\mathbf{u}_9 = [1.2100\ 0.6987\ 0.9418\ -1.8010]^{\mathrm{T}}$ and $\mathbf{u}_{10} = [1.2142\ 0.7006\ 0.9435\ -1.8016]^{\mathrm{T}}$. Using (7.12), make an estimate of the major principal curvature at the design point. Compare this result with the major principal curvature obtained in Example 7.2.

7.4 Use CalREL to obtain a SORM solution of Problem 6.7. Use both curvature-fitting and point-fitting methods and for the former use both the classical approach and the gradient-based approach. Compare the results with each other and with the FORM result. The "exact" Monte Carlo solution for this problem yields the generalized reliability index $\beta_g = 2.812$ with 2% c.o.v.

7.5 Use CalREL to obtain a SORM solution of Problem 6.5. Use both curvature-fitting and point-fitting methods and for the former use both the classical approach and the gradient-based approach. Compare the results with each other and with the FORM result. The "exact" Monte Carlo solution of this problem yields the generalized reliability index $\beta_g = 3.127$ with 2% c.o.v.

7.6 Using FORM and SORM, compute the CDF of $Y = X_1 + X_2 + X_3$, where X_1, X_2, and X_3 are statistically independent random variables having gamma distributions with (k, λ) parameters $(1, 1)$, $(2, 1)$, and $(5, 1)$, respectively. Note that the exact distribution of Y is gamma with parameters $(8, 1)$. Plot and compare the FORM and SORM approximations of the CDF with the exact result. Also compute the FORM approximation of the PDF of Y and compare it with the exact result.

8 System Reliability

8.1 Introduction

A system is a collection of possibly interacting or interdependent components each performing a certain function, so that the functioning state of the system depends on the states of its constituent components. The system is described in terms of its configuration, i.e., the positioning of the components and their interactions with each other, by the demands placed on the system and its components, and by the definition of its intended function. Examples of engineering systems include a power distribution system with generation plants, substations, transmission lines, and consumption nodes; a water distribution system with reservoirs, filtration plants, pumps, pipelines, and consumption nodes; a transportation system with roadways, bridges, embankments, and tunnels; and a building structure with its foundation, beams, columns, walls, and slabs. At a more granular level, some of the mentioned components may themselves be considered as systems. For example, a power substation is a system of components such as buses, transformers, circuit breakers, switches, etc., and a column of a building is a system if it has multiple failure modes such as failure due to bending, shear, torsion, or buckling. Our primary interest in this chapter is in determining the reliability of a system when the reliabilities of its constituent components are known.

There are different graphical ways of describing a system. These include the physical diagram, the block diagram, the event tree, the fault tree, and the Bayesian network (see, e.g., Rausand and Høyland, 2004). This chapter mostly uses the block diagram, where components are shown as blocks with links describing their interconnections, and also develops a method employing event-tree diagrams. Chapter 15 develops the Bayesian network approach.

The chapter starts with some elementary definitions and notions of classical system theory. Then it specializes to what are called *structural systems*, meaning systems for which the states of the components are defined in terms of limit-state functions. Much of the classical systems reliability theory, starting from the pioneering work of Barlow and Proschan (1975), is built upon the notion that component reliabilities are expressed in terms of failure rates, which are obtained from available failure or life-testing data. This is a purely statistical approach. Defining component states through limit-state functions brings the physics of the component behavior into the analysis. Furthermore, the need for data is at a more basic level, e.g., for basic demand and capacity random variables. More specifically, the chapter considers capacity and demand variables for each component and defines the failure event following the rules of physics through limit-state functions, as already described for structural component problems in Chapters 6 and 7.

8.2 Representation of Systems

Consider a system of N components and define the component state variables s_i, $i = 1, \ldots, N$, and the system state variable s_{sys}. A system component may have discrete or continuous states. For example, a power transmission line has two states: either it is capable of transmitting power (working state) or it is unable to transmit power (fail state). On the other hand, a pipe in a water distribution system may be capable of transmitting water in a range of values from zero to its full capacity, depending on the condition of the pipe (blockage, leakage, etc.) – hence, a continuum of performance states. Similarly, the system state can be discrete or continuous. For example, a power system may or may not be able to supply electricity at a critical consumption node – hence, a discrete system state – while a water distribution system may provide a range of water flow and pressure values at a consumption node – hence, a continuous system state. Of course, constinuous states can be discretized by dividing the range of values into intervals. In the following we assume that such discretization has been carried out, and the discussion will be limited to systems and components with discrete states.

Let component i of the system have $m_i + 1$ states designated by the numerical values $s_i = (0, 1, \ldots, m_i)$. For example, for a water pipe component, $s_i = 0$ may indicate that the pipe is entirely broken and cannot transmit any water, $s_i = 1$ may indicate that the pipe can transmit only 10% of its capacity, and $s_i = m_i$ may indicate that the pipe is capable of transmitting full capacity. Also let $s_{\text{sys}} = (0, 1, \ldots, M)$ denote the $M + 1$ states of the system, with increasing values of s_{sys} indicating better system performance. By definition, s_{sys} depends on s_i, $i = 1, \ldots, N$. This dependence is defined through a *system function*

$$s_{\text{sys}} = \phi(s_1, \ldots, s_N) = \phi(\mathbf{s}), \tag{8.1}$$

where $\mathbf{s} = [s_1, s_2, \ldots, s_n]^{\mathrm{T}}$ is the vector of component state variables. A system is called *coherent* when ϕ is non-decreasing in each s_i.

Much of the system theory has been developed for two-state systems with two-state components, i.e., systems with component states defined by Boolean variables $s_i = (0, 1)$, where $s_i = 0$ indicates component i is in the failed state and $s_i = 1$ indicates component i is in the working state, and system states defined by $s_{\text{sys}} = (0, 1)$, where $s_{\text{sys}} = 0$ indicates the system is in the failed state (it fails to perform its intended function) and $s_{\text{sys}} = 1$ indicates the system is in the working state, i.e., capable of performing its intended function. Next, we develop notions specific to such two-state systems with two-state components.

The number of distinct configurations of a two-state system with N two-state components, i.e., distinct configurations with failed and working components, is 2^N. For $N = 100$, this is 1.27×10^{30}. This is an enormously large number of configurations and shows one of the challenges of systems analysis. A system with N m-state components has m^N distinct system configurations.

A *series system* is a system that fails if any of its components fails. Figure 8.1 shows a block diagram of such a system, where the system performance criterion is connection between points A and B. The system is like a chain, where the failure of any link results in the failure of the chain. For this reason, a series system is often called a *chain-like system*. It is also called a *weakest-link system* because the safety of the system depends on the safety of the weakest link in the chain. It should be clear that for such a system, the system function can be described as the product of the component state variables,

$$\phi(\mathbf{s}) = s_1 s_2 \cdots s_N. \tag{8.2}$$

Note that $\phi(\mathbf{s}) = 0$ if any one $s_i = 0$, and $\phi(\mathbf{s}) = 1$ if all $s_i = 1$, i.e., if all components survive.

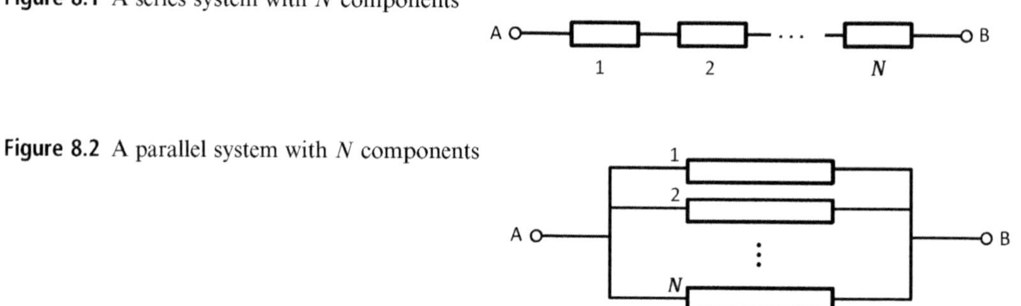

Figure 8.1 A series system with N components

Figure 8.2 A parallel system with N components

A *parallel system* is a system that fails if all of its components fail. Figure 8.2 shows a block diagram of such a system, where the system performance criterion again is connection between points A and B. The system is like a cable built of a bundle of wires, where the failure occurs when all wires fail. For such a system, the system function can be described as

$$\phi(\mathbf{s}) = 1 - (1 - s_1)(1 - s_2) \cdots (1 - s_N). \tag{8.3}$$

Note that if all $s_i = 0$, then we have $\phi(\mathbf{s}) = 0$, but if one or more $s_i = 1$, then $\phi(\mathbf{s}) = 1$.

Another special system is the so-called *k-out-of-N system*. This is a system that survives if at least k out of N components survive. The system function of such a system is defined as

$$\begin{aligned} \phi(\mathbf{s}) &= 1, \quad \text{if } k \leq \textstyle\sum_{i=1}^{N} s_i \\ &= 0, \quad \text{otherwise.} \end{aligned} \tag{8.4}$$

It should be clear that $k = N$ corresponds to a series system while $k = 1$ corresponds to a parallel system.

Now consider the simple but more general system with four components shown in Figure 8.3. This system cannot be categorized as one of the special systems described above.

To formulate the system function, the following definitions are introduced (Rausand and Høyland, 2004):

Cut set: Any set of components whose joint failure constitutes failure of the system.

Minimum cut set: A cut set that ceases to be a cut set if any one component is removed from the set. From here on, this is referred to as a *min-cut set*.

Disjoint cut sets: Disjoint sets of component states each constituting failure of the system.

Link set: Any set of components whose joint survival constitutes survival (working condition) of the system.

Minimum link set: A link set that ceases to be a link set if any one component is removed from the set. From here on, this is referred to as a *min-link set*.

Disjoint link sets: Disjoint sets of component states each constituting survival of the system.

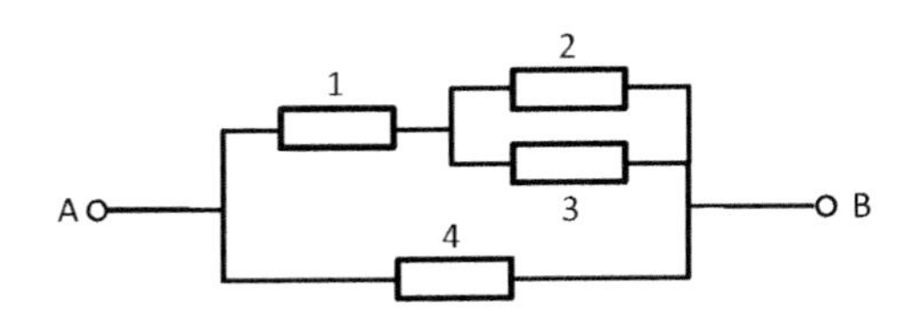

Figure 8.3 General system with four components

For the system in Figure 8.3, we have

$$\text{Cut sets} = \{(1,2,3,4),(1,2,4),(1,3,4),(2,3,4),(1,4)\}, \tag{8.5}$$

$$\text{Min-cut sets} = \{(2,3,4),(1,4)\}, \tag{8.6}$$

$$\text{Disjoint cut sets} = \left\{(1,2,3,4),\left(\overline{1},2,3,4\right),\left(1,\overline{2},3,4\right),\left(1,2,\overline{3},4\right),\left(1,\overline{2},\overline{3},4\right)\right\}, \tag{8.7}$$

$$\text{Link sets} = \left\{\begin{matrix}(1,2,3,4),(1,2,3),(1,2,4),(1,3,4),(2,3,4),\\(1,2),(1,3),(1,4),(2,4),(3,4),(4)\end{matrix}\right\}, \tag{8.8}$$

$$\text{Min-link sets} = \{(1,2),(1,3),(4)\}, \tag{8.9}$$

$$\text{Disjoint link sets} = \left\{\begin{matrix}(1,2,3,4),\left(\overline{1},2,3,4\right),\left(1,\overline{2},3,4\right),\left(1,2,\overline{3},4\right),\left(1,2,3,\overline{4}\right),\\\left(\overline{1},\overline{2},3,4\right),\left(\overline{1},2,\overline{3},4\right),\left(1,\overline{2},\overline{3},4\right),\left(1,\overline{2},3,\overline{4}\right),\\\left(1,2,\overline{3},\overline{4}\right),\left(\overline{1},\overline{2},\overline{3},4\right)\end{matrix}\right\}. \tag{8.10}$$

In the case of disjoint sets, a superposed bar above a component number indicates the complementary state of the component. For example, the disjoint cut set $\left(\overline{1},2,3,4\right)$ indicates that component 1 is in the working state while components 2, 3, and 4 are in the failed state, whereas in the disjoint link set $\left(\overline{1},2,3,4\right)$ component 1 is in the failed state while components 2, 3, and 4 are in the working state.

In formulating the system function, only the min-cut/link sets need be considered because the complete or disjoint cut/link sets contain information that is redundant. Naturally, if any of the min-cut sets is realized, the system is in the failed state, and if any of the min-link sets is realized, the system is in the working state. Thus, a system can be represented as a series system of its min-cut sets, or a parallel system of its min-link sets. As a cut set itself can be considered a parallel system (all components must fail for the cut set to fail), its system function can be formulated as in (8.3). Thus, for a general two-state system we can write

$$\phi(\mathbf{s}) = \prod_{j=1}^{N_C}\left[1 - \prod_{i\in C_j}(1 - s_i)\right], \tag{8.11}$$

where $i \in C_j$ indicates components i that are members of the jth min-cut set, and N_C is the number of min-cut sets. Note that the bracket on the right represents the system function for the jth min-cut set. Alternatively, using the min-link sets, the system function is

$$\phi(\mathbf{s}) = 1 - \prod_{j=1}^{N_L}\left[1 - \prod_{i\in L_j} s_i\right], \tag{8.12}$$

where $i \in L_j$ indicates components i that are members of the jth min-link set, and N_L is the number of min-link sets. Note that the bracket on the right represents the system function for the jth min-link set.

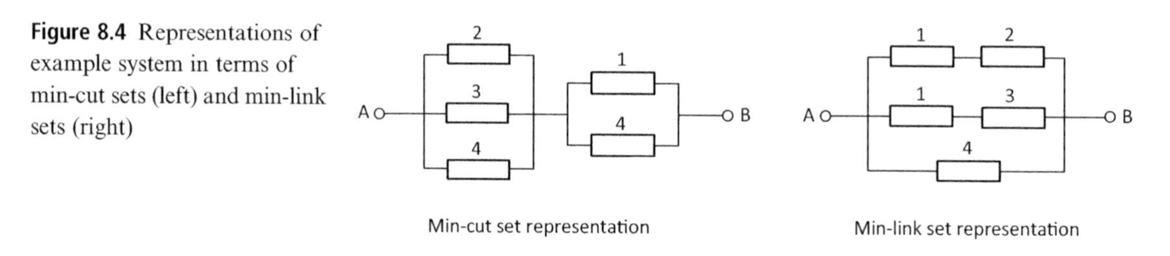

Figure 8.4 Representations of example system in terms of min-cut sets (left) and min-link sets (right)

Based on the above definitions, the example system in Figure 8.3 can be represented in alternative block diagrams in terms of min-cut sets or min-link sets, as shown in Figure 8.4. On the left, the system is represented as a series system of its min-cut sets, each min-cut set represented as a parallel subsystem. On the right, the system is represented as a parallel system of its min-link sets, each min-link set represented as a series subsystem.

Using the min-cut set formulation in (8.11), the system function for the example system is

$$\phi(\mathbf{s}) = [1 - (1 - s_2)(1 - s_3)(1 - s_4)][1 - (1 - s_1)(1 - s_4)], \tag{8.13}$$

and using the min-link set formulation in (8.12), it is

$$\phi(\mathbf{s}) = 1 - (1 - s_1 s_2)(1 - s_1 s_3)(1 - s_4). \tag{8.14}$$

Expansion of (8.13) and (8.14), together with the Boolean property $s_i^m = s_i$ for any $0 < m$ of the state variables, reveals that these expressions are identical, and both reduce to

$$\phi(\mathbf{s}) = s_4 + s_1 s_2 + s_1 s_3 - s_1 s_2 s_3 - s_1 s_2 s_4 - s_1 s_3 s_4 + s_1 s_2 s_3 s_4. \tag{8.15}$$

Example 8.1 – A Truss as a System

Consider the truss system consisting of 10 members shown in Figure 8.5. Note that members 1 and 2 overlap and are not connected. Loads are applied only at the nodes of the truss. It is well known that a simply supported planar truss needs at least $2j - 3$ members to be stable, where j is the number of joints (nodes) of the truss. Here, $j = 6$ so that nine members are needed for stability and there is one extra member. Recall that a truss remains stable as long as each of its members forms a triangle with two other members. For the example truss, failure of members 7, 8, 9, or 10 constitutes failure of the system, as the remaining members do not form triangles. For example, upon failure of member 7, member 9 does not form a triangle. As a result, it will rotate around its end nodes, resulting in the collapse of the truss. On the other hand, failure of member 3 does not constitute failure of the truss because all remaining members form triangles with other remaining members and rotations at nodes are not possible. It follows that members 7, 8, 9, and 10 individually represent min-cut sets. The truss needs at least nine members to remain stable, so all pairs of the remaining members constitute additional min-cut sets. Thus, we have

$$\text{Min-cut sets} = \left\{ \begin{array}{c} (7),(8),(9),(10),(1,2),(1,3),(1,4)(1,5),(1,6),(2,3), \\ (2,4),(2,5),(2,6),(3,4),(3,5),(3,6),(4,5),(4,6),(5,6) \end{array} \right\}. \quad \text{(E1)}$$

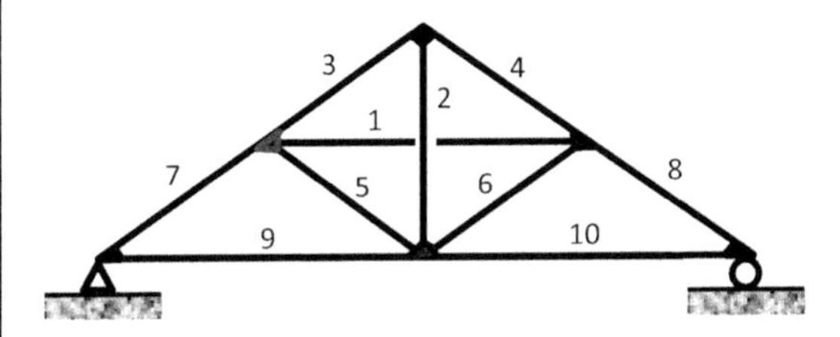

Figure 8.5 Example truss with 10 members

For the min-link sets, it is clear that each link set must include at least nine members and that members 7, 8, 9, and 10 must be present in all link sets. Only one of the remaining members can be removed. Hence, the min-link set are

$$\text{Min-link sets} = \left\{ \begin{array}{c} (2,3,4,5,6,7,8,9,10),(1,3,4,5,6,7,8,9,10),(1,2,4,5,6,7,8,9,10), \\ (1,2,3,5,6,7,8,9,10),(1,2,3,4,6,7,8,9,10),(1,2,3,4,5,7,8,9,10) \end{array} \right\}. \quad \text{(E2)}$$

There are too many disjoint cut sets of the truss to list. However, there are only seven disjoint link sets, as follows:

$$\text{Disjoint link sets} = \left\{ \begin{array}{c} (1,2,3,4,5,6,7,8,9,10),(\overline{1},2,3,4,5,6,7,8,9,10), \\ (1,\overline{2},3,4,5,6,7,8,9,10),(1,2,\overline{3},4,5,6,7,8,9,10), \\ (1,2,3,\overline{4},5,6,7,8,9,10),(1,2,3,4,\overline{5},6,7,8,9,10) \\ (1,2,3,4,5,\overline{6},7,8,9,10) \end{array} \right\}. \quad \text{(E3)}$$

8.3 Definition of System Reliability

The system reliability problem arises when the states of the components and, therefore, the state of the system are uncertain. Often the component states and, therefore, the state of the system stochastically evolve with time. Under these conditions, several measures of reliability are of interest, including:

Availability: The probability that the system is available (is functioning) at a given time.
Mean failure rate: The mean number of system failures per unit time.
First-passage probability: Probability distribution of the time to the first failure of the system.
Mean time to failure: Mean of the time to the first failure of the system.
Mean downtime: The mean duration of system downtime upon failure.
Reliability: The probability that the system continuously functions during a specified interval of time.

This chapter focuses on the availability problem and considers it as the point-in-time reliability of the system. The topic of time-variant reliability analysis and some of the other quantities defined above are addressed in Chapters 11 and 13.

To signify the random nature of component and system states, upper-case letters are used: S_i, $i = 1, \ldots, N$, for the component states and S_{sys} for the system state. Thus, $p_i = \Pr(S_i = 0)$ denotes the probability of failure of component i and $p_F = \Pr(S_{\text{sys}} = 0)$ denotes the probability of failure of the system. The following sections describe various methods for computing p_F or bounds for it.

8.4 System Reliability by Expectation

It is well known that for Boolean variables, probability and expectation are interchangeable. To see this, observe that

$$\begin{aligned} \mathrm{E}[S_i] &= 0 \times \Pr(S_i = 0) + 1 \times \Pr(S_i = 1) \\ &= \Pr(S_i = 1) \\ &= 1 - \Pr(S_i = 0) \\ &= 1 - p_i. \end{aligned} \tag{8.16}$$

In a similar way,

$$\begin{aligned} \mathrm{E}[\phi(\mathbf{S})] &= \Pr(S_{\text{sys}} = 1) \\ &= 1 - \Pr(S_{\text{sys}} = 0) \\ &= 1 - p_F. \end{aligned} \tag{8.17}$$

Thus, the expectation of the system function equals the system survival probability, which is the complement of the system failure probability.

Assume the component states of the system are statistically independent. This may happen in systems where the component failures are due to random wear and tear, or to local environmental effects that independently affect each component. Using (8.2) and (8.3), we have the following results for series and parallel systems, respectively:

$$\begin{aligned} p_{F,\text{series}} &= 1 - \mathrm{E}[S_1 S_2 \cdots S_N] \\ &= 1 - \mathrm{E}[S_1]\mathrm{E}[S_2] \cdots \mathrm{E}[S_N] \\ &= 1 - (1 - p_1)(1 - p_2) \cdots (1 - p_N), \end{aligned} \tag{8.18}$$

$$\begin{aligned} p_{F,\text{parallel}} &= 1 - \mathrm{E}[1 - (1 - S_1)(1 - S_2) \cdots (1 - S_N)] \\ &= \mathrm{E}[1 - S_1]\mathrm{E}[1 - S_2] \cdots \mathrm{E}[1 - S_N] \\ &= p_1 p_2 \cdots p_N. \end{aligned} \tag{8.19}$$

For general systems, using (8.11) and (8.12), we can write

$$p_F = 1 - \mathrm{E}\left\{\prod_{j=1}^{N_C}\left[1 - \prod_{i \in C_j}(1 - S_i)\right]\right\} \tag{8.20a}$$

$$= \mathrm{E}\left\{\prod_{j=1}^{N_L}\left[1 - \prod_{i \in L_j} S_i\right]\right\}. \tag{8.20b}$$

Because of the possibility of an individual S_i appearing in more than one of the square-bracketed terms (see 8.13 and 8.14 for the example problem), the expectation operation cannot be taken past the outside product operator. One option is to expand the system function, as in (8.15) for the example problem, and then take the expectation of the resulting expression that only involves sums of products of individual S_i. However, this can be extremely tedious for a large system. Another approach is to simplify the system function by conditioning on component states, as described below.

Using the total probability rule, we write

$$\mathrm{E}[\phi(\mathbf{S})] = (1 - p_i)\mathrm{E}[\phi(\mathbf{S})|\, S_i = 1] + p_i \mathrm{E}[\phi(\mathbf{S})|\, S_i = 0], \tag{8.21}$$

where $\mathrm{E}[\phi(\mathbf{S})|\, S_i = 1]$ is the expectation of the system function given that component i is in state 1, and $\mathrm{E}[\phi(\mathbf{S})|\, S_i = 0]$ is the expectation of the system function given that component i is in state 0. These conditioned system functions are simpler because S_i appearing in them are replaced by the numerical values 1 or 0. Obviously, it is advantageous to select a component i that appears in many min-cut sets or min-link sets. This conditioning on components is repeated until the reduced system functions are sufficiently simple and their expectations are easy to compute. The Boolean property $S_i^2 = S_i$ also proves useful in this analysis. The final result is a function of individual component failure probabilities, p_i.

If the component states are statistically dependent, then the expectation calculations in (8.20) must account for the dependence between the random variables S_i, $i = 1, \ldots, N$. The formulation presented in this section is not practical for that case.

Example 8.2 – Reliability of the Truss System

For the truss example in Figure 8.5, assume the member states are statistically independent with p_i denoting the failure probability of member i. (This assumption makes sense only if the uncertainty is in the member capacities, say due to deterioration. Uncertainty in the loads would induce statistical dependence among the member states because of the effect of load sharing.) Using the min-cut set formulation in (8.20a) and the identified min-cut sets in Example 8.1, we can write

$$\begin{aligned}
\mathrm{E}[\phi(\mathbf{S})] &= \mathrm{E}\{[1-(1-S_7)]\cdots[1-(1-S_{10})][1-(1-S_1)(1-S_2)]\cdots \\
&\quad [1-(1-S_5)(1-S_6)]\} \\
&= (1-p_7)\cdots(1-p_{10})\times \qquad \text{(E1)} \\
&\quad [(1-p_1)\mathrm{E}\{[1-(1-S_2)(1-S_3)]\cdots[1-(1-S_5)(1-S_6)]\}+ \\
&\quad p_1\,\mathrm{E}\{S_2\cdots S_6[1-(1-S_2)(1-S_3)]\cdots[1-(1-S_5)(1-S_6)]\}],
\end{aligned}$$

where we have conditioned on component 1. Note that the first four terms are for the single-component min-cut sets, which are independent of the other terms and are readily evaluated. Also observe that $S_2[1-(1-S_2)(1-S_3)] = S_2(S_2+S_3-S_2S_3) = S_2$, where the Boolean property $S_2^2 = S_2$ has been used. Thus, the expectation in the last line above reduces to $\mathrm{E}\{S_2\cdots S_6\} = (1-p_2)\cdots(1-p_6)$. Now, conditioning on component 2, the expectation in the penultimate line above reduces to

$$\begin{aligned}
&\mathrm{E}\{[1-(1-S_2)(1-S_3)]\cdots[1-(1-S_5)(1-S_6)]\} \\
&\quad = (1-p_2)\mathrm{E}\{[1-(1-S_3)(1-S_4)]\cdots[1-(1-S_5)(1-S_6)]\} \qquad \text{(E2)} \\
&\quad + p_2(1-p_3)\cdots(1-p_6),
\end{aligned}$$

where again we have used the Boolean property mentioned above to obtain the last line. Continuing this process eventually leads to

$$\begin{aligned}
\mathrm{E}[\phi(\mathbf{S})] &= 1-p_F \\
&= (1-p_7)(1-p_8)(1-p_9)(1-p_{10}) \\
&\{(1-p_1)\{(1-p_2)\{(1-p_3)[(1-p_4)(1-p_5p_6)+p_4(1-p_5)(1-p_6)] \\
&\quad + p_3(1-p_4)(1-p_5)(1-p_6)\} \qquad \text{(E3)} \\
&\quad + p_2(1-p_3)(1-p_4)(1-p_5)(1-p_6)\} \\
&\quad + p_1(1-p_2)(1-p_3)(1-p_4)(1-p_5)(1-p_6)\}.
\end{aligned}$$

For $p_i = p$, $i = 1, \ldots, 10$, the above expression reduces to

$$p_F = 1-(1+5p)(1-p)^9. \qquad \text{(E4)}$$

Example 8.2 (cont.)

For $p = 0.01$, $p_F = 0.0408$. An obvious lower bound for the system failure probability is obtained by considering only the failure of components 7, 8, 9, and 10. In that case, the system is a series system and $p_F \geq 1 - (1 - p_7)(1 - p_8)(1 - p_9)(1 - p_{10})$. For the equal probability case, $p_F \geq 1 - (1 - p)^4$ and for $p = 0.01$, $p_F \geq 0.0394$. Hence, the difference $0.0408 - 0.0394 = 0.0014$ is due to the contribution of failures by pairs of members 1 to 6.

8.5 System Reliability by the Inclusion–Exclusion Rule

Let $\mathcal{C}_j$ denote the event of occurrence of the jth min-cut set of the system. Because the system is a series system of its min-cut sets, we have

$$p_F = \Pr(\mathcal{C}_1 \cup \mathcal{C}_2 \cup \cdots \cup \mathcal{C}_{N_C}). \tag{8.22}$$

Expanding the union of events by the inclusion–exclusion rule (2.11), we have

$$p_F = \sum_{j=1}^{N_C} \Pr(\mathcal{C}_j) - \sum_{j=1}^{N_C-1} \sum_{k=j+1}^{N_C} \Pr(\mathcal{C}_j\mathcal{C}_k) + \sum_{j=1}^{N_C-2} \sum_{k=j+1}^{N_C-1} \sum_{l=k+1}^{N_C} \Pr(\mathcal{C}_j\mathcal{C}_k\mathcal{C}_l) - \cdots + (-1)^{N_C-1}\Pr(\mathcal{C}_1 \cdots \mathcal{C}_{N_C}). \tag{8.23}$$

Each of the intersecting events $\mathcal{C}_j\mathcal{C}_k$, $\mathcal{C}_j\mathcal{C}_k\mathcal{C}_l$, etc., describes the event of joint failure of all components that are members of the min-cut sets included in the intersection. Thus, each probability term in the above expression is the failure probability of a parallel system having components that are members of all the included min-cut sets. Of course, if a system has a large number of min-cut sets, then the above expression has many terms and it would be difficult to compute them all. However, as mentioned following (2.12), truncating the inclusion–exclusion series at successive terms provides bounds on the probability of the union. Thus, computing only the first-sum term provides an upper bound to the failure probability; including the double-sum term provides a lower bound; terminating after the triple-sum term provides a tighter upper bound, and so on. It follows that computing these terms until the difference between the successive truncations is sufficiently small is one way of reducing the computational effort.

If the component states are statistically independent, each probability term in (8.23) is simply the product of the failure probabilities of all the components that are members of the intersecting min-cut sets. If the component states are statistically dependent, then one needs to compute the joint probability of failure of the components in the intersection set, accounting for their dependence.

A similar formulation is available with the min-link sets. Let $\mathcal{L}_j$ denote the jth min-link set event, i.e., the survival of all components in L_j. Then,

$$\begin{aligned} p_F &= 1 - \Pr(\mathcal{L}_1 \cup \mathcal{L}_2 \cup \cdots \cup \mathcal{L}_{N_L}) \\ &= 1 - \sum_{j=1}^{N_L} \Pr(\mathcal{L}_j) + \sum_{j=1}^{N_L-1} \sum_{k=j+1}^{N_L} \Pr(\mathcal{L}_j \mathcal{L}_k) \\ &\quad - \sum_{j=1}^{N_L-2} \sum_{k=j+1}^{N_L-1} \sum_{l=k+1}^{N_L} \Pr(\mathcal{L}_j \mathcal{L}_k \mathcal{L}_l) + \cdots + (-1)^{N_L} \Pr(\mathcal{L}_1 \cdots \mathcal{L}_{N_L}). \end{aligned} \tag{8.24}$$

Each of the intersecting events $\mathcal{L}_j\mathcal{L}_k$, $\mathcal{L}_j\mathcal{L}_k\mathcal{L}_l$, etc., now describes the event of joint survival of all the components that are members of the min-link sets included in the intersection. Thus, each probability term in the expression is the survival probability of a series system having components that are members of all the min-link sets included in the intersection. Again, bounds on the system failure probability are obtained by terminating the series at successive sum terms. However, for highly reliable systems, these bounds often are too wide to be useful.

When the component states are statistically independent, each of the probability terms is the product of the complements of component failure probabilities of all the components that are members of the min-link sets included in the intersection. For dependent component states, one will need to compute the joint survival probability of all components within each intersection.

Another approach for computing the system failure probability uses the disjoint cut sets. Let $\mathcal{DC}_j$ denote the jth disjoint cut-set event. Because the probability of the union of disjoint events is the sum of their probabilities, we have

$$p_F = \sum_{j=1}^{N_{DC}} \Pr(\mathcal{DC}_j), \tag{8.25}$$

where N_{DC} is the number of disjoint cut sets. Note that each $\mathcal{DC}_j$ includes one of the min-cut sets with the remaining components being in a distinct combination of fail and survival states. Naturally, the number of terms here is greater than the number of min-cut sets. The advantage, on the other hand, is that no intersections of cut sets need to be considered. In the case of statistically independent component states, the probability terms can be computed as the product of the failure or survival probabilities of the components. For dependent component states, probabilities of joint occurrences of distinct combinations of component failure and survival events need to be computed.

A similar formulation is available in terms of the disjoint link sets. Let $\mathcal{DL}_j$ denote the jth disjoint link set. We can write

$$p_F = 1 - \sum_{j=1}^{N_{DL}} \Pr(\mathcal{DL}_j), \tag{8.26}$$

where N_{DL} is the number of disjoint link sets. Note that each $\mathcal{DL}_j$ includes one of the min-link sets, with the remaining components being in a distinct combination of fail and survival states. In the case of statistically independent component states, the probability terms can be computed as the product of component failure or survival probabilities. For dependent component states, probabilities of joint occurrences of distinct combinations of component failure and survival events need to be computed.

For a system with many components, the number of disjoint cut or link sets can be very large. It is noted, however, that excluding any disjoint cut set from (8.25) provides a lower bound to the system failure probability while excluding any disjoint link set from (8.26) provides an upper bound. Thus, working with both terms helps bound the system failure probability.

Example 8.3 – Bounds on the Reliability of the Truss System

Consider the truss example in Example 8.1 with statistically independent component states. The 19 min-cut sets of this system were identified in Example 8.1. For the case with all member failure probabilities equal to p, using (8.23), the first summation term gives the upper bound of the system failure probability as

$$p_F \leq \sum_{j=1}^{19} \Pr(\mathcal{C}_j) = 4p + 15p^2. \tag{E1}$$

Using the first two summation terms gives the lower bound

$$p_F \geq \sum_{j=1}^{N_C} \Pr(\mathcal{C}_j) - \sum_{j=1}^{N_C-1} \sum_{k=j+1}^{N_C} \Pr(\mathcal{C}_j \mathcal{C}_k) = 4p + 15p^2 - 6p^2 - 120p^3 - 45p^4. \tag{E2}$$

For $p = 0.01$, (E1) yields $p_F \leq 0.0415$ while (E2) yields $p_F \geq 0.04078$. These values bound the correct system failure probability, $p_F = 0.0408$, obtained in Example 8.2.

As shown in Example 8.1, the example truss has six min-link sets. Using (8.24) for the case with equal component failure probabilities, we obtain the lower bound

$$p_F \geq 1 - \sum_{j=1}^{6} \Pr(\mathcal{L}_j) = 1 - 6(1-p)^9. \tag{E3}$$

If we include the second summation term in the series, the upper bound is

$$p_F \leq 1 - \sum_{j=1}^{6} \Pr(\mathcal{L}_j) + \sum_{j=1}^{5} \sum_{k=j+1}^{6} \Pr(\mathcal{L}_j \mathcal{L}_k) = 1 - 6(1-p)^9 + 15(1-p)^{10}. \tag{E4}$$

For $p = 0.01$, (E3) gives $p_F \geq -4.481$ and (E4) gives $p_F \leq 9.085$, both of which are useless bounds. It is evident that bounding the failure probability by truncating the series in (8.24) is not effective. This has to do with the fact that probabilities of min-link sets tend to be close to 1 for reliable systems of interest.

Now consider the disjoint cut set formulation in (8.25). As mentioned in Example 8.1, the number of disjoint cut sets of the truss system are too many to be enumerated. However, including any partial set of the disjoint cut sets, (8.25) will provide a lower bound to the failure probability. Consider the disjoint cut sets formed by the min-cut sets identified in Example 8.1 with all remaining members in the safe state. That is, for example, for the min-cut set (7), only

Example 8.3 (cont.)

consider the disjoint cut set $(\bar{1},\bar{2},\bar{3},\bar{4},\bar{5},\bar{6},7,\bar{8},\bar{9},\bar{10})$. This results in 19 disjoint cut sets, which is a small fraction of all the disjoint cut sets. Equation (8.25) then gives the lower bound

$$p_F \geq \sum_{j=1}^{19} \Pr(\mathcal{DC}_j) = 4p(1-p)^9 + 15p^2(1-p)^8. \tag{E5}$$

For $p = 0.01$, the above gives $p_F \geq 0.0379$.

As shown in Example 8.1, the truss system has seven disjoint link sets. The analysis yields,

$$p_F = 1 - \sum_{j=1}^{7} \Pr(\mathcal{DL}_j) = 1 - (1-p)^{10} - 6p(1-p)^9. \tag{E6}$$

For $p = 0.01$, this gives $p_{sys} = 0.0408$, which is the exact solution obtained in Example 8.2.

8.6 Bounds on Series-System Reliability

Series systems play a significant role in structural reliability analysis. A structure that can fail in one of several failure mechanisms or failure modes is a series system. Furthermore, a general system is a series system of its min-cut sets. Hence, there is special interest in developing bounds on series systems with dependent component states. The following derivation of bounds for series system is due to Ditlevsen (1979b). Earlier versions of these bounds were derived by Kounias (1968) and Hunter (1976).

Let S_i denote the state of the ith component and $\bar{S}_i$ denote its complement. As S_i is a binary (0,1) variable, $\bar{S}_i = 1 - S_i$. Furthermore, $p_i = \mathrm{E}[\bar{S}_i]$ is the failure probability of component i. Also let $\bar{S}_{\text{sys}}$ denote the complement of S_{sys} so that $p_F = \mathrm{E}[\bar{S}_{\text{sys}}]$ is the failure probability of the system.

For a series system with N components, from (8.2) we have $S_{\text{sys}} = S_1 S_2 \cdots S_N$. We can write

$$\begin{aligned} S_{\text{sys}} &= S_1 S_2 \cdots S_N \\ &= S_1 S_2 \cdots S_{N-1}(1 - \bar{S}_N) \\ &= S_1 S_2 \cdots S_{N-1} - S_1 S_2 \cdots S_{N-1} \bar{S}_N. \end{aligned} \tag{8.27}$$

Repeating this formulation in turn for S_{N-1}, S_{N-2}, and eventually for S_1, results in

$$S_{\text{sys}} = 1 - \bar{S}_1 - S_1 \bar{S}_2 - \cdots - S_1 S_2 \cdots S_{N-1} \bar{S}_N \tag{8.28}$$

or

$$\bar{S}_{\text{sys}} = \bar{S}_1 + S_1 \bar{S}_2 + \cdots + S_1 S_2 \cdots S_{N-1} \bar{S}_N. \tag{8.29}$$

The above expression essentially says that the series system fails if component 1 fails, or if component 1 does not fail but component 2 fails, or if components 1 and 2 do not fail but component 3 fails, etc., all the way to components 1 to $N-1$ not failing but component N failing.

Observe that $\bar{S}_{\text{sys}} \geq \bar{S}_i$ for any i. Taking expectation of both sides, we have

$$p_F \geq \max_{1 \leq i \leq N} p_i. \tag{8.30}$$

It should be clear from (8.29) that $\bar{S}_{\text{sys}} \leq \bar{S}_1 + \bar{S}_2 + \cdots + \bar{S}_N$. Taking expectation of both sides, we have

$$p_F \leq \sum_{i=1}^{N} p_i. \tag{8.31}$$

Equations (8.30) and (8.31) are known as *uni-modal bounds* for series systems. They are given in terms of the individual component failure probabilities and disregard any dependence that may exist between the component failure events.

Now observe that

$$S_1 S_2 \cdots S_{i-1} \geq 1 - \left(\bar{S}_1 + \cdots + \bar{S}_{i-1}\right). \tag{8.32}$$

This can be proven by inspection. Note that the left-hand side is either 0 or 1. It is 0 if any of $S_1, S_2, \ldots, S_{i-1}$ equals 0, in which case the corresponding complement variable equals 1 and the right-hand side is either 0 or a negative number. The left-hand side equals 1 if every one of the variables $S_1, S_2, \ldots, S_{i-1}$ equals 1, in which case the right-hand side also equals 1 and the equality between the two sides holds. Multiplying both sides of (8.32) by $\bar{S}_i$, one obtains

$$S_1 S_2 \cdots S_{i-1} \bar{S}_i \geq \bar{S}_i - \sum_{j=1}^{i-1} \bar{S}_i \bar{S}_j. \tag{8.33}$$

Taking expectation of both sides, we have

$$\mathrm{E}\left[S_1 S_2 \cdots S_{i-1} \bar{S}_i\right] \geq \max\left\{0, p_i - \sum_{j=1}^{i-1} p_{ij}\right\}, \tag{8.34}$$

where $p_{ij} = \mathrm{E}\left[\bar{S}_i \bar{S}_j\right]$ is the probability of joint failure of components i and j. The max operator is used on the right-hand side because the left-hand side cannot be negative. Now taking expectation of both sides of (8.29) and using the lower bound in (8.34) for the terms on the right-hand side, we obtain

$$p_F \geq p_1 + \sum_{i=2}^{N} \max\left\{0, p_i - \sum_{j=1}^{i-1} p_{ij}\right\}. \tag{8.35}$$

This is known as the *bi-modal lower bound* for a series system because it accounts for the pairwise joint component failure probabilities. Note that the above result depends on the ordering of the components. Usually the best result is obtained if components are ordered so that $p_1 \geq p_2 \geq \cdots \geq p_N$.

Now observe that

$$S_1 S_2 \cdots S_{i-1} \leq S_j = 1 - \bar{S}_j, \quad \text{for } j < i. \tag{8.36}$$

Multiplying both sides by $\bar{S}_i$,

$$S_1 S_2 \cdots S_{i-1} \bar{S}_i \leq \bar{S}_i - \bar{S}_i \bar{S}_j, \quad \text{for } j < i. \tag{8.37}$$

Substituting these upper bounds for each term in (8.29) and taking expectations of both sides, we obtain

$$p_F \leq \sum_{i=1}^{N} p_i - \sum_{i=2}^{N} \max_{j<i} p_{ij}, \tag{8.38}$$

where we have used the max operator because for any i we can select j to maximize p_{ij}. This is known as the *bi-modal upper bound* for a series system.

It is possible to refine the above bounds by including the joint failure probabilities of triple sets of the components. Zhang (1993) derived the following *tri-modal bounds*:

$$p_F \geq p_1 + p_2 - p_{12} + \sum_{i=3}^{N} \max \left\{ 0, p_i - \sum_{j=1}^{i-1} p_{ij} + \max_{k<i} \sum_{\substack{j=1 \\ j \neq k}}^{i-1} p_{ijk} \right\}, \tag{8.39}$$

$$p_F \leq \sum_{i=1}^{N} p_i - p_{12} - \sum_{i=3}^{N} \max_{j<k<i} \left(p_{ik} + p_{ij} - p_{ijk} \right), \tag{8.40}$$

where $p_{ijk} = \mathrm{E}\left[\bar{S}_i \bar{S}_j \bar{S}_k\right]$ is the probability of joint failure of components i, j, and k. However, experience has shown that for most systems the improvement obtained by including tri-modal probabilities is modest.

The above bounds are also applicable to general systems. As indicated in (8.22), a general system is a series system of its min-cut sets. It follows that the above bounds are applicable to such a system if p_i, p_{ij}, and p_{ijk} are interpreted as the probabilities of realization of the individual, pairwise, and triple min-cut sets, respectively. Note that in that case, N in the bounding formulas should be replaced by N_C, the number of min-cut sets. This approach for the truss system example is demonstrated below.

Example 8.4 – Bounds on the Reliability of the Truss System

In this example, we compute the uni-modal and bi-modal bounds for the truss system with independent component states with equal failure probabilities. Note that, although the component states are statistically independent, the min-cut sets are statistically dependent.

The 19 min-cut sets of the truss system were identified in Example 8.1. For the case considered, the probabilities of the first four min-cut sets are p and those of the remaining 15 are p^2. Using the uni-modal bounds in (8.30) and (8.31),

$$p \leq p_F \leq 4p + 15p^2. \tag{E1}$$

For $p = 0.01$, the above gives $0.01 \leq p_F \leq 0.0415$. Recall that the exact failure probability obtained in Example 8.2 is $p_F = 0.0408$.

To apply the bi-modal bounds, we need to compute the pairwise min-cut set probabilities. We have 6 such terms with joint probability p^2 (intersections of the first four min-cut sets),

Example 8.4 (cont.)

120 such terms with joint probability p^3 (intersections of the first 4 min-cut sets with the remaining 15, plus intersections of the remaining 15 min-cut sets with common components), and 45 such terms with probability p^4 (intersections of the remaining 15 min-cut-sets without common components). Using (8.35) and (8.38), we obtain

$$p_F \geq p + \left(3p + 15p^2 - 6p^2 - 120p^3 - 45p^4\right), \tag{E2}$$

$$p_F \leq 4p + 15p^2 - \left(3p^2 + 15p^3\right). \tag{E3}$$

For $p = 0.01$, this gives $0.04079 \leq p_F \leq 0.0412$. Observe that the bi-modal bounds provide a major improvement over the uni-modal bounds.

8.7 Bounds on System Reliability by Linear Programming

In real-world system reliability problems, the probabilistic information available about the components is often incomplete or imprecise. For example, one may entirely lack information about the failure probabilities of a subset of the components, or have bounding information only. It would be useful in such cases if bounds on the system failure probability could be developed based on the available information. This section describes a linear programming (LP) approach originally suggested by Hailperin (1965) and further developed and extended by Song and Der Kiureghian (2003a) for this purpose. Although the approach is applicable to systems having components with any number of states, our presentation is limited to systems with two-state components only.

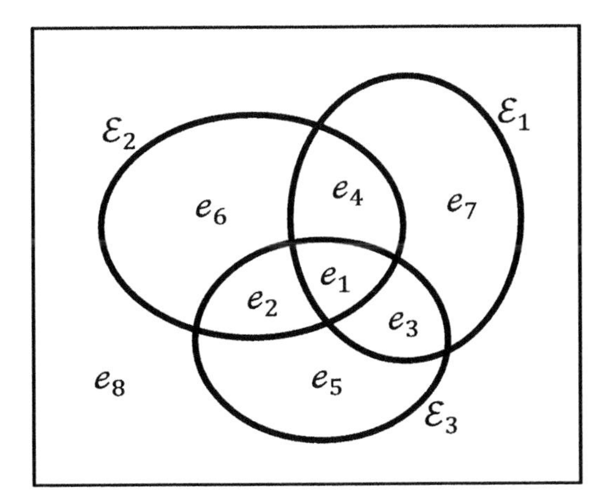

Figure 8.6 Venn diagram of a system with three components

Consider the outcome space of the states of the system components and let $\mathcal{E}_i$ denote the event of failure of component i. For a two-state system with N two-state components, this space consists of 2^N distinct system realizations, i.e., configurations of the system with distinct combinations of working and failed components. Let e_l, $l = 1, 2, \ldots, 2^N$, denote these distinct events and $\pi_l = \Pr(e_l)$ their respective probabilities. Figure 8.6 shows the Venn diagram for a system with three components and $2^3 = 8$ events formed by the distinct intersections of events $\mathcal{E}_i$, $i = 1, 2, 3$, and their complements: $e_1 = \mathcal{E}_1\mathcal{E}_2\mathcal{E}_3$, $e_2 = \bar{\mathcal{E}}_1\mathcal{E}_2\mathcal{E}_3$, $e_3 = \mathcal{E}_1\bar{\mathcal{E}}_2\mathcal{E}_3$, $e_4 = \mathcal{E}_1\mathcal{E}_2\bar{\mathcal{E}}_3$, $e_5 = \bar{\mathcal{E}}_1\bar{\mathcal{E}}_2\mathcal{E}_3$, $e_6 = \bar{\mathcal{E}}_1\mathcal{E}_2\bar{\mathcal{E}}_3$, $e_7 = \mathcal{E}_1\bar{\mathcal{E}}_2\bar{\mathcal{E}}_3$, and $e_1 = \bar{\mathcal{E}}_1\bar{\mathcal{E}}_2\bar{\mathcal{E}}_3$. Observe that e_l are mutually exclusive and collectively exhaustive events. Thus, their probabilities π_l must satisfy the following conditions:

$$0 \leq \pi_l, l = 1, 2, \ldots, 2^N, \tag{8.41}$$

$$\sum_{l=1}^{2^N} \pi_l = 1. \tag{8.42}$$

Furthermore, because any logical function of events $\mathcal{E}_i$ can be written as the union of the disjoint events e_l contained in it, the probability of the logical function can be written as a sum of probabilities π_l. For example,

$$\Pr(\mathcal{E}_i) = p_i = \sum_{l:\, e_l \subset \mathcal{E}_i} \pi_l, \tag{8.43}$$

$$\Pr(\mathcal{E}_i\mathcal{E}_j) = p_{ij} = \sum_{l:\, e_l \subseteq \mathcal{E}_i\mathcal{E}_j} \pi_l, \tag{8.44}$$

$$\Pr(\mathcal{E}_i\mathcal{E}_j\mathcal{E}_k) = p_{ijk} = \sum_{l:\, e_l \subseteq \mathcal{E}_i\mathcal{E}_j\mathcal{E}_k} \pi_l, \tag{8.45}$$

where p_i, p_{ij}, and p_{ijk} are the probabilities of failure of component i, joint failure of components i and j, and joint failure of components i, j, and k, respectively, and the summations are over the values of index l such that e_l is contained in the event of interest. Similar expressions can be written for other logical functions of component events, such as $\mathcal{E}_i \cup \mathcal{E}_j \cup \mathcal{E}_k$ or $\mathcal{E}_i\bar{\mathcal{E}}_j\bar{\mathcal{E}}_k$. Importantly, because the system failure event is also a logical function of the component failure and survival events, we can write

$$p_F = \sum_{l:\, e_l \subseteq \mathcal{E}_{\text{sys}}} \pi_l, \tag{8.46}$$

where $\mathcal{E}_{\text{sys}}$ denotes the event of failure of the system. Observe that all the above expressions are linear functions of probabilities π_l.

We now formulate the following LP problem:

$$\text{Minimize or Maximize: } p_F = \mathbf{a}^{\mathrm{T}}\boldsymbol{\pi} \tag{8.47a}$$

$$\text{Subject to: } \mathbf{A}_1\boldsymbol{\pi} = \mathbf{b}_1 \tag{8.47b}$$

$$\mathbf{A}_2\boldsymbol{\pi} \leq \mathbf{b}_2 \tag{8.47c}$$

$$\mathbf{A}_3\boldsymbol{\pi} \geq \mathbf{b}_3. \tag{8.47d}$$

In the above, $\boldsymbol{\pi} = \left[\pi_1\, \pi_2 \cdots \pi_{2^N}\right]^{\mathrm{T}}$ is the vector of non-negative decision variables of the problem, $\mathbf{a}$, $\mathbf{A}_1$, $\mathbf{A}_2$, and $\mathbf{A}_3$ are coefficient vector and matrices, respectively, and $\mathbf{b}_1$, $\mathbf{b}_2$, and $\mathbf{b}_3$ are vectors of known probabilities, upper bound probabilities, and lower bound probabilities, respectively. Equation (8.47a) defines the objective function of the LP problem, (8.47b) defines the set of equality constraints including (8.42), and (8.47c) and (8.47d) define inequality constraints, including (8.41). Note that the inequalities must be interpreted row-wise. The lth element of the coefficient vector $\mathbf{a}$ is 1 or 0, depending on whether e_l is contained in the system failure event or not. The (l, m) element of $\mathbf{A}_1$ is 1 or 0, depending on whether e_m is contained in the event defined by the logical expression in the lth row or not. The lth row of $\mathbf{b}_1$ is the known probability for the event defined by the lth row of $\mathbf{A}_1$. To represent (8.42), the corresponding row of $\mathbf{A}_1$ and element of $\mathbf{b}_1$ are all 1s. Coefficient matrices $\mathbf{A}_2$ and $\mathbf{A}_3$ are similarly constructed to represent events for which probability bounds are known. Minimizing the objective function provides a lower bound for the system failure probability while maximizing it leads to an upper bound.

In addition to any logical function of component events, the LP formulation can account for known conditional probabilities of component states. Suppose we know $\Pr(\mathcal{E}_i|\mathcal{E}_j) = p_{i|j}$. Because $\Pr(\mathcal{E}_i|\mathcal{E}_j) = \Pr(\mathcal{E}_i\mathcal{E}_j)/\Pr(\mathcal{E}_j)$, we can write the equality constraint

$$\sum\nolimits_{l:\, e_l \subseteq \mathcal{E}_i\mathcal{E}_j} \pi_l - p_{i|j} \sum\nolimits_{l:\, e_l \subseteq \mathcal{E}_j} \pi_l = 0. \tag{8.48}$$

Observe that this is again a linear function of the variables $\boldsymbol{\pi}$ and can be included in the set of equality constrains (8.47b).

The LP formulation presented above has a number of important advantages: It offers a unified approach for series, parallel, or general systems; it can incorporate any kind of information, including marginal, joint, or conditional component probabilities; and it finds the narrowest possible bounds on the system failure probability for the given information. Later in this chapter it is shown that this approach can also be used for probability updating as well as for computing component importance and sensitivity measures. On the other hand, the size of the problem is 2^N, which can be extremely large for systems with many components. For this reason, the LP bounds approach is limited by the available computer memory size.

Several approaches have been proposed to alleviate the size issue of the LP bounds formulation. Usually they involve some relaxation scheme in decomposing the event space and accepting a trade-off between the size of the optimization problem and the distance between the bounds. Der Kiureghian and Song (2008) proposed a multi-scale approach, where "super-components" representing subsystems within the system and composed of subsets of the components are introduced. The LP bounds approach is applied to each subsystem and the computed subsystem bounds are then used as the available information to compute probability bounds for the whole system. This approach inevitably leads to wider bounds than the original LP formulation. Wang et al. (2011), Wei et al. (2013), and Chang and Mori (2013) proposed alternative decomposition strategies instead of using the basic mutually exclusive and collectively exhaustive events e_i. Recently, Byun and Song (2020) proposed the delayed column generation method to address the LP size issue. Their approach eliminates the need for explicit enumeration of all the variables when it is possible to obtain an efficient description of the solution set. The original LP problem is reformulated as an iteration of smaller problems of binary integer programming. Consequently, the memory required to formulate the problem is significantly reduced with a trade-off of additional computational cost to repeatedly solve smaller problems.

Example 8.5 – Seismic Reliability of Electrical Substation System

Consider the two-transmission-line electrical substation system shown in Figure 8.7, which is adopted from Song and Der Kiureghian (2003b). The system has 12 components, including 3 disconnect switches (DS), 2 circuit breakers (CB), 2 power transformers (PT), 2 drawout breakers (DB), 1 tie breaker (TB), and 2 feeder breakers (FB). We are interested in the reliability of the system against an impending earthquake. Let A denote the uncertain peak ground acceleration in units of gravity acceleration, g, and assume each component is affected by a local site effect factor, S_i, such that the intensity experienced by the ith component is AS_i. Also let R_i denote the capacity of the ith component in terms of the intensity of ground acceleration in g. Assume these random variables are all lognormal with the means and coefficients of variation (c.o.v.) listed in Table 8.1. Also assume all random variables are statistically independent, except that the capacities of pairs of similar equipment are jointly lognormal with the correlation coefficients listed in Table 8.1.

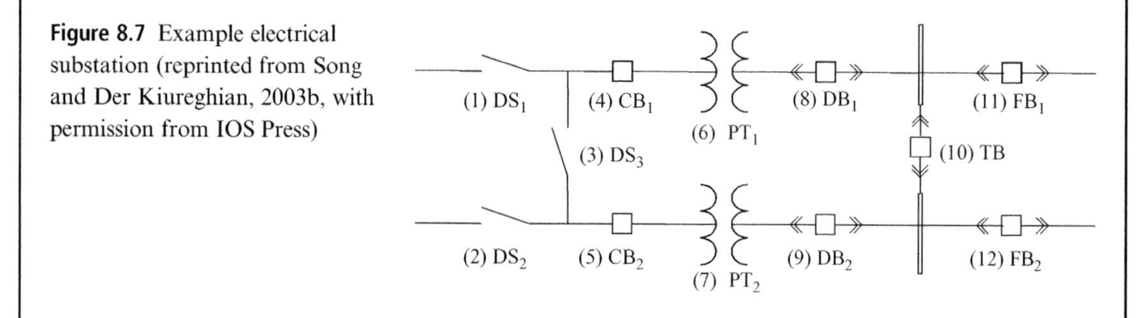

Figure 8.7 Example electrical substation (reprinted from Song and Der Kiureghian, 2003b, with permission from IOS Press)

It should be clear that component i fails if $R_i \leq AS_i$. This can be stated as the event

$$\mathcal{E}_i = \{\ln R_i - \ln A - \ln S_i \leq 0\}. \tag{E1}$$

Owing to the lognormal distribution of the random variables,

$$Z_i = \ln R_i - \ln A - \ln S_i, i = 1, \ldots, 12 \tag{E2}$$

have the multinormal distribution, which facilitates computation of the component marginal and joint failure probabilities. Note that $\{Z_i \leq 0\}$ is tantamount to the failure of component i.

Example 8.5 (cont.)

Table 8.1 Statistics of random variables for example electrical substation

Variable description	Mean	c.o.v.	Correlation between pairs of capacities for similar equipment
Peak ground acceleration, A	0.15	0.5	–
Site effect factors, S_i	1.0	0.2	–
Capacities of DS_1, DS_2, and DS_3, R_1, R_2, R_3	0.4	0.3	0.3
Capacities of CB_1 and CB_2, R_4, R_5	0.3	0.3	0.3
Capacities of PT_1 and PT_2, R_6, R_7	0.5	0.5	0.5
Capacities of DB_1 and DB_2, R_8, R_9	0.4	0.3	0.3
Capacity of TB, R_{10}	1.0	0.3	–
Capacities of FB_1 and FB_2, R_{11}, R_{12}	1.0	0.3	0.3

The 21 min-cut sets of the system have been identified in Song and Der Kiureghian (2003b) as

$$\begin{aligned}\text{Min-cut sets} = \{&(1,2),(4,5),(4,7),(4,9),(5,6),(6,7),(6,9),(5,8),(7,8),\\ &(8,9),(11,12),(1,3,5),(1,3,7),(1,3,9),(2,3,4),(2,3,6),\\ &(2,3,8),(4,10,12),(6,10,12),(8,10,12),(5,10,11),\ (7,10,11),\\ &(9,10,11),(1,3,10,12),(2,3,10,11)\}.\end{aligned} \tag{E3}$$

For example, the 1st cut set event is $\mathcal{C}_1 = \mathcal{E}_1\mathcal{E}_2$ and the 25th min-cut set event is $\mathcal{C}_{25} = \mathcal{E}_2\mathcal{E}_3\mathcal{E}_{10}\mathcal{E}_{11}$. The system failure event is formulated as the union of the min-cut sets. This allows determination of the coefficient vector **a** in the objective function of the LP problem. This vector has length $2^{12} = 4,096$, which determines the size of the LP problem.

Table 8.2 summarizes the results of the LP analyses for four different cases. Listed for each case are the LP bounds using only the marginal component failure probabilities (denoted as "uni-comp" bounds), using marginal and pairwise joint-component failure probabilities ("bi-comp" bounds), and using marginal, pairwise, and triple joint-component failure probabilities ("tri-comp" bounds). In Case 1, we assume these probabilities are available for all the components; hence, there are 12 equality constraints for the uni-comp bounds, 66 for the bi-comp bounds, and an additional 55 for the tri-comp bounds. For this case, the "exact" solution obtained by Monte Carlo analysis (with 1% c.o.v.) is also shown. Of course, the full probability information was used for Monte Carlo analysis. In Case 2, we assume no information is available on TB (component 10). Hence, all equality constraints involving this component are removed in the LP formulation. In Case 3, we assume no information is available on CB_1

Example 8.5 (cont.)

(component 4) and remove all equality constraints involving this component. Finally, in Case 4, we assume only the upper bound $p_4 \leq 0.01$ is available on CB_1. Note that for the information provided in Table 8.1, the probability of failure of CB_1 is $p_4 = 0.0929$. So, the known upper bound suggests a much higher reliability than the full information case for this component.

Table 8.2 Results of analysis for substation system with LP bounds method (reprinted from Song and Der Kiureghian, 2003b, with permission from IOS Press)

Case	LP bounds on p_F			Monte Carlo	Case description
	Uni-comp	Bi-comp	Tri-comp		
1	1.13×10^{-12} ~0.202	0.0436~0.146	0.0616~0.0942	0.0752	Information on all components
2	1.82×10^{-11} ~0.202	0.0436~0.146	0.0615~0.0943	–	No information on TB
3	1.26×10^{-9} ~0.202	0.0267~0.147	0.0395~0.136	–	No information on CB_1
4	5.19×10^{-9} ~0.120	0.0267~0.0995	0.0395~0.0701	–	Upper bound on CB_1, $p_4 \leq 0.01$

The following observations about the results reported in Table 8.2 are noteworthy: (a) In general, the uni-comp bounds are found to be too wide to be of practical use. The lower bound is virtually 0 and the upper bound is much greater than the “exact” result obtained by Monte Carlo simulation (with a c.o.v. of 0.01); (b) Accounting for pairwise joint component failure probabilities significantly narrows the bounds, but they are still quite wide; (c) Accounting for triple joint component failure probabilities further improves the bounds; (d) For Case 1, the “exact” Monte Carlo result is enveloped by all three sets of LP bounds; (e) In Case 2, where it is assumed that no information on TB (component 10) is available, little change in the bounds is observed. This is because this component has a relatively insignificant influence on the reliability of the system. This is apparent from the fact that this component appears only in min-cut sets that have three or four components (see E3); (f) For Case 3, where it is assumed that no information is available on CB_1 (component 4), significant changes, particularly in the lower bounds, are observed. This is an indication that this component has a significant influence on the system reliability. Indeed, this component is present in 3 out of 11 two-component min-cut sets (see E3); (g) For Case 4 with a favorable upper bound failure probability information on CB_1, significant changes (reductions) in all the upper bounds are observed.

8.8 Matrix-Based System Reliability Method

Song and Kang (2009) used the idea behind the LP formulation to develop an efficient matrix-based system reliability analysis method. This method is described first for the case of statistically independent component states. Later an extension of the method to account for a special type of dependence between the component states is described. This method requires complete probability information about the component states. Although the approach is applicable to systems having components with any number of states, the presentation here is limited to systems with two-state components only.

We have seen that the system failure probability can be written as the inner product of two vectors, $p_F = \mathbf{a}^{\mathrm{T}}\boldsymbol{\pi}$, where $\mathbf{a}$ is a coefficient vector defining the system event of interest and $\boldsymbol{\pi}$ is the vector of probabilities of the mutually exclusive and collectively exhaustive events e_i formed by distinct combinations of the component states. The size of these vectors is 2^N, where N denotes the number of system components. Song and Kang (2009) developed efficient ways of forming these vectors. Let $\mathbf{a}^{\mathcal{E}}$ denote the coefficient vector for an event $\mathcal{E}$. Song and Kang show that

$$\mathbf{a}^{\bar{\mathcal{E}}} = \mathbf{1} - \mathbf{a}^{\mathcal{E}}, \tag{8.49}$$

$$\mathbf{a}^{\mathcal{E}_1\mathcal{E}_2\cdots\mathcal{E}_n} = \mathbf{a}^{\mathcal{E}_1}.*\mathbf{a}^{\mathcal{E}_2}.*\cdots.*\mathbf{a}^{\mathcal{E}_n}, \tag{8.50}$$

$$\mathbf{a}^{\mathcal{E}_1\cup\mathcal{E}_2\cup\cdots\cup\mathcal{E}_n} = \mathbf{1} - \left(\mathbf{1} - \mathbf{a}^{\mathcal{E}_1}\right).*\left(\mathbf{1} - \mathbf{a}^{\mathcal{E}_2}\right).*\cdots.*\left(\mathbf{1} - \mathbf{a}^{\mathcal{E}_n}\right), \tag{8.51}$$

where a superposed bar indicates the complement of an event, $\mathbf{1}$ is a 2^N vector of 1s, and $.*$ denotes element by element multiplication. Because the event vector of the system can be written as the union of intersections of component state events (see the min-cut set and min-link set formulations in 8.22 and 8.24), the above operations can be used to efficiently construct the event coefficient vector $\mathbf{a}$ for the system. For example, (8.50) can be used to compute the coefficient vector for each min-cut set and then (8.51) can be used to compute the coefficient vector for the union of min-cut sets. Matrix operations in many codes can perform this kind of calculations in a few lines.

Recall that, in the case of statistically independent component states, each element of $\boldsymbol{\pi}$ is the product of a distinct combination of failure and survival probabilities of the system components. Let $\boldsymbol{\pi}_{[1]} = [p_1 \; 1 - p_1]^{\mathrm{T}}$. Song and Kang (2009) showed that, for systems with statistically independent component states, $\boldsymbol{\pi}$ can be constructed by the following iterative operation

$$\boldsymbol{\pi}_{[i]} = \begin{bmatrix} p_i\boldsymbol{\pi}_{[i-1]} \\ (1 - p_i)\boldsymbol{\pi}_{[i-1]} \end{bmatrix}, i = 2, 3, \cdots, N, \tag{8.52}$$

with $\boldsymbol{\pi} = \boldsymbol{\pi}_{[N]}$. This matrix operation is computationally much faster than individually computing the elements of $\boldsymbol{\pi}$.

In many systems, statistical dependence between component states arises from "common-source" effects, e.g., a demand variable that is common to all components. For such cases, Song and Kang (2009) suggest using the total probability rule

$$p_F = \int_{\mathbf{x}} \mathbf{a}^{\mathrm{T}} \boldsymbol{\pi}(\mathbf{x}) f_{\mathbf{X}}(\mathbf{x}) d\mathbf{x}, \tag{8.53}$$

where $\mathbf{X}$ is the set of common-source random variables with probability density function (PDF) $f_{\mathbf{X}}(\mathbf{x})$, and $\boldsymbol{\pi}(\mathbf{x})$ is the probability vector for $\mathbf{X} = \mathbf{x}$. Provided $\mathbf{X}$ does not have too many elements, the above integral can be computed numerically to obtain the unconditional system failure probability. As the coefficient vector $\mathbf{a}$ is independent of $\mathbf{X}$, an alternative form of the integral is

$$p_F = \mathbf{a}^{\mathrm{T}} \int_{\mathbf{x}} \boldsymbol{\pi}(\mathbf{x}) f_{\mathbf{X}}(\mathbf{x}) d\mathbf{x} = \mathbf{a}^{\mathrm{T}} \bar{\boldsymbol{\pi}}, \tag{8.54}$$

where $\bar{\boldsymbol{\pi}}$ is the mean of the uncertain conditional probability vector $\boldsymbol{\pi}(\mathbf{X})$. This requires integrating all elements of the vector $\boldsymbol{\pi}(\mathbf{x})$, which can be time consuming. However, the formulation in (8.54) is advantageous if one wishes to study several different system events with different coefficient vectors $\mathbf{a}$, because $\bar{\boldsymbol{\pi}}$ remains the same for all system events.

In some problems, the statistical dependence between the random variables (or component limit-state functions) is of a form that can be described in terms of a small number of common-source random variables. This is the case, for example, for Nataf-distributed random variables if the elements of the correlation matrix $\mathbf{R}_0$ can be written in the so-called Dunnett-Sobel (Dunnett and Sobel, 1955) form, $\rho_{0,ij} = r_i r_j$ for $i \neq j$, where $|r_i| < 1$. In that case, the transformed correlated standard normal random variables can be written as

$$Z_i = \sqrt{1 - r_i^2}\, U_i + r_i X, i = 1, 2, \ldots, n, \tag{8.55}$$

where U_i and X are statistically independent standard normal random variables and X represents a common-source variable for all Z_i. One important case is that in which pairs of random variables Z_i and Z_j, $i \neq j$, have identical correlations $\rho_{ij} = \rho$, in which case $r_i = \sqrt{\rho}$. For a general correlation matrix, Song and Kang (2009) suggest finding an approximating Dunnett-Sobel-class correlation matrix by minimizing a measure of the difference between the two correlation matrices. For an improved approximation, they suggest introducing more than one common-source variable by writing Z_i in the form

$$Z_i = \sqrt{1 - \sum_{k=1}^{m} r_{ik}^2}\, U_i + \sum_{k=1}^{m} r_{ik} X_k, i = 1, 2, \ldots, n, \tag{8.56}$$

where m common-source standard normal random variables, X_k, $k = 1, \ldots, m$, are introduced. The correlation coefficient between Z_i and Z_j is now given by $\sum_{k=1}^{m} r_{ik} r_{jk}$ for $i \neq j$, where r_{ij} satisfy the conditions $-1 < r_{ik} < 1$ and $0 < 1 - \sum_{k=1}^{m} r_{ik}^2$. The following example demonstrates an application of this method.

Example 8.6 – Seismic Reliability of Electrical Substation by Matrix-Based System Reliability Method

The author is grateful to J.-E. Byun and J. Song for providing this example. Consider the two-transmission-line electrical substation system of Example 8.5. As noted there, the system has 12 components with capacities R_i, $i = 1, \ldots, 12$, and it is subject to an impending earthquake of random peak ground acceleration A and statistically independent site effects S_i at the component locations. Among the components, three disconnect switches (DSs), and pairs of circuit breakers (CBs), power transformers (PTs), drawout breakers (DBs), and feeder breakers (FBs) have correlated capacities (see Table 8.1). Furthermore, the peak ground acceleration A is a common demand variable to all components. As a result, random variables Z_i, $i = 1, \ldots, 12$, in (E2) of Example 8.5 are correlated. Let $\mathbf{R}$ denote the 12×12 correlation matrix of these random variables and $\mathbf{R}_{\text{DS}}(\mathbf{r})$ denote the approximating Dunnett–Sobel-class correlation matrix with $\mathbf{r}$ representing the collection of r_{ik}, $k = 1, \ldots, m$, values, where m denotes the number of common-source variables to be introduced (see 8.56). The latter matrix is constructed by solving the optimization problem

$$\begin{aligned} &\min_{\mathbf{r}} \|\mathbf{R} - \mathbf{R}_{\text{DS}}(\mathbf{r})\| \\ &\text{Subject to: } -1 < r_{ik} < 1, 0 < 1 - \sum_{k=1}^{m} r_{ik}^2, \end{aligned} \tag{E1}$$

where $\|\cdot\|$ denotes the Euclidean norm of the matrix. Now, using (8.56) to represent each of the random variables Z_i, the system reliability problem is defined by $12 + m$ statistically independent random variables U_i, $i = 1, \ldots, 12$, and X_k, $k = 1, \ldots, m$. Conditioned on the m common-source random variables, $\mathbf{X} = (X_1, \cdots, X_m)$, the component states are now statistically independent and the iterative algorithm in (8.52) can be used to compute the probability vector $\boldsymbol{\pi}(\mathbf{x})$ as a function of each outcome $\mathbf{X} = \mathbf{x}$. This is then used in (8.53) to compute the system failure probability by integration over the common-source variables. The results of the analysis are summarized in Table 8.3. Three cases with $m = 1$, 2, and 3 are considered and comparisons are made with the "exact" solution $p_{F,\text{MC}} \cong 0.0752$ obtained by Monte Carlo simulation, as reported in Song and Der Kiureghian (2003b). Also listed are the measures of error $\|\mathbf{R} - \mathbf{R}_{\text{DS}}(\mathbf{r})\| / \|\mathbf{R}\|$ in representing the correlation matrix.

Table 8.3 Summary of results and comparison with Monte Carlo solution

	Number of common-source random variables		
Description	$m = 1$	$m = 2$	$m = 3$
p_F	0.0688	0.0729	0.0779
$\|\mathbf{R} - \mathbf{R}_{\text{DS}}(\mathbf{r})\| / \|\mathbf{R}\|$	2.83%	2.82%	0.953%
$\lvert p_{F,\text{MC}} - p_F \rvert / p_{F,\text{MC}}$	−8.51%	−3.06%	+3.59%

Example 8.6 (cont.)

Observe in Table 8.3 that increasing the number of common-source random variables from one to three significantly improves the accuracy in representing the correlation matrix. Furthermore, with two or three common-source random variables, the error in the estimate of the system failure probability is of the order of 3%. Of course, the Monte Carlo solution itself is prone to error (1% c.o.v.), so the error estimates relative to the Monte Carlo solution should be examined with this caveat in mind.

8.9 Formulation of Structural System Reliability

The system reliability methods described in the preceding sections do not explicitly consider the physics of the component failure events. They assume that component failure probabilities are somehow available. One exception was Example 8.5, where component failure events were modeled in terms of limit-state functions. This section introduces methods to account for the physics of the component events through limit-state functions. The word "structural" used in naming the formulation is indicative of this approach.

Let $\mathbf{X} = (X_1, \cdots, X_n)$ denote the collection of all basic random variables that describe the performance of the system components and $g_i = g_i(\mathbf{X})$, $i = 1, \ldots, N$, denote the limit-state function for the ith two-state component such that

$$\{S_i = 0\} \equiv \{g_i(\mathbf{X}) \leq 0\} \tag{8.57}$$

describes the failure event of the component. In this formulation, dependence between the component states arises from random variables that are common in their limit-state functions, or from statistical dependence between random variables that appear in different limit-state functions.

It is instructive to see the above formulation in graphical form. Figure 8.8 depicts representations of series, parallel, and general systems with $N = 3$ components in the outcome space of two random variables $\mathbf{X} = (X_1, X_2)$. Dark gray bands depict the negative sides of the component limit-state surfaces, thus indicating component failure events, while light gray areas marked as domain $\mathcal{F}$ depict the failure domain of the system. The figure on the left depicts a series system, the one in the middle depicts a parallel system, and the one on the right depicts a general system with $N_C = 2$ min-cut sets, one including component 1 alone and one including components 2 and 3. As can be seen, the series system is in the failed state if any of its components fails, so that its failure domain is defined as

$$\mathcal{F} = \bigcup_{i=1}^{N} \{g_i(\mathbf{X}) \leq 0\}, \tag{8.58}$$

The parallel system is in the failed state when all components fail so that its failure domain is defined as

$$\mathcal{F} = \bigcap_{i=1}^{N} \{g_i(\mathbf{X}) \le 0\}. \tag{8.59}$$

Finally, the general system is in the failed state when any of its min-cut set fails so that the failure domain is

$$\mathcal{F} = \bigcup_{j=1}^{N_C} \bigcap_{i \in C_j} g_i(\mathbf{X}) \le 0, \tag{8.60}$$

where C_j is the index set for the jth min-cut set. The probability of failure of the system is given by the integral over the domain $\mathcal{F}$,

$$p_F = \int_{\mathcal{F}} f_{\mathbf{X}}(\mathbf{x}) \mathrm{d}\mathbf{x}. \tag{8.61}$$

For a large number of random variables, direct evaluation of this integral is clearly impractical. Next, first-order approximations for these types of systems are developed.

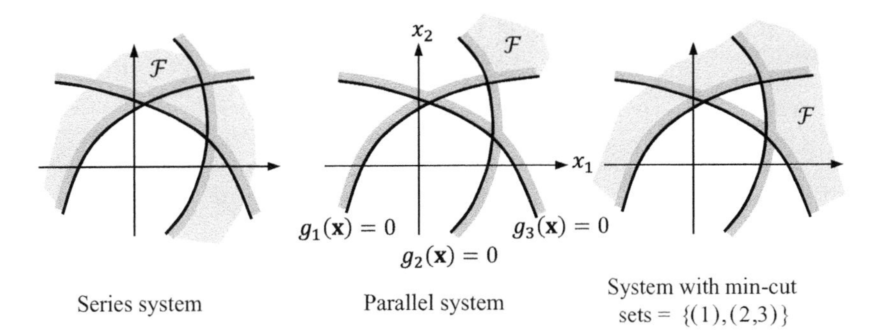

Figure 8.8 Depictions of a three-component structural system in the outcome space of two random variables

8.10 First-Order Approximations for Series and Parallel Systems

The formulation in this section is based on the pioneering work of Hohenbichler and Rackwitz (1983). First consider a series system of N components and let n denote the number of random variables. We transform the set of random variables to the standard normal space, $\mathbf{U} = \mathbf{T}(\mathbf{X})$, with the corresponding transformed limit-state functions denoted $G_i(\mathbf{U}) = g_i\left[\mathbf{T}^{-1}(\mathbf{U})\right]$, $i = 1, \ldots, N$. Figure 8.9 exemplifies the component and system states in an $(n = 2)$-dimensional space of standard normal random variables for a system with $N = 3$ components. Let $\mathbf{u}_i^*$ denote the design point on the limit-state surface of the ith component and $\widehat{\boldsymbol{\alpha}}_i$ denote the corresponding unit normal vector so that $\beta_i = \widehat{\boldsymbol{\alpha}}_i \mathbf{u}_i^*$ is the component reliability index. Linearizing the component limit-state functions at the corresponding design points (shown as dashed lines in Figure 8.9), the series system failure probability is written as

$$
\begin{aligned}
p_{F,\text{series}} &= \Pr\left[\bigcup_{i=1}^{N} \{g_i(\mathbf{X}) \leq 0\}\right] \\
&= \Pr\left[\bigcup_{i=1}^{N} \{G_i(\mathbf{U}) \leq 0\}\right] \\
&\cong \Pr\left[\bigcup_{i=1}^{N} \{\beta_i - \widehat{\boldsymbol{\alpha}}_i \mathbf{U} \leq 0\}\right],
\end{aligned}
\tag{8.62}
$$

where the approximation in the last line is due to the linearization $G_i(\mathbf{U}) \cong \|\nabla G_i(\mathbf{u}_i^*)\|(\beta_i - \widehat{\boldsymbol{\alpha}}_i \mathbf{U})$ in which we have removed the positive-valued scale factor $\|\nabla G_i(\mathbf{u}_i^*)\|$. Observe that the functions $\beta_i - \widehat{\boldsymbol{\alpha}}_i \mathbf{U}$ are linear in the standard normal random variables $\mathbf{U}$ and, therefore, are normally distributed. Let

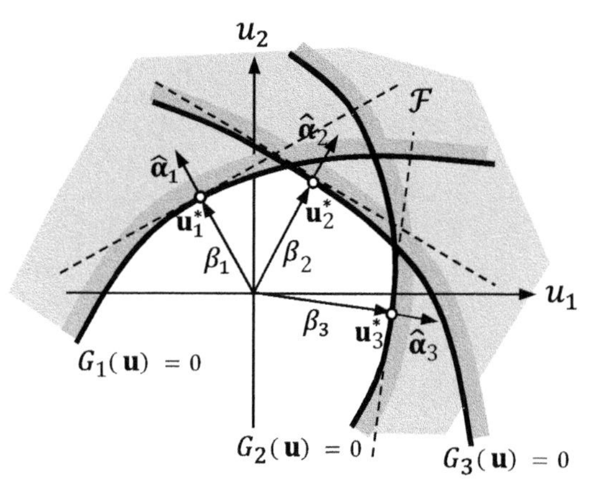

Figure 8.9 FORM approximation of a series system

$$
\mathbf{V} = \mathbf{A}\mathbf{U},
\tag{8.63}
$$

where $\mathbf{A}$ is an $N \times n$ matrix with $\widehat{\boldsymbol{\alpha}}_i$ as its ith row. It should be clear that $\mathbf{V}$ is an N-vector of standard normal random variables with correlation matrix $\mathbf{R} = \mathbf{A}\mathbf{A}^{\mathrm{T}}$ having the elements $\rho_{ij} = \widehat{\boldsymbol{\alpha}}_i \widehat{\boldsymbol{\alpha}}_j^{\mathrm{T}}$. Replacing $\widehat{\boldsymbol{\alpha}}_i \mathbf{U}$ in (8.62) by the ith element V_i of $\mathbf{V}$, we have

$$
\begin{aligned}
p_{F,\text{series}} &\cong \Pr\left[\bigcup_{i=1}^{N} \{V_i \geq \beta_i\}\right] \\
&= 1 - \Pr\left[\bigcap_{i=1}^{N} \{V_i \leq \beta_i\}\right],
\end{aligned}
\tag{8.64}
$$

where we have used De Morgan's rule (2.7) to arrive at the last expression. The probability term in the last expression is the joint CDF of a set of correlated standard normal random variables evaluated at thresholds β_i, $i = 1, \ldots, N$. Thus, in first-order approximation,

$$p_{F,\text{series}} \cong 1 - \Phi_N(\mathbf{B}, \mathbf{R}), \tag{8.65}$$

where $\Phi_N(\mathbf{B}, \mathbf{R})$ is the N-variate standard normal CDF with correlation matrix $\mathbf{R}$, evaluated at thresholds $\mathbf{B} = (\beta_1, \beta_2, \ldots, \beta_N)$ (see (3.9) for the definition of this function). Observe that in the case of a single component ($N = 1$), (8.65) reduces to $p_{F1} = 1 - \Phi(\beta) = \Phi(-\beta)$, which is the first-order reliability method (FORM) approximation for a component. It is evident that the FORM approximation of the probability of failure of a series system requires computing the N-variate standard normal CDF with correlation matrix $\mathbf{R}$. There is no closed-form solution of this function. However, several methods for its numerical computation are available, which are cited in the paragraph following (3.9). In CalREL, the sequential conditioned importance sampling method by Ambartzumian et al. (1998) is implemented.

Now consider a parallel system in the standard normal space, as depicted in Figure 8.10. Note that for this system, component 3 is inactive as it does not border the failure domain $\mathcal{F}$. Define the joint design point $\mathbf{u}^*$ according to

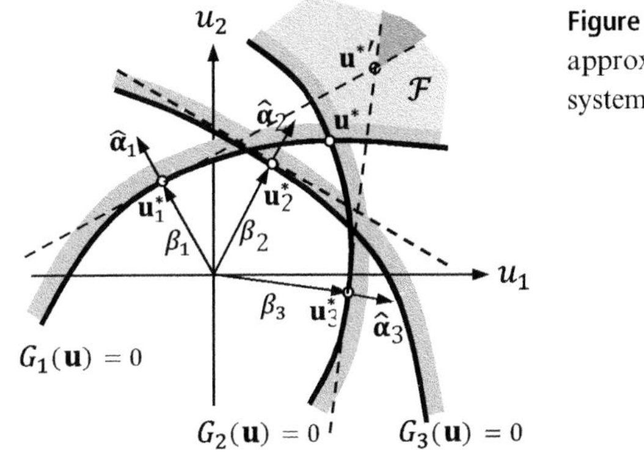

Figure 8.10 FORM approximation of a parallel system

$$\mathbf{u}^* = \min\{\|\mathbf{u}\| \, | G_i(\mathbf{u}) \leq 0, i = 1, \ldots, N\}. \tag{8.66}$$

This is an optimization problem with N inequality constraints. A FORM approximation can be developed by linearizing each active limit-state surface at this point. However, finding this point is a lot harder than finding the design point for a single component. A simpler alternative is to linearize the limit-state surface for each component at its respective design point. The failure domain of this approximation is depicted in Figure 8.10 with a slightly darker gray shade. It can be seen that this approximation can be crude when the individual design points are far from the failure domain $\mathcal{F}$. A simple, yet much better alternative is to linearize the limit-state functions at the joint design point of the linearized component limit-state surfaces (Der Kiureghian, 2006). That is, at the point obtained by solving (8.66) with the linearized constraints

$$\mathbf{u}^{*\prime} = \min\{\|\mathbf{u}\| \mid \beta_i - \widehat{\boldsymbol{\alpha}}_i \mathbf{u} \leq 0, i = 1, \ldots, N\}. \tag{8.67}$$

The solution of (8.67) is much simpler than that of (8.66). Of course, this point may not lie on any of the component limit-state surfaces. As a result, the linearized hyperplanes will not be tangent to the actual limit-state surfaces. But provided $\mathbf{u}^{*\prime}$ is not too far from $\mathbf{u}^*$, the approximation works well. In any case, this alternative is much better than using the linearized limit-state surfaces at the individual component design points. CalREL has options for using either $\mathbf{u}^*$ or $\mathbf{u}^{*\prime}$.

Linearizing each limit-state function at point $\mathbf{u}^{*\prime}$ (or $\mathbf{u}^*$), we have

$$\begin{aligned} G_i(\mathbf{u}) &\cong G_i(\mathbf{u}^{*\prime}) + \nabla G_i(\mathbf{u}^{*\prime})(\mathbf{u} - \mathbf{u}^{*\prime}) \\ &= \|\nabla G_i(\mathbf{u}^{*\prime})\| \left[\frac{G_i(\mathbf{u}^{*\prime})}{\|\nabla G_i(\mathbf{u}^{*\prime})\|} + \widehat{\boldsymbol{\alpha}}_i' \mathbf{u}^{*\prime} - \widehat{\boldsymbol{\alpha}}_i' \mathbf{u} \right] \\ &= \|\nabla G_i(\mathbf{u}^{*\prime})\| (\beta_i' - \widehat{\boldsymbol{\alpha}}_i' \mathbf{u}), \end{aligned} \tag{8.68}$$

where we have introduced $\widehat{\boldsymbol{\alpha}}_i' = -\nabla G_i(\mathbf{u}^{*\prime})/\|\nabla G_i(\mathbf{u}^{*\prime})\|$ and $\beta_i' = G_i(\mathbf{u}^{*\prime})/\|\nabla G_i(\mathbf{u}^{*\prime})\| + \widehat{\boldsymbol{\alpha}}_i' \mathbf{u}^{*\prime}$. Now, the failure probability of the parallel system is written as

$$\begin{aligned} p_{F,\text{parallel}} &= \Pr\left[\bigcap_{i=1}^{N} \{G_i(\mathbf{U}) \leq 0\}\right] \\ &\cong \Pr\left[\bigcap_{i=1}^{N} \{\beta_i' - \widehat{\boldsymbol{\alpha}}_i' \mathbf{U} \leq 0\}\right]. \end{aligned} \tag{8.69}$$

We now introduce

$$\mathbf{V}' = \mathbf{A}'\mathbf{U}, \tag{8.70}$$

where $\mathbf{A}'$ is an $N \times n$ matrix with $\widehat{\boldsymbol{\alpha}}_i'$ as its ith row. It should be clear that $\mathbf{V}'$ is an N-vector of standard normal random variables with correlation matrix $\mathbf{R}' = \mathbf{A}'\mathbf{A}'^{\mathrm{T}}$ having elements $\rho_{ij} = \widehat{\boldsymbol{\alpha}}_i' \widehat{\boldsymbol{\alpha}}_i'^{\mathrm{T}}$. Replacing $\widehat{\boldsymbol{\alpha}}_i' \mathbf{U}$ in (8.69) by the ith element V_i' of $\mathbf{V}'$, we have

$$p_{F,\text{parallel}} \cong \Pr\left[\bigcap_{i=1}^{N} \{V_i' \geq \beta_i'\}\right]. \tag{8.71}$$

Owing to the rotational symmetry property of the standard normal space, $\{V_i' \geq \beta_i'\} = \{V_i' \leq -\beta_i'\}$. Hence, the FORM approximation of the parallel system probability is

$$p_{F,\text{parallel}} \cong \Phi_N(-\mathbf{B}', \mathbf{R}'), \tag{8.72}$$

where $\mathbf{B}' = (\beta_1', \beta_2', \ldots, \beta_N')$ is the set of thresholds. Observe again that for $N = 1$, the above expression reduces to the FORM solution for a single component.

As we have seen, the failure event of a general system can be expressed as the union of its min-cut set events. Writing the union of the min-cut set in the form of the inclusion–exclusion rule as in (8.23), the system probability is expressed as a summation of probability terms, each representing the failure probability of a parallel system. Each of these probability terms can be computed in first-order approximation by use of (8.72), thus providing a FORM approximation for the general system failure probability. Similarly, the survival of a general system can be expressed as a union of its min-link set events. Expressing the union in the form of the inclusion–exclusion rule as in (8.24), the complement of the system failure probability is obtained as a summation of probability terms, each representing the survival probability of a series system. The complement of each of these survival probabilities can be computed in first-order approximation by use of (8.65), thus providing a FORM approximation for the general system failure probability. However, because the survival probabilities are close to 1 for practical structural systems, this second approach tends not to provide good accuracy. For that reason, the min-cut set formulation is normally used. The following example demonstrates one such application.

Example 8.7 – Reliability Analysis of Ductile Frame

Consider the one-bay frame in Figure 8.11 under horizontal and vertical loads H and V, respectively. Assume the beam and columns have elastoplastic bending behavior and let M_1, $M_2, \ldots, M_5$ denote the plastic moment capacities at the joints and the middle of the beam, which are potential locations for formation of plastic hinges, as shown in the figure. Under the applied loads, the frame may experience any of the three plastic mechanisms shown in Figure 8.11. Let $\mathbf{X} = (H, V, M_1, M_2, M_3, M_4, M_5)$ denote the set of random variables of the problem with the distributions and second moments listed in Table 8.4.

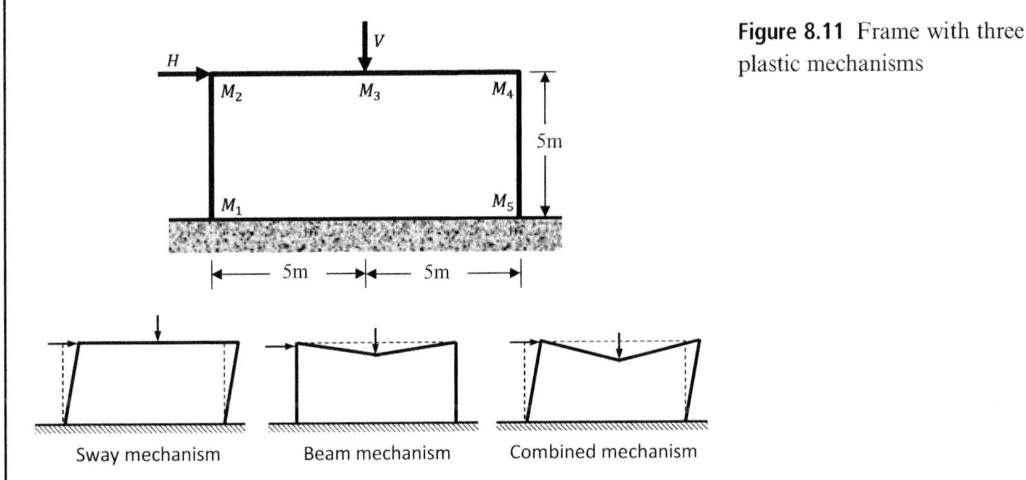

Figure 8.11 Frame with three plastic mechanisms

Equating the virtual work of internal and external forces (see Neal, 1985), the following limit-state functions are formulated to describe each mechanism:

$$\text{Sway mechanism: } g_1(\mathbf{X}) = M_1 + M_2 + M_4 + M_5 - 5H \tag{E1}$$

Example 8.7 (cont.)

Table 8.4 Distributions and second moments of random variables

Variable	Distribution	Mean	c.o.v.	Correlation
M_i, $i = 1, \ldots, 5$, kNm	Joint lognormal	150	0.2	$\rho_{ij} = 0.3, i \neq j$
H, kN	Gumbel	50	0.4	Independent
V, kN	Gamma	60	0.2	Independent

$$\text{Beam mechanism: } g_2(\mathbf{X}) = M_2 + 2M_3 + M_4 - 5V, \tag{E2}$$

$$\text{Combined mechanism: } g_3(\mathbf{X}) = M_1 + 2M_3 + 2M_4 + M_5 - 5H - 5V. \tag{E3}$$

Solving for the design point of each limit-state function, we obtain the following results for the product $\beta_i \widehat{\boldsymbol{\alpha}}_i$:

$$\beta_1 \widehat{\boldsymbol{\alpha}}_1 = 2.29[-0.238 - 0.174 - 0.044 - 0.131 - 0.112\ 0.939\ 0.000], \tag{E4}$$

$$\beta_2 \widehat{\boldsymbol{\alpha}}_2 = 2.87[-0.263 - 0.356 - 0.425 - 0.204 - 0.000\ 0.000\ 0.763], \tag{E5}$$

$$\beta_3 \widehat{\boldsymbol{\alpha}}_3 = 2.00[-0.313 - 0.137 - 0.291 - 0.240 - 0.113\ 0.792\ 0.317]. \tag{E6}$$

The above is a convenient form for presenting FORM results for systems analysis. We have

$$\mathbf{B} = (2.29, 2.87, 2.00) \text{ and } \mathbf{R} = \begin{bmatrix} 1 & 0.170 & 0.899 \\ 0.170 & 1 & 0.545 \\ 0.899 & 0.545 & 1 \end{bmatrix}, \tag{E7}$$

where the elements of $\mathbf{R}$ are obtained as $\rho_{ij} = \widehat{\boldsymbol{\alpha}}_i \cdot \widehat{\boldsymbol{\alpha}}_j$. Using (8.65) with $N = 3$, we obtain the FORM approximation of the frame system as

$$p_{F1} = 1 - \Phi_3(\mathbf{B}, \mathbf{R}) = 0.0264. \tag{E8}$$

From the β values in (E4) – (E6), it is clear that the combined mechanism is the most likely failure mode of the frame. The probability of failure due to that mode alone is $p_{F1} = \Phi(-2.00) = 0.0227$. The difference, $0.0264 - 0.0227 = 0.0037$, is the contribution of the sway and beam failure modes.

Now, suppose the frame is proof tested under the load values $h_0 = 70$ kN and $v_0 = 72$ kN, and survives. We are interested in updating our estimate of the system failure probability in the light of this information. For this purpose, we introduce the following additional limit-state functions that define the survival of the frame under the three mechanisms:

$$\text{Survival in sway mechanism: } g_4(\mathbf{X}) = 5h_0 - M_1 - M_2 - M_4 - M_5, \tag{E9}$$

Example 8.7 (cont.)

$$\text{Survival in beam mechanism: } g_5(\mathbf{X}) = 5v_0 - M_2 - 2M_3 - M_4, \tag{E10}$$

$$\text{Survival in combined mechanism: } g_6(\mathbf{X}) = 5h_0 + 5v_0 - M_1 - 2M_3 - 2M_4 - M_5. \tag{E11}$$

Let $\mathcal{E}_i \equiv \{g_i(\mathbf{X}) \leq 0\}$, $i = 1, \ldots, 6$, denote the component events. The probability that the frame will fail under the loads H and V, given that it has survived under the loads h_0 and v_0, is given by

$$\begin{aligned} p_{F|\text{proof test}} &= \Pr(\mathcal{E}_1 \cup \mathcal{E}_2 \cup \mathcal{E}_3 | \mathcal{E}_4\mathcal{E}_5\mathcal{E}_6) \\ &= \frac{\Pr(\mathcal{E}_1\mathcal{E}_4\mathcal{E}_5\mathcal{E}_6 \cup \mathcal{E}_2\mathcal{E}_4\mathcal{E}_5\mathcal{E}_6 \cup \mathcal{E}_3\mathcal{E}_4\mathcal{E}_5\mathcal{E}_6)}{\Pr(\mathcal{E}_4\mathcal{E}_5\mathcal{E}_6)}. \end{aligned} \tag{E12}$$

It is seen that the numerator is the probability of a general system with six components and three min-cut sets, while the denominator is the probability of a parallel system with three components. Defining $\mathcal{C}_1 = \mathcal{E}_1\mathcal{E}_4\mathcal{E}_5\mathcal{E}_6$, $\mathcal{C}_2 = \mathcal{E}_2\mathcal{E}_4\mathcal{E}_5\mathcal{E}_6$, and $\mathcal{C}_3 = \mathcal{E}_3\mathcal{E}_4\mathcal{E}_5\mathcal{E}_6$ and using the inclusion–exclusion rule in (8.23), the numerator is written as

$$\begin{aligned} \Pr(\mathcal{C}_1 \cup \mathcal{C}_2 \cup \mathcal{C}_3) &= \Pr(\mathcal{C}_1) + \Pr(\mathcal{C}_2) + \Pr(\mathcal{C}_3) - \Pr(\mathcal{C}_1\mathcal{C}_2) \\ &\quad - \Pr(\mathcal{C}_1\mathcal{C}_3) - \Pr(\mathcal{C}_2\mathcal{C}_3) + \Pr(\mathcal{C}_1\mathcal{C}_2\mathcal{C}_3). \end{aligned} \tag{E13}$$

Observe that each of the above probability terms represents a parallel system problem. The first three terms involve four components each, the next three terms involve five components each, and the last term involves all six components.

To obtain a FORM solution, we first obtain the design points for the limit-state functions 4, 5, and 6. The results are:

$$\beta_4\widehat{\boldsymbol{\alpha}}_4 = -3.79[0.692\ 0.505\ 0.127\ 0.381\ 0.325\ 0.000\ 0.000], \tag{E14}$$

$$\beta_5\widehat{\boldsymbol{\alpha}}_5 = -3.32[0.408\ 0.559\ 0.645\ 0.324\ 0.000\ 0.000\ 0.000], \tag{E15}$$

$$\beta_6\widehat{\boldsymbol{\alpha}}_6 = -1.56[0.602\ 0.261\ 0.556\ 0.459\ 0.220\ 0.000\ 0.000]. \tag{E16}$$

Note that it is important to maintain the size of the vector of random variables, even though H and V are not involved in these limit-state functions (thus 0s appear in the last two elements of the $\widehat{\boldsymbol{\alpha}}_i$ vectors). For each of the parallel systems, linearization is carried out at the joint design point of the hyperplanes tangent to the limit-state surfaces at their individual design points, obtained by solving (8.67). The results of the analyses, using the multinormal probability formulation for parallel systems in (8.72), are $\Pr(\mathcal{C}_1) \cong 0.00867$, $\Pr(\mathcal{C}_2) \cong 0.00074$, $\Pr(\mathcal{C}_3) \cong 0.01447$, $\Pr(\mathcal{C}_1\mathcal{C}_2) \cong 0.00001$, $\Pr(\mathcal{C}_1\mathcal{C}_3) \cong 0.00607$, $\Pr(\mathcal{C}_2\mathcal{C}_3) \cong 0.00018$, and $\Pr(\mathcal{C}_1\mathcal{C}_2\mathcal{C}_3) \cong 0.00001$, so that we obtain the FORM approximation $\Pr(\mathcal{C}_1 \cup \mathcal{C}_2 \cup \mathcal{C}_3) \cong 0.0176$. The parallel-system probability in the denominator is similarly solved, resulting in the FORM approximation $\Pr(\mathcal{E}_4\mathcal{E}_5\mathcal{E}_6) \cong 0.941$. Thus, the updated failure probability of the frame after survival under the proof-load test is

Example 8.7 (cont.)

$$p_{F|\text{proof test}} \cong \frac{0.0176}{0.941} = 0.0187. \tag{E17}$$

In the light of the frame's survival under the proof-load test, the estimate of the failure probability of the system has gone down from 0.0264 to 0.0187.

8.11 Bi-Component Bounds for FORM Approximation of Series Systems

In (8.35) and (8.38), bi-component bounds were developed involving the pairwise joint component failure probabilities $p_{ij} = \mathrm{E}\left[\bar{S}_i\bar{S}_j\right] = \Pr\left(\mathcal{E}_i\mathcal{E}_j\right)$. When using the FORM approximation, this probability reduces to the bivariate standard normal CDF, $p_{ij} \cong \Phi_2\left(-\beta_i, -\beta_j, \rho_{ij}\right)$. In this expression, the β_i and the unit normal vectors that define the correlation coefficient as $\rho_{ij} = \widehat{\boldsymbol{\alpha}}_i \cdot \widehat{\boldsymbol{\alpha}}_j$ can be those associated with the individual limit-state surfaces, or, for a better approximation, the primed ones associated with the joint design point of the linearized surfaces (see (8.65)–(8.70)). Because of the importance of this probability term in series-system reliability analysis, this section presents an efficient method for computing this bivariate probability.

The bivariate standard normal CDF is defined as

$$\Phi_2\left(-\beta_i, -\beta_j, \rho\right) = \int_{-\infty}^{-\beta_i}\int_{-\infty}^{-\beta_j} \phi_2(x, y, \rho)\mathrm{d}x\mathrm{d}y, \tag{8.73}$$

where

$$\phi_2(x, y, \rho) = \frac{1}{2\pi\sqrt{1-\rho^2}} \exp\left[-\frac{x^2 + y^2 - 2\rho xy}{2(1-\rho^2)}\right] \tag{8.74}$$

is the bivariate standard normal PDF with correlation coefficient ρ. Taking derivatives of the above function, it is easily found that

$$\frac{\partial^2\phi_2(x, y, \rho)}{\partial x \partial y} = \frac{\partial\phi_2(x, y, \rho)}{\partial\rho}. \tag{8.75}$$

Furthermore,

$$\phi_2(x, y, \rho) = \phi_2(x, y, 0) + \int_0^{\rho} \frac{\partial\phi_2(x, y, r)}{\partial r}\mathrm{d}r. \tag{8.76}$$

Substituting this last expression in (8.71), we can write

$$\begin{aligned}\Phi_2\left(-\beta_i,-\beta_j,\rho\right) &= \Phi(-\beta_i)\Phi\left(-\beta_j\right)+\int_{-\infty}^{-\beta_i}\int_{-\infty}^{-\beta_j}\int_0^{\rho}\frac{\partial\phi_2(x,y,r)}{\partial r}\mathrm{d}r\mathrm{d}x\mathrm{d}y\\ &= \Phi(-\beta_i)\Phi\left(-\beta_j\right)+\int_{-\infty}^{-\beta_i}\int_{-\infty}^{-\beta_j}\int_0^{\rho}\frac{\partial^2\phi_2(x,y,r)}{\partial x\partial y}\mathrm{d}r\mathrm{d}x\mathrm{d}y\\ &= \Phi(-\beta_i)\Phi\left(-\beta_j\right)+\int_0^{\rho}\int_{-\infty}^{-\beta_i}\int_{-\infty}^{-\beta_j}\frac{\partial^2\phi_2(x,y,r)}{\partial x\partial y}\mathrm{d}x\mathrm{d}y\mathrm{d}r\\ &= \Phi(-\beta_i)\Phi\left(-\beta_j\right)+\int_0^{\rho}\phi_2\left(-\beta_i,-\beta_j,r\right)\mathrm{d}r.\end{aligned} \tag{8.77}$$

Hence, the double integral in (8.73) is reduced to a single-fold integral of the bivariate PDF over the correlation coefficient. Given its finite limits, this integral is easily computed numerically.

Example 8.8 – Bounds on the Reliability of Ductile Frame

For the elastoplastic frame in Example 8.7, using (8.35) and (8.38), the bi-component bounds of the system failure probability are

$$P_F \geq p_1 + \max(0, p_2 - p_{12}) + \max(0, p_3 - p_{13} - p_{23}) \cong 0.02639, \tag{E1}$$

$$P_F \leq p_1 + p_2 + p_3 - p_{12} - \max(p_{13}, p_{23}) \cong 0.02647. \tag{E2}$$

The results are shown as approximations because the FORM is used to compute the marginal and joint probabilities. The bi-component probabilities are computed using (8.77) with β_i and $\widehat{\boldsymbol{\alpha}}_i$ corresponding to the individual design points. It is seen that these bounds narrowly bracket the exact FORM result, $p_{F1} = 0.0264$, given in (E8) of Example 8.7.

Example 8.9 – Bounds on Reliability of Brittle Structure with Load-Path Dependence

The elastoplastic behavior of the frame in Example 8.7 is a special structural attribute that simplifies the reliability analysis. Specifically, the description of the failure mechanisms is not affected by the order of formation of plastic hinges or by the manner in which the loads are applied. This is not the case with structures having other kinds of behavior, such as gradual or abrupt loss of strength. This is demonstrated through a simple example.

Consider the simple structural system in Figure 8.12(a), consisting of two vertical bars holding a horizontal beam that carries loads W_1 and W_2. The vertical bars have tensile capacities R_1 and R_2, while the horizontal beam has a large capacity and is not subject to failure. Consider two different

Example 8.9 (cont.)

load-deformation behaviors of the bars: ductile, shown in Figure 8.12(b), and brittle, shown in Figure 8.12(c). The random variables, $\mathbf{X} = (W_1, W_2, R_1, R_2)$, are assumed to be statistically independent with the distributions listed in Table 8.5. As shown in this table, two cases for the second moments are considered: Case 1 represents a statistically "symmetric" system, and Case 2 represents a statistically "asymmetric" system.

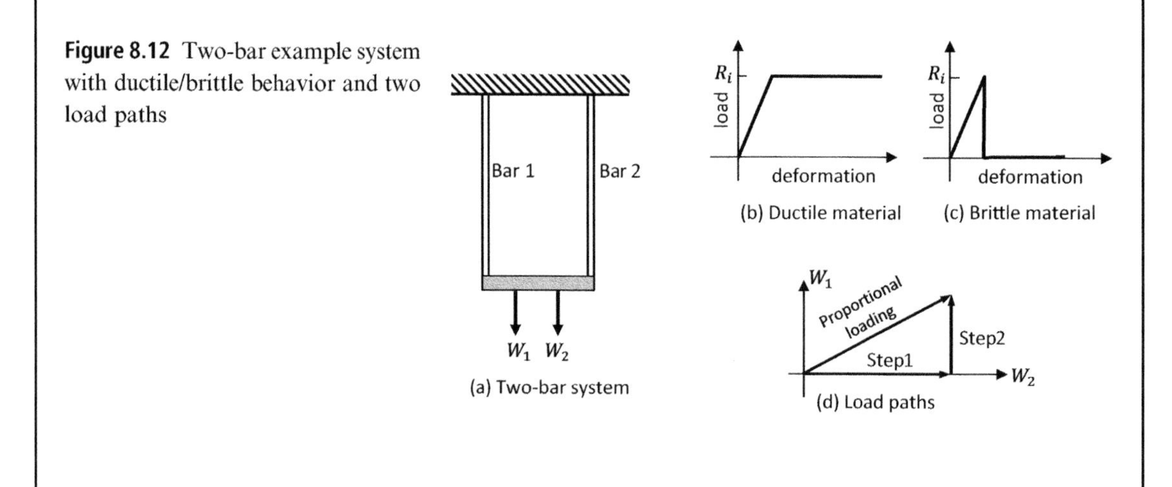

Figure 8.12 Two-bar example system with ductile/brittle behavior and two load paths

Table 8.5 Description of random variables

Variable	Distribution	Case 1		Case 2	
		Mean	Standard deviation	Mean	Standard deviation
R_1	Lognormal	100	30	120	35
R_2	Lognormal	100	30	90	25
W_1	Gamma	50	20	40	18
W_2	Gamma	50	20	60	25

First consider the case of ductile behavior. As the loads are applied and increased to their final value, one of the two bars may reach its plastic yield level, in which case it freely elongates but continues to carry a load equal to its plastic capacity. Any further increase in the loads must be carried by the second bar. Failure occurs when the second bar also reaches its yield limit, in which case it also freely elongates while carrying a load equal to its plastic capacity.

Example 8.9 (cont.)

The two-bar system then elongates without limit, which constitutes failure of the system. The limit-state function that describes this event is

$$g_1(\mathbf{X}) = R_1 + R_2 - W_1 - W_2. \tag{E1}$$

Now assume the bars have the brittle behavior shown in Figure 8.12(c). As can be seen, when the internal (tension) load in a bar reaches its brittle capacity, the bar fractures and carries no load. The acting loads and any increase in them must be carried by the surviving bar. Failure of the surviving bar under these loads constitutes failure of the system. Clearly, the reliability formulation of the problem should account for the order of failures of the bars. Furthermore, the path through which the loads are applied may have an influence and should be considered. Here, we consider the two load paths shown in Figure 8.12(d). Path 1 describes proportional loading, while Path 2 is a case where we first apply load W_1 up to its final value (step 1 of loading), then apply load W_2 (step 2 of loading). To do the analysis, we also need to know the proportions of the loads that are carried by each member in each configuration of the system. As the structure behaves linearly in each configuration between member failures, we can use linear structural analysis to determine these internal loads. For the intact structure, let c_{ij} be an influence coefficient such that $c_{ij}W_j$ denotes the internal force in member i resulting from load W_j. These influence coefficients, which represent the internal force in each member for a unit value of the load, are determined by simple linear structural analysis. For the system under consideration, their values are $c_{11} = c_{22} = 2/3$ and $c_{12} = c_{21} = 1/3$. After the failure of one bar, all the load must be carried by the surviving bar, which means the influence coefficient then is 1. Below, we consider four distinct cases:

Load path 1, failure sequence 1 → 2:
Under proportional loading, three events must occur for bar 1 to fail first, followed by the failure of bar 2. These are enumerated below together with the corresponding limit-state functions. Note that in each case the event of interest occurs when the limit-state function takes on a negative value:

$$\text{Bar 1 fails before bar 2: } g_2 = (c_{21}W_1 + c_{22}W_2)/R_2 - (c_{11}W_1 + c_{12}W_2)/R_1 \tag{E2}$$

$$\text{Bar 1 fails: } g_3 = R_1 - c_{11}W_1 - c_{12}W_2 \tag{E3}$$

$$\text{Bar 2 fails after bar 1 has failed: } g_4 = R_2 - W_1 - W_2 \tag{E4}$$

The limit-state function in (E2) describes the condition that the ratio of the internal force in bar 1 to its capacity is larger than the same ratio for bar 2. This is the condition for bar 1 to fail before bar 2 as the loads are increased. (E3) describes the condition for bar 1 to fail under its internal load, and (E4) describes the condition for bar 2 to fail under the total load after bar 1 has failed.

Example 8.9 (cont.)

Load path 1, failure sequence 2 → 1:
This case is similar to the above case with the indices exchanged:

$$\text{Bar 2 fails before bar 1: } g_5 = -g_2 = (c_{11}W_1 + c_{12}W_2)/R_1 - (c_{21}W_1 + c_{22}W_2)/R_2, \quad \text{(E5)}$$

$$\text{Bar 2 fails: } g_6 = R_2 - c_{21}W_1 - c_{22}W_2, \quad \text{(E6)}$$

$$\text{Bar 1 fails after bar 2 has failed: } g_7 = R_1 - W_1 - W_2. \quad \text{(E7)}$$

Note that the limit-state function in (E5) is the complement of that in (E2). Only one of the failure sequences can happen, so the above sets of three limit-state functions for each sequence constitute disjoint cut sets of the system under load path 1. Thus, the system has the disjoint cut sets $\mathcal{DC} = \{(2,3,4),(5,6,7)\} = \{(2,3,4),(\overline{2},6,7)\}$, where we have replaced component 5 with its equivalent, the complement of component 2.

For load path 2, the first failure of one of the bars can happen under load W_1 alone, i.e., during step 1 of the loading, or under the application of load W_2 with the full value of W_1 acting, i.e., during step 2 of the loading, followed by the failure of the second bar. Note that these are disjoint events. They are formulated as follows:

Load path 2, step 1, failure sequence 1 → 2:

$$\text{Bar 1 fails before bar 2 in step 1 of loading: } g_8 = (c_{21}W_1)/R_2 - (c_{11}W_1)/R_1, \quad \text{(E8)}$$

$$\text{Bar 1 fails in step 1 of loading: } g_9 = R_1 - c_{11}W_1, \quad \text{(E9)}$$

$$\text{Bar 2 fails after bar 1 has failed: } g_{10} = g_4 = R_2 - W_1 - W_2. \quad \text{(E10)}$$

Load path 2, step 2, failure sequence 1 → 2:

$$\text{Bar 1 does not fail in step 1 of loading: } g_{11} = -g_9 = c_{11}W_1 - R_1, \quad \text{(E11)}$$

$$\text{Bar 2 does not fail in step 1 of loading: } g_{12} = c_{21}W_1 - R_2, \quad \text{(E12)}$$

$$\begin{gathered}\text{Bar 1 fails before bar 2 in step 2 of loading:}\\ g_{13} = (c_{22}W_2)/(R_2 - c_{21}W_1) - (c_{12}W_2)/(R_1 - c_{11}W_1),\end{gathered} \quad \text{(E13)}$$

$$\text{Bar 1 fails in step 2 of loading: } g_{14} = g_3 = R_1 - c_{11}W_1 - c_{12}W_2, \quad \text{(E14)}$$

$$\text{Bar 2 fails after bar 1 has failed: } g_{15} = g_4 = R_2 - W_1 - W_2. \quad \text{(E15)}$$

The limit-state function in (E13) describes the condition where the ratio of the internal force in bar 1 due to the application of W_2 in the second loading step to its residual capacity is greater than the same ratio for bar 2. This condition ensures that bar 1 is the one to fail first. Note that $g_{10} = g_4, g_{11} = -g_9, g_{14} = g_3$, and $g_{15} = g_4$. The case for the failure sequence *2 → 1* is obtained from the above by exchanging indices:

Example 8.9 (cont.)

Load path 2, step 1, failure sequence 2 → 1:

$$\text{Bar 2 fails before bar 1 in step 1 of loading:} \quad g_{16} = -g_8 = (c_{11}W_1)/R_1 - (c_{21}W_1)/R_2, \tag{E16}$$

$$\text{Bar 2 fails in step 1 of loading: } g_{17} = -g_{12} = R_2 - c_{21}W_1, \tag{E17}$$

$$\text{Bar 1 fails after bar 2 has failed: } g_{18} = g_7 = R_1 - W_1 - W_2. \tag{E18}$$

Load path 2, step 2, failure sequence 2 → 1:

$$\text{Bar 2 does not fail in step 1 of loading: } g_{19} = g_{12} = c_{21}W_1 - R_2, \tag{E19}$$

$$\text{Bar 1 does not fail in step 1 of loading: } g_{20} = -g_9 = c_{11}W_1 - R_1, \tag{E20}$$

$$\text{Bar 2 fails before bar 1 in step 2 of loading:} \quad g_{21} = -g_{13} = (c_{12}W_2)/(R_1 - c_{11}W_1) - (c_{12}W_2)/(R_2 - c_{21}W_1), \tag{E21}$$

$$\text{Bar 2 fails in step 2 of loading: } g_{22} = g_6 = R_2 - c_{21}W_1 - c_{22}W_2, \tag{E22}$$

$$\text{Bar 1 fails after bar 2 has failed: } g_{23} = g_7 = R_1 - W_1 - W_2. \tag{E23}$$

Note that $g_{16} = -g_8$, $g_{17} = -g_{12}$, $g_{18} = g_7$, $g_{19} = g_{12}$, $g_{20} = -g_9$, $g_{21} = -g_{13}$, $g_{22} = g_6$, and $g_{23} = g_7$. As only one of the failure sequences can occur, the above set of 16 limit-state functions (E8)–(E23) for four failure sequences constitute disjoint cut sets of the system under load path 2. Thus, under the two-step loading path, the system has the disjoint cut sets $\mathcal{DC} = \{(8,9,10),(11,12,13,14,15),(16,17,18),(19,20,21,22,23)\} = \{(8,9,4),(\overline{9},12,13,3,4),(\overline{8},\overline{12},7),(12,\overline{9},\overline{13},6,7)\}$, where several components have been replaced with their equivalents.

In total, one limit-state function defines the ductile failure event as a component problem, five distinct limit-state functions define the disjoint cut sets for the case of the brittle system under proportional loading, and eight distinct limit-state functions define the disjoint cut sets for the case of the brittle system under two-step loading. For the two cases of brittle material, we use the FORM approximation for parallel systems (employing the joint design point of the linearized limit-state surfaces) to compute the probability for each disjoint cut set. These are then added to obtain the system failure probability. The results are summarized in Table 8.6.

The following observations in Table 8.6 are noteworthy: (a) For both load paths, the probability of failure of the brittle system is much higher than that of the ductile system. This is a clear demonstration of the advantage of ductile behavior versus brittle behavior in structures. (b) In the case of the statistically symmetric system, the load path has no effect on the system failure probability, whereas in the case of the statistically asymmetric system, there is a small load-path dependence. (c) In the case of the two-step load path, the probability of

Example 8.9 (cont.)

Table 8.6 Reliability analysis results for two-bar system

Material behavior, load path, failure sequence	Cut sets	System failure probability Case 1	Case 2
Ductile	(1)	**0.0192**	**0.0154**
Brittle, proportional loading			
Failure sequence 1 → 2	$(2, 3, 4)$	0.0363	0.00720
Failure sequence 2 → 1	$(\overline{2}, 6, 7)$	0.0363	0.07162
Total		**0.0726**	**0.0788**
Brittle, two-step loading			
Failure sequence 1 → 2, during step 1 of loading	$(8, 9, 4)$	0.00751	0.00066
Failure sequence 1 → 2, during step 2 of loading	$(\overline{9}, 12, 13, 3, 4)$	0.03043	0.00735
Failure sequence 2 → 1, during step 1 of loading	$(\overline{8}, \overline{12}, 7)$	0.00002	0.00001
Failure sequence 2 → 1, during step 2 of loading	$(12, \overline{9}, \overline{13}, 6, 7)$	0.03467	0.07015
Total		**0.0726**	**0.0782**

failure during step 1 of loading is much smaller than that during the second step of loading. (d) For the statistically symmetric case, the probabilities for the two sequences of bar failures under the two-step loading are different. This is because the two-step loading is more favorable to bar 2 than bar 1.

8.12 Event-Tree Approach for Modeling Sequential Failures

Structural reliability problems such as the above example are often characterized by sequences of events that lead to the failure of the system. An event tree is a convenient graphical means for analyzing such problems. Consisting of nodes and branches, the event tree helps in codifying the sequence of events and identifying the corresponding cut sets. The following description is motivated by Der Kiureghian (2006) and Lee and Song (2011).

The event-tree concept is demonstrated for a structure consisting of n two-state members under proportional loading, assuming that the structure behaves linearly between member state transitions and that member state transitions are irreversible., The tree starts at the root node, which represents the structure in its intact state, as shown in Figure 8.13. Disregarding the unlikely event that two or more members simultaneously make state transitions, one of

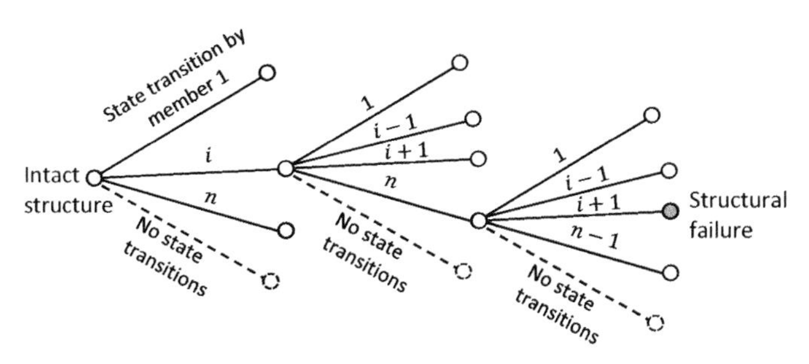

Figure 8.13 Event tree for a structure with n two-state members

$n + 1$ events will occur as loads are proportionally increased: Either no member will make a state transition, or member i, $i = 1, \ldots, n$, will make a state transition with all other members remaining in their original states. These are shown in Figure 8.13 as $n + 1$ branches emanating from the root node. The branch with no member state transition is shown by a dashed line and those with state transitions are shown as solid lines. The nodes with solid borders at the end of the solid branches now define new states of the structure, while the node at the end of the dashed branch is the end of that branch and is shown with a dashed border. At each solid line node, analysis is performed for the new configuration of the structure and new member state transitions are formulated. From each such node, n branches emanate, one of which represents the case of no member transition and is a terminating branch with a dashed node, while the remaining $n - 1$ represent transitions by the remaining active members. This process is continued until an unstable configuration of the structure, indicated by a gray node, is arrived at. All white nodes with solid borders remain active and can produce additional branches.

The event associated with each branch of the tree is formulated in terms of limit-state functions involving the loads, the member capacities, and the influence coefficients for the particular configuration of the structure. These functions essentially describe the conditions for transition of one member before all other members or, in the case of a dashed branch, non-transition of all members. As we saw for the two-bar system, the former event is formulated by writing the conditions that the particular member is the first to make a transition and that it makes the transition. The latter event is formulated by stating the condition that none of the active members makes a transition.

With the above formulation, each continuously solid sequence of branches leading to a gray node represents a disjoint cut set of the structure and each sequence of branches leading to a dashed node represents a disjoint link set of the structure. Each disjoint cut or link set is formulated as the intersection of all the conditioning events of all the branches in the sequence. The corresponding probability is then computed as that of a parallel system.

For a structure with a high degree of indeterminacy and a large number of members, the above formulation could lead to an enormously large tree. To reduce the number of computations, Lee and Song (2011) proposed a bounding method. Observe that disregarding any continuously solid sequence of branches will result in underestimating the failure probability, while disregarding any sequence of branches that ends in a dashed node will result in underestimating the survival probability or overestimating the failure probability. It follows that the sum of probabilities of any subset of the continuously solid sequence of branches ending in gray nodes provides a lower bound to the failure probability, while the sum of probabilities of any subset of the sequence of branches ending in dashed nodes provides a lower bound to the survival probability or the complement of an upper bound to the failure probability. In the simplest case, at each solid white node, one may expand only the solid branch that has the highest probability and the corresponding dashed branch that leads to survival. The latter is necessary so that the upper bound to the failure probability can be calculated. A simpler approach for selecting the branch would be to select the one that has the highest marginal probability of member transition, i.e., disregard the conditions for other members not to make transitions. This is a simple component problem and much easier to compute. Including additional branches, e.g., the branch with the second, third, etc., highest probability, will of course improve the bounds.

A seemingly simpler approach in formulating the event for each solid branch is to include the condition that a member makes a state transition but disregard the condition that this happens before other members make state transitions. This renders each sequence of solid branches a min-cut set, not a disjoint cut set. As a result, the events for different branches are dependent. The sum of the branch probabilities then renders the first-sum term in (8.23), which can be an excessively large upper bound to the system failure probability.

The above analysis can be extended to structures with multi-state members as long as the structural behavior between member state transitions is linear, proportional loading is assumed, and member transitions are irreversible. For example, a member may have three states: linear elastic, linear hardening, and fracture when an ultimate capacity is reached. The analysis for such a system naturally would require additional branches and nodes to capture the successive transitions of each member and the corresponding configurations of the system.

Example 8.10 – Reliability of Truss under Proportional Loading

The author is grateful to E.-H. Choi for assistance in developing this example. Consider the truss structure under horizontal and vertical loads shown in Figure 8.14. The members of the truss are made of S355 steel and have square hollow sections with the dimensions listed in Table 8.7. Let R_i, $i = 1, \dots, 10$, denote the capacity of the ith member of the truss in tension or compression. If the member fails in tension, then it carries a load equal to its yield capacity R_i. If the member fails in compression, i.e. by buckling, it carries no load and is effectively removed from the truss. The set of basic random variables of the problem is $\mathbf{X} = (R_1, \dots, R_{10}, H, V)$. Table 8.8 lists the distributions and second moments of the random variables. Note that R_i are assumed to be equicorrelated, while H and V are assumed to be statistically independent of the other random variables. We aim to determine the reliability of the truss under proportional loading, i.e., when the loads are proportionally increased from 0 to their final values.

Example 8.10 (cont.)

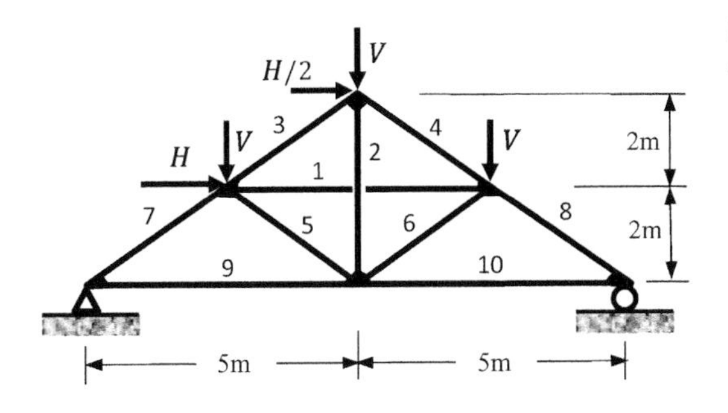

Figure 8.14 Truss system under horizontal and vertical loads

The truss system is statically one-degree indeterminate. However, as we saw in Example 8.1, the failure of any of the members 7, 8, 9, or 10, individually and either in tension or compression, leads to the failure of the truss. Of the remaining members, two must fail for the truss to fail. The event tree for this system is shown in Figure 8.15, expanded only for node 6. Nodes 1 to 5 have similar expansions. As demonstrative examples, the events for four sequences of branches are formulated below.

Table 8.7 Cross-sectional dimensions of truss members

Member	Section (mm × mm × mm)	Area (cm^2)
1	100×100×8	27.24
2, 5, 6	90×90×3	10.21
3, 4	90×90×6	19.23
7, 8, 9, 10	160×160×10	56.57

Table 8.8 Description of random variables

Variable	Distribution	Mean (kN)	Standard deviation (kN)	Correlation
R_1	Lognormal	900	270	Equicorrelated $\rho_{ij} = 0.15$, $i \neq j$
R_2, R_5, R_6	Lognormal	350	105	
R_3, R_4	Lognormal	700	210	
R_7, R_8, R_9, R_{10}	Lognormal	2000	600	
H	Gamma	200	60	Independent
V	Gamma	150	45	Independent

Let $c_{iH}^{(0)}$ denote the influence coefficient for member i relative to the horizontal loads in the intact state of the truss. In essence, $c_{iH}^{(0)}$ is the internal force in member i for $H = 1$. Similarly, let $c_{iV}^{(0)}$ denote the influence coefficient for member i relative to the vertical loads in the intact state

Example 8.10 (cont.)

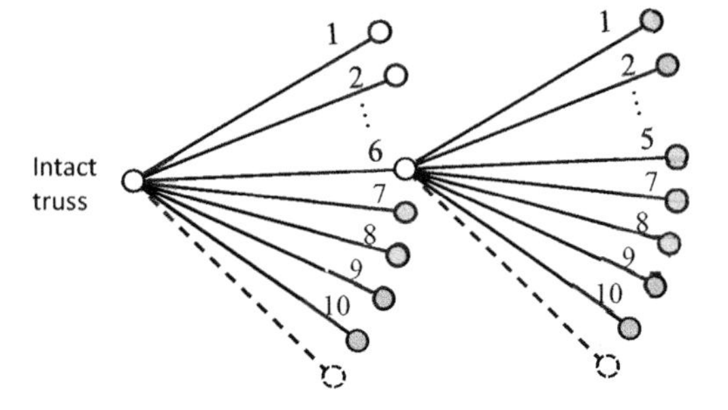

Figure 8.15 Event tree for truss system

of the truss. Then $c_{iH}^{(0)}H + c_{iV}^{(0)}V$ is the internal force in member i in the intact state of the truss. We determine the influence coefficients though linear structural analysis by applying the two load cases $H = 1, V = 0$ and $H = 0, V = 1$. The influence coefficients for the intact state of the truss are collected in the 10×2 matrix

$$\mathbf{c}^{(0)} = \begin{bmatrix} -0.3891 & -1.1790 \\ 0.0887 & 0.0569 \\ 0.2492 & -0.8459 \\ -0.3911 & -0.8459 \\ -0.3911 & -0.0455 \\ 0.2492 & -0.0455 \\ 0.6403 & -2.4012 \\ -0.6403 & -2.4012 \\ 1.0000 & 1.8750 \\ 0.5000 & 1.8750 \end{bmatrix}. \tag{E1}$$

Hence, the matrix product $\mathbf{c}^{(0)}[H \;\; V]^{\mathrm{T}}$ provides the internal forces in the members. Observe from the signs of the influence coefficients that members 1, 4, 5, and 8 are in compression while members 2, 9, and 10 are in tension. Given the magnitudes of the influence coefficients and the mean lead values, we assume that members 3 and 7 are also in compression and member 6 is in tension.

Let $\mathbf{s} = [-1 \;\; 1 \;\; -1 \;\; -1 \;\; -1 \;\; 1 \;\; -1 \;\; -1 \;\; 1 \;\; 1]$ be a vector with elements s_i, $i = 1, \ldots, 10$, indicating the nature of the internal forces in the members: $s_i = 1$ indicates that member i is in tension and $s_i = -1$ indicates that member i is in compression.

Now consider the solid branch in the event tree corresponding to the state transition of member 6. Limit-state functions defining the event that member 6 is the first to make a state transition in the intact state of the truss are

Example 8.10 (cont.)

$$g_i(\mathbf{X}) = s_i \frac{c_{iH}^{(0)} H + c_{iV}^{(0)} V}{R_i} - s_6 \frac{c_{6H}^{(0)} H + c_{6V}^{(0)} V}{R_6}, i = 1, \ldots, 5, 7, \ldots, 10. \quad \text{(E2)}$$

The limit-state function that defines the event that member 6 indeed makes a transition is

$$g_6(\mathbf{X}) = R_6 - s_6\left(c_{6H}^{(0)} H + c_{6V}^{(0)} V\right). \quad \text{(E3)}$$

The above 10 limit-state functions define the solid branch indicating the state transition by member 6. This event leads to a revised configuration of the truss, as indicated by the white node with solid border at the end of the branch. As member 6 is in tension, it is removed to arrive at the revised configuration of the truss with a pair of collinear tension forces of magnitude R_6 applied at the corresponding nodes. Similar limit-state functions apply to the solid branches for the other members. However, for members 1, 3, 4, and 5, which fail in compression (buckling), the member is simply removed to arrive at the revised configuration of the system with no loads added at the corresponding nodes. Furthermore, for members 7, 8, 9, and 10, the state transitions lead to the failure of the truss, as indicated by the gray nodes at the end of the corresponding branches.

The limit-state functions that define the dashed survival branch from the intact state of the truss (see Figure 8.15) are

$$g_{10+i}(\mathbf{X}) = s_i\left(c_{iH}^{(0)} H + c_{iV}^{(0)} V\right) - R_i, i = 1, \ldots, 10. \quad \text{(E4)}$$

This branch represents a link set of the system. The complement of the probability of this branch is an upper bound to the system failure probability.

Now consider the state of the truss after member 6 makes its state transition. Let $c_{iH}^{(-6)}$ and $c_{iV}^{(-6)}$ denote the influence coefficients in the revised truss configuration, where the superscript (-6) indicates the absence of member 6. Because member 6 failed in tension, a pair of collinear tension forces of magnitude R_6 are applied at the end nodes of that member. Let $c_{iR_6}^{(-6)}$, $i = 1, \ldots, 5, 7, \ldots, 10$, denote the influence coefficients relative to the tension forces R_6 in the revised truss configuration. The set of limit-state functions that define the solid branch indicating a state transition by member 5 in the revised truss configuration are

$$\begin{aligned} g_{20+i}(\boldsymbol{X}) = {} & s_i^{(-6)} \frac{c_{iH}^{(-6)} H + c_{iV}^{(-6)} V + c_{iR_6}^{(-6)} R_6}{R_i} \\ & - s_5^{(-6)} \frac{c_{5H}^{(-6)} H + c_{5V}^{(-6)} V + c_{5R_6}^{(-6)} R_6}{R_5}, i = 1, \ldots, 4, 7, \ldots, 10 \end{aligned} \quad \text{(E5)}$$

$$g_{25}(\mathbf{X}) = R_5 - s_5^{(-6)}\left(c_{5H}^{(-6)} H + c_{5V}^{(-6)} V + c_{5R_6}^{(-6)} R_6\right). \quad \text{(E6)}$$

Example 8.10 (cont.)

The eight limit-state functions in (E5) state the conditions for member 5 to be the first to make a state transition, and (E6) indicates the condition for member 5 to indeed make a transition. With member 5 making the transition, the truss system fails, as indicated by the gray node at the end of the branch in the event tree (see Figure 8.15). A similar formulation applies to the remaining members 1–4, and 7–10.

Finally, the set of limit-state functions defining the dashed survival branch after member 6 makes its state transition are

$$g_{30+i}(\mathbf{X}) = s_i^{(-6)}\left(c_{iH}^{(-6)}H + c_{iV}^{(-6)}V + c_{iR_6}^{(-6)}R_6\right) - R_i, i = 1, \ldots, 5, 7, \ldots, 10. \quad \text{(E7)}$$

This essentially says that no other member makes a state transition and, therefore, this sequence of branches represents a link set that is disjoint of the earlier link set. The complement of the probability of this sequence of branches can be deducted from the previous upper bound to obtain a tighter upper bound to the truss failure probability.

The probability for each sequence of branches leading to failure or survival of the truss system is computed as a parallel system probability using the FORM approximation in (8.72). All the limit-state functions describing the events along the sequence of branches define the "components" of the parallel system. The joint design point of the component limit-state surfaces linearized at the respective design points is used. The "exact" solution for each sequence of branches is also obtained by Monte Carlo simulation using a sample of size 5×10^8. Many of the branch sequences provide virtually zero probabilities. Table 8.9 lists the results for the branch sequences that have dominant contributions. These include 4 sequences that lead to the survival of the truss and that represent disjoint link sets, and 10 sequences that lead to the failure of the truss and represent disjoint cut sets. The sum of the disjoint-link-set probabilities gives a lower bound to the survival probability, and its complement provides an upper bound to the failure probability. The sum of the disjoint-cut-set probabilities gives a lower bound to the system failure probability. The results based on the FORM approximation and the Monte Carlo simulation are:

$$\text{By FORM: } 7.78 \times 10^{-4} \leq p_F \leq 7.99 \times 10^{-4}, \quad \text{(E8a)}$$

$$\text{By Monte Carlo: } 7.18 \times 10^{-4} \leq p_F \leq 8.27 \times 10^{-4}. \quad \text{(E8b)}$$

Observe that accounting for only 4 disjoint link sets and 10 disjoint cut sets provides narrow bounds on the system failure probability. Furthermore, the FORM provides a good approximation of the failure probability bounds. It is also worth noting that this approach provides information about the sequences of member failures that dominantly contribute to the failure of the system. From Table 8.9, it is evident that the failure of member 4 followed by the failure of member 6 is the most likely sequence of member failures giving rise to the failure of the truss.

Example 8.10 (cont.)

Table 8.9 Critical survival and failure sequences and probability estimates

Survival sequence	Disjoint link-set probability		Failure sequence	Disjoint cut-set probability ($\times 10^{-5}$)	
	by FORM	by Monte Carlo		by FORM	by Monte Carlo
0	0.9990688	0.9990276	$4 \to 6$	13.92	14.38
$5 \to 0$	0.0000863	0.0000945	$1 \to 5$	12.62	13.14
$4 \to 0$	0.0000246	0.0000248	$1 \to 2$	11.02	11.65
$1 \to 0$	0.0000217	0.0000257	$4 \to 2$	10.13	11.55
			8	9.21	7.33
			$4 \to 1$	8.97	9.75
			$5 \to 6$	3.76	3.44
			$1 \to 4$	3.18	3.36
			9	2.98	4.00
			$5 \to 1$	1.99	2.16

8.13 Component Importance Measures

It is often of interest to determine the order of importance of the components of a system relative to the performance of the system. Over the years, a number of such measures have been introduced with different definitions of "importance" (see, e.g., Anders, 1990; Henley and Kumamoto, 1981; Song and Der Kiureghian, 2005). This section presents several such measures for two-state systems with two-state components.

Structural Important Measure

Recall that the binary function $\phi(s_1, \ldots, s_N)$ defines the state of the system as a function of the component state variables s_i, $i = 1, \ldots, N$, where $s_i = 1$ $(= 0)$ indicates that component i is in the working (failed) state. Specifically, the system is in the functioning state when $\phi = 1$ and the failed state when $\phi = 0$. Define

$$\Delta_i = \phi(s_1, \ldots, s_{i-1}, 1, s_{i+1}, \ldots, s_N) - \phi(s_1, \ldots, s_{i-1}, 0, s_{i+1}, \ldots, s_N). \tag{8.78}$$

Clearly, when $\Delta_i = 1$, the state of component i alters the state of the system for the particular selection of other component states. The realization $(s_1, \ldots, s_{i-1}, s_{i+1}, \ldots, s_N)$ of the other components is said to be a *critical path* for component i. When $\Delta_i = 0$, the state of component i does not alter the state of the system and $(s_1, \ldots, s_{i-1}, s_{i+1}, \ldots, s_N)$ is not a critical path of component i. The *structural importance* measure of component i is defined as

$$\mathrm{SI}_i = \frac{1}{2^{N-1}} \sum\nolimits_{(s_1,\ldots,s_{i-1},s_{i+1},\ldots,s_N)} \Delta_i(s_1,\ldots,s_{i-1},s_{i+1},\ldots,s_N), \tag{8.79}$$

where the summation is over all the state values of components other than component i. Essentially, SI_i provides the fraction of realizations of component states that are critical paths for component i. The higher this fraction, the higher the importance of component i. Note that this measure has to do only with the configuration of the system. It has no relation to the reliability of the components or the system.

Marginal Important Measure

Consider a system with statistically independent component states and let p_i denote the probability of failure of component i. We define the *marginal importance* measure for component i as

$$\mathrm{MI}_i = \frac{\partial p_F}{\partial p_i}. \tag{8.80}$$

Recall from Section 8.4 that $p_F = 1 - \mathrm{E}[\phi(\mathbf{s})] = 1 - p_i \mathrm{E}[\phi(s_1,\ldots,s_{i-1},0,s_{i+1},\ldots,s_N)] - (1-p_i)\mathrm{E}[(s_1,\ldots,s_{i-1},1,s_{i+1},\ldots,s_N)]$. It follows that

$$\mathrm{MI}_i = \mathrm{E}[\phi(s_1,\ldots,s_{i-1},1,s_{i+1},\ldots,s_N)] - \mathrm{E}[\phi(s_1,\ldots,s_{i-1},0,s_{i+1},\ldots,s_N)]. \tag{8.81}$$

Let $\mathcal{E}_{F+i} = \{\phi(s_1,\ldots,s_{i-1},1,s_{i+1},\ldots,s_N) = 0\}$ be the event of failure of the system when component i is replaced with a perfectly reliable substitute and $\mathcal{E}_{F-i} = \{\phi(s_1,\ldots,s_{i-1},0,s_{i+1},\ldots,s_N) = 0\}$ be the event of failure of the system when component i is removed. Noting that the probabilities of these events are the respective complements of the expectations in (8.81), the above measure can be written as

$$\mathrm{MI}_i = \Pr[\mathcal{E}_{F-i}] - \Pr[\mathcal{E}_{F+i}]. \tag{8.82}$$

So, MI_i measures the difference in the system failure probabilities when component i is removed and when it is replaced by a perfectly reliable substitute. This form of the measure is also known as the *boundary probability* measure (Anders, 1990), denoted $\mathrm{BP}_i = \Pr[\mathcal{E}_{F-i}] - \Pr[\mathcal{E}_{F+i}]$. In this form, the measure also applies to systems with dependent components.

Risk Achievement Worth Measure

The *risk achievement worth* measure is defined as

$$\mathrm{RAW}_i = \frac{\Pr(\mathcal{E}_{F-i})}{\Pr(\mathcal{E}_F)}. \tag{8.83}$$

This measure is the ratio of the failure probability of the system with component i removed to the failure probability of the original system. In essence, it represents a measure of how much the system failure probability increases when component i is removed.

Risk Reduction Worth Measure

The risk reduction worth measure is defined as

$$\text{RRW}_i = \frac{\Pr(\mathcal{E}_F)}{\Pr(\mathcal{E}_{F+i})}. \tag{8.84}$$

This is the ratio of the failure probability of the original system to the failure probability of the system with component i replaced by a perfectly reliable substitute. In essence, it represents a measure of how much the system failure probability is reduced when component i is replaced by a perfectly reliable substitute.

Fussell–Vesely Measure

Recall that the failure event of a system can be written as the union of its min-cut sets, each min-cut set representing the intersection of a subset of component failure events whose joint occurrence constitutes the failure of the system (see 8.22). The *Fussell-Vesely* measure (Fussell, 1973) is defined as

$$\text{FV}_i = \frac{\Pr\left[\bigcup_{k:\, i\in C_k}\bigcap_{j\in C_k}\mathcal{E}_j\right]}{\Pr\left[\bigcup_k\bigcap_{j\in C_k}\mathcal{E}_j\right]}, \tag{8.85}$$

where $\mathcal{E}_i$ is the failure event of component i. The numerator is the probability of the union of all the min-cut sets that include the ith component, while the denominator is the probability of the union of all the min-cut sets. Hence, according to this measure, a component that is a member of min-cut sets with high probability or a member of many min-cut sets tends to be more important.

Conditional Probability Measure

Song and Kang (2009) proposed the *conditional probability* importance measure. This is the conditional probability of component i having failed, given that the system has failed. Noting that $\Pr(\mathcal{E}_i|\mathcal{E}_F) = \Pr(\mathcal{E}_i\mathcal{E}_F)/\Pr(\mathcal{E}_F)$, this measure is defined as

$$\text{CP}_i = \frac{\Pr(\mathcal{E}_i\mathcal{E}_F)}{\Pr(\mathcal{E}_F)}, \tag{8.86}$$

where the numerator is the probability of the intersection of the failure events of component i and the system. This event is formed by intersecting all the min-cut sets of the system with the event $\mathcal{E}_i$.

Aside from the structural importance measure that requires evaluations of the system function only, the above measures all require computation of the failure probability of the system or its modified configurations. This can be done by methods described earlier in this chapter. When the available information is limited to marginal or low-order joint component probabilities, the LP formulation described in Section 8.7 can be used to obtain bounds on these importance measures. For the measure MI_i or BP_i, the computation is straightforward. We just need to define the coefficient vector $\mathbf{a}$ in (8.47a) for the configurations of the system with component i removed or replaced with a perfectly reliable substitute. However, the measures defined in (8.83)–(8.86) represent ratios of probabilities of the form

$$\mathrm{IM}_i = \frac{\Pr(\mathcal{A})}{\Pr(\mathcal{B})}, \tag{8.87}$$

where IM_i may represent any of RAW_i, RRW_i, FV_i, or CP_i. Let $\mathbf{a}_{\mathcal{A}}$ and $\mathbf{a}_{\mathcal{B}}$ define the coefficient vectors for events $\mathcal{A}$ and $\mathcal{B}$ so that $\Pr(\mathcal{A}) = \mathbf{a}_{\mathcal{A}}^{\mathrm{T}}\boldsymbol{\pi}$ and $\Pr(\mathcal{B}) = \mathbf{a}_{\mathcal{B}}^{\mathrm{T}}\boldsymbol{\pi}$. Substituting these relations in (8.85), we obtain a nonlinear function of $\boldsymbol{\pi}$, which cannot be directly handled by LP. Instead, we formulate the iterative LP problem

$$\text{Minimize or Maximize: } p = \left(\mathbf{a}_{\mathcal{A}}^{\mathrm{T}} - \lambda\mathbf{a}_{\mathcal{B}}^{\mathrm{T}}\right)\boldsymbol{\pi} \tag{8.88a}$$

$$\text{Subject to: } \mathbf{A}_1\boldsymbol{\pi} = \mathbf{b}_1 \tag{8.88b}$$

$$\mathbf{A}_2\boldsymbol{\pi} \leq \mathbf{b}_2 \tag{8.88c}$$

$$\mathbf{A}_3\boldsymbol{\pi} \geq \mathbf{b}_3, \tag{8.88d}$$

where λ is a parameter to be adjusted. One starts with a trial value of λ, usually $\lambda = 1$ for minimization and $\lambda = 0$ for maximization. The value of λ is then adjusted according to the rule

$$\lambda = \frac{\mathbf{a}_{\mathcal{A}}^{\mathrm{T}}\widehat{\boldsymbol{\pi}}}{\mathbf{a}_{\mathcal{B}}^{\mathrm{T}}\widehat{\boldsymbol{\pi}}}, \tag{8.89}$$

where $\widehat{\boldsymbol{\pi}}$ is the solution of (8.88) for the current value of λ. Provided the objective function in (8.88a) converges to 0 at the solution point, the fraction in (8.89) provides the desired bound. According to Dinkelbach (1967), convergence in this algorithm is guaranteed and is usually achieved within a few iterations. An example in the next section demonstrates this approach for computing the component importance measures.

The matrix-based reliability method by Song and Kang (2009) offers another alternative for computing the component importance measures. Recall that in this method the system failure probability is computed as $\mathbf{a}^{\mathrm{T}}\boldsymbol{\pi}$, where for the case of statistically independent component states efficient matrix manipulation methods are developed for constructing the vectors $\mathbf{a}$ and $\boldsymbol{\pi}$. For each of the measures in (8.81)–(8.86), one just needs to formulate the coefficient vector $\mathbf{a}$ for each of the events defining the importance measure. For example, to compute the RAW_i measure, we need to construct $\mathbf{a}$ for the event $\mathcal{E}_F$ and for the event $\mathcal{E}_{F-i}$ for each component of interest. Note that the vector $\boldsymbol{\pi}$ remains unchanged for all events defining the importance measure.

In the case of statistically dependent component states with a small number of "common-source" random variables, (8.53) can be used. However, if the coefficient vector $\mathbf{a}$ needs to be constructed for many different events, e.g., $\mathcal{E}_{F-i}$ for many components, then the form in (8.54) is advantageous as the integration is performed once for all events.

Upgrade Worth Measure

The importance measures described in this section are often used to identify the best component to upgrade in order to enhance the reliability of the system. Any of the above measures can be used to identify the importance ordering of components in influencing the system reliability. However, in practice, different components can have vastly different costs to upgrade. The component that has the highest influence may not be the most appropriate for an upgrade because of the cost.

Let θ_i denote a capacity measure of component i, e.g., the mean value if the capacity is uncertain and is modeled as a random variable. Also let $\Delta\theta_i$ denote a positive increase in this capacity value if a fixed amount of resources is expended. The *upgrade worth* measure introduced by Song and Der Kiureghian (2005) is defined as

$$\mathrm{UW}_i = -\frac{\partial p_F}{\partial \theta_i} \Delta\theta_i, \tag{8.90}$$

where the partial derivative represents the sensitivity of the system failure probability with respect to θ_i. The measure represents the positive reduction in the system failure probability for a fixed-cost increase in the capacity of component i. It is a product of two factors: the rate of reduction in the system failure probability for a unit increase in the capacity of component i, and the amount of increase in the capacity of the component that can be achieved for the fixed amount of resources. UW_i can be high for a component that has a significant influence on the system failure probability or for one whose capacity can be increased by a larger amount than those of other components with the same amount of resources.

The above measure requires computing the sensitivity of the system failure probability with respect to parameter θ, which could be a distribution parameter or a parameter in a component limit-state function. This topic is addressed in Section 8.14.

Example 8.11 – Component Importance Measures for a Simple System

Consider the simple system with four components in Figure 8.3. Assume the component states are statistically independent, with p_i denoting the failure probability of component i. Table 8.10 lists the component importance measures relative to the system performance criterion of connection between nodes A and B. The last two columns in the table show the results when the component failure probabilities are identical, $p_i = p$, $i = 1, \ldots, 4$, and the numerical results for $p = 0.01$, respectively. For the upgrade worth measure, we consider the amount that the failure probability of each component can be reduced by a fixed amount of resources. Hence, in applying (8.90), $\theta_i = p_i$ and $\Delta\theta_i = \Delta p_i$. Note that in this case the partial derivative in (8.90) is identical to the marginal importance measure.

Example 8.11 (cont.)

The structural importance measure shows that component 4 has five critical paths (i.e., distinct realizations of system states for which component 4 working or not working changes the state of the system), whereas component 1 has three critical paths and components 2 and 3 each have one critical path. Note that these measures only have to do with the layout of the system and do not involve the component probabilities. For the other measures, the relative importance of the components depends on the component failure probabilities. It is useful, therefore, to examine the case where the component failure probabilities are identical, as listed in the last two columns of Table 8.9. For the marginal importance or boundary probability measures, MI_i and BP_i, we clearly have $(1+p-p^2)p > (1-p^2)p > (1-p)p^2$. Hence, component 4 is the most important, next is component 1, and components 2 and 3 are equally least important. The same order of importance of the components is evident from the results for the RAW_i, RRW_i, FV_i, and CP_i measures. However, the scales of the importance measures are vastly different. For example, for MI_i the measures for components 4 and 1 are 0.010099 and 0.009999, for RAW_i they are 100 and 99.01970, for RRW_i they are ∞ and 100.9900, for FV_i they are 1 and 0.009902, and for CP_i they are 1 and 0.990197, respectively. For these measures, therefore, it is the order of the importance measures that is relevant, not their numerical values.

Table 8.10 Component importance measures for example system

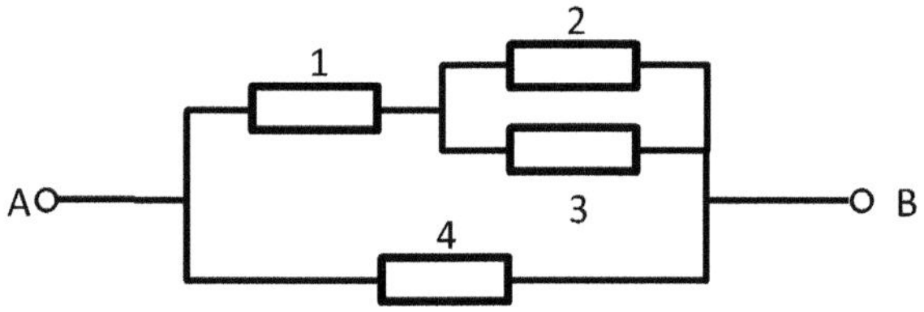

Component number i	Component importance measures		
	Component probabilities p_i	$p_i = p,\ i = 1, \ldots, 4$	$p = 0.01$
	SI_i		
1	3/8	3/8	
2	1/8	1/8	
3	1/8	1/8	
4	5/8	5/8	
	$MI_i = BP_i$		
1	$(1-p_2p_3)p_4$	$(1-p^2)p$	0.009999
2	$(1-p_1)p_3p_4$	$(1-p)p^2$	0.000099
3	$(1-p_1)p_2p_4$	$(1-p)p^2$	0.000099
4	$p_1+(1-p_1)p_2p_3$	$(1+p-p^2)p$	0.010099

Example 8.11 (cont.)

Table 8.10 (cont.)

Component number i	Component importance measures		
	Component probabilities p_i	$p_i = p,\ i = 1, \ldots, 4$	$p = 0.01$
	RAW_i		
1	$1/(p_1 + p_2p_3 - p_1p_2p_3)$	$1/(p + p^2 - p^3)$	99.01970
2	$(p_1 + p_3 - p_1p_3)/(p_1 + p_2p_3 - p_1p_2p_3)$	$(2 - p)/(1 + p - p^2)$	1.970492
3	$(p_1 + p_2 - p_1p_2)/(p_1 + p_2p_3 - p_1p_2p_3)$	$(2 - p)/(1 + p - p^2)$	1.970492
4	$1/p_4$	$1/p$	100
	RRW_i		
1	$(p_1 + p_2p_3 - p_1p_2p_3)/(p_2p_3)$	$(1 + p - p^2)/p$	100.9900
2	$(p_1 + p_2p_3 - p_1p_2p_3)/p_1$	$1 + p - p^2$	1.009900
3	$(p_1 + p_2p_3 - p_1p_2p_3)/p_1$	$1 + p - p^2$	1.009900
4	∞	∞	∞
	FV_i		
1	$p_1p_4/(p_1 + p_2p_3 - p_1p_2p_3)$	$p/(1 + p - p^2)$	0.009902
2	$p_2p_3p_4/(p_1 + p_2p_3 - p_1p_2p_3)$	$p^2/(1 + p - p^2)$	0.000099
3	$p_2p_3p_4/(p_1 + p_2p_3 - p_1p_2p_3)$	$p^2/(1 + p - p^2)$	0.000099
4	1	1	1
	CP_i		
1	$p_1/(p_1 + p_2p_3 - p_1p_2p_3)$	$1/(1 + p - p^2)$	0.990197
2	$(p_1 + p_3 - p_1p_4)p_2/(p_1 + p_2p_3 - p_1p_2p_3)$	$(2p - p^2)/(1 + p - p^2)$	0.019705
3	$(p_1 + p_2 - p_1p_4)p_3/(p_1 + p_2p_3 - p_1p_2p_3)$	$(2p - p^2)/(1 + p - p^2)$	0.019705
4	1	1	1
	UW_i		
1	$(1 - p_2p_3)p_4\Delta p_1$	$(1 - p^2)p\Delta p_1$	$0.009999\Delta p_1$
2	$(1 - p_1)p_3p_4\Delta p_2$	$(1 - p)p^2\Delta p_2$	$0.000099\Delta p_2$
3	$(1 - p_1)p_2p_4\Delta p_3$	$(1 - p)p^2\Delta p_3$	$0.000099\Delta p_3$
4	$(p_1 + p_2p_3 - p_1p_2p_3)\Delta p_4$	$(1 + p - p^2)p\Delta p_4$	$0.010099\Delta p_4$

Example 8.11 (cont.)

For the UW_i measure, the order of importance depends on the incremental reduction in the failure probability of each component that can be achieved by a fixed cost. If the costs are equal, then the order is the same as for the other measures and the numerical values are proportional to those of MI_i. But suppose $\Delta p_1 = 2\Delta p_2$, that is, for the same cost, the failure probability of component 1 can achieve double the reduction that can be achieved for component 4. In that case, it is more profitable to use the available resources to upgrade component 1 and not component 4.

8.14 System Sensitivity Measures

Parameters in system reliability analysis may appear in the distribution of random variables or in the limit-state functions defining component states. Let θ define one such parameter. The system sensitivity with respect to θ is defined as the derivative $\mathrm{d}p_F/\mathrm{d}\theta$. These sensitivities have many useful applications in such areas as optimal design, uncertainty management, reliability upgrading, and reliability analysis under parameter uncertainties. This section describes several methods for computing these sensitivities. Examples in this section and in later chapters demonstrate various applications of parameter sensitivities for systems.

First consider the case of a system with statistically independent component states. As shown in Section 8.4, the system failure probability then is a function of the individual component failure probabilities. Let $p_F = Q(p_1, \ldots, p_N)$ define this function. Then, by the chain rule of differentiation,

$$\frac{\mathrm{d}p_F}{\mathrm{d}\theta} = \sum_{i=1}^{N} \frac{\partial Q(p_1, \ldots, p_N)}{\partial p_i} \frac{\mathrm{d}p_i}{\mathrm{d}\theta}, \tag{8.91}$$

where $\mathrm{d}p_i/\mathrm{d}\theta$ is the sensitivity of the ith component failure probability with respect to θ. This derivative is computed based on the formulation of the component reliability problem. In a FORM approximation of the component problem, the component sensitivity is computed by the methods described in Section 6.6. Note that the above formulation allows for the possibility of a single parameter influencing more than one component. This may happen, for example, when the mean capacity values of several components (a distribution parameter) or their safe demand thresholds (a limit-state function parameter) are identical.

For the case of systems with statistically dependent component states, various methods of reliability analysis were described in Sections 8.4–8.12. Exact solutions of the system failure probability are given through the inclusion–exclusion rule involving the min-cut sets as in (8.23) or the min-link sets as in (8.24), or as a summation of probabilities of the disjoint cut sets as in (8.25) or the disjoint link sets as in (8.26). Each term in these formulations represents the probability of joint failure/survival of a set of components. Hence, for a direct differentiation of the terms in these equations, it is necessary to compute derivatives of the form $\mathrm{dPr}\left(\bigcap_{i\in I}\mathcal{E}_i\bigcap_{j\in J}\bar{\mathcal{E}}_j\right)/\mathrm{d}\theta$, where I and J are index sets of subsets of the components and a superposed bar indicates a survival event. This is essentially the sensitivity of a parallel system. Hence, exact evaluation of the system sensitivities requires computing the sensitivities of parallel systems. This is not an easy task.

In the FORM approximation of parallel systems, the failure probability is given in terms of a multinormal probability function involving a set of thresholds $\mathbf{B} = (\beta_1, \ldots, \beta_N)$ and a correlation matrix $\mathbf{R}$ with elements $\rho_{ij} = \widehat{\boldsymbol{\alpha}}_i \cdot \widehat{\boldsymbol{\alpha}}_j$, where β_i represent distances (reliability indices) and $\widehat{\boldsymbol{\alpha}}_i$ represent unit normal vectors of hyperplanes obtained from linearization of component limit-state functions (see (8.70)). Thus, differentiation of this probability requires not only the derivatives of β_i, which were developed in Section 6.6, but also those of $\widehat{\boldsymbol{\alpha}}_i$. Bjerager and Krenk (1989) and Ditlevsen and Madsen (1996) have shown that

$$\frac{\mathrm{d}\widehat{\boldsymbol{\alpha}}_i}{\mathrm{d}\theta} = -(\mathbf{I} + \beta_i \mathbf{H}_i)^{-1} \frac{1}{\nabla_{\mathbf{u}} G_i(\mathbf{u}_i^*, \theta)} \frac{\partial \nabla_{\mathbf{u}} G_i(\mathbf{u}_i^*, \theta)}{\partial \theta}, \tag{8.92}$$

where $G_i(\mathbf{u}, \theta)$ is the limit-state function for component i in the standard normal space, $\mathbf{I}$ is an identity matrix, and $\mathbf{H}_i$ is the Hessian of $G_i(\mathbf{u}, \theta)$ with respect to $\mathbf{u}$ evaluated at the linearization point $\mathbf{u}_i^*$. As $\nabla_{\mathbf{u}} G_i(\mathbf{u}_i^*, \theta) = \nabla_{\mathbf{x}} g_i(\mathbf{x}_i^*, \theta)\mathbf{J}_{\mathbf{x},\mathbf{u}}(\mathbf{x}_i^*, \theta)$, where $\mathbf{J}_{\mathbf{x},\mathbf{u}}(\mathbf{x}_i^*, \theta)$ is the Jacobian of the probability transformation to the standard normal space, the last term in (8.92) is

$$\frac{\partial \nabla_{\mathbf{u}} G_i(\mathbf{u}_i^*, \theta)}{\partial \theta} = \frac{\partial \nabla_{\mathbf{x}} g_i(\mathbf{x}_i^*, \theta)}{\partial \theta} \mathbf{J}_{\mathbf{x},\mathbf{u}}(\mathbf{x}_i^*, \theta) + \nabla_{\mathbf{x}} g_i(\mathbf{x}_i^*, \theta) \frac{\partial \mathbf{J}_{\mathbf{x},\mathbf{u}}(\mathbf{x}_i^*, \theta)}{\partial \theta}. \tag{8.93}$$

Note that when θ is a distribution parameter, the first term on the right-hand side of the above expression is 0, and when θ is a limit-state function parameter, the last term on the right-hand side vanishes.

It is evident that in order to compute the sensitivity of the FORM approximation of a parallel system, we need to compute the Hessians of the component limit-state functions at their respective linearization points as well as the derivatives of the gradient vector of each limit-state function and of the Jacobian of the probability transformation with respect to the parameter of interest, both of which involve mixed second derivatives. These second derivatives require rather tedious numerical analysis. Hence, they are seldom used.

An alternative for system sensitivity analysis is the LP formulation described in Section 8.7. Recall that this approach provides bounds on the system failure probability when the available probability information is incomplete, e.g., only marginal and low-order joint component-failure probabilities or bounds on these probabilities are available. Because the system failure probability in general is a function of the marginal and joint component-failure probabilities, using the chain rule of differentiation, one can write

$$\frac{\mathrm{d}p_F}{\mathrm{d}\theta} = \sum\nolimits_{i\in I_1} \frac{\partial p_F}{\partial p_i}\frac{\mathrm{d}p_i}{\mathrm{d}\theta} + \sum\nolimits_{(i,j)\in I_2} \frac{\partial p_F}{\partial p_{ij}}\frac{\mathrm{d}p_{ij}}{\mathrm{d}\theta} + \sum\nolimits_{(i,j,k)\in I_3} \frac{\partial p_F}{\partial p_{ijk}}\frac{\mathrm{d}p_{ijk}}{\mathrm{d}\theta} + \cdots, \tag{8.94}$$

where p_i, p_{ij}, p_{ijk}, etc., are the marginal, pairwise joint, triple-wise joint, etc., component failure probabilities and I_1, I_2, I_3, etc. are index sets of components for which these probabilities or their bounds are known. Most LP algorithms provide the sensitivities of the optimal solution with respect to the constraints of the problem as a by-product of the solution (Bertsimas and Tsitsiklis, 1997). Thus, the terms $\partial p_F/\partial p_i$, $\partial p_F/\partial p_{ij}$, $\partial p_F/\partial p_{ijk}$, etc. are readily available as by-products of the LP solution for the lower- and upper-bound estimates of p_F. We only need to compute the terms $\mathrm{d}p_i/\mathrm{d}\theta$, $\mathrm{d}p_{ij}/\mathrm{d}\theta$, and $\mathrm{d}p_{ijk}/\mathrm{d}\theta$. These are computed based on the type of probability information that is available on component failure events. Naturally, this approach can only provide bounds on the system sensitivity measures.

Another alternative for computing system sensitivities is the matrix-based method of Song and Kang (2009). As described in Section 8.8, in this method the system failure probability is computed as $\mathbf{a}^{\mathrm{T}}\boldsymbol{\pi}$, where, for the case of statistically independent component states, efficient matrix manipulation methods are available for constructing the vectors $\mathbf{a}$ and $\boldsymbol{\pi}$. Observe that the variation in a parameter θ may affect $\boldsymbol{\pi}$ but does not affect the coefficient vector $\mathbf{a}$. Hence, $\mathrm{d}p_F/\mathrm{d}\theta = \mathbf{a}^{\mathrm{T}}\mathrm{d}\boldsymbol{\pi}/\mathrm{d}\theta$. Song and Kang (2009) have shown that $\mathrm{d}\boldsymbol{\pi}/\mathrm{d}\theta$ can be efficiently computed by the matrix operation

$$\frac{\mathrm{d}\boldsymbol{\pi}}{\mathrm{d}\theta} = [\boldsymbol{\pi}_1\ \boldsymbol{\pi}_2 \cdots \boldsymbol{\pi}_N]\frac{\mathrm{d}\mathbf{p}}{\mathrm{d}\theta}. \tag{8.95}$$

In the above, $\boldsymbol{\pi}_i$, $i = 1, \ldots, N$, is the probability vector computed by the algorithm in (8.52) with the probabilities p_i and $1 - p_i$ of component i replaced by 1 and -1, respectively, and $\mathbf{p} = [p_1\ p_2 \cdots p_N]^{\mathrm{T}}$ is the vector of component failure probabilities.

In the case of statistically dependent component states with a small number of "common-source" random variables, based on (8.53), the derivative is computed as

$$\frac{\mathrm{d}p_F}{\mathrm{d}\theta} = \int_{\mathbf{x}} \mathbf{a}^{\mathrm{T}}\left[\frac{\mathrm{d}\boldsymbol{\pi}(\mathbf{x})}{\mathrm{d}\theta}f_{\mathbf{X}}(\mathbf{x}) + \boldsymbol{\pi}(\mathbf{x})\frac{\mathrm{d}f_{\mathbf{X}}(\mathbf{x})}{\mathrm{d}\theta}\right]\mathrm{d}\mathbf{x}. \tag{8.96}$$

If θ is a parameter of the distribution of the common-source random variables $\mathbf{X}$, the first term inside the square brackets drops, and if the distribution of the common-source random variables does not depend on θ, then the second term drops.

Example 8.12 – Component Importance and System Sensitivity Measures for an Electrical Substation System

Consider again the two-transmission-line electrical substation system with 12 components shown in Figure 8.7. We compute bounds on the component importance measures defined in Section 8.13 by the LP formulation. For the marginal importance or boundary probability (BP) measure in (8.82), the original LP formulation with the objective function $\Pr[\mathcal{E}_{F-i}] - \Pr[\mathcal{E}_{F+i}]$ is used. Note that this function can be formulated as $(\mathbf{a}_{-i}^{\mathrm{T}} - \mathbf{a}_{+i}^{\mathrm{T}})\boldsymbol{\pi}$, where $\mathbf{a}_{-i}$ and $\mathbf{a}_{+i}$ are the coefficient vectors for the two events $\mathcal{E}_{F-i}$ and $\mathcal{E}_{F+i}$, respectively. For the

Example 8.12 (cont.)

risk achievement worth (RAW), risk reduction worth (RRW), and Fussell–Veseley (FV) measures, the iterative LP formulation (8.88)–(8.89) is used. Up to tri-component probabilities are included in the analysis. The results for the lower and upper bounds of these measures, as reported in Song and Der Kiureghian (2005), are summarized in Figure 8.16. All but the RAW measure predict circuit breakers CB_1 and CB_2, followed by PT_1 and PT_2, to be the most important components, whereas RAW predicts power transformers PT_1 and PT_2 and drawout breakers DB_1 and DB_2 to be equally important as the two circuit breakers. All measures predict negligible importance for TB, FB_1, and FB_2.

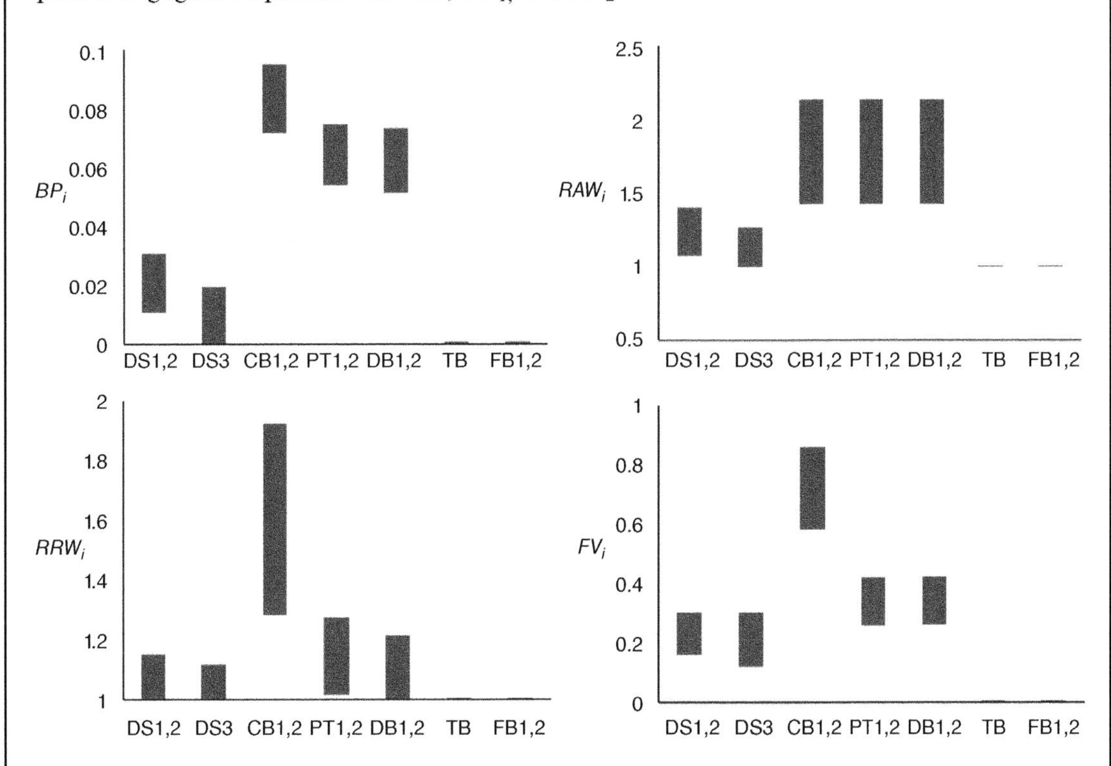

Figure 8.16 Bounds on component importance measures (reprinted from Song and Der Kiureghian, 2005, with permission from IOS Press)

To demonstrate an application of the upgrade worth measure (UW), we consider the relative importance of the system components for a fixed-cost increase in the logarithmic mean $\theta_i = \mathrm{E}[\ln R_i]$, $i = 1, \ldots, 12$, of the component capacities. We assume, for a fixed cost, the incremental increases in the component capacities, $\Delta\theta_i$, are 1.0 for DS_1, DS_2, and DS_3, 0.9 for CB_1 and CB_2, 0.3 for PT_1 and PT_2, 1.2 for DB_1 and DB_2, 0.8 for TB, and 1.0 for FB_1 and FB_2.

Example 8.12 (cont.)

As defined in (8.90), the upgrade worth measure requires the derivatives of the system failure probability with respect to θ_i. Here, we use the LP bounds method to compute these sensitivities. Up to tri-component probabilities are included as constraints and, owing to the jointly lognormal distribution of the capacities, the partial derivatives $\partial p_i/\partial\theta_i$, $\partial p_{ij}/\partial\theta_i$, and $\partial p_{ijk}/\partial\theta_i$ in (8.94) are computed exactly in terms of the uni-, bi-, and tri-normal cumulative probability functions. The terms $\partial p_F/\partial p_i$, $\partial p_F/\partial p_{ij}$, and $\partial p_F/\partial p_{ij}$ are obtained as by-products of the LP solution. The resulting sensitivities $\partial p_F/\partial\theta_i$ are then multiplied by the corresponding $\Delta\theta_i$ to obtain UW_i. Details are given in Song and Der Kiureghian (2005) and the results are plotted in Figure 8.17. Note that the ranges shown do not represent bounds on the UW_i values; rather, they represent UW_i values obtained for the lower and upper bounds of p_F. For DS_1, DS_2, DS_3, PT_1, PT_2, DB_1, and DB_2, which are shown in dark shade, the lower values represent the UW_i values based on the lower bound of p_F and the upper values represent the values based on the upper bound of p_F. This is reversed for the remaining components, CB_1, CB_2, TB, FB_1, and FB_2, which are shown in lighter shade (the latter three hardly visible in the figure). It is observed that CB_1 and CB_2 are still the most important components, but PT_1 and PT_2 have lost their importance because of their high upgrade cost while DS_1 and DS_2 have gained more importance because of the low cost of their upgrading.

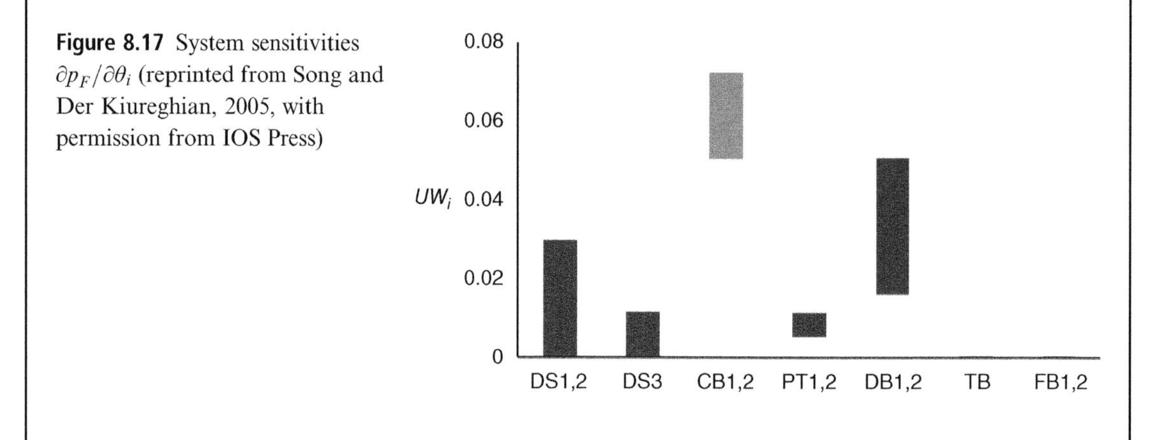

Figure 8.17 System sensitivities $\partial p_F/\partial\theta_i$ (reprinted from Song and Der Kiureghian, 2005, with permission from IOS Press)

8.15 Concluding Remarks

This chapter described the topic of system reliability, starting from the classical concepts developed for general systems that are primarily statistics based and assume component failure probabilities are estimated from life-testing data (Barlow and Proschan, 1975). It then focused on structural systems, where predictive models of system components are employed in the form of limit-state functions of basic random variables. The available data are for the basic random variables, and the limit-state functions bring in the physics of component behavior.

It was shown that the concepts of cut sets and link sets (minimum or disjoint) play important roles in the formulation and calculation of system reliability. Various methods were developed to assess the system reliability exactly, in approximation, or by bounding. These included methods that assume component states are statistically independent as well as methods that account for statistical dependence between components, which may arise from statistical dependence between the random variables or from random variables common to the limit-state functions defining different components. The chapter ends with the introduction of importance measures for system components and system sensitivity analysis.

Admittedly, application of the methods developed in this chapter to large systems can be computationally demanding. For this reason, the field is still evolving with researchers developing newer and computationally more efficient methods. For structural systems, one promising approach is the integration of reliability methods with general structural-analysis software, such as a finite-element code. As an example, consider the probability of failure of an elastoplastic frame structure under lateral loads. Rather than considering the system as a series system of its multiple modes of plastic mechanisms, one can formulate a limit-state function for the drift at the top of the structure exceeding a safe threshold. Because the drift of the structure is theoretically unbounded when a plastic mechanism forms, the reliability problem formulated in terms of the drift provides an upper bound to the probability of the frame structure forming a plastic mechanism. Chapter 12 describes this approach to structural reliability analysis.

PROBLEMS

8.1 The truss shown is subject to a variety of loads. It is considered failed if it cannot carry any of the load applications.

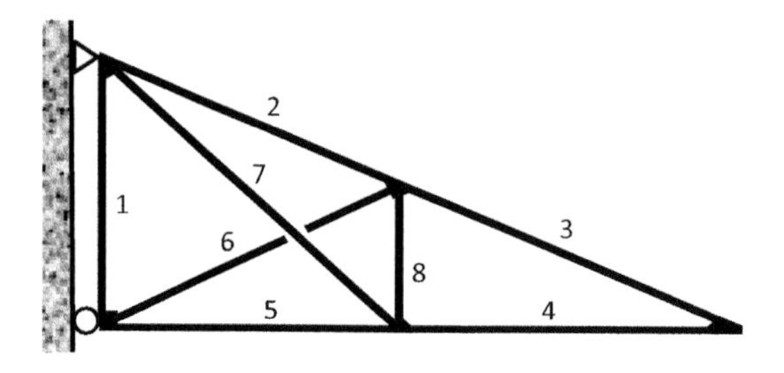

1) Determine the min-cut sets and min-link sets of the system.
2) Suppose the member failures are statistically independent (say, they are due to corrosion of members) and the probability of failure of each member is p. Using the expectation method described in Section 8.4, determine the failure probability of the truss as a function of p and evaluate it for $p = 0.01$.
3) Use the inclusion–exclusion rule described in Section 8.5 to obtain bounds on the failure probability of the truss and compare with the result in (2). Include at least two sum terms.

8.2 The structure shown in the figure consists of a propped cantilever beam and a bar supporting a random load P. The beam is made of a ductile material and has a random plastic moment capacity M. The bar is made of a brittle material and has infinite stiffness and a random tensile capacity T. The random variables $\mathbf{X} = (P, M, T)$ are statistically independent and have lognormal distributions with the second moments listed in the table below. Failure of the beam to carry load P constitutes failure of the system.

Variable	Mean	c.o.v.
P, kN	100	0.40
M, kNm	500	0.15
T, kN	40	0.25

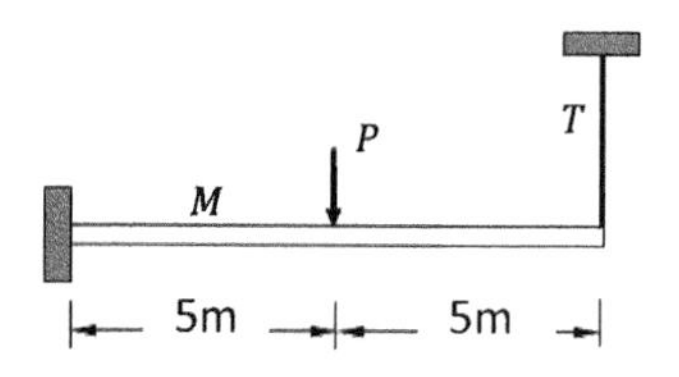

1) Draw an event tree showing sequential member failures leading to the failure of the system.
2) Formulate limit-state functions describing the disjoint cut sets and min-cut sets of the system.
3) Determine the FORM approximation of the system failure probability using (a) the min-cut sets and (b) the disjoint cut sets. Compare the results and comment on any differences. You may use CalREL or a similar program.

8.3 The two-story, one-bay frame structure shown in the figure has elastoplastic columns and rigid beams. The plastic moment capacities of the first-story columns are M_1 and those of the second-story columns are M_2. The structure is subjected to earthquake loads P_1 and P_2, as shown in the figure. The four random variables $\mathbf{X} = (M_1, M_2, P_1, P_2)$ have a Nataf joint distribution with the marginal distributions and second moments listed in the table below.

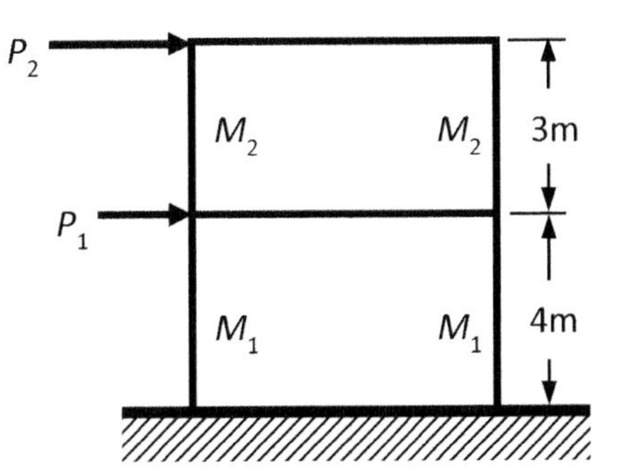

Variable	Marginal distribution	Mean	Standard deviation	Correlation coefficients			
				M_1	M_2	P_1	P_2
M_1, kNm	lognormal	300	60	1	0.3	0	0
M_2, kNm	lognormal	200	40	0.3	1	0	0
P_1, kN	gamma	50	20	0	0	1	0.5
P_2, kN	gamma	100	40	0	0	0.5	1

1) Determine the generalized reliability index of the structure against formation of a plastic mechanism.
2) Suppose that during its lifetime the structure is subject to three earthquakes with the forces P_1 and P_2 renewing at each earthquake, i.e., for each earthquake they have the same joint distribution as described above, but they are statistically independent from earthquake to earthquake. Determine the generalized reliability index of the structure during its lifetime.
3) Suppose the structure survives a static horizontal force of $f = 250$ kN applied at the roof level. Update the reliability estimate in part (1) in the light of this information.

8.4 The gabled frame shown below is subjected to random loads X_1 and X_2. The members have elastoplastic flexural behavior. Note that the columns are hinged at their bottoms. The plastic moment capacities at locations 1–5 are random variables M_i, $i = 1, \ldots, 5$. All random variables are independent lognormals with the means and standard deviations listed in the table below.

Variable	Mean	Standard deviation
X_1, kN	100	20
X_2, kN	60	30
$M_1, \ldots, M_5$, kNm	1500	400

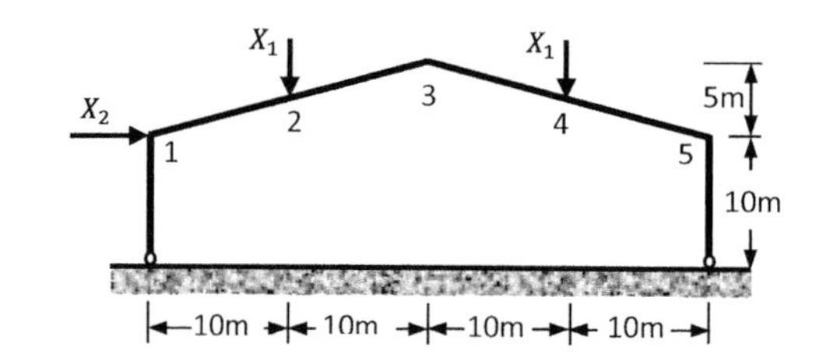

1) Using the method of virtual work, formulate limit-state functions for the failure mechanisms of the frame. Note that formation of two plastic hinges will lead to a failure mechanism. Disregard any mechanisms that involve negative work done by external forces.
2) Compute uni- and bi-component bounds on the failure probability of the frame.
3) Compute the FORM approximation of the failure probability of the frame and compare with the bounds computed in part (2).

8.5 Consider the rectangular plate in Problem 6.4. Suppose that in addition to E, ν, and K, the applied stress, denoted S, is also random, having a normal distribution with mean 8 ksi and standard deviation 1 ksi. FORM analysis with the four random variables $\mathbf{X} = (E, \nu, K, S)$ yields the result $\beta\hat{\boldsymbol{\alpha}} = 2.408\,[-0.492 \ \ -0.118 \ \ -0.698 \ \ 0.507]$. Now suppose the plate is subjected to repeated load applications with the resulting stresses S_i, $i = 1, \ldots, n$, being statistically independent and having the normal distribution just mentioned. The plate is considered to have failed if it fails under any of the applied loads. Determine the FORM approximation of the failure probability of the plate for $n = 1, 2, 5$, and 10. Hint: You do not need to perform additional reliability analyses but should consider the expanded space of the random variables $\mathbf{X} = (E, \nu, K, S_1, \ldots, S_n)$.

8.6 Reconsider the spacecraft with debris impact described in Problem 6.8. Suppose that during the service period of the spacecraft, each meter square of its outer shield will be subjected to five projectile impacts. Suppose the random variables $\mathbf{X} = (R, V, \gamma_P)$ are statistically independent and identically distributed from projectile to projectile.

1) Determine the probability that one or more holes will puncture through a 1 m^2 panel of the spacecraft shield.
2) Suppose the shield is composed of four panels, as shown in the figure below. Panel 1 protects the solid fuel, whereas panels 2–4 protect the oxygen tanks for the crew. Define the indicator variables s_i, $i = 1, \ldots, 4$, where $s_i = 0$ if panel i is punctured and $s_i = 1$ if panel i is not punctured. A catastrophic failure of the spacecraft will occur if panel 1 is punctured or any two of panels 2, 3, and 4 are punctured. Carry out the following analysis:
   a) Identify the minimum cut sets of the system of panels of the spacecraft.
   b) Formulate the system function $\phi(\mathbf{s})$ in terms of the indicator variables s_i, $i = 1, \ldots, 4$.
   c) Using the result in part (1), estimate the probability of catastrophic failure of the spacecraft during its service period.

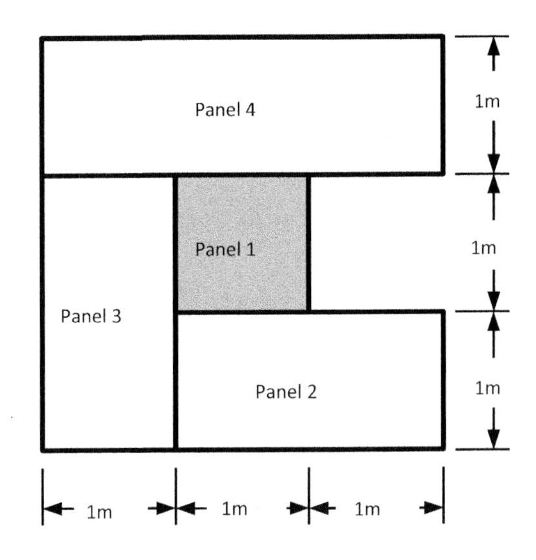

9 Simulation Methods

9.1 Introduction

Chapters 6 and 7 developed first- and second-order reliability methods (FORM and SORM) as approximate methods for assessing structural reliability. In Chapter 8, these methods were further developed for reliability assessment of structural systems. This chapter introduces simulation methods as an alternative to FORM and SORM. The word *simulation* is used because these methods simulate the real world by generating artificial outcomes of random variables and events. The task of computing probabilities then becomes a statistical problem with the artificially generated data. Some of these methods are asymptotically exact as the number of simulations grows, provided the random variable outcomes are properly generated. As a result, a major issue is the number of simulations, i.e., the size of the generated sample of artificial numbers to obtain accurate estimates of the probability. These methods are often computationally demanding, yet they offer a powerful tool to obtain "exact" solutions in order to check the accuracy of approximate methods, such as FORM and SORM. Furthermore, they are less restricted. For example, differentiability of the limit-state functions or one-to-one transformability of random variables are not necessary, though some simulation methods do take advantage of such properties.

The chapter starts with a description of the seemingly simple problem of simulating a random number uniformly distributed between 0 and 1. It turns out that this is not a simple matter. In fact, one could write a whole book about this topic. Here, only a brief introduction is provided but some of the challenges that require close attention are discussed. Next, methods for generating artificial samples of random variables with specified distributions are discussed. This is followed by the presentation of the ordinary Monte Carlo simulation method. This approach does not take shortcuts but usually is computationally demanding when the probability to be computed is small, which is often the case in structural-reliability applications. Next, the importance-sampling method that aims to employ fewer but a more effective sample of random variables is presented. This is followed by a host of other simulation methods that use various means to reduce the size of the needed sample. These include directional sampling, orthogonal-plane sampling, and subset simulation. The chapter ends with a presentation of methods for computing reliability sensitivities by simulation. Numerical examples carried out by use of CalREL or MATLAB are presented throughout the chapter.

9.2 Generation of Pseudorandom Numbers with Uniform Distribution

Our aim here is to generate an artificial sample of numbers that fits a Uni(0,1) distribution defined by the probability density function (PDF)

$$f_Y(y) = \begin{cases} 1, & 0 \le y \le 1, \\ 0, & \text{otherwise.} \end{cases} \tag{9.1}$$

It is not only necessary that the normalized frequency diagram of the sample fit the above distribution. We must also ascertain that the members of the sample are truly random in nature. This requires that the members of the sample be statistically independent of one another. As an example, imagine the set of numbers $y_i = i/1000$, $i = 1, 2, \ldots, 1000$. If we plot the normalized frequency diagram of these numbers, it will perfectly fit the above PDF. However, this is not a proper random sample from the uniform distribution because there is perfect dependence between the generated numbers. In fact, if we know the ith number in the sequence, we can determine the values of all the preceding and subsequent numbers. Generating a sequence of numbers that not only fits the distribution shape but is also random in nature is the main challenge in generating artificial samples.

There are of course ways of physically generating random numbers. The most common examples are throwing a dice and examining the sequence of a deck of cards after a thorough shuffling. However, it is difficult to use such methods when a large sample of random numbers is needed. In numerical applications, the most common approach is to use a pseudorandom number generator, which is typically an algorithm that generates a sequence of numbers that have the desired properties. One of the oldest and best known of such algorithms is the linear congruential generator (Rubinstein, 1981; Knuth, 1998), which is defined by the recursive relation

$$x_{i+1} = (ax_i + c) \bmod m, i = 0, 1, \ldots, \tag{9.2a}$$

$$y_{i+1} = \frac{x_{i+1}}{m}, \tag{9.2b}$$

where m is the modulus, a the multiplier, c the increment, and “mod” refers to the modulo operation, which in this case produces the remainder of the division of $(ax_i + c)$ by m. One must select a large value for m, $0 < a < m$, and $0 \le c < m$. The calculation starts with an initial “seed” x_0, which must be selected so that $0 \le x_0 < m$. The division of the result of the operation by m, as shown in (9.2b), produces a number that is between 0 and 1.

It should be clear that the sequence of numbers generated by the above algorithm is completely predictable if m, a, c, and x_0 are known. Furthermore, if any generated x_{i+1} matches x_0, then the sequence repeats itself. It follows that the sequence of pseudorandom numbers generated by this algorithm has a period and it turns out that the period of the sequence is closely related to the modulus m. For that reason, it is important to select as large m as possible. Various implementers of this algorithm have selected values for the constants based on empirical tests of randomness, e.g., test for lack of correlation between pairs (y_i, y_{i+k}), $i = 1, 2, \ldots$, for different k values. For example, Numerical Recipes (Press et al., 1986) recommends $m = 2^{32}$, $a = 1{,}664{,}525$, and $c = 1{,}013{,}904{,}223$. To further randomize the sequence, the initial seed is often selected by the computer using the current time or an atmospheric value, e.g., current ambient temperature or pressure.

There are many other pseudorandom number generators that use more sophisticated algorithms in order to assure stronger statistical independence for the members of the sequence. For example, some algorithms use two or more pseudorandom numbers to generate one sample outcome. As mentioned above, this is a vast field and well beyond our scope. The interested reader is referred to Knuth (1998) for a more in-depth treatment of the subject.

9.3 Generation of Pseudorandom Numbers with Specified Distribution

Suppose we wish to generate pseudorandom numbers according to the cumulative distribution function (CDF) $F_X(x)$. Let y_i, $i = 1, 2, \ldots$, be a sequence of pseudorandom numbers generated according to the algorithm in (9.2) that fits the Uni(0,1) distribution. Then, provided $F_X(x)$ is a strictly increasing function of x, the sequence

$$x_i = F_X^{-1}(y_i), i = 1, 2, \ldots \tag{9.3}$$

is the desired result. This is because for $Y = \text{Uni}(0,1)$, $X = F_X^{-1}(Y)$ has the PDF $f_X(x) = \mathrm{d}F_X(x)/\mathrm{d}x$. This can be verified by using the relation $f_X(x) = f_Y(y)|\mathrm{d}y/\mathrm{d}x|$ (see 2.87) and noting that $f_Y(y) = 1$ in the range of interest and $\mathrm{d}y/\mathrm{d}x = f_X(x)$.

The above approach requires inverting the CDF. For some important distributions, such as normal and lognormal, a closed-form solution of the inverse CDF is not available and the computation can be time consuming. For the case of these two distributions, an alternative solution is as follows: Select a pair (y_i, y_{i+1}) of the standard uniformly distributed pseudorandom numbers and compute a corresponding pair from the relations

$$u_i = \sqrt{-2\ln y_i}\sin(2\pi y_{i+1}), \tag{9.4a}$$

$$u_{i+1} = \sqrt{-2\ln y_i}\cos(2\pi y_{i+1}). \tag{9.4b}$$

One can show that u_i and u_{i+1} have the standard normal distribution. (The reader can easily verify this by using (2.86) in Chapter 2. Also see Problem 2.9.) This process is repeated with pairs of independent, standard uniformly distributed pseudorandom numbers to obtain the desired number of standard normal pseudorandom numbers. A linear transformation of the form $x_i = \mu + \sigma u_i$ can then be used to obtain a sample of pseudorandom numbers having a normal distribution with mean μ and standard deviation σ. To obtain a sample of pseudorandom numbers having a lognormal distribution with parameters λ and ζ (see Table A2.2), one should use the transformation $x_i = \exp(\lambda + \zeta u_i)$.

9.4 Generation of Pseudorandom Numbers for Dependent Random Variables

First consider the case of an n-vector of multinormal random variables $\mathbf{X} = \mathrm{N}(\mathbf{M}, \boldsymbol{\Sigma})$ having the $(n \times 1)$ mean vector $\mathbf{M}$ and the $(n \times n)$ covariance matrix $\boldsymbol{\Sigma}$. We assume $\boldsymbol{\Sigma}$ is a positive-definite matrix. Let $\mathbf{D} = \mathrm{diag}[\sigma_i]$ be the diagonal matrix of the standard deviations and $\mathbf{R} = \mathbf{D}^{-1}\boldsymbol{\Sigma}\mathbf{D}^{-1}$ be the correlation matrix. Also let $\mathbf{L}$ denote the lower-triangular Cholesky decomposition of the correlation matrix so that $\mathbf{R} = \mathbf{L}\mathbf{L}^{\mathrm{T}}$. To generate a pseudorandom sample of $\mathbf{X}$, first generate an n-vector of independent, standard normal pseudorandom numbers $\mathbf{u}$ according to the algorithm described in the preceding section. The desired sample of $\mathbf{X}$ then is computed as $\mathbf{x} = \mathbf{M} + \mathbf{L}\mathbf{D}\mathbf{u}$. Repeat this to obtain as many samples of $\mathbf{X}$ as needed.

Next consider the n-vector of random variables $\mathbf{X}$ having the Nataf distribution with marginals $F_j(x_j)$, $j = 1, \ldots, n$, and correlation matrix $\mathbf{R}$. Let $\mathbf{Z}$ denote the vector of correlated standard normal random variables defined by the element-to-element transformations $Z_j = \Phi^{-1}[F_j(X_j)], j = 1, \ldots, n$. Also let $\mathbf{R}_0$ denote the correlation matrix of random variables $\mathbf{Z}$, which is obtained in terms of $\mathbf{R}$ and the marginal distributions using the relations given in Appendix 3A, and $\mathbf{L}_0$ as its Cholesky decomposition so that $\mathbf{R}_0 = \mathbf{L}_0\mathbf{L}_0^{\mathrm{T}}$. To generate a pseudorandom sample of $\mathbf{X}$, we proceed as follows: (a) Using the algorithms described in the preceding section, generate a pseudorandom n-vector $\mathbf{u}$ having the standard normal distribution; (b) Compute the corresponding sample of $\mathbf{Z}$ by using the relation $\mathbf{z} = \mathbf{L}_0\mathbf{u}$; (c) Compute the elements of the desired sample $\mathbf{x}$ of $\mathbf{X}$ by performing element-to-element transformations $x_j = F_j^{-1}[\Phi(z_j)]$, $j = 1, \ldots, n$. This process can be repeated to obtain as many pseudorandom samples of $\mathbf{X}$ as desired.

Finally, consider the case of an n-vector of dependent random variables $\mathbf{X}$ having a non-Nataf joint distribution. Depending on the type of the joint distribution, a variety of methods are available. Here, a method that presumes availability of the conditional CDFs $F_{j|1,\ldots,j-1}(x_j|x_1, \ldots, x_{j-1}) = \Pr(X_j \leq x_j|X_1 = x_1, \ldots, X_{j-1} = x_{j-1})$, $j = 1, \ldots, n$ is described. (See (3.60)–(3.62) for formulas for obtaining these conditional CDFs.) To generate a pseudorandom sample of $\mathbf{X}$, generate an n-vector of standard uniformly distributed pseudorandom numbers $\mathbf{y}$. Using (9.3), the first element of $\mathbf{x}$ is obtained as $x_1 = F_1^{-1}(y_1)$. The second element is obtained as $x_2 = F_{2|1}^{-1}(y_2|x_1)$, where x_1 is the sample of X_1 just generated. This process is continued so that the jth element of $\mathbf{x}$ is computed as $x_j = F_{j|1,\ldots,j-1}^{-1}(|y_j|x_1, \ldots, x_{j-1})$. Note that each of these equations requires inverting the conditional CDF to obtain its argument x_j. A standard root-finding method, such as the Newton-Raphson algorithm or the secant method, can be used for this purpose. Because FORM and SORM employ the conditional CDFs for general joint distributions, this simulation method is convenient for structural-reliability analysis.

For implementation in a software code that includes capabilities for FORM analysis, it is more convenient to perform the sampling in the standard normal space. One generates a sample of standard normal vectors $\mathbf{u}_i$, $i = 1, 2, \ldots$, and for each sample computes the corresponding point in the original space using the inverse transform $\mathbf{x}_i = \mathbf{T}^{-1}(\mathbf{u}_i)$. This then allows computing the limit-state function(s) in the original space. This approach only requires independent standard normal variables to be generated, as described in the preceding section. The CalREL program employs this approach for its simulation options.

9.5 Monte Carlo Simulation

Our aim is to compute the probability of failure of a structural system defined by the integral

$$p_F = \int_{\mathcal{F}} f_{\mathbf{X}}(\mathbf{x})\mathrm{d}\mathbf{x}, \tag{9.5}$$

where $\mathcal{F} = \bigcup_{k=1}^{N_C} \bigcap_{j\in\mathcal{C}_k} \{g_j(\mathbf{x}) \leq 0\}$ is the failure domain. This expression is for a general structural system problem. In it, $g_j(\mathbf{x})$ denotes the limit-state function for the jth component, $\mathcal{C}_k$ denotes the kth min-cut set describing the kth minimum set of components whose joint failure constitutes the failure of the system, and the union is over all the N_C min-cut sets of the system. Let $\mathbf{x}$ be an outcome of $\mathbf{X}$ and define the indicator function

$$I(\mathbf{x}) = \begin{cases} 1, & \text{if } \mathbf{x} \in \mathcal{F}, \\ 0, & \text{otherwise.} \end{cases} \tag{9.6}$$

Equation (9.5) can be written as

$$\begin{aligned} p_F &= \int_{\mathbf{x}} I(\mathbf{x}) f_{\mathbf{X}}(\mathbf{x})\mathrm{d}\mathbf{x} \\ &= \mathrm{E}[I(\mathbf{X})]. \end{aligned} \tag{9.7}$$

Hence, the failure probability equals the expectation of the indicator function defined in (9.6) relative to the distribution of $\mathbf{X}$.

Let $q(\mathbf{X}) = I(\mathbf{X})$ so that $p_F = \mathrm{E}[q(\mathbf{X})]$. Also let $\mathbf{x}_i$, $i = 1, \ldots, N$, be a sample of size N of $\mathbf{X}$, artificially generated using one of the methods described in the preceding section, and define $q_i = q(\mathbf{x}_i)$ for the ith sample outcome. A natural *estimator* of p_F then is

$$\begin{aligned} \widehat{p}_F &= \frac{1}{N}\sum_{i=1}^{N} q_i \\ &= \bar{q}, \end{aligned} \tag{9.8}$$

where $\bar{q}$ denotes the sample mean of q_i. Naturally, a different sample of $\mathbf{X}$ will lead to a different estimate of the failure probability, likely with a small difference if N is large. For this reason, we are interested in the statistical properties of the estimator $\widehat{p}_F$.

To examine the statistical properties of $\widehat{p}_F$, we regard $\mathbf{x}_i$ as a random vector, representing the ith statistically independent realization of $\mathbf{X}$. The mean of the estimator then is

$$\begin{aligned} \mathrm{E}[\widehat{p}_F] &= \frac{1}{N}\sum_{i=1}^{N}\mathrm{E}[q_i] \\ &= \frac{1}{N}N\mathrm{E}[q(\mathbf{X})] \\ &= p_F. \end{aligned} \tag{9.9}$$

Hence, the mean of the estimator is identical to the true failure probability. That is, in the average, the estimator correctly predicts the failure probability. For this reason, we regard the estimator in (9.8) as being *unbiased*. Next, we examine the variance of the estimator, keeping in mind that the sample $\mathbf{x}_i$ and, therefore, the sample q_i, are randomly selected so that no statistical dependence exists between any pairs q_i and q_j, $i \neq j$. Furthermore, as $q(\mathbf{X})$ is a Bernoulli variable with $\Pr[q(\mathbf{X}) = 1] = p_F$ and $\Pr[q(\mathbf{X}) = 0] = 1 - p_F$, its variance is given by $\mathrm{Var}[q(\mathbf{X})] = p_F(1 - p_F)$. Therefore,

$$\begin{aligned} \mathrm{Var}\,[\widehat{p}_F] &= \frac{1}{N^2}\sum_{i=1}^{N}\mathrm{Var}[q_i] \\ &= \frac{1}{N}\mathrm{Var}[q(\mathbf{X})] \\ &= \frac{1}{N}p_F(1 - p_F). \end{aligned} \tag{9.10}$$

A good measure of the accuracy of the estimator is its coefficient of variation (c.o.v.), which is obtained by dividing the square root of the variance by the mean value. The result is

$$\begin{aligned} \delta_{\widehat{p}_F} &= \frac{\sqrt{\frac{1}{N}p_F(1 - p_F)}}{p_F} \\ &= \sqrt{\frac{1 - p_F}{Np_F}}. \end{aligned} \tag{9.11}$$

It is apparent that the c.o.v. of the estimator decreases with increasing N. Furthermore, for small failure probabilities, the numerator is essentially 1 so that the c.o.v. increases with decreasing p_F in proportion to the reciprocal of the square root of the failure probability. The number of samples needed to achieve a specified c.o.v. is

$$N = \frac{1 - p_F}{p_F \delta_{\widehat{p}_F}^2}. \tag{9.12}$$

Figure 9.1 shows plots of the relations in (9.11) and (9.12) for selected p_F values as a function of N and $\delta_{\hat{p}_F}$, respectively. Observe, for example, that when $p_F = 0.0001$, with a simulation sample of size $N = 10^6$, the c.o.v. of the estimator is around $\delta_{p_F} = 0.10$ (left chart). For the same probability level, to achieve a c.o.v. of $\delta_{\hat{p}_F} = 0.05$ a sample of size $N = 3,999,600$ (right chart) is needed.

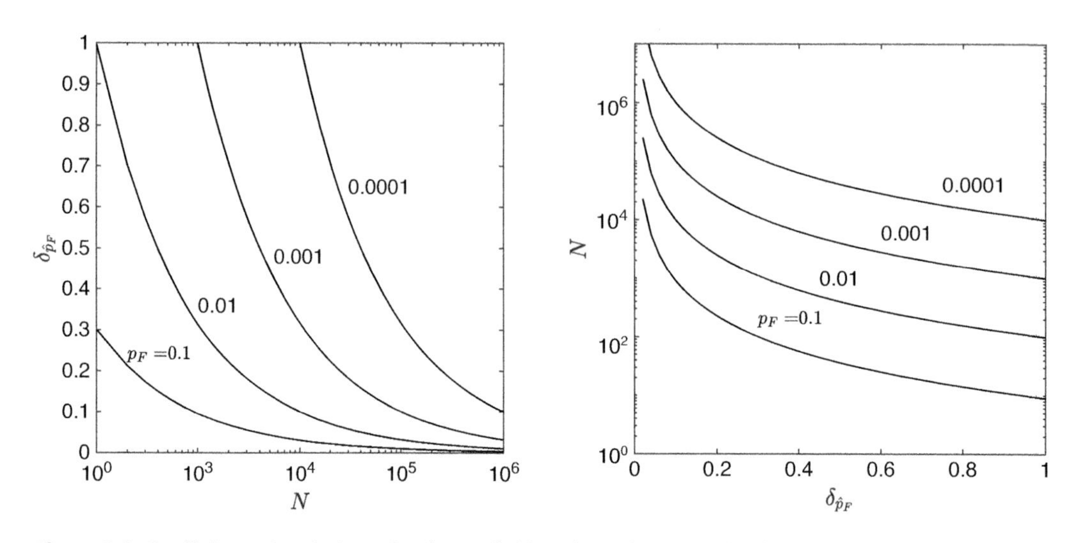

Figure 9.1 Coefficient of variation of estimate (left) and required sample size (right) for selected failure probabilities levels

The above expression for the c.o.v. of the estimator involves the failure probability, which is of course unknown and the subject of the simulation exercise itself. To find an expression for the c.o.v. that is computable, we estimate the variance based on the sample itself. Expressing the variance as the difference between the mean-square and the square of the mean of the sample values q_i, we obtain

$$\begin{aligned} \text{Var}[\widehat{p}_F] &= \frac{1}{N}\text{Var}[q(\mathbf{X})] \\ &\cong \frac{1}{N}\left(\frac{1}{N}\sum_{i=1}^{N} q_i^2 - \bar{q}^2\right). \end{aligned} \tag{9.13}$$

Taking the square root and dividing by the sample mean $\bar{q}$, we have the sample c.o.v. of the estimator as

$$\begin{aligned} \delta_{\hat{p}_F} &\cong \frac{\sqrt{\frac{1}{N}\left(\frac{1}{N}\sum_{i=1}^{N} q_i^2 - \bar{q}^2\right)}}{\bar{q}} \\ &= \frac{\sqrt{\sum_{i=1}^{N} q_i^2 - N\bar{q}^2}}{N\bar{q}}. \end{aligned} \tag{9.14}$$

In computer implementation, it is more efficient to write the above expression in the form

$$\delta_{\hat{p}_F} \cong \frac{\sqrt{\sum_{i=1}^{N} q_i^2 - \left(\sum_{i=1}^{N} q_i\right)^2 / N}}{\sum_{i=1}^{N} q_i}. \tag{9.15}$$

One stores the current values of the two sums $\sum_{i=1}^{N} q_i$ and $\sum_{i=1}^{N} q_i^2$. For each new simulation, one then only needs to add the values q_i and q_i^2 to the stored sums to update the above expression.

In typical Monte Carlo simulation of a reliability problem, one monitors the c.o.v. based on the above expression and stops the calculations when it falls below an acceptable level. CalREL allows specification of a c.o.v. threshold, but also the specification of a maximum sample size so that the computations are not continued indefinitely if the specified c.o.v. threshold is not reached.

It is apparent from the above analysis that a straight Monte Carlo simulation approach to compute the failure probability may require a large number of simulations, which means a large number of computations of the limit-state function(s). This is especially exacerbated by the fact that typical structural-reliability problems involve small failure probabilities. When the limit-state function is expensive to compute, as happens for example when it is an implicit function involving algorithmic calculations or numerical solution of differential equations, the Monte Carlo simulation approach may not be practical. The question then is what can be done to improve the efficiency of the simulation approach.

Many alternatives are available to improve the efficiency of the Monte Carlo simulation approach. These methods essentially aim to increase the mean of $q(\mathbf{X})$ or reduce its variance through reformulation. However, while the Monte Carlo simulation approach is generally applicable and does not involve restrictions, some of the alternative methods are problem-specific and may involve additional computational requirements or assumptions. The following sections present several efficient simulation methods specifically designed for structural-reliability problems. But first an example demonstrating the application of the ordinary Monte Carlo approach is presented.

Example 9.1 – Reliability of a Column by Monte Carlo Simulation

The column reliability problem in Example 6.5 is solved by the Monte Carlo simulation method. Figure 9.2 shows plots of $\widehat{p}_F$ (left) and $\delta_{\hat{p}_F}$ (right) versus the number of simulations, N.

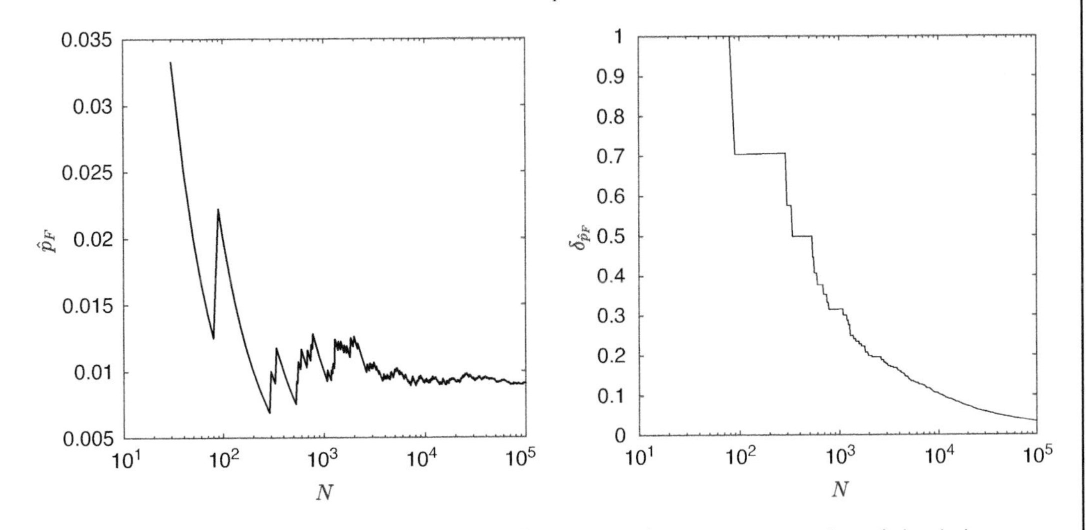

Figure 9.2 Failure probability and corresponding c.o.v. estimates versus number of simulations

Example 9.1 (cont.)

Observe that the probability estimate rapidly fluctuates for small N and then stabilizes as N grows. Also observe that the c.o.v. of the probability estimate steadily decreases with increasing N. To achieve a c.o.v. of 0.05, we need $N = 42{,}090$ samples.

9.6 Use of Antithetic Variates

The idea behind antithetic variates is to use the antithetic of each sample point in order to introduce a negative covariance between the outcomes for pairs of sample values and, hence, reduce the variance of the estimator. Suppose we are performing simulations in the standard normal space and let $\mathbf{u}_i$, $i = 1, 2, \ldots$, be the artificially generated sample of the standard normal random variables for a structural-reliability problem. As mentioned at the end of Section 9.4, these sample points are transformed to the original space using $\mathbf{x}_i = \mathbf{T}^{-1}(\mathbf{u}_i)$ so that the limit-state function(s) can be computed and $q_i = q(\mathbf{x}_i)$ can be determined. In the antithetic variates approach, for each $\mathbf{u}_i$, we also use $-\mathbf{u}_i$. Obviously these two sample points are strongly dependent. In particular, if $\mathbf{u}_i$ is in the failure domain, $-\mathbf{u}_i$ is likely to be in the safe domain and vice versa. It follows that the corresponding q_i values, i.e., $q_{i+} = q\left[\mathbf{T}^{-1}(\mathbf{u}_i)\right]$ and $q_{i-} = q\left[\mathbf{T}^{-1}(-\mathbf{u}_i)\right]$, are likely to be negatively correlated. (If q_{i+} equals 1, it is likely that q_{i-} equals 0.) Therefore, each of these sample pairs introduces a negative covariance term in the total variance of the estimator $\widehat{p}_F$. Specifically, if we generate $N/2$ pairs of antithetic variates,

$$\begin{aligned} \mathrm{Var}[\widehat{p}_F] &= \mathrm{Var}\left[\frac{1}{N}\sum_{i=1}^{N/2}(q_{i+} + q_{i-})\right] \\ &= \frac{1}{N^2}\sum_{i=1}^{N/2}\left\{\mathrm{Var}\left[q_{i+}\right] + \mathrm{Var}[q_{i-}] + 2\mathrm{Cov}\left[q_{i+}, q_{i-}\right]\right\}. \end{aligned} \tag{9.16}$$

As we saw before, $\mathrm{Var}\left[q_{i+}\right] = \mathrm{Var}[q_{i-}] = p_F(1 - p_F)$. Furthermore, $\mathrm{Cov}\left[q_{i+}, q_{i-}\right] = \mathrm{E}\left[q_{i+}q_{i-}\right] - \mathrm{E}\left[q_{i+}\right]E[q_{i-}] = \mathrm{E}\left[q_{i+}q_{i-}\right] - p_F^2$. In the best case, the mean of the product $\mathrm{E}\left[q_{i+}q_{i-}\right]$ is 0. This is because when either q_{i+} or q_{i-} equals 1, it is likely that the other is 0 so that the product remains 0. Thus, we can say $\mathrm{Cov}\left[q_{i+}, q_{i-}\right] \geq -p_F^2$ and (9.16) simplifies to

$$\mathrm{Var}[\widehat{p}_F] \geq \frac{1}{N}p_F(1 - 2p_F). \tag{9.17}$$

We have used the greater than or equal sign instead of an equality sign on account of the possibility that $\mathrm{E}\left[q_{i+}q_{i-}\right]$ is greater than 0. Compared to (9.10), the gain is the factor 2 of p_F inside the parentheses on the right-hand side. Clearly, this benefit is insignificant when the failure probability is small so that p_F^2 is negligible compared to p_F. Nevertheless, the option of using antithetic variates has the added advantage of needing to generate only half the sample, as the other half is obtained by changing the signs of the sample points. Furthermore, it can be effective when the failure probability is relatively large. Unfortunately, an exact estimate of the variance of $\widehat{p}_F$ from the simulated sample is impractical because the value of $\mathrm{Cov}\left[q_{i+}, q_{i-}\right]$ is difficult to compute.

Example 9.2 – Reliability of a Column by Monte Carlo Simulation using Antithetic Variates

Figure 9.3 shows two plots of $\widehat{p}_F$ and $\delta_{\hat{p}_f}$ versus the number of simulations for the column problem of Example 9.1. The dashed lines are the results obtained from ordinary Monte Carlo simulation whereas the solid lines are the results obtained using antithetic variates. It can be seen that using antithetic variates reduces the c.o.v. of the probability estimate by a small amount.

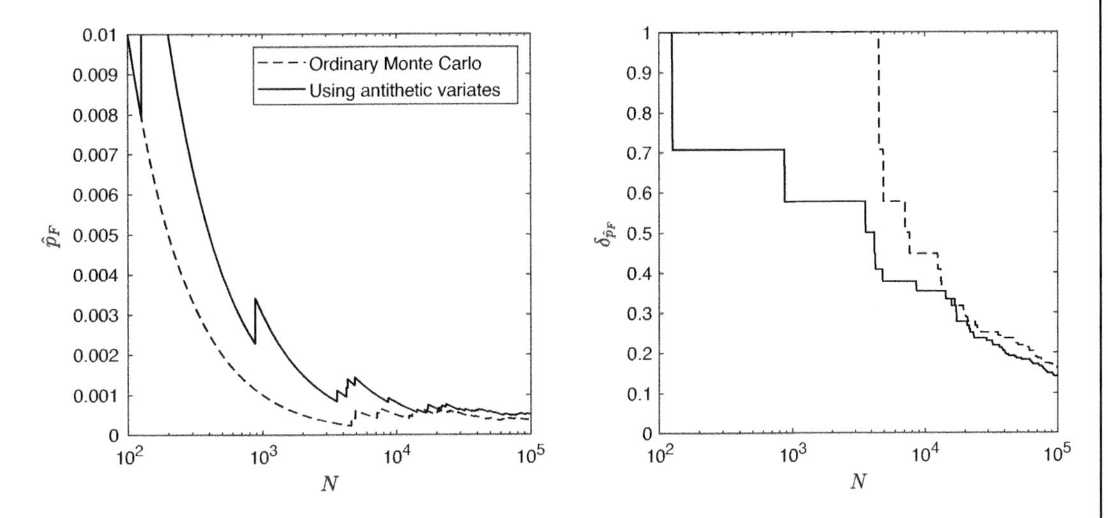

Figure 9.3 Monte Carlo solution with (thick line) and without (dashed line) antithetic variates

9.7 Importance Sampling

We rewrite (9.7) in the standard normal space

$$\begin{aligned} p_F &= \int_{\mathbf{u}} I(\mathbf{u})\phi(\mathbf{u})d\mathbf{u}, \\ &= \mathrm{E}[I(\mathbf{U})], \end{aligned} \tag{9.18}$$

where $I(\mathbf{u}) = 1$ if $\mathbf{x} = \mathbf{T}^{-1}(\mathbf{u}) \in \mathcal{F}$; otherwise $I(\mathbf{u}) = 0$. Let $h(\mathbf{u})$ be a PDF such that $0 < h(\mathbf{u})$ for any $\mathbf{u}$ in the failure domain, i.e., wherever $I(\mathbf{u}) = 1$. We rewrite (9.18) in the form

$$p_F = \int_{\mathbf{u}} I(\mathbf{u}) \frac{\phi(\mathbf{u})}{h(\mathbf{u})} h(\mathbf{u}) d\mathbf{u}$$
$$= \mathrm{E}_h \left[I(\mathbf{U}) \frac{\phi(\mathbf{U})}{h(\mathbf{U})} \right], \tag{9.19}$$

where $\mathrm{E}_h[\cdot]$ denotes the expectation with respect to the distribution $h(\mathbf{u})$. It follows that instead of sampling according to the PDF $\phi(\mathbf{u})$ as in (9.18) or according to the PDF $f_{\mathbf{X}}(\mathbf{x})$ as in (9.7), we can sample according to the PDF $h(\mathbf{u})$ and compute the estimate of the failure probability as the expectation on the right-hand side of (9.19). Let

$$q_i = q(\mathbf{u}_i)$$
$$= I(\mathbf{u}_i) \frac{\phi(\mathbf{u}_i)}{h(\mathbf{u}_i)}, \quad i = 1, 2, \ldots, N, \tag{9.20}$$

be the sample values. Advantage is gained if q_i has a small variance. The ideal choice for the sampling density is

$$h(\mathbf{u}) \propto I(\mathbf{u})\phi(\mathbf{u}), \tag{9.21}$$

in which case $q_i = 1$ for all i and the variance is zero. It should be clear from (9.18) that the normalizing factor for the sampling PDF then is $1/p_F$ so that

$$h(\mathbf{u}) = \frac{I(\mathbf{u})\phi(\mathbf{u})}{p_F}. \tag{9.22}$$

However, p_F is unknown and the choice in (9.21) is not practical. Therefore, we try to select a sampling density that is close to (9.21). Obviously, it is desirable for the PDF to have large values in the failure domain and small or zero values in the safe domain. In other words, to sample more in the important region – hence the name *importance sampling*.

The first suggestion of using importance sampling for structural-reliability assessment came from Shinozuka (1983), who suggested sampling uniformly within a box around the design point(s). However, this choice violates the condition that the sampling density have non-zero values in the entire failure domain, as specified following (9.18), and may lead to biased estimates of the failure probability. Harbitz (1986) suggested the sampling density

$$h(\mathbf{u}) = \frac{\phi(\mathbf{u})}{1 - \mathrm{X}_n^2(\beta^2)}, \quad \beta \leq \|\mathbf{u}\|,$$
$$= 0, \quad \text{otherwise}, \tag{9.23}$$

where β is the distance from the origin to the nearest point on the limit-state surface and $\mathrm{X}_n^2(\cdot)$ the chi-square CDF with n degrees of freedom, n being the number of random variables. The above sampling density is proportional to a standard normal PDF with a hypersphere of radius β centered in the origin cut out from it. In essence, Harbitz suggests not sampling within the hypersphere of radius β because we know all points there are safe.

However, we saw in Section 6.4 that the probability content in a hypersphere centered at the origin in the standard normal space quickly diminishes as the dimension n increases. That is, the complementary CDF in the denominator of (9.23) quickly approaches 1 as n increases. Therefore, the sampling density in (9.23) is ineffective in high-dimension problems.

The most common choice in the current practice of importance sampling is a normal PDF centered at the design point(s) (Melchers, 1989). For a structural component problem with design point $\mathbf{u}^*$, the recommended sampling density is

$$h(\mathbf{u}) = \mathrm{N}\left(\mathbf{u}^*, \sigma^2\mathbf{I}\right), \tag{9.24}$$

where $\mathbf{I}$ is the identity matrix, i.e., $h(\mathbf{u})$ is a multinormal distribution with mean $\mathbf{u}^*$ and a diagonal covariance matrix with equal variances σ^2. Based on numerical tests, Melchers (1989) recommends using $\sigma = 1$.

For a series-system reliability problem with m design points $\mathbf{u}_i^*$, $i = 1, \ldots, m$, a weighted sum of normal PDFs may be used, i.e.,

$$h(\mathbf{u}) = \sum\nolimits_{i=1}^{m} w_i \mathrm{N}\left(\mathbf{u}_i^*, \sigma^2\mathbf{I}\right), \tag{9.25}$$

where w_i, satisfying the condition $\sum_{i=1}^{m} w_i = 1$, are the weights. It is desirable to give more weight to design points that are closer to the origin and, hence, contribute more to the failure probability. Various choices for this purpose are available, e.g., $w_i \propto \Phi(-\beta_i)$ or $w_i \propto 1/\beta_i^k$, where β_i is the reliability index corresponding to design point $\mathbf{u}_i^*$ and k is a selected positive constant.

For parallel-system reliability problems, a normal PDF centered at the joint design point obtained according to (8.66) can be used. CalREL offers this choice, but also the choice of using the joint design point of the linearized limit-state surfaces at their respective design points, see (8.67) and Figure 8.10. The latter approach avoids the difficult problem of finding the exact joint design point, which is an optimization problem with multiple nonlinear inequality constraints. In some problems, the failure domain of the parallel system is a narrow wedge. One such case arises in determining the mean out-crossing rate of a structure under stochastic loading, as described in Section 11.5.1 and Figure 11.9 of Chapter 11. Regardless of the choice of the center of the sampling density, importance sampling for such a failure domain is not effective as few samples fall within the wedge-like failure domain. For such a problem, the orthogonal-plane sampling approach described later in this chapter is more effective.

Chapter 8 showed that a general system can be formulated as a series system of its min-cut sets, with each min-cut set representing a parallel system of components whose joint failure constitutes the failure of the system. Thus, for such a problem, (9.25) can be used with $\mathbf{u}_i^*$ representing the joint design point of the ith min-cut set. Again, either the exact joint design points or the ones obtained for the component-wise linearized limit-state surfaces may be used.

The above formulations of the sampling density all require knowledge of the design point(s). This means that a considerable amount of preparatory calculations are necessary before importance sampling is conducted. It also imposes certain conditions on the reliability problem, specifically that the limit-state functions be continuously differentiable so that the design points can be found, which is not necessary for the importance sampling itself. To overcome these issues, some researchers have suggested using an adaptive importance sampling, see, e.g., Bucher (1988), Karamchandani et al. (1989), Melchers (1990), and Au and Beck (1999). One starts with an ordinary Monte Carlo sampling, i.e., with the standard multinormal density centered at the origin as the sampling distribution, until a sample point in the failure domain is found. One then places a sampling PDF centered at the newly found failure point. The sampling density may be moved around as failure points closer to the origin are found, or additional sampling densities may be added as distinct failure points far from the earlier points are found. In this kind of approach, there is always a danger that one may miss important segments of the failure domain, in which case the estimate of the failure probability will be biased. To avoid this problem, it is best to include a "background" sampling density centered at the origin to capture potential undiscovered areas of the failure domain. For this purpose, the sampling density is written as

$$h(\mathbf{u}) = w_0 \mathrm{N}(\mathbf{0}, \mathbf{I}) + \sum\nolimits_{i=1}^{m} w_i \mathrm{N}\left(\mathbf{u}_i^*, \sigma^2 \mathbf{I}\right), \tag{9.26}$$

where w_0 is the weight for the background sampling. Of course, we must now have $\sum_{i=0}^{m} w_i = 1$. CalREL gives the user the option of selecting w_0 with a default value of $w_0 = 0.5$.

Some researchers have suggested conducting importance-sampling simulation in the original space so as to avoid transforming random variables (Melchers, 1992; Maes et al., 1993; Melchers and Beck, 2018). This can be done, but several issues need to be addressed: (a) One needs to select centers for the sampling densities; the adaptive approach described above can be used for this purpose; (b) The sampling density needs to be scaled in each direction, depending on the standard deviations of the basic random variables; and (c) The sampling density may need to be rotated in order to account for correlations between the basic random variables. The choices to be made are not unique and their effectiveness may depend on the form of the joint PDF $f_{\mathbf{X}}(\mathbf{x})$ of the basic random variables. For these reasons, sampling in the standard normal space is generally preferred if that choice is available. Benchmark studies with various importance-sampling methods are reported in Schueller and Stix (1987) and Engelund and Rackwitz (1993).

Example 9.3 – Reliability of Ductile Frame by Importance Sampling

Consider the reliability of the ductile frame in Example 8.7. The problem is a series-system reliability problem with three components. We solve this problem by three simulation methods: (a) ordinary Monte Carlo, (b) importance sampling using Harbitz's sampling density in (9.23), where $\beta = 2.00$ is the reliability index for the design point closest to the origin (component 3, corresponding to the combined failure mechanism), and (c) importance sampling using design points with the sampling density described in (9.25) with w_i, $i = 1, 2, 3$, taken as proportional to

Example 9.3 (cont.)

$\Phi(-\beta_i)$. See Example 8.7 for the coordinates of the design points, which are used as the centers for the importance sampling density, and the corresponding β_i values. Figure 9.4 shows plots of $\hat{p}_F$ (left) and $\delta_{\hat{p}_F}$ (right) as functions of the number of simulations. It can be seen that Harbitz's sampling density offers practically no advantage over the ordinary Monte Carlo method. This is because the radius of the hypersphere is relatively small ($\beta = 2$) and the dimension of the space relatively large ($n = 7$). Importance sampling using design points, on the other hand, offers a significant advantage. If we set the required c.o.v. threshold as $\delta_{\hat{p}_F} \leq 0.05$, then the importance sampling method requires less than 900 samples, whereas the other two methods require about 15,500 samples. At this c.o.v. threshold, the estimated failure probabilities by the three simulation methods are $\hat{p}_F = 0.0278, 0.0255,$ and 0.0292, respectively. In comparison, the FORM approximation obtained in Example 8.7 is $p_{F1} = 0.0264$.

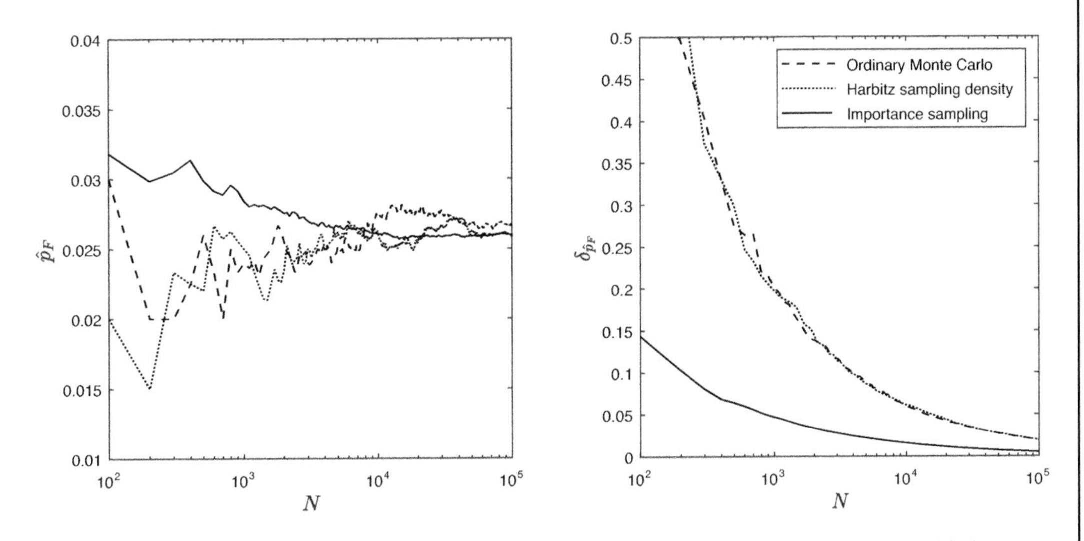

Figure 9.4 Comparison of three sampling methods: ordinary Monte Carlo (dashed), Harbitz's sampling density (thin solid), and importance sampling using design points

If the failure domain is known to lie on one side of a hyperplane having the unit normal $\hat{\boldsymbol{\alpha}}$ and distance β from the origin, as in Figure 9.5, then a highly effective importance sampling method is to sample within the half-space $\beta - \hat{\boldsymbol{\alpha}}\mathbf{u} \leq 0$. The existence of this condition, at least in the neighborhood of the design point, can be verified by the point-fitting SORM. Referring to the notation used in Section 7.4, the condition is satisfied if the u'_n coordinates of all the fitting points are greater than β. The importance-sampling density in this case is given by

$$h(\mathbf{u}) = \begin{cases} \dfrac{\phi(\mathbf{u})}{\Phi(-\beta)}, & \beta - \hat{\boldsymbol{\alpha}}\mathbf{u} \leq 0, \\ 0 & \text{Otherwise.} \end{cases} \tag{9.27}$$

To simulate points with the above density, proceed as follows:

1) Generate a standard normal n-vector $\mathbf{v} = \mathrm{N}(\mathbf{0}, \mathbf{I})$.
2) Project $\mathbf{v}$ on the plane $\widehat{\boldsymbol{\alpha}}\mathbf{u} = 0$, which is orthogonal to $\widehat{\boldsymbol{\alpha}}$ and passes through the origin; the result is $\mathbf{v}_0 = \mathbf{v} - \widehat{\boldsymbol{\alpha}}\mathbf{v}\widehat{\boldsymbol{\alpha}}^{\mathrm{T}}$.
3) Generate a standard uniform variate $a = \mathrm{Uni}(0, 1)$.
4) Compute $b = \Phi^{-1}[a + (1 - a)\Phi(\beta)]$. This simulates b as a truncated standard normal variate satisfying the condition $\beta \leq b$.
5) Compute the desired random point in the half-space $\beta - \widehat{\boldsymbol{\alpha}}\mathbf{u} \leq 0$ as $\mathbf{u} = \mathbf{v}_0 + b\widehat{\boldsymbol{\alpha}}^{\mathrm{T}}$.

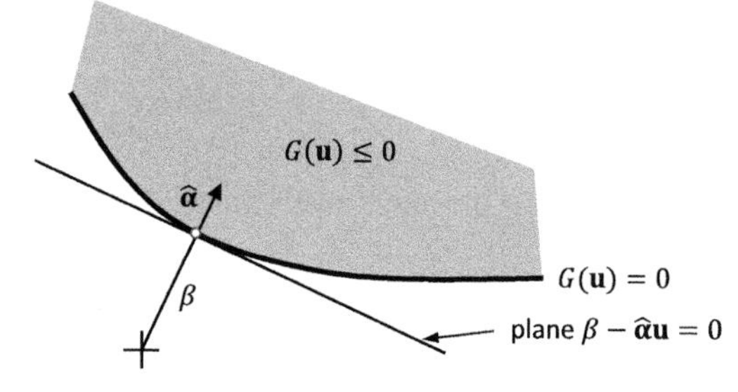

Figure 9.5 Failure domain contained in a half-space

Example 9.4 – Reliability of a Two-Degree-of-Freedom Primary-Secondary System

Consider the two-degree-of-freedom primary-secondary system in Figure 9.6, where the mass of the secondary oscillator is much smaller than that of the primary oscillator. When the frequencies of the two oscillators match, there is a tuning effect in the combined system whereby the response of the secondary oscillator is strongly amplified. Suppose the frequencies ω_p and ω_s and the damping ratios ζ_p and ζ_s of the primary and secondary oscillators are uncertain and modeled as independent lognormal random variables with the means and c.o.v. listed in Table 9.1. Note that the mean system is perfectly tuned. Igusa and Der Kiureghian (1985) showed that, when subjected to a broadband (white noise) base acceleration, the displacement response of the secondary oscillator relative to the primary has the expected peak value

$$\mathrm{E}[X_{\max}] = p\sqrt{\frac{\pi S_0}{4\zeta_s\omega_s^3}\left[\frac{\zeta_a\zeta_s}{\zeta_p\zeta_s(4\zeta_a^2 + \theta^2) + \gamma\zeta_a^2}\frac{\left(\zeta_p\omega_p^3 + \zeta_s\omega_s^3\right)\omega_p}{4\zeta_a\omega_a^4}\right]}, \tag{E1}$$

where p is a peak factor dependent on the duration of the excitation, S_0 the intensity of the white noise, $\gamma = m_s/m_p$ the ratio of masses, $\omega_a = (\omega_p + \omega_s)/2$ and $\zeta_a = (\zeta_p + \zeta_s)/2$ the averages of the frequencies and damping ratios, respectively, and $\theta = (\omega_p - \omega_s)/\omega_a$ a "tuning" parameter. This expression includes the effect of interaction between the primary and secondary oscillators, which is important in the case of tuning when θ is near zero. Owing to the tuning and interaction effects, the response of the secondary oscillator is strongly sensitive to the frequency and damping properties of the system. Hence, uncertainties in these values can

Example 9.4 (cont.)

have a profound effect on the reliability of the secondary oscillator. For this example we assume $p = 3$, $S_0 = 1\ \text{cm}^2/\text{s}^3$, and $\gamma = 0.05$. Because the variance of the peak response is small, we ignore the variability in the peak response and adopt the limit-state function

$$g(\omega_p, \omega_s, \zeta_p, \zeta_s) = r - \text{E}[X_{\max}], \tag{E2}$$

where $r = 6$ cm is the acceptable threshold. Reliability analysis with FORM yields the reliability index $\beta = 3.77$, the failure probability estimate $p_{F1} = 8.04 \times 10^{-5}$, the design point $\mathbf{u}^* = [-2.17 - 2.46 - 1.24 - 1.39]^{\text{T}}$, and unit normal vector $\widehat{\boldsymbol{\alpha}} = [-0.578 - 0.650 - 0.328 - 0.368]$. As the random variables are independent, the elements of $\widehat{\boldsymbol{\alpha}}$ indicate their relative importances. It is seen that the frequencies are more important than the damping ratios, even though the latter have larger c.o.v. This is because of the effect of tuning. The mapping of the design point in the original space is $\mathbf{x}^* = [4.01\ 3.78\ 0.0333\ 0.0318]^{\text{T}}$. It is seen that at the most likely failure point, both frequencies are lower than their mean values (the system is softer than the mean system) and they are close to each other, indicating tuning. Additionally, both damping ratios are smaller than their mean values.

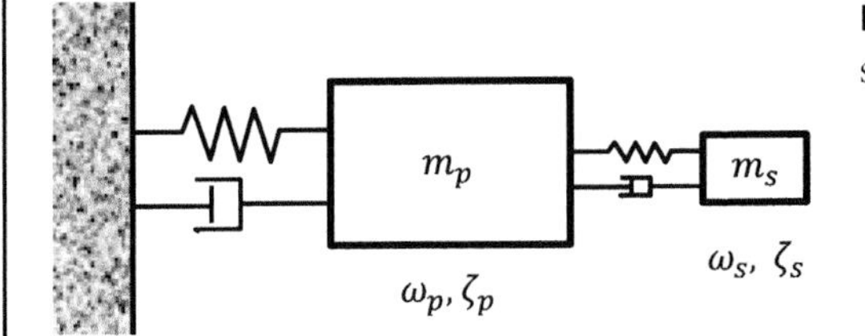

Figure 9.6 Two-degree-of-freedom primary-secondary system

Table 9.1 Description of random variables

Variable	Distribution	Mean	c.o.v.
ω_p, rad/s	Lognormal	2π	20%
ω_s, rad/s	Lognormal	2π	20%
ζ_p	Lognormal	0.05	30%
ζ_s	Lognormal	0.05	30%

The curvature-fitting SORM reveals the principal curvatures as $\kappa_1 = 4.89$, $\kappa_2 = 0.0736$, and $\kappa_3 = 0.0184$. All three are positive, indicating that the surface bends away from the origin, at least near the design point. The point-fitting SORM reveals fitting points that have coordinates u'_n ranging from 3.96 to 4.80, all greater than β, indicating that the surface does indeed bend

Example 9.4 (cont.)

away from the origin in a bigger neighborhood of the design point. Further exploration reveals that the major principal curvature κ_1 is in the plane of the transformed frequencies ω_p and ω_s and, therefore, is due to the effect of tuning. This can be seen in Figure 9.7, where we show the limit-state surface in the plane of u_1 and u_2 (one-to-one transformations of ω_p and ω_s, respectively) and going through the design point. Observe that the surface has a strong curvature in this plane but remains on the far side of the tangent plane at the design point. The SORM analysis yields the failure probability estimate $p_{F2} = 1.47 \times 10^{-5}$ and the generalized reliability index $\beta_g = 4.18$. The latter is significantly greater than that of the FORM estimate, again suggesting that the surface bends away from the origin.

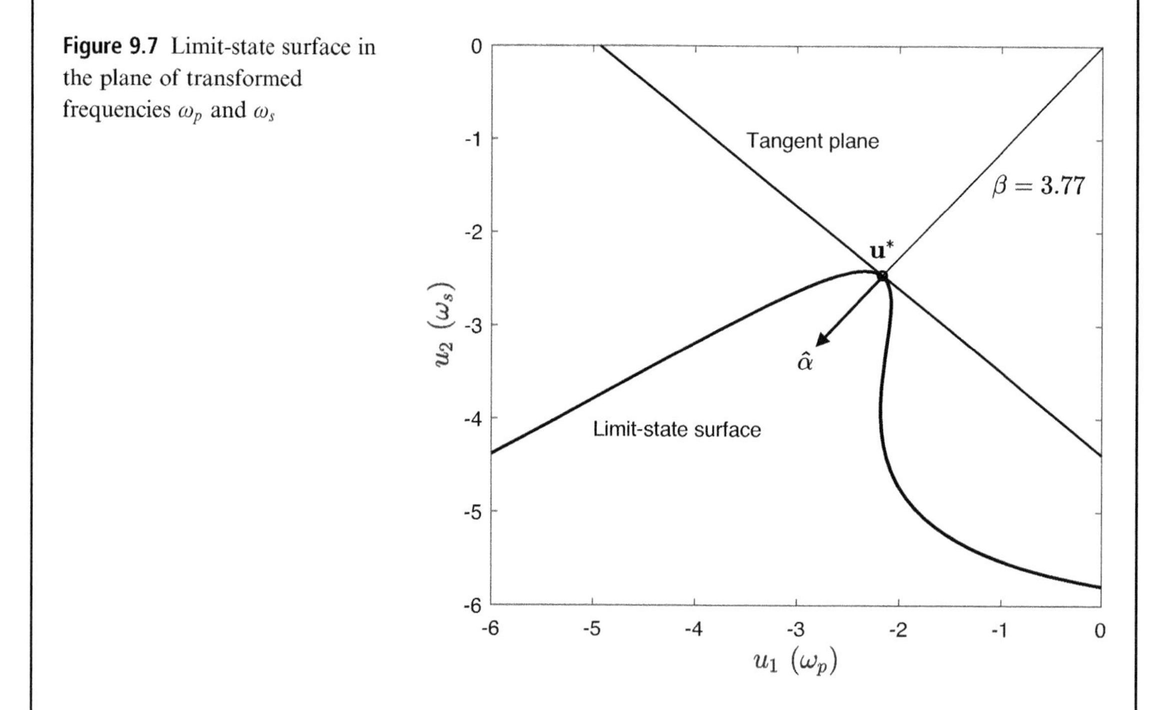

Figure 9.7 Limit-state surface in the plane of transformed frequencies ω_p and ω_s

Figure 9.8 shows plots of $\widehat{p}_F$ and $\delta_{\hat{p}_F}$ versus the number of simulations for ordinary Monte Carlo simulation and importance sampling in the half-space using the PDF in (9.27). Owing to the small failure probability, ordinary sampling does not produce a point in the failure domain in the initial 40,000 samples and remains highly fluctuating even after 1,000,000 samples with the c.o.v. of the probability estimate exceeding 20% at that sample size. On the other hand, the importance sampling in the half-space is highly efficient with a c.o.v. of the probability estimate around 6% after only 1,000 samples. The estimated failure probability, $\widehat{p}_F = 1.64 \times 10^{-5}$, is fairly close to the SORM estimate mentioned above.

Example 9.4 (cont.)

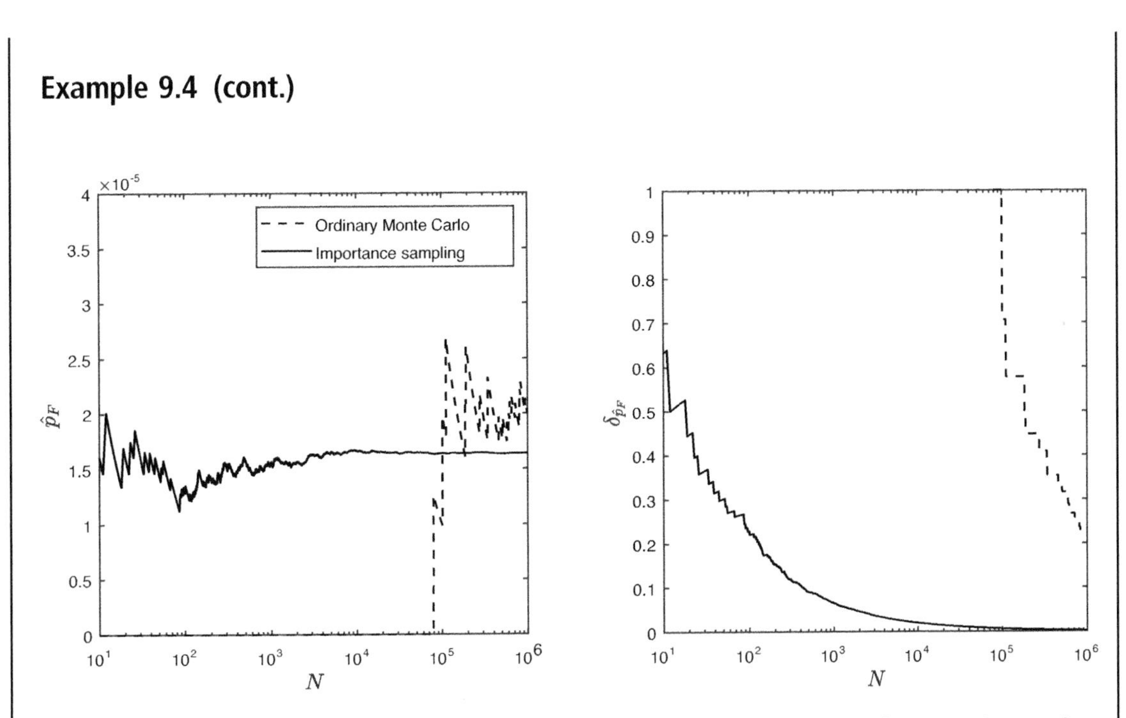

Figure 9.8 Comparison of ordinary Monte Carlo and importance sampling for a two-degree-of-freedom primary-secondary system

9.8 Numerical Integration by Importance Sampling

Suppose we wish to numerically compute the multifold integral

$$I = \int_{\mathbf{x}} g(\mathbf{x})\mathrm{d}\mathbf{x} \tag{9.28}$$

over the space of an n-vector of variables $\mathbf{x}$. Let $h(\mathbf{x})$ be the PDF of an n-vector of random variables $\mathbf{X}$ with non-zero values within the domain of the integration where $g(\mathbf{x}) \neq 0$. Let Ω denote that domain. We can write

$$\begin{aligned} I &= \int_{\Omega} \frac{g(\mathbf{x})}{h(\mathbf{x})} h(\mathbf{x})\mathrm{d}\mathbf{x} \\ &= \mathrm{E}_h\left[\frac{g(\mathbf{X})}{h(\mathbf{X})}\right], \end{aligned} \tag{9.29}$$

where $\mathrm{E}_h[\cdot]$ is the expectation with respect to the distribution $h(\mathbf{x})$. It follows that the integral can be approximately computed by sampling from $h(\mathbf{x})$ and estimating the mean of the ratio $g(\mathbf{X})/h(\mathbf{X})$. Specifically, for the ith sample, we set $q_i = g(\mathbf{x}_i)/h(\mathbf{x}_i)$, $i = 1, \ldots, N$, and compute the "mean" estimate of I as

$$\begin{aligned} \mathrm{E}[\hat{I}] &= \frac{1}{N}\sum\nolimits_{i=1}^{N} q_i \\ &= \bar{q} \end{aligned} \tag{9.30}$$

and its c.o.v. (a measure of accuracy of the integration) as

$$\delta_{\hat{I}} \cong \frac{\sqrt{\sum_{i=1}^{N} q_i^2 - \left(\sum_{i=1}^{N} q_i\right)^2 / N}}{\sum_{i=1}^{N} q_i}. \tag{9.31}$$

To obtain good accuracy with this method, variations in q_i must be small. In the case where $g(\mathbf{x})$ is non-negative within the domain of integration, the ideal choice of sampling density is $h(\mathbf{x}) \propto g(\mathbf{x})$, in which case q_i have equal values and the result in (9.30) is exact. However, this choice is not possible because one needs the value of I to scale the sampling density as $h(\mathbf{x}) = g(\mathbf{x})/I$. This observation suggests that the sampling density should be selected such that, as much as possible, it mimics the shape of the function $g(\mathbf{x})$, i.e., have large (small) values where $g(\mathbf{x})$ is large (small) and a zero value where $g(\mathbf{x})$ is zero. It also suggests that the method would not be as effective for an integrand $g(\mathbf{x})$ that has both positive and negative values, as q_i would then widely fluctuate.

An important application of the above method is in computing integrals for Bayesian updating of the distribution of uncertain parameters. As will be seen in Chapter 10, for a vector of uncertain parameters $\boldsymbol{\Theta}$, the Bayesian updating involves integrals of the form

$$I = \int_{\boldsymbol{\theta}} K(\boldsymbol{\theta}) L(\boldsymbol{\theta}) f'_{\boldsymbol{\Theta}}(\boldsymbol{\theta}) \mathrm{d}\boldsymbol{\theta}, \tag{9.32}$$

where $K(\boldsymbol{\theta})$ is the Bayesian kernel and depends on what we wish to compute (i.e., the normalizing constant of the distribution, the posterior means or variance/covariances, posterior marginal distributions, or the predictive distribution of the underlying random variable), $L(\boldsymbol{\theta})$ is the likelihood function, and $f'_{\boldsymbol{\Theta}}(\boldsymbol{\theta})$ is the prior distribution of $\boldsymbol{\Theta}$ (see Section 10.3). In some cases, $K(\boldsymbol{\theta})$ is a vector or matrix, and in other cases the integration should be performed for a range of parameter values. In such cases, one can use the same simulated sample of $\boldsymbol{\Theta}$ for evaluating the multiple integrals involved, thus reducing the computational task of generating random numbers.

As shown in Chapter 10, the integrand $L(\boldsymbol{\theta}) f'_{\boldsymbol{\Theta}}(\boldsymbol{\theta})$ is proportional to the posterior distribution of $\boldsymbol{\Theta}$. It follows that it would be desirable to select that distribution as the sampling density. However, that is not possible because the posterior distribution is not readily available. This observation suggests that one should select a sampling distribution that is close to the posterior. One can show (see Beck and Katafygiotis, 1998) that, under certain conditions, the shape of the likelihood function asymptotically approaches a multinormal distribution with the mean located where the likelihood function takes its maximum value (the maximum likelihood estimate, MLE) and a covariance matrix that is equal to the inverse of the Hessian of the negative log-likelihood function at its maximum point. Thus, finding the maximum point of the negative log-likelihood function and then computing the inverse of the Hessian at that point provides an approximation of the posterior mean and covariance matrix. Adopting a sampling distribution with this mean vector and covariance matrix usually proves effective. One convenient choice is a multivariate Nataf distribution with normal marginals for parameters that are unbounded and lognormal marginals for non-negative parameters. For parameters bounded on both sides, a uniform or beta marginal may be used. With such a choice, the sampling density has the form (see 3.37)

$$h_{\boldsymbol{\Theta}}(\boldsymbol{\theta}) = f_{\Theta_1}(\theta_1)\cdots f_{\Theta_n}(\theta_n)\frac{\phi_n(\mathbf{z}, \mathbf{R}_0)}{\phi(z_1)\cdots\phi(z_n)}, \tag{9.33}$$

where n is the number of variables, $f_{\Theta_j}(\theta_j), j = 1, \ldots, n$, are the assumed marginal sampling distributions (normal, lognormal, beta, etc.), $z_j = \Phi^{-1}\left[F_{\Theta_j}(\theta_j)\right]$ are marginally transformed samples of standard but correlated normal random variables, $\phi(z_j)$ is the standard normal PDF of Z_j, and $\phi_n(\mathbf{z}, \mathbf{R_0})$ is the n-variate standard normal distribution with correlation matrix $\mathbf{R}_0$. The means of the marginal distributions are set equal to the MLE values and the variances and the correlation matrix $\mathbf{R}_0$ are obtained from the covariance matrix computed as the inverse of the Hessian. The following example illustrates an application of this approach.

Example 9.5 – Bayesian Updating by Importance Sampling

The sampling method described above is used to compute the posterior statistics for the problem to be described as Case 5 in Example 10.3. The unknown parameters are the mean and variance, $\boldsymbol{\Theta} = (\mu, \sigma^2)$, of the normal distribution of a random variable X. The available data consist of six statistically independent equality observations $x_1 = 208$, $x_2 = 220$, $x_3 = 201$, $x_4 = 175$, $x_5 = 190$, and $x_6 = 245$, and one lower-bound observation $x_7 = 250 < X_7$. The likelihood function is given by

$$L(\mu, \sigma^2) \propto \frac{1}{\sigma^6}\exp\left[-\frac{1}{2\sigma^2}\sum_{i=1}^{6}(x_i - \mu)^2\right] \times \left[1 - \Phi\left(\frac{x_7 - \mu}{\sigma}\right)\right], \tag{E1}$$

and the prior distribution is specified as

$$f'(\mu, \sigma^2) \propto \frac{1}{\sigma^2}. \tag{E2}$$

Three kernels are used in (9.32): $K(\boldsymbol{\theta}) = 1$ in order to compute the normalizing factor c of the posterior distribution $f''_{\boldsymbol{\Theta}}(\boldsymbol{\theta}) = cL(\boldsymbol{\theta})f'_{\boldsymbol{\Theta}}(\boldsymbol{\theta})$ from

$$c = \left[\int_{\boldsymbol{\theta}} L(\boldsymbol{\theta})f'_{\boldsymbol{\Theta}}(\boldsymbol{\theta})\mathrm{d}\boldsymbol{\theta}\right]^{-1}, \tag{E3}$$

$\mathbf{K}(\boldsymbol{\theta}) = c\boldsymbol{\theta}$ to compute the posterior mean vector $\mathbf{M}''_{\boldsymbol{\Theta}}$, and $\mathbf{K}(\boldsymbol{\theta}) = c\left(\boldsymbol{\theta}\boldsymbol{\theta}^{\mathrm{T}} - \mathbf{M}''_{\boldsymbol{\Theta}}\mathbf{M}''^{\mathrm{T}}_{\boldsymbol{\Theta}}\right)$ to compute the posterior covariance matrix.

Example 9.5 (cont.)

To formulate the sampling density, we use MATLAB's `fminunc` function to determine the MLE (the maximum point) of the negative log-likelihood function and the inverse of the Hessian at that point. For the above data and likelihood function, the results are

$$\mathbf{M}_{\Theta} = \begin{Bmatrix} 215 \\ 852 \end{Bmatrix} \text{ and } \boldsymbol{\Sigma}_{\Theta\Theta} = \begin{bmatrix} 126 & 464 \\ 464 & 262{,}186 \end{bmatrix}, \tag{E4}$$

from which standard deviations are computed as 11.2 and 512 and the correlation coefficient is computed as 0.0809. For the sampling PDF, we use the bivariate Nataf distribution

$$h_{\Theta}(\boldsymbol{\theta}) = f_{\Theta_1}(\theta_1) f_{\Theta_2}(\theta_2) \frac{\phi_2(\mathbf{z}, \mathbf{R}_0)}{\phi(z_1)\phi(z_2)}, \tag{E5}$$

where $f_{\Theta_1}(\theta_1)$ is normal with mean 215 and standard deviation 11.2 and, in view of the fact that σ^2 must be positive, $f_{\Theta_2}(\theta_2)$ is lognormal with mean 852 and standard deviation 512. The correlation coefficient between Z_1 and Z_2 is only slightly different from that between Θ_1 and Θ_2 (see Section 3.5.2 and Figure 3.7). Hence, for the sake of simplicity, we use 0.0809 as the correlation coefficient between Z_1 and Z_2.

The results of the calculations, which are also reported in Case 5 of Example 10.3, are $\mu''_{\mu} = 215.5$, $\mu''_{\sigma^2} = 1{,}482$, $\sigma''_{\mu} = 13.2$, $\sigma''_{\sigma^2} = 1{,}113$, and $\rho''_{\mu\sigma^2} = 0.089$. These are obtained with a sample of size 10^5 with the c.o.v of the posterior mean estimates being less than 5%. The c.o.v. in the other estimated values are larger. It is noted that, although the posterior mean and variance of μ are not far from the MLE estimates in (E4), those for the variance σ^2 are significantly different. This has to do with the fact that the distribution of σ^2 is far from being normal, which is the assumed asymptotic distribution in computing the Hessian.

In Chapter 10, the *predictive* distribution of X, which incorporates the uncertainty in the parameters μ and σ^2 together with its inherent variability, is also described. This distribution is obtained through the integral

$$\tilde{f}_X(x) = \int_{\boldsymbol{\theta}} c f_X(x|\boldsymbol{\theta}) L(\boldsymbol{\theta}) f'_{\Theta}(\boldsymbol{\theta}) d\boldsymbol{\theta}, \tag{E6}$$

where $f_X(x|\boldsymbol{\theta})$ denotes the PDF of X conditioned on the parameter values. The importance-sampling method described in this section is used to compute this integral. For that purpose, we use the kernel $K(\boldsymbol{\theta}, x) = c f_X(x|\boldsymbol{\theta})$ for a series of selected x values in the anticipated range of outcomes of X. Note that the same set of random samples are used to compute the kernel values for all selected x. This saves on generating samples and computing the corresponding likelihood function. The resulting PDF is shown in Figure 10.3 and is further discussed as part of Example 10.3.

9.9 Directional Sampling

Several simulation methods employ conditioning on a subset of the random variables and use sampling on the conditioned variables. Directional sampling described in this section is one such method. The orthogonal-plane sampling method described in Section 9.10 also belongs to this group of simulation methods.

Consider the partitioning $\mathbf{X} = (\mathbf{Y}, \mathbf{Z})$ of the vector of basic random variables in a reliability problem. Letting $\mathcal{F}$ denote the failure domain, we can write

$$\begin{aligned} p_F &= \int_{\mathcal{F}} f_{\mathbf{X}}(\mathbf{x})\mathrm{d}\mathbf{x} \\ &= \int_{\mathbf{y}} \Pr[(\mathbf{Y}, \mathbf{Z}) \in \mathcal{F} | \mathbf{Y} = \mathbf{y}] f_{\mathbf{Y}}(\mathbf{y})\mathrm{d}\mathbf{y} \\ &= \int_{\mathbf{y}} \Pr[(\mathbf{y}, \mathbf{Z}) \in \mathcal{F}] \frac{f_{\mathbf{Y}}(\mathbf{y})}{h(\mathbf{y})} h(\mathbf{y})\mathrm{d}\mathbf{y}. \end{aligned} \quad (9.34)$$

In the second line, we have conditioned on the set of random variables $\mathbf{Y}$, and in the last line we have introduced the sampling density $h(\mathbf{y})$. Obviously, in order to obtain an unbiased result, $h(\mathbf{y})$ must be non-zero wherever the product $\Pr[(\mathbf{y}, \mathbf{Z}) \in \mathcal{F}] f_{\mathbf{Y}}(\mathbf{y})$ is non-zero. It follows that the probability of failure equals the expectation

$$p_F = \mathrm{E}_h\left[\Pr[(\mathbf{y}, \mathbf{Z}) \in \mathcal{F}] \frac{f_{\mathbf{Y}}(\mathbf{y})}{h(\mathbf{y})}\right] \quad (9.35)$$

relative to the sampling distribution. Thus, the sampling method described in Section 9.5 can be used with

$$q_i = \Pr[(\mathbf{y}_i, \mathbf{Z}) \in \mathcal{F}] \frac{f_{\mathbf{Y}}(\mathbf{y}_i)}{h(\mathbf{y}_i)}, \quad (9.36)$$

where $\mathbf{y}_i$ are samples generated according to the distribution $h(\mathbf{y})$. More specifically, (9.8) provides the mean estimate $\widehat{p}_F$ and (9.15) provides the c.o.v. of the estimate, $\delta_{\widehat{p}_F}$. The above approach is effective if the conditional probability $\Pr[(\mathbf{y}_i, \mathbf{Z}) \in \mathcal{F}]$ can be easily computed for each generated sample $\mathbf{y}_i$.

In the directional sampling approach (Bjerager, 1988), the random variables in the standard normal space are partitioned according to $\mathbf{U} = \left(\widehat{\mathbf{A}}, R\right)$, where $\widehat{\mathbf{A}} = \mathbf{U}^{\mathrm{T}}/\|\mathbf{U}\|$ is the unit directional row-vector along $\mathbf{U}$ and R is the distance from the origin. Note that the number of random variables on the two sides of the partition equation is the same because $\widehat{\mathbf{A}}$ is normalized to have a unit length and, therefore, has one fewer independent random variable than its size. Also note that $\mathbf{U} = R\widehat{\mathbf{A}}^{\mathrm{T}}$. Owing to the rotational symmetry property of the standard normal space (see Section 6.2), $\widehat{\mathbf{A}}$ has a uniform distribution on the sphere of unit radius (unit sphere) centered at the origin of the standard normal space. Furthermore, as $R^2 = U_1^2 + \cdots + U_n^2$, i.e., R^2 is the sum of squares of n standard normal random variables, it has the Chi-square distribution with n degrees of freedom. Following the approach described in the preceding paragraph, we condition on the random variables $\widehat{\mathbf{A}}$ so that

$$q_i = \Pr(R\widehat{\mathbf{a}}_i \in \mathcal{F}) \frac{f_{\widehat{\mathbf{A}}}(\widehat{\mathbf{a}}_i)}{h(\widehat{\mathbf{a}}_i)}, \tag{9.37}$$

where $f_{\widehat{\mathbf{A}}}(\cdot)$ is the PDF of $\widehat{\mathbf{A}}$ (uniform on the unit sphere) and $\widehat{\mathbf{a}}_i$ is the ith sample of $\widehat{\mathbf{A}}$ according to the sampling density $h(\widehat{\mathbf{a}})$.

For a structural system with component limit-state functions $G_k(\mathbf{U})$, $k = 1, \ldots, K$, and min-cut sets $\mathcal{C}_l$, $l = 1, \ldots, L$, the failure domain is defined as $\mathcal{F} = \left\{\bigcup_{l=1}^{L} \bigcap_{k\in\mathcal{C}_l} G_k\left(R\widehat{\mathbf{A}}^{\mathrm{T}}\right) \le 0\right\}$. For a sample $\widehat{\mathbf{A}} = \widehat{\mathbf{a}}_i$, the conditional probability in the above expression is given by

$$\Pr(R\widehat{\mathbf{a}}_i \in \mathcal{F}) = 1 - \mathrm{X}_n^2(r_i^2), \tag{9.38}$$

where $\mathrm{X}_n^2(\cdot)$ is the CDF of the Chi-square distribution with n degrees of freedom and r_i is the solution to

$$r_i = \min_r \left\{\bigcup_{l=1}^{L} \bigcap_{k\in\mathcal{C}_l} G_k\left(r\widehat{\mathbf{a}}_i^{\mathrm{T}}\right) \le 0\right\}. \tag{9.39}$$

This is a root-finding problem: Along the direction $\widehat{\mathbf{a}}_i$, find the distance from the origin to the nearest point on the failure domain. Usually a secant method with a few evaluations of the limit-state function(s) is sufficient to find r_i. One should note, however, that in some problems, not the entire domain beyond r_i is a failure domain. For example, for a parallel system with a wedge-like failure domain as exemplified in Figure 9.9, there could be two roots along direction $\widehat{\mathbf{a}}_i$ such that failure occurs when $r_i \le r \le r_{i+1}$. In that case, (9.38) should be modified to read

$$\Pr(R\widehat{\mathbf{a}}_i \in \mathcal{F}) = \mathrm{X}_n^2(r_{i+1}^2) - \mathrm{X}_n^2(r_i^2). \tag{9.40}$$

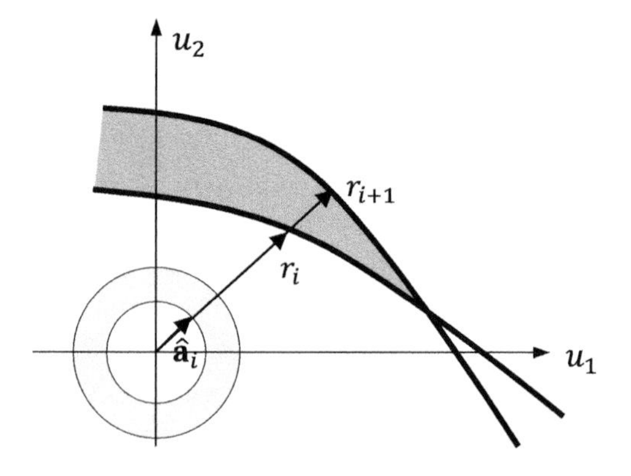

Figure 9.9 Directional sampling for a wedge-like failure domain

This problem may also arise for component problems if the limit-state surface is strongly nonlinear. The analyst must be aware of the possibility of the second root from the context of the problem and look for it.

For a component reliability problem with reliability index β and unit normal vector $\widehat{\boldsymbol{\alpha}}$ at the design point, Bjerager (1988) suggests sampling points in the half-space $\beta - \widehat{\boldsymbol{\alpha}}\mathbf{u} \leq 0$ and using the corresponding unit direction vectors. However, he suggests also sampling according to the uniform distribution on the unit sphere to capture failure points that may fall outside the half-space. The corresponding importance-sampling density is written as

$$h(\widehat{\mathbf{a}}) = (1 - w)f_{\widehat{\mathbf{A}}}(\widehat{\mathbf{a}}) + wh_{\widehat{\mathbf{A}}1}(\widehat{\mathbf{a}}), \tag{9.41}$$

where $0 \leq w \leq 1$ is a weight factor and $h_{\widehat{\mathbf{A}}1}(\widehat{\mathbf{a}})$ the importance sampling PDF for the unit direction vector of points falling in the half-space $\beta - \widehat{\boldsymbol{\alpha}}\mathbf{u} \leq 0$. Hence, for $1 - w$ fraction of the time one samples according to the uniform density over the unit sphere and for w fraction of the time one samples according to the PDF $h_{\widehat{\mathbf{A}}1}(\widehat{\mathbf{a}})$. The procedure described following (9.27) can be used to sample from the latter PDF. Figure 9.10 illustrates the sampling density for a two-dimensional problem with $w = 0.5$.

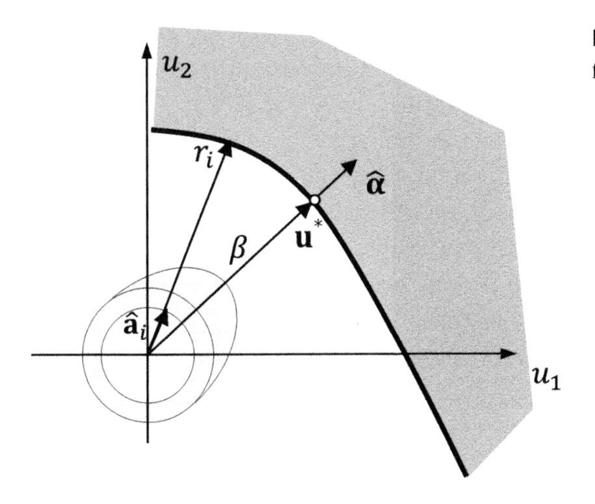

Figure 9.10 Directional importance sampling for a component problem

For a series-system problem with K components, one can sample in the direction of the half-space for each component. Extending (9.41), the importance-sampling density takes the form

$$h(\widehat{\mathbf{a}}) = w_0 f_{\widehat{\mathbf{A}}}(\widehat{\mathbf{a}}) + \sum_{k=1}^{K} w_k h_{\widehat{\mathbf{A}}k}(\widehat{\mathbf{a}}), \tag{9.42}$$

where $\sum_{k=0}^{K} w_k = 1$ are the weights and $h_{\widehat{\mathbf{A}}k}(\widehat{\mathbf{a}})$ is the importance sampling density associated with the half-space $\beta_k - \widehat{\boldsymbol{\alpha}}_k\mathbf{u} \leq 0$ of the kth system component. For the weights, values suggested following (9.25) may be used.

For a parallel system, the sampling density in (9.41) can be used with β and $\widehat{\boldsymbol{\alpha}}$ of the joint design point as defined in (8.66). Alternatively, the joint design point of the linearized component limit-state surfaces, as defined in (8.67), may be used to avoid the difficult problem of finding the joint design point of nonlinear limit-state surfaces. In all cases, it is prudent to include sampling in all directions to capture failure subdomains that fall outside the half-space $\beta - \hat{\boldsymbol{\alpha}}\mathbf{u} \leq 0$. As mentioned earlier, (9.40) should be used when a safe domain exists beyond a wedge-like failure domain, as illustrated in Figure 9.9.

A general system is solved in a manner similar to a series system with each "component" representing a min-cut set with β_k and $\hat{\boldsymbol{\alpha}}_k$ corresponding to its joint design point. The sampling density in (9.42) is used, including the background sampling with $f_{\hat{\mathbf{A}}}(\hat{\mathbf{a}})$.

The main computational effort in directional sampling (aside from finding the design points when importance sampling is employed) is finding the roots r_i in (9.39). As mentioned earlier, using a secant method, this essentially requires multiple evaluations of the limit-state functions for each generated directional sample. This may require an enormous effort if the limit-state functions involve computationally expensive algorithms, such as finite-element analysis. To reduce the amount of computation, CalREL offers two approximate alternatives: (a) use of limit-state surfaces linearized at their respective design points; or (b) use of point-fitted second-order limit-state surfaces. In both cases, the root in a given direction to the approximated surface is obtained in closed form without any iterations (Lin and Der Kiureghian, 1987). This greatly reduces the computational effort when the limit-state functions are computationally demanding, but also introduces a bias in the simulation result by replacing the exact surfaces with approximate ones. Naturally, FORM/SORM analysis must precede this type of directional sampling.

Finally, the idea of antithetic variates can be employed to increase the efficiency of the directional sampling method. This idea is useful when no information about the failure domain, such as component design points, is available and sampling is done uniformly in all directions. For each simulated direction $\hat{\mathbf{a}}_i$, one also explores the opposite direction, $-\hat{\mathbf{a}}_i$. The chances are that the two corresponding roots, r_i and r_{-i}, are negatively correlated, i.e., if r_i is small, then r_{-i} is likely to be large or infinite (no failure domain along that direction) and vice versa. This introduces negative covariance between the corresponding q_i and q_{-i} values and, hence, a smaller variability in the probability estimate. Unfortunately, it is not easy to evaluate this reduced c.o.v. Hence, one must be content with a more stable estimate of $\hat{p}_F$ without knowing its true c.o.v.

Example 9.6 – Comparison of Four Sampling Methods

Consider the linear limit-state function $G(\mathbf{U}) = 3 - \left(1/\sqrt{10}\right)(U_1 - \cdots - U_{10})$ in the standard normal space. The reliability index for this problem is $\beta = 3$, the failure probability is $p_F = \Phi(-3) = 0.00135$, and the design point is at $u_1^* = \cdots = u_{10}^* = 3/\sqrt{10}$. The failure probability of this problem is estimated by four simulation methods: ordinary Monte Carlo simulation, importance sampling using the design point (ISUDP), directional sampling with uniform density over the unit sphere (DS), and directional importance sampling (DIS) using the sampling density in (9.41) with $w = 0.5$. Figure 9.11 shows plots of the probability estimate $\hat{p}_F$ and its c.o.v. $\delta_{\hat{p}_F}$ as a function of the number of samples N. It is observed that all four simulation methods converge to the correct solution; however, their rates of convergence are vastly different. The least-efficient method is Monte Carlo, which achieves a c.o.v. of 8.60%

Example 9.6 (cont.)

after $N = 100{,}000$ simulations. Next is DS, which achieves a c.o.v. of 3.13%, followed by ISUDP with a c.o.v. of 0.58% after the same number of simulations. The most-efficient method is DIS, which achieves a c.o.v. of 0.31% after $N = 100{,}000$ simulations. Indeed, the last two methods achieve c.o.v. of 5.89% and 3.22% after only 100 simulations. It is noted that the efficiencies of the four methods remain essentially unchanged if the dimension of the problem is varied, as long as the reliability index remains the same.

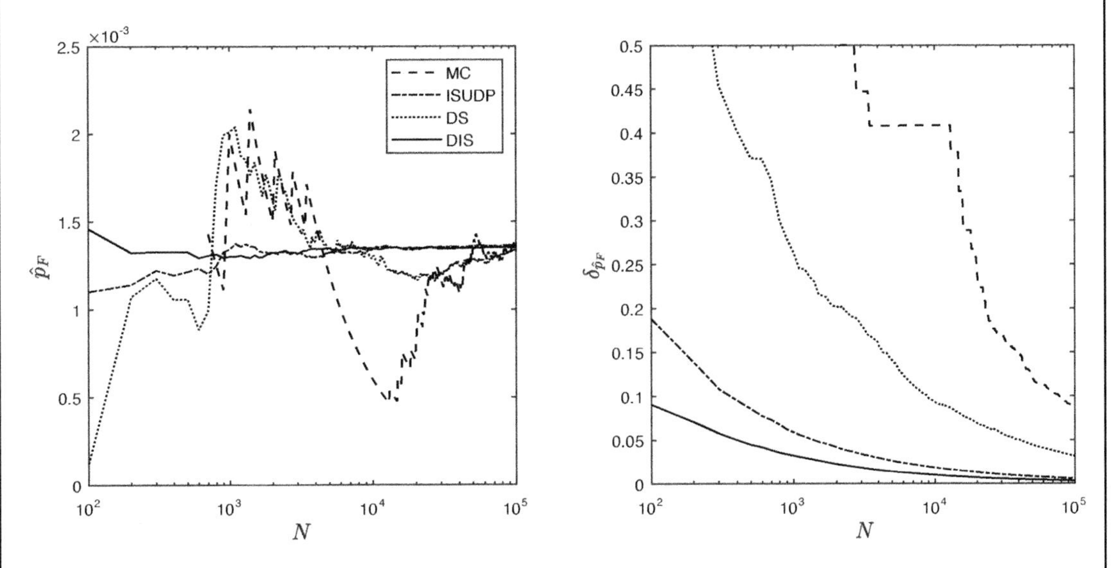

Figure 9.11 Comparison of simulation methods: ordinary Monte Carlo (MC), ISUDP, DS, and DIS

Example 9.7 – Reliability of Ductile Frame by Directional Sampling

Consider the ductile frame in Example 8.7 with three plastic failure mechanisms. The limit-state functions describing the three failure mechanisms and the statistics of the random variables are given in (E1)–(E3) and Table 8.4 of Example 8.7, respectively. Here, we compute the failure probability of the series system with directional sampling using three methods: (a) solving the exact roots according to (9.39) (involving the union of three component events), (b) solving the roots with the limit-state surfaces replaced by their first-order approximations at the respective design points, and (c) solving the roots with the limit-state surfaces replaced by their point-fitted, second-order approximations at the respective design points. Whereas the first method requires iterative calculations to find the distance r_i for each simulated direction,

Example 9.7 (cont.)

the second and third methods employ closed-form solutions of the distance and no iterations are necessary; hence, the latter methods require less computation but produce approximate values of the distance r_i.

Table 9.2 Results of analysis by directional sampling

Method	$\hat{p}_F$	N for $\delta_{\hat{p}_F} = 0.02$
Directional sampling with exact limit-state surfaces	0.0260	13,600
Directional sampling with first-order approximate surfaces	0.0265	13,000
Directional sampling with second-order approximate surfaces	0.0261	13,500

The results of the analysis with a target c.o.v. of 2% for the probability estimate are summarized in Table 9.2. The estimate based on the first-order approximate surfaces is practically identical to the result obtained by FORM in (E8) of Example 8.7. The results based on the exact limit-state surfaces and the second-order approximate surfaces also practically coincide. However, one must admit that the extent of nonlinearity in the limit-state surfaces in this problem, which arises only from the non-normal distribution of the random variables, is relatively mild – hence the close agreement between the three results.

Next, we apply the directional sampling method to compute the updated failure probability of the frame in light of its survival under proof-load tests, as described in Example 8.7. This probability is given as the ratio of a general system failure probability to that of a parallel system – see (E12) of Example 8.7. Directional sampling using exact surfaces provides the result $0.0179/0.947 = 0.0188$ and directional sampling using point-fitted second-order surfaces provides the result $0.0177/0.947 = 0.0186$, where all probability estimates have c.o.v. equal to or less than 0.02. These results closely agree with the result obtained in Example 8.7 using the multinormal probability.

9.10 Orthogonal-Plane Sampling

Hohenbichler and Rackwitz (1988) introduced the idea of orthogonal-plane sampling as an improvement to the second-order reliability approximation. This method can be used as a simulation method for component-reliability problems as long as the design point is available.

Consider the limit-state function $G(\mathbf{U})$ in the standard normal space and assume the design point $\mathbf{u}^*$ is known. Consider a rotation of the axes according to the transformation $\mathbf{u}' = \mathbf{P}\mathbf{u}$ as described following (7.1), where $\mathbf{P}$ is an orthonormal matrix with the unit normal vector $\hat{\boldsymbol{\alpha}}$ as its last row. As shown in Section 7.2, this puts the design point on the u'_n axis with

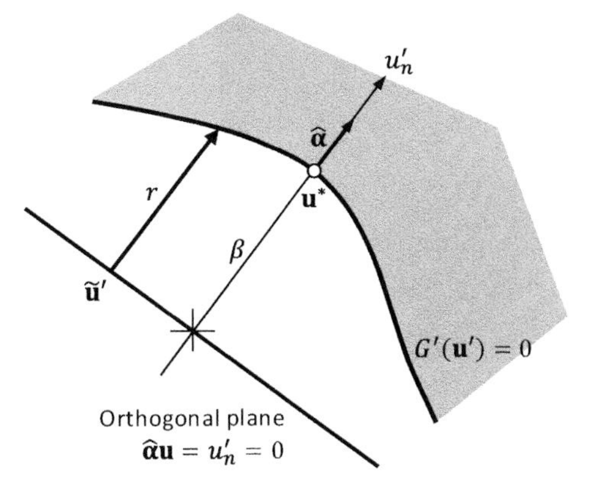

Figure 9.12 Sampling in the orthogonal plane

coordinates $(0, \ldots 0, \beta)$, where $\beta = \widehat{\boldsymbol{\alpha}}\mathbf{u}^*$ is the reliability index. Furthermore, $\widehat{\boldsymbol{\alpha}}\mathbf{u} = u'_n = 0$ denotes the plane that passes through the origin and is orthogonal to the u'_n axis, see Figure 9.12. Let $\widetilde{\mathbf{u}}' = \left(u'_1, \ldots, u'_{n-1}, 0\right)$ be a point on the orthogonal plane. We can write the failure probability in the standard normal space as

$$
\begin{aligned}
p_F &= \int_{G(\mathbf{u}) \leq 0} \phi(\mathbf{u}) \mathrm{d}\mathbf{u} \\
&= \int_{G'(\mathbf{u}') \leq 0} \phi(\mathbf{u}') \mathrm{d}\mathbf{u}' \qquad (9.43) \\
&= \int_{\widetilde{\mathbf{u}}'} \int_{G'(\mathbf{u}') \leq 0} \phi\left(u'_n | \widetilde{\mathbf{u}}'\right) \phi_{n-1}(\widetilde{\mathbf{u}}') \mathrm{d}u'_n \mathrm{d}\widetilde{\mathbf{u}}',
\end{aligned}
$$

where $G'(\mathbf{u}') = G\left(\mathbf{P}^{\mathrm{T}}\mathbf{u}'\right)$ is the limit-state function in the rotated space and in the last line we have conditioned on the point in the orthogonal plane. Let $r(\widetilde{\mathbf{u}}')$ denote the distance from the point in the orthogonal plane to the limit-state surface in the direction parallel to u'_n, as shown in Figure 9.12. This distance is the solution to

$$
G'(\widetilde{\mathbf{u}}' + r\hat{\mathbf{i}}) = 0. \qquad (9.44)
$$

where $\hat{\mathbf{i}}$ is a unit vector along the u'_n axis. We clearly have

$$
\int_{G'(\mathbf{u}') \leq 0} \phi\left(u'_n | \widetilde{\mathbf{u}}'\right) \mathrm{d}u'_n = \Phi[-r(\widetilde{\mathbf{u}}')]. \qquad (9.45)
$$

It follows that

$$
p_F = \int_{\widetilde{\mathbf{u}}'} \Phi[-r(\widetilde{\mathbf{u}}')] \phi_{n-1}(\widetilde{\mathbf{u}}') \mathrm{d}\widetilde{\mathbf{u}}'. \qquad (9.46)
$$

Thus, the probability of failure is the expectation of $\Phi\left[-r\left(\widetilde{\mathbf{U}}'\right)\right]$ relative to the $(n-1)$-dimensional standard normal PDF $\phi_{n-1}(\widetilde{\mathbf{u}}')$,

$$p_F = \mathrm{E}\left\{\Phi\left[-r\left(\widetilde{\mathbf{U}}'\right)\right]\right\}. \tag{9.47}$$

We compute the above expectation by sampling in the orthogonal plane. Let $\widetilde{\mathbf{u}}_i'$ be the ith sample and $r_i = r(\widetilde{\mathbf{u}}_i')$ be the corresponding solution obtained from (9.44). Then setting $q_i = \Phi[-r_i]$, we use (9.8) and (9.15) to compute the probability estimate $\widehat{p}_F$ and its c.o.v.

Recall from the discussion following (9.27) that points in the orthogonal plane can be easily simulated by generating a random point $\mathbf{u}$ in the standard normal space and projecting it on the plane $\widehat{\boldsymbol{\alpha}}\mathbf{u} = 0$. The result is $\widetilde{\mathbf{u}}'^{\mathrm{T}} = \mathbf{u} - \widehat{\boldsymbol{\alpha}}\mathbf{u}\widehat{\boldsymbol{\alpha}}^{\mathrm{T}}$. It is noted that this sampling method is primarily suitable for component-reliability problems.

Example 9.8 – Reliability of Two-Degree-of-Freedom Primary-Secondary System by Different Sampling Methods

Consider the two-degree-of-freedom primary-secondary system of Example 9.4 with random frequencies and damping ratios. We saw in Figure 9.7 that the limit-state surface of this problem is highly nonlinear. In Example 9.4 we obtained the solution of this problem by importance sampling in the half-space defined by the design point. Here, solutions are obtained and compared using several methods, as described in Table 9.3. Given in the table are the probability estimate $\widehat{p}_F$ and the number of samples needed to achieve a 5% c.o.v. in the probability estimate. It is seen that orthogonal-plane sampling is the most-efficient simulation method for this problem, followed by importance sampling in the half-space and DIS. Of the four methods used, importance sampling using the design point is the least-efficient method for this problem.

Table 9.3 Comparison of four sampling methods

Method	$\widehat{p}_F \times 10^5$	N for $\delta_{\widehat{p}_F} = 0.05$
Importance sampling using design point	1.63	6,200
Importance sampling in half-space	1.66	1,536
Directional importance sampling with $w = 0.5$	1.57	2,100
Orthogonal-plane sampling	1.57	800

9.11 Subset Simulation

We have seen that the main drawback of the ordinary Monte Carlo simulation method is that it needs a large number of samples when the probability to be estimated is small. As shown in (9.11), for small values of the failure probability, the coefficient of variation of the probability estimate is approximately equal to $1/\sqrt{Np_F}$, where N is the number of samples. As an example, for $p_F = 10^{-4}$ we need a sample of 4,000,000 to achieve a probability estimate with a c.o.v. of 0.05. In subset simulation, introduced in the field of structural reliability by Au and Beck (2001b), the failure probability is formulated as the product of several conditional probabilities, each having a fairly large value so that it can be estimated with a relatively small sample. A more recent treatment of the subject in the broader context of engineering risk assessment is given in Au and Wang (2014). The following description of the method is based on the later work of Papaioannou et al. (2015).

Let $\mathcal{F}$ denote the failure event. Formulating the problem in the standard normal space for a general system, this event can be expressed as $\mathcal{F} = \{\min_{1\le l \le L} \max_{k\in \mathcal{C}_l} G_k(\mathbf{U}) \le 0\}$, where $G_k(\mathbf{U})$ is the limit-state function of the kth component and $\mathcal{C}_l$, $l = 1, \ldots, L$, are the L min-cut sets of the system. We regard $\mathcal{F}$ as a subset in the outcome space of the standard normal random variables $\mathbf{U}$. Now consider a series of nested subsets $\mathcal{F}_0, \mathcal{F}_1, \ldots, \mathcal{F}_M$ such that $\mathcal{F}_0 \supset \mathcal{F}_1 \supset \cdots \supset \mathcal{F}_M$ with $\mathcal{F}_0$ being the certain event (the entire outcome space of $\mathbf{U}$) and $\mathcal{F}_M = \mathcal{F}$. Clearly $\bigcap_{m=0}^{M} \mathcal{F}_m = \mathcal{F}_M = \mathcal{F}$. Thus, by conditioning,

$$\begin{aligned} p_F &= \Pr(\mathcal{F}) \\ &= \Pr\left(\bigcap\nolimits_{m=0}^{M} \mathcal{F}_m\right) \\ &= \prod\nolimits_{m=1}^{M} \Pr(\mathcal{F}_m | \mathcal{F}_{m-1}). \end{aligned} \tag{9.48}$$

Note that $\Pr(\mathcal{F}_m|\mathcal{F}_{m-1}) \ge \Pr(\mathcal{F})$ for any m. Hence, the last line above expresses the failure probability as the product of larger probabilities. By proper definition of the subsets $\mathcal{F}_m$ and their number M, the conditional probabilities can be made sufficiently large for their estimation to require a small number of samples. The subsets are formulated as

$$\mathcal{F}_m = \left\{ \min_{1\le l\le L} \max_{k\in\mathcal{C}_l} G_k(\mathbf{U}) \le b_m \right\}, m = 1, \ldots, M-1, \tag{9.49}$$

with $b_1 > b_2 > \cdots > b_{M-1} > 0$. In practice, these thresholds are set adaptively so that the conditional probabilities $\Pr(\mathcal{F}_m|\mathcal{F}_{m-1})$ are around 0.1–0.2.

Of course, the conditional probabilities $\Pr(\mathcal{F}_m|\mathcal{F}_{m-1})$ can be estimated by Monte Carlo simulation. This would require counting the samples that fall in $\mathcal{F}_m$ and dividing that number by the number of samples that fall in $\mathcal{F}_{m-1}$. But as m increases and approaches M, few samples will fall in the subset of interest. Hence, a different approach is needed. Au and Beck (2001b) proposed the Markov chain Monte Carlo (MCMC) simulation method with a modified Metropolis algorithm. Papaioannou et al. (2015) reviewed this and several other MCMC algorithms and proposed one that is conducted in the standard normal space. Their algorithm, which was earlier used by Neal (1998) for regression and classification and has the advantage of not having to deselect candidate samples, is described below.

A Markov chain consists of a discrete sequence of events with the property that the probability of the current event being in a particular state conditioned on the sequence history depends only on the state of the directly preceding event. Suppose $\mathbf{u}_1, \mathbf{u}_2, \ldots, \mathbf{u}_i$ are outcomes of a sequence of random n-vectors $\mathbf{U}_1, \mathbf{U}_2, \ldots, \mathbf{U}_i$ and let $\mathcal{A}$ be a set in their outcome space. The sequence constitutes a Markov chain if

$$Pr(\mathbf{U}_{i+1} \in \mathcal{A} \,|\mathbf{U}_1 = \mathbf{u}_1, \ldots, \mathbf{U}_i = \mathbf{u}_i) = \Pr(\mathbf{U}_{i+1} \in \mathcal{A} \,|\mathbf{U}_i = \mathbf{u}_i). \tag{9.50}$$

That is, the sequence is a Markov chain if the probability that the vector in the $(i+1)$th step falls in $\mathcal{A}$, given the outcomes in all previous steps, depends only on the outcome of the last step. $\Pr(\mathbf{U}_{i+1} \in \mathcal{A} \,|\mathbf{U}_i = \mathbf{u}_i)$ is known as the transition probability. The Markov chain is stationary if the marginal probability $\Pr(\mathbf{U}_i \in \mathcal{A})$ and the transition probability are invariant of i. The method described below employs a stationary Markov chain.

In the subset simulation method with the MCMC algorithm, one first simulates N points in $\mathcal{F}_0$ (the entire outcome space) according to the n-variate standard normal PDF $\phi_n(\mathbf{u})$. One then computes and orders the set of values $\{\min_{1\le k\le K} \max_{i\in\mathcal{C}_k} G_i(\mathbf{u})\}$, $i = 1, \ldots, N$, and selects the threshold b_1 as the p_0-percentile value so that $\Pr\{\min_{1\le k\le K} \max_{i\in\mathcal{C}_k} G_i(\mathbf{u}) \le b_1\} = p_0$. This defines the subset $\mathcal{F}_1$ and we have $\Pr(\mathcal{F}_1|\mathcal{F}_0) = p_0$. Next, each of the Np_0 sample points in $\mathcal{F}_1$ is used as an initial seed to generate a Markov chain of length $(1/p_0 - 1)$ so that the total number of sample points in $\mathcal{F}_1$, including the starting seed points, is $Np_0(1/p_0 - 1) + Np_0 = N$. (One should choose p_0 such that Np_0 and $1/p_0$ are both integers.) These points are generated according to a multinormal conditional PDF $\phi_n(\mathbf{u}\,|\mathcal{F}_1)$. To achieve this, Papaioannou et al. (2015) suggest using a multinormal distribution of the seed, $\mathbf{s}$, and a new candidate vector $\mathbf{v}$ with zero means, unit variances, and correlation coefficient ρ between pairwise elements of $\mathbf{s}$ and $\mathbf{v}$. Given $\mathbf{s}$, the conditional distribution of $\mathbf{v}$ then is normal with mean vector $\mathbf{Rs}$ and covariance matrix $\mathbf{I} - \mathbf{RR}$, where $\mathbf{R} = \text{diag}[\rho]$ and $\mathbf{I}$ is the identity matrix. Of course, it is not guaranteed that the generated sample $\mathbf{v}$ will fall in $\mathcal{F}_1$. Hence, the new sample is selected as $\mathbf{u} = \mathbf{v}$ if $\mathbf{v} \in \mathcal{F}_1$ and $\mathbf{u} = \mathbf{s}$ if $\mathbf{v} \notin \mathcal{F}_1$. The new sample point, $\mathbf{u}$, is now considered to be the seed for generating the next sample and these steps are repeated until all the $(1/p_0 - 1)$ samples associated with the initial seed are generated. This process is repeated for each of the Np_0 initial seeds in $\mathcal{F}_1$ to achieve N sample points, including the initial seed points, as mentioned above. The generated samples in $\mathcal{F}_1$ are now used to compute and order the values $\{\min_{1\le l\le L} \max_{k\in\mathcal{C}_l} G_k(\mathbf{u})\}$ so that the p_0-percentile threshold b_2 defining $\mathcal{F}_2$ can be determined. This process is repeated for each subset until the value of the p_0-percentile threshold b_m for the subset $\mathcal{F}_m$ turns out to be negative. This indicates that we have reached the subset $\mathcal{F}_M$. Let $N_{\mathcal{F}}$ denote the number of points among the N generated in $\mathcal{F}_{m-1}$ that fall within $\mathcal{F}$. The estimate of the failure probability then is

$$\widehat{p}_F = p_0^M \frac{N_{\mathcal{F}}}{N}. \tag{9.51}$$

The main drawback of the MCMC method is that the generated samples are correlated because they depend on seeds that are samples themselves. A lower bound to the coefficient of variation of the probability estimate is given by (Au and Beck 2001a, Papaioannou et al. 2015)

$$\delta_{\hat{p}_F} \leqq \sqrt{\sum_{m=1}^{M} \delta_m^2}, \tag{9.52}$$

where

$$\delta_m = \sqrt{\frac{1 - p_m}{N p_m}(1 + \gamma_m)} \tag{9.53}$$

is the coefficient of variation of the probability estimate of the mth subset and

$$\gamma_m = 2\sum_{k=1}^{1/p_0 - 1} (1 - k p_0)\rho_m(k), \tag{9.54}$$

in which $\rho_m(k)$ is a measure of the correlation between the samples. Equation (9.52) assumes independence between the estimates for the different conditional probabilities and (9.54) assumes independence between the different chains. As a result, (9.52) may underestimate the true c.o.v. of the failure probability estimate.

Comparing (9.53) with (9.11), we see that positive correlation between the samples decreases the efficiency of the MCMC simulation relative to the ordinary Monte Carlo. In particular, a large ρ would lead to candidate samples $\mathbf{v}$ that are highly correlated with the seed, while a small ρ would lead to candidate samples $\mathbf{v}$ that have lower correlation with the seed but are more likely to fall outside $\mathcal{F}_m$, in which case the previous sample is utilized and the chain does not move. Hence, a trade-off value of ρ must be found. Papaioannou et al. (2015) do this adaptively, such that the rate of accepting a candidate sample $\mathbf{v}$ is in the order of 0.44, which gives a near-optimal performance. In most previous applications of this method, the values $p_0 = 0.1$ and $\rho = 0.8$ have been used.

The main advantage of the subset simulation method is that it is not affected by the dimension of the problem. In particular, as shown by Papaioannou et al. (2015) through example applications, the coefficient of variation of the probability estimate remains fairly constant with increasing dimension. On the other hand, the c.o.v. tends to slowly increase with decreasing failure probability because more subsets need to be used.

Example 9.9 – Reliability of Two-Degree-of-Freedom Primary-Secondary System by Subset Simulation

We solve the two-degree-of-freedom primary-secondary system by subset simulation. $N = 1{,}000$ samples are used in each subset with $p_0 = 0.10$. Five subsets are generated with threshold values $b_1 = 3.24$, $b_2 = 2.31$, $b_3 = 1.48$, $b_4 = 0.662$, and $b_5 = -0.098$, with a total of 5,000 samples. The resulting failure probability estimate is $\widehat{p}_f = 1.47 \times 10^{-5}$, which favorably compares with the results in Table 9.3 of Example 9.8. However, the c.o.v. of the estimate is 39%, which is much larger than those of the methods listed in Example 9.8. It is noted that the relative advantage of the subset simulation method becomes more revealing in problems with large numbers of random variables.

9.12 Reliability Sensitivities by Simulation

Section 6.6 presented formulas for computing the sensitivities of the reliability index and the FORM approximation of the failure probability with respect to parameters in the limit-state function or in the probability distribution of the basic random variables. The question arises whether these sensitivities can be computed by simulation. A naive answer is that one can perform multiple simulations while perturbing each parameter and then compute the sensitivities by finite differences. However, we have seen that simulation results are not exact, and every simulation run produces a different estimate of the failure probability. When a parameter is perturbed, the variation in the probability estimate will be due both to the change in the parameter value and to normal variation due to the random sampling. To reduce the latter variation, one can use the same set of random numbers in the simulations for the central and perturbed parameter values. Nevertheless, variations in sampling will occur depending on where the sample points fall relative to the limit-state surface. As it is not possible to differentiate the variations due to parameter perturbations and those due to random sampling, this type of finite-difference approach is not reliable. Furthermore, multiple simulation runs can be costly when there are many parameters. The following presents alternative approaches for computing parameter sensitivities by simulation. An advantage of these approaches is that one can obtain sensitivities for structural system problems, whereas the FORM results in Section 6.6 are limited to component-reliability problems.

First consider the case with limit-state function parameters $\boldsymbol{\theta}_g$ in a component reliability problem. Let $G(\mathbf{U}, \boldsymbol{\theta}_g)$ denote the limit-state function in the n-dimensional standard normal space. Using DIS with $\mathbf{U} = R\widehat{\mathbf{A}}^{\mathrm{T}}$ (see Section 9.9), the probability of failure is

$$\begin{aligned} p_F &= \int_{\hat{\mathbf{a}}} \Pr\left[G\left(R\widehat{\mathbf{A}}^{\mathrm{T}}, \boldsymbol{\theta}_g\right) \leq 0 \mid \widehat{\mathbf{A}} = \widehat{\mathbf{a}}\right] \frac{f_{\hat{\mathbf{A}}}(\widehat{\mathbf{a}})}{h(\widehat{\mathbf{a}})} h(\widehat{\mathbf{a}}) \mathrm{d}\widehat{\mathbf{a}} \\ &= \int_{\hat{\mathbf{a}}} \Pr\left[r(\widehat{\mathbf{a}}, \boldsymbol{\theta}_g) \leq R\right] \frac{f_{\hat{\mathbf{A}}}(\widehat{\mathbf{a}})}{h(\widehat{\mathbf{a}})} h(\widehat{\mathbf{a}}) \mathrm{d}\widehat{\mathbf{a}}, \end{aligned} \tag{9.55}$$

where $r(\widehat{\mathbf{a}}, \boldsymbol{\theta}_g)$ is the solution to

$$G\left(r\widehat{\mathbf{a}}^{\mathrm{T}}, \boldsymbol{\theta}_g\right) = 0. \tag{9.56}$$

We differentiate (9.55) with respect to $\boldsymbol{\theta}_g$ to obtain

$$\nabla_{\boldsymbol{\theta}_g} p_F = \int_{\hat{\mathbf{a}}} \nabla_{\boldsymbol{\theta}_g} \Pr\left[r(\widehat{\mathbf{a}}, \boldsymbol{\theta}_g) \leq R\right] \frac{f_{\hat{\mathbf{A}}}(\widehat{\mathbf{a}})}{h(\widehat{\mathbf{a}})} h(\widehat{\mathbf{a}}) \mathrm{d}\widehat{\mathbf{a}}. \tag{9.57}$$

Using (9.38), the gradient of the probability term inside the integrand is written as

$$\begin{aligned} \nabla_{\boldsymbol{\theta}_g} \Pr\left[r(\widehat{\mathbf{a}}, \boldsymbol{\theta}_g) \leq R\right] &= \nabla_{\boldsymbol{\theta}_g} \left\{1 - \mathrm{X}_n^2\left[r^2(\widehat{\mathbf{a}}, \boldsymbol{\theta}_g)\right]\right\} \\ &= -2r(\widehat{\mathbf{a}}, \boldsymbol{\theta}_g) \nabla_{\boldsymbol{\theta}_g} r(\widehat{\mathbf{a}}, \boldsymbol{\theta}_g) \chi_n^2\left[r^2(\widehat{\mathbf{a}}, \boldsymbol{\theta}_g)\right], \end{aligned} \tag{9.58}$$

where $\mathrm{X}_n^2[\cdot]$ is the CDF and $\chi_n^2[\cdot]$ the PDF of the chi-square distribution with n degrees of freedom. To obtain the gradient $\nabla_{\boldsymbol{\theta}_g} r(\widehat{\mathbf{a}}, \boldsymbol{\theta}_g)$, we differentiate the sides of (9.56):

$$\nabla_{\mathbf{u}} G\left(r\widehat{\mathbf{a}}^{\mathrm{T}}, \boldsymbol{\theta}_g\right)\widehat{\mathbf{a}}^{\mathrm{T}} \nabla_{\boldsymbol{\theta}_g} r(\widehat{\mathbf{a}}, \boldsymbol{\theta}_g) + \nabla_{\boldsymbol{\theta}_g} G\left(r\widehat{\mathbf{a}}^{\mathrm{T}}, \boldsymbol{\theta}_g\right) = 0. \tag{9.59}$$

It follows that

$$\nabla_{\boldsymbol{\theta}_g} r(\widehat{\mathbf{a}}, \boldsymbol{\theta}_g) = -\frac{1}{\nabla_{\mathbf{u}} G(r\widehat{\mathbf{a}}^{\mathrm{T}}, \boldsymbol{\theta}_g)\widehat{\mathbf{a}}^{\mathrm{T}}} \nabla_{\boldsymbol{\theta}_g} G\left(r\widehat{\mathbf{a}}^{\mathrm{T}}, \boldsymbol{\theta}_g\right). \tag{9.60}$$

Substituting the above expression in (9.58) and the result in (9.57), we have, for the gradient vector of the failure probability,

$$\nabla_{\boldsymbol{\theta}_g} p_F = \int_{\widehat{\mathbf{a}}} \frac{2r(\widehat{\mathbf{a}}, \boldsymbol{\theta}_g)}{\nabla_{\mathbf{u}} G(r\widehat{\mathbf{a}}^{\mathrm{T}}, \boldsymbol{\theta}_g)\widehat{\mathbf{a}}^{T}} \nabla_{\boldsymbol{\theta}_g} G\left(r\widehat{\mathbf{a}}^{\mathrm{T}}, \boldsymbol{\theta}_g\right) \chi_n^2\left[r^2(\widehat{\mathbf{a}}, \boldsymbol{\theta}_g)\right] \frac{f_{\widehat{\mathbf{A}}}(\widehat{\mathbf{a}})}{h(\widehat{\mathbf{a}})} h(\widehat{\mathbf{a}}) \mathrm{d}\widehat{\mathbf{a}}. \tag{9.61}$$

It is immediately apparent that the above integral can be computed by DIS using the outcome row vectors

$$\mathbf{q}_i = \frac{2r_i}{\nabla_{\mathbf{u}} G(\mathbf{u}_i, \boldsymbol{\theta}_g)\widehat{\mathbf{a}}_i^{\mathrm{T}}} \nabla_{\boldsymbol{\theta}_g} G(\mathbf{u}_i, \boldsymbol{\theta}_g) \chi_n^2(r_i^2) \frac{f_{\widehat{\mathbf{A}}}(\widehat{\mathbf{a}}_i)}{h(\widehat{\mathbf{a}}_i)}, \tag{9.62}$$

where $\mathbf{u}_i = r_i \widehat{\mathbf{a}}_i$ has been used. Observe that the above formulation requires computing the gradient vectors $\nabla_{\mathbf{u}} G(\mathbf{u}_i, \boldsymbol{\theta}_g)$ and $\nabla_{\boldsymbol{\theta}_g} G(\mathbf{u}_i, \boldsymbol{\theta}_g)$ at each sampling point. Both are required in FORM sensitivity analysis as well – the latter only at the design point. For convenience, we recall that $\nabla_{\mathbf{u}} G(\mathbf{u}, \boldsymbol{\theta}_g) = \nabla g(\mathbf{x}, \boldsymbol{\theta}_g)\mathbf{J}_{\mathbf{x},\mathbf{u}}$ (see 6.18e) and $\nabla_{\boldsymbol{\theta}_g} G(\mathbf{u}, \boldsymbol{\theta}_g) = \nabla_{\boldsymbol{\theta}_g} g(\mathbf{x}, \boldsymbol{\theta}_g)$ (see 6.45), where $\mathbf{x}$ is the mapping of $\mathbf{u}$ in the original space and $\mathbf{J}_{\mathbf{x},\mathbf{u}}$ is the inverse of the Jacobian of the probability transformation. With $\mathbf{q}_i$ computed for a sample of size N, the mean estimate of the probability sensitivity vector is obtained as

$$\begin{aligned} \nabla_{\boldsymbol{\theta}_g} \widehat{p}_F &= \frac{1}{N} \sum\nolimits_{i=1}^{N} \mathbf{q}_i \\ &= \bar{\mathbf{q}} \end{aligned} \tag{9.63}$$

and the element-wise c.o.v. are (see (9.15))

$$\delta_{\nabla_{\boldsymbol{\theta}_g} \hat{p}_F} \cong \frac{\sqrt{\sum\nolimits_{j=1}^{N} q_{i,j}^2 - \left(\sum\nolimits_{j=1}^{N} q_{i,j}\right)^2 / N}}{\sum\nolimits_{j=1}^{N} q_{i,j}}. \tag{9.64}$$

In general, the convergence rate of the probability sensitivities may not be the same as the convergence rate of the probability estimate itself. Furthermore, an importance-sampling density that is effective for estimating the failure probability may not be as effective for estimating the probability sensitivities. Nevertheless, it is convenient to use the same sample for computing the probability estimate as well as the estimates of probability sensitivities. The extra effort for sensitivity analysis then is that of computing the gradients $\nabla_{\mathbf{u}} G(\mathbf{u}_i, \boldsymbol{\theta}_g)$ and $\nabla_{\boldsymbol{\theta}_g} G(\mathbf{u}_i, \boldsymbol{\theta}_g)$ at each sampling point. It is important to monitor the c.o.v. values for all desired estimates to make sure that adequate accuracy is achieved.

The above directional sampling approach can be extended to general systems problems. Let $\mathcal{F} = \left\{ \bigcup_{1 \le l \le L} \bigcap_{k \in C_l} G_k(\mathbf{U}, \boldsymbol{\theta}_g) \le 0 \right\}$ define the failure domain in the standard normal space. For each directional sample $\widehat{\mathbf{a}}_i$, let r_i be the solution of r in

$$\min_{1 \le l \le L} \max_{k \in C_l} G_k\left(r\widehat{\mathbf{a}}_i^{\mathrm{T}}, \boldsymbol{\theta}_g \right) = 0. \tag{9.65}$$

This solution must fall on one of the limit-state surfaces $G_k\left(r\widehat{\mathbf{a}}_i^{\mathrm{T}}, \boldsymbol{\theta}_g \right) = 0$, $k = 1, 2, \ldots, K$, say $k = k'$. Then expressions (9.62)–(9.64) remain valid with $G(\mathbf{u}_i, \boldsymbol{\theta}_g)$ in (9.62) replaced by $G_{k'}(\mathbf{u}_i, \boldsymbol{\theta}_g)$. This assumes that a small perturbation in $\boldsymbol{\theta}_g$ does not change the active limit-state surface on which the root r_i falls. Such a change will occur only when the root is located at the intersection of two or more limit-state surfaces. The probability of such an event is of lower order than that of falling in the interior of the limit-state surface and, hence, can be neglected.

The orthogonal-plane sampling method described in Section 9.10 can also be used to compute the sensitivities with respect to the limit-state function parameters for a component-reliability problem. Without providing detailed derivations, we state that the only change is that the statistic $\mathbf{q}_i$ of the ith sample should be computed as

$$\mathbf{q}_i = -\phi(r_i) \nabla_{\boldsymbol{\theta}_g} r_i, \tag{9.66}$$

where $\nabla_{\boldsymbol{\theta}_g} r_i$ is the gradient of the root of (9.44) for the ith sample with respect to $\boldsymbol{\theta}_g$. It can be shown that

$$\nabla_{\boldsymbol{\theta}_g} r_i = -\frac{1}{\partial G / \partial r_i} \nabla_{\boldsymbol{\theta}_g} g(\mathbf{x}_i, \boldsymbol{\theta}_g), \tag{9.67}$$

where $\partial G / \partial r_i$ is the derivative of the limit-state function along the search direction. This derivative is computed in the unrotated $\mathbf{u}$ space along the vector $\mathbf{u}_{0,i} + r_i \widehat{\boldsymbol{\alpha}}$, where $\mathbf{u}_{0,i}$ is the sample point on the orthogonal plane. The derivation of these results is left as an exercise for the student.

Now consider the sensitivities of the failure probability with respect to distribution parameters, $\boldsymbol{\theta}_f$. Directional sampling is not effective in computing sensitivities with respect to these parameters. This is because a change in the distribution parameters changes the probability transformation and, hence, the configuration of the limit-state surfaces in the standard normal space. For these parameters, Monte Carlo importance sampling can be used, as described below.

Let $f_{\mathbf{X}}(\mathbf{x}, \boldsymbol{\theta}_f)$ denote the PDF of the basic random variables in a structural system reliability problem, where $\boldsymbol{\theta}_f$ denotes the set of distribution parameters, e.g., means, variances, correlation coefficients, bounds, or scale and shape parameters. Using Monte Carlo importance sampling in the original space, the probability of failure is written as

$$p_F = \int_{\mathbf{x}} I(\mathbf{x}) \frac{f_{\mathbf{X}}(\mathbf{x}, \boldsymbol{\theta}_f)}{h(\mathbf{x})} h(\mathbf{x}) \mathrm{d}\mathbf{x}, \tag{9.68}$$

where $I(\mathbf{x}) = 1$ if $\mathbf{x} \in \mathcal{F}$ and $I(\mathbf{x}) = 0$ otherwise, and $h(\mathbf{x})$ is the sampling density in the original space. Differentiating with respect to $\boldsymbol{\theta}_f$, we have

$$\nabla_{\boldsymbol{\theta}_f} p_F = \int_{\mathbf{x}} I(\mathbf{x}) \frac{\nabla_{\boldsymbol{\theta}_f} f_{\mathbf{X}}(\mathbf{x}, \boldsymbol{\theta}_f)}{h(\mathbf{x})} h(\mathbf{x}) d\mathbf{x}. \tag{9.69}$$

It follows that we can sample according to $h(\mathbf{x})$, $\mathbf{x}_i$, $i = 1 \ldots N$, to compute the values

$$\mathbf{q}_i = I(\mathbf{x}_i) \frac{\nabla_{\boldsymbol{\theta}_f} f_{\mathbf{X}}(\mathbf{x}_i, \boldsymbol{\theta}_f)}{h(\mathbf{x}_i)}, \tag{9.70}$$

and obtain the estimate $\nabla_{\boldsymbol{\theta}_f} \widehat{p}_F$ and its c.o.v. from formulas analogous to (9.63) and (9.64), respectively.

A critical step in the above approach is the selection of the importance-sampling density $h(\mathbf{x})$; this choice should consider the dimensional properties and correlation structure of the basic random variables in the original space. If FORM analysis has been carried out, one good choice is to sample according to (9.26) in the standard normal space and transform the sample points $\mathbf{u}_i$ back into the original space to obtain the sample points $\mathbf{x}_i$. The corresponding sampling density $h(\mathbf{x})$ is then obtained by transforming the random variables $\mathbf{U}$ having the distribution in (9.26) into the original space. As mentioned before, the same set of samples can be used to estimate the failure probability as well as the sensitivities with respect to the distribution parameters. Observe that with this approach, no additional calculations of the limit-state surface are necessary to compute the sensitivities. The only major additional calculation is that of the gradient $\nabla_{\boldsymbol{\theta}_f} f_{\mathbf{X}}(\mathbf{x}_i, \boldsymbol{\theta}_f)$ at each sampling point.

Example 9.10 – Sensitivities of the Failure Probability of a Two-Degree-of-Freedom Primary-Secondary System

Consider the two-degree-of-freedom primary-secondary system introduced in Example 9.4. The limit-state function is given in (E1)–(E2) and the probability distributions of the random variables are listed in Table 9.1. We wish to compute the sensitivities of the failure probability with respect to the limit-state function parameters $\boldsymbol{\theta}_g = (r, S_0, \gamma)$ and the probability-distribution parameters $\boldsymbol{\theta}_f = \left(\mu_{\omega_p}, \mu_{\omega_s}, \mu_{\zeta_p}, \mu_{\zeta_s}\right)$, where μ denotes the mean, with the corresponding c.o.v. fixed.

For the sensitivities with respect to the limit-state function parameters, the orthogonal-plane sampling method described in Section 9.10 and the paragraph following (9.65) is used. The statistics $\mathbf{q}_i$ are computed using (9.66) and (9.67), and (9.63) and (9.64) are used to compute the mean and c.o.v. of the sensitivity estimates, respectively. Using a sample of size $N = 10{,}000$, the results are $\widehat{p}_f = 1.62 \times 10^{-5}$ for the failure probability and $\nabla_{\boldsymbol{\theta}_g} \widehat{p}_F = 10^{-5}[-4.61\ 13.8 - 219]$ for its sensitivities with all the estimates having c.o.v. less than 2%. It is apparent that increasing the threshold r or the mass ratio γ reduces the failure probability, while increasing the amplitude of the white-noise excitation S_0 increases it. These are of course all as expected. As an example, reducing the mass ratio by 2%, i.e., from $\gamma = 0.05$ to $\gamma = 0.0490$, results in an increase of $\Delta p_f \cong 0.00219 \times 0.0010 = 0.0000022$ in the failure probability, i.e., to $\widehat{p}_f \cong 1.84 \times 10^{-5}$. On the other hand, increasing the allowable threshold

Example 9.10 (cont.)

by 2%, i.e., from $r = 6$ to $r = 6.12$, results in a decrease in the failure probability of $\Delta p_f \cong -0.0000461 \times 0.12 = -0.0000055$, i.e., to $\widehat{p}_f \cong 1.07 \times 10^{-5}$.

For the sensitivities with respect to $\boldsymbol{\theta}_f$, we know the failure domain is located in the half-space defined by the tangent hyperplane at the design point, so we use the importance sampling in the half-space described following Example 9.3. As defined in (9.27), the sampling density is the scaled, truncated normal PDF in the half-space $\beta - \widehat{\boldsymbol{\alpha}}\mathbf{u} \leq 0$. Hence, random sampling is done in that half-space in the standard normal space, and the generated points are mapped back onto the original space to compute the gradient $\nabla_{\boldsymbol{\theta}_f} f_{\mathbf{X}}(\mathbf{x}_i, \boldsymbol{\theta}_f)$. Because all the generated sample points fall in the half-space, the sampling density in the original space is given by $h(\mathbf{x}) = f_{\mathbf{X}}(\mathbf{x})/\Phi(-\beta)$. A sample of size $N = 10{,}000$ is used to compute the failure probability as well as the sensitivities with respect to all four means. The results are $\widehat{p}_f = 1.65 \times 10^{-5}$ and $\nabla_{\boldsymbol{\theta}_f} \widehat{p}_F = 10^{-5}[-2.97 - 3.43 - 144 - 172]$. The corresponding c.o.v. are all around 2%. Recall that these sensitivities describe the change in the failure probability value for a unit change in each parameter. Of course, a unit change in the mean damping ratios is an enormously large change – hence, the large sensitivity values with respect to the mean damping ratios. It is evident that increasing any of the mean values results in a reduction in the failure probability. For the frequencies, this is because with increasing fundamental frequency the system becomes stiffer and has a smaller deformation; for the damping ratios, it is because the system with a higher damping ratio dissipates more energy and has a smaller response. It is also noteworthy that the sensitivities are higher for the mean properties of the secondary oscillator relative to the corresponding mean properties of the primary system. This is because the failure criterion is specified in terms of the secondary response.

As an example, a reduction of 0.1 rad/s in the primary frequency, an increase of 0.1 rad/s in the secondary frequency, a reduction of 0.01 in the primary damping ratio, and an increase of 0.01 in the secondary damping ratio result in a change of approximately

$$\begin{aligned}\Delta p_f &\cong \left[-0.1(-2.97) + 0.1(-3.43) - 0.01(-144) + 0.01(-172) \times 10^{-5}\right] \\ &= -0.321 \times 10^{-5}\end{aligned} \tag{E1}$$

in the failure probability so that the revised failure probability estimate is $\widehat{p}_f \cong (1.65 - 0.32)\times 10^{-5} = 1.33 \times 10^{-5}$.

9.13 Concluding Remarks

This chapter presents a number of methods for structural and system reliability analysis by simulation. Methods for generation of pseudorandom numbers with prescribed distributions were described first. Although random number generators are readily available in most software used by engineers, knowing how these numbers are generated provides insight into some of the limitations that sampling methods may encounter. These include the finite period of the sequence, after which the sample of pseudorandom numbers repeats, and possible statistical dependence between pairs of sample values. The presentation also describes methods for generating dependent vectors of random variables with a prescribed joint distribution, which are not readily available in standard software.

The Monte Carlo simulation method is the backbone of all simulation methods. It provides an unbiased estimate of the failure probability as long as the generated sample of pseudorandom numbers is statistically independent and accurately reflects the joint distribution of the random variables. This method is also free of restrictions on the form of the limit-state function(s) and is equally applicable to component as well as systems problems. On the other hand, this method is inefficient for estimating small failure probabilities, with the c.o.v. of the estimate being inversely proportional to the square root of the failure probability. For ordinary structural-reliability problems, typically hundreds of thousands of simulations must be employed to achieve adequate accuracy. This is equal to the number of times that the limit-state function(s) must be computed. When the limit-state function requires costly computations, such as finite-element analysis or numerical integration, obviously the Monte Carlo solution is not practical. In practice, the Monte Carlo solution is often used as a benchmark for assessing the accuracy of other approximate methods.

A number of alternative simulation methods are presented, aimed at enhancing the efficiency of the ordinary Monte Carlo method. These include the use of antithetic variates, importance sampling, directional sampling, orthogonal-plane sampling, and subset simulation. Several of these methods require knowledge of the design point(s). These include importance-sampling methods using design points and the orthogonal-plane sampling method. Naturally, these methods impose certain restrictions on the form of the limit-state function, e.g., continuous differentiability, so that the design points can be found. Depending on the form of the reliability problem, i.e., component, series, parallel, or general system, these alternative methods can provide significant reduction in the required computational effort compared to the ordinary Monte Carlo method.

Other topics presented in this chapter include an importance-sampling method for multifold numerical integration, which is particularly useful for Bayesian updating (the topic of Chapter 10), and simulation methods for computing the sensitivities of the failure probability with respect to parameters in the limit-state function(s) or the probability distribution of random variables. Several numerical examples are used to demonstrate the presented methods and their limitations and advantages.

Sampling methods for structural-reliability analysis remain an active topic of research and development. A particularly active area of current research is the construction of surrogate models for implicit limit-state functions, i.e., those involving algorithmic solutions such as finite-element analysis, with a limited number of adaptively selected sample realizations. These surrogate models are then used as explicit mathematical models of limit-state functions in FORM, SORM, or simulation methods without engaging costly algorithmic calculations. This topic is further developed in Chapter 12.

PROBLEMS

9.1 Consider Problem 6.4 concerning the reliability of a plate against buckling. Carry out the following analysis:

1) Estimate the probability of failure by: (a) ordinary Monte Carlo simulation, (b) ISUDP, (c) DIS using the design point, and (d) orthogonal-plane sampling. Plot the mean probability estimates and their c.o.v. as functions of the number of simulations. Comment on the relative efficiency and accuracy of the four simulation methods.
2) Using directional sampling, determine the sensitivity of the failure probability estimate with respect to the plate thickness, t.
3) Using an importance sampling method, determine the sensitivities of the failure probability with respect to the mean and standard deviation of E.

9.2 Consider Problem 6.5 concerning the reliability of an oscillator subjected to harmonic excitation. The design point for this problem is at $\mathbf{u}^* = [-2.502 \quad -0.956 \ 0.936]^{\mathrm{T}}$ and the FORM reliability index is $\beta = 2.837$. Owing to the resonance effect, the limit-state surface for this problem is strongly nonlinear. Analysis by the point-fitting SORM reveals that all fitting points have coordinates u_n' greater than β, which suggests that the failure domain for this problem lies in the half-space defined by the tangent plane at the design point. Carry out the following analysis:

1) Estimate the probability of failure by (a) ordinary Monte Carlo simulation, (b) ISUDP, and (c) importance sampling in the half-space. Plot the mean probability estimates and their c.o.v. as functions of the number of simulations for the three methods. Comment on the accuracy and efficiency of the three methods and compare the results with the FORM approximation obtained in Problem 6.5.
2) Using importance sampling in the half-space, determine the sensitivities of the failure probability with respect to the means of Ω_0 and Ω.

9.3 Consider Problem 8.5 concerning the reliability of a plate against buckling under repeated load applications. Employing ISUDP, determine an estimate of the failure probability for the case of $n = 5$ load applications and compare it with the FORM approximation obtained in Problem 8.5. Use the same sampling method to compute the sensitivities of the failure probability with respect to the mean and standard deviation of E.

9.4 Determine the failure probability of the system defined in Problem 8.2 by (a) directional sampling, (b) DIS. Make plots of the probability estimate and its c.o.v. versus the number of simulations. Compare the results with the FORM approximation obtained in Problem 8.2.

10 Bayesian Parameter Estimation and Reliability Updating

10.1 Introduction

Formulation of the structural-reliability problem invariably involves parameters. These appear in two places: In the probability-distribution models that are used to describe the basic random variables, and in the limit-state functions that define component states. Specifically, let $f_{\mathbf{X}}(\mathbf{x}, \boldsymbol{\theta}_f)$ be the joint probability density function (PDF) of the basic random variables $\mathbf{X}$, where $\boldsymbol{\theta}_f = (\theta_{f1}, \theta_{f2}, \ldots,)$ is the set of distribution parameters, and let $g_i(\mathbf{X}, \boldsymbol{\theta}_g)$, $i = 1, \ldots, N$, be the set of limit-state functions for N structural system components, where $\boldsymbol{\theta}_g = (\theta_{g1}, \theta_{g2}, \ldots,)$ is the set of limit-state function parameters. The parameters $\boldsymbol{\theta}_f$ usually include the means, standard deviations, bounds, shape, scale, or other parameters that define the distribution function. These typically need to be estimated based on available observational data of the basic variables, and such estimation invariably involves uncertainties. The set $\boldsymbol{\theta}_g$ usually involves parameters that are present in mathematical models of capacities or demands that enter into the limit-state functions. These parameters need to be estimated by calibration of models to observed data and often also involve uncertainties.

This chapter addresses two topics related to parameter uncertainties: How to assess the uncertainties in the estimation of distribution and model parameters based on observed data, and how to account for these uncertainties in the assessment of structural reliability. Throughout, a Bayesian approach is used. Toward that end, this chapter starts with a description of the Bayesian statistical approach to parameter estimation. Methods for both distribution and mathematical model parameters are developed. Analysis of structural reliability under parameter uncertainties is then developed. Several methods are presented that implicitly or explicitly account for the effects of parameter uncertainties in the assessment of probability of failure or the reliability index. Examples throughout the chapter demonstrate these ideas and methods.

A third, related topic is the updating of structural reliability in light of observations. Bayesian techniques again prove to be the logical framework for this purpose. With the advent of sensor technology, measurements of structural properties and observation of structural behavior are easy and inexpensive. However, such measurements and observations are often indirect and affected by errors and uncertainties. The last section of this chapter presents methods for updating the reliability estimate of a structural system in the light of such indirect, error-prone, and uncertain information.

As mentioned in Section 1.3, there is a distinct difference between the classical (frequentist) and Bayesian notions of probability and treatment of uncertainties. Whereas the former sees the probability of an event as a property of the event itself, the Bayesian approach sees the probability of an event as the degree of belief of the person who assigns the probability value, subject to the available information. The Bayesian probability estimate, therefore, may change as the information available to the person who assigns it evolves. Furthermore, while the classical approach treats inherent variabilities and statistical uncertainties differently, in the Bayesian approach all types of uncertainties are modeled and analyzed by the same rules of probability and predictions are made by combining their effects. This Bayesian philosophical approach underpins the presentation in this chapter.

10.2 Sources and Types of Uncertainties

A variety of sources give rise to uncertainty in engineering analysis and decision making. These include the following:

Inherent variability or randomness: This is the kind of uncertainty that is inherent or intrinsic to a phenomenon. It cannot be predicted, but it can be measured through observations. Furthermore, it cannot be influenced or reduced without altering the phenomenon itself. Examples include the wind pressure on the façade of a building, the intensity of ground motion for a future earthquake, or the compressive strength of a cylinder of concrete. Note, for example, that the variability in the compressive strength of concrete cylinders cannot be altered without changing the concrete mix or the curing process. However, one may argue that some phenomena that appear inherently uncertain are in fact predictable if we can develop the correct scientific models. For example, perhaps the wind pressure on the façade of a building can be predicted if one has a sophisticated meteorological model for the region that predicts wind velocities. Similarly, for the strength of the concrete cylinder, perhaps one can predict it if one has precise information about the positioning of sand and gravel particles, and their respective strengths. Not having such models, we choose to see certain phenomena as inherently uncertain. So, the choice to regard a phenomenon as inherently random is to some extent ours. We do it to simplify our analysis, or because we do not possess the prerequisite scientific models.

Statistical uncertainty: This is the uncertainty that arises from the estimation of parameters, fitting of distributions, or assessment of probabilities from observational data, when such data are limited. This type of uncertainty is reducible because, at least theoretically, one always has the option of collecting more data.

Model error: Use of mathematical models is common in all aspects of engineering. The model describing the strength of a reinforced-concrete beam and the finite element model describing the stress at a critical joint in a structure or equations describing the contaminant transported in the ground-water flow to a consumption node are examples of such models. Mathematical models are invariably idealizations of more complex reality and, therefore, entail uncertain errors. Two aspects contribute to this error: an incorrect model form, e.g., a linear model when the true behavior is nonlinear, and "missing variables," i.e., variables that affect the phenomenon of interest but are not included in the mathematical model either for the sake of simplicity or because we are not aware of their influence. Additional uncertainty of the statistical type arises when the model parameters are estimated based on limited data.

Measurement error: When measuring observational data, errors are often encountered that arise from imperfection of the measuring devices or because of incorrect reading of the measurement. A measurement is called unbiased when the mean of the measurement error is zero. The standard deviation of this error is a measure of accuracy of the measurement. In most cases a normal distribution for the measurement error is assumed.

Human error: Humans are a source of uncertainty in many aspects of engineering. Mistakes or omissions in the design, analysis, construction, or operation of facilities could lead to uncertain variability in their performance. This kind of error is difficult to model and assess mathematically, though there are ways of controlling and managing its effects (Grigoriu, 1984a; Nowak, 1986). For example, independent review of design and analysis documents and construction practices is an effective way to identify and remedy human error in structural engineering.

It is common to categorize uncertainties as either *epistemic* or *aleatory*. An epistemic uncertainty is one that arises from lack of knowledge or data. A good example is statistical uncertainty in the estimation of parameters, which arises from limited data and possible measurement errors. An important characteristic of this type of uncertainty is that, at least theoretically speaking, it can be reduced. Specifically, it can be reduced by gathering more data. An aleatory uncertainty is one that is intrinsic to the phenomenon. It can be measured, but it cannot be reduced without altering the phenomenon itself. Der Kiureghian and Ditlevsen (2009) argue that, fundamentally, all uncertainties are epistemic in nature. This is because a phenomenon that appears to be intrinsically random may not be so if predictive scientific models are developed. Our state of knowledge is continuously evolving, and predictive models may be developed in the future for phenomena for which models are not available today. The uncertainty in the wind pressure on the façade of a building mentioned above is an example. For this reason, Der Kiureghian and Ditlevsen (2009) accept the epistemic versus aleatory categorization only within the universe of models that are adopted for an assessment project. For example, if a meteorological model is not available to predict the wind pressure on the façade of the building then the variability in the wind pressure may be considered an aleatory uncertainty within the confines of the project. This kind of separation of uncertainties is useful because it then becomes clear which uncertainties are reducible and which are not, within a given project.

10.3 Bayesian Parameter Estimation

In the Bayesian statistical approach, the rules of probability are used to represent and analyze all types of uncertainties. Furthermore, all uncertain quantities are modeled as random variables. This is not the case in classical statistics, where alternative measures, e.g., confidence interval, are used to represent the effect of statistical uncertainties and model parameters are not considered as random variables. Here, we follow the Bayesian approach. In a later section, we discuss the advantages of this approach. With this in mind, from here on we use upper-case letters to denote uncertain parameters and lower-case letters to represent their outcomes. Hence, Θ may represent an uncertain parameter modeled as a random variable, while θ denotes its outcome. Using the rules of probability, we define the probability distribution of Θ in terms of its PDF, $f_\Theta(\theta)$. This distribution characterizes the uncertainty in the parameter Θ.

Let $\mathcal{A}$ be an event of interest and $\mathcal{E}$ be an observed event. According to Bayes' rule in (2.19), the updated probability of $\mathcal{A}$ in light of observation of $\mathcal{E}$ is

$$\Pr(\mathcal{A}|\mathcal{E}) = \frac{\Pr(\mathcal{E}|\mathcal{A})}{\Pr(\mathcal{E})}\Pr(\mathcal{A}). \tag{10.1}$$

Now, if we let $\mathcal{A} = \{\theta < \Theta \leq \theta + \Delta\theta\}$, it follows that $\Pr(\mathcal{A}) = \Pr(\theta < \Theta \leq \theta + \Delta\theta) = f_\Theta(\theta)\Delta\theta$ and $\Pr(\mathcal{A}|\mathcal{E}) = \Pr(\theta < \Theta \leq \theta + \Delta\theta|\mathcal{E}) = f_\Theta(\theta|\mathcal{E})\Delta\theta$, where $f_\Theta(\theta|\mathcal{E})$ is the conditional PDF of Θ given event $\mathcal{E}$. Substituting these relations in (10.1) and dropping $\Delta\theta$ from both sides, we obtain

$$f_\Theta(\theta|\mathcal{E}) = \frac{\Pr(\mathcal{E}|\Theta = \theta)}{\Pr(\mathcal{E})}f_\Theta(\theta), \tag{10.2}$$

where in $\Pr(\mathcal{E}|\Theta = \theta)$ we have replaced the event $\{\theta < \Theta \leq \theta + \Delta\theta\}$ with $\{\Theta = \theta\}$, as $\Delta\theta$ can be arbitrarily small. Note that this is the probability of observing event $\mathcal{E}$ had we known that $\{\Theta = \theta\}$. The above is a rule for updating the distribution of Θ in the light of observing event $\mathcal{E}$.

Before going further, we make two generalizations for the above formulation: First, the rule in (10.2) equally applies to a vector of parameters, denoted as $\boldsymbol{\Theta}$ with outcomes $\boldsymbol{\theta}$. Second, the rule applies if we observe a sequence of events $\mathcal{E}_1, \ldots, \mathcal{E}_m$. Thus, more generally, (10.2) is written as

$$f_{\boldsymbol{\Theta}}(\boldsymbol{\theta}|\mathcal{E}_1\mathcal{E}_2\cdots\mathcal{E}_m) = \frac{\Pr(\mathcal{E}_1\mathcal{E}_2\cdots\mathcal{E}_m|\boldsymbol{\Theta} = \boldsymbol{\theta})}{\Pr(\mathcal{E}_1\mathcal{E}_2\cdots\mathcal{E}_m)}f_{\boldsymbol{\Theta}}(\boldsymbol{\theta}). \tag{10.3}$$

It can be shown that the order of the sequence of observations does not affect the final updated distribution.

The rule in (10.3) has four elements: $f_{\boldsymbol{\Theta}}(\boldsymbol{\theta})$ is the distribution of $\boldsymbol{\Theta}$ before information about the observed events becomes available. It is known as the *prior* distribution and reflects our prior knowledge about the uncertainty in $\boldsymbol{\Theta}$. From here on, we denote it with a prime, i.e., $f'_{\boldsymbol{\Theta}}(\boldsymbol{\theta})$. $f_{\boldsymbol{\Theta}}(\boldsymbol{\theta}|\mathcal{E}_1\mathcal{E}_2\cdots\mathcal{E}_m)$ is the updated distribution of $\boldsymbol{\Theta}$ that incorporates the information gained from observing the events $\mathcal{E}_1, \mathcal{E}_2, \ldots, \mathcal{E}_m$ together with the prior information. It is known as the *posterior* distribution and we denote it $f''_{\boldsymbol{\Theta}}(\boldsymbol{\theta})$. The probability term $\Pr(\mathcal{E}_1\mathcal{E}_2\cdots\mathcal{E}_m|\boldsymbol{\Theta}=\boldsymbol{\theta})$ describes the likelihood of observing the events $\mathcal{E}_1, \mathcal{E}_2, \ldots, \mathcal{E}_m$ if $\boldsymbol{\Theta}$ were equal to $\boldsymbol{\theta}$. This is known as the *likelihood* function and we denote it $L(\boldsymbol{\theta})$. Finally, the term $\Pr(\mathcal{E}_1\mathcal{E}_2\cdots\mathcal{E}_m)$ in the denominator, which is independent of $\boldsymbol{\theta}$, is essentially a scale factor so that the expression on the right-hand side is a valid PDF, i.e., so that it integrates to 1. With these considerations, we write (10.3) in the form

$$f''_{\boldsymbol{\Theta}}(\boldsymbol{\theta}) = cL(\boldsymbol{\theta})f'_{\boldsymbol{\Theta}}(\boldsymbol{\theta}), \tag{10.4}$$

where c is a normalizing factor so that

$$c = \left[\int_{\boldsymbol{\theta}} L(\boldsymbol{\theta})f'_{\boldsymbol{\Theta}}(\boldsymbol{\theta})\mathrm{d}\boldsymbol{\theta}\right]^{-1}. \tag{10.5}$$

Figure 10.1 illustrates the elements of the updating formula in (10.4) for the case of a single parameter. Observe that the prior is a broad distribution, the likelihood has a shape similar to a PDF but is not normalized, and the posterior is proportional to the product of the two so that its shape is influenced by both the prior and the likelihood shapes.

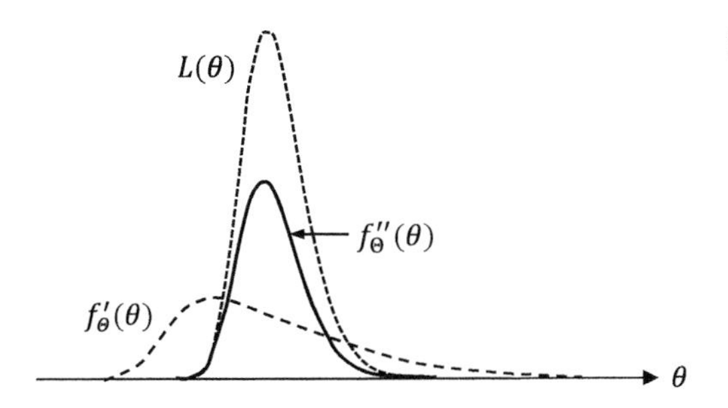

Figure 10.1 Bayesian updating of one parameter

Often the posterior statistics of $\boldsymbol{\Theta}$ are of interest. In particular, the posterior mean vector and covariance matrix are obtained by integration as

$$\mathbf{M}''_{\boldsymbol{\Theta}} = \int_{\boldsymbol{\theta}} c\boldsymbol{\theta}L(\boldsymbol{\theta})f'_{\boldsymbol{\Theta}}(\boldsymbol{\theta})\mathrm{d}\boldsymbol{\theta}, \tag{10.6}$$

$$\boldsymbol{\Sigma}''_{\boldsymbol{\Theta\Theta}} = \int_{\boldsymbol{\theta}} c\left(\boldsymbol{\theta} - \mathbf{M}''_{\boldsymbol{\Theta}}\right)\left(\boldsymbol{\theta} - \mathbf{M}''_{\boldsymbol{\Theta}}\right)^{\mathrm{T}}L(\boldsymbol{\theta})f'_{\boldsymbol{\Theta}}(\boldsymbol{\theta})\mathrm{d}\boldsymbol{\theta}. \tag{10.7}$$

For the latter, it is often more convenient to first compute the posterior mean-square matrix $\mathrm{E}''\left[\boldsymbol{\Theta}\boldsymbol{\Theta}^{\mathrm{T}}\right]$ and then compute the covariance matrix as $\boldsymbol{\Sigma}''_{\boldsymbol{\Theta\Theta}} = \mathrm{E}''\left[\boldsymbol{\Theta}\boldsymbol{\Theta}^{\mathrm{T}}\right] - \mathbf{M}''_{\boldsymbol{\Theta}}\mathbf{M}''^{\mathrm{T}}_{\boldsymbol{\Theta}}$.

Furthermore, if the marginal distribution of one of the random variables, say Θ_1, is required, then integration over all the other parameters is carried out:

$$f''_{\Theta_1}(\theta_1) = \int_{\theta_2} \cdots \int_{\theta_m} cL(\boldsymbol{\theta}) f'_{\boldsymbol{\Theta}}(\boldsymbol{\theta}) \mathrm{d}\theta_2 \cdots \mathrm{d}\theta_m. \tag{10.8}$$

We see that computations of the normalizing factor, posterior statistics, and marginal distributions require multifold integrations. As we will show later, for certain forms of the prior distribution and likelihood function, closed-form solutions exist. But more generally, the above integrals may need to be evaluated numerically. In Section 9.8, we described a sampling method to compute the above type of integrals and Example 9.5 demonstrated the approach. This example is further developed later in this chapter.

When the object of the updating is the set of parameters of the probability distribution of random variables **X**, it becomes desirable to obtain the so-called *predictive* distribution. This is the distribution of **X** that incorporates the uncertainty in the parameters through the total-probability rule:

$$\begin{aligned} \widetilde{f}_{\mathbf{X}}(\mathbf{x}) &= \int_{\boldsymbol{\theta}} f_{\mathbf{X}}(\mathbf{x}|\boldsymbol{\theta}) f''_{\boldsymbol{\Theta}}(\boldsymbol{\theta}) \mathrm{d}\boldsymbol{\theta} \\ &= \int_{\boldsymbol{\theta}} c f_{\mathbf{X}}(\mathbf{x}|\boldsymbol{\theta}) L(\boldsymbol{\theta}) f'_{\boldsymbol{\Theta}}(\boldsymbol{\theta}) \mathrm{d}\boldsymbol{\theta} \\ &= \mathrm{E}_{\boldsymbol{\theta}}[f_{\mathbf{X}}(\mathbf{x}|\boldsymbol{\Theta})]. \end{aligned} \tag{10.9}$$

In the above, $f_{\mathbf{X}}(\mathbf{x}|\boldsymbol{\theta})$ denotes the distribution of **X**, where we have shown the conditioning on the parameters to emphasize that this distribution is for specified values of the parameters. As indicated in the last line, the predictive distribution is essentially the expectation of the conditional distribution relative to the uncertain parameters $\boldsymbol{\Theta}$. Therefore, it is free of these parameters. Furthermore, it is apparent that the calculation of this distribution also requires integration over the $\boldsymbol{\theta}$ values. Therefore, the sampling method described in Section 9.8 is also useful for this purpose.

As the predictive distribution incorporates the uncertainty in its parameters, it is appropriate to use it when predictions involving **X** are to be made. It is also the appropriate distribution to use in decision-making based on the maximum expected utility criterion (von Neumann and Morgenstern, 1947). Examples later in this chapter will demonstrate the use of this distribution.

10.3.1 Formulation of the Likelihood Function

As mentioned earlier, the likelihood function is the conditional probability of the observed events given $\boldsymbol{\Theta} = \boldsymbol{\theta}$. This probability should depend on $\boldsymbol{\theta}$; otherwise, the observed events do not provide information about $\boldsymbol{\Theta}$. This section considers the case where the dependence arises because the observations are related to the outcomes of a set of random variables **X** with a joint distribution parametrized by $\boldsymbol{\theta}$ so that these parameters show up when we formulate the likelihood probability. We use $f_{\mathbf{X}}(\mathbf{x}|\boldsymbol{\theta})$ and $F_{\mathbf{X}}(\mathbf{x}|\boldsymbol{\theta})$ to denote the conditional PDF and conditional cumulative distribution function (CDF) of $\mathbf{X}$, given $\boldsymbol{\Theta} = \boldsymbol{\theta}$, respectively. Below, we present the formulation of the likelihood function for different types of observations.

Random sample of observations with exact measurements: Suppose we have a random sample of observations of the random variables, $\mathbf{x}_i$, $i = 1, \ldots, n$, with exact measurement. By "random sample" we mean that the samples are selected so that the observations are statistically independent of one another. The likelihood function then is

$$\begin{aligned} L(\boldsymbol{\theta}) &= \Pr(\mathbf{X}_1 = \mathbf{x}_1 \cap \cdots \cap \mathbf{X}_n = \mathbf{x}_n|\boldsymbol{\theta}) \\ &= \prod\nolimits_{i=1}^{n} \Pr(\mathbf{X}_i = \mathbf{x}_i|\boldsymbol{\theta}) \\ &= \lim_{\Delta\mathbf{x}_i \to \mathbf{0}} \prod\nolimits_{i=1}^{n} f_{\mathbf{X}}(\mathbf{x}_i|\boldsymbol{\theta})\Delta\mathbf{x}_i \\ &\propto \prod\nolimits_{i=1}^{n} f_{\mathbf{X}}(\mathbf{x}_i|\boldsymbol{\theta}), \end{aligned} \tag{10.10}$$

where $\mathbf{X}_i = \mathbf{x}_i$ denotes the ith instance of observing $\mathbf{X}$. In the second line, we have used the product of probabilities because the observations are statistically independent; the third line is based on the definition of the joint PDF; and in the last line we have dropped $\Delta\mathbf{x}_i$, which is independent of $\boldsymbol{\theta}$, and replaced the equality with proportionality. Note that any constant factor of the likelihood function can be incorporated into the normalizing factor c in (10.4), so we need to determine the likelihood function only in proportion. Thus, in the case under consideration, the likelihood function is proportional to the product of the joint PDFs of the random variables evaluated at the observed measurements and expressed as a function of the distribution parameters.

Random sample of observations with inexact measurements: Suppose we have a random sample of observations of the random variables, but the measurements are inexact. Let $\widehat{\mathbf{x}}_i$, $i = 1, \ldots, n$, denote the measured values and $\boldsymbol{\epsilon}_i$, $i = 1, \ldots, n$, denote the associated measurement errors, which are, of course, random. The superposed hat is used to indicate that the measured values are inexact. It follows that the exact measured instances are $\mathbf{X}_i = \widehat{\mathbf{x}}_i - \boldsymbol{\epsilon}_i$, or that we are actually observing the random variables $\mathbf{X}_i + \boldsymbol{\epsilon}_i$ with outcomes $\widehat{\mathbf{x}}_i$. Assuming the measurement errors in different observations to be statistically independent, it follows from the derivation of (10.10) that the likelihood function in this case is

$$L(\boldsymbol{\theta}) \propto \prod\nolimits_{i=1}^{n} f_{\mathbf{X}+\boldsymbol{\epsilon}_i}(\widehat{\mathbf{x}}_i|\boldsymbol{\theta}), \tag{10.11}$$

where $f_{\mathbf{X}+\boldsymbol{\epsilon}_i}(\cdot|\boldsymbol{\theta})$ denotes the joint PDF of random variables $\mathbf{X} + \boldsymbol{\epsilon}_i$. It is clear that the formulation of the likelihood function in this case requires determining the joint PDF of the function of random variables $\mathbf{X} + \boldsymbol{\epsilon}_i$.

Observation of bounds: Suppose we have a random sample of observations $\mathbf{X}_i = \mathbf{x}_i$, $i = 1, \ldots, n$, but also a set of lower-bound observations $\mathbf{X}_j \geq \mathbf{x}_j$, $j = 1, \ldots, l$, and a set of upper-bound observations $\mathbf{X}_k \leq \mathbf{x}_k$, $k = 1, \ldots, m$, where the inequalities are to be interpreted element-wise. Assuming the observations are statistically independent with exact measurements, the likelihood function takes the form

$$L(\boldsymbol{\theta}) \propto \prod_{i=1}^{n} f_{\mathbf{X}}(\mathbf{x}_i|\boldsymbol{\theta}) \prod_{j=1}^{l} \left[1 - F_{\mathbf{X}}\left(\mathbf{x}_j|\boldsymbol{\theta}\right)\right] \prod_{k=1}^{m} F_{\mathbf{X}}(\mathbf{x}_k|\boldsymbol{\theta}). \tag{10.12}$$

The second product involving the complementary joint CDF is the probability of observing the lower bounds, while the third product involving the joint CDF is the probability of observing the upper bounds. If measurement errors are present, the PDF and CDF of $\mathbf{X}$ should be replaced with those of $\mathbf{X} + \boldsymbol{\epsilon}$ and the observed values $\mathbf{x}_i$, $\mathbf{x}_j$, and $\mathbf{x}_k$ become $\widehat{\mathbf{x}}_i$, $\widehat{\mathbf{x}}_j$, and $\widehat{\mathbf{x}}_k$.

Observation of indirect events: Sometimes we make observations that are not measurements of the random variables $\mathbf{X}$ but events that involve these random variables. Examples are the survival of a structure under proof loading, which gives information about the capacity of the structure, or measurement of the drift of a structure under loads, which provides information about the stiffness characteristics of the structure and the applied loads. Mathematically speaking, such events can be expressed in terms of functions of the random variables, either in an equality form or an inequality form. For example, the information that the structure has survived a proof loading implies that the capacity is greater than the applied load effect. This can be expressed as an event of the form $\{g(\mathbf{X}|\boldsymbol{\theta}) \leq 0\}$, where $g(\mathbf{X}|\boldsymbol{\theta})$ denotes the load effect minus the capacity of the structure as a function of a set of basic random variables $\mathbf{X}$ describing member sizes and properties. The parameters $\boldsymbol{\theta}$ may appear in these functions explicitly as parameters of the model that predicts the structural capacity, or implicitly as parameters in the distribution of $\mathbf{X}$. In the case of measurement of the structural draft, the event of interest can be expressed as $\{h(\mathbf{X}|\boldsymbol{\theta}) = 0\}$, where $h(\mathbf{X}|\boldsymbol{\theta})$ denotes the difference between the drift of the structure, expressed as a function of structural properties and loads, and the measured drift. Hence, in a general form, the likelihood function can be expressed as

$$L(\boldsymbol{\theta}) = \Pr\left[\bigcap_{i=1}^{n} \{g_i(\mathbf{X}|\boldsymbol{\theta}) \leq 0\} \bigcap_{j=1}^{m} \left\{h_j(\mathbf{X}|\boldsymbol{\theta}) = 0\right\}\right], \tag{10.13}$$

where we have included n events of inequality type and m events of equality type. If the observed events are statistically independent, the probability of intersections can be written as the product of the probabilities of the individual events and the likelihood function reduces to

$$L(\boldsymbol{\theta}) = \prod_{i=1}^{n} \Pr[g_i(\mathbf{X}|\boldsymbol{\theta}) \leq 0] \prod_{j=1}^{m} \Pr\left[h_j(\mathbf{X}|\boldsymbol{\theta}) = 0\right]. \tag{10.14}$$

Computation of the above probabilities would require determining the joint or marginal CDFs and PDFs of the functions $g_i(\mathbf{X}|\boldsymbol{\theta})$ and $h_j(\mathbf{X}|\boldsymbol{\theta})$, respectively. This is not an easy task and may require introducing approximations. Specific examples later in this chapter will demonstrate this approach.

The power of the Bayesian approach should be evident from the broad nature of the type of observational data that can be considered in estimating model parameters. The fact that we are able to account for incorrectly measured data, information on bounds, and even observation of events that indirectly provide information about the parameters is a remarkable advantage of the Bayesian approach.

10.3.2 Selection of Prior Distribution

As defined earlier, the prior distribution $f'_{\boldsymbol{\Theta}}(\boldsymbol{\theta})$ reflects our state of knowledge about the parameters $\boldsymbol{\Theta}$ before observations are made. As Figure 10.1 illustrates, the selection of the prior distribution is important only when the likelihood curve is not sharp, i.e., when the observational data are not rich in information content. This usually happens when the observational data sample is small. The prior is a controversial part of the Bayesian statistical approach, as its selection is not unique and is subject to the analyst's choice. In this section, we briefly review various available options for selecting the prior distribution.

Subjective prior: In many engineering-decision problems, reliance on expert judgment and opinion cannot be avoided. This is the case when the available objective information is limited, and additional observational data cannot be gathered because of time or cost constraints. For example, suppose we are to estimate the mean rate of occurrence of earthquakes in a region with a short historical record. If our interest is in large-magnitude earthquakes, the chances are that very few would have been recorded. Obviously, additional observational data cannot be gathered without waiting a very long time. In such a case, the expert opinion of a seismologist based on indirect information, such as carbon dating of fault trenches, comparison with other regions of similar seismicity, etc., can be valuable. One important advantage of the Bayesian approach is that it provides a means to incorporate subjective information of experts in the estimation of the model parameters.

In selecting a subjective prior distribution, it is best to first determine the prior mean and standard deviation and then select a distribution type that is computationally convenient and satisfies the applicable constraints, e.g., boundedness of a parameter. The mean should represent a central estimate of the parameter, whereas the standard deviation should represent a measure of the expert's uncertainty about the true value of the parameter. The interval of the mean plus and minus one standard deviation should correspond roughly to a 70% confidence interval, that is, the expert should feel about 70% certain that the interval includes the unknown parameter. If possible, a conjugate distribution shape, as described later in this chapter, can be selected to simplify the analysis.

Previous posterior as prior: The posterior distribution from a previous updating can be used as the prior for updating with new information. In fact, this procedure can be repeatedly applied as many times as new observational data become available. One can show that the order in which information is received does not influence the final posterior distribution. Furthermore, one can perform the updating for each event separately as it is observed (using the previous posterior as prior), or update for all the observed events at once, as indicated in (10.3). The two methods will yield the same posterior distribution.

Diffuse prior: When the prior distribution is flat in the non-zero interval of the likelihood function (see Figure 10.1), it has practically no influence on the posterior. One can in fact assume it to be a constant, or a "locally" uniform distribution, i.e., a distribution that is uniform over the interval where the likelihood is non-zero. For updating purposes, the prior is then considered a constant and incorporated into the normalizing factor c. This kind of prior is known as a *diffuse* prior.

Non-informative prior: In some situations, no prior information (objective or subjective) is available and one must entirely rely on observational data to estimate the parameters of a model. In such cases, a prior that has minimal influence on the shape of the posterior is desired. Such a prior is called *non-informative*. A diffuse prior may appear to be non-informative and is often used in practice with that in mind. However, the choice of the distribution parameter is often not unique, and a diffuse prior for the parameter in one formulation of the model may, in fact, be informative for the parameter in an alternative formulation of the model. Consider, for example, the exponential PDF defined as $f_X(x|\lambda) = \lambda \exp(-\lambda x)$, where $0 < \lambda$ is the sole parameter. An alternative formulation of this PDF is $f_X(x|\theta) = (1/\theta) \exp(-x/\theta)$, where $\theta = 1/\lambda$ is now the parameter. A locally uniform (diffuse) distribution for λ, say over $[0, \infty)$, implies a PDF for θ that is proportional to $1/\theta^2$; conversely, a locally uniform distribution for θ over the same interval implies a PDF for λ that is proportional to $1/\lambda^2$. Here lies the paradox in assigning prior distributions: A seemingly non-informative prior for a model parameter may imply an informative prior for the parameter in an alternative formulation of the same model.

To overcome the above difficulty, Box and Tiao (1973) define a non-informative prior in the following way: Let $\lambda = \lambda(\theta)$ be a one-to-one transformation of the parameter θ of a distribution model and $L(\lambda)$ be the likelihood function expressed in terms of the transformed parameter λ. If the likelihood function $L(\lambda)$ is data translated, i.e., if the data serve to change the location of $L(\lambda)$ along the λ axis but not its shape, then a locally uniform prior for λ is non-informative. The corresponding non-informative prior for θ is proportional to $|\mathrm{d}\lambda/\mathrm{d}\theta|$. Box and Tiao (1973) also show that the non-informative prior for the mean μ of a normal distribution with known standard deviation is locally uniform, whereas for the standard deviation σ of a normal distribution with known mean the non-informative prior is locally uniform for the transformed parameter $\lambda = \ln \sigma$, which implies a density proportional to $|\mathrm{d}\lambda/\mathrm{d}\sigma| = 1/\sigma$ for σ itself. Unfortunately, it is not always possible to find the transformation $\lambda = \lambda(\theta)$ that makes the likelihood function data translated. In such cases, numerical tests with different transformed parameters can be conducted to see which choice yields a likelihood function that is at least approximately data translated. The following example illustrates such a case.

A method to derive a parameter transformation that has an approximately data-translated likelihood is described by Box and Tiao (1973). When such detailed analysis is unavailable, a good practical choice is to use a diffuse prior for an unbounded parameter and a prior proportional to the reciprocal of the parameter for a non-negative parameter.

Example 10.1 – Non-Informative Prior for the Parameter of the Exponential Distribution

Consider the exponential distribution $f_X(x|\lambda) = \lambda \exp(-\lambda x), 0 \leq x$, with unknown parameter $0 < \lambda$. For a sample x_i, $i = 1, \ldots, n$, of n observations, using (10.10), the likelihood function is

$$L(\lambda) = \lambda^n \exp(-n\lambda\bar{x}), 0 < \lambda, \tag{E1}$$

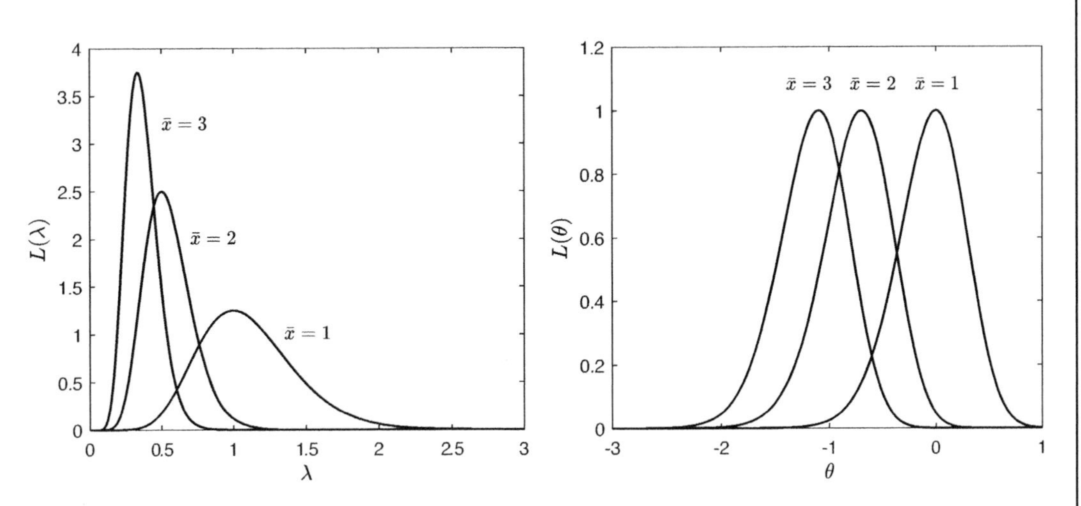

Figure 10.2 Likelihood functions for two parametrizations of the exponential distribution

where $\bar{x} = \sum_{i=1}^{n} x_i/n$ is the sample mean. For $n = 10$, the left chart in Figure 10.2 shows plots of the normalized likelihood function for $\bar{x} = 1, 2$, and 3. Observe that this likelihood function is not data translated, as its shape as well as its location depends on the data, i.e., on $\bar{x}$. Now suppose we were to express the exponential distribution in terms of the transformed parameter $\theta = \ln \lambda$, i.e., as $f_X(x, \lambda) = \exp(\theta) \exp[-\exp(\theta)x], 0 \leq x$. The likelihood function for θ is obtained by substituting $\lambda = \exp(\theta)$ in the preceding equation, which yields

$$L[\theta] = \exp(n\theta) \exp[-n \exp(\theta)\bar{x}]. \tag{E2}$$

The right-hand chart in Figure 10.2 shows plots of the likelihood function in terms of θ for the same values of n and $\bar{x}$ as used for the chart on the left. Observe that this likelihood function is approximately data translated, in the sense that its shape is nearly invariant of the data and only its location depends on $\bar{x}$. Obviously, if a locally uniform prior distribution is assumed for θ, it will have virtually no influence on the posterior distribution. Thus, the non-informative prior for θ is locally uniform. This implies a non-informative prior for λ that is proportional to $|d\theta/d\lambda| = 1/\lambda$.

Conjugate prior: For certain distributions of the underlying random variable X and certain types of observational data, there exist prior distributions such that the posterior of Θ is of the same distribution type as the prior. Such prior/posterior pairs are known as conjugate distributions. The selection of a conjugate prior, if it exists, greatly simplifies the updating process, as well as the calculation of posterior statistics of the parameter Θ and the predictive distribution of X. Table A10.1 in Appendix 10A lists the conjugate prior distributions and the updating rules for several commonly used distribution models. The following example demonstrates the derivation of some of the results in that table.

Example 10.2 – Conjugate Prior of Exponential Distribution

Consider the exponential distribution $f_X(x|\lambda) = \lambda \exp(-\lambda x), 0 \leq x$, with unknown parameter $0 < \lambda$. Assume the available observation is a random sample of exact measurements x_i, $i = 1, \ldots, n$. We have seen that the likelihood function for this case is $L(\lambda) = \lambda^n \exp(-n\lambda\bar{x}), 0 < \lambda$. Let us adopt the gamma distribution as the prior distribution of λ. This is given as $f'_\Lambda(\lambda) = v'(v'\lambda)^{k'-1} \exp(-v'\lambda)/\Gamma(k'), 0 < \lambda$, where $0 < v'$ and $1 \leq k'$ are the distribution parameters and $\Gamma(\cdot)$ is the gamma function. Using the updating formula in (10.4), we have

$$\begin{aligned} f''_\Lambda(\lambda) &= c\,\lambda^n \exp(-n\lambda\bar{x}) \frac{v'(v'\lambda)^{k'-1} \exp(-v'\lambda)}{\Gamma(k')} \\ &\propto \lambda^{n+k'-1} \exp[-(v' + n\bar{x})\lambda]. \end{aligned} \tag{E1}$$

The last expression is proportional to the gamma PDF with parameters

$$v'' = v' + n\bar{x} \tag{E2}$$

$$k'' = k' + n. \tag{E3}$$

It follows that the posterior distribution is also gamma but with parameters v'' and k''. Hence, gamma is the conjugate prior for the parameter λ of the exponential distribution for the particular type of observation considered, and (E2) and (E3) describe the updating rule. Furthermore, the predictive distribution of X is given by

$$\begin{aligned} \tilde{f}_X(x) &= \int_0^\infty \lambda \exp(-\lambda x) \frac{v''(v''\lambda)^{k''-1} \exp(-v''\lambda)}{\Gamma(k'')} \mathrm{d}\lambda \\ &= \frac{v''^{k''}}{\Gamma(k'')} \int_0^\infty \lambda^{k''} \exp[-(v'' + x)\lambda] \mathrm{d}\lambda \\ &= \frac{v''^{k''}}{\Gamma(k'')} \frac{\Gamma(k''+1)}{(v''+x)^{k''+1}} \\ &= \frac{k'' v''^{k''}}{(v''+x)^{k''+1}}, 0 \leq x. \end{aligned} \tag{E4}$$

Example 10.2 (cont.)

It is evident that in the case of a conjugate prior we are also able to obtain a closed-form solution for the predictive distribution of the underlying random variable.

Now suppose in addition to the above observations, we have m observations of lower bounds, $y_j \leq X_j, j = 1, \ldots, m$. Using (10.12), the likelihood function then is

$$\begin{aligned} L(\lambda) &= \lambda^n \exp(-n\bar{x}\lambda) \exp(-m\bar{y}\lambda) \\ &= \lambda^n \exp[-(n\bar{x} + m\bar{y})\lambda], \end{aligned} \tag{E5}$$

where $\bar{y}$ is the sample mean of the lower bounds. The posterior distribution is given by

$$\begin{aligned} f''_\Lambda(\lambda) &= c\,\lambda^n \exp[-(n\bar{x} + m\bar{y})\lambda] \frac{\nu'(\nu'\lambda)^{k'-1} \exp(-\nu'\lambda)}{\Gamma(k')} \\ &\propto \lambda^{n+k'-1} \exp[-(\nu' + n\bar{x} + m\bar{y})\lambda]. \end{aligned} \tag{E6}$$

It is clear that the posterior distribution is gamma with parameters

$$\nu'' = \nu' + n\bar{x} + m\bar{y} \tag{E7}$$

$$k'' = k' + n. \tag{E8}$$

Furthermore, it is easily shown that the predictive distribution is the same as in (E4), albeit with ν'' computed according to (E7).

The above results are summarized in Table A10.1 along with results for other known conjugate distribution pairs.

10.3.3 Conjugate Priors for the Normal Distribution

Considering the importance of the normal distribution, in this section we present updating formulas for the parameters of the normal distribution. We present cases where the variance is known and the mean is to be estimated, and where both the mean and variance are unknown and need to be estimated from observational data. An example demonstrates the use of these formulas. The formulas are also summarized in Table A10.1.

Consider the normal distribution $f_X(x|\mu, \sigma) = (1/\sqrt{2\pi}\sigma) \exp\left[-(x-\mu)^2/2\sigma^2\right]$ with mean μ and standard deviation σ. Assume the available observation is a random sample of exact measurements x_i, $i = 1, \ldots, n$. We first consider the case where μ is unknown while σ is known. This situation may arise in applications where the variability in the data is independent of the sample mean and can be estimated from a large population while the mean needs to be estimated for smaller subsets of the population. The production of concrete by batch plants is one possible example: The variability remains nearly the same for all mixes and can be estimated from long-term production results, while the mean for each mix of concrete is different and needs to be estimated from a smaller sample.

One can show that, under the above assumptions, the conjugate prior of μ is normal with mean μ'_μ and standard deviation σ'_μ. The updating formulas for the mean and standard deviation of the normal posterior distribution are

$$\mu''_\mu = \frac{\bar{x}\sigma'^2_\mu + \mu'_\mu \dfrac{\sigma^2}{n}}{\sigma'^2_\mu + \dfrac{\sigma^2}{n}} \tag{10.15}$$

$$\sigma''_\mu = \sqrt{\frac{\sigma'^2_\mu \dfrac{\sigma^2}{n}}{\sigma'^2_\mu + \dfrac{\sigma^2}{n}}}, \tag{10.16}$$

where $\bar{x} = \sum_{i=1}^{n} x_i/n$ is the sample mean. It can be seen that the posterior mean is the weighted average of the sample mean and the prior mean, with the weights being the counter-respective variances. (We note that σ^2/n is the variance of the sample mean.) As $n \to \infty$ or $\sigma'_\mu \to \infty$ (i.e., a diffuse prior), we have $\mu''_\mu \to \bar{x}$, which is the data-based estimate of the mean. On the other hand, as $n \to 0$, i.e., no observed sample, we have $\mu''_\mu \to \mu'_\mu$, which is logical because the prior is the only information available. For the posterior standard deviation, we can easily verify that σ''_μ is smaller than both σ'_μ and $\sigma/\sqrt{n}$. This is because the posterior incorporates the information contained both in the prior and in the observed data. Hence, the posterior distribution of μ is less dispersed than either the prior distribution or the distribution of the sample mean.

It can be shown that, under the above assumptions, the predictive distribution of X is also normal with mean equal to the posterior mean, $\tilde{\mu}_X = \mu''_\mu$, and variance $\tilde{\sigma}^2_X = \sigma^2 + \sigma''^2_\mu$. It is apparent that the variance of the predictive distribution is increased on account of the statistical uncertainty in the mean. This last expression clearly shows the amalgamation of two types of uncertainties: the uncertainty due to the inherent variability in X, represented by σ^2, and the statistical uncertainty in the estimation of the mean of the distribution, represented by σ''^2_μ. This amalgamation of uncertainties is a fundamental notion of Bayesian analysis and is essential for any decision making that involves predictions of the random variable X.

Now consider the case where both μ and σ are unknown and need to be estimated from the sample of n observations. We need a bivariate prior distribution. It can be shown that the conjugate prior for μ and σ^2 is the normal-inverse-gamma distribution. This distribution is defined by the product of two distributions: the normal conditional distribution of μ given σ^2, denoted $\mathrm{N}\left(\mu'_\mu, \sigma^2/n'\right)$, where parameter n' can be interpreted as an equivalent sample size for the prior information, and the marginal distribution of σ^2, which is inverse-gamma with parameters k' and ν'. Specifically, the joint PDF has the form

$$f'\left(\mu,\sigma^2\right)=\frac{\sqrt{n'}}{\sqrt{2\pi}\sigma}\exp\left[-\frac{n'}{2\sigma^2}\left(\mu-\mu'_\mu\right)^2\right]\frac{\nu'^{k'}}{\Gamma(k')}\left(\frac{1}{\sigma^2}\right)^{k'+1}\exp\left(-\frac{\nu'}{\sigma^2}\right)$$
$$=\frac{\nu'^{k'}}{\Gamma(k')}\frac{\sqrt{n'}}{\sqrt{2\pi}\sigma}\left(\frac{1}{\sigma^2}\right)^{k'+1}\exp\left[-\frac{2\nu'+n'\left(\mu-\mu'_\mu\right)^2}{2\sigma^2}\right]. \tag{10.17}$$

One can show that the marginal distribution of $\left(\mu-\mu'_\mu\right)\sqrt{n'k'/\nu'}$ is the standard Student's t-distribution with $2k'$ degrees of freedom so that

$$f'(\mu)=\frac{\Gamma(k'+1/2)}{\Gamma(k')}\sqrt{\frac{n'}{2\pi\nu'}}\exp\left[1+\frac{n'\left(\mu-\mu'_\mu\right)^2}{2\nu'}\right]^{-\left(k'+\frac{1}{2}\right)}. \tag{10.18}$$

Furthermore, the updating rules for the parameters are

$$\mu''_\mu=\frac{\bar{x}n+\mu'_\mu n'}{n+n'}, \tag{10.19}$$

$$n''=n'+n, \tag{10.20}$$

$$k''=k'+\frac{n}{2}, \tag{10.21}$$

$$\nu''=\nu'+\frac{1}{2}\left[\sum_{i=1}^{n}(x_i-\bar{x})^2+\frac{nn'}{n+n'}\left(\bar{x}-\mu'_\mu\right)^2\right]. \tag{10.22}$$

The predictive distribution of X obtained according to (10.9) is

$$\tilde{f}_X(x)=\frac{\Gamma(k''+1/2)}{\Gamma(k'')}\sqrt{\frac{n''}{2\pi\nu''(n''+1)}}\left[1+\frac{n''\left(x-\mu''_\mu\right)^2}{2\nu''(n''+1)}\right]^{-\left(k''+\frac{1}{2}\right)}, \tag{10.23}$$

which corresponds to the standard Student's t-distribution with $2k''$ degrees of freedom for the shifted and scaled variable $\left(X-\mu''_\mu\right)\sqrt{n''k''/\nu''(n''+1)}$. This distribution has the mean μ''_μ and variance $\tilde{\sigma}_X^2=\nu''(n''+1)/n''(k''-1)$. The above formulas are summarized in Table A10.1.

Example 10.3 – Parameter Estimation with the Normal Distribution

Suppose six samples of a material specimen have been tested for strength and the measured values (in an appropriate unit) are $x_1=208$, $x_2=220$, $x_3=201$, $x_4=175$, $x_5=190$, and $x_6=245$. We have $n=6$, the sample mean $\bar{x}=\sum_{i=1}^{n}x_i/n=206.5$, and the sample variance

Example 10.3 (cont.)

$s^2 = \sum_{i=1}^{n}(x_i - \bar{x})^2/(n-1) = 24.3$. We assume the distribution of the strength is normal with mean μ and standard deviation σ. Below, we analyze these data under various assumptions.

Case 1: Suppose the standard deviation of the strength is known to be $\sigma = 25$. Assuming a diffuse prior for μ ($\sigma'_\mu = \infty$) and using (10.15) and (10.16), the posterior statistics of the unknown mean are $\mu''_\mu = \bar{x} = 206.5$ and $\sigma''_\mu = \sigma/\sqrt{n} = 10.2$. The corresponding predictive distribution of the strength is normal with mean $\tilde{\mu}_X = 206.5$ and standard deviation $\tilde{\sigma}_X = \sqrt{25^2 + 10.2^2} = 27.0$.

Case 2: Now suppose that we have information from prior experience that suggests the mean is around 200 with a coefficient of variation (c.o.v.) of 10%. Using a conjugate normal prior with $\mu'_\mu = 200$ and $\sigma'_\mu = 200 \times 0.10 = 20$ in (10.15) and (10.16), we obtain $\mu''_\mu = 205.2$ and $\sigma''_\mu = 9.09$. In comparison to Case 1, the posterior mean of μ has decreased because of the lower prior mean estimate relative to the sample mean, but only slightly because of the rather large prior standard deviation $\sigma'_\mu = 20$ in comparison to the standard deviation of the sample mean, $\sigma/\sqrt{n} = 8.16$. The posterior standard deviation has decreased because of the added information. The predictive distribution of X now is normal with mean $\tilde{\mu}_X = 205.2$ and standard deviation $\tilde{\sigma}_X = \sqrt{25^2 + 9.09^2} = 26.6$.

Case 3: Now suppose the sample measurements include an error ϵ that is normally distributed with mean $\mu_\epsilon = 0$ (unbiased measurements) and standard deviation $\sigma_\epsilon = 5$ and that the errors for different measurements are statistically independent. This implies that the observations are for the normal random variable $X + \epsilon$, which has an unknown mean $\mu_{X+\epsilon} = \mu$ and standard deviation $\sigma_{X+\epsilon} = \sqrt{25^2 + 5^2} = 25.5$. Using (10.15) and (10.16) with a diffuse prior, we obtain $\mu''_\mu = 206.5$ and $\sigma''_\mu = 10.4$. The increase in the standard deviation relative to Case 1 is a result of the deteriorated quality of the available information, due to measurement errors. The predictive distribution of $X + \epsilon$ is normal with mean $\tilde{\mu}_{X+\epsilon} = \mu''_\mu = 206.5$ and standard deviation $\tilde{\sigma}_{X+\epsilon} = \sqrt{25.5^2 + 10.4^2} = 27.5$. It follows that the predictive distribution of X is normal with mean $\tilde{\mu}_X = \tilde{\mu}_{X+\epsilon} - \mu_\epsilon = 206.5$ and standard deviation $\tilde{\sigma}_X = \sqrt{27.5^2 - 5^2} = 27.1$. Compared to Case 1 with exact measurements, we see that the predictive standard deviation has increased only slightly (from 26.6 to 27.1). This is because some of the observed variability in the data is being "explained away" as measurement error. If we consider the normal prior distribution with mean $\mu'_\mu = 200$ and standard deviation $\sigma'_\mu = 20$, the results are $\mu''_\mu = 205.1$, $\sigma''_\mu = 9.23$, $\tilde{\mu}_{X+\epsilon} = 205.1$, $\tilde{\sigma}_{X+\epsilon} = 27.1$, $\tilde{\mu}_X = 205.1$, and $\tilde{\sigma}_X = 26.7$. Observe that, relative to Case 2, the predictive mean has slightly decreased because more weight is being given to the prior mean, due to the deteriorated quality of the measured data. The predictive standard deviation has increased because of the deteriorated quality of the data, but only slightly (from 26.6 to 26.7) because of the "explaining away" phenomenon.

Example 10.3 (cont.)

Case 4: Next, consider the case where both μ and σ are unknown. We assume non-informative priors, which are diffuse for μ and proportional to $1/\sigma^2$ for σ^2. This is equivalent to the normal-inverse-gamma prior with $n' = 0$, $\nu' = 0$, and $k' = 0$. Assuming exact measurements and using (10.19)–(10.22), the posterior joint distribution of μ and σ^2 is normal-inverse-gamma with $\mu''_\mu = 206.5$, $n'' = 6$, $k'' = 3$, and $\nu'' = 1{,}481$. Using the expressions given in Table A10.1, the posterior mean of σ^2 is $\mu''_{\sigma^2} = \nu''/(k'' - 1) = 740$, which is greater than the "known" value $25^2 = 625$ of Case 1. The posterior variance of μ is $\sigma''^2_\mu = \nu''/[n''(k'' - 1)] = 123$, which results in the standard deviation of $\sigma''_\mu = 11.1$. The increase in the posterior standard deviation of μ relative to Case 1 (from 10.2 to 11.1) is due to the larger expected variance σ^2 than assumed known in Case 1. The posterior variance of σ^2 is $\sigma''^{\,2}_{\sigma^2} = \nu''^2/\left[(k'' - 1)^2(k'' - 2)\right] = 548{,}160$, which results in the standard deviation of $\sigma''_{\sigma^2} = 740$. This indicates a large uncertainty in the estimation of the variance, which results from the rather small sample size ($n = 6$) and the non-informative prior. It is evident that the uncertainty in estimating the mean is much smaller than that in estimating the variance. In terms of the posterior c.o.v., they are $11.1/206.5 = 0.0538$ for the mean versus $740/740 = 1.00$ for the variance. It is noted that, for the normal-inverse-gamma distribution, the posterior correlation coefficient between μ and σ^2 is 0, even though the two parameters are statistically dependent. Finally, the predictive distribution of X in this case is the shifted and scaled student's t-distribution given in (10.23) with the mean $\tilde{\mu}_X = \mu''_\mu = 206.5$ and the standard deviation $\tilde{\sigma}_X = \sqrt{\nu''(n'' + 1)/[n''(k'' - 1)]} = 29.4$. The standard deviation is larger than in the previous cases because of the uncertainty in σ^2.

Case 5: Finally, consider a case where, in addition to the six exact measurements, a seventh observation indicates $x_7 < X_7$, where $x_7 = 250$. Note that this observed lower bound is larger than all the measured values, so it represents an extreme case. We assume both μ and σ^2 are unknown and employ the non-informative prior $f'(\mu, \sigma^2) \propto 1/\sigma^2$ as in the previous case. With this observation, the normal-inverse-gamma distribution is no longer a conjugate prior. Hence, we must resort to numerical integration to determine the posterior statistics. The likelihood function in this case is

$$L(\mu, \sigma^2) \propto \frac{1}{\sigma^6} \exp\left[-\frac{1}{2\sigma^2} \sum\nolimits_{i=1}^{6} (x_i - \mu)^2\right] \times \left[1 - \Phi\left(\frac{x_7 - \mu}{\sigma}\right)\right], \tag{E1}$$

where $\Phi(\cdot)$ is the standard normal CDF and the last bracketed term indicates the probability of the event $\{x_7 < X_7\}$. Thus, the posterior distribution is

$$f''(\mu, \sigma^2) = c \frac{1}{\sigma^2} L(\mu, \sigma^2), \tag{E2}$$

Example 10.3 (cont.)

where c denotes the normalizing factor obtained from

$$c = \left\{ \int_0^\infty \int_{-\infty}^\infty \frac{1}{\sigma^7} \exp\left[-\frac{1}{2\sigma^2} \sum_{i=1}^5 (x_i - \mu)^2 \right] \times \left[1 - \Phi\left(\frac{x_6 - \mu}{\sigma} \right) \right] d\mu d\sigma^2 \right\}^{-1}. \tag{E3}$$

Numerical integration by importance sampling (see Example 9.5) yields $\mu''_\mu = 215.5$, $\mu''_{\sigma^2} = 1,482$, $\sigma''_\mu = 13.2$, $\sigma''_{\sigma^2} = 1,113$, and $\rho''_{\mu\sigma^2} = 0.089$. The posterior means of μ and σ^2 are larger than in the previous cases in light of the rather large lower-bound observation. Note, also, that the two parameters are now weakly correlated. The coefficients of variation in the estimates of the posterior means are less than 5%; those in the estimates of the other statistics are larger.

Figure 10.3 compares the predictive distributions of X for Cases 1, 4, and 5, the latter computed by importance sampling as described in Example 9.5. Recall that in Cases 1 and 4 only six exact measurements are available, whereas in Case 5 a high lower bound is additionally observed. Furthermore, Case 1 assumes that σ^2 is known, whereas Cases 4 and 5 assume that this parameter is unknown. Observe that the added uncertainty from not knowing the variance in Cases 4 and 5 broadens the predictive distribution of X. Furthermore, when a high lower bound is observed in Case 5, the distribution shifts toward higher values and further broadens on account of the added uncertainty.

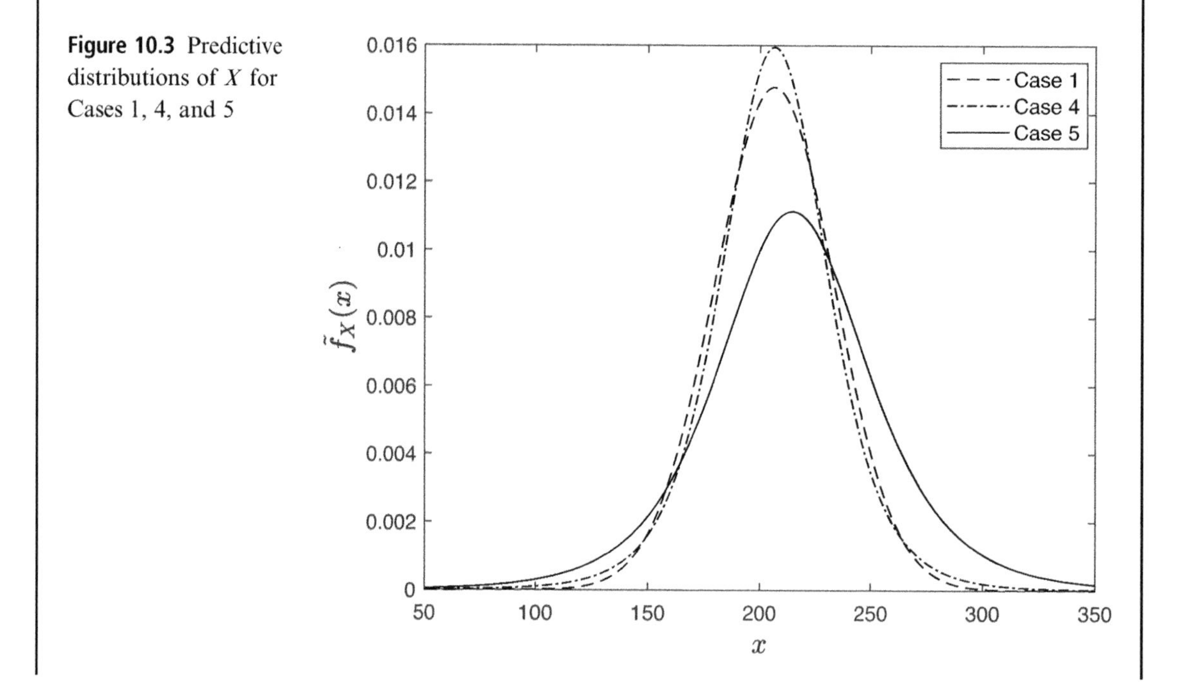

Figure 10.3 Predictive distributions of X for Cases 1, 4, and 5

10.4 Assessing Mathematical Models of Physical Phenomena

Use of idealized mathematical models of physical phenomena is widespread in all aspects of engineering. We use them in design, analysis, assessment, and decision making. Virtually all mathematical models are idealizations of more complex reality and, therefore, entail errors. Typically, the model includes a number of parameters, which are estimated by fitting the model to observational data. The process of model assessment also entails estimating the model error. Because of the uncertainties involved, a probabilistic method must be used for model assessment. This section describes the Bayesian approach to model assessment. A different approach, based on the pragmatic needs of a probabilistic design code, is presented in Ditlevsen (1982).

In the most general case, a mathematical model can be written in the form

$$h(\mathbf{x}, \boldsymbol{\theta}) = 0, \tag{10.24}$$

where $\mathbf{x}$ is a set of observable variables, $\boldsymbol{\theta}$ is a set of unobservable model parameters that are introduced in the process of constructing the model, and $h(\cdot, \cdot)$ is a function or mathematical operation that describes the relation between the variables with the aid of the parameters. The latter can be an algebraic expression, a differential equation, an integral equation, or an algorithmic process such as a finite element code. Often, a reduced form of the model is used, where the aim is to predict one of the variables in terms of the other variables. This form of the model is written as

$$y = h(\mathbf{x}, \boldsymbol{\theta}), \tag{10.25}$$

where y is the response variable to be predicted. The following development focuses on this reduced form of the model as it is the form most often used in engineering. In practice, one often deals with more than one model. In that case, the set of models is written in the form of a vector of functions

$$\mathbf{y} = \mathbf{h}(\mathbf{x}, \boldsymbol{\theta}), \tag{10.26}$$

where $\mathbf{h}(\mathbf{x}, \boldsymbol{\theta}) = [h_1(\mathbf{x}, \boldsymbol{\theta}), h_2(\mathbf{x}, \boldsymbol{\theta}), \ldots]^{\mathrm{T}}$ is the vector of predictive models and $\mathbf{y} = [y_1, y_2, \ldots]^{\mathrm{T}}$ is the vector of variables to be predicted. To be general, we assume that these models share variables and parameters. Naturally, each model may contain only subsets of the variables and parameters.

There are several sources that give rise to uncertainty in the model. They include:

Imperfect model form: The form of the model as defined by the function $h(\cdot, \cdot)$ may not be perfect. For example, a linear relation may be used when the actual relationship is nonlinear. Often, out of necessity or for practical expediency, we employ simplifications in formulating the model, which result in an imperfect model form. In other cases, the exact form of the relationship is unknown to us so there is no choice but to resort to idealized forms. The uncertainty in the component of the model error that arises from the imperfect model form is largely epistemic in nature as, by acquiring knowledge, one can develop a more exact model form.

Missing variables: The model may be missing variables that have influence on y. Variables may be excluded from the model for the sake of simplicity, because they are difficult to measure, or because the model builder is not aware of their existence and influence. The component of model uncertainty arising from missing variables is aleatory in nature if the missing variables have inherent variability.

Statistical uncertainty: Model assessment involves the estimation of model parameters based on observational data on the variables. Statistical uncertainty arises when the observational data are limited. Naturally, this component of model uncertainty is epistemic in nature because, at least theoretically, one has the option of increasing the amount of data.

Measurement error: Errors in measuring the observational data give rise to additional uncertainties in assessing the model.

10.4.1 Formulation of Likelihood Function for Model Assessment

First consider the case where the only source of model uncertainty is the imperfect model form. Let $\widehat{h}(\mathbf{x}, \boldsymbol{\theta})$ denote the imperfect model. One can write

$$y = \widehat{h}(\mathbf{x}, \boldsymbol{\theta}) - E, \tag{10.27}$$

where E represents the model error. We assume the model error is normal with mean μ_E and standard deviation σ_E. For this assumption to be valid, it may be necessary to transform the model. For example, if y is a non-negative quantity, it might be appropriate to make a logarithmic transformation so that the transformed response variable is unbounded. The validity of this assumption is verified by examining the distribution of the residuals, as described below.

The mean μ_E of the model error defines the bias in the model. In most cases an unbiased model is desired. In that case, we set $\mu_E = 0$. This essentially causes an adjustment in the parameter values so that $\widehat{h}(\mathbf{x}, \boldsymbol{\theta})$ with the estimated parameters becomes an unbiased model.

A second assumption, known as homoscedasticity, is that the standard deviation σ_E of the model error is independent of variables $\mathbf{x}$. Again, a transformation of the model may be necessary to satisfy this requirement. Tests with model residuals can be conducted to examine the validity of this assumption. σ_E can be considered as a measure of the quality of the model.

Let $\mathbf{x}_i$, $i = 1, \ldots, n$, be a set of observations of the model variables for which the corresponding responses, y_i, are measured. It follows that $e_i = \widehat{h}(\mathbf{x}_i, \boldsymbol{\theta}) - y_i$ are the corresponding observations of the model error. These are known as the model residuals. Assuming these observations are statistically independent and normally distributed, the likelihood function takes the form

$$L(\boldsymbol{\theta}, \sigma_E) \propto \frac{1}{\sigma_E^n} \exp\left\{-\frac{1}{2\sigma_E^2} \sum\nolimits_{i=1}^{n} \left[\widehat{h}(\mathbf{x}_i, \boldsymbol{\theta}) - y_i\right]^2\right\}. \tag{10.28}$$

Note that the error variance is considered to be identical for all observations and has been factored out of the summation in account of the homoscedasticity assumption. The set of parameters to be estimated is $(\boldsymbol{\theta}, \sigma_E)$. In conjunction with an assumed prior distribution, the above likelihood function is used to estimate this set of parameters. Once the parameters $\boldsymbol{\theta}$ are estimated, their posterior mean values are used to compute the residuals e_i. A frequency diagram or q-q plot of the residuals is then used to ascertain the validity of the normal distribution. Furthermore, plots of the residuals against observed variables are used to verify the assumption of homoscedasticity. If necessary, alternative transformations of the model may be explored to make sure that the normality and homoscedasticity assumptions are valid. Examples below demonstrate this type of analysis.

Now consider the case where the measurement of the response variable y entails an error ϵ. Let $\widehat{y}_i$ denote the ith measured value so that we have $y_i = \widehat{y}_i - \epsilon_i$, where ϵ_i is the corresponding outcome of the error. We assume the measurements are unbiased (the mean of ϵ is zero) and errors at different observations are statistically independent and identically normal with standard deviation σ_ϵ. We can write $\widehat{y}_i - \epsilon_i = \widehat{h}(\mathbf{x}_i, \boldsymbol{\theta}) - e_i$ or $e_i - \epsilon_i = \widehat{h}(\mathbf{x}_i, \boldsymbol{\theta}) - \widehat{y}_i$. Thus, $\widehat{h}(\mathbf{x}_i, \boldsymbol{\theta}) - \widehat{y}_i$, $i = 1, \ldots, n$, are observations of the random variable $E - \epsilon$, which is normal with zero mean and variance $\sigma^2_{E-\epsilon} = \sigma^2_E + \sigma^2_\epsilon$. The likelihood function now takes the form

$$L(\boldsymbol{\theta}, \sigma_E, \sigma_\epsilon) \propto \frac{1}{\left(\sigma_E^2 + \sigma_\epsilon^2\right)^{n/2}} \exp\left\{-\frac{1}{2\left(\sigma_E^2 + \sigma_\epsilon^2\right)} \sum\nolimits_{i=1}^{n} \left[\widehat{h}(\mathbf{x}_i, \boldsymbol{\theta}) - \widehat{y}_i\right]^2\right\}. \tag{10.29}$$

Usually the variance of the measurement error is known from calibration of the measuring device. If it is not known then the above likelihood function will allow estimation of the sum $\sigma_E^2 + \sigma_\epsilon^2$ but not the individual error variances.

Now suppose there are also errors in measuring the input variables $\mathbf{x}_i$. Let $\widehat{\mathbf{x}}_i$ be the measured values and $\boldsymbol{\epsilon}_i$ be the associated errors so that we have $\mathbf{x}_i = \widehat{\mathbf{x}}_i - \boldsymbol{\epsilon}_i$, $i = 1, \ldots, n$. Including the model error and the error in measuring the response variable, for each observation we have the equality $\widehat{y}_i - \epsilon_i = \widehat{h}(\widehat{\mathbf{x}}_i - \boldsymbol{\epsilon}_i, \boldsymbol{\theta}) - e_i$. We assume the errors in measuring the input variables are statistically independent of other variables and normally distributed with zero means (unbiased measurements) and covariance matrix $\boldsymbol{\Sigma}_{\boldsymbol{\epsilon\epsilon}}$, which is diagonal if the measurement errors for different variables are statistically independent. For a general, nonlinear model function $\widehat{h}(\cdot, \cdot)$, the joint distribution of all the error terms would be difficult to obtain. Hence, assuming measurement errors are small, we use a first-order approximation of the model function. Expanding around the zero mean of $\boldsymbol{\epsilon}_i$, we have

$$\widehat{h}(\widehat{\mathbf{x}}_i - \boldsymbol{\epsilon}_i, \boldsymbol{\theta}) \cong \widehat{h}(\widehat{\mathbf{x}}_i, \boldsymbol{\theta}) - \nabla_{\mathbf{x}} \widehat{h}(\widehat{\mathbf{x}}_i, \boldsymbol{\theta}) \boldsymbol{\epsilon}_i. \tag{10.30}$$

Thus, for each observation we have $\widehat{y}_i - \epsilon_i \cong \widehat{h}(\widehat{\mathbf{x}}_i, \boldsymbol{\theta}) - \nabla_{\mathbf{x}} \widehat{h}(\widehat{\mathbf{x}}_i, \boldsymbol{\theta}) \boldsymbol{\epsilon}_i - e_i$, where $\nabla_{\mathbf{x}} \widehat{h}(\widehat{\mathbf{x}}_i, \boldsymbol{\theta})$ is the gradient row-vector of the model function with respect to the input variables. Taking all the error terms to the left-hand side, we obtain $e_i - \epsilon_i - \nabla_{\mathbf{x}} \widehat{h}(\widehat{\mathbf{x}}_i, \boldsymbol{\theta}) \boldsymbol{\epsilon}_i \cong \widehat{h}(\widehat{\mathbf{x}}_i, \boldsymbol{\theta}) - \widehat{y}_i$. The left-hand side now is a realization of a linear function of normally distributed random variables, which is normal with zero mean and variance $\sigma^2_{E+\epsilon+\boldsymbol{\epsilon}} = \sigma^2_E + \sigma^2_\epsilon + \nabla_{\mathbf{x}}\widehat{h}(\widehat{\mathbf{x}}_i, \boldsymbol{\theta})\boldsymbol{\Sigma}_{\boldsymbol{\epsilon\epsilon}}\nabla_{\mathbf{x}}\widehat{h}(\widehat{\mathbf{x}}_i, \boldsymbol{\theta})^{\mathrm{T}}$. It follows that the likelihood function takes the form

$$L(\boldsymbol{\theta}, \sigma_E, \sigma_\epsilon, \boldsymbol{\Sigma}_{\boldsymbol{\epsilon\epsilon}}) \propto \frac{1}{\sigma^n_{E+\epsilon+\boldsymbol{\epsilon}}} \exp\left\{-\frac{1}{2\sigma^2_{E+\epsilon+\boldsymbol{\epsilon}}}\sum\nolimits_{i=1}^n \left[\widehat{h}(\widehat{\mathbf{x}}_i, \boldsymbol{\theta}) - \widehat{y}_i\right]^2\right\}. \tag{10.31}$$

As mentioned earlier, in most cases the variances of measurement errors are known from calibration of the measuring devices. If this is not the case and these variances are unknown, then the above likelihood function can provide information about the total variance $\sigma^2_{E+\epsilon+\boldsymbol{\epsilon}}$ only and not the individual variances.

Now suppose the measurements of the response variable y include bounds. For example, if y denotes the displacement capacity of a reinforced concrete column, it is possible that for the observed input variables the column does not reach its failure capacity. In that case, the measured displacement value y_i is a lower bound to the true displacement capacity y. To formulate the likelihood function for such a case, let there be n equality measurements $y = y_i, i = 1, \ldots, n$, l lower-bound measurements $y > y_j, j = 1, \ldots, l$, and m upper-bound measurements $y < y_k, k = 1, \ldots, m$. The likelihood function then takes the form

$$\begin{aligned} L(\boldsymbol{\theta}, \sigma_E, \sigma_\epsilon, \boldsymbol{\Sigma}_{\boldsymbol{\epsilon\epsilon}}) \propto \frac{1}{\sigma^n_{E+\epsilon+\boldsymbol{\epsilon}}} \exp\left\{-\frac{1}{2\sigma^2_{E+\epsilon+\boldsymbol{\epsilon}}}\sum\nolimits_{i=1}^n \left[\widehat{h}(\widehat{\mathbf{x}}_i, \boldsymbol{\theta}) - \widehat{y}_i\right]^2\right\} \times \\ \prod\nolimits_{j=1}^l \left\{1 - \Phi\left[\frac{\widehat{h}(\widehat{\mathbf{x}}_j, \boldsymbol{\theta}) - \widehat{y}_j}{\sigma_{E+\epsilon+\boldsymbol{\epsilon}}}\right]\right\} \times \prod\nolimits_{k=1}^m \left\{\Phi\left[\frac{\widehat{h}(\widehat{\mathbf{x}}_k, \boldsymbol{\theta}) - \widehat{y}_k}{\sigma_{E+\epsilon+\boldsymbol{\epsilon}}}\right]\right\}, \end{aligned} \tag{10.32}$$

where $\Phi[\cdot]$ is the standard normal CDF.

In some cases, it is desirable to develop a correction or improvement to an existing model rather than developing a new model. This is particularly the case when the model has been in use for a long time and there would be resistance to changing it. Let $y = \widehat{h}(\mathbf{x})$ be such a model. The chances are that this model is biased and the bias depends on $\mathbf{x}$. Gardoni et al. (2002) formulated the improved model in the form

$$y = \widehat{h}(\mathbf{x}) + \mu(\mathbf{x}, \boldsymbol{\theta}) + E, \tag{10.33}$$

where $\mu(\mathbf{x}, \boldsymbol{\theta})$ is the correction term for the model bias and E is the zero-mean model error. In their study of capacity models for bridge columns, Gardoni et al. (2002) employed bias correction models of the form

$$\mu(\mathbf{x}, \boldsymbol{\theta}) = \sum\nolimits_{j=1}^m h_j(\mathbf{x})\theta_j, \tag{10.34}$$

where $h_j(\mathbf{x})$ are "explanatory" functions and θ_j are coefficients to be estimated. Note that this model is linear in the unknown parameters θ_j, which simplifies the Bayesian updating calculations. The explanatory functions are selected based on knowledge about the subject phenomenon, an exercise that is part of the art of model building and requires deep knowledge of the subject area. In general, it is good practice to formulate the model so that the response variable as well as the explanatory functions and, therefore, the coefficients θ_j are dimensionless. An estimated coefficient θ_j that has a mean near 0 or has a large variance indicates that the corresponding explanatory function does not improve the model. Gardoni et al. (2002) start the modeling process with a large number of explanatory functions, gradually dropping those that do not provide significant improvement in order to arrive at a simpler model. Furthermore, a strong correlation between any pair of coefficients θ_j and θ_k, say $\rho_{jk} > 0.7$, is an indication that the two explanatory functions work in tandem. In such cases it is reasonable to replace one of the coefficients with a linear function of the other coefficient, thus further simplifying the model. The likelihood function for the above model for each of the cases described earlier is developed by replacing $\widehat{h}(\mathbf{x}_i, \boldsymbol{\theta})$ in (10.28)–(10.32) with $\widehat{h}(\mathbf{x}_i) + \sum_{j=1}^{m} h_j(\mathbf{x}_i)\theta_j$.

Finally, consider the case of assessing a vector of models. We will only consider the case with an imperfect model form with exact measurements. Extensions to the cases with measurement error should be clear from the preceding analysis. For a vector of imperfect models $\widehat{\mathbf{h}}(\mathbf{x}, \boldsymbol{\theta})$, we can write

$$\mathbf{y} = \widehat{\mathbf{h}}(\mathbf{x}, \boldsymbol{\theta}) - \mathbf{E}, \tag{10.35}$$

where $\mathbf{E}$ is the vector of model errors. For observations $\mathbf{x}_i$ and $\mathbf{y}_i$, $i = 1, \ldots, n$, we can write $\mathbf{e}_i = \widehat{\mathbf{h}}(\mathbf{x}_i, \boldsymbol{\theta}) - \mathbf{y}_i$, where $\mathbf{e}_i$ is the outcome of $\mathbf{E}$ for the ith observation. Note that it is reasonable to assume that errors at different observation instances are statistically independent; however, it is quite possible that there will be correlation between the errors for different models at each observation. That is, $\mathbf{e}_i$ are statistically independent for different i, but the elements of $\mathbf{e}_i$ for each i might be correlated. Hence, we assume that $\mathbf{E}$ has the multinormal distribution with zero means and covariance matrix $\boldsymbol{\Sigma}_{\mathbf{EE}}$, which may not be diagonal. The likelihood function then takes the form

$$L(\boldsymbol{\theta}, \boldsymbol{\Sigma}_{\mathbf{EE}}) \propto \frac{1}{(\det\boldsymbol{\Sigma}_{\mathbf{EE}})^{n/2}} \exp\left\{-\frac{1}{2}\sum_{i=1}^{n}\left[\widehat{\mathbf{h}}(\mathbf{x}_i, \boldsymbol{\theta}) - \mathbf{y}_i\right]^{\mathrm{T}} \boldsymbol{\Sigma}_{\mathbf{EE}}^{-1}\left[\widehat{\mathbf{h}}(\mathbf{x}_i, \boldsymbol{\theta}) - \mathbf{y}_i\right]\right\}. \tag{10.36}$$

Note that the object of this estimation is not only $\boldsymbol{\theta}$ and the error variances but also the covariances of the errors in different models.

Example 10.4 – Capacity Model for Reinforced-Concrete Columns

Gardoni et al. (2002) developed probabilistic deformation and shear capacity models for reinforced-concrete bridge columns with circular cross sections using observations from 106 cyclic load tests. The experiments included cases where the column failed in deformation, in shear, or in a mixed mode. For developing the deformation-capacity model, cases in which the column failed in deformation were considered as equality observations (the applied deformation at failure is considered as the deformation capacity of the column), while the measured deformation for columns failing in shear or mixed mode were considered as lower-bound observations (censored data) for the deformation capacity. Likewise, for developing the

Example 10.4 (cont.)

shear-capacity model, tests in which the column failed in shear were considered to provide equality observations, while those failing in deformation or mixed mode were considered to provide lower-bound observations. Consequently, the likelihood function in (10.32) with the first two terms was used. Owing to long-standing existing models, Gardoni et al. (2002) used the model correction method described in (10.33)–(10.34) rather than developing new models.

The existing reinforced-concrete column deformation-capacity model is written in the dimensionless form

$$\widehat{d}(\mathbf{x}) = \frac{1}{H}\left(\widehat{\Delta}_y + \widehat{\Delta}_p\right) = \frac{1}{H}\left(\widehat{\Delta}_f + \widehat{\Delta}_{sh} + \widehat{\Delta}_{sl} + \widehat{\Delta}_p\right), \tag{E1}$$

where $\mathbf{x}$ is the set of material and geometry variables, H is the column height, $\widehat{\Delta}_y$ is the deformation (displacement at the top of the column) at the onset of yield and is composed of the flexural component $\widehat{\Delta}_f$, the shear component $\widehat{\Delta}_{sh}$, and a component $\widehat{\Delta}_{sl}$ due to local rotation at the base caused by slippage of longitudinal reinforcement, and $\widehat{\Delta}_p$ is the inelastic deformation due to plastic flow. Gardoni et al. (2002) present detailed state-of-the-art formulas for each of these components of deformation. To satisfy normality and homoscedasticity requirements, they formulate the corrected model in the form

$$\ln D(\mathbf{x}) = \ln \widehat{d}(\mathbf{x}) + \sum\nolimits_{j=1}^{m} h_j(\mathbf{x})\theta_j + E, \tag{E2}$$

where E is the model error for the deformation capacity. Starting with 11 explanatory functions, they eliminate 7 non-informative functions and express θ_8 as a linear function of θ_{11} due to high correlation between the two, to arrive at the reduced model

$$\ln D(\mathbf{x}) = \ln \widehat{d}(\mathbf{x}) + \theta_1 + \frac{4V_I}{\pi D_g^2 f'_t}\theta_7 + \frac{\rho_s f_{yh} D_c}{f'_c D_g}(-0.6035 - 1.034\theta_{11}) + \epsilon_{cu}\theta_{11} + E, \tag{E3}$$

in which $V_I = M_I/H$ is the shear force corresponding to the moment capacity at yield curvature, D_g the gross cross-sectional diameter, f'_t the tensile strength of concrete, ρ_s the volumetric ratio of the confining steel, f_{yh} the yield stress of the transverse reinforcement, D_c the core diameter, f'_c the compressive strength of concrete, and ϵ_{cu} the ultimate concrete strain. Table 10.1 lists the posterior statistics of the parameters θ_1, θ_7, θ_{11} and the standard deviation of the model error, σ_E. The positive mean of θ_1 indicates that, overall, the existing model tends to underestimate the deformation capacity. The explanatory function for θ_7, $4V_I/\left(\pi D_g^2 f'_t\right)$, is intended to correct for the effect of the so-called idealized elastic-perfectly plastic shear force. The positive mean value of θ_7 indicates that the existing model tends to underestimate this effect. The next explanatory function, $\rho_s f_{yh} D_c/\left(f'_c D_g\right)$, is intended to correct for the effect of the confining transverse reinforcement. The original coefficient for this function is θ_8, which turns out to be closely correlated with θ_{11} and, hence, is replaced by a linear function of the

Example 10.4 (cont.)

latter. For the negative mean value of θ_{11}, the positive coefficient of this explanatory term indicates that the existing model also tends to underestimate the effect of the confining reinforcement. Finally, the negative mean value of θ_{11} indicates that the existing model tends to overestimate the contribution of the ultimate concrete strain, ϵ_{cu}, to the deformation capacity of the column. Owing to the logarithmic form of the model, the mean of σ_E is approximately equal to the c.o.v. of the deformation capacity. Hence, for a given set of material and geometry variables, we are able to predict the deformation capacity of a reinforced-concrete column of circular cross section with a 38% c.o.v.

Table 10.1 Posterior statistics of parameters in the deformation-capacity model (after Gardoni et al., 2002, with permission from ASCE)

			Correlation coefficient			
Parameter	Mean	Standard deviation	θ_1	θ_7	θ_{11}	σ_E
θ_1	0.531	0.119	1			
θ_7	0.701	0.204	−0.37	1		
θ_{11}	−48.4	13.6	−0.59	−0.38	1	
σ_E	0.383	0.050	−0.04	0.14	0.20	1

For the shear capacity, Gardoni et al. (2002) examined three existing models and selected the one that had the smallest error variance. The selected model was developed by Moehle et al. (2000) and is a refinement of an earlier model. The shear capacity is written in the form

$$\widehat{v}(\mathbf{x}) = \widehat{v}_c + \widehat{v}_s, \tag{E4}$$

where $\widehat{v}_c$ is the contribution from concrete and $\widehat{v}_s$ is the contribution from transverse steel. Details of the two components are given in Gardoni et al. (2002). For this model, they selected four explanatory functions, two of which turned out to be significant. The reduced corrected model has the form

$$\ln V(\mathbf{x}) = \ln \widehat{v}(\mathbf{x}) + \rho_l\theta_2 + \frac{A_v f_{yh} D_g}{A_g f'_t S}\theta_4 + E, \tag{E5}$$

where ρ_l is the longitudinal reinforcement ratio, A_v the area of the transverse reinforcement, A_g the gross cross-sectional area, S the spacing of transverse reinforcement, and f_{yh}, D_g, and f'_t are as defined earlier. Table 10.2 lists the posterior statistics of the model parameters. The positive mean of θ_2 indicates that the existing model tends to underestimate the contribution of the longitudinal reinforcement to the shear capacity. The explanatory function associated with coefficient θ_4, $A_v f_{yh} D_g / (A_g f'_t S)$, is intended to examine the effect of the transverse reinforcement. The fact that the mean of θ_4 is negative is an indication that the existing model

Example 10.4 (cont.)

tends to overestimate this effect. Finally, the mean value of σ_E indicates that the corrected model is capable of predicting the shear capacity of a column with a 15% c.o.v.

Table 10.2 Posterior statistics of parameters in the shear-capacity model (after Gardoni et al., 2002, with permission from ASCE)

			Correlation coefficient		
Parameter	Mean	Standard deviation	θ_2	θ_4	σ_E
θ_2	23.1	1.20	1		
θ_4	−0.614	0.120	−0.87	1	
σ_E	0.153	0.013	−0.13	0.06	1

A given column may of course fail in either deformation or shear mode, or in a mixed mode. Given that these capacity models share many property and geometric variables, the errors in the two models are expected to be dependent. In order to assess this dependence, it is necessary to develop a bivariate model. Gardoni et al. (2002) used the reduced model forms in (E3) and (E5) to develop a bivariate model. They used the likelihood function in (10.36) but also included terms for the censored data. The posterior statistics of the parameters are listed in Table 10.3, where a subscript D indicates a deformation capacity parameter and a subscript V indicates a shear capacity parameter. It is noted that the estimated means and variances of the parameters are nearly the same as those obtained for the marginal models. What is new here is the correlation coefficient ρ of the two model errors, which has a posterior mean of −0.535 and a standard deviation of 0.166.

Table 10.3 Posterior statistics of parameters in the bivariate deformation-shear capacity model (after Gardoni et al., 2002, with permission from ASCE)

			Correlation coefficient							
Parameter	Mean	Standard deviation	θ_{1D}	θ_{7D}	θ_{11D}	σ_D	θ_{2V}	θ_{4V}	σ_V	ρ
θ_{1D}	0.512	0.054	1							
θ_{7D}	0.828	0.198	−0.38	1						
θ_{11D}	−50.8	11.5	−0.60	−0.51	1					
σ_D	0.383	0.048	−0.41	0.36	0.07	1				
θ_{2V}	21.6	1.40	0.02	−0.10	0.07	−0.06	1			
θ_{4V}	−0.584	0.180	0.12	0.00	−0.11	−0.01	−0.84	1		
σ_V	0.189	0.019	0.06	0.05	−0.10	0.06	0.02	0.03	1	
ρ	−0.535	0.166	0.14	−0.29	0.12	−0.26	0.12	0.01	−0.15	1

Example 10.4 (cont.)

The negative value of the mean of the correlation coefficient indicates that the deformation and shear capacities tend to be negatively correlated: if a column has a larger than expected deformation capacity then it is likely that its shear capacity is below its mean value, and vice versa.

Finally, as a demonstration of the effect of model correction, Figure 10.4, taken from Gardoni et al. (2002), compares the observed shear-capacity data with predictions by the existing deterministic model and the median predictions of the probabilistic model. Failure (equality) data are shown as solid dots and lower-bound data are shown as open triangles. It is seen that the existing deterministic model is strongly biased on the conservative side, whereas the probabilistic model corrects for this bias and has about 80% of the failure data within plus/minus one standard deviation interval (dotted lines) of the median.

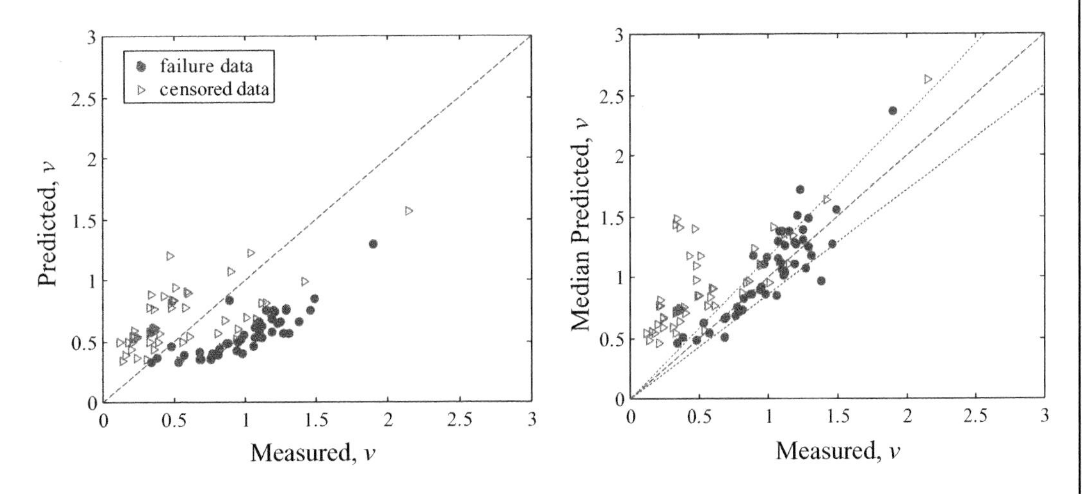

Figure 10.4 Measured and predicted column shear capacities by existing deterministic (left) and corrected probabilistic (right) models (after Gardoni et al., 2002, with permission from ASCE)

10.5 Analysis of Structural Reliability under Statistical and Model Uncertainties

Previous chapters of this book developed methods for reliability analysis of structural components and systems. These methods employed probability distributions of the basic random variables and limit-state functions defining the states of the structural components. These probability distributions of course need to be selected and their parameters estimated based on observational data, and the limit-state functions need to be formulated using mathematical models of structural capacities and demands. However, as we have seen already in this chapter, significant uncertainties may be present in the estimated distribution parameters. Furthermore, models used to formulate limit-state functions are likely to be imperfect and prone to uncertain errors. The focus of this section is on ways to account for these uncertainties in the estimation of structural reliability. The approach is based on the earlier work reported in Der Kiureghian (2008).

Let $f_{\mathbf{X}}(\mathbf{x}|\boldsymbol{\theta}_f)$ define the joint PDF of the basic random variables with $\boldsymbol{\theta}_f$ being the distribution parameters and $g_i = g_i(\mathbf{X}, \boldsymbol{\theta}_g, \mathbf{E})$, $i = 1, \ldots, N$, denoting the set of component limit-state functions with parameters $\boldsymbol{\theta}_g$. As just mentioned, usually these functions are formulated from mathematical models of capacities and demands that are imperfect and, therefore, associated with errors. The vector $\mathbf{E}$ collects all these model error terms. Assuming unbiased models are used, $\mathbf{E}$ has zero means and we define its joint PDF as $f_{\mathbf{E}}(\mathbf{e}|\boldsymbol{\Sigma}_{\mathbf{EE}})$, where $\boldsymbol{\Sigma}_{\mathbf{EE}}$ is the covariance matrix. Because the elements of $\boldsymbol{\Sigma}_{\mathbf{EE}}$ are estimated along with the model parameters, for the sake of brevity of notation we choose to include $\boldsymbol{\Sigma}_{\mathbf{EE}}$ as a part of $\boldsymbol{\theta}_g$. The collection of parameters of the reliability problem then is $\boldsymbol{\theta} = (\boldsymbol{\theta}_f, \boldsymbol{\theta}_g)$. For clarification, it is noted that the measurement errors discussed in the preceding section do not directly appear in the reliability problem. When included in the process of parameter estimation, they affect the posterior statistics of the parameters but do not appear in the formulation of limit-state functions. It is also noted that, whereas the uncertainty in parameters $\boldsymbol{\theta}$ is completely epistemic in nature, the uncertainty in $\mathbf{E}$, particularly the component arising from missing model variables having inherent variability, is aleatory in nature. For this reason, it makes sense to treat $\mathbf{E}$ in the same manner as the basic variables $\mathbf{X}$.

We have seen that, in the most general case, the failure domain of a structural system is defined in terms of the union of min-cut sets, each min-cut set defining the joint failure of a set of components. Hence, in a compact form, the failure probability of a structural system can be written as

$$p_F(\boldsymbol{\theta}) = \int_{\mathcal{F}} f_{\mathbf{X}}(\mathbf{x}|\boldsymbol{\theta}_f) f_{\mathbf{E}}(\mathbf{e}|\boldsymbol{\Sigma}_{\mathbf{EE}}) \mathrm{d}\mathbf{x}\mathrm{d}\mathbf{e}, \tag{10.37}$$

where $\mathcal{F} = \bigcup_k \bigcap_{i \in \mathcal{C}_k} \{g_i(\mathbf{X}, \boldsymbol{\theta}_g, \mathbf{E}) \leq 0\}$ is the failure domain, in which $\mathcal{C}_k$ denotes the kth cut set, the intersection operation is over component events that are included in the kth cut set, and the union operation is over all the min-cut sets. The corresponding generalized reliability index is defined by

$$\beta(\boldsymbol{\theta}) = \Phi^{-1}[1 - p_F(\boldsymbol{\theta})]. \tag{10.38}$$

It is seen that the failure probability and the generalized reliability index of the structural system depend on the set of parameters $\boldsymbol{\theta}$. The question to ask is how to account for the uncertainty that is inherent in these parameters. Below, we offer several approaches. For the subsequent discussion, let $f''_{\boldsymbol{\Theta}}(\boldsymbol{\theta}) = f''_{\boldsymbol{\Theta}_f}(\boldsymbol{\theta}_f) f''_{\boldsymbol{\Theta}_g}(\boldsymbol{\theta}_g)$ denote the posterior PDFs of the parameters, where we have assumed that $\boldsymbol{\Theta}_f$ and $\boldsymbol{\Theta}_g$ are statistically independent. Considering that the estimations of these parameters involve entirely separate processes, this assumption is reasonable.

Point estimates: The simplest approach is to use central measures of the posterior distributions of the parameters, at which point estimates of the failure probability and reliability index are obtained. These are denoted as

$$\widehat{p}_F = p_F\left(\widehat{\boldsymbol{\theta}}\right) \tag{10.39}$$

$$\widehat{\beta} = \beta\left(\widehat{\boldsymbol{\theta}}\right) \tag{10.40}$$

where $\widehat{\boldsymbol{\theta}}$ denotes central measures of the parameters, such as the posterior means or maximum likelihood estimates. This approach clearly disregards the uncertainties in the parameters. However, it uses the available observational data to select updated central measures of the parameters.

Predictive estimate: In a Bayesian context, one treats the uncertainties in $\mathbf{X}$, $\mathbf{E}$, and $\boldsymbol{\Theta}$ similarly. The predictive measure of the failure probability is obtained through the total probability rule as

$$\begin{aligned}\widetilde{p}_F &= \int_{\boldsymbol{\theta}} p_F(\boldsymbol{\theta}) f''_{\boldsymbol{\Theta}}(\boldsymbol{\theta}) \mathrm{d}\boldsymbol{\theta} \\ &= \mathrm{E}_{\boldsymbol{\Theta}}[p_F(\boldsymbol{\Theta})]\end{aligned} \tag{10.41}$$

and the corresponding predictive generalized reliability index from

$$\widetilde{\beta} = \Phi^{-1}\left(1 - \widetilde{p}_F\right). \tag{10.42}$$

The predictive failure probability is essentially the expected value of the failure probability relative to the posterior distribution of the parameters $\boldsymbol{\Theta}$. $\widetilde{p}_F$ and $\widetilde{\beta}$ are our best estimates of the failure probability and the reliability index for the state of information we have, respectively. They incorporate the uncertainties in the basic variables, the model errors, and the epistemic uncertainties in the distribution and limit-state function parameters. As such, they are appropriate predictions given the available information. Naturally, if more data become available, the posterior distribution of $\boldsymbol{\Theta}$ may change and these predictive estimates may also change. So, the values of $\widetilde{p}_F$ and $\widetilde{\beta}$ reflect our state of information. Computation of these values clearly poses a challenge because the conditional failure probability $p_F(\boldsymbol{\theta})$ in the integrand of (10.41) must be computed for each outcome $\boldsymbol{\Theta} = \boldsymbol{\theta}$ of the parameters. Below, four approaches for computing the predictive failure probability are presented.

The first approach for computing the predictive failure probability is to treat $\boldsymbol{\Theta} = \left(\boldsymbol{\Theta}_f, \boldsymbol{\Theta}_g\right)$ as additional random variables. The failure-probability integral takes the form

$$\widetilde{p}_F = \int_{\mathcal{F}} f_{\mathbf{X}}(\mathbf{x}|\boldsymbol{\theta}_f) f_{\mathbf{E}}(\mathbf{e}|\boldsymbol{\Sigma}_{\mathbf{EE}}) f''_{\boldsymbol{\Theta}_f}(\boldsymbol{\theta}_f) f''_{\boldsymbol{\Theta}_g}(\boldsymbol{\theta}_g) \mathrm{d}\mathbf{x}\, \mathrm{d}\mathbf{e}\, \mathrm{d}\boldsymbol{\theta}_f\, \mathrm{d}\boldsymbol{\theta}_g. \tag{10.43}$$

This is essentially reliability analysis with the set of basic random variables $\left(\mathbf{X}, \mathbf{E}, \boldsymbol{\Theta}_f, \boldsymbol{\Theta}_g\right)$. Among these variables, $\boldsymbol{\Theta}_f$ and $\boldsymbol{\Sigma}_{\mathbf{EE}}$, which is included in $\boldsymbol{\Theta}_g$, do not explicitly appear in the limit-state functions, but they affect the probability integral through the distributions of $\mathbf{X}$

and $\mathbf{E}$, respectively. The solution of (10.43) by first- or second-order reliability methods (FORM or SORM) is fairly straightforward. However, the number of random variables is now greater and there is the possibility of stronger nonlinearity if, for example, the dependence of $\mathbf{X}$ on parameters $\mathbf{\Theta}_f$ is strongly nonlinear. It is important to note that because of the dependence of $\mathbf{X}$ on $\mathbf{\Theta}_f$ and $\mathbf{E}$ on $\mathbf{\Theta}_g$ (through $\mathbf{\Sigma}_{\mathbf{EE}}$), the use of the Rosenblatt transformation (see Section 3.6.5) is unavoidable. In formulating this transformation, the order of random variables must be $\mathbf{\Theta}_f$ before $\mathbf{X}$ and $\mathbf{\Theta}_g$ before $\mathbf{E}$ so that the conditional distributions $f_{\mathbf{X}}(\mathbf{x}|\mathbf{\theta}_f)$ and $f_{\mathbf{E}}(\mathbf{e}|\mathbf{\Sigma}_{\mathbf{EE}})$ can be properly used.

A somewhat simpler solution can be formulated if the predictive distribution of $\mathbf{X}$, denoted $\widetilde{f}_{\mathbf{X}}(\mathbf{x})$ and defined in (10.9), is available. Carrying out the integral over $\mathbf{\theta}_f$ in (10.43), one can write

$$\widetilde{p}_F = \int_{\mathcal{F}} \widetilde{f}_{\mathbf{X}}(\mathbf{x}) f_{\mathbf{E}}(\mathbf{e}|\mathbf{\Sigma}_{\mathbf{EE}}) f''_{\mathbf{\Theta}_g}(\mathbf{\theta}_g) \mathrm{d}\mathbf{x}\, \mathrm{d}\mathbf{e}\, \mathrm{d}\mathbf{\theta}_g. \tag{10.44}$$

The set of random variables is now reduced to $(\mathbf{X}, \mathbf{E}, \mathbf{\Theta}_g)$. This approach is particularly effective when a conjugate-pair distribution is used for $\mathbf{\Theta}_f$, in which case a closed-form solution of $\widetilde{f}_{\mathbf{X}}(\mathbf{x})$ is typically available (see Table A10.1).

The third approach is nested reliability analysis. As shown by Wen and Chen (1987), if $\beta(\mathbf{\Theta})$ is the solution of a reliability problem conditioned on a set of random variables $\mathbf{\Theta}$, then the unconditional solution is obtained by solving a reliability problem defined by the limit-state function $\widetilde{g}(\mathbf{\Theta}, U) = U + \beta(\mathbf{\Theta})$, where U is a standard normal random variable independent of $\mathbf{\Theta}$. This is verified as follows:

$$\begin{aligned}
\int_{\{u+\beta(\mathbf{\theta})\le 0\}} \phi(u) f''_{\mathbf{\Theta}}(\mathbf{\theta}) \mathrm{d}u \mathrm{d}\mathbf{\theta} &= \int_{\mathbf{\theta}} \left[\int_{-\infty}^{-\beta(\mathbf{\theta})} \phi(u) \mathrm{d}u \right] f''_{\mathbf{\Theta}}(\mathbf{\theta}) \mathrm{d}\mathbf{\theta} \\
&= \int_{\mathbf{\theta}} \Phi[-\beta(\mathbf{\theta})] f''_{\mathbf{\Theta}}(\mathbf{\theta}) \mathrm{d}\mathbf{\theta} \\
&= \int_{\mathbf{\theta}} p_F(\mathbf{\theta}) f''_{\mathbf{\Theta}}(\mathbf{\theta}) \mathrm{d}\mathbf{\theta} \\
&= \widetilde{p}_F.
\end{aligned} \tag{10.45}$$

This approach essentially requires solving the seemingly simple outer-reliability problem with the limit-state function $\widetilde{g}(\mathbf{\Theta}, U) = U + \beta(\mathbf{\Theta})$ and involving random variables $(U, \mathbf{\Theta})$. However, each computation of this limit-state function requires the solution of an inner-reliability problem for fixed values $\mathbf{\Theta} = \mathbf{\theta}$ of the uncertain parameters, hence the name *nested reliability analysis*. Therefore, this approach is useful when a closed-form solution of the inner-reliability problem is available.

The above formulation offers yet another solution approach, which is the simplest of all though it entails an approximation. Recall that in the mean-centered, first-order, second-moment (MCFOSM) reliability method (Section 5.2.1), the reliability index is given by

$\beta = \mu_g / \sigma_g$, where μ_g and σ_g are the mean and standard deviation, respectively, of the limit-state function. Employing this approximation for the outer reliability problem, we have $\mu_{\tilde{g}} = 0 + \mu_\beta$ and $\sigma_{\tilde{g}}^2 = 1 + \sigma_\beta^2$, where μ_β and σ_β are the mean and standard deviation of the conditional reliability index $\beta(\boldsymbol{\Theta})$. Thus, within a second-moment context, we have

$$\tilde{\beta} \cong \frac{\mu_\beta}{\sqrt{1 + \sigma_\beta^2}}. \tag{10.46}$$

As expected, this expression shows that any uncertainty in the parameters $\boldsymbol{\Theta}$ decreases the reliability index relative to the mean conditional reliability index, provided the latter is positive. The next question to ask is how one gets hold of μ_β and σ_β. Considering the fact that the epistemic uncertainties in parameters $\boldsymbol{\Theta}$ are of secondary importance relative to the aleatory uncertainties in $\mathbf{X}$ and $\mathbf{E}$, it is reasonable to use an approximation method to compute μ_β and σ_β. Using the first-order approximation method described in Section 2.9, we have

$$\mu_\beta \cong \beta\left(\mathbf{M}_{\boldsymbol{\Theta}}''\right) \tag{10.47}$$

and

$$\sigma_\beta^2 \cong \nabla_{\boldsymbol{\theta}}\beta\left(\mathbf{M}_{\boldsymbol{\Theta}}''\right) \boldsymbol{\Sigma}_{\boldsymbol{\Theta\Theta}}'' \nabla_{\boldsymbol{\theta}}\beta\left(\mathbf{M}_{\boldsymbol{\Theta}}''\right)^{\mathrm{T}}, \tag{10.48}$$

where $\mathbf{M}_{\boldsymbol{\Theta}}''$ is the posterior mean vector of $\boldsymbol{\Theta}$, $\boldsymbol{\Sigma}_{\boldsymbol{\Theta\Theta}}''$ its posterior covariance matrix, and $\nabla_{\boldsymbol{\theta}}\beta\left(\mathbf{M}_{\boldsymbol{\Theta}}''\right)$ the gradient (sensitivity) row-vector of the conditional reliability index with respect to the parameters $\boldsymbol{\theta}$, evaluated at the posterior mean values $\mathbf{M}_{\boldsymbol{\Theta}}''$. A single reliability analysis using the posterior mean values of parameters $\boldsymbol{\Theta}$ together with the parameter sensitivities is sufficient to approximately compute the mean and standard deviation of the conditional reliability index from (10.47) and (10.48) and the predictive reliability index from (10.46). It is noted that this calculation of the predictive reliability index entails two approximations: (a) The formula in (10.46) is based on the second-moment approximation of the reliability index. This expression is accurate only when $\beta(\boldsymbol{\Theta})$ has the normal distribution. (b) The expressions in (10.47) and (10.48) employ a first-order approximation. These expressions would be exact if $\beta(\boldsymbol{\Theta})$ were a linear function of $\boldsymbol{\Theta}$, which is not likely to be the case. Nevertheless, (10.46)–(10.47) offer a simple and quick way of computing the predictive reliability index that accounts for parameter uncertainties. As long as the uncertainties in the parameters are of secondary importance relative to those in $\mathbf{X}$ and $\mathbf{E}$, this approach provides adequate results.

Confidence bounds: A philosophically different approach insists on treating the epistemic and aleatory uncertainties separately. This approach is prevalent in the United States nuclear industry (Vesely and Rasmuson, 1984). Although the author is a strong believer in the Bayesian approach of treating all uncertainties similarly and combining them to arrive at predictive estimates, methods for this different treatment are presented below for those who wish to pursue that approach. In this alternative approach, confidence intervals are computed for the failure probability or the reliability index, reflecting the effect of epistemic uncertainties. Let $F_B(b)$ denote the CDF of $B = \beta(\mathbf{\Theta})$. Using a nested reliability analysis, $F_B(b)$ can be computed as

$$F_B(b) = \int_{\{\beta(\boldsymbol{\theta})-b\leq 0\}} f''_{\mathbf{\Theta}}(\boldsymbol{\theta})d\boldsymbol{\theta}. \tag{10.49}$$

The corresponding PDF is $f_B(b) = dF_B(b)/db$, which is the sensitivity with respect to the limit-state function parameter b. The $(1-\alpha)$-confidence interval of B is the difference between $b_{1-\alpha/2} = F_B^{-1}(1-\alpha/2)$ and $b_{\alpha/2} = F_B^{-1}(\alpha/2)$, where $F_B^{-1}(\cdot)$ is the inverse of the above CDF. This is conceptually shown in Figure 10.5. Of course, this requires nested reliability analysis, which can be quite cumbersome unless a closed-form solution of $\beta(\mathbf{\Theta})$ is available. A similar formulation can be developed for $p_F(\mathbf{\Theta})$. More expeditiously, the confidence interval for the failure probability can be computed as the difference between $\Phi(-b_{1-\alpha/2})$ and $\Phi(-b_{\alpha/2})$.

A simple but approximate method for computing the confidence interval of the reliability index is to assume a distribution for $\beta(\mathbf{\Theta})$ with the mean and standard deviation approximately computed according to (10.47) and (10.48). Considering that $\beta(\mathbf{\Theta})$ is unbounded (as opposed to $p_F(\mathbf{\Theta})$, which is bounded from below at 0 and typically has a strongly skewed distribution), the normal distribution is a reasonable choice. Numerical investigations largely support this assertion. In that case, the interval $(\mu_\beta \pm \sigma_\beta)$ contains the true reliability index with approximately 68% confidence. The corresponding interval of the failure probability is given by $\Phi\left[-(\mu_\beta \pm \sigma_\beta)\right]$.

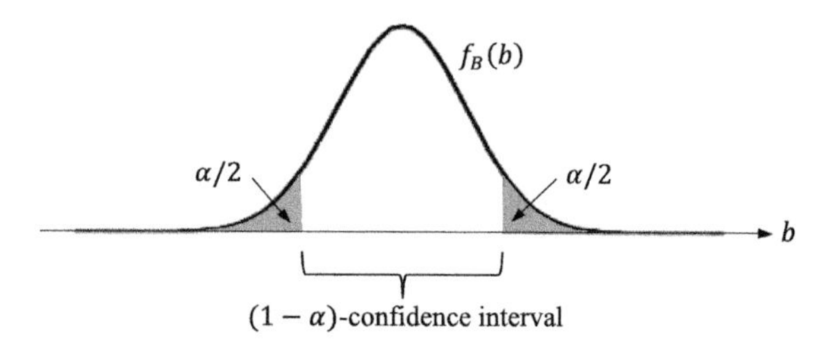

Figure 10.5 Confidence interval of the reliability index

Example 10.5 – Effect of Parameter and Model Uncertainties on Reliability Estimate

We consider a simple example, adopted from Der Kiureghian (2008), for which some closed-form solutions can be obtained. This allows a more explicit examination of the effect of uncertainties on the reliability measure and a closer assessment of the accuracy of the various approximation methods described above.

Consider the limit-state function

$$g(X_1, X_2) = X_1 - X_2, \tag{E1}$$

Example 10.5 (cont.)

where X_i are statistically independent normal random variables with unknown means M_i and known standard deviations σ_i, $i = 1, 2$. The set of unknown parameters of the problem is $\boldsymbol{\Theta}_f = (M_1, M_2)$. Suppose we have a sample of size n for each random variable with sample means $\bar{x}_i$, $i = 1, 2$. Using Bayesian updating with a diffuse prior, we obtain the normal posterior distributions $f''_{M_i}(\mu_i) \sim \mathrm{N}(\bar{x}_i, \sigma_i/\sqrt{n})$, $i = 1, 2$, for M_1 and M_2.

Given that the limit-state function is linear in normal random variables, the conditional reliability index is given by the mean of g divided by its standard deviation, or

$$\beta(\boldsymbol{\Theta}_f) = \frac{M_1 - M_2}{\sqrt{\sigma_1^2 + \sigma_2^2}}. \tag{E2}$$

As $\beta(\boldsymbol{\Theta}_f)$ is a linear function of the uncertain parameters, its mean and variance are obtained exactly as

$$\mu_\beta = \frac{\bar{x}_1 - \bar{x}_2}{\sqrt{\sigma_1^2 + \sigma_2^2}}, \tag{E3}$$

$$\begin{aligned} \sigma_\beta^2 &= \frac{\sigma_1^2/n + \sigma_2^2/n}{\sigma_1^2 + \sigma_2^2} \\ &= \frac{1}{n}. \end{aligned} \tag{E4}$$

It is seen that $\mu_\beta = \beta\left(\mathbf{M}''_{\boldsymbol{\Theta}_f}\right)$, where $\mathbf{M}''_{\boldsymbol{\Theta}_f} = (\bar{x}_1, \bar{x}_2)$ is the posterior mean of the unknown parameters. Thus, in this case, (10.47) provides an exact solution. It is noteworthy that the variance of the reliability index is inversely proportional to the sample size so that it decreases with increasing n, indicating the effect of a larger sample size in reducing the uncertainties in the distribution parameters and, therefore, in the reliability index. The value in (E3) is a natural point estimate of the reliability index. Note, however, that this measure does not account for the uncertainties in the parameters.

To obtain the predictive reliability index, we solve the nested reliability problem defined by the limit-state function

$$\tilde{g}(M_1, M_2, U) = U + \frac{M_1 - M_2}{\sqrt{\sigma_1^2 + \sigma_2^2}}, \tag{E5}$$

where U is a standard normal random variable independent of M_1 and M_2. The above is a linear function of normal random variables with mean $\mu_{\tilde{g}} = 0 + (\bar{x}_1 - \bar{x}_2)/\sqrt{\sigma_1^2 + \sigma_2^2} = (\bar{x}_1 - \bar{x}_2)/\sqrt{\sigma_1^2 + \sigma_2^2}$ and variance $\sigma_{\tilde{g}}^2 = 1 + (\sigma_1^2/n)/(\sigma_1^2 + \sigma_2^2) + (\sigma_2^2/n)/(\sigma_1^2 + \sigma_2^2) = 1 + 1/n$. Thus, the predictive reliability index is

Example 10.5 (cont.)

$$\begin{aligned}\widetilde{\beta} &= \frac{\mu_{\tilde{g}}}{\sigma_{\tilde{g}}} \\ &= \frac{\bar{x}_1 - \bar{x}_2}{\sqrt{\sigma_1^2 + \sigma_2^2}} \frac{1}{\sqrt{1 + 1/n}}.\end{aligned} \tag{E6}$$

This result is identical to that obtained from the approximate formula in (10.46) by substituting for μ_β and σ_β from (E3) and (E4), respectively. It is noteworthy that the predictive reliability index is smaller than the mean reliability index and approaches it as $n \to \infty$. This is essentially a penalty for the added uncertainty present in the distribution parameters.

Next, consider the confidence interval of the reliability index. Because the conditional reliability index in (E2) is a linear function of normal random variables, $B = \beta(\boldsymbol{\Theta}_f)$ has the normal distribution with the mean and variance given in (E3) and (E4), respectively. Therefore, the $(1 - \alpha)$-confidence interval of B is given by

$$\langle B \rangle_{1-\alpha} = \left\{ \frac{\bar{x}_1 - \bar{x}_2}{\sqrt{\sigma_1^2 + \sigma_2^2}} - \frac{1}{\sqrt{n}} k_{\alpha/2}, \frac{\bar{x}_1 - \bar{x}_2}{\sqrt{\sigma_1^2 + \sigma_2^2}} + \frac{1}{\sqrt{n}} k_{\alpha/2} \right\}, \tag{E7}$$

where $k_{\alpha/2} = \Phi^{-1}(1 - \alpha/2)$. Furthermore, the one standard deviation (68%) confidence interval is

$$\langle B \rangle_{0.68} = \left\{ \frac{\bar{x}_1 - \bar{x}_2}{\sqrt{\sigma_1^2 + \sigma_2^2}} - \frac{1}{\sqrt{n}}, \frac{\bar{x}_1 - \bar{x}_2}{\sqrt{\sigma_1^2 + \sigma_2^2}} + \frac{1}{\sqrt{n}} \right\}. \tag{E8}$$

Figure 10.6 shows plots of the above results for $(\bar{x}_1 - \bar{x}_2)/\sqrt{\sigma_1^2 + \sigma_2^2} = 3$ as a function of the sample size n. Shown are the mean and predictive reliability indices, the mean plus or minus one standard deviation indices, and the 95% confidence interval. Note that the predictive reliability index always stays below the mean reliability index but approaches it asymptotically as $n \to \infty$, and that the confidence intervals narrow as n grows.

Now assume that the limit-state function is an imperfect model and add a random correction term so that it reads

$$g(X_1, X_2, E) = X_1 - X_2 - E, \tag{E9}$$

where E is a normal random variable with zero mean, i.e., the imperfect model is unbiased, and an unknown variance $\Sigma = \sigma_E^2$. The set of unknown parameters of the reliability problem is now $\boldsymbol{\Theta} = (M_1, M_2, \Sigma)$. Suppose the posterior distribution of Σ is gamma with mean $\mu_\Sigma'' = r(\sigma_1^2 + \sigma_2^2)$ and c.o.v. $\delta_\Sigma'' = \delta$. Observe that r is a measure of the model error, while δ

Example 10.5 (cont.)

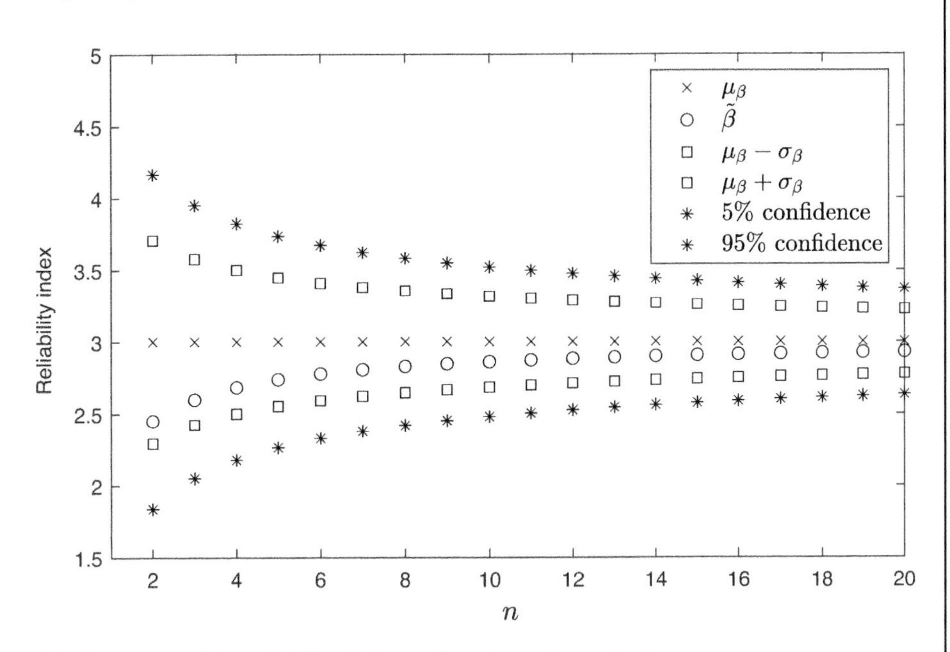

Figure 10.6 Estimates of the reliability index as functions of sample size

represents the variability in this measure. Both are dimensionless quantities. The conditional reliability index now is

$$\beta(\Theta) = \frac{M_1 - M_2}{\sqrt{\sigma_1^2 + \sigma_2^2 + \Sigma}}, \tag{E10}$$

which is a nonlinear function of the uncertain parameters. Exact analytical results for this case cannot be derived and we resort to approximations. Using the first-order approximations in (10.47) and (10.48), the mean and variance of the conditional reliability index are

$$\mu_\beta \cong \frac{\bar{x}_1 - \bar{x}_2}{\sqrt{\sigma_1^2 + \sigma_2^2}} \frac{1}{\sqrt{1+r}}, \tag{E11}$$

$$\sigma_\beta^2 \cong \frac{1}{n(1+r)} + \frac{(\bar{x}_1 - \bar{x}_2)^2}{\sigma_1^2 + \sigma_2^2} \frac{(r\delta)^2}{4(1+r)^3}. \tag{E12}$$

The point estimate at the posterior mean values of the parameters is $\widehat{\beta} = \mu_\beta$, as given in (E11). It can be seen that this estimate decreases with increasing r, i.e., increasing mean of the error variance, but is not affected by the variability in the error variance (as represented by δ). On the

Example 10.5 (cont.)

other hand, the variance of β is affected by both the mean and variability of the model error variance.

Using the approximate formula in (10.46) and the above estimates of the mean and variance of the conditional reliability index, the predictive reliability index is obtained as

$$\tilde{\beta} \cong \frac{\bar{x}_1 - \bar{x}_2}{\sqrt{\sigma_1^2 + \sigma_2^2}} \frac{1}{\sqrt{1 + r + \frac{1}{n} + \frac{(\bar{x}_1 - \bar{x}_2)^2}{\sigma_1^2 + \sigma_2^2}\left[\frac{r\delta}{2(1+r)}\right]^2}}. \tag{E13}$$

The first term in the above is the mean reliability index without accounting for the model error. The second, dimensionless term is a factor representing the effects of the uncertainties in the distribution parameters and the model error. It can be seen that this term is always smaller than unity, i.e., the predictive reliability index is always smaller than the mean reliability index without the model error. Furthermore, the term increases with increasing n (larger sample size) and decreases with increasing r (larger mean of the error variance) and increasing δ (larger variability of the error variance). These trends are all reasonable and as expected.

Because we have a closed-form solution of the conditional reliability index, the predictive failure probability can be obtained by nested reliability analysis using the limit-state function $\tilde{g}(M_1, M_2, \Sigma, U) = U + (M_1 - M_2)/\sqrt{\sigma_1^2 + \sigma_2^2 + \Sigma}$, i.e.,

$$\tilde{p}_F = \int_{\left\{u + (\mu_1 - \mu_2)/\sqrt{\sigma_1^2 + \sigma_2^2 + \sigma_E^2} \leq 0\right\}} \phi(u) f''_{\boldsymbol{\Theta}}(\boldsymbol{\theta}) \mathrm{d}u \mathrm{d}\boldsymbol{\theta}, \tag{E14}$$

where U is a standard normal random variable independent of $\boldsymbol{\Theta} = (M_1, M_2, \Sigma)$. Furthermore, the CDF of $B = \beta(\boldsymbol{\Theta})$ can be obtained by nested reliability analysis using

$$F_B(b) = \int_{\left\{(\mu_1 - \mu_2)/\sqrt{\sigma_1^2 + \sigma_2^2 + \sigma_E^2} - b \leq 0\right\}} f''_{\boldsymbol{\Theta}}(\boldsymbol{\theta}) \mathrm{d}\boldsymbol{\theta} \tag{E15}$$

with the PDF obtained as the sensitivity $f_B(b) = \mathrm{d}F_B(b)/\mathrm{d}b$ with respect to the limit-state function parameter b. The corresponding PDF of the failure probability is obtained as $f_{P_F}(p) = f_B(b)/\phi(b)$, where $b = \Phi^{-1}(1 - p)$. Performing this analysis for a range of b values, we are able to compute the full distribution of the conditional reliability index and the failure probability.

Figure 10.7 shows a plot of the predictive reliability index $\tilde{\beta} = -\Phi^{-1}(\tilde{p}_F)$ computed by FORM using the nested reliability formulation in (E14) together with the approximation in (E13) for $(\bar{x}_1 - \bar{x}_2) = 3$, $\sqrt{\sigma_1^2 + \sigma_2^2} = 1$, $r = 0.2$, and $\delta = 0.3$ as a function of the sample size n. Also shown is the approximate mean in (E11). Observe that the two estimates of the predictive reliability index are almost indistinguishable and that both lie below the mean estimate. This is

Example 10.5 (cont.)

an indication that the approximate formulas in (10.46)–(10.48) work well, at least for this example problem. Also observe that, for a given sample size, the predictive reliability index in Figure 10.7 is smaller than that in Figure 10.6. This is of course due to the added uncertainty arising from the model error.

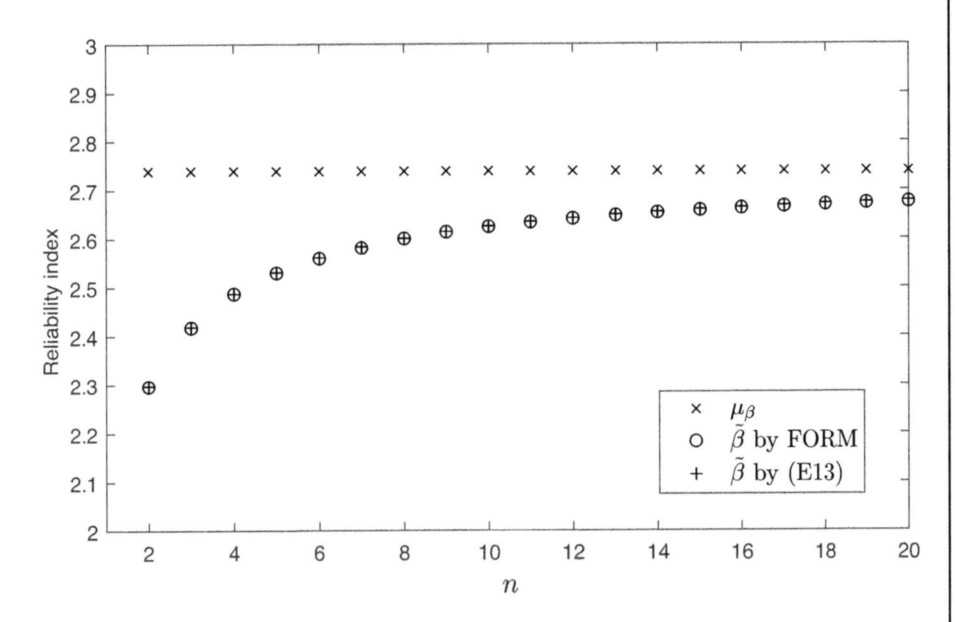

Figure 10.7 Estimates of the reliability index by different methods for the case with model error

Next, we solve the problem in (E15) by FORM for a range of values of the threshold b to compute the CDF of the predictive reliability index and also compute the FORM sensitivities with respect to b, which yields the corresponding PDF $f_B(b)$. Figure 10.8 shows plots of the CDF and PDF of the predictive reliability index together with those of a normal distribution fitted to the mean and variance computed from (E11)–(E12) for $n = 10$, $r = 0.2$, and $\delta = 0.3$. It can be seen that the two approximations are virtually indistinguishable. This suggests that the assumption of a normal distribution for the conditional reliability index is good for this example problem.

Finally, consider a k-out-of-N system with exchangeable components, each characterized by the limit-state function in (E9). Such a system survives if k or more of its N components survive. Conditioned on the parameters $\mathbf{\Theta} = (M_1, M_2, \Sigma)$, the random variables X_1, X_2, and E for different components are statistically independent. However, all components share the same set of uncertain parameters, so that their predictive states are statistically dependent. This dependence can have a significant influence on the reliability of the system. In the following, we

Example 10.5 (cont.)

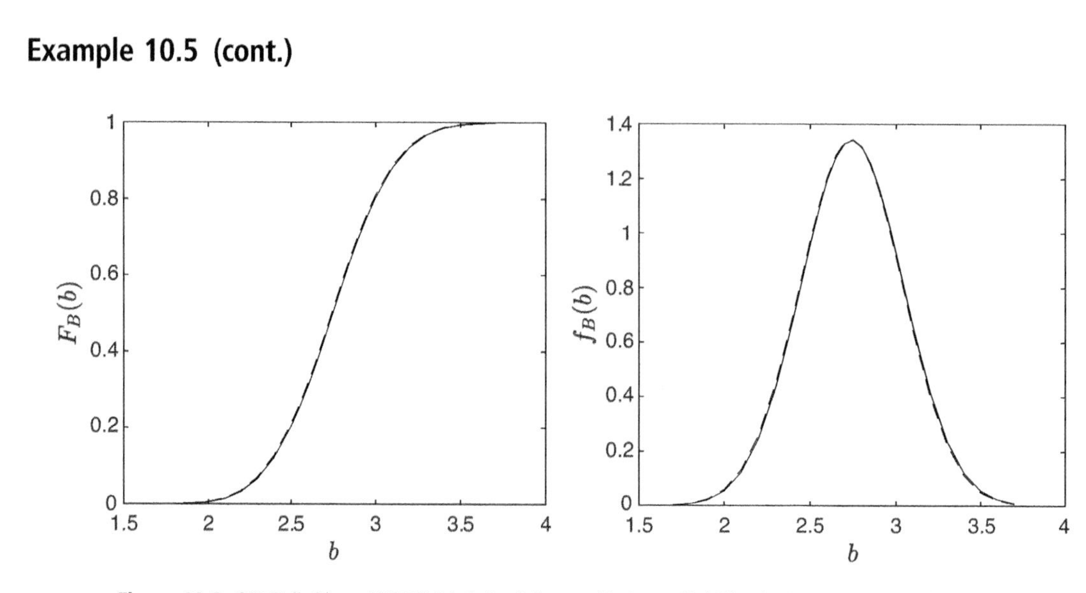

Figure 10.8 CDF (left) and PDF (right) of the predictive reliability index by FORM (solid line) and a normal distribution fitted to the approximate mean and variance (dashed line)

study systems with $N = 5$ components with $k = 1, 3,$ and 5. Note that the case $k = 1$ corresponds to a parallel system while $k = 5$ represents a series system. The case $k = 3$ is in between and represents a system with redundant components.

Owing to the conditional independence of the component states, the conditional failure probability of the system for given $\mathbf{\Theta} = (M_1, M_2, \Sigma)$ is described by the binomial cumulative distribution function

$$p_F(\mathbf{\Theta}) = \sum\nolimits_{i=N-k+1}^{N} \frac{N!}{i!(N-i)!} [p(\mathbf{\Theta})]^i [1 - p(\mathbf{\Theta})]^{N-i}, \tag{E16}$$

where

$$p(\mathbf{\Theta}) = \Phi\left(-\frac{M_1 - M_2}{\sqrt{\sigma_1^2 + \sigma_2^2 + \Sigma}}\right) \tag{E17}$$

is the conditional failure probability of each component. The corresponding conditional generalized reliability index for the system is given by $\beta_g(\mathbf{\Theta}) = \Phi^{-1}[1 - p_F(\mathbf{\Theta})]$. The predictive failure probability of the system is now computed by the nested reliability formulation

$$\widetilde{p}_F = \int_{\{u+\beta_g(\boldsymbol{\theta})\leq 0\}} \phi(u) f''_{\mathbf{\Theta}}(\boldsymbol{\theta}) \mathrm{d}u \mathrm{d}\boldsymbol{\theta}, \tag{E18}$$

Example 10.5 (cont.)

and the corresponding predictive reliability index is computed as $\widetilde{\beta} = \Phi^{-1}(1 - \widetilde{p}_F)$. For numerical analysis, we consider the parameter values $(\bar{x}_1 - \bar{x}_2) = 3$, $\sqrt{\sigma_1^2 + \sigma_2^2} = 1$, $\delta = 0.3$, $n = 10$ and 20, and a range of r values representing the magnitude of the variance of the model error.

The nested reliability problem in (E18) is solved by FORM and by Monte Carlo simulation, the latter employing importance sampling using the design point and having less than 1% c.o.v. The results are presented in Figure 10.9 in terms of the predictive reliability index, where the left chart is for a sample of size $n = 10$ and the right chart is for $n = 20$. Also shown in this figure is the predictive reliability index computed from the approximate formula in (10.46), which employs the approximate estimates of the mean and variance of the conditional reliability index in (10.47) and (10.48), respectively. Computation of the variance requires the sensitivities of $\beta_g(\boldsymbol{\theta})$ with respect to $\boldsymbol{\theta}$ at the posterior mean point. These are obtained by applying the chain rule of differentiation to the expression in (E16) and using the component sensitivities computed using (E17). Also shown are the mean estimates of the reliability index in both charts in Figure 10.9. Note that the mean estimates do not depend on the sample size n.

The following observations in Figure 10.9 are noteworthy: (a) On account of the parameter uncertainties, all estimates of the predictive reliability index are below the respective mean estimates of the conditional reliability index. (b) Both the mean and predictive reliability indices decrease with increasing r, i.e., increasing variance of the model error. (c) The predictive reliability indices are smaller for $n = 10$ than for $n = 20$, reflecting the effect of more

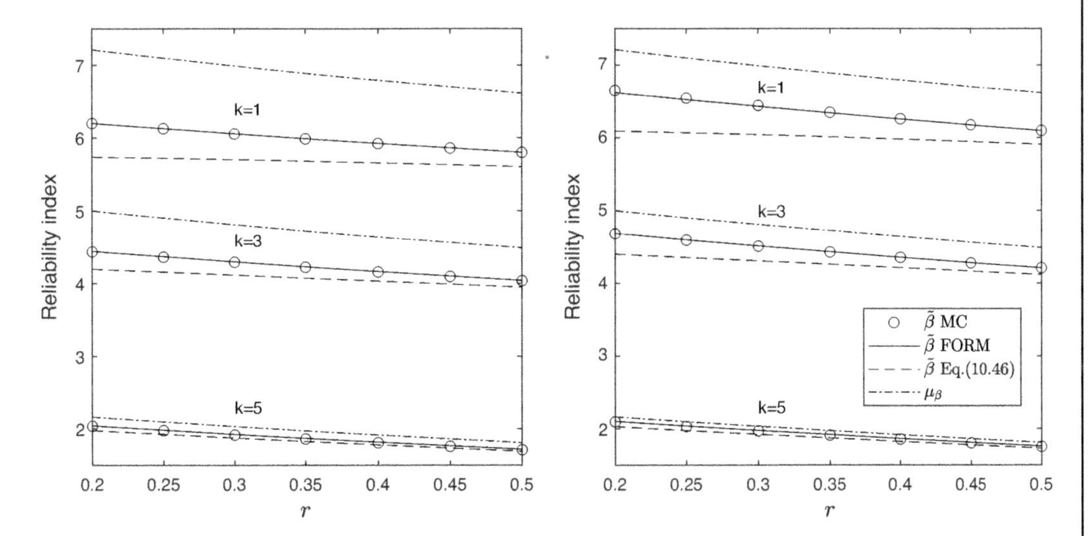

Figure 10.9 Predictive and mean estimates of the reliability index for sample sizes $n = 10$ (left) and $n = 20$ (right)

Example 10.5 (cont.)

uncertainty in the estimation of the means of X_1 and X_2 with a smaller sample size. (d) The "exact" predictive-reliability indices computed by Monte Carlo importance sampling practically coincide with the FORM results. Given the strongly nonlinear dependence of the system-failure probability on the parameters $\mathbf{\Theta}$, as evident in (E16), this level of accuracy by FORM is quite remarkable. The approximate formula in (10.46) is fairly accurate but tends to underestimate the predictive reliability index. (e) It is also worth observing the vast difference in the reliability levels of the three systems. Naturally, the series system with $k = 5$ has the lowest reliability, the parallel system with $k = 1$ has the highest reliability, and the system with $k = 3$ is in between. It is also noteworthy that the redundant systems are more influenced by the parameter uncertainties than the series system.

10.6 Updating of Structural Reliability

There are fundamental differences between the reliability assessment of a planned structure (in the design stage) and that of an existing structure. In the reliability assessment of a planned structure, the uncertainties in the structural properties, dimensions, and loads are aleatory in nature, as none of these have yet materialized. One can imagine that the structure to be built will be a random selection from an infinite sample of possible realizations. On the other hand, for an existing structure, the uncertainties in the properties and dimensions are epistemic in nature, as one can determine them if one has the time and resources to make extensive measurements. Furthermore, although future loads remain aleatory in nature, the past load history of an existing structure may be used to estimate future load values. Additionally, the performance of the structure during its existence gives indirect information about its properties and future performance. Having said this, we remind the reader that the distinction between aleatory and epistemic uncertainties is blurred, as what is purely random depends on the universe of predictive models we choose to use (see Section 10.2). Nevertheless, the possibility of making measurements and observations on an existing structure to learn about its current and future states is a distinct advantage that is not available for planned structures.

Naturally, for an existing structure, it is desirable to update the reliability estimate in the light of information gained from observations and measurements (Madsen, 1987; Jiao and Moan, 1990, Straub, 2011; Straub and Papaioannou, 2015; Luque and Straub, 2016). Example 8.7 in Chapter 8 provided one such example, where the reliability of a ductile frame structure was updated in the light of information about its survival under proof loading. In that example, we used the conditional-probability concept. In this section, we further elaborate on that approach and consider different types of observations and present specific methods for reliability updating.

Let the limit-state functions $g_i(\mathbf{X})$, $i = 1, \ldots, N$, and the min-cut sets $\mathcal{C}_k$, $k = 1, \ldots, K$, define the reliability problem for a structural system such that the failure event is described as $\mathcal{F} = \bigcup_{k=1}^{K} \left\{ \bigcap_{i \in \mathcal{C}_k} [g_i(\mathbf{X}) \leq 0] \right\}$. Observations about the structure in general may involve the basic random variables $\mathbf{X}$ but possibly also additional random variables $\mathbf{Y}$. For example, when a property of the structure (say one of the random variables $\mathbf{X}$) is measured, the measurement error is an additional random variable that needs to be considered. We collect such additional random variables into the vector $\mathbf{Y}$.

In general, observations about a structure can be formulated as inequality or equality events in the form $\{h_j(\mathbf{X}, \mathbf{Y}) < 0\}$ or $\{h_j(\mathbf{X}, \mathbf{Y}) = 0\}$, $j = 1, \ldots, M$, respectively, in which $h_j(\mathbf{X}, \mathbf{Y})$ can be seen as analogous to a limit-state function. An example of the inequality-type observation is that component 1 of the structure has survived a load X_1 that was measured as $\widehat{x}_1$ with measurement error Y_1. In that case, $h_1(\mathbf{X}, \mathbf{Y}) = -g_1(\widehat{x}_1 - Y_1, X_2, \ldots, X_n) < 0$. Note that we have replaced X_1 with its measured value minus the measurement error, and that the negative sign in the front is to indicate the survival event. An example of the equality-type observation is that we measure property X_2 of the structure as $\widehat{x}_2$ with measurement error Y_2, in which case $h_2(\mathbf{X}, \mathbf{Y}) = X_2 + Y_2 - \widehat{x}_2 = 0$. For the following analysis, let $\mathcal{H}_j = \{h_j(\mathbf{X}, \mathbf{Y}) < 0\}$ or $\mathcal{H}_j = \{h_j(\mathbf{X}, \mathbf{Y}) = 0\}$ define the jth inequality- or equality-type observed event.

When M observations are made, the updated probability of failure of the structural system is given by the conditional probability

$$\Pr(\mathcal{F}|\mathcal{H}_1 \cdots \mathcal{H}_M) = \frac{\Pr(\mathcal{F}\mathcal{H}_1 \cdots \mathcal{H}_M)}{\Pr(\mathcal{H}_1 \cdots \mathcal{H}_M)}. \tag{10.50}$$

When all the observations $\mathcal{H}_j$ are of the inequality type, both the numerator and the denominator in the right-hand side of the above expression are analogous to a system-reliability formulation, albeit with additional random variables $\mathbf{Y}$, and can be solved by FORM, SORM, or simulation methods, as described in the previous chapters of this book. Indeed, Example 8.7, where the reliability of a ductile frame was updated in the light of its survival under proof loading, represented one such case with the numerator being a system with three min-cut sets and the denominator being a parallel system – see (E12) in Example 8.7. However, when one or more of the $\mathcal{H}_i$ represent equality-type observations, difficulty arises as the established methods of structural reliability (FORM, SORM, and most simulation methods described in Chapter 9) cannot handle equality events. Next, we describe two methods to solve this problem. For the sake of simplicity of notation, we describe the problem for a structure with only one component and for only one equality-type observation. The extension to more general problems is straightforward, and is briefly described for each method below.

For a single component and one equality observation, the updated probability takes the form

$$\begin{aligned}\Pr(\mathcal{F}|\mathcal{H}) &= \frac{\Pr(\mathcal{F}\mathcal{H})}{\Pr(\mathcal{H})} \\ &= \frac{\Pr[g\{(\mathbf{X}) \leq 0\} \cap \{h(\mathbf{X}, \mathbf{Y}) = 0\}]}{\Pr[h(\mathbf{X}, \mathbf{Y}) = 0]}.\end{aligned} \tag{10.51}$$

For continuous random variables, the probability of an event of the form $\{h(\mathbf{X}, \mathbf{Y}) = 0\}$ is 0; hence, the above ratio is indefinite. Madsen (1987) showed that the above expression can be converted into the form

$$\Pr(\mathcal{F}|\mathcal{H}) = \frac{\frac{\partial}{\partial \delta}\Pr[\{g(\mathbf{X}) \leq 0\} \cap \{h(\mathbf{X}, \mathbf{Y}) - \delta \leq 0\}]_{\delta=0}}{\frac{\partial}{\partial \delta}\Pr[h(\mathbf{X}, \mathbf{Y}) - \delta \leq 0]_{\delta=0}}, \tag{10.52}$$

where δ is an auxiliary parameter. It is seen that the problem is now defined in terms of inequality events and can be solved by existing structural-reliability methods as the ratio of probability sensitivities of a parallel system and a component. More generally, for a system problem and multiple observations, the solution is in the form of the ratio of probability sensitivities of a general system and a parallel system. These system probability sensitivities can be computed by FORM or SORM techniques and finite differences, or by the directional or orthogonal plane sampling methods described in Section 9.12.

Straub (2011) used the likelihood function from Bayesian analysis to recast the problem with equality-type observations into a form with inequality events so that existing structural-reliability methods can be used. Observe that the probability of the observation $\mathcal{H}$ can be written as

$$\Pr(\mathcal{H}) = \int_x \Pr(\mathcal{H}|\mathbf{X}=\mathbf{x}) f_{\mathbf{X}}(\mathbf{x})\mathrm{d}\mathbf{x}. \tag{10.53}$$

The conditional probability $\Pr(\mathcal{H}|\mathbf{X}=\mathbf{x})$ can be interpreted as the likelihood function relative to the random variables $\mathbf{X}$. Allowing for a proportionality constant, we write the likelihood function as $L(\mathbf{x}) \propto \Pr(\mathcal{H}|\mathbf{X}=\mathbf{x})$. As an example, if the observation is a measurement $\widehat{x}_2$ of the basic random variable X_2 with measurement error Y_2, then

$$L(\mathbf{x}) \propto f_{Y_2}(\widehat{x}_2 - x_2), \tag{10.54}$$

where $f_{Y_2}(\cdot)$ is the PDF of the measurement error (typically a normal distribution with zero mean if the measurement is unbiased). Motivated by the nested reliability formulation of Wen and Chen (1987), Straub (2011) found that the likelihood function can be written in the form

$$L(\mathbf{x}) = \frac{1}{c}\Pr\{U - \Phi^{-1}[cL(\mathbf{x})] \leq 0\}, \tag{10.55}$$

where U is a standard normal random variable independent of $\mathbf{X}$ and c is a positive constant such that $cL(\mathbf{x}) \leq 1$. This identity is easily proven by observing that

$$\begin{aligned}\Pr\{U - \Phi^{-1}[cL(\mathbf{x})] \le 0\} &= \int_{-\infty}^{\Phi^{-1}[cL(\mathbf{x})]} \phi(u)\mathrm{d}u \\ &= \Phi\{\Phi^{-1}[cL(\mathbf{x})]\} \\ &= cL(\mathbf{x}).\end{aligned} \tag{10.56}$$

It follows that the observed event can be described equivalently as

$$\mathcal{H}_e = \{h_e(\mathbf{X}, U) \le 0\}, \tag{10.57}$$

where

$$h_e(\mathbf{x}, u) = u - \Phi^{-1}[cL(\mathbf{x})]. \tag{10.58}$$

In essence, random variable U represents the set of additional random variables $\mathbf{Y}$. Equation (10.53) now takes the form

$$\Pr(\mathcal{H}) = \frac{\alpha}{c}\int_{(\mathbf{x},u)\in\mathcal{H}_e} \phi(u) f_{\mathbf{X}}(\mathbf{x})\mathrm{d}u\mathrm{d}\mathbf{x}, \tag{10.59}$$

where α is the proportionality constant. In a similar way,

$$\Pr(\mathcal{F}\cap\mathcal{H}) = \frac{\alpha}{c}\int_{(\mathbf{x},u)\in\{\mathcal{F}\cap\mathcal{H}_e\}} \phi(u) f_{\mathbf{X}}(\mathbf{x})\mathrm{d}u\mathrm{d}\mathbf{x}. \tag{10.60}$$

Inserting (10.59) and (10.60) in (10.51), we see that the normalizing factor α vanishes; however, we still need to have the constant c as it enters the definition of h_e in (10.58). Observe that the probability integrals in (10.59) and (10.60) are over inequality domains. Hence, methods of structural reliability can be used to compute these probabilities.

The above formulation is easily extended to multiple observations. If the observations are statistically independent, the overall likelihood function is the product of the likelihood functions of the individual observations. A single observation event $\mathcal{H}_e$ is then defined by (10.57) and (10.58) through the overall likelihood function. In this case, a single auxiliary random variable U accounts for all the additional random variables $\mathbf{Y}$ in all the equality observation events. Alternatively, for each observation, a separate event $\mathcal{H}_{ej}$ is defined as in (10.57) and (10.58) with the corresponding auxiliary variable U_j. The updated probability is then computed according to (10.50), with each $\mathcal{H}_j$ replaced by the corresponding $\mathcal{H}_{ej}$. Note that in this case one needs to augment $\mathbf{X}$ with M auxiliary standard normal random variables $U_1, \ldots, U_M$, one variable per observed equality event. Straub (2011) cautions that the limit-state function defined by (10.58) can be strongly nonlinear and that one must be careful in using FORM or SORM for computing probabilities with this approach.

Example 10.6 – Updating of Column Reliability

Reconsider the reliability of the short column in Example 6.5 under random bi-axial bending moments M_1 and M_2 and axial force P, with an uncertain yield strength Y. In that example,

Example 10.6 (cont.)

the FORM approximation of the failure probability was obtained as $p_{F1} = 0.00682$, and the mapping of the design point in the original space was $\mathbf{x}^* = [341\ 170\ 3{,}223\ 31{,}770]^{\mathrm{T}}$. The fourth element of this vector is Y, which has a Weibull distribution with a mean of 40,000 kPa and a standard deviation of 4,000 kPa. We can see that the mapping of the design point into the original space suggests that, at the most likely failure point, the yield strength is far below its mean value. Furthermore, Example 6.6 shows that the most important random variable in this problem is Y. Obviously, measuring Y would be a good way to update the reliability of the column.

Suppose the column yield strength is measured, which produces the result $\widehat{y} = 38{,}000$ kPa. This is half a standard deviation below the mean. Assume the measurement is associated with an error, ϵ, that is normally distributed with zero mean (unbiased measurement) and a standard deviation of 1,000 kPa. Hence, we have $h(Y, \epsilon, \widehat{y}) = Y + \epsilon - \widehat{y} = 0$ as an equality-type observation. We use (10.52) to update the column reliability in the light of this information. According to this formula, the updated probability of failure is the ratio of the probability sensitivity with respect to δ of a parallel system with two components, one having the failure domain $\{g(M_1, M_2, P, Y) \leq 0\}$ (see (E1) in Example 6.5) and the other having the failure domain $\{h(Y, \epsilon, \widehat{y}) - \delta \leq 0\}$, to the probability sensitivity with respect to δ of a component problem with the latter failure domain, both evaluated for $\delta = 0$. The latter sensitivity is readily available from FORM component-sensitivity analysis. For this example, the sensitivity of the parallel system is obtained by finite differences of FORM results. Considering the magnitude of $\widehat{y}$, the value $\Delta\delta = 100$ kPa is used. For $\delta = 0$, the FORM approximation of the failure probability of the parallel system is obtained as 0.00663867, and for $\delta = 100$ it is obtained as 0.00665067. The FORM sensitivity of the component probability with respect to δ (at $\delta = 0$) is obtained as 0.0000752. Thus, the FORM approximation of the updated failure probability of the column is

$$p_{F1|\widehat{y}=38{,}000} \cong \frac{0.00665067 - 0.00663867}{100 \times 0.00007522} = 0.00160. \tag{E1}$$

We see that the updated FORM approximation of probability of failure is smaller than the FORM approximation of the unconditional probability of failure ($p_{F1} = 0.00682$). Two factors influence this outcome: the sharp reduction in the uncertainty of Y (standard deviation has gone down from 4,000 kPa to 1,000 kPa), and the specific measurement outcome that is below the mean value. Evidently, the first factor has a greater influence here. The fact that the design-point value of the unconditional problem, $y^* = 31{,}770$ kPa, is more than two standard deviations below the mean is an indication that the measured value of 38,000 kPa is not a pessimistic outcome.

To further explore this problem, we consider the case where $\widehat{y} = 42{,}000$ kPa is the measured value. The result is

Example 10.6 (cont.)

$$p_{F1|\hat{y}=42,000} \cong \frac{0.00681637 - 0.00681586}{100 \times 0.00001019} = 0.000501. \quad \text{(E2)}$$

The updated probability of failure is now much smaller than the unconditional probability of failure. This is because not only has the uncertainty in Y decreased but also the measured value is above the mean.

We now solve this problem using Straub's approach. We formulate the observation likelihood function as $L(y) = \exp\left[-0.5(y - \hat{y})^2/\sigma_\epsilon^2\right]$. As $\max L(y) = 1$, we can set $c = 1$ in (10.58). However, to avoid having the limit-state surface go to infinity in its interior, we select $c = 0.9$. We need to solve (10.51) with $\mathcal{F} = \{g(M_1, M_2, P, Y) \le 0\}$ and $\mathcal{H}$ replaced by $\mathcal{H}_e = \left\{U - \Phi^{-1}[0.9L(Y)] \le 0\right\}$. The results for the two cases of the measured value computed by Monte Carlo simulation (with c.o.v. equal to or less than 8%) are

$$p_{F1|\hat{y}=38,000} \cong \frac{0.000581}{0.149} = 0.00390, \quad \text{(E3)}$$

$$p_{F1|\hat{y}=42,000} \cong \frac{0.000142}{0.204} = 0.000698. \quad \text{(E4)}$$

As mentioned earlier, Straub cautions against the use of FORM and SORM for his approach because the observation limit-state function can be strongly nonlinear. To demonstrate this, in Figure 10.10 we show a plot of the surface $h_e(y, u) = u - \Phi^{-1}[cL(y)] = 0$ for the case with

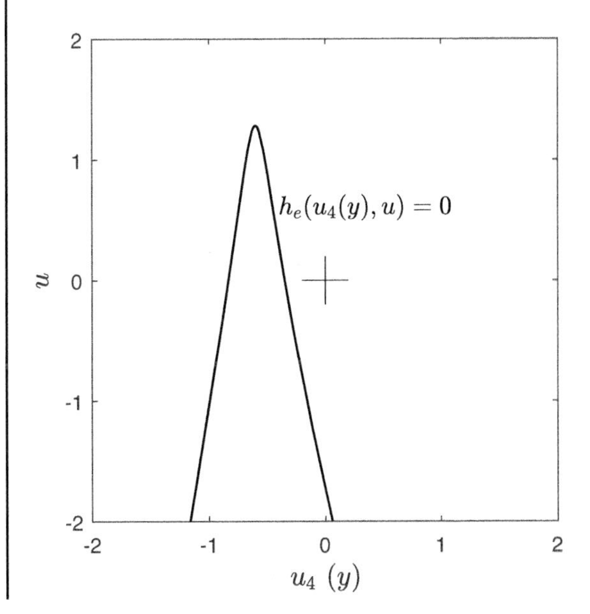

Figure 10.10 Limit-state surface of the observation event in the standard normal space

Example 10.6 (cont.)

$\widehat{y} = 38{,}000$ kPa in the transformed standard normal space. It can be seen that the surface is indeed strongly nonlinear and not well suited for a FORM approximation.

The discrepancy between the results in (E1)–(E2) based on Madsen's approach and those in (E3)–(E4) based on Straub's approach is due to the error in the FORM approximation used to compute the probability sensitivities in (10.52). Indeed, if the Monte Carlo approach is used to solve (10.52), the results closely match those in (E3) and (E4).

10.7 Updating the Distribution of Basic Random Variables

Observations giving information about a structural-reliability problem, e.g., measurements of structural dimensions, material properties, or response, or observation of events such as survival under proof loading, can be used to learn about the basic random variables and update their probability distributions. Let $f'_{\mathbf{X}}(\mathbf{x})$ denote the prior distribution of the basic random variables and $L(\mathbf{x})$ denote the likelihood function of observations made about the structure. As defined earlier, the latter function is proportional to the probability of making the observations given $\mathbf{X} = \mathbf{x}$. According to (10.4) and (10.5), the posterior distribution of the basic random variables is given by

$$f''_{\mathbf{X}}(\mathbf{x}) = \frac{L(\mathbf{x})f'_{\mathbf{X}}(\mathbf{x})}{\int_{\mathbf{x}} L(\mathbf{x})f'_{\mathbf{X}}(\mathbf{x})\mathrm{d}\mathbf{x}}. \tag{10.61}$$

The posterior distribution can be used to update the reliability estimate of the structure in the light of the available information. The result should be the same as directly updating the structural reliability, as described in the preceding section. However, there are benefits to updating the distribution of the basic random variables, as this provides updated information that can also be used in other applications.

As we have seen, updating requires the calculation of multifold integrals to compute the normalizing constant (the reciprocal of the denominator in (10.61)) and the posterior statistics, such as means, variance/covariances, and marginal distributions. Conjugate priors are unlikely to exist for the types of likelihood functions that arise in such applications, so one must invariably resort to numerical integration. When there are more than two or three random variables, conventional numerical integration becomes impractical. The importance-sampling method described in Section 9.8 is one viable approach. Another approach is the Markov chain Monte Carlo sampling method, which has the advantage that it can work with a distribution that is not normalized, i.e., samples of the posterior distribution can be generated by use of the product $L(\mathbf{x})f'_{\mathbf{X}}(\mathbf{x})$ without knowing the normalizing factor. A description of this method and example applications to structural and geotechnical problems were given by Straub and Papaioannou (2015). In particular, they showed that methods of structural-reliability analysis can be used to carry out the numerical integrations necessary for updating the distribution of basic parameters, even when equality-type observations are involved.

A somewhat simpler formulation results if one works with the CDF rather than the PDF. Consider updating the distribution of random variable X_1 in light of the set of observations $\mathcal{H}_i$, $i = 1, \ldots, M$. The updated CDF is

$$F_{X_1}(x_1|\mathcal{H}_1 \cdots \mathcal{H}_M) = \frac{\Pr[\{X_1 \leq x_1\} \cap \mathcal{H}_1 \cdots \mathcal{H}_M]}{\Pr(\mathcal{H}_1 \cdots \mathcal{H}_M)}. \tag{10.62}$$

Observe that the first event in the numerator is a simple component-reliability problem with the limit-state function $g(X_1) = X_1 - x_1$. If the observations are of inequality type, then both the numerator and denominator represent parallel-system reliability problems, which can be solved by standard structural-reliability methods. If equality-type observations are involved, then the methods by Madsen (1987) or Straub (2011) described in the preceding section can be used to convert them to inequality-type events. Naturally, one needs to solve the above problem for a range of x_1 values to obtain a full description of the CDF. Furthermore, the corresponding PDF is obtained as the sensitivity with respect to the limit-state function parameter x_1.

The above formulation can be extended to the joint CDF of several random variables. The updating formula takes the form

$$F_{\mathbf{X}}(\mathbf{x}|\mathcal{H}_1 \cdots \mathcal{H}_M) = \frac{\Pr[\{X_1 \leq x_1\} \cdots \{X_n \leq x_n\} \cap \mathcal{H}_1 \cdots \mathcal{H}_M]}{\Pr(\mathcal{H}_1 \cdots \mathcal{H}_M)}. \tag{10.63}$$

Observe that the numerator is still a parallel-system reliability problem, albeit with a larger number of components.

Example 10.7 – Updating the Distribution of Yield Strength in a Column

Reconsider the column-reliability problem introduced in Example 6.5 and further analyzed in Example 10.6. Recall that the column is under random bi-axial bending moments M_1 and M_2 and axial force P, with an uncertain yield strength Y. In Figure 10.11, we show updated PDFs of Y conditioned on the events of failure and survival of the column under the applied loads. Also shown is the unconditional (prior) PDF of Y. The updated PDFs are obtained as sensitivities of the updated CDF

$$F_Y(y|\mathcal{E}) = \frac{\Pr[\{Y \leq y\} \cap \mathcal{E}]}{\Pr(\mathcal{E})} \tag{E1}$$

with respect to parameter y, where $\mathcal{E} = \mathcal{F}$ for the failure event and $\mathcal{E} = \bar{\mathcal{F}}$ for the survival event with $\mathcal{F} = \{g(M_1, M_2, P, Y) \leq 0\}$, in which $g(\cdot)$ is the limit-state function in (E1) of

Example 10.7 (cont.)

Example 6.5. The results shown are computed by FORM analyses for a range of values of y. Observe that the updated PDF given failure of the column is sharply shifted towards lower values of y, while the updated PDF given survival of the column virtually coincides with the prior PDF. This is because the survival event does not offer an unexpected outcome and has a probability close to 1.

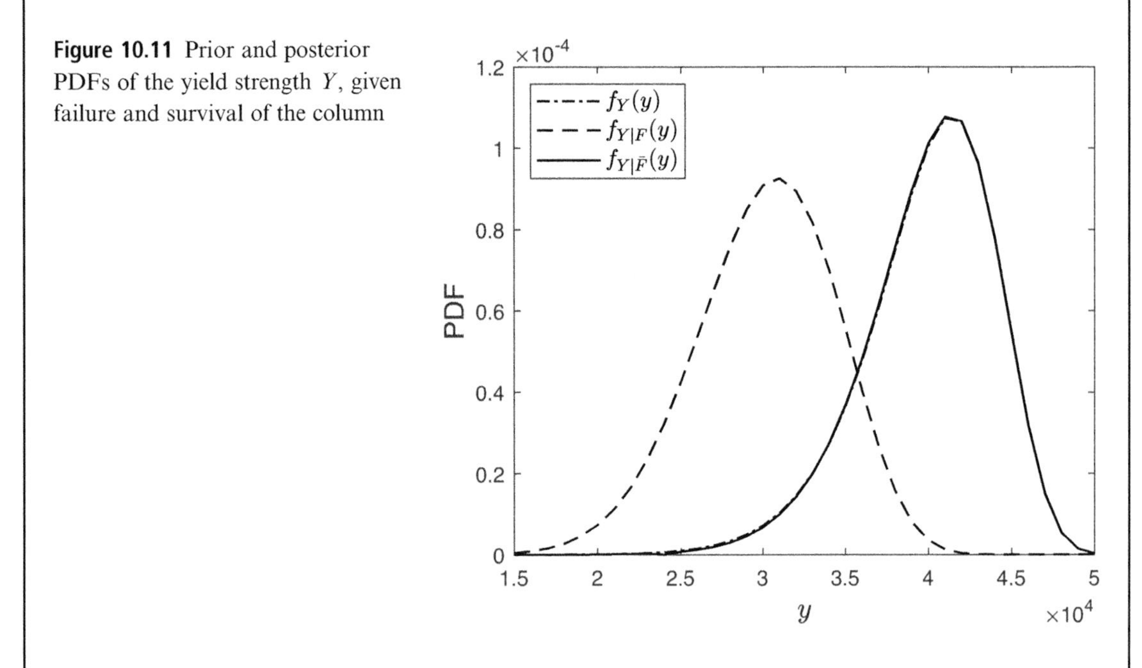

Figure 10.11 Prior and posterior PDFs of the yield strength Y, given failure and survival of the column

10.8 Concluding Remarks

This chapter introduced the Bayesian approach for the analysis uncertainties present in the estimation of model parameters. Two kinds of models were considered: probability-distribution models and mathematical models of physical phenomena. Sources of uncertainty considered were inherent variability, statistical error, model error, and measurement error. These were categorized as either intrinsic or epistemic. It was argued, however, that the distinction between these two types of uncertainties is blurry and makes sense only within the universe of models used in the reliability problem.

The Bayesian updating formula involves the prior distribution reflecting our prior state of information about the uncertain parameters and the likelihood function that represents the information gained from a set of observations. The product of the two is proportional to the posterior (updated) distribution of the parameters, which incorporates both sets of information. Consistent with the Bayesian philosophy, we treated parameter uncertainties similarly to inherent variabilities represented by random variables and combined the two through the total probability rule to obtain the predictive distribution of the underlying random variables. This distribution incorporates both types of uncertainties. Details were given on the formulation of the prior distribution and the likelihood function under different conditions, and methods were described for computing the posterior statistics and the predictive distribution. The class of conjugate-pair distributions presents a particularly convenient case, where closed-form solutions of the posterior and predictive distributions are available. A selection of conjugate-pair distributions is listed in Appendix 10A.

The next topic considered was how to account for parameter uncertainties in the assessment of structural reliability. Three measures were introduced: the point estimate, the predictive estimate, and confidence bounds. Philosophically, the predictive estimate, which incorporates the parameter uncertainties into the reliability estimate, is consistent with the Bayesian philosophy and is the measure advocated by the author. It is also the proper measure in the context of decision making. Nevertheless, the other measures were presented for the reader's information and as a contrast with the predictive measure. Several methods were presented for computing the three measures.

When observations are made that relate to the random variables affecting the reliability of a structure, the information contained in them can be used to update the reliability estimate. The Bayesian updating formula provides the proper framework for this purpose. Methods for such updating, involving both inequality- and equality-type observations, were described. Finally, observations about the state of a structure can be used to update our information about the basic random variables involved in the reliability problem. In particular, the distribution of random variables can be updated in the light of information about the state of a structure.

With the methods for sensing and measuring rapidly advancing, collecting observational data is becoming cheap and omnipresent. The real challenge is how to make use of this information in making decisions. As we have seen, the Bayesian updating formulation provides the proper mathematical framework for analyzing observational data and for incorporating them in predictive models. However, computational challenges for large-scale problems remain. In Chapter 15, we present the topic of Bayesian networks, which offers a different perspective and computational approach to Bayesian analysis.

APPENDIX 10A
Conjugate-Pair Distributions

Table A10.1 Conjugate-pair distributions

Distribution of X	Unknown parameters	Observation	Prior and posterior distribution of parameters	Mean and standard deviation	Updating rule	Predictive distribution of X
Discrete random variables						
Binomial: $p_X(x\|\theta) = \binom{n}{x}\theta^x(1-\theta)^{n-x}$, $x = 0, \ldots, n$	θ	x occurrences in n trials	Beta: $f_\Theta(\theta) = \frac{\Gamma(q+r)}{\Gamma(q)\Gamma(r)}\theta^{q-1}(1-\theta)^{r-1}$, $0 < \theta < 1$	$\mu_\Theta = \frac{q}{q+r}$ $\sigma_\Theta = \frac{1}{q+r}\sqrt{\frac{qr}{q+r+1}}$	$q'' = q' + x$ $r'' = r' + n - x$	$\widetilde{p}_X(x) = \binom{n}{x} \frac{\Gamma(q''+r'')\Gamma(q''+x)\Gamma(r''+n-x)}{\Gamma(q'')\Gamma(r'')\Gamma(q''+r''+1)}$, $x = 0, \ldots, n$
Geometric: $p_X(x\|\theta) = \theta(1-\theta)^{x-1}$, $x = 1, 2, \ldots$	θ	x trials to first occurrence	Beta: $f_\Theta(\theta) = \frac{\Gamma(q+r)}{\Gamma(q)\Gamma(r)}\theta^{q-1}(1-\theta)^{r-1}$, $0 < \theta < 1$	$\mu_\Theta = \frac{q}{q+r}$ $\sigma_\Theta = \frac{1}{q+r}\sqrt{\frac{qr}{q+r+1}}$	$q'' = q' + 1$ $r'' = r' + x - 1$	$\widetilde{p}_X(x) = \frac{q''\Gamma(q''+r'')\Gamma(r''+x-1)}{\Gamma(r'')\Gamma(q''+r''+x)}$, $x = 1, 2, \ldots$
Negative binomial: $p_X(x\|\theta) = \binom{x-1}{k-1}\theta^k(1-\theta)^{x-k}$, $x = k, k+1, \ldots$	θ (k known)	x trials to kth occurrence	Beta: $f_\Theta(\theta) = \frac{\Gamma(q+r)}{\Gamma(q)\Gamma(r)}\theta^{q-1}(1-\theta)^{r-1}$, $0 < \theta < 1$	$\mu_\Theta = \frac{q}{q+r}$ $\sigma_\Theta = \frac{1}{q+r}\sqrt{\frac{qr}{q+r+1}}$	$q'' = q' + k$ $r'' = r' + x - k$	$\widetilde{p}_X(x) = \binom{x-1}{k-1} \frac{\Gamma(q''+r'')\Gamma(q''+k)\Gamma(r''+x-k)}{\Gamma(q'')\Gamma(r'')\Gamma(q''+r''+x)}$, $x = k, k+1, \ldots$
Poisson process: $p_X(x\|\lambda) = \frac{(\lambda t)^x}{x!}\exp(-\lambda t)$, $x = 0, 1, \ldots$	λ	x occurrences in $(0, t)$	Gamma: $f_\Lambda(\lambda) = \frac{\nu(\nu\lambda)^{k-1}}{\Gamma(k)}\exp(-\nu\lambda)$, $0 < \lambda$	$\mu_\Lambda = \frac{k}{\nu}$ $\sigma_\Lambda = \frac{\sqrt{k}}{\nu}$	$k'' = k' + x$ $\nu'' = \nu' + t$	$\widetilde{p}_X(x) = \frac{\nu''^{k''} t^x \Gamma(k''+x)}{(\nu''+t)^{k''+x}\Gamma(k'')x!}$, $x = 0, 1, \ldots$
Continuous random variables						
Exponential: $f_X(x\|\lambda) = \lambda\exp(-\lambda x)$, $0 < x$	λ	Sample x_i, $i = 1, \ldots, n$	Gamma: $f_\Lambda(\lambda) = \frac{\nu(\nu\lambda)^{k-1}}{\Gamma(k)}\exp(-\nu\lambda)$, $0 < \lambda$	$\mu_\Lambda = \frac{k}{\nu}$ $\sigma_\Lambda = \frac{\sqrt{k}}{\nu}$	$k'' = k' + n$ $\nu'' = \nu' + n\bar{x}$, $\bar{x} = \sum_{i=1}^{n} x_i / n$	$\widetilde{f}_X(x) = \frac{k''\nu^{k''}}{(\nu''+x)^{k''+1}}$, $0 < x$
Exponential: $f_X(x\|\lambda) = \lambda\exp(-\lambda x)$, $0 < x$	λ	Sample x_i, $i = 1, \ldots, n$; Lower bounds y_j, $j = 1, \ldots, m$	Gamma: $f_\Lambda(\lambda) = \frac{\nu(\nu\lambda)^{k-1}}{\Gamma(k)}\exp(-\nu\lambda)$, $0 < \lambda$	$\mu_\Lambda = \frac{k}{\nu}$ $\sigma_\Lambda = \frac{\sqrt{k}}{\nu}$	$k'' = k' + n$ $\nu'' = \nu' + n\bar{x} + m\bar{y}$, $\bar{x} = \frac{1}{n}\sum_{i=1}^{n} x_i$, $\bar{y} = \frac{1}{m}\sum_{j=1}^{m} y_j$	$\widetilde{f}_X(x) = \frac{k''\nu''k''}{(\nu''+x)^{k''+1}}$, $0 < x$

Table A10.1 (cont.)

Distribution of X	Unknown parameters	Observation	Prior and posterior distribution of parameters	Mean and standard deviation	Updating rule	Predictive distribution of X
Gamma: $f_X(x\|\lambda) = \frac{\lambda(\lambda x)^{\kappa-1}}{\Gamma(\kappa)} \exp(-\lambda x)$, $0 < x$	λ (κ known)	Sample $x_i, i = 1, \ldots, n$	Gamma: $f_\Lambda(\lambda) = \frac{\nu(\nu\lambda)^{k-1}}{\Gamma(k)} \exp(-\nu\lambda)$, $0 < \lambda$	$\mu_\Lambda = \frac{k}{\nu}$ $\sigma_\Lambda = \frac{\sqrt{k}}{\nu}$	$k'' = k' + n\kappa$ $\nu'' = \nu' + n\bar{x}$ $\bar{x} = \frac{1}{n}\sum_{i=1}^n x_i$	$\tilde{f}_X(x) = \frac{\Gamma(k''+\kappa)}{\Gamma(k'')\Gamma(\kappa)} \frac{\nu''^{k''} x^{\kappa-1}}{(\nu''+x)^{k''+\kappa}}$, $0 < x$
Rayleigh: $f_X(x\|\lambda) = 2\lambda x \exp(-\lambda x^2)$, $0 < x$	λ	Sample $x_i, i = 1, \ldots, n$	Gamma: $f_\Lambda(\lambda) = \frac{\nu(\nu\lambda)^{k-1}}{\Gamma(k)} \exp(-\nu\lambda)$, $0 < \lambda$	$\mu_\Lambda = \frac{k}{\nu}$ $\sigma_\Lambda = \frac{\sqrt{k}}{\nu}$	$k'' = k' + n$ $\nu'' = \nu' + \sum_{i=1}^n x_i^2$	$\tilde{f}_X(x) = \frac{2k''\nu''^{k''} x}{(\nu''+x^2)^{k''+1}}$, $0 < x$
Normal: $f_X(x\|\mu) = \frac{1}{\sqrt{2\pi}\sigma} \exp\left[-\frac{(x-\mu)^2}{2\sigma^2}\right]$	μ (σ known)	Sample $x_i, i = 1, \ldots, n$	Normal: $f_M(\mu) = \frac{1}{\sqrt{2\pi}\sigma_M} \exp\left[-\frac{(x-\mu_M)^2}{2\sigma_M^2}\right]$	μ_M σ_M	$\mu''_M = \frac{\bar{x}\sigma'^2_M + \mu'_M\left(\frac{\sigma^2}{n}\right)}{\sigma'^2_M + \mu'_M\left(\frac{\sigma^2}{n}\right)}$ $\sigma''_M = \sqrt{\frac{\sigma'^2_M \sigma^2/n}{\sigma'^2_M + \sigma^2/n}}$, $\bar{x} = \frac{1}{n}\sum_{i=1}^n x_i$	$\tilde{f}_X(x) = \frac{1}{\sqrt{2\pi(\sigma^2 + \sigma''^2_M)}} \exp\left[-\frac{(x-\mu''_M)^2}{2(\sigma^2+\sigma''^2_M)}\right]$
Normal: $f_X(x\|\mu, \sigma^2) = \frac{1}{\sqrt{2\pi}\sigma} \exp\left[-\frac{(x-\mu)^2}{2\sigma^2}\right]$	μ, σ^2	Sample $x_i, i = 1, \ldots, n$	Normal-inverse-gamma: $f_{M,\Sigma^2}(\mu, \sigma^2) = \frac{\nu^k}{\Gamma(k)} \frac{\sqrt{\eta}}{\sqrt{2\pi}\sigma} \left(\frac{1}{\sigma^2}\right)^{k+1} \exp\left[-\frac{2\nu + \eta(\mu-\mu_M)^2}{2\sigma^2}\right]$, $0 < \sigma^2$ Marginals: $f_M(\mu) = \frac{\Gamma(k+1/2)}{\Gamma(k)} \sqrt{\frac{\eta}{2\pi\nu}} \exp\left[1 + \frac{\eta(\mu-\mu_M)^2}{2\nu}\right]^{-\left(k+\frac{1}{2}\right)}$ $f_{\Sigma^2}(\sigma^2) = \frac{\nu^k}{\Gamma(k)} \left(\frac{1}{\sigma^2}\right)^{k+1} \exp\left(-\frac{\nu}{\sigma^2}\right)$	μ_M $\sigma_M = \sqrt{\frac{\nu}{\eta(k-1)}}$ $\mu_{\Sigma^2} = \frac{\nu}{k-1}$ $\sigma_{\Sigma^2} = \sqrt{\frac{\nu^2}{(k-1)^2(k-2)}}$	$\mu''_M = \frac{\bar{x}n + \mu'_M\eta'}{n+\eta'}$ $\eta'' = \eta' + n$ $k'' = k' + \frac{n}{2}$ $\nu'' = \nu' + \frac{1}{2}\left[\sum_{i=1}^n (x_i - \bar{x})^2 + \frac{n\eta'}{n+\eta'}(\bar{x} - \mu'_M)^2\right]$, $\bar{x} = \frac{1}{n}\sum_{i=1}^n x_i$	$\tilde{f}_X(x) = \frac{\Gamma(k''+1/2)}{\Gamma(k'')} \sqrt{\frac{\eta''}{2\pi\nu''(\eta''+1)}} \left[1 + \frac{\eta''(x-\mu''_M)^2}{2\nu''(\eta''+1)}\right]^{-\left(k''+\frac{1}{2}\right)}$

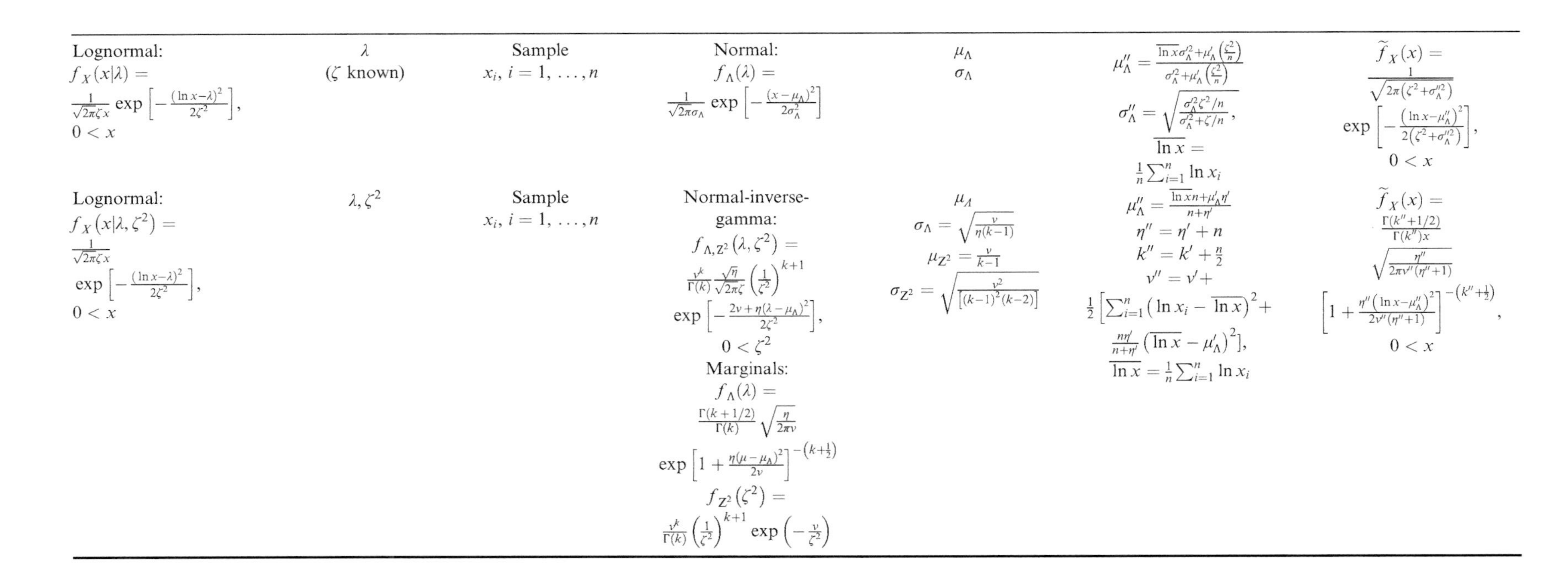

Lognormal: $f_X(x \vert \lambda) = \frac{1}{\sqrt{2\pi}\zeta x} \exp\left[-\frac{(\ln x-\lambda)^2}{2\zeta^2}\right]$, $0 < x$	λ (ζ known)	Sample $x_i,\ i = 1, \ldots, n$	Normal: $f_\Lambda(\lambda) = \frac{1}{\sqrt{2\pi}\sigma_\Lambda} \exp\left[-\frac{(x-\mu_\Lambda)^2}{2\sigma_\Lambda^2}\right]$	μ_Λ σ_Λ	$\mu''_\Lambda = \frac{\overline{\ln x}\sigma'^2_\Lambda + \mu'_\Lambda\left(\frac{\zeta^2}{n}\right)}{\sigma'^2_\Lambda + \mu'_\Lambda\left(\frac{\zeta^2}{n}\right)}$ $\sigma''_\Lambda = \sqrt{\frac{\sigma'^2_\Lambda\zeta^2/n}{\sigma'^2_\Lambda + \zeta/n}}$, $\overline{\ln x} = \frac{1}{n}\sum_{i=1}^n \ln x_i$	$\widetilde{f}_X(x) = \frac{1}{\sqrt{2\pi\left(\zeta^2+\sigma''^2_\Lambda\right)}} \exp\left[-\frac{\left(\ln x-\mu''_\Lambda\right)^2}{2\left(\zeta^2+\sigma''^2_\Lambda\right)}\right]$, $0 < x$
Lognormal: $f_X\left(x \vert \lambda, \zeta^2\right) = \frac{1}{\sqrt{2\pi}\zeta x} \exp\left[-\frac{(\ln x-\lambda)^2}{2\zeta^2}\right]$, $0 < x$	λ, ζ^2	Sample $x_i,\ i = 1, \ldots, n$	Normal-inverse-gamma: $f_{\Lambda,Z^2}\left(\lambda, \zeta^2\right) = \frac{\nu^k}{\Gamma(k)} \frac{\sqrt{\eta}}{\sqrt{2\pi}\zeta} \left(\frac{1}{\zeta^2}\right)^{k+1} \exp\left[-\frac{2\nu+\eta(\lambda-\mu_\Lambda)^2}{2\zeta^2}\right]$, $0 < \zeta^2$ Marginals: $f_\Lambda(\lambda) = \frac{\Gamma(k+1/2)}{\Gamma(k)} \sqrt{\frac{\eta}{2\pi\nu}} \exp\left[1 + \frac{\eta(\mu-\mu_\Lambda)^2}{2\nu}\right]^{-\left(k+\frac{1}{2}\right)}$ $f_{Z^2}\left(\zeta^2\right) = \frac{\nu^k}{\Gamma(k)} \left(\frac{1}{\zeta^2}\right)^{k+1} \exp\left(-\frac{\nu}{\zeta^2}\right)$	μ_Λ $\sigma_\Lambda = \sqrt{\frac{\nu}{\eta(k-1)}}$ $\mu_{Z^2} = \frac{\nu}{k-1}$ $\sigma_{Z^2} = \sqrt{\frac{\nu^2}{\left[(k-1)^2(k-2)\right]}}$	$\mu''_\Lambda = \frac{\overline{\ln x}n + \mu'_\Lambda\eta'}{n+\eta'}$ $\eta'' = \eta' + n$ $k'' = k' + \frac{n}{2}$ $\nu'' = \nu' + \frac{1}{2}\left[\sum_{i=1}^n\left(\ln x_i - \overline{\ln x}\right)^2 + \frac{n\eta'}{n+\eta'}\left(\overline{\ln x} - \mu'_\Lambda\right)^2\right]$, $\overline{\ln x} = \frac{1}{n}\sum_{i=1}^n \ln x_i$	$\widetilde{f}_X(x) = \frac{\Gamma(k''+1/2)}{\Gamma(k'')x} \sqrt{\frac{\eta''}{2\pi\nu''(\eta''+1)}} \left[1 + \frac{\eta''\left(\ln x-\mu''_\Lambda\right)^2}{2\nu''(\eta''+1)}\right]^{-\left(k''+\frac{1}{2}\right)}$, $0 < x$

PROBLEMS

10.1 Consider the Rayleigh distribution with the CDF defined as $F_X(x) = 1 - \exp(-\lambda x^2)$, where $0 < \lambda$ is the unknown parameter. Suppose the available information consists of n equality observations x_i, $i = 1, \ldots, n$, and m lower-bound observations, x_i, $i = n+1, \ldots, n+m$.

1) Show that the gamma distribution is the conjugate prior for the above distribution and observations. Determine the updating rules for the parameters of the gamma prior.
2) Derive an expression for the predictive distribution of X that incorporates the uncertainty in the parameter λ.
3) Suppose we have the $n = 8$ equality observations 4.5, 5.1, 3.7, 4.7, 5.4, 5.2, 4.9, and 3.9 and the $m = 2$ lower-bound observations 5.5 and 6.0. Also suppose our prior information about λ is a gamma distribution with mean 0.04 and standard deviation 0.02. Compute and plot the prior and posterior PDFs of λ.
4) Compute and plot the predictive PDF of X and compare it with the PDF of the Rayleigh distribution using the posterior mean of λ. Note that the latter distribution uses the updated mean of the parameter but does not incorporate the uncertainty in the parameter in the distribution of X.

10.2 Reconsider Problem 6.8 concerning a spacecraft subject to debris impact. As stated there, the marginal distribution of the radii of the projectiles impacting the spacecraft is exponential, $f_R(r) = \lambda \exp(-\lambda r)$, with the parameter λ having the value $\lambda = 1/2.5 = 0.4$ 1/mm. Now assume the value of this parameter is uncertain. Suppose prior information suggests a mean of $\mu'_\Lambda = 0.4$ 1/mm and a c.o.v. of $\delta'_\Lambda = 0.50$. Employ the conjugate prior distribution.

1) Determine the predictive reliability index, accounting for the uncertainty in parameter λ.
2) Suppose that upon landing of the spacecraft, careful observation of the shield reveals 10 projectile impacts with a mean radius of 1.5 mm. Determine the posterior distribution of λ and the predictive distribution of R. Make plots of and compare the prior and posterior distributions of λ. Also, make plots of and compare the distribution of R with λ set to its posterior mean value with the predictive distribution of R.
3) Update the probability that a projectile impacting the spacecraft will puncture a hole in the shield during a future mission, accounting for the above information on λ. Compare this result with that obtained in part (1).

10.3 For Problem 6.8 concerning a spacecraft subject to debris impact, suppose it is known that a projectile has punctured a hole in the shield. Determine the updated distribution of the radius of the projectile that caused the hole. Make plots of and compare the distributions of the radius of the projectile before and after the above observation of the punctured hole.

10.4 Reconsider the oscillator reliability problem (Problem 6.5) with the limit-state function

$$g(\mathbf{X}, \theta) = \left(\Omega_0^2 - \Omega^2\right)^2 + 4Z^2\Omega_0^2\Omega^2 - \frac{1}{\theta^2},$$

where $\mathbf{X} = (\Omega_0, Z, \Omega)$ is the set of basic random variables and $\theta = 0.3$. Suppose Ω_0 is normally distributed with unknown mean μ_{Ω_0} and standard deviation $\sigma_{\Omega_0} = 0.9$ rad/s, Z is lognormally distributed with an unknown logarithmic mean λ_Z and c.o.v. $\delta_Z = 0.5$, and Ω is normally distributed with an unknown mean μ_Ω and standard deviation $\sigma_\Omega = 0.6$ rad/s. Suppose the available information indicates that $\mu_{\Omega_0} \sim \mathrm{N}(6, 0.6)$, $\lambda_Z \sim \mathrm{N}(-1.721, 0.5)$, and $\mu_\Omega \sim \mathrm{N}(3, 0.3)$, where $\mathrm{N}(\mu, \sigma)$ denotes the normal distribution with mean μ and standard deviation σ. Using this information, compute

1) The first-order approximation of the mean and standard deviation of the reliability index.
2) The predictive reliability index. Apply the three formulations described in (10.43), (10.44), and (10.46) and compare the results.
3) The mean plus or minus one standard deviation confidence interval of the conditional reliability index.

10.5 A widely used model describing the flexural capacity of a reinforced concrete beam is

$$\widehat{M}_n = \rho f_y b d^2 \left(1 - 0.59 \rho f_y / f'_c\right),$$

where ρ is the reinforcement ratio, f_y the yield strength of the reinforcing bars, f'_c the compressive strength of concrete, and b and d the effective width and depth, respectively, of the beam. The table below lists the predicted and measured values of the flexural capacity for 10 beams that were tested in a laboratory. Note that 1 MPa is equivalent to 1,000 kN/m^2. To account for the model bias and error, consider the following "corrected" model form

$$\ln M = \ln \widehat{M}_n + E,$$

where M is the measured (true) capacity, $\widehat{M}_n$ the predicted (nominal) capacity using the above idealized model, and E the model error. We have used the logarithmic transform of the capacity model to satisfy the normality and homoscedasticity requirements. Assume E is a normal random variable with mean μ and variance σ^2, both unknown. Note that μ describes the bias in the model.

					Flexural capacity in kNm	
f'_c, MPa	f_y, MPa	ρ	b, m	d, m	predicted	measured
28	320	0.012	0.35	0.50	309	372
28	410	0.012	0.35	0.50	386	444
28	500	0.015	0.40	0.50	631	699
32	320	0.015	0.35	0.60	551	698

f'_c, MPa	f_y, MPa	ρ	b, m	d, m	Flexural capacity in kNm	
					predicted	measured
32	410	0.018	0.40	0.60	918	1109
32	500	0.018	0.35	0.60	946	1031
36	410	0.015	0.30	0.50	415	393
36	500	0.018	0.35	0.60	967	922
40	410	0.020	0.40	0.60	1038	1305
40	410	0.015	0.40	0.70	1096	905

1) Employing a non-informative prior, use the normal-inverse-gamma conjugate distribution (see Section 10.3.3 and Example 10.3) to determine the posterior joint distribution and second moments of the unknown parameters μ and σ^2.
2) Consider a beam with dimensions $b = 0.40$ m and $d = 0.60$ m and reinforcement ratio $\rho = 0.015$. Let f_y be a lognormal random variable with mean 430 MPa and c.o.v. equal to 0.10, and f'_c be a normal random variable with mean 42 MPa and c.o.v. equal to 0.20. Accounting for the model error and parameter uncertainties, determine the first-order approximations of the mean and standard deviation of the conditional reliability index. Also determine the predictive reliability index of the beam to carry a bending moment of 600 kNm that accounts for the uncertainties in the distribution parameters.

10.6 Reconsider Problem 8.4 concerning the reliability of a gabled frame under vertical and horizontal loads. Suppose it is known that the frame has failed in a sway mode. Update the probability distribution of the horizontal load X_2 that caused the failure.

11 Time- and Space-Variant Reliability Analysis

11.1 Introduction

The methods developed in the previous chapters of this book presented the structural and system reliability problem without consideration of variations in time or space. In the real world, variations in time or space are inevitable. Most demands on structures, e.g., loads, temperature, base motion, foundation settlement, etc., fluctuate in time and space; furthermore, structural properties often vary randomly in space and properties such as stiffness and strength may evolve in time due to wear and tear, deterioration, fatigue, and aging. This chapter addresses this class of problems in a broad sense. Chapter 12 introduces methods for finite element reliability analysis, which includes consideration of random fields, and Chapter 13 focuses on the specific problem of the dynamic response of structures subjected to stochastic excitations. The topic of stochastic processes and random vibrations is a very rich one on which volumes have been written. Here, this chapter and Chapters 12 and 13 approach the topic narrowly, from the viewpoint of reliability analysis, especially in the context of the methods developed in the earlier chapters of this book. For a broader treatment of stochastic processes and random vibrations, the reader should consult one of the many texts available on those subjects.

Most of the discussion presented below is written in the context of a component reliability problem. The extension to systems problems, at least conceptually, is straightforward. This will be shown through example applications to systems problems.

Let t denote the time coordinate and $\mathbf{s} = (s_1, s_2, s_3)$ denote the space coordinates. In most cases we will focus on a single spatial coordinate, denoted s, along a line. In that case, the time- and space-variant reliability problems are mathematically similar and t and s can be exchanged without altering the nature of the problem or the solution approach. The extension to multi-dimensional space-variant problems is not trivial. Example 12.4 in Chapter 12 demonstrates one such application.

One can envision the following categories of time- and space-variant structural reliability problems:

Encroaching problem: Consider a structural component reliability problem defined by a limit-state function of the form $g = g(\mathbf{X}, t)$ or $g = g(\mathbf{X}, \mathbf{s})$, where $\mathbf{X}$ are the basic random variables. Note that the coordinates t or $\mathbf{s}$ enter as parameters in the limit-state function. The reliability problem is still defined in terms of time- and space-invariant random variables, but the limit-state function varies with time or space. Examples of such problems include structures with capacities that vary in time or space according to a deterministic rule. A structure having a capacity defined as $R = R_0 \exp(-vt)$, where R_0 is the random initial capacity and v is a deterministic decay rate, is one example. For a space-variant example, consider a shell structure under random external loads $\mathbf{X}$, where the thickness of the shell varies deterministically in space. The location of the potential failure obviously depends on the internal forces and the local thickness of the shell. Figure 11.1 shows a conceptual portrayal of this category of problems for the time-variant case. The outcome of the basic random variables is an unknown but fixed point $\mathbf{x}$, while the limit-state surface evolves in time. Failure happens when the limit-state surface *encroaches* on the outcome point $\mathbf{x}$. The failure event for this category of problems is defined as

$$\mathcal{F} = \left\{ \min_{t \in T} g(\mathbf{X}, t) \leq 0 \right\} \text{ or } \mathcal{F} = \left\{ \min_{\mathbf{s} \in \Omega} g(\mathbf{X}, \mathbf{s}) \leq 0 \right\}, \tag{11.1}$$

where T is a time interval within which the structure operates and Ω is a spatial domain within which the structure is located.

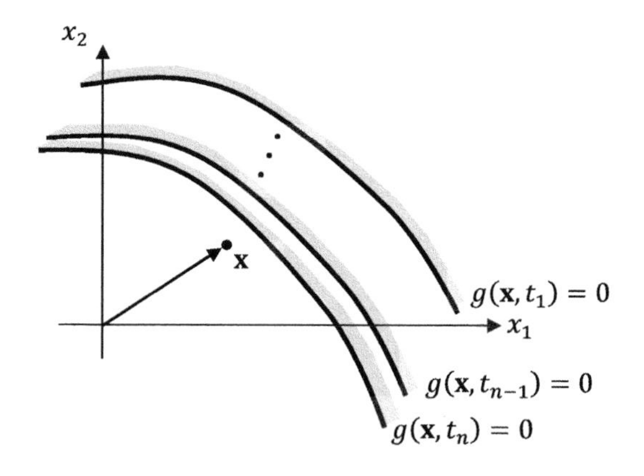

Figure 11.1 The time-variant encroaching problem

Outcrossing problem: Now consider a structural reliability problem defined by a limit-state function of the form $g = g[\mathbf{X}(t)]$ or $g = g[\mathbf{X}(\mathbf{s})]$, where $\mathbf{X}(t)$ and $\mathbf{X}(\mathbf{s})$ are random processes. In this case, the basic random quantities vary with time or space coordinates; hence, they are random processes. Examples include structures under stochastic loads or structures with spatially varying random properties. Figure 11.2 shows a conceptual portrayal of this category of problems for the time-variant case. Note that in both time- and space-variant cases the limit-state surface $g(\mathbf{x}) = 0$ is invariant in time or space. It is

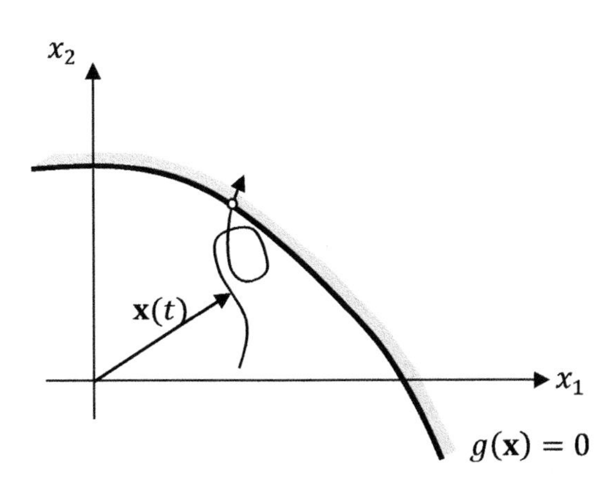

Figure 11.2 The time-variant outcrossing problem

the vector of random quantities that evolves in time or space and failure occurs when it *outcrosses* the limit-state surface. In this case the failure event is defined by

$$\mathcal{F} = \left\{ \min_{t \in T} g[\mathbf{X}(t)] \leq 0 \right\} \text{ or } \mathcal{F} = \left\{ \min_{\mathbf{s} \in \Omega} g[\mathbf{X}(\mathbf{s})] \leq 0 \right\}. \tag{11.2}$$

Encroaching-outcrossing problem: Now consider problems defined by limit-state functions of the form $g = g[\mathbf{X}(t), t]$ or $g = g[\mathbf{X}(\mathbf{s}), \mathbf{s}]$. In this case, the time or space coordinates appear not only through the random processes but also as parameters in the limit-state function. Thus, both the realization of the basic random quantities and the limit-state surface evolve in time or space. A structure under stochastic loads and having a capacity that evolves in time according to a deterministic rule belongs to this category of problems. Figure 11.3 shows a conceptual portrayal of this category of problems for the time-variant case. Failure occurs when the outcome vector $\mathbf{x}(t)$ outcrosses the evolving limit-state surface $g(\mathbf{x}, t) = 0$. In the example figure, this happens at time t_{n-1}.The failure event is now defined by

$$\mathcal{F} = \left\{ \min_{t \in T} g[\mathbf{X}(t), t] \leq 0 \right\} \text{ or } \mathcal{F} = \left\{ \min_{\mathbf{s} \in \Omega} g[\mathbf{X}(\mathbf{s}), \mathbf{s}] \leq 0 \right\}. \tag{11.3}$$

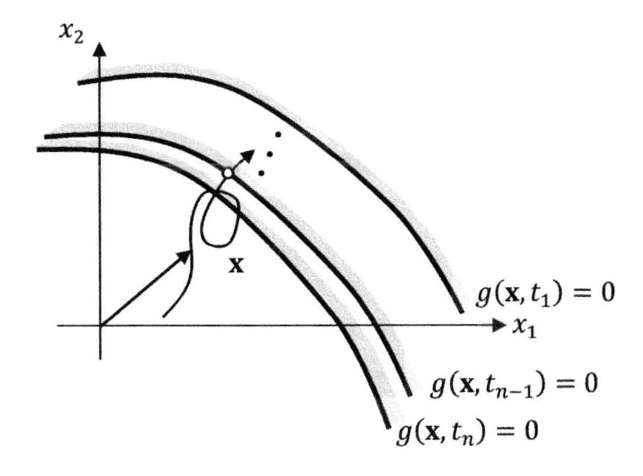

Figure 11.3 The time-variant encroaching-outcrossing problem

Time- and space-variant problems: In the most general case, the limit-state function is defined by $g = g[\mathbf{X}(t, \mathbf{s}), t, \mathbf{s}]$. The basic random quantities now are random processes dependent on both time and space coordinates; these coordinates also appear as parameters in the limit-state function. The failure event is now defined by

$$\mathcal{F} = \left\{ \min_{t \in T, \mathbf{s} \in \Omega} g[\mathbf{X}(t, \mathbf{s}), t, \mathbf{s}] \leq 0 \right\}. \tag{11.4}$$

Another way to visualize the above problems is to monitor the limit-state function as a function of the coordinates t and/or $\mathbf{s}$. As an example, Figure 11.4 shows a realization of $g[\mathbf{X}(t), t]$ in time for $T = (0, T_0]$. Failure occurs when the limit-state function down-crosses the zero level. This is a scalar random process problem; however, when more than one time or space coordinates are involved, the realizations of $g[\mathbf{X}(t, \mathbf{s}), t, \mathbf{s}]$ appear as surfaces evolving in the $(t, \mathbf{s})$ coordinate space. Failure then occurs when the surface penetrates below the zero level. One simple case is when $g[\mathbf{X}(t, \mathbf{s}), t, \mathbf{s}]$ monotonically decreases with the time and special coordinates, i.e., the gradient $\nabla_{(t,\mathbf{s})} g$ has all negative elements regardless of the outcome of $\mathbf{X}(t, \mathbf{s})$. In that case the problem reduces to that of solving the reliability problem defined by the limit-state function $g = g[\mathbf{X}(T_0, \mathbf{S}_0), T_0, \mathbf{S}_0]$, where $(T_0, \mathbf{S}_0)$ are the coordinates where the limit-state function takes its minimum value. One such example is when the strength of a structure under constant loads monotonically decreases in time. In that case, one only needs to perform reliability analysis at the projected end of the time period of interest.

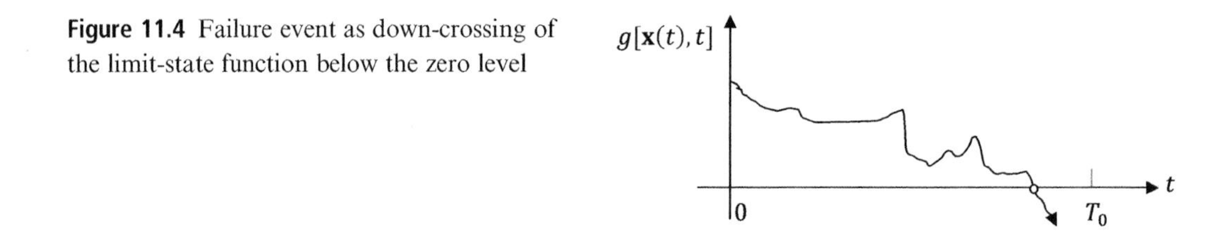

Figure 11.4 Failure event as down-crossing of the limit-state function below the zero level

The crossing of a random process or a vector of random processes outside a safe domain has been of interest for a long time. Analytical solutions exist for Gaussian processes or non-Gaussian processes obtained by translation of Gaussian processes (see, e.g., Rice, 1944, 1945; Belyaev, 1968; Veneziano et al., 1977; Ditlevsen, 1983; Grigoriu, 1984b; and Hohenbichler and Rackwitz, 1986b). This chapter focuses on numerical methods that employ first- or second-order reliability method (FORM/SORM) approximations. The well-known analytical solution for a scalar Gaussian process obtained by Rice (1944, 1945) is presented in Chapter 13.

Before proceeding to develop solution methods for time- and space-variant structural reliability problems, brief reviews of random processes are presented. Special attention is given to Gaussian processes. The Poisson process is developed later. These processes are subsequently employed in developing load models and used in example applications.

11.2 Review of Random Processes

A random process $X(t)$ is a sequence of parametrized random variables. For any selected set $t_1 < t_2 < \cdots < t_n$ of parameter t, the sequence $X(t_1), \ldots, X(t_n)$ is a collection of dependent random variables. The random process is completely defined by specifying the joint distribution of $X(t_1), \ldots, X(t_n)$ for any selection of t_i, $i = 1, \ldots, n$, and n. Depending on whether t is discrete or continuous and whether $X(t)$ has discrete or continuous outcomes, the process is a discrete- or continuous-parameter discrete or continuous random process. Later examples demonstrate these four types of random processes. In the following, we let $\mu_X(t) = \mathrm{E}[X(t)]$, $\sigma_X^2(t) = \mathrm{Var}[X(t)]$, $f_X(x,t)$, and $F_X(x,t)$ define the mean, variance, probability density function (PDF), and cumulative distribution function (CDF) of $X(t)$, respectively, all expressed as functions of t. In addition, we define $\Sigma_{XX}(t_1,t_2) = \mathrm{E}[X(t_1)X(t_2)] - \mu_X(t_1)\mu_X(t_2)$ as the *auto-covariance* and $\rho_{XX}(t_1,t_2) = \Sigma_{XX}(t_1,t_2)/\sigma_X(t_1)\sigma_X(t_2)$ as the *auto-correlation coefficient* functions of the process, respectively. Note that $\Sigma_{XX}(t,t) = \sigma_X^2(t)$ and $\rho_{XX}(t,t) = 1$. Also note that the above PDF and CDF are marginal distributions. As just mentioned, complete characterization of a random process requires specification of the joint distribution of all the random variables for any set of the parameter values.

The auto-covariance and auto-correlation coefficient functions have several important properties. Here, these properties for $\Sigma_{XX}(t_1,t_2)$ are summarized without proof. The corresponding properties of $\rho_{XX}(t_1,t_2)$ should be evident from the relation between the two.

Symmetry: $\Sigma_{XX}(t_1,t_2) = \Sigma_{XX}(t_2,t_1)$.
Boundedness: $|\Sigma_{XX}(t_1,t_2)| \le \sqrt{\Sigma_{XX}(t_1,t_1)\Sigma_{XX}(t_2,t_2)} = \sigma_X(t_1)\sigma_X(t_2)$.
Continuity: If $\Sigma_{XX}(t_1,t_2)$ is continuous at (t_1,t_1) and at (t_2,t_2), then it must be continuous at (t_1,t_2). Continuity here means that, regardless of how $|\epsilon_1|$ and $|\epsilon_2|$ approach zero, $\Sigma_{XX}(t_1+\epsilon_1, t_2+\epsilon_2)$ must approach a unique value, $\Sigma_{XX}(t_1,t_2)$.
Loss of correlation: For a random process not containing periodic components, correlation between $X(t_1)$ and $X(t_2)$ diminishes as t_1 and t_2 depart so that $\lim_{|t_2-t_1|\to\infty}\Sigma_{XX}(t_1,t_2) = 0$.
Positive semi-definiteness: The auto-covariance matrix formed for any discrete set of parameter values $t_1, t_2, \ldots, t_n$ must be positive semi-definite. Furthermore, in the case of a continuous-parameter process, the quantity $\int_{-\infty}^{+\infty}\int_{-\infty}^{+\infty}\Sigma_{XX}(t_1,t_2)\exp[\mathrm{i}\omega(t_1-t_2)]\mathrm{d}t_1\mathrm{d}t_2$ must be non-negative for any ω, where $\mathrm{i} = \sqrt{-1}$ denotes the imaginary unit. Note that the double integral is a bivariate Fourier transform.

Often, one must work with more than one random process, say $X(t)$ and $Y(t)$. It is then useful to define the *cross-covariance* function $\Sigma_{XY}(t_1,t_2) = \mathrm{E}[X(t_1)Y(t_2)] - \mu_X(t_1)\mu_Y(t_2)$ and the *cross-correlation coefficient* function $\rho_{XY}(t_1,t_2) = \Sigma_{XY}(t_1,t_2)/\sigma_X(t_1)\sigma_Y(t_2)$. These functions possess some of the properties described above. In particular, $\Sigma_{XY}(t_1,t_2) = \Sigma_{YX}(t_2,t_1)$ and $|\Sigma_{XY}(t_1,t_2)| \le \sigma_X(t_1)\sigma_Y(t_2)$. Furthermore, if $\Sigma_{XY}(t_1,t_2)$ is continuous at (t_1,t_1) and at (t_2,t_2), then it must be continuous at (t_1,t_2).

As with deterministic functions, the continuity, differentiability, and integrability of continuous-parameter processes are of interest. However, because a random process has multiple (often infinite) possible realizations, the conditions for possessing these properties must be stated in terms of the ensemble of outcomes. Without going into the details, here we state the conditions for these properties based on the so-called *mean-square convergence criterion* (Lin, 1967).

A continuous-parameter process $X(t)$ is mean-square continuous at t if and only if $\Sigma_{XX}(t_1, t_2)$ is finite and continuous at $t_1 = t_2 = t$. A continuous-parameter process $X(t)$ is mean-square differentiable at t if and only if the second derivative $\partial^2\Sigma_{XX}(t_1, t_2)/\partial t_1 \partial t_2$ exists and is unique at $t_1 = t_2 = t$. The integral $Y(u) = \int_a^b X(t)h(t, u)\mathrm{d}t$, where $h(t, u)$ is a deterministic, possibly complex-valued function, is finite in the mean-square sense if $\left|\int_a^b \int_a^b \Sigma_{XX}(t_1, t_2)h(t_1, u_1)h^*(t_2, u_2)\mathrm{d}t_1\mathrm{d}t_2\right|$ is finite for $u_1 = u_2$, where the asterisk denotes the complex conjugate. Furthermore, if the differential $\dot{X}(t) = \mathrm{d}X(t)/\mathrm{d}t$ and the integral $Y(u) = \int_a^b X(t)h(t, u)\mathrm{d}t$ exist, then one can show that the corresponding mean, auto-, and cross-covariance functions are given by

$$\mu_{\dot{X}}(t) = \frac{\mathrm{d}\mu_X(t)}{\mathrm{d}t}, \tag{11.5}$$

$$\Sigma_{X\dot{X}}(t_1, t_2) = \frac{\partial\Sigma_{XX}(t_1, t_2)}{\partial t_2}, \Sigma_{\dot{X}X}(t_1, t_2) = \frac{\partial\Sigma_{XX}(t_1, t_2)}{\partial t_1}, \tag{11.6}$$

$$\Sigma_{\dot{X}\dot{X}}(t_1, t_2) = \frac{\partial^2\Sigma_{XX}(t_1, t_2)}{\partial t_1 \partial t_2}, \tag{11.7}$$

$$\mu_Y(t) = \int_a^b \mu_X(t)h(t, u)\mathrm{d}t, \tag{11.8}$$

$$\Sigma_{YY}(u_1, u_2) = \int_a^b \int_a^b \Sigma_{XX}(t_1, t_2)h(t_1, u_1)h^*(t_2, u_2)\mathrm{d}t_1\mathrm{d}t_2. \tag{11.9}$$

As is evident, the auto-covariance function plays an important role in determining the mean-square properties of a random process and its derivatives and integrals.

An important subclass of random processes is the stationary or homogeneous processes. The random process $X(t)$ is said to be *stationary* or *homogeneous* if its probabilistic description is invariant of a shift in the parameter origin, i.e., if the probabilistic properties of $X(t)$ and $X(t + \tau)$ are identical for any t and τ. It follows that for such a process we must have constant values of the mean and variance, μ_X and σ_X^2, and that the marginal PDF and CDF should be independent of t, i.e., $f_X(x)$, and $F_X(x)$. Furthermore, the auto-covariance and auto-correlation coefficient functions must be functions of the parameter lag $\tau = t_2 - t_1$, i.e., $\Sigma_{XX}(\tau)$ and $\rho_{XX}(\tau)$. Additionally, two processes $X(t)$ and $Y(t)$ are jointly stationary if their joint probabilistic description is invariant of a shift in the parameter origin. The corresponding cross-covariance and cross-correlation coefficient functions are then functions of τ only, i.e., $\Sigma_{XY}(\tau)$ and $\rho_{XY}(\tau)$.

The above-stated properties of the auto- and cross-covariance functions take simpler forms for stationary processes:

Symmetry: $\Sigma_{XX}(\tau) = \Sigma_{XX}(-\tau)$ and $\Sigma_{XY}(\tau) = \Sigma_{YX}(-\tau)$.
Boundedness: $|\Sigma_{XX}(\tau)| \leq \Sigma_{XX}(0) = \sigma_X^2$ and $|\Sigma_{XY}(\tau)| \leq \Sigma_{XY}(0) = \sigma_X\sigma_Y$.
Continuity: $\Sigma_{XX}(\tau)$ and $\Sigma_{XY}(\tau)$ must be continuous at all τ if they are continuous at $\tau = 0$.
Loss of correlation: $\lim_{|\tau|\to\infty}\Sigma_{XX}(\tau) = 0$, provided the process does not contain periodic components.
Positive semi-definiteness: $\int_{-\infty}^{+\infty} \Sigma_{XX}(\tau)\exp(-i\omega\tau)d\tau \geq 0$.

Furthermore, a continuous-parameter stationary process $X(t)$ is continuous at t if and only if $\Sigma_{XX}(\tau)$ is finite and continuous at $\tau = 0$; it is mean-square differentiable if and only if the second derivative $d^2\Sigma_{XX}(\tau)/d\tau^2$ exists and is unique at $\tau = 0$; and the integral $Y(u) = \int_a^b X(t)h(t,u)dt$ is finite in the mean-square sense if $\left|\int_a^b\int_a^b \Sigma_{XX}(t_2 - t_1)h(t_1,u_1)h^*(t_2,u_2)\, dt_1dt_2\right|$ is finite. Also, if $X(t)$ is stationary, then its derivative $\dot{X}(t)$ must also be stationary. Additionally, if the derivative and integral processes exist, then the corresponding mean, auto-, and cross-covariance functions are given by

$$\mu_{\dot{X}} = 0, \tag{11.10}$$

$$\Sigma_{X\dot{X}}(\tau) = -\Sigma_{\dot{X}X}(\tau) = \frac{d\Sigma_{XX}(\tau)}{d\tau}, \tag{11.11}$$

$$\Sigma_{\dot{X}\dot{X}}(\tau) = -\frac{d^2\Sigma_{XX}(\tau)}{d\tau^2}, \tag{11.12}$$

$$\mu_Y(t) = \mu_X\int_a^b h(t,u)dt, \tag{11.13}$$

$$\Sigma_{YY}(u_1,u_2) = \int_a^b\int_a^b \Sigma_{XX}(t_2 - t_1)h(t_1,u_1)h^*(t_2,u_2)dt_1dt_2. \tag{11.14}$$

Based on the symmetry property, we must have $\Sigma_{X\dot{X}}(0) = \Sigma_{\dot{X}X}(0)$. It follows from (11.11) that the only possible solution is $\Sigma_{X\dot{X}}(0) = \Sigma_{\dot{X}X}(0) = 0$. Hence, a stationary process and its derivative are uncorrelated at the same time instant.

Example 11.1 – Candidates for Auto-Covariance Function

Consider the following three functions:

1) $\Sigma(\tau) = \begin{cases} \sigma^2(1 - |\tau|/a), 0 < a, & \text{for } |\tau| \leq a/2 \\ 0, & \text{otherwise,} \end{cases}$
2) $\Sigma(\tau) = \sigma^2\exp(-|\tau|/a), 0 < a,$
3) $\Sigma(\tau) = \sigma^2\exp(-\tau^2/a^2), 0 < a.$

Example 11.1 (cont.)

Figure 11.5 shows plots of these functions. We want to determine whether each function is a valid candidate to serve as the auto-covariance function of a stationary process, and, if so, whether the process is mean-square continuous and mean-square differentiable.

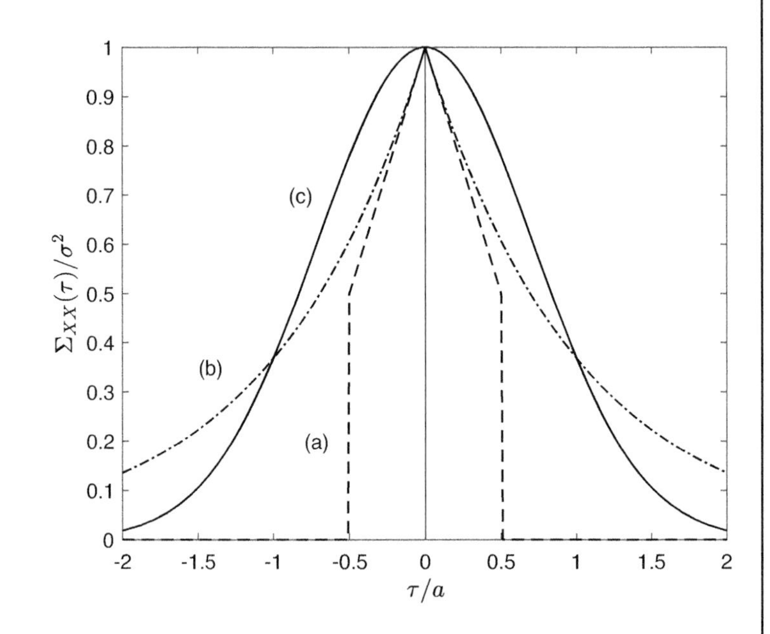

Figure 11.5 Candidate auto-covariance functions

Although candidate function (1) is symmetric with respect to τ, satisfies the property $|\Sigma(\tau)| \leq \Sigma(0)$, and has zero value as $|\tau| \rightarrow \infty$, it is not valid as an auto-covariance function because, although it is continuous at $\tau = 0$, it is discontinuous at $\tau = \pm a/2$. Hence, it violates the continuity property of an auto-covariance function. One can also show that this function is not positive semi-definite.

Candidate function (2) is symmetric in τ, satisfies the boundedness condition $|\Sigma_{XX}(\tau)| \leq \Sigma_{XX}(0)$, is continuous for all τ, and approaches zero as $|\tau| \rightarrow \infty$. Furthermore, one can show that

$$\int_{-\infty}^{+\infty} \sigma^2 \exp\left(-\frac{|\tau|}{a}\right) \exp(-i\omega\tau) d\tau = \frac{2a\sigma^2}{1 + a^2\omega^2}, \tag{E1}$$

which is positive for any value of ω, thus satisfying the positive semi-definiteness property. Hence, function (2) is a valid candidate as the auto-covariance function of a stationary process. A process having this auto-covariance function is mean-square continuous at all points because $\Sigma(\tau)$ is continuous at $\tau = 0$. However, the process is not mean-square differentiable because $\Sigma(\tau)$ is not differentiable at $\tau = 0$.

Example 11.1 (cont.)

Candidate function (3) is symmetric in τ, satisfies the boundedness condition $|\Sigma_{XX}(\tau)| \leq \Sigma_{XX}(0)$, is continuous for all τ, and approaches zero as $|\tau| \to \infty$. Furthermore, one can show that

$$\int_{-\infty}^{+\infty} \sigma^2 \exp\left(-\frac{\tau^2}{a^2}\right) \exp(-\mathrm{i}\omega\tau)\mathrm{d}\tau = \sqrt{\pi} a \sigma^2 \exp\left(-\frac{a^2\omega^2}{4}\right), \tag{E2}$$

which is positive for any value of ω, thus satisfying the positive semi-definiteness property. Hence, function (3) is a valid candidate as the auto-covariance function of a stationary process. A stationary process having this auto-covariance function is mean-square continuous at all points because $\Sigma(\tau)$ is continuous at $\tau = 0$. Furthermore, it is differentiable because $\Sigma(\tau)$ is twice differentiable at $\tau = 0$. The mean of the derivative process is 0 and its cross- and auto-covariance functions are

$$\Sigma_{X\dot{X}}(\tau) = -\Sigma_{\dot{X}X}(\tau) = \frac{\mathrm{d}}{\mathrm{d}\tau}\left[\sigma^2 \exp\left(-\frac{\tau^2}{a^2}\right)\right] = -2\left(\frac{\sigma^2}{a}\right)\frac{\tau}{a} \exp\left(-\frac{\tau^2}{a^2}\right), \tag{E3}$$

$$\Sigma_{\dot{X}\dot{X}}(\tau) = -\frac{\mathrm{d}^2}{\mathrm{d}\tau^2}\left[\sigma^2 \exp\left(-\frac{\tau^2}{a^2}\right)\right] = 2\left(\frac{\sigma^2}{a^2}\right)\left(1 - \frac{\tau^2}{a^2}\right) \exp\left(-\frac{\tau^2}{a^2}\right). \tag{E4}$$

Figure 11.6 shows plots of the above cross- and auto-covariance functions. Observe that the cross-covariance is an odd function of τ with a value of 0 at $\tau = 0$ as expected, and the auto-covariance function is an even function of τ approaching 0 as $\tau \to \infty$.

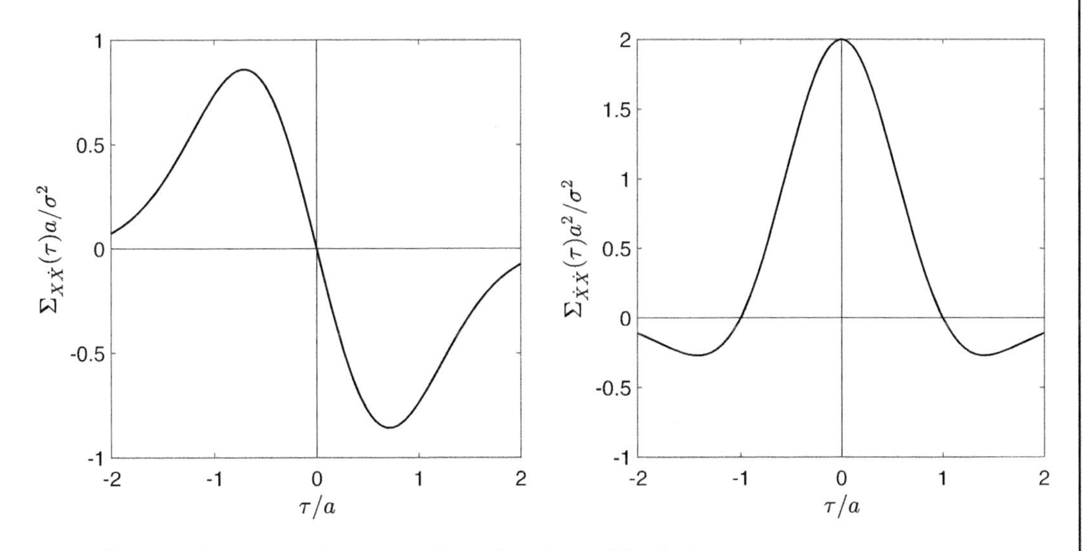

Figure 11.6 Cross- and auto-covariance functions of the derivative process

11.3 Power-Spectral Density of a Stationary Process

With interest in the frequency content of a stationary process $X(t)$, we examine the Fourier transform of the process, defined as

$$\bar{X}(\omega) = \lim_{T\to\infty} \frac{1}{2\pi} \int_{-T}^{T} X(t) \exp(-\mathrm{i}\omega t) \mathrm{d}t, \tag{11.15}$$

where, for convenience, we have used the factor $1/2\pi$ instead of the usual $1/\sqrt{2\pi}$ in the deterministic Fourier transform. As described in the preceding section, the above integral is finite in the mean-square sense if the integral

$$I(\omega) = \lim_{T\to\infty} \frac{1}{(2\pi)^2} \int_{-T}^{T} \int_{-T}^{T} \Sigma_{XX}(t_2 - t_1) \exp[-\mathrm{i}\omega(t_2 - t_1)] \mathrm{d}t_1 \mathrm{d}t_2 \tag{11.16}$$

is finite. Letting $\tau = t_2 - t_1$, after some derivations we arrive at

$$I(\omega) = \lim_{T\to\infty} \frac{1}{(2\pi)^2} \int_{-2T}^{2T} (2T - |\tau|) \Sigma_{XX}(\tau) \exp(-\mathrm{i}\omega\tau) \mathrm{d}\tau. \tag{11.17}$$

Because of the term $2T$ inside the integrand, this integral does not have a finite value. Hence, the Fourier transform of a stationary process does not exist. This is not surprising, because a stationary process evolves with an unchanging probabilistic character and constant power from $-\infty$ to $+\infty$ so that its total power is infinite. With this finding in mind and noting that the integral in (11.16) represents the expectation $\mathrm{E}\left[|\bar{X}(\omega)|^2\right]$, we consider the normalized integral

$$\begin{aligned}
\lim_{T\to\infty} \frac{\pi}{T} \mathrm{E}\left[|\bar{X}(\omega, T)|^2\right] &= \lim_{T\to\infty} \frac{1}{2\pi} \int_{-2T}^{2T} \left(1 - \frac{|\tau|}{2T}\right) \Sigma_{XX}(\tau) \exp(-\mathrm{i}\omega\tau) \mathrm{d}\tau \\
&= \frac{1}{2\pi} \int_{-\infty}^{\infty} \Sigma_{XX}(\tau) \exp(-\mathrm{i}\omega\tau) \mathrm{d}\tau \\
&= \Phi_{XX}(\omega).
\end{aligned} \tag{11.18}$$

We define this as the *power-spectral density* (PSD) function of the stationary process $X(t)$. As $\Sigma_{XX}(\tau)$ is an even function of τ, using $\exp(-\mathrm{i}\omega\tau) = \cos(\omega\tau) + \mathrm{i}\sin(\omega\tau)$, one can write

$$\begin{aligned}
\Phi_{XX}(\omega) &= \frac{1}{2\pi} \int_{-\infty}^{\infty} \Sigma_{XX}(\tau) \exp(-\mathrm{i}\omega\tau) \mathrm{d}\tau \\
&= \frac{1}{\pi} \int_{-\infty}^{\infty} \Sigma_{XX}(\tau) \cos(\omega\tau) \mathrm{d}\tau,
\end{aligned} \tag{11.19}$$

which shows that $\Phi(\omega)$ is real-valued, even function of ω. Furthermore, owing to the positive semi-definite property of the auto-covariance function, $\Phi_{\mathrm{XX}}(\omega)$ is non-negative. Finally, employing the inverse Fourier transform, we have

$$\Sigma_{XX}(\tau) = \int_{-\infty}^{\infty} \Phi_{XX}(\omega) \exp(\mathrm{i}\omega\tau)\mathrm{d}\tau$$
$$= 2\int_{0}^{\infty} \Phi_{XX}(\omega) \cos(\omega\tau)\mathrm{d}\tau. \tag{11.20}$$

The auto-covariance and PSD functions are alternatives for defining the second-moment properties of a stationary random process. Having one, the other can be derived from one of the above relations. The formulas in (11.19) and (11.20) are known as Wiener–Khintchine relations (Lin, 1967). It is noted that in many texts the PSD is defined as the Fourier transform of the auto-correlation function, which is related to the auto-covariance function through $\mathrm{E}[X(t)X(t+\tau)] = \Sigma_{\mathrm{XX}}(\tau) + \mu_X^2$. The difference between the two PSD functions is that a Dirac delta function (spike) of magnitude μ_X^2 appears at $\omega = 0$ when the auto-correlation function is used to define the PSD.

It is instructive to examine (11.20) for $\tau = 0$. Because $\Sigma_{XX}(0) = \sigma_X^2$,

$$\sigma_X^2 = \int_{-\infty}^{\infty} \Phi_{XX}(\omega)\mathrm{d}\tau. \tag{11.21}$$

It follows that the area underneath the PSD function equals the variance of the process. This implies that, as $|\omega| \to \infty$, the tails of the PSD should approach zero faster than $1/\omega$ for the process to have a finite variance. Furthermore, the above relation shows that the PSD is a decomposition of the variance (power) of the process in the frequency domain, thus the name *power-spectral density*.

Taking the derivative with respect to τ from both sides of (11.20) and noting (11.11), one has

$$\Sigma_{X\dot{X}}(\tau) = \int_{-\infty}^{\infty} \mathrm{i}\omega\Phi_{XX}(\omega) \exp(\mathrm{i}\omega\tau)\mathrm{d}\tau. \tag{11.22}$$

It follows that $\Phi_{X\dot{X}}(\omega) = -\Phi_{\dot{X}X}(\omega) = \mathrm{i}\omega\Phi_{XX}(\omega)$ is the cross-PSD of the process and its derivative process. Note that this is an imaginary and odd function of ω. Taking one more derivative with respect to τ from both sides of (11.22) and using (11.12), one obtains

$$\Sigma_{\dot{X}\dot{X}}(\tau) = \int_{-\infty}^{\infty} \omega^2\Phi_{XX}(\omega) \exp(\mathrm{i}\omega\tau)\mathrm{d}\tau. \tag{11.23}$$

Hence, the auto-PSD of the derivative process is $\Phi_{\dot{X}\dot{X}}(\omega) = \omega^2\Phi_{XX}(\omega)$. The derivative process exists, i.e., $X(t)$ is differentiable, if the area underneath $\omega^2\Phi(\omega)$ is finite.

Example 11.2 – PSD of a Stationary Process

In Example 11.1, we saw that two functions (2) and (3) below are valid as the auto-covariance function of a stationary process:

2) $\Sigma(\tau) = \sigma^2 \exp(-|\tau|/a), 0 < a,$
3) $\Sigma(\tau) = \sigma^2 \exp\left(-\tau^2/a^2\right), 0 < a.$

Example 11.2 (cont.)

Here we derive the corresponding PSDs and examine their properties.

For function (2), from the result in (E1) of Example 11.1, it is clear that

$$\Phi_{XX}(\omega) = \frac{1}{2\pi}\frac{2a\sigma^2}{1 + a^2\omega^2} = \frac{a\sigma^2}{\pi(1 + a^2\omega^2)}. \tag{E1}$$

A plot of this function is shown in Figure 11.7. Observe that the function is symmetric with respect to ω. Furthermore, as $|\omega| \rightarrow \infty$, $\Phi_{XX}(\omega)$ approaches 0 at the rate of $1/\omega^2$. It follows that the variance of this process is finite. In fact, the area underneath the PSD is σ^2, which equals the variance of the process. If this process were differentiable, the PSD of the process would be the function in (E1) multiplied by ω^2. As $|\omega| \rightarrow \infty$, that function approaches a non-zero value and, hence, has infinite area underneath it. It follows that the derivative process does not exist.

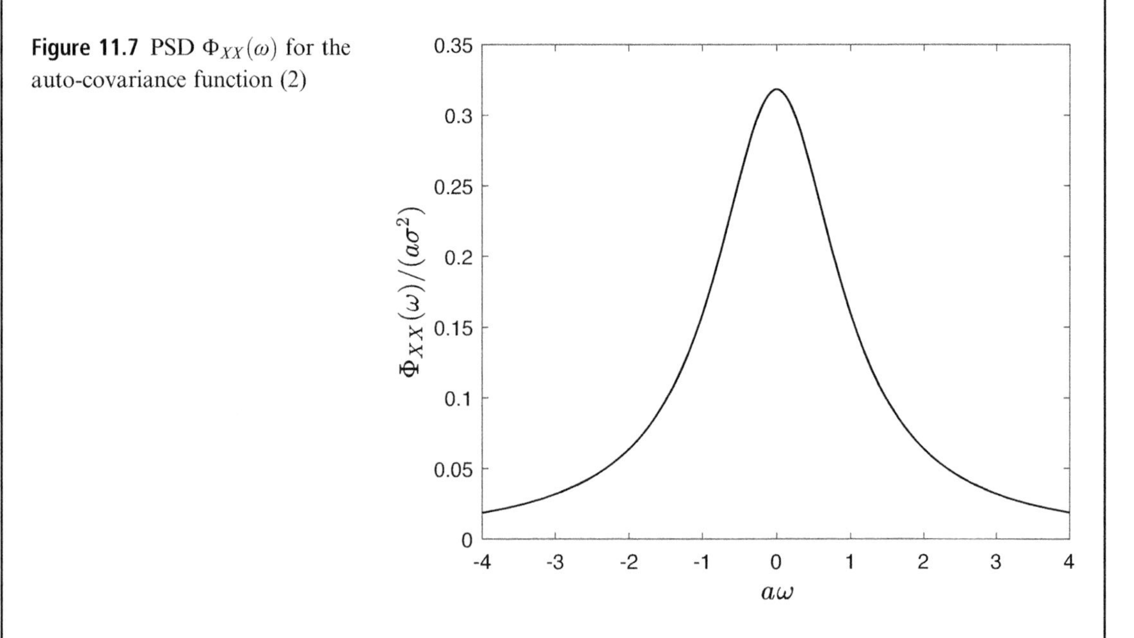

Figure 11.7 PSD $\Phi_{XX}(\omega)$ for the auto-covariance function (2)

For function (3), from the result in (E2) of Example 11.1, it is clear that

$$\Phi_{XX}(\omega) = \frac{a\sigma^2}{2\sqrt{\pi}} \exp\left(-\frac{a^2\omega^2}{4}\right). \tag{E2}$$

Observe again that this is an even function of ω and that, as $|\omega| \rightarrow \infty$, the tails of the function approach 0 exponentially so that the area underneath it is finite and equals σ^2. Furthermore, if the function is multiplied by ω^2, the exponential decay rate still governs, so the area

Example 11.2 (cont.)

underneath the multiplied function remains finite. Hence, the process is differentiable and its cross- and auto-PSDs are given by

$$\Phi_{X\dot{X}}(\omega) = -\Phi_{\dot{X}X}(\omega) = \frac{\mathrm{i}a\omega\sigma^2}{2\sqrt{\pi}} \exp\left(-\frac{a^2\omega^2}{4}\right) \tag{E3}$$

$$\Phi_{\dot{X}\dot{X}}(\omega) = \frac{a\omega^2\sigma^2}{2\sqrt{\pi}} \exp\left(-\frac{a^2\omega^2}{4}\right). \tag{E4}$$

Figure 11.8 shows plots of the PSD functions (E2)–(E4).

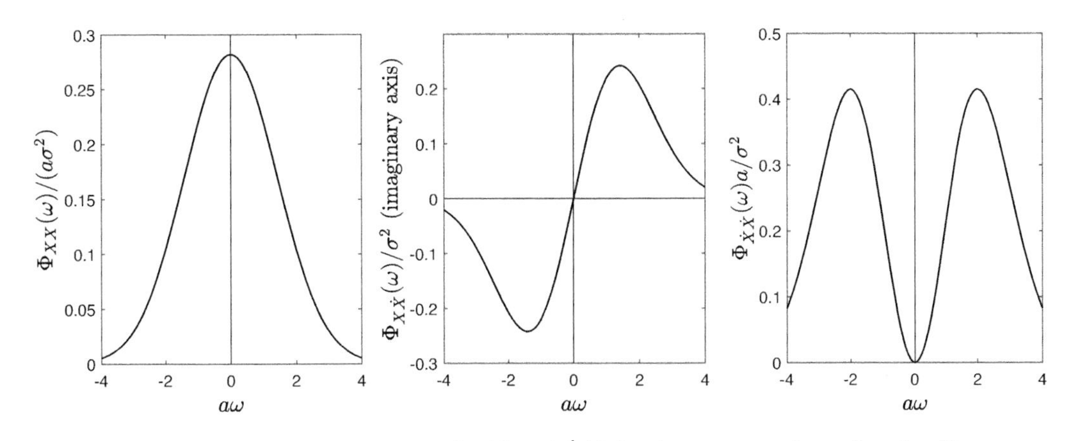

Figure 11.8 Auto- and cross-PSDs of $X(t)$ and $\dot{X}(t)$ for the auto-covariance function (3)

11.4 The Gaussian Process

A random process $X(t)$ is said to be Gaussian if for any selected parameter set $t_1, \ldots, t_n$ and any n, the random variables $X(t_1), \ldots, X(t_n)$ have the multinormal distribution (see Section 3.2). Because the multinormal distribution is defined in terms of the mean vector and the covariance matrix, the Gaussian process is completely defined by specifying the mean and auto-covariance functions, $\mu_X(t)$ and $\Sigma_{XX}(t, t')$, respectively. Furthermore, a set of random processes $X_1(t), \ldots, X_m(t)$ are jointly Gaussian if for any selected parameter set $t_1, \ldots, t_n$ and any n, the random variables $X_j(t_i)$, $i = 1, \ldots, n$, $j = 1, \ldots, m$, have the multinormal distribution. Specification of the mean, auto-, and cross-covariance functions, $\mu_{X_j}(t)$, $\Sigma_{X_jX_j}(t, t')$, and $\Sigma_{X_jX_k}(t, t')$, respectively, completely defines the set of jointly Gaussian processes.

Let $\mathbf{X}(t) = [X_1(t) \cdots X_m(t)]^{\mathrm{T}}$ be a vector of jointly Gaussian processes. Then, owing to the fact that linear functions of multinormal random variables are also multinormal (see property 5 in Section 3.2), the linearly transformed processes

$$\mathbf{Y}(t) = \mathbf{A}(t)\mathbf{X}(t) + \mathbf{b}(t), \tag{11.24}$$

where $\mathbf{A}(t)$ is an $l \times m$ deterministic coefficient matrix and $\mathbf{b}(t)$ is an $l \times 1$ deterministic vector, is an l-vector of jointly Gaussian processes. The vector of mean functions and the matrix of auto- and cross-covariance functions of $\mathbf{Y}(t)$ are given by (see 2.91 and 2.92)

$$\mathbf{M_Y}(t) = \mathbf{A}(t)\mathbf{M_X}(t) + \mathbf{b}(t), \tag{11.25}$$

$$\mathbf{\Sigma_{YY}}(t, t') = \mathbf{A}(t)\mathbf{\Sigma_{XX}}(t, t')\mathbf{A}(t')^{\mathrm{T}}, \tag{11.26}$$

where $\mathbf{M_X}(t) = \left[\mu_{X_1}(t) \cdots \mu_{X_m}(t)\right]^{\mathrm{T}}$ is the vector of mean functions of $\mathbf{X}(t)$ and $\mathbf{\Sigma_{XX}}(t, t')$ is an $m \times m$ matrix collecting the auto-covariance functions $\Sigma_{X_jX_j}(t, t')$, $j = 1, \ldots, m$, along its diagonal and the cross-covariance functions $\Sigma_{X_jX_k}(t, t')$, $j, k = 1, \ldots, m$, $j \neq k$, off its diagonal.

Observe that the derivative $\dot{X}(t) = \lim_{\Delta t \to 0}[X(t + \Delta t) - X(t)]/\Delta t$ of a process is a linear function of the process values at times t and $t + \Delta t$. It follows that the derivative of a Gaussian process, if it exists, is also Gaussian. Furthermore, an integral of the form $Y(u) = \int_a^b X(t)h(t, u)\mathrm{d}t$, where $h(t, u)$ is a deterministic function, is also linear in $X(t)$. Therefore, such an integral of a Gaussian process is also a Gaussian process. These properties enormously simplify stochastic analysis of the response of linear structures subjected to Gaussian excitations. Because the stochastic responses of such a system are all Gaussian, one only needs to determine the mean and auto-/cross-covariance functions to completely characterize the response processes.

11.5 Solution Approaches for Reliability Analysis

This section presents several general solution approaches for time- and space-variant structural reliability problems. Derivations are presented for the structural-component-reliability problem; applications to systems are demonstrated through examples. Additional solution methods are presented in later sections of this chapter and in Chapters 12 and 13.

11.5.1 Upper-Bound Solution

As seen in Section 11.1, in time- and space-variant structural reliability problems, the event of failure occurs when the minimum of the limit-state function over the time interval of operation and spatial domain of occupation takes on a negative value. In the case of a one-dimensional process with t denoting either the time or space coordinate and $t \in [0, T]$ denoting the interval of interest, an upper bound to the failure probability can be derived as follows. We first write the identity

$$\Pr\left\{\min_{0\le t\le T} g[\mathbf{X}(t), t] \le 0\right\} = \Pr\{g[\mathbf{X}(0), 0] \le 0 \cup 0 < N(T)\}, \tag{11.27}$$

where $N(T)$ is the number of down-crossings of the limit-state function below the level zero during the interval $(0, T]$. The right-hand side of the equation essentially says that the component fails if it starts from a failed state at $t = 0$ or if the limit-state function down-crosses the zero level one or more times during the considered interval. Disregarding possible dependence between the two events, we can write

$$\Pr\left\{\min_{0\le t\le T} g[\mathbf{X}(t), t] \le 0\right\} \le \Pr\{g[\mathbf{X}(0), 0] \le 0\} + \Pr[0 < N(T)]. \tag{11.28}$$

Note that the first probability term on the right-hand side is a time-invariant structural-reliability problem and can be solved by methods developed in the earlier chapters with $\mathbf{X}(0)$ as the set of basic random variables. For the second term, we write

$$\begin{aligned}\Pr[0 < N(T)] &= \sum\nolimits_{n=1}^{\infty} \Pr[N(T) = n] \\ &\le \sum\nolimits_{n=0}^{\infty} n\Pr[N(T) = n] \\ &= \mathrm{E}[N(T)] \\ &= \int_0^T v(t)\mathrm{d}t.\end{aligned} \tag{11.29}$$

In the second line, we have multiplied each probability term by n and changed the lower bound of the summation to $n = 0$. This does not affect the terms for $n = 0$ and $n = 1$. However, it amplifies the probability terms for $n = 2$ and higher n values – hence the replacement of the equality sign with an upper-bound sign. As structures are expected to be highly reliable, the probability that $N(T)$ will be equal to or greater than 2 is expected to be much smaller than the probability of $N(T)$ being equal to 1. Therefore, the upper bound is expected to be good for highly reliable systems. The third line merely states that the summation in the second line equals the mean of $N(T)$. In the last line, we have introduced $v(t)$ as the mean rate of zero-level down-crossings of the limit-state function. The integral is over the duration of operation of the structure and equals the mean of $N(T)$. Thus, one can write (11.28) as

$$\Pr\left\{\min_{0\le t\le T} g[\mathbf{X}(t), t] \le 0\right\} \le \Pr\{g[\mathbf{X}(0), 0] \le 0\} + \int_0^T v(t)\mathrm{d}t. \tag{11.30}$$

It is clear that the mean rate of down-crossings of the limit-state function is a key ingredient in time- and space-variant reliability analysis. We now focus on developing a method to compute this term.

Provided $N(t)$ is a differentiable process, we can write

$$v(t) = \lim_{\Delta t\to 0} \frac{\mathrm{E}[N(t + \Delta t) - N(t)]}{\Delta t}. \tag{11.31}$$

For a sufficiently small Δt, the probability of two or more zero-level down-crossings of the limit-state function is of higher order in relation to the probability of one down-crossing. Hence, one can write,

$$\begin{aligned}
\mathrm{E}[N(t+\Delta t)-N(t)] &= 0\times \Pr\{0 \text{ down crossings in } (t,t+\Delta t]\} \\
&\quad + 1\times \Pr\{1 \text{ down crossing in } (t,t+\Delta t]\} \\
&\quad + 2\times \Pr\{2 \text{ down crossings in } (t,t+\Delta t]\} \\
&\quad + \cdots \\
&= \Pr\{1 \text{ down crossing in } (t,t+\Delta t]\} + \mathrm{O}(\Delta t).
\end{aligned} \tag{11.32}$$

where $\mathrm{O}(\Delta t)$ denotes terms of order higher than Δt. On the other hand, the probability of a zero-level down-crossing in $(t, t+\Delta t]$ can be written as

$$\Pr\{1 \text{ down crossing in } (t,t+\Delta t]\} = \Pr\{-g[\mathbf{X}(t),t]\leq 0 \cap g[\mathbf{X}(t+\Delta t),t+\Delta t]\leq 0\}. \tag{11.33}$$

In the probability term on the right-hand side, the first event states the condition that the limit-state function has a positive value at t, and the second event states the condition that the limit-state function has a negative value at $t+\Delta t$. Surely, one or more zero-level down-crossings must have occurred during the interval $(t, t+\Delta t]$. Because for small Δt the likelihood of two or more down-crossings is of order higher than that of one down-crossing, as $\Delta t \to 0$, the expression provides the probability of one down-crossing in $(t, t+\Delta t]$.

For a small but finite Δt, (11.33) represents a parallel-system reliability problem with the limit-state functions $-g[\mathbf{X}(t),t]$ and $g[\mathbf{X}(t+\Delta t),t+\Delta t]$ and $[\mathbf{X}(t),\mathbf{X}(t+\Delta t)]$ as the set of basic random variables. In general, the set of random variables $\mathbf{X}(t)$ and $\mathbf{X}(t+\Delta t)$ are strongly correlated for small Δt and this may cause numerical instability in computing the parallel-system reliability. For this reason, and provided $\mathbf{X}(t)$ is differentiable, it is advantageous to replace the latter vector with its approximation for small Δt, $\mathbf{X}(t+\Delta t) \cong \mathbf{X}(t)+\dot{\mathbf{X}}(t)\Delta t$. The second limit-state function now takes the form $g\left[\mathbf{X}(t)+\dot{\mathbf{X}}(t)\Delta t, t+\Delta t\right]$ and the set of basic random variables is $\left[\mathbf{X}(t),\dot{\mathbf{X}}(t)\right]$. Note that the processes $\mathbf{X}(t)$ and $\dot{\mathbf{X}}(t)$ are usually weakly correlated (in the case of stationary processes, they are uncorrelated) and this formulation does not pose numerical problems. Thus, using (11.32) and (11.33) in (11.31), the latter is written as

$$v(t) = \lim_{\Delta t\to 0} \frac{\Pr\left\{-g[\mathbf{X}(t),t]\leq 0\cap g\left[\mathbf{X}(t)+\dot{\mathbf{X}}(t)\Delta t, t+\Delta t\right]\leq 0\right\}}{\Delta t}, \tag{11.34}$$

where calculations are performed for a small Δt.

Figure 11.9 portrays the parallel-system event in the standard normal space, where $G(\cdot)$, $\mathbf{U}$, and $\dot{\mathbf{U}}$ represent transformations of $g[\cdot]$, $\mathbf{X}(t)$, and $\dot{\mathbf{X}}(t)$, respectively. It is seen that the event of interest is defined by a wedge-like domain. Also shown in the figure are linearizations of the limit-state surfaces at their respective design points. The mean crossing rate as a parallel-system reliability problem was first formulated by Hagen and Tvedt (1991).

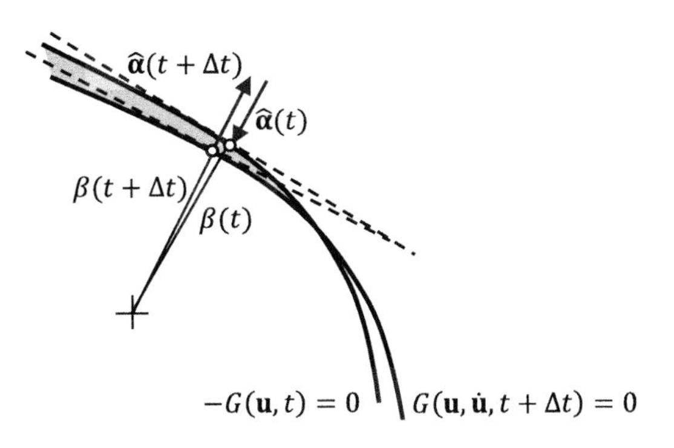

Figure 11.9 Parallel-system event defining the mean outcrossing rate

As seen in Chapter 8, the FORM approximation of the probability of a parallel system with two components is given by the bi-normal probability $\Phi_2[-\beta(t), -\beta(t+\Delta t), \rho(t,\Delta t)]$ where $\beta(t)$ and $\beta(t+\Delta t)$ are the component reliability indices and $\rho(t,\Delta t) = \widehat{\boldsymbol{\alpha}}(t)\cdot\widehat{\boldsymbol{\alpha}}(t+\Delta t)$ is the correlation coefficient, in which $\widehat{\boldsymbol{\alpha}}(t)$ and $\widehat{\boldsymbol{\alpha}}(t+\Delta t)$ are the unit normal vectors at the respective design points, as shown in Figure 11.9. Note that here $\beta(t)$ has a negative value because $\widehat{\boldsymbol{\alpha}}(t)$ is directed toward the origin while $\beta(t+\Delta t)$ has a positive value. Thus, using (3.10), we have, in first-order approximation,

$$\begin{aligned}&\Pr\{-g[\mathbf{X}(t),t]\leq 0\cap g[\mathbf{X}(t)+\dot{\mathbf{X}}(t)\Delta t, t+\Delta t]\leq 0\}\\ &\cong \Phi[-\beta(t)]\Phi[-\beta(t+\Delta t)]+\int_0^{\rho}\phi_2[-\beta(t), -\beta(t+\Delta t), r]dr.\end{aligned} \tag{11.35}$$

For small Δt, the two design points tend to be close to each other so that $\beta(t)\cong-\beta(t+\Delta t)$. Furthermore, ρ is near -1 because the two unit vectors are nearly coinciding but have opposite directions. Using these properties, Koo and Der Kiureghian (2003) derived a closed-form approximation for the right-hand side of (11.35). In Appendix 11A, a slightly improved version of their expression is obtained as

$$\nu(t)\cong\frac{1}{\Delta t}\exp\left[\frac{\beta(t)\beta(t+\Delta t)}{2}\right]\left\{\frac{1}{4}+\frac{\sin^{-1}[\rho(t,\Delta t)]}{2\pi}\right\}, \tag{11.36}$$

where the β values and $\rho(t,\Delta t)$ are computed for small Δt. The appropriate value for Δt depends on the scale of fluctuations of $\mathbf{X}(t)$, which in turn depends on the auto-covariance or PSD functions of the processes involved. Examples 11.3 and 11.4 below illustrate the selection of Δt.

An alternative solution of the mean crossing rate was derived by Sudret (2008) through a different formulation. Starting from the observation that the numerator in (11.34) is 0 for $\Delta t = 0$ (because the two intersecting events are then disjoint), the ratio in (11.34) is interpreted as the derivative of the parallel-system probability. Using the FORM approximation of (11.33) and differentiating with respect to t, Sudret (2008) arrived at the following expression for the mean down-crossing rate:

$$\nu(t) \cong \phi[\beta(t)] \frac{\|\widehat{\boldsymbol{\alpha}}(t+\Delta t) - \widehat{\boldsymbol{\alpha}}(t)\|}{\Delta t} \Psi\left[\frac{\beta(t+\Delta t) - \beta(t)}{\|\widehat{\boldsymbol{\alpha}}(t+\Delta t) - \widehat{\boldsymbol{\alpha}}(t)\|}\right], \tag{11.37}$$

where $\Psi(x) = \phi(x) - x\Phi(-x)$, in which $\phi[\cdot]$ and $\Phi(\cdot)$ are the standard normal PDF and CDF, respectively. We note that $\beta(t)$ and $\widehat{\boldsymbol{\alpha}}(t)$ in Sudret's formulation are the reliability index and unit normal vector for the point-in-time reliability problem $\{g[\mathbf{X}(t), t] \leq 0\}$ at time t (i.e., without the negative sign as in (11.33)). Similar to (11.36), the above formula requires computing the β and $\widehat{\boldsymbol{\alpha}}$ values at t and $t + \Delta t$ for a small Δt.

The above formulas take simplified forms when the process $g[\mathbf{X}(t), t]$ is stationary. This happens when $\mathbf{X}(t)$ is composed of jointly stationary processes and $g[\cdot]$ is not an explicit function of t. In that case $\beta(t) = \beta$ and $\rho(t, \Delta t) = \rho(\Delta t)$. Furthermore, the mean down-crossing rate is a constant. Consequently, (11.36) simplifies to

$$\nu \cong \frac{1}{\Delta t} \exp\left(-\frac{\beta^2}{2}\right)\left[\frac{1}{4} + \frac{\sin^{-1}[\rho(\Delta t)]}{2\pi}\right], \tag{11.38}$$

and, noting that $\Psi(0) = 1/\sqrt{2\pi}$, (11.37) simplifies to

$$\nu \cong \frac{1}{2\pi} \exp\left(-\frac{\beta^2}{2}\right) \frac{\|\widehat{\boldsymbol{\alpha}}(t+\Delta t) - \widehat{\boldsymbol{\alpha}}(t)\|}{\Delta t}. \tag{11.39}$$

Note again that the $\widehat{\boldsymbol{\alpha}}$ vectors in the latter formula are the unit-normal vectors for the point-in-time reliability problems at times t and $t + \Delta t$. As the mean down-crossing rate is a constant, the upper bound in (11.30) takes the form

$$\Pr\left\{\min_{0 \leq t \leq T} g[\mathbf{X}(t)] \leq 0\right\} \leq \Pr\{g[\mathbf{X}(0)] \leq 0\} + \nu T, \tag{11.40}$$

where T is the duration of operation of the structure.

The above approaches employ the FORM or SORM approximation at the design point to obtain an approximation of the mean down-crossing rate. However, as shown by Pearce and Wen (1985) and Ditlevsen (1987), for a time-variant reliability problem, this might not be the best choice of linearization point. They showed that a better approximation is obtained if the surface is linearized at the stationary point of the mean crossing rate out of the tangent plane. The differences, however, are relatively small for most problems.

Example 11.3 – Mean Crossing Rate of a Rod with Stochastic Cross-Sectional Area

Consider a rod of length l having a randomly varying cross-sectional area $A(s)$ under a deterministic axial load q, as shown in Figure 11.10. We assume $A(s)$ is a homogeneous lognormal random process with mean μ_A, coefficient of variation δ_A, and auto-correlation coefficient function $\rho_{AA}(s_2 - s_1) = \exp\left[-(|s_1 - s_2|/a)^2\right]$, where a is the correlation length, a measure of the rate of fluctuations of the random process. The stress at location $0 \leq s \leq l$ is given by $X(s) = q/A(s)$. We are interested in the mean rate of up-crossings of the stress above

Example 11.3 (cont.)

the level σ_0, which is the same as the mean rate of zero-level down-crossings of the limit-state function $g[A(s)] = \sigma_0 - q/A(s)$.

The reciprocal of a lognormal process is also lognormal, so one can easily show that $X(s)$ is a lognormal random process with mean $\mu_X = q(1+\delta_A^2)/\mu_A$ and coefficient of variation (c.o.v.) and auto-correlation function identical to those of $A(s)$. Furthermore, $Z(s) = \ln X(s)$ is a homogeneous Gaussian random process with mean $\mu_Z = \ln \mu_X - 0.5\sigma_Z^2$, variance $\sigma_Z^2 = \ln(1+\delta_A^2)$, and auto-correlation coefficient function $\rho_{ZZ}(s_2 - s_1) = \ln\left[1 + \rho_{AA}(s_2 - s_1)\delta_A^2\right]/\ln(1+\delta_A^2)$. Because the up-crossings of $X(s)$ above the level σ_0 are identical to the up-crossings of $Z(s)$ above the level $\ln \sigma_0$, we can use the well-known formula by Rice (1944, 1945) for homogeneous Gaussian processes to obtain the desired up-crossing rate as

$$\nu = \frac{\sigma_{\dot{Z}}}{2\pi\sigma_Z} \exp\left[-\frac{(\ln \sigma_0 - \mu_Z)^2}{2\sigma_Z^2}\right], \tag{E1}$$

where $\sigma_{\dot{Z}}^2 = 2\delta_A^2/\left[a^2(1+\delta_A^2)\right]$ is the variance of the derivative process $\dot{Z} = \mathrm{d}Z(s)/\mathrm{d}s$. This exact solution gives us an opportunity to examine the accuracy of the approximate formulas in (11.38) and (11.39).

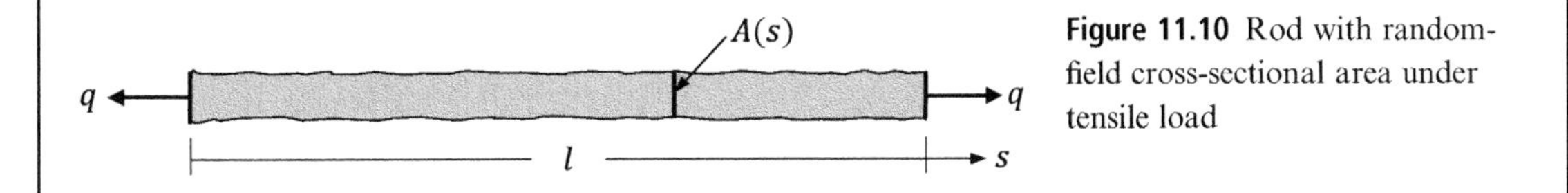

Figure 11.10 Rod with random-field cross-sectional area under tensile load

Table 11.1 lists the exact and approximate results for the dimensionless quantity νa for selected values of the scaled threshold $\sigma_0\mu_A/q$, for two different values $\Delta s/a = 1/4$ and $1/8$ of the distance $\Delta s = s_2 - s_1$ between the two selected points (indicated as Δt in (11.38) and (11.39)) and for $\delta_A = 0.20$. Also shown are the correlation coefficients computed using the normal vectors at the design points of the two limit-state surfaces for the selected $\Delta s/a$ values. Observe that the limit formulas (11.38) and (11.39) provide accurate estimates of the up-crossing rate as long as the selected Δs is small in relation to the scale of fluctuation of the random process. It appears that a value of Δs about 1/8 the correlation length of the random process provides good accuracy.

Having said the above, it is important to note that, for the above problem, the limit-state surfaces in the standard normal space are linear and the FORM provides an exact solution. Therefore, any errors in the results in Table 11.1 are due to the approximations involved in deriving the limit formulas in (11.38) and (11.39). Thus, any conclusions derived from this comparison are relevant to these limit formulas rather than the accuracy of the FORM approximation.

Example 11.3 (cont.)

Table 11.1 Comparison of results for scaled up-crossing rate ***va***

		Formula in (11.38)		Formula in (11.39)	
Threshold $\sigma_0\mu_A/q$	Exact solution	$\Delta s/a = 1/4$ $\rho = -0.941$	$\Delta s/a = 1/8$ $\rho = -0.985$	$\Delta s/a = 1/4$ $\rho = 0.941$	$\Delta s/a = 1/8$ $\rho = 0.985$
1.0	0.222	0.220	0.221	0.218	0.220
1.5	0.0334	0.0331	0.0333	0.0328	0.0331
2.0	6.86×10^{-4}	6.79×10^{-4}	6.84×10^{-4}	6.73×10^{-4}	6.79×10^{-4}

Example 11.4 – Reliability of a Column under Stochastic Loads – Upper-Bound Solution

Reconsider the column in Example 6.5, but assume the applied loads are stochastic in nature so that the point-in-time limit-state function is described as

$$g[M_1(t), M_2(t), P(t), Y] = 1 - \frac{M_1(t)}{s_1 Y} - \frac{M_2(t)}{s_2 Y} - \left[\frac{P(t)}{aY}\right]^2. \tag{E1}$$

Let the set of load processes $\mathbf{X}(t) = [M_1(t), M_2(t), P(t)]$ be jointly stationary and Gaussian with mean vector $\mathbf{M_X}$ and auto-/cross-covariance matrix $\mathbf{\Sigma_{XX}}(\tau)$ with elements $\Sigma_{X_iX_j}(\tau) = \sigma_{ij}\exp\left(-\tau^2/a_{ij}^2\right)$, $i,j = 1,2,3$, where σ_{ij} denotes the point-in-time covariance between $X_i(t)$ and $X_j(t)$, and a_{ij} are positive constants having units of time. At a given time, i.e., for $\tau = 0$, the standard deviations are $\sigma_{X_i} = \sqrt{\sigma_{ii}}$, $i = 1,2,3$, and the correlation coefficients are $\rho_{X_iX_j} = \sigma_{ij}/\sqrt{\sigma_{ii}\sigma_{jj}}$. We assign values to σ_{ij} so that the standard deviations and correlation coefficients are identical to those listed in Table 6.1. We also assume the process mean values are as listed in Table 6.1. However, as we saw in Example 6.7, the dominant random variable in this problem was the yield strength Y. To highlight the stochastic nature of the loads, we reduce the coefficient of variation of Y from 10% to 2.5% while maintaining the same mean value as in Table 6.1.

For the assumed auto-covariance functions, all the processes are mean-square differentiable. The cross-covariance function between $X_i(t)$ and the derivative process $\dot{X}_j(t)$ is given by

$$\Sigma_{X_i\dot{X}_j}(\tau) = \frac{\mathrm{d}}{\mathrm{d}\tau}\left[\sigma_{ij}\exp\left(-\frac{\tau^2}{a_{ij}^2}\right)\right] = -2\sigma_{ij}\frac{\tau}{a_{ij}^2}\exp\left(-\frac{\tau^2}{a_{ij}^2}\right), \tag{E2}$$

Example 11.4 (cont.)

which is 0 for $\tau = 0$. Hence, at any given time t, each load process is uncorrelated with its own derivative process, as expected, as well as the derivatives of the other load processes. The cross-covariance function between pairs of the derivative processes is

$$\begin{aligned}\Sigma_{\dot{X}_i\dot{X}_j}(\tau) &= -\frac{\mathrm{d}}{\mathrm{d}\tau}\left[-2\sigma_{ij}\frac{\tau}{a_{ij}^2}\exp\left(-\frac{\tau^2}{a_{ij}^2}\right)\right] \\ &= 2\sigma_{ij}\frac{1}{a_{ij}^2}\left(1-2\frac{\tau^2}{a_{ij}^2}\right)\exp\left(-\frac{\tau^2}{a_{ij}^2}\right).\end{aligned} \tag{E3}$$

For $\tau = 0$, the above gives $\Sigma_{\dot{X}_i\dot{X}_j}(0) = 2\sigma_{ij}/a_{ij}^2$. Thus, the standard deviations of the derivative processes are $\sigma_{\dot{X}_i} = \sqrt{2}\sigma_{X_i}/a_{ii}$ and the correlation coefficients are $\rho_{\dot{X}_i\dot{X}_j} = \rho_{X_iX_j}\left(a_{ii}a_{jj}/a_{ij}^2\right)$. The assigned values for a_{ij} and the corresponding point-in-time standard deviations and correlation coefficients of the derivative processes are listed in Table 11.2. Note that the derivative processes all have zero means because of the stationarity assumption. These assumed numerical values are representative of random fluctuations that may occur in load processes during the strong-motion phase of an earthquake. In particular, we note that $\sigma_{\dot{X}_i}/2\pi\sigma_{X_i} = 1/\sqrt{2}\pi a_{ii}$ represents the mean frequency of the ith process in Hz and its reciprocal represents the predominant period of the process. Of the three processes, the shortest period is that of the axial load $P(t)$, which is $\sqrt{2}\pi a_{33} = 0.471$ s. We assume a $T = 10$ s duration of loading.

With the means, variances, and covariances of the normally distributed basic random variables $\mathbf{X}(t)$ and $\dot{\mathbf{X}}(t)$ determined, we are now ready to perform reliability analysis. But first we need to select an appropriate value for Δt. One measure of the scale of fluctuations of a random process is the time between two instances of the process for the auto-correlation coefficient to reduce to $\exp(-1)$, which is commonly known as the correlation length. In our case, the auto-correlation coefficient functions are given by $\rho_{X_iX_i} = \exp\left(-\tau^2/a_{ii}^2\right)$. Thus, the correlation length is $\tau_i = a_{ii}$. The process $P(t)$ has the shortest correlation length, i.e., $\tau_3 = 1/3\pi = 0.106$ s. It is necessary that we select a Δt that is much smaller than this time

Table 11.2 Statistics of derivative processes

	a_{ij}, s				$\rho_{\dot{X}_i\dot{X}_j}$		
Process	$\dot{M}_1(t)$	$\dot{M}_2(t)$	$\dot{P}(t)$	$\sigma_{\dot{X}_i}$	$\dot{M}_1(t)$	$\dot{M}_2(t)$	$\dot{P}(t)$
$\dot{M}_1(t)$, kNm/s	$1/2\pi$			$150\sqrt{2}\pi$	1		
$\dot{M}_2(t)$, kNm/s	$2/3\pi$	$1/2\pi$		$75\sqrt{2}\pi$	0.281	1	
$\dot{P}(t)$, kN/s	$1/2\pi$	$1/2\pi$	$1/3\pi$	$1{,}500\sqrt{2}\pi$	0.200	0.200	1

Example 11.4 (cont.)

interval. On this basis, we select $\Delta t = 0.01$ s. Note that this value is also much shorter than the predominant period of the process described above.

Now we perform FORM analysis for the two limit-state functions $-g[M_1(t), M_2(t), P(t), Y]$ and $g[M_1(t) + \dot{M}_1(t)\Delta t, M_2(t) + \dot{M}_2(t)\Delta t, P(t) + \dot{P}(t)\Delta t, Y]$ with the set of basic random variables $[M_1(t), M_2(t), P(t), Y, \dot{M}_1(t), \dot{M}_2(t), \dot{P}(t)]$. The resulting β and $\widehat{\boldsymbol{\alpha}}$ values are as follows:

$$\beta\widehat{\boldsymbol{\alpha}}(t) = -3.43\,[-0.759 - 0.438 - 0.408\ 0.257\ 0.000\ 0.000\ 0.000] \tag{E4}$$

$$\beta\widehat{\boldsymbol{\alpha}}(t + \Delta t) = 3.43\,[0.755\ 0.436\ 0.407 - 0.255\ 0.059\ 0.044\ 0.056]. \tag{E5}$$

The correlation coefficient between the linearized limit-states is $\rho(\Delta t) = \widehat{\boldsymbol{\alpha}}(t) \cdot \widehat{\boldsymbol{\alpha}}(t + \Delta t) = -0.9957$. Using this in (11.38), we obtain $= 0.00415\ \text{s}^{-1}$. Thus, in first-order approximation, the upper bound of the failure probability according to (11.40) is

$$\begin{aligned}\Pr\left\{\min_{0\le t\le T} g[\mathbf{X}(t)] \le 0\right\} &\le \Phi(-3.43) + 0.00415 \times 10 \\ &= 0.000304 + 0.0415 \\ &= 0.0418.\end{aligned} \tag{E6}$$

It is evident that the major contribution to the failure probability comes from the out-crossings of the safe domain, not the point-in-time failure probability at time zero.

11.5.2 Lower-Bound Solution

Let $t_i, i = 1, \ldots, n$, be a set of discrete points within the interval $[0, T]$. A lower bound to the probability of failure is given by

$$\Pr\left\{\min_{0\le t\le T} g[\mathbf{X}(t), t] \le 0\right\} \ge \Pr\left\{\bigcup_{i=1}^{n} \{g[\mathbf{X}(t_i), t_i] \le 0\}\right\}. \tag{11.41}$$

Naturally, the more points we select within the interval, the tighter the lower-bound estimate will be. The right-hand side of the above equation is a series-system reliability problem with identical component limit-state functions and the set of random variables $\{\mathbf{X}(t_1), \cdots, \mathbf{X}(t_n)\}$. A FORM approximation of the lower bound is obtained by linearizing each limit-state surface at its design point. Thus, based on (8.65), in first-order approximation, the right-hand side of (11.41) is

$$\Pr\left\{\bigcup_{i=1}^{n} \{g[\mathbf{X}(t_i), t_i] \le 0\}\right\} \cong 1 - \Phi_n(\mathbf{B}, \mathbf{R}), \tag{11.42}$$

where $\Phi_n(\cdot,\cdot)$ is the n-variate multinormal CDF, $\mathbf{B} = (\beta_1, \beta_2, \ldots, \beta_n)$ is the set of reliability indices for the point-in-time reliability problems at t_i, $i = 1, \ldots, n$, and $\mathbf{R}$ is the $n \times n$ correlation matrix with elements $\rho_{ij} = \widehat{\boldsymbol{\alpha}}_i \cdot \widehat{\boldsymbol{\alpha}}_j$, $i, j = 1, \ldots, n$. Thus, we need only determine the reliability indices and the unit normal vectors at time points t_i to compute the lower bound in first-order approximation. In the special case when $g[\mathbf{X}(t), t]$ is stationary, the reliability indices are identical and the correlation coefficients depend only on the time lags $t_j - t_i$, which means that the diagonals of $\mathbf{R}$ have identical values. In that case it is necessary to compute only one row of the correlation matrix.

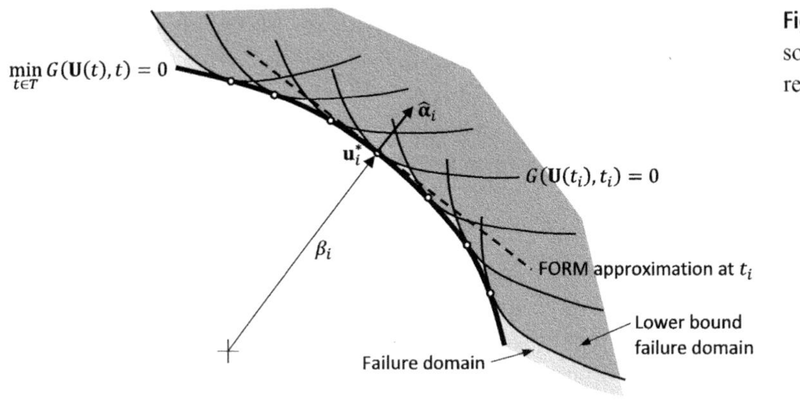

Figure 11.11 Lower-bound solution of time-variant reliability problem

Figure 11.11 portrays the series-system lower-bound solution of the time-variant reliability problem together with its FORM approximation. Details are given for one time instant. The exact failure domain is shown in light gray and the lower-bound failure domain is shown in darker gray. Observe that each point-in-time limit-state surface is active over a small segment near its design point. As a result, the first-order approximation normally provides good accuracy for this problem. However, depending on the rate of fluctuations of the process $g[\mathbf{X}(t), t]$, a large number of time points may need be used to obtain a good lower-bound solution. The following example demonstrates this shortcoming of this approach.

Example 11.5 – Reliability of a Column under Stochastic Loads – Lower-Bound Solution

We use the formulation in (11.41)–(11.42) to obtain a lower bound to the probability of failure of the column under stochastic loads. Suppose we select $n = 11$ equally spaced time points within the 10 s duration of the strong-motion phase of the earthquake, i.e., at $t_0 = 0$, $t_1 = 1$ s,…, $t_{10} = 10$ s with $\Delta t = 1$ s. The correlation coefficient between $M_1(t)$ and $M_1(t + \Delta t)$ is $\exp\left[-1^2/(1/2\pi)^2\right] = 7.16 \times 10^{-18}$, which is practically 0. Correlation coefficients between all other pairs of load processes at times t and $t + \Delta t$ are equally small.

Example 11.5 (cont.)

Hence, for all practical purposes, the process values at time instances separated by 1 s are statistically independent. Naturally, for time points further apart, the correlations are even smaller. However, correlation between the $\widehat{\boldsymbol{\alpha}}$ vectors at different time points arises because the random variable Y is common to all the time points. Performing FORM analysis, the result for the reliability index at all time points is $\beta = 3.43$, and the correlation coefficient between the linearized limit-state functions (arising from the common random variable Y) is $\widehat{\boldsymbol{\alpha}}(t_i) \cdot \widehat{\boldsymbol{\alpha}}(t_j) = 0.0660$, $i \neq j$. For these values, (11.42) yields the lower bound as 0.00333. As the correlation between the linearized limit-state functions is small, assuming independence among the component failure events, a quick approximation to this probability is obtained using (8.18) as $1 - [1 - \Phi(-3.43)]^{11} = 0.00334$, only slightly greater than that obtained from (11.42). We can see that these estimates of the lower bound are much smaller than the upper bound found in (E6) of the preceding example.

Suppose we increase the number of time points to $n = 21$ by using $\Delta t = 0.5$ s. The correlation coefficient between $M_1(t)$ and $M_1(t + \Delta t)$ now is $\exp\left[-0.5^2/(1/2\pi)^2\right] = 5.17 \times 10^{-5}$, which is still virtually 0. The same is true of the correlation coefficients between the other pairs of load processes at the two time points. FORM analysis yields the same values of β and the correlation coefficient between the linearized limit-state functions as for the previous case. The lower-bound estimate according to (11.42) for $n = 21$ now is 0.00634. If we assume statistical independence of the component failure events, (8.18) yields $1 - [1 - \Phi(-\beta)]^{21} = 0.00637$, which is only slightly greater than the result based on (11.42). These are still much smaller than the upper bound obtained in Example 11.3. If we use $\Delta t = 0.25$ s with $n = 41$ points, the correlation coefficient between $M_1(t)$ and $M_1(t + \Delta t)$ is $\exp\left[-0.25^2/(1/2\pi)^2\right] = 0.085$ and similarly small correlation values are obtained for the other pairs of load processes. The correlation coefficient between the linearized limit-state functions now is 0.159 and the result based on (11.42) is 0.0112, whereas assuming statistically independent component failure events we obtain 0.0124. It is clear that the lower-bound estimate improves as we consider more points within the 10 s duration of the load processes. The proper number of points to use depends on the scale of fluctuation of the load processes. For the present example, n may have to be as large as 100 or 200 to achieve convergence of the lower bound. The large number of points needed to achieve a good lower bound is a main drawback of this approach.

11.6 The Poisson Process

The Poisson process is a counting process. It counts the number of events randomly occurring in time or along a spatial coordinate. As we will see shortly, it can be used to obtain an approximate solution to time- or space-variant reliability problems.

Furthermore, it provides a building block for several stochastic load models that are commonly used in structural reliability analysis. In this section, mathematical expressions describing important properties of this process are derived. For this purpose, time, t, is used as the parameter. But all expressions are also valid if the parameter is a spatial coordinate, s. In a subsequent section, the Poisson process is used to construct various stochastic load models.

The Poisson process, denoted $N(t)$, is a continuous-parameter, discrete process that counts the number of events randomly occurring in the interval $(0, t]$. (The interval includes the instant t, but not the instant 0.) This process is formed based on three fundamental assumptions:

1) The probability of occurrence of an event in a small interval $(t, t + \Delta t]$ is proportional to the length of the interval, i.e.,

$$\lim_{\Delta t \to 0} \frac{\Pr[\text{occurrence in } (t, t + \Delta t)]}{\Delta t} = \nu(t), \tag{11.43}$$

   where $\nu(t)$ is called the intensity function.
2) The probability of two or more events in $(t, t + \Delta t]$ is of an order higher than Δt, i.e., it is negligible in comparison to the probability of one or zero events in the same interval as $\Delta t \to 0$.
3) The numbers of events in disjoint intervals are statistically independent. This assumption is based on the idea that the events are occurring entirely randomly.

A typical realization of the process $N(t)$ may appear as in Figure 11.12.

We now proceed to derive the first-order (marginal) distribution of the Poisson process based on the above assumptions. Let $p_N(n, t) = \Pr[N(t) = n]$ denote the probability mass function (PMF) of the process at time t, i.e., the probability that $N(t) = n$ events occur in the interval $(0, t]$. Considering the small interval $(t - \Delta t, t]$ preceding t, there are $n + 1$ different ways that n events can occur within the two non-overlapping intervals $(t, t - \Delta t]$ and $(t - \Delta t, t]$. These are $(n, 0)$, $(n - 1, 1)$, $(n - 2, 2)$,..., $(0, n)$. Hence, using the total probability rule and considering the third property above concerning the statistical independence of the number of events in non-overlapping intervals, one can write

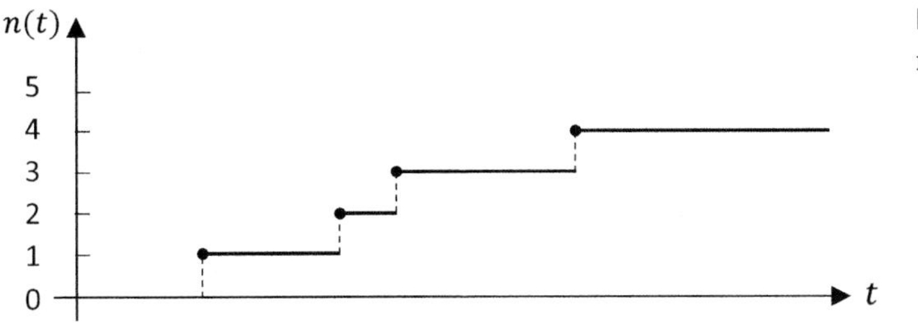

Figure 11.12 Typical realization of a Poisson process

$$\begin{aligned} p_N(n,t) &= p_N(n,t-\Delta t)\Pr\{0 \text{ events in } (t-\Delta t, t]\} \\ &\quad + p_N(n-1,t-\Delta t)Pr\{1 \text{ event in } (t-\Delta t, t]\} \\ &\quad + p_N(n-2,t-\Delta t)\Pr\{2 \text{ events in } (t-\Delta t, t]\} \\ &\quad + \cdots \\ &\quad + p_N(0,t-\Delta t)\Pr\{n \text{ events in } (t-\Delta t, t]\} \\ &= p_N(n,t-\Delta t)[1-\nu(t)\Delta t] + p_N(n-1,t-\Delta t)\nu(t)\Delta t + \mathrm{O}(\Delta t). \end{aligned} \tag{11.44}$$

In the last line we have used the second assumption above, i.e., that the probability of two or more events in a small interval of length Δt is of an order higher than Δt. Note also that we have used $\nu(t)\Delta t$ as the probability of an event occurring in the interval $(t-\Delta t, t]$ and its complement as the probability of zero events in the same interval. Rearranging terms and dividing by Δt, we arrive at

$$\frac{p_N(n,t)-p_N(n,t-\Delta t)}{\Delta t} + \nu(t)p_N(n,t-\Delta t) = \nu(t)p_N(n-1,t-\Delta t) + \frac{\mathrm{O}(\Delta t)}{\Delta t}. \tag{11.45}$$

Taking the limit as $\Delta t \to 0$, we obtain

$$\frac{\mathrm{d}}{\mathrm{d}t}p_N(n,t) + \nu(t)p_N(n,t) = \nu(t)p_N(n-1,t). \tag{11.46}$$

The above is a recursive differential equation for the first-order PMF of the Poisson process. For $n=0$, noting that the right-hand side is 0, the above reduces to

$$\frac{\mathrm{d}}{\mathrm{d}t}p_N(0,t) + \nu(t)p_N(0,t) = 0. \tag{11.47}$$

Solving this first-order differential equation with the initial condition $p_N(0,0)=1$, we obtain

$$p_N(0,t) = \exp\left[-\int_0^t \nu(t)\mathrm{d}t\right] = \exp[-m(t)], \tag{11.48}$$

where

$$m(t) = \int_0^t \nu(t)\mathrm{d}t. \tag{11.49}$$

This result is now placed in (11.46) to obtain the differential equation for $n=1$, which is solved with the initial condition $p_N(1,0)=0$. Repeating this process recursively, we obtain the general solution for the first-order PMF of the Poisson process

$$p_N(n,t) = \frac{[m(t)]^n \exp[-m(t)]}{n!}, n = 0,1,2,\ldots. \tag{11.50}$$

The second moments of the Poisson process are:

$$\mu_N(t) = \mathrm{E}[N(t)] = m(t) \text{ mean function,} \tag{11.51}$$

$$E\left[N^2(t)\right] = m(t) + m^2(t) \text{ mean-square function,} \tag{11.52}$$

$$\sigma_N^2(t) = m(t) \text{ variance function.} \tag{11.53}$$

As $m(t)$ is the mean function, $\nu(t)$ can be interpreted as the mean rate of occurrences. It is also clear that the c.o.v. at any given time is given by $\delta_N(t) = 1/\sqrt{m(t)}$.

The second-order PMF of the Poisson process, representing the joint distribution of $N(t_1)$ and $N(t_2)$, is derived by using the third assumption described above. For $t_1 \le t_2$, we write

$$\begin{aligned}
p_{NN}(n_1, t_1; n_2, t_2) &= \Pr[N(t_1) = n_1 \cap N(t_2) = n_2] \\
&= \Pr[N(t_1) = n_1 \cap N(t_2) - N(t_1) = n_2 - n_1] \\
&= \Pr[N(t_1) = n_1]\Pr[N(t_2) - N(t_1) = n_2 - n_1] \\
&= \frac{[m(t_1)]^{n_1} \exp[-m(t_1)]}{n_1!} \\
&\quad \times \frac{[m(t_2) - m(t_1)]^{n_2-n_1} \exp[-m(t_2) + m(t_1)]}{(n_2 - n_1)!} \\
&= \frac{[m(t_1)]^{n_1}[m(t_2) - m(t_1)]^{n_2-n_1} \exp[-m(t_2)]}{n_1!(n_2 - n_1)!}, \\
n_1 &= 0, 1, \ldots; n_2 = n_1, n_1 + 1, \ldots.
\end{aligned} \tag{11.54}$$

In going from the second to the third line above, we have used the fact that the numbers of occurrences within the non-overlapping intervals $(0, t_1]$ and $(t_1, t_2]$ are statistically independent.

The above joint PMF can be used to compute the auto-covariance function of the process. However, it is easier to derive that function by directly taking expectations:

$$\begin{aligned}
\Sigma_{NN}(t_1, t_2) &= \mathrm{E}[N(t_1)N(t_2)] - \mathrm{E}[N(t_1)]\mathrm{E}[N(t_2)] \\
&= \mathrm{E}\{N(t_1)[N(t_2) - N(t_1) + N(t_1)]\} - m(t_1)m(t_2) \\
&= \mathrm{E}\left[N^2(t_1)\right] + \mathrm{E}[N(t_1)]\mathrm{E}[N(t_2) - N(t_1)] - m(t_1)m(t_2) \\
&= m(t_1) + m^2(t_1) + m(t_1)[m(t_2) - m(t_1)] - m(t_1)m(t_2) \\
&= m(t_1) \\
&= \sigma_N^2(t_1).
\end{aligned} \tag{11.55}$$

Note that in going from the second to the third line we have again used the statistical independence of the numbers of events in the non-overlapping intervals $(0, t_1]$ and $(t_1, t_2]$.

We see that the auto-covariance function of the Poisson process at two time points is equal to the variance at the earlier time. Furthermore, the auto-correlation coefficient function is

$$\begin{aligned}\rho_{NN}(t_1,t_2) &= \frac{\Sigma_{NN}(t_1,t_2)}{\sigma_N(t_1)\sigma_N(t_2)} \\ &= \sqrt{\frac{m(t_1)}{m(t_2)}}.\end{aligned} \tag{11.56}$$

In the special case when the mean rate of occurrences, ν, is constant, the process is named the homogeneous Poisson process. In that case $m(t) = \nu t$ and the above results simplify. In particular, the first-order PMF and the mean, auto-covariance, and auto-correlation coefficient functions take the form:

$$p_N(n,t) = \frac{(\nu t)^n \exp(-\nu t)}{n!}, n = 0, 1, 2, \ldots, \tag{11.57}$$

$$\mu_N(t) = \nu t, \tag{11.58}$$

$$\Sigma_{NN}(t_1,t_2) = \nu t_1, t_1 \leq t_2, \tag{11.59}$$

$$\rho_{NN}(t_1,t_2) = \sqrt{t_1/t_2}, t_1 \leq t_2. \tag{11.60}$$

Furthermore, the increment process $\Delta N(t,\theta) = N(t+\theta) - N(t)$ formed by a homogeneous Poisson process for a fixed θ is a stationary process with the properties

$$p_{\Delta N}(n,t,\theta) = \frac{(\nu\theta)^n \exp(-\nu\theta)}{n!}, n = 0, 1, 2, \ldots, \tag{11.61}$$

$$\mu_{\Delta N} = \mathrm{E}[\Delta N(t,\theta)] = \nu\theta, \tag{11.62}$$

$$\Sigma_{\Delta N \Delta N}(t_1,t_2;\theta) = \begin{cases} \nu(\theta - |t_2 - t_1|), & \theta \leq t_2 - t_1, \\ 0, & \theta \geq t_2 - t_1. \end{cases} \tag{11.63}$$

Note that the auto-covariance function depends on the time lag $\tau = t_2 - t_1$ rather than the individual time points. Figure 11.13 shows a plot of the auto-covariance function. Because this is continuous at $\tau = 0$, the increment process is mean-square continuous. This might be surprising, as realizations of the process include jumps at occurrence points. Observe, however, that these jumps occur at random points in time. For a given time instance t, the likelihood of a jump to occur is zero; hence, the reason for the continuity. On the other hand, because the auto-covariance function is not differentiable at $\tau = 0$, the process is not mean-square differentiable. This again could be surprising as the process is continuous. Loosely speaking, this has to do with the fact that, although the probability of an occurrence at a time instance is zero, the corresponding change in the value of the process is infinite (1 divided by Δt as $\Delta t \to 0$) so that the two effects result in non-differentiability.

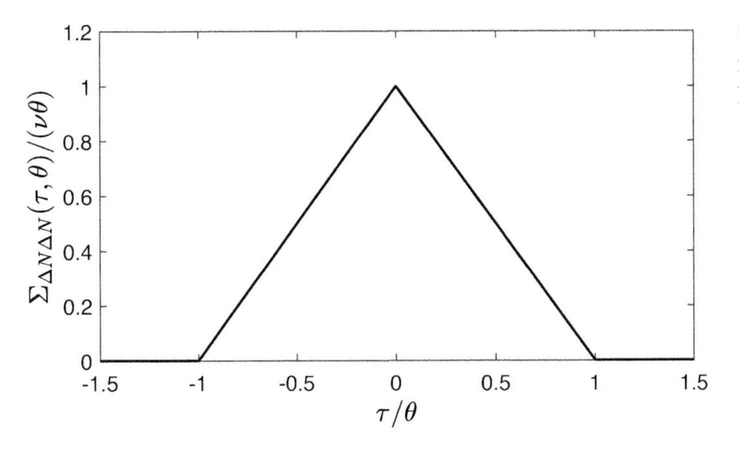

Figure 11.13 Auto-covariance function of the increment of a homogeneous Poisson process

11.6.1 Poisson Process with Random Selections

Suppose at each occurrence of a Poisson process we randomly select or deselect the occurred event with probabilities p and $1-p$, respectively. We assume the selections of successive events are statistically independent. As shown below, the number of selected events is a Poisson process with the mean rate $p\nu(t)$.

Let $M(t)$ be the number of events selected over the time interval $(0, t]$ out of $N(t)$ events of the parent Poisson process. Obviously, $M(t) \leq N(t)$. Furthermore, because the selections are statistically independent and have identical probabilities, the distribution of $M(t)$ for $N(t) = n$ is binomial with parameters (n, p). Therefore, using the total probability rule, one can write

$$\begin{aligned} p_M(m,t) &= \Pr\{M(t) = m \text{ in } (0,t]\} \\ &= \sum_{n=m}^{\infty} \binom{n}{m} p^m (1-p)^{n-m} \frac{[m(t)]^n \exp[-m(t)]}{n!} \\ &= \frac{[pm(t)]^m \exp[-m(t)]}{m!} \sum_{n=m}^{\infty} \frac{[(1-p)m(t)]^{n-m}}{(n-m)!}. \end{aligned} \tag{11.64}$$

The summation in the last expression is identical to the Maclaurin series expansion of $\exp[(1-p)m(t)]$. Hence, the final result is

$$p_M(m,t) = \frac{[pm(t)]^m \exp[-pm(t)]}{m!}, m = 0, \ldots, n, \tag{11.65}$$

which is identical to the Poisson distribution with the mean rate $p\nu(t)$. As shown later, this result is useful in deriving the probabilistic characteristics of stochastic load models.

11.6.2 Waiting and Interarrival Times in a Poisson Process

Let W_n denote the time to the nth occurrence and T_n denote the time between the $(n-1)$th and nth occurrences in a Poisson process, W_n is also called the *waiting time* to the nth occurrence and T_n is called the nth *interarrival time*. We are interested in probabilistic characterization of these random variables.

The PDF of W_n is obtained as follows:

$$\begin{aligned} f_{W_n}(t) &= \lim_{\Delta t \to 0} \frac{\Pr(t < W_n \le t + \Delta t)}{\Delta t} \\ &= \lim_{\Delta t \to 0} \frac{p_N(n-1, t)\nu(t)\Delta t + O(\Delta t)}{\Delta t} \\ &= p_N(n-1, t)\nu(t) \\ &= \frac{\nu(t)[m(t)]^{n-1} \exp[-m(t)]}{(n-1)!}, 0 < t. \end{aligned} \tag{11.66}$$

In the second line of the above derivation, the first term in the numerator represents the product of the probabilities of $n-1$ events in $(0, t]$ and one event in the non-overlapping interval $(t, t + \Delta t]$. The last term in the numerator represents the higher-order contribution of other combinations of events that involve two or more occurrences in the latter interval. This term vanishes as $\Delta t \to 0$. In the case of a homogeneous Poisson process, the above distribution simplifies to

$$f_{W_n}(t) = \frac{\nu(\nu t)^{n-1} \exp(-\nu t)}{(n-1)!}, 0 < t, \tag{11.67}$$

which is the gamma distribution with shape parameter n and scale parameter ν. It is also noted that, in this case, the time to the first occurrence has the exponential distribution

$$f_{W_1}(t) = \nu \exp(-\nu t), 0 < t. \tag{11.68}$$

The mean time to the first occurrence in a homogeneous Poisson process is $\mu_{W1} = 1/\nu$, while the mean time to the nth occurrence is $\mu_{W_n} = n/\nu$.

For T_n, first consider the conditional probability

$$\Pr(t < T_n \mid W_{n-1} = w) = \Pr\{\text{zero events in } (w, w + t]\} = \exp\{-[m(w + t) - m(w)]\}. \tag{11.69}$$

The unconditional probability is obtained by the total probability rule,

$$\begin{aligned} \Pr(t < T_n) &= \int_0^\infty \Pr(t < T_n \mid W_{n-1} = w) f_{W_{n-1}}(w) \mathrm{d}w \\ &= \int_0^\infty \exp\{-[m(w + t) - m(w)]\} \frac{\nu(w)[m(w)]^{n-2} \exp[-m(w)]}{(n-2)!} \mathrm{d}w \\ &= \int_0^\infty \frac{\nu(w)[m(w)]^{n-2} \exp[-m(w + t)]}{(n-2)!} \mathrm{d}w. \end{aligned} \tag{11.70}$$

It follows that the CDF of T_n is given by

$$F_{T_n}(t) = 1 - \int_0^\infty \frac{v(w)[m(w)]^{n-2} \exp[-m(w+t)]}{(n-2)!} \mathrm{d}w, 0 < t \tag{11.71}$$

and the PDF is obtained by taking the derivative with respect to t, resulting in

$$f_{T_n}(t) = \int_0^\infty \frac{v(w)v(w+t)[m(w)]^{n-2} \exp[-m(w+t)]}{(n-2)!} \mathrm{d}w, 0 < t. \tag{11.72}$$

Note that the above result depends on n, i.e., the interarrival times between successive occurrences in the general case are not identically distributed. For the case of a homogeneous Poisson process, it is easily shown that the above result reduces to

$$f_{T_n}(t) = v \exp(-vt), 0 < t, \tag{11.73}$$

which is the exponential distribution with v as the scale parameter. This result is independent of n, i.e., the interarrival times between successive events have identical exponential distributions. Because of the random nature of event occurrences in the Poisson process, the interarrival times $T_1, T_2, \ldots$ are statistically independent random variables.

11.6.3 Poisson Approximation for Time- and Space-Variant Reliability Problems

The Poisson process can be used to obtain an approximation to the time- or space-variant structural reliability problem. In the previous section we derived an approximation to the mean rate of down-crossings of the limit-state function below the zero level, which mark the failure events of the structure. If these down-crossing events can be assumed to be statistically independent, then it is reasonable to assume that they occur according to a Poisson process. Thus, under this assumption and including the probability of failure at the initial point, the failure probability over the interval $[0, T]$ is approximated as

$$\Pr\left\{\min_{0 \le t \le T} g[\mathbf{X}(t), t] \le 0\right\} \cong \Pr\{g[\mathbf{X}(0), 0] \le 0\} + 1 - \exp\left[-\int_0^T v(t)\mathrm{d}t\right], \tag{11.74}$$

where the last two terms represent the complement of the probability of zero Poisson events in $(0, T]$. Furthermore, using (11.66) for $n = 1$,

$$f_{W_1}(t) = v(t) \exp\left[-\int_0^t v(\tau)\mathrm{d}\tau\right], 0 < t, \tag{11.75}$$

describes the PDF of the time to the first failure of the structure if it operates indefinitely.

Example 11.6 – Reliability of a Column under Stochastic Loads – Poisson Approximation

In Example 11.4 for the column under stationary stochastic loads, we obtained $v = 0.00415\ \mathrm{s}^{-1}$ as the mean rate of down-crossings of the limit-state function below the zero level. The crossing events are expected to be rare, so they are likely to occur far apart, so the load processes are

Example 11.6 (cont.)

effectively statistically independent at successive crossing events. However, the yield strength random variable Y is common to all the crossing events and introduces dependence among them. Therefore, the assumption of statistically independent crossings is not correct in this case. Nevertheless, as we saw in Example 11.5, the common random variable Y induces a small correlation coefficient of only 0.0660 between the linearized limit-state functions in the standard normal space. Thus, the dependence between the crossing events introduced by the common random variable Y is minimal. On this basis, it is reasonable to use the Poisson approximation. As v is a constant, we have a homogeneous Poisson process and (11.74) yields

$$\begin{aligned}\Pr\left\{\min_{0\leq t\leq T} g[\mathbf{X}(t),t]\leq 0\right\} &\cong \Phi(-3.43)+1-\exp(-0.00415\times 10)\\ &=0.000304+0.0407\\ &=0.0410.\end{aligned} \tag{E1}$$

This result compares well with the upper bound 0.0418 obtained in Example 11.5. Furthermore, if the stochastic loads were to continue indefinitely, according to (11.75) the time to failure would have the exponential PDF

$$f_{W_1}(t)=0.00415\exp(-0.00415t), 0<t \tag{E2}$$

with the mean time to failure $\mu_{W_1}=1/0.00415=241$ s.

One way to properly account for the dependence arising from the common random variable Y is to use conditioning. Let $\beta(y)$ and $v(y)$ be the solutions of the reliability index and the mean crossing rate for a given outcome y of Y. Determining these values for the full range of outcomes of Y, we obtain the unconditional failure probability estimate by use of the total probability rule

$$\Pr\left\{\min_{0\leq t\leq T} g[\mathbf{X}(t),t]\leq 0\right\}\cong\int_0^\infty\{\Phi[-\beta(y)]+10v(y)\}f_Y(y)\mathrm{d}y \tag{E3}$$

where $f_y(y)$ denotes the PDF of Y. Another approach is the nested reliability analysis described in Section 10.5. We will leave the pursuit of these approaches to the student's initiative.

11.7 Stochastic Load Models and Load Combination

Virtually all loads acting on structures are stochastic in nature, randomly fluctuating in not only time but also space. Gravity loads change in time and space because of reconfiguration of space or change in the occupancy of a building. Live loads fluctuate in time and space because of movement of humans and movable objects, as do environmental loads such as snow, wind, waves, and earthquakes. If the predominant period of temporal fluctuations of the load is close to the fundamental period of the structure, significant dynamic effects may occur. In that case, the micro-scale model of the fluctuating load is important. Otherwise, a macro-scale load model showing only significant changes in time is sufficient for most analysis. Chapter 13 describes macro-scale modeling of stochastic loads and corresponding reliability analysis methods under dynamic effects. In this section are described several macro-scale models of stochastic loads that do not introduce dynamic effects and corresponding reliability analysis methods. Furthermore, because structures are typically subjected to multiple loads, the combined load effect should be considered. The latter is the demand placed on the structure, such as the stress, strain, or displacement at a particular point in time and space. When the structure behaves linearly, the combined load effect arising from multiple loads is the sum of the load amplitudes, each multiplied by an influence coefficient that represents the demand value for a unit value of the load. Much of the literature on macro-scale load modeling and load combination assumes this linear case. The stochastic dynamic problem described in Chapter 13 considers nonlinear structural behavior.

In analyzing loads and load combinations, two quantities are of particular interest: (a) the mean rate of a critical load effect up-crossing a specified threshold, and (b) the probability distribution of the maximum load effect over a specified duration of time. The following sections present solutions for these quantities for specific load models and their combinations.

11.7.1 Ferry Borges–Castanheta Load Model

One of the earliest stochastic load models is due to Ferry Borges and Castanheta (1971). Their objective was to develop a load model and combination rule that could be used in codified design. Each load-effect process (load magnitude multiplied by the influence coefficient for a response quantity of interest), denoted $S(t,\tau)$, is assumed to have statistically independent and identically distributed (SIID) amplitudes over fixed intervals of time, τ, as exemplified in Figure 11.14. Letting $F_S(s)$ denote the common CDF of the load effect amplitudes, the CDF of the maximum load effect $S_{\max} = \max_{0\le t\le T} S(t,\tau)$ over the interval $[0, T]$, for $\tau \le T$, is given by

$$F_{S_{\max}}(s) = [F_S(s)]^{T/\tau}, \tag{11.76}$$

where T/τ, assumed to be an integer, is the number of distinct load amplitudes during $[0, T]$.

Now suppose we have several statistically independent load effects $S_i(t,\tau_i)$, $i = 1, \ldots, n$, with CDFs $F_{S_i}(s)$, where $\tau_n \le \tau_{n-1} \le \cdots \le \tau_1$. Our aim is to determine the distribution of the maximum of the combined load effect $S(t) = S_1(t,\tau_1) + \cdots + S_n(t,\tau_n)$ over a time interval $[0, T]$ with $\tau_1 \le T$. We make the simplifying assumption that $n_i = \tau_i/\tau_{i-1}$ are integers for all $i = 2, \ldots, n$. In that case, n_i SIID occurrences of $S_i(t,\tau_i)$ fall within one interval τ_{i-1} of $S_{i-1}(t,\tau_{i-1})$, see Figure 11.15. To obtain the maximum of the sum of these two load effects within a τ_{i-1} interval, one needs to add the maximum of the n_i SIID occurrences of $S_i(t,\tau_i)$ to the random-point-in-time occurrence of $S_{i-1}(t,\tau_{i-1})$. We perform this analysis starting with

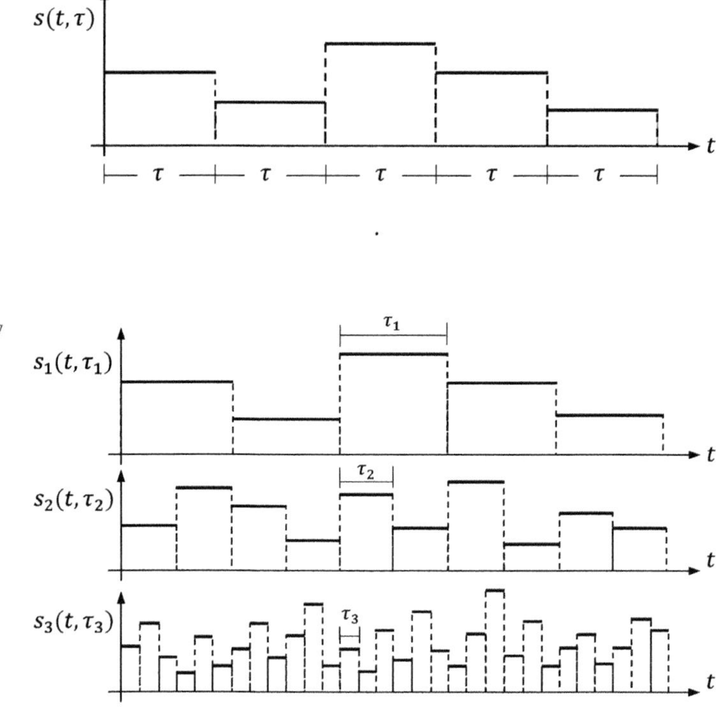

Figure 11.14 Sample realization of a Ferry Borges-Castanheta load model

Figure 11.15 Combination of three Ferry Borges-Castanheta load effects

the two load effects with the smallest fixed intervals, $S_n(t, \tau_n)$ and $S_{n-1}(t, \tau_{n-1})$. Let $Z_n(\tau_{n-1})$ denote the maximum of $S_n(t, \tau_n)$ over the interval τ_{n-1} having the CDF $[F_{S_n}(z)]^{n_n}$. The CDF of the sum $Z_n(\tau_{n-1}) + S_{n-1}(t, \tau_{n-1})$ is given by the convolution integral (see Example 2.9)

$$F_{Z_n+S_{n-1}}(z) = \int_{-\infty}^{+\infty} [F_{S_n}(z-s)]^{n_n} f_{S_{n-1}}(s) \mathrm{d}s. \tag{11.77}$$

It should be clear that the resulting sum process complies with the definition of a Ferry Borges-Castanheta process with the above CDF and the fixed time interval τ_{n-1}. We now replace $S_{n-1}(t, \tau_{n-1})$ with the sum process $Z_n(\tau_{n-1}) + S_{n-1}(t, \tau_{n-1})$ and its CDF $F_{S_{n-1}}(s)$ with the CDF $F_{Z_n+S_{n-1}}(s)$. We have thus eliminated one of the load processes. Next, we proceed with the revised $S_{n-1}(t, \tau_{n-1})$ and the load process $S_{n-2}(t, \tau_{n-2})$. The result is the elimination of $S_{n-1}(t, \tau_{n-1})$ and revision of $S_{n-2}(t, \tau_{n-2})$. This process is continued until we arrive at a revised version of the load effect process $S_1(t, \tau_1)$. Equation (11.76) is then used to obtain the distribution of the maximum of the combined load effects during the interval $[0, T]$.

The above analysis requires repeated computations of the integral in (11.77) to determine the revised CDF that incorporates the contribution of the eliminated load effect. This can be cumbersome because analytical solutions are unlikely to exist and the integral needs to be evaluated for the full range of values of z. Rackwitz and Fiessler (1978) developed a special algorithm to address this computational problem when one is interested in the tail of the distribution of the maximum combined load effect. This algorithm was the precursor to the development of the FORM. Since then, more realistic load models and simpler load combination rules have been developed, as described in the remainder of this chapter. For that reason, we will not provide more details of the Ferry Borges-Castanheta load model and the specific algorithm developed by Rackwitz and Fiessler. The load models and load combination analysis presented in the following sections are mostly based on the work of Wen (1977, 1990, 1993).

11.7.2 The Filtered Poisson Model

Consider a Poisson process $N(t)$ with the mean occurrence rate $\nu(t)$. At the kth occurrence of the Poisson process we assign a *response function* $w(t, t_k, \mathbf{X}_k)$, which starts at the time of occurrence, t_k, of the event and is shaped by a set of random variables $\mathbf{X}_k$. Shape variables at different occurrences are assumed to be statistically independent. The *filtered Poisson process* is defined as

$$S(t) = \sum_{k=0}^{N(t)} w(t, t_k, \mathbf{X}_k). \tag{11.78}$$

Figure 11.16 shows a realization of one such process, where the response function is a right triangle with height $X_{k,1}$ and base $X_{k,2}$ so that $\mathbf{X}_k = (X_{k,1}, X_{k,2})$. Note that it is possible for the response functions of two or more events to overlap, in which case the response functions are added. One such occurrence is shown in Figure 11.16. This feature of the filtered Poisson process is not realistic for most loads, except for loads such as snow on the roof of a building where accumulation can indeed occur at successive events. In fact, the triangular response function is meant to model the snow load on the roof of a building, which gradually decays because of melting. Unfortunately, the probabilistic characteristics of this process, such as the maximum value during a specified interval, are difficult to derive analytically because of the overlapping events. For this reason, the filtered Poisson process is seldom used in structural reliability analysis.

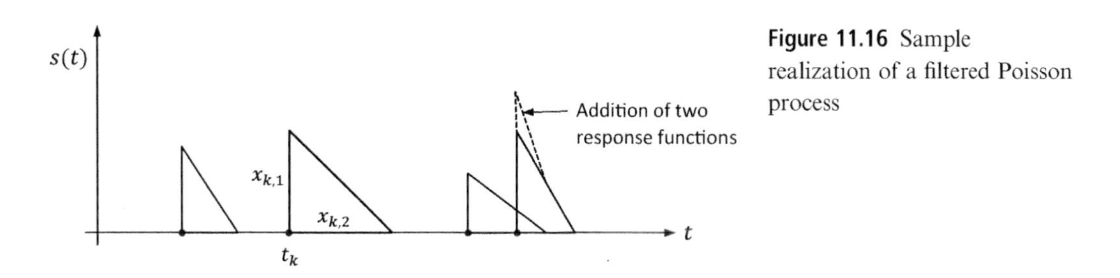

Figure 11.16 Sample realization of a filtered Poisson process

One special case of the filtered Poisson process that is useful for modeling extraordinary loads is the *Poisson spike process*. The response function for this process is defined as

$$\begin{aligned} w(t, t_k, X_k) &= X_k, t = t_k, \\ &= 0, \text{elsewhere}, \end{aligned} \tag{11.79}$$

where X_k, $k = 1, 2, \ldots$, are SIID random variables. The load events in this model obviously cannot overlap. A typical realization of this process is shown in Figure 11.17. This process is useful for modeling extraordinary loads, such as the passage of extraordinarily large vehicles over a bridge or the gathering of a large number of people in a hall, which occur randomly and typically over infinitesimal periods of time.

Let the load magnitudes X_k in the Poisson spike process have the common CDF $F_X(x)$. Suppose we are interested in those loads that exceed a specified magnitude s. The probability of a given load occurrence exceeding this level is $1 - F_X(s)$. Therefore, based on the result derived in Section 11.6.1, the load events with magnitudes greater than s are Poisson with the mean rate equal to $[1 - F_X(s)]\nu(t)$. Furthermore, the distribution of the maximum load $S_{\max} = \max_{0<t\leq T} S(t)$ over the interval $[0, T]$ is given by

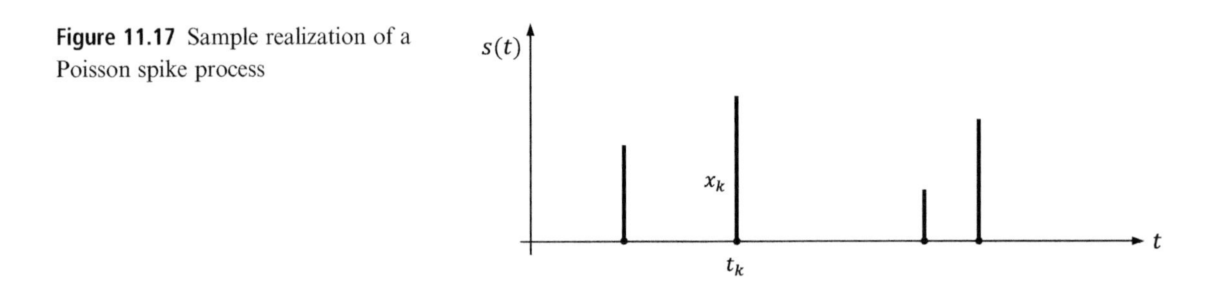

Figure 11.17 Sample realization of a Poisson spike process

$$\begin{aligned} F_{S_{\max}}(s) &= \Pr\{\text{zero occurrences of } s \leq X_k \text{ in } [0, T]\} \\ &= \exp\{-[1 - F_X(s)]m(T)\}, \end{aligned} \tag{11.80}$$

where $m(T) = \int_0^T \nu(t)dt$. The PDF of the above distribution is

$$f_{S_{\max}}(s) = \delta(s)\exp[-m(T)] + f_X(s)m(T)\exp\{-[1 - F_X(s)]m(T)\}, \tag{11.81}$$

where $\delta(s)$ is the Dirac delta function on account of the fact that there is probability $\exp[-m(T)]$ that no load events will occur during the considered time interval, in which case $S_{\max} = 0$. Figure 11.18 depicts such a mixed PDF with a probability mass at zero.

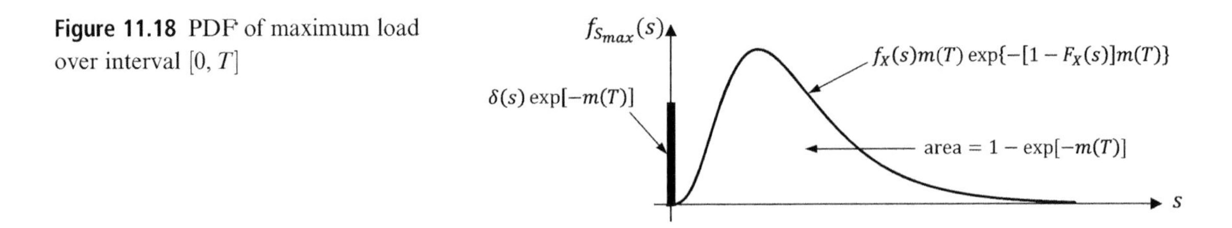

Figure 11.18 PDF of maximum load over interval $[0, T]$

In the case of a homogeneous Poisson process, the above formulas are simplified by replacing $m(T)$ with νT. The interarrival times between the load applications have the distribution in (11.71) and, in the case when the parent Poisson process is homogeneous, the interarrival times are exponentially distributed according to (11.73).

11.7.3 The Poisson Square-Wave Process

Consider again a Poisson process $N(t)$ with the mean occurrence rate $\nu(t)$ and let X_i, $i = 0, 1, \ldots,$ be a set of SIID random variables having the common CDF $F_X(x)$. The *Poisson square-wave process* is defined as

$$S(t) = X_{N(t)}. \tag{11.82}$$

A typical realization of this process may appear as in Figure 11.19. The process starts with the constant load value X_0 at $t = 0$ and, then, at the first occurrence of the Poisson process, the load value changes to X_1 and stays constant until the next occurrence. This process is good for modeling live loads, which tend to experience major changes when the type of operation or occupancy of a building is changed. The Poisson points here model the significant load change points in time; between these points the load is assumed to remain constant.

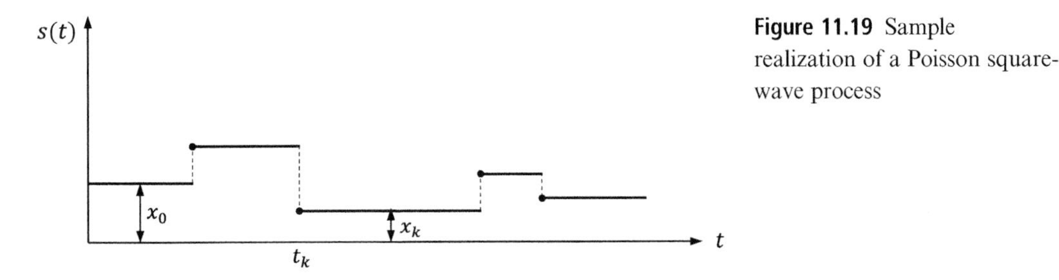

Figure 11.19 Sample realization of a Poisson square-wave process

From what we have seen in the preceding section, load-change events with the load value exceeding a specified threshold s are Poisson with the mean rate $[1 - F_X(s)]\nu(t)$. However, this assumes that we start below the threshold, i.e., $X_0 < s$. Thus, accounting for the starting load value, the CDF of the maximum load over the interval $[0, T]$ for this model is

$$F_{S_{\max}}(s) = F_X(s)\exp\{-[1 - F_X(s)]m(T)\} \tag{11.83}$$

and the corresponding PDF is

$$f_{S_{\max}}(s) = f_X(s)[1 + F_X(s)m(T)]\exp\{-[1 - F_X(s)]m(T)\}. \tag{11.84}$$

The distribution of each load duration is given by the interarrival time of the parent Poisson process, as specified in (11.71). In the case of a homogeneous Poisson process, the load durations are identically exponentially distributed according to (11.73), with a mean value of $1/\nu$.

The Poisson spike process and the Poisson square-wave process are the extreme cases of load duration, the first having loads acting only over infinitesimal intervals whereas the latter has loads acting at all times. The next model allows loads to act intermittently over finite durations.

11.7.4 The Poisson Pulse Process

Consider a Poisson square-wave process where the load amplitudes X_i, $i = 0, 1, \ldots$, have a common PDF of the form

$$f_X(x) = (1 - p)\delta(x) + p\tilde{f}_X(x). \tag{11.85}$$

Here, p is the probability that the load is "on," in which case the PDF of the load magnitude is $\tilde{f}_X(x)$; otherwise, the load is "off" with probability $1 - p$ and a value of 0. The above mixed PDF has a shape similar to that shown in Figure 11.18. This process is called the *Poisson pulse process*. Figure 11.20 shows a sample realization of such a process, where the solid dots indicate "on" events of the load and the open dots indicate "off" events.

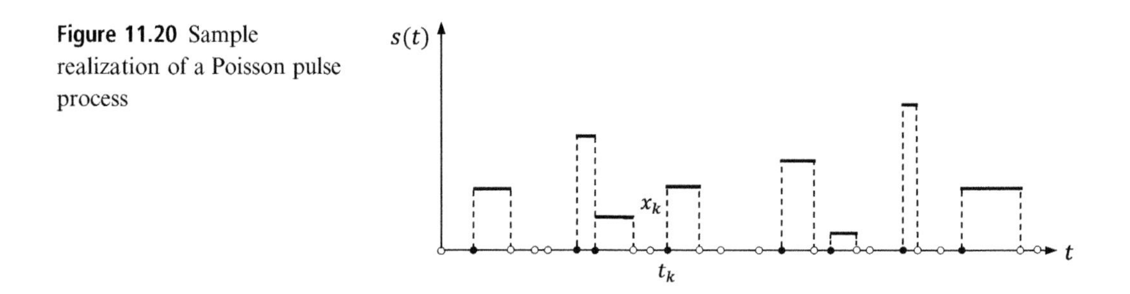

Figure 11.20 Sample realization of a Poisson pulse process

Because the on/off load events occur randomly, it follows that the "on" load events are Poisson distributed with the mean rate being $p\nu(t)$. Furthermore, the duration of each "on" load event is distributed according to (11.71) and, in the case of the homogeneous Poisson process, is exponentially distributed according to (11.73) with a mean duration of $1/\nu$. Thus, the mean rate of occurrence of the "on" events and their mean durations can be controlled by adjusting p and ν. The mean rate of an "on" load event exceeding the threshold s is given by $\left[1 - \tilde{F}_X(s)\right]p\nu(t)$, where $\tilde{F}_X(x)$ is the CDF of the load magnitude when it is "on." Furthermore, the CDF of the maximum load over the interval $[0, T]$ is given by

$$F_{S_{\max}}(s) = \left\{1 - p\left[1 - \tilde{F}_X(s)\right]\right\}\exp\left\{-p\left[1 - \tilde{F}_X(s)\right]m(T)\right\}. \tag{11.86}$$

The corresponding PDF is

$$\begin{aligned} f_{S_{\max}}(s) &= (1 - p)\exp\{-pm(T)\}\delta(s) \\ &\quad + p\tilde{f}_X(s)\left\{1 + \left[1 - p + p\tilde{F}_X(s)\right]m(T)\right\}\exp\left\{-p\left[1 - \tilde{F}_X(s)\right]m(T)\right\}. \end{aligned} \tag{11.87}$$

The first term represents the probability mass corresponding to there being no "on" load events during the specified interval, including at the starting point.

The Poisson spike process and the Poisson square-wave process are limiting cases of the Poisson pulse process. The former model is reached when $p \to 0$ and $\nu \to \infty$ in such a way that $p\nu$ remains finite, and the latter model is reached when $p \to 1$.

When the structure is subjected to a single load effect, the above distribution of the maximum load effect along with the distributions of other random quantities, e.g., capacity values, can be used to perform reliability analyses. The real challenge is when the structure is subject to multiple stochastic loads. This topic is addressed in the following section.

11.8 Combination of Homogeneous Poisson Pulse Load Processes

Suppose a structure is subjected to multiple stochastic load effects $S_i(t)$, $i = 1, 2, \ldots$, each modeled as a Poisson pulse process, including the limiting cases of Poisson spike and Poisson square-wave processes. Assume all loads are statistically independent and their parent Poisson processes are homogeneous. Let v_i and p_i denote the mean rate of occurrence and the probability of "on" events of the load effect process $S_i(t)$ and denote $\lambda_i = p_i v_i$ as the mean rate of the "on" events and $\mu_i = 1/v_i$ as the mean duration of an "on" event.

A structure or system may fail under any combination of the applied load coincidences. Hence, we first focus on the rates of load coincidences. Observe that, on average, load effect $S_i(t)$ is "on" during $\lambda_i \mu_i$ fraction of the time. During this fraction of time, load effect $S_j(t)$, on average, will have $\lambda_j \lambda_i \mu_i$ "on" occurrences. These are coincidences of the two loads that occur when $S_j(t)$ is triggered, while $S_i(t)$ is "on." There are, on average, $\lambda_i \lambda_j \mu_j$ additional coincidences that occur when $S_i(t)$ is triggered, while $S_j(t)$ is "on." Adding the two, we have

$$\lambda_{i,j} \cong \lambda_i \lambda_j \left(\mu_i + \mu_j \right) \tag{11.88}$$

as the mean rate of coincidences of "on" events of load effects $S_i(t)$ and $S_j(t)$, $i \neq j$. Furthermore, observe that these coincidences occur, on average, over $\lambda_i \mu_i \lambda_j \mu_j$ fraction of the time. Dividing by the mean rate of coincidences, we obtain the mean duration of each coincidence as

$$\mu_{i,j} \cong \frac{\mu_i \mu_j}{\mu_i + \mu_j}. \tag{11.89}$$

In a similar way, the mean rate of coincidence of three load effects $S_i(t)$, $S_j(t)$, $S_k(t)$, $i \neq j \neq k$, and their mean duration of coincidence are obtained as

$$\lambda_{i,j,k} \cong \lambda_i \lambda_j \lambda_k \left(\mu_i \mu_j + \mu_j \mu_k + \mu_k \mu_i \right) \tag{11.90}$$

$$\mu_{i,j,k} \cong \frac{\mu_i \mu_j \mu_k}{\mu_i \mu_j + \mu_j \mu_k + \mu_k \mu_i}. \tag{11.91}$$

The extension to four or more loads should be obvious.

Under the combined load effects, the structure or system may fail under the action of a single load, under the action of pairwise load coincidences, under the action of triple load coincidences, etc. Hence, the mean rate of failures can be conservatively approximated as

$$\lambda_F \cong \sum_i \lambda_i P_i + \sum_{i<j} \lambda_{i,j} P_{i,j} + \sum_{i<j<k} \lambda_{i,j,k} P_{i,j,k} + \cdots, \tag{11.92}$$

where P_i is the probability of failure under an "on" event of load effect $S_i(t)$ alone, $P_{i,j}$ is the probability of failure under a coincidence event of load effects $S_i(t)$ and $S_j(t)$, $P_{i,j,k}$ is the probability of failure under a triple coincidence event of load effects $S_i(t)$, $S_j(t)$, and $S_k(t)$, etc. The above expression is conservative because the single load occurrences are also present in the pairwise and higher order load coincidences, the pairwise load coincidences are also present in the triple and higher order load coincidences, etc. However, because these coincidence events are typically rare, the added conservatism is usually insignificant. It is important to note that if a load process $S_i(t)$ is always "on," i.e., has $p_i = 1$, then that load should be present in all the terms in (11.92). Any term not including that load should be dropped from the equation. It is also noted that typically, but not always, $\lambda_i > \lambda_{i,j} > \lambda_{i,j,k} > \cdots$, while $P_i < P_{i,j} < P_{i,j,k} < \cdots$. Hence, the term that contributes most to the mean rate of failures depends on the magnitudes of the product terms $\lambda_i P_i$, $\lambda_{i,j} P_{i,j}$, $\lambda_{i,j,k} P_{i,j,k}$, etc.

If the load effects are the only random quantities involved in the reliability problem, then the probabilities P_i, $P_{i,j}$, $P_{i,j,k}$, etc., are computed by time-invariant reliability analysis using the "on" distributions of the load effects. In this case, the failure events under various load coincidences can be conservatively considered to be statistically independent. (Dependence between these failure events may exist when a load value is shared between two or more load coincidences. This happens, for example, when load $S_j(t)$ is triggered two or more times during one occurrence of load $S_i(t)$.) Under this assumption, the failure events under any one of the load coincidence terms constitute a Poisson process with random selections (see Section 11.6.1). Because the sum of Poisson processes is also a Poisson process, the failure events constitute a Poisson process and we can write the failure probability over the interval $(0, T]$ as

$$p_F(T) \cong 1 - \exp(-\lambda_F T). \tag{11.93}$$

To account for the possibility of failure at the initial time, the time-invariant probability of failure at $t = 0$ may be added to the right-hand side. It is important to note that the above load-coincidence formulation applies even when the load effects are nonlinearly combined. For that case, the probability of failure for each load-coincidence term should account for the nonlinear behavior of the structure.

The presence of other, time-invariant random variables, e.g., capacity variables, induces statistical dependence between the failure events under various load coincidences. The failure events then are not statistically independent, and the Poisson assumption is not valid. One recourse in that case is to use the total probability rule

$$p_F(T) = \int_{\mathbf{x}} p_F(T|\mathbf{x}) f_{\mathbf{X}}(\mathbf{x}) \mathrm{d}\mathbf{x}, \tag{11.94}$$

where $\mathbf{X}$ is the set of time-invariant random variables with joint PDF $f_{\mathbf{X}}(\mathbf{x})$ and $p_F(T|\mathbf{x})$ is the conditional probability of failure obtained from (11.93) for $\mathbf{X} = \mathbf{x}$. This approach is viable when the number of time-invariant random variables is no more than one or two.

An alternative is the nested reliability approach described in Section 10.5, provided that an analytical solution can be obtained for the conditional failure probability $p_F(T|\mathbf{x})$. Example 11.8 demonstrates this approach. Another alternative is to numerically determine the distribution of the maximum combined load effect and use it together with the other time-invariant random variables to determine the failure probability by ordinary FORM, SORM, or simulation methods. The complementary CDF of the maximum combined load effect is determined using (11.92) and (11.93) with $P_i = \Pr[s < S_i(t)]$, $P_{i,j} = \Pr\left[s < S_i(t) + S_j(t)\right]$, $P_{i,j,k} = \Pr\left[s < S_i(t) + S_j(t) + S_k(t)\right]$, etc., where s is the selected threshold. Naturally, a range of thresholds must be considered to determine the full distribution of the maximum combined load effect.

Example 11.7 – Reliability Analysis under Combination of Poisson Pulse Loads

Consider a structure under the combined action of two statistically independent load effects $S_1(t)$ and $S_2(t)$, which are modeled as Poisson pulse processes with $\lambda_1 = \lambda_2 = 1\ \text{yr}^{-1}$ and $\mu_1 = \mu_2 = 0.01$ yr. These parameter values correspond to $\nu_1 = \nu_2 = 1/0.01 = 100\ \text{yr}^{-1}$ and $p_1 = p_2 = 1/100 = 0.01$. Furthermore, assume the "on" events of load 1 have the normal distribution with mean $\mu_1 = 100$ and standard deviation $\sigma_1 = 45$, while the "on" events of load 2 have the normal distribution with mean $\mu_2 = 100$ and standard deviation $\sigma_2 = 60$. Suppose the limit-state function of the structure is defined as

$$g[r, S_1(t), S_2(t)] = r - S_1(t) - S_2(t), \tag{E1}$$

where r is a deterministic capacity value. We are interested in the failure probability over a 10-year period for $r = 250, 300,$ and 350.

Using (11.88) and (11.89), we obtain $\lambda_{1,2} = 1 \times 1 \times (0.01 + 0.01) = 0.02\ \text{yr}^{-1}$ and $\mu_{1,2} = 0.01 \times 0.01/(0.01 + 0.01) = 0.005$ yr. Furthermore,

$$P_1 = \Pr(r - X_1 \le 0) = \Pr(r \le X_1) = 1 - \Phi\left(\frac{r - 100}{45}\right) \tag{E2}$$

$$P_2 = \Pr(r - X_2 \le 0) = \Pr(r \le X_2) = 1 - \Phi\left(\frac{r - 100}{60}\right) \tag{E3}$$

$$P_{1,2} = \Pr(r - X_1 - X_2 \le 0) = \Pr(r \le X_1 + X_2) = 1 - \Phi\left(\frac{r - 100 - 100}{\sqrt{45^2 + 60^2}}\right). \tag{E4}$$

In the last line, we have used the property that the sum of two normal random variables is also normal. For $r = 250$, we obtain $P_1 = 0.000429$, $P_2 = 0.00621$, and $P_{1,2} = 0.252$. Using (11.92), one has

$$\begin{aligned}\lambda_F &= 1 \times 0.000429 + 1 \times 0.00621 + 0.02 \times 0.252 \\ &= 0.000429 + 0.00621 + 0.00505 \\ &= 0.0117.\end{aligned} \tag{E5}$$

Example 11.7 (cont.)

Observe that for this value of r the dominant contribution to the failure probability comes from individual applications of load 2. Using (11.93), the failure probability estimate in this case is $p_F(10\,\text{yr}) = 0.110$. Similar analysis for $r = 300$ leads to $\lambda_F = 4.41 \times 10^{-6} + 4.29 \times 10^{-4} + 1.82 \times 10^{-3} = 0.00226$ with $p_F(10\,\text{yr}) = 0.0223$ and for $r = 350$ leads to $\lambda_F = 1.38 \times 10^{-8} + 1.55 \times 10^{-5} + 4.55 \times 10^{-4} = 0.000470$ with $p_F(10\,\text{yr}) = 0.00469$. Observe that in the latter cases the dominant contribution to the failure probability comes from the coincidence of the two load effects.

Repeating the above analysis for a range of thresholds $r = s$, we obtain the complementary CDF of the maximum combined load effect from (11.93). Numerically differentiating with respect to s, the PDF is obtained. The results are shown in Figure 11.21. These numerical results can now be used to perform reliability analysis with a random capacity.

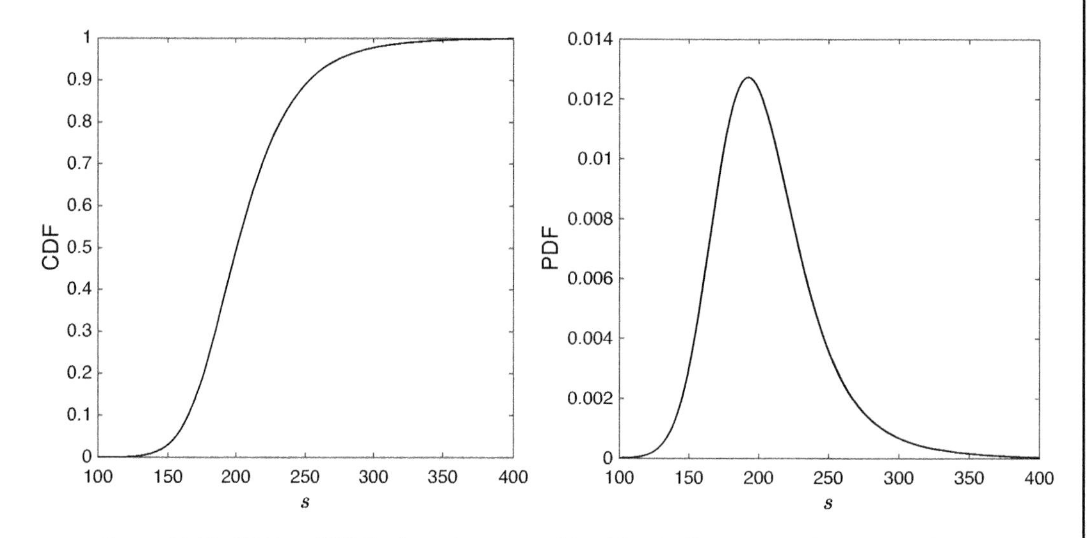

Figure 11.21 CDF and PDF of maximum combined load effect over a 10-year duration

Example 11.8 – Reliability of Ductile Frame under Poisson Pulse Loads

Consider the ductile frame in Example 8.7 under horizontal and vertical loads. Assume the horizontal load, $H(t)$, representing the static equivalent of earthquake loads, is modeled as a Poisson spike process with mean rate $\lambda_H = 0.2\,\text{yr}^{-1}$ with a magnitude at each occurrence that is normally distributed with mean 0 kN and standard deviation 25 kN, and the vertical load, representing a live load on the roof, is modeled as a Poisson pulse process with mean rate of

Example 11.8 (cont.)

"on" events $\lambda_V = 10\ \text{yr}^{-1}$, mean duration $\mu_V = 0.05$ yr, and a magnitude at each "on" event that is normally distributed with mean 40 kN and standard deviation 16 kN. In addition, we assume the plastic-moment capacities at the potential locations of forming plastic hinge are random variables M_1, M_2, M_3, M_4, and M_5, having a joint lognormal distribution with identical means 150 kNm, identical standard deviations 15 kNm, and pairwise correlation coefficients 0.3. We are interested in the reliability of the frame over a $T = 50$ yr lifetime.

Using (11.88), we first compute $\lambda_{H,V} = 0.2 \times 10 \times (0 + 0.05) = 0.1\ \text{yr}^{-1}$ as the mean rate of load coincidences. Obviously $\mu_{H,V} = 0$, as the Poisson spike loads have zero duration. Using (11.92), the mean rate of failure is computed as

$$\lambda_F \cong \lambda_H P_H + \lambda_V P_V + \lambda_{H,V} P_{H,V}, \tag{E1}$$

where P_H is the probability of failure under a random action of the horizontal load alone, P_V is the probability of failure under an "on" event of the vertical load alone, and $P_{H,V}$ is the probability of failure under a coincidence event of the two loads.

As we saw in Example 8.7, the frame may fail by one of three plastic mechanisms. Under the horizontal load alone, only the sway mode shown in the bottom left of Figure 8.11 is applicable; however, since $H(t)$ has a zero mean, we need to consider the possibility of failure in the sway mode in either direction. These failure events are defined by the limit-state functions

$$g_1(\mathbf{M}, H) = M_1 + M_2 + M_4 + M_5 - 5H \tag{E2}$$

$$g_2(\mathbf{M}, H) = M_1 + M_2 + M_4 + M_5 + 5H, \tag{E3}$$

where $\mathbf{M} = (M_1, M_2, M_3, M_4, M_5)$. Under the vertical load alone, only the beam mechanism shown in the bottom middle of Figure 8.11 is applicable. This is described by the limit-state function

$$g_3(\mathbf{M}, V) = M_2 + 2M_3 + M_4 - 5V. \tag{E4}$$

Finally, under the joint action of the horizontal and vertical loads, all three mechanisms apply but, on account of the zero mean of the horizontal load, both directions should be considered. Thus, for this case, we need to consider the following two limit-state functions, in addition to the three given above:

$$g_4(\mathbf{M}, H, V) = M_1 + 2M_3 + 2M_4 + M_5 - 5H - 5V \tag{E5}$$

$$g_5(\mathbf{M}, H, V) = M_1 + 2M_3 + 2M_4 + M_5 + 5H - 5V. \tag{E6}$$

Where H and V are normally distributed, so the conditional failure probabilities given $\mathbf{M}$ are easily computed. Denoting $q_i(\mathbf{m}) = \Pr(g_i \leq 0 \mid \mathbf{M} = \mathbf{m})$, they are

$$q_1(\mathbf{m}) = q_2(\mathbf{m}) = 1 - \Phi\left(\frac{m_1 + m_2 + m_4 + m_5}{5 \times 25}\right) \tag{E7}$$

Example 11.8 (cont.)

$$q_3(\mathbf{m}) = 1 - \Phi\left(\frac{m_2 + 2m_3 + m_4 - 40}{5 \times 16}\right) \tag{E8}$$

$$q_4(\mathbf{m}) = q_5(\mathbf{m}) = 1 - \Phi\left(\frac{m_1 + 2m_3 + 2m_4 + m_5 - 40}{5\sqrt{25^2 + 16^2}}\right). \tag{E9}$$

We use the nested reliability analysis method to account for the uncertainty in the plastic-moment capacities. For this purpose, it is advantageous to have an analytical solution for the conditional failure probability. With this in mind, we opt to use the uni-modal upper bound for series systems (see Section 8.4). The probability terms in (E1) for $\mathbf{M} = \mathbf{m}$ are computed as follows:

$$P_H(\mathbf{m}) = q_1(\mathbf{m}) + q_2(\mathbf{m}), \tag{E10}$$

$$P_V(\mathbf{m}) = q_3(\mathbf{m}), \tag{E11}$$

$$P_{H,V}(\mathbf{m}) = q_1(\mathbf{m}) + q_2(\mathbf{m}) + q_3(\mathbf{m}) + q_4(\mathbf{m}) + q_5(\mathbf{m}). \tag{E12}$$

Now, using the Poisson approximation in (11.93), the conditional probability of failure is obtained as

$$p_F(\mathbf{m}, T) = 1 - \exp[-\lambda_F(\mathbf{m})T], \tag{E13}$$

where $\lambda_F(\mathbf{m})$ is obtained by substituting the conditional probabilities in (E10)–(E12) in (E1). The corresponding reliability index is

$$\beta(\mathbf{m}, T) = \Phi^{-1}[1 - p_F(\mathbf{m}, T)]. \tag{E14}$$

The nested reliability problem is formulated through its limit-state function

$$\tilde{g}(U, \mathbf{M}, T) = U + \beta(\mathbf{M}, T), \tag{E15}$$

where U is a standard normal random variable, independent of $\mathbf{M}$. As we saw in Section 10.5, the solution of this problem is identical to the solution of the reliability problem with the complete set of random variables. Note that a FORM solution of this problem requires repeated evaluations of $\beta(\mathbf{m}, T)$ and its gradient with respect to $\mathbf{m}$ for selected values of $\mathbf{m}$ as determined by a search algorithm such as the iHL-RF algorithm. This is easily done because an analytical solution of $\beta(\mathbf{m}, T)$ is available as described above.

The solution of the above problem by the FORM for $T = 50$ yr yields $p_{F1} = 0.00239$ and the generalized reliability index $\beta = 2.82$. Virtually identical results are obtained by SORM analysis. A Monte Carlo solution with 100,000 samples yields $\hat{p}_F = 0.00249$ and $\beta = 2.81$ with a 6.33% c.o.v.

11.9 Concluding Remarks

Time- and space-variant reliability problems are the ultimate in structural reliability analysis. When these problems involve stochastic processes, multidimensional random fields, dynamic effects, and nonlinear structural behavior then they present serious challenges in finding viable solution approaches. This chapter presented several approximate solution approaches for certain classes of these problems. Chapters 12 and 13 offer additional solution approaches involving finite element methods and stochastic structural dynamics. Nevertheless, the field is still evolving and is wide open for further development and research.

The chapter started with a brief outline of random process theory. This topic is essential for modeling stochastic processes and random fields. The brief outline is sufficient as an introduction to the topic, but those aspiring to perform research in this field will need more in-depth knowledge of that topic. Specific attention was paid to two families of processes, Gaussian processes and the Poisson and related processes, which offer convenient means for modeling and analysis.

Structural reliability problems are categorized as encroaching, outcrossing, or encroaching-outcrossing, depending on how the temporal or spatial variability enters the problem. Encroaching problems arise when the limit-state functions involve time or spatial coordinates as parameters; outcrossing problems arise when the limit-state functions involve random processes; and encroaching-outcrossing problems arise when the limit-state functions involve both time or spatial coordinates as parameters as well as random processes. It was shown that a key quantity in all three problems is the mean rate at which the limit-state function down-crosses the zero level.

An upper-bound solution to the time/space variant component reliability problem is obtained in terms of the mean rate of zero-level down-crossings of the limit-state function. The mean rate is formulated as a parallel-system reliability problem involving the limit-state function values at two closely spaced time/space coordinates. A lower-bound solution is obtained as a series-system reliability problem defined by the limit-state function values at a series of time/space coordinates within the temporal/spatial domain of operation of the structure. Finally, an approximate solution was presented by modeling the failure events in time or space as Poisson events. Additional solution approaches are presented in Chapters 12 and 13. Although the solutions presented were for a component reliability formulation, the extension to systems problems should be self-evident.

The last part of the chapter presented several macro-level stochastic load models. These models employ the Poisson process to describe random changes in load values in time. Both sustained and intermittent load models were developed. The important topic of load combination was then presented, which aims to determine the probability distribution of the maximum combined load effect resulting from the combination of multiple stochastic loads. Finally, the load-coincidence method for the analysis of structural reliability under a combination of loads was presented. Several numerical examples illustrate the models and methods developed in this chapter.

APPENDIX 11A

Derivation of Limit Formula for Mean Down-Crossing Rate

We wish to compute the parallel-system probability

$$\Phi_2[-\beta(t),\,-\beta(t+\Delta t),\rho]$$
$$=\Phi[-\beta(t)]\Phi[-\beta(t+\Delta t)]+\int_0^{\rho}\phi_2[-\beta(t),\,-\beta(t+\Delta t),r]\mathrm{d}r \tag{A10.1}$$

for a small Δt and ρ near -1. Let $\beta(t)=-b$, $\beta(t+\Delta \mathrm{t})=b+\Delta b$, and $\rho=-1+\epsilon$.

First observe that $\Phi_2(b,-b+\Delta b,\,-1)=0$. It follows that

$$\lim_{\epsilon\to 0}\int_0^{-1+\epsilon}\phi_2(b,-b-\Delta b,r)\mathrm{d}r=-\Phi(b)\Phi(-b-\Delta b). \tag{A10.2}$$

Therefore,

$$\begin{aligned}\Phi_2(b,-b-\Delta b,\rho)&=\Phi(b)\Phi(-b-\Delta b)\\&\quad+\int_0^{-1}\phi_2(b,-b-\Delta b,r)\mathrm{d}r+\int_{-1}^{-1+\epsilon}\phi_2(b,-b-\Delta b,r)\mathrm{d}r\\&=\int_{-1}^{-1+\epsilon}\phi_2(b,-b-\Delta b,r)\mathrm{d}r\\&=\frac{1}{2\pi}\int_{-1}^{-1+\epsilon}\frac{1}{\sqrt{1-r^2}}\exp\left[-\frac{b^2+(b+\Delta b)^2+2b(b+\Delta b)r}{2(1-r^2)}\right]\mathrm{d}r.\end{aligned} \tag{A10.3}$$

Expanding the numerator of the exponential term and dropping Δb^2 terms, we obtain

$$\Phi_2(b,-b-\Delta b,\rho)=\frac{1}{2\pi}\int_{-1}^{-1+\epsilon}\frac{1}{\sqrt{1-r^2}}\exp\left[-\frac{b(b+\Delta b)}{1-r}\right]\mathrm{d}r. \tag{A10.4}$$

Observe that for $-1<r<-1+\epsilon$, the exponential term is virtually constant for small ϵ and can be taken out of the integral. Hence,

$$\Phi_2(b,-b-\Delta b,\rho)=\frac{1}{2\pi}\exp\left[-\frac{b(b+\Delta b)}{2}\right]\int_{-1}^{-1+\epsilon}\frac{1}{\sqrt{1-r^2}}\mathrm{d}r, \tag{A10.5}$$

which leads to the solution in (11.36).

PROBLEMS

11.1 Determine the admissible values of parameter a, for which the following functions are valid candidates for the auto-covariance function of a stationary process:

1) $\Sigma_{XX}(\tau) = |\tau|^{-a}$.
2) $\Sigma_{XX}(\tau) = \frac{\cosh(a\tau)}{\cosh(\pi\tau)}$.

11.2 Show that the function $\Sigma(\tau) = \frac{a^2}{1+\tau^4}$ is not a valid candidate as the auto-covariance function of a stationary process.

11.3 For each of the auto-covariance functions listed below, determine whether the process $X(t)$ is mean-square continuous and whether it is mean-square differentiable. If the process is mean-square differentiable, determine and plot $\Sigma_{X\dot{X}}(\tau)$, $\Sigma_{\dot{X}X}(\tau)$, and $\Sigma_{\dot{X}\dot{X}}(\tau)$.

1) $\Sigma_{XX}(\tau) = \left(1 - \frac{a|\tau|}{2}\right)\exp(-a|\tau|), 0 < a$.
2) $\Sigma_{XX}(\tau) = \frac{\sin(a\tau)}{\tau}, 0 < a$.

11.4 Determine the power-spectral density function of a zero-mean stationary process defined by the auto-covariance function $\Sigma_{XX}(\tau) = \exp(-a|\tau|)\cos(\omega_0\tau)$, $0 < a$. Plot the auto-covariance and PSD functions for $a = 1$ 1/s and $\omega_0 = \pi$ rad/s.

11.5 The limit-state function of a deteriorating element of an offshore structure is defined as $g[A, R, S(t), t] = (1 - At)R - S(t)$, where R is the initial capacity, A the rate of deterioration of the capacity, t the time in years, and $S(t)$ the stochastic wave load effect, which is modeled as a zero-mean stationary Gaussian process. The random variables R and A are statistically independent of each other and of $S(t)$ and have the distributions listed below. $S(t)$ has the auto-covariance function $\Sigma_{SS}(\tau) = 15^2\cos(\pi\tau)\exp\left[-(\pi\tau)^2\right]$, where τ is in seconds.

Variable	Distribution	Mean	c.o.v.
R	Normal	100	0.20
A	Lognormal	0.004 yr^{-1}	0.25

1) Determine the mean zero-level down-crossing rate of the limit-state function for $t = [0, 50]$ years.
2) Compute an upper bound to the failure probability of the structural element over a 50-year lifetime.

11.6 Reconsider the spacecraft with debris impact in Problems 6.8 and 8.6. Suppose the projectile impact events occur according to a Poisson process with the mean rate of 0.1 per day per square meter surface area. The random variables $\mathbf{X} = (R, V, \gamma_P)$ are statistically independent and identically distributed from projectile to projectile with the distributions listed in the table in Problem 6.8. Determine the probability that one or more holes will puncture through a 2m^2 panel of the spacecraft shield if the spacecraft stays in orbit for 20 days.

11.7 A structure is subjected to the combined stochastic load effects $S_1(t)$, $S_2(t)$, and $S_3(t)$, as well as a dead load effect S_0. The capacity of the structure, R, is also a random variable so that the limit-state function is formulated as

$$g[R, S_0, S_1(t), S_2(t), S_3(t)] = R - S_0 - S_1(t) - S_2(t) - S_3(t).$$

Suppose $S_1(t)$ is a square-wave process with mean rate of load changes $\nu_1 = 2\ \text{yr}^{-1}$, $S_2(t)$ is a Poisson pulse process with mean rate of "on" events $\lambda_2 = 5\ \text{yr}^{-1}$ and a mean duration of $\mu_2 = 0.1$ yr at each occurrence, and $S_3(t)$ is a Poisson spike process with the mean rate of occurrence $\lambda_3 = 0.2\ \text{yr}^{-1}$. The distributions of the load effects at each occurrence are listed in the following table together with those of the dead load and the capacity.

Variable	Distribution	Mean	c.o.v.
R	Lognormal	275	0.10
S_0	Normal	50	0.10
$S_1(t)$	Normal	50	0.20
$S_2(t)$	Normal	25	0.20
$S_3(t)$	Normal	50	0.30

1) Compute and plot the distribution of the maximum combined load effect over a $T = 20$ yr duration.
2) Determine the probability of failure and the reliability index of the structure over the $T = 20$ yr period.

12 Finite-Element Reliability Methods

12.1 Introduction

Today, structural, mechanical, geotechnical, and environmental problems are almost invariably solved numerically, by finite element (FE) or similar methods. These include static and dynamic problems and involve stress/strain, displacement, flow, conduction, and stability analysis. Failure criteria in such systems are usually formulated in terms of local or global response quantities, which are themselves implicit functions of input variables such as geometric dimensions, material properties, loads, and boundary conditions. These input quantities are often uncertain and, hence, give rise to uncertainty in the computed response quantities. From a probabilistic point of view, two problems of interest arise: (a) analysis of uncertainty propagation, and (b) reliability analysis. The first aims at assessing the uncertainty in response quantities, which is usually expressed in terms of first and second moments, i.e., means, variances, and covariances of the response quantities. The second aims at estimating the probability of occurrence of a failure event, such as the excursion of a response quantity above a safe threshold, as the complement of structural reliability. Because failure events usually occur under extreme conditions where the behavior of the system is nonlinear, reliability assessment often involves nonlinear FE analysis.

A variety of methods have been developed for second-moment analysis of FE solutions. These include the perturbation method developed by Handa and Anderson (1981), Hisada and Nakagiri (1981), and Kleiber and Hien (1992) for structural mechanics problems, Beacher and Ingra (1981) and Phoon et al. (1990) for geotechnical problems, and Liu et al. (1986a) for nonlinear dynamic problems. The approach involves a Taylor-series expansion of the governing equilibrium equations around the mean values of the random variables and evaluation of the coefficients in the expansion by perturbation analysis. In their analysis, Liu et al. (1986b) considered random fields of material properties and used shape functions to discretize them (see Section 12.5). Deodatis (1990, 1991), Deodatis and Shinozuka (1991), and Takada (1990) developed the weighted-integral method for the case where the elasticity matrix can be written as the product of a deterministic matrix and a scalar random field. The stiffness matrix can then be formulated as a linear function of deterministic matrices with random coefficients. Coupled with a first-order Taylor-series expansion of the response vector, this then leads to expressions for the second moments of the response quantities. Deodatis (1990) also derived an upper bound to the variance of a response quantity that is independent of the correlation structure of the underlying random field. Another method developed by Yamazaki et al. (1988) uses the Neumann expansion in conjunction with Monte Carlo simulation. The spectral stochastic FE method developed by Ghanem and Spanos (1991a) and described later in this chapter is also essentially a second-moment method, although it has been used for reliability analysis as well (Ghanem and Spanos 1991b). As this book focuses on reliability analysis, we will refrain from going into further details about second-moment FE methods. Furthermore, while most of the mentioned methods are limited to linear problems or have been applied to linear problems, our interest is in nonlinear problems because that is the regime under which structural failures occur and reliability analysis is of interest. Hence, this chapter focuses mostly on nonlinear problems, including structures with inelastic behavior and undergoing large deformations.

The chapter addresses a number of issues that arise when reliability methods are used in conjunction with nonlinear FE to assess the reliability of complex systems. We have seen that structural reliability methods, such as the first-order reliability method (FORM), second-order reliability method (SORM), and simulation, require repeated computations of limit-state functions and, in the case of the former two methods, their gradients and possibly the Hessians with respect to selected realizations of the uncertain quantities. Furthermore, to assess reliability sensitivities or importance measures, one needs to compute gradients of limit-state functions with respect to parameters present in these functions or in the probability distributions of uncertain quantities. One approach is to compute these gradients by finite differences. However, nonlinear FE solutions can be computationally time-consuming and costly. Hence, reducing the number of repeated computations becomes an important objective. Furthermore, because of the iterative numerical algorithms used in nonlinear FE analysis, which involve error tolerances, computation of gradients by finite differences is not only expensive but also unreliable. A host of other issues arise when dealing with FE representation of uncertain systems, such as discretization of random fields, definition of failure events, differentiability of limit-state functions, and the intrusive versus non-intrusive nature of the approach.

The chapter starts with a brief review of the FE formulation for nonlinear structural dynamic problems. This serves as an example for FE formulations for other problems. The next section presents the direct-differentiation method (DDM) for computing response gradients as an integral part of the FE analysis. It is shown that, at the expense of implementing derivative routines in the FE code, the needed gradients can be computed accurately and efficiently through linear analysis at the convergence of each iteration step in solving the nonlinear response. Next, the discretization of random fields and their representation in terms of random variables are discussed. The subsequent section presents a brief review of the spectral stochastic FE method, which has gained wide popularity. This is followed by a section on response-surface or metamodel methods. Several illustrative examples are presented.

It is natural that there will be notational conflicts between the two distinct fields of FE and reliability analyses. For example, x denotes coordinates in FE but is commonly used to denote the outcome of a basic random variable in reliability analysis. Similarly, σ, ϵ, and u denote stress, strain, and nodal displacement, respectively, in the FE formulation, but standard deviation, random model error, and the outcome of a standard normal random variable in the reliability formulation. We will not attempt to invent new notation to avoid these conflicts, with only one exception: In this chapter, $\mathbf{Y}$ will denote a vector of random variables with $\mathbf{y}$ as its outcome, so that $\mathbf{x}$ can be used to denote coordinates in the FE formulation. The reader should be able to understand the meaning of other conflicting symbols from their context.

12.2 Brief Review of the Finite-Element Formulation

In a standard FE formulation, the time- and space-discretized equation of motion of a structure under dynamic loads is written as (Zienkiewicz et al., 2013a)

$$\mathbf{M}\ddot{\mathbf{u}}_{n+1} + \mathbf{C}_0\dot{\mathbf{u}}_{n+1} + \mathbf{P}^{\text{int}}_{n+1} = \mathbf{P}^{\text{ext}}_{n+1}, \tag{12.1}$$

where $\mathbf{M}$ is the mass matrix, $\mathbf{C}_0$ the linear viscous damping matrix, $\mathbf{P}^{\text{int}}$ the vector of internal forces, $\mathbf{P}^{\text{ext}}$ the vector of external forces, and $\mathbf{u}$ the unknown vector of nodal displacements. A superposed dot indicates the derivative with respect to time and the subscript $n+1$ is used to denote quantities at discrete time t_{n+1}. For a nonlinear structure, $\mathbf{P}^{\text{int}}$ in general depends on the deformation of the structure; in the case of an inelastic material, it depends on the entire deformation history. (More generally, $\mathbf{P}^{\text{int}}$ also depends on the velocities if viscous effects are present.) For the sake of simplicity, we assume the external forces do not depend on the deformation of the structure.

In the FE method, the terms $\mathbf{P}^{\text{int}}$ and $\mathbf{P}^{\text{ext}}$ are computed at the nodal points of the structure by assembling over the finite elements. Using isoparametric elements and numerical quadrature (Zienkiewicz et al., 2013a, 2013b), the integration is performed in a normalized "parent domain" with coordinates χ, which are related to the original coordinates $\mathbf{x}$ through a transformation of the form $\mathbf{x} = \mathbf{N}(\chi)\hat{\mathbf{x}}$, with $\mathbf{N}(\chi)$ being the matrix of shape functions and $\hat{\mathbf{x}}$ the nodal coordinates of the element. The typical integral is of the form

$$\int_{\Omega_{\text{el}}} F(\mathbf{x})\text{d}\mathbf{x} \cong \sum_{i=1}^{np} \omega_i F(\chi_i)\left|\mathbf{J}_{\mathbf{x},\chi}\right|, \tag{12.2}$$

where χ_i are quadrature points, ω_i are corresponding weights, np is the number of integration points, $F(\cdot)$ is the integrand, and $\left|\mathbf{J}_{\mathbf{x},\chi}\right|$ is the determinant of the Jacobian of the coordinate transformation. In particular, the integrand for assembling the internal force vector is

$$F(\mathbf{x}) = \mathbf{B}^{\text{T}}\boldsymbol{\sigma}, \tag{12.3}$$

where $\boldsymbol{\sigma}$ is the stress vector and $\mathbf{B}$ the strain-displacement matrix defined by the incremental kinematic relation

$$d\boldsymbol{\epsilon} = \mathbf{B}d\mathbf{u}, \tag{12.4}$$

in which $\boldsymbol{\epsilon}$ is the strain vector. $\mathbf{B}$ depends on the element shape functions and nodal coordinates. When geometric nonlinearities are included, $\mathbf{B}$ may also depend on the displacements $\mathbf{u}$.

The set of equations in (12.1) is typically solved by assuming that the acceleration and velocity at time t_{n+1} are given by the displacement at that time step and the displacement, velocity, and acceleration values in the previous time step in the form

$$\ddot{\mathbf{u}}_{n+1} = a_1\mathbf{u}_{n+1} + a_2\mathbf{u}_n + a_3\dot{\mathbf{u}}_n + a_4\ddot{\mathbf{u}}_n \tag{12.5}$$

$$\dot{\mathbf{u}}_{n+1} = a_5\mathbf{u}_{n+1} + a_6\mathbf{u}_n + a_7\dot{\mathbf{u}}_n + a_8\ddot{\mathbf{u}}_n, \tag{12.6}$$

respectively, where a_i are constants depending on the type of integration rule used. This formulation includes well-known integration algorithms such as Newmark's β-method (Newmark, 1959), for which the coefficients are $a_1 = -a_2 = 1/\left(\beta h^2\right)$, $a_3 = -1/(\beta h)$, $a_4 = 1 - 1/(2\beta)$, $a_5 = -a_6 = \gamma/(\beta h)$, $a_7 = 1 - \gamma/\beta$, and $a_8 = h[1 - \gamma/(2\beta)]$, in which h is the time increment and β and γ are integration parameters. Among other options, for $\gamma = 1/2$ and $\beta = 1/4$ the integration scheme is second-order accurate and unconditionally stable. Stability and accuracy analysis of this and other integration methods are reported in Goudreau and Taylor (1972) and Hilber et al. (1977).

Substituting (12.5) and (12.6) in (12.1) and rearranging terms, one obtains

$$a_1\mathbf{M}\mathbf{u}_{n+1} + a_5\mathbf{C}_0\mathbf{u}_{n+1} + \mathbf{P}^{\text{int}}_{n+1}(\mathbf{u}_{n+1}) = \widetilde{\mathbf{P}}^{\text{ext}}_{n+1}, \tag{12.7}$$

where

$$\widetilde{\mathbf{P}}^{\text{ext}}_{n+1} = \mathbf{P}^{\text{ext}}_{n+1} - \mathbf{M}(a_2\mathbf{u}_n + a_3\dot{\mathbf{u}}_n + a_4\ddot{\mathbf{u}}_n) - \mathbf{C}_0(a_6\mathbf{u}_n + a_7\dot{\mathbf{u}}_n + a_8\ddot{\mathbf{u}}_n). \tag{12.8}$$

Equation (12.7) represents a nonlinear equation for the unknown displacements at time step t_{n+1}. This is solved by an iterative scheme, such as the Newton–Raphson algorithm, which amounts to iteratively solving the set of equations

$$\left[a_1\mathbf{M} + a_5\mathbf{C}_0 + \mathbf{K}\left(\mathbf{u}^i_{n+1}\right)\right]\Delta\mathbf{u}^i_{n+1} = \Delta\mathbf{P}^{\text{int}}\left(\mathbf{u}^i_{n+1}\right) \tag{12.9a}$$

$$\Delta\mathbf{P}^{\text{int}}\left(\mathbf{u}^i_{n+1}\right) = \widetilde{\mathbf{P}}^{\text{ext}}_{n+1} - \left[a_1\mathbf{M}\mathbf{u}^i_{n+1} + a_5\mathbf{C}_0\mathbf{u}^i_{n+1} + \mathbf{P}^{\text{int}}_{n+1}\left(\mathbf{u}^i_{n+1}\right)\right] \tag{12.9b}$$

$$\mathbf{u}^{i+1}_{n+1} = \mathbf{u}^i_{n+1} + \Delta\mathbf{u}^i_{n+1}, \tag{12.9c}$$

where $\mathbf{K} = \partial\mathbf{P}^{\text{int}}/\partial\mathbf{u}$ is the tangent stiffness matrix.

The above provides a brief account of the main steps in a standard nonlinear FE formulation. In the following sections, we discuss a number of issues that relate to reliability analysis using a formulation as described above.

12.3 Formulation of the Finite-Element Reliability Problem

In a FE formulation, the basic random variables $\mathbf{Y}$ are typically such quantities as material properties, geometric dimensions, and variables that define load values. On the other hand, failure criteria are usually defined in terms of response quantities such as strains, stresses, and displacements. Examples include classical failure criteria such as the maximum shear stress failure criterion, fracture due to excessive strain, inter-story drift for multistory buildings, excessive foundation settlement, and local or global instability (unbounded displacement). These quantities are implicit functions of the basic variables and, as described above, are computed through iterative numerical algorithms.

For the sake of illustration, let the limit-state function in a reliability problem be defined as

$$g(\mathbf{Y}) = g[\mathbf{S}(\mathbf{Y}), \mathbf{Y}], \tag{12.10}$$

where $\mathbf{S}(\mathbf{Y})$ is a vector of responses quantities obtained from a FE code as an implicit function of the basic random variables $\mathbf{Y}$. We have allowed explicit dependence of g on $\mathbf{Y}$, which may include such quantities as allowable thresholds or capacity values.

As we saw in the previous chapter, methods of structural reliability require repeated computations of the limit-state function for selected realizations $\mathbf{Y} = \mathbf{y}$. Each such computation requires a FE solution. Because the realization is usually generated in the standard normal space as $\mathbf{u}$, the inverse transform $\mathbf{y} = \mathbf{T}^{-1}(\mathbf{u})$ is used to obtain the realization in the original space, which then allows the response outcome $\mathbf{s}(\mathbf{y})$ and, therefore, $g(\mathbf{y})$ to be computed by the FE code. As we have also seen, the FORM requires the gradients in the standard normal space to be computed as $\nabla G(\mathbf{u}) = \nabla g(\mathbf{y})\mathbf{J}_{\mathbf{y},\mathbf{u}}$, which involves the gradient vector $\nabla g(\mathbf{y})$ and the Jacobian of the inverse transform. The latter depends on the probability distribution of the basic random variables and does not involve FE computations. However, the gradient $\nabla g(\mathbf{y})$ must be computed by FE analysis. For the particular limit-state function in (12.10), we have

$$\nabla g(\mathbf{y}) = \nabla_{\mathbf{s}} g|_{\mathbf{y}} \mathbf{J}_{\mathbf{s},\mathbf{y}} + \nabla_{\mathbf{y}} g|_{\mathbf{s}}, \tag{12.11}$$

where $\nabla_{\mathbf{s}} g|_{\mathbf{y}}$ and $\nabla_{\mathbf{y}} g|_{\mathbf{s}}$ are the gradient vectors of $g(\mathbf{s}, \mathbf{y})$ with respect to $\mathbf{s}$ and $\mathbf{y}$, respectively, with the other variables fixed, and $\mathbf{J}_{\mathbf{s},\mathbf{y}}$ is the Jacobian matrix of the response vector with respect to the vector of basic variables. To form the latter, we need to compute the partial derivatives of the response quantities with respect to the basic random variables at the realization point $\mathbf{y}$. A naïve approach would suggest computing the derivative with respect to an element y_i of $\mathbf{y}$ by finite differences, such as the forward-difference formula $\partial\mathbf{s}/\partial y_i \cong [\mathbf{s}(y_i + \Delta y_i) - \mathbf{s}(y_i)]/\Delta y_i$, where Δy_i is a small perturbation of y_i. This approach poses two problems: First, one extra FE solution is required for each perturbed variable, which significantly increases the computational effort if the number of random variables is large; second, because the FE solution requires iterative numerical computations that invariably involve convergence errors, a part of the variation $\mathbf{s}(y_i + \Delta y_i) - \mathbf{s}(y_i)$ would be due to these numerical errors, thus making the derivative computation by finite differences unreliable. To address this problem, in the next section we present the DDM to compute the response gradients as an integral part of the FE formulation. As shown there, DDM requires the solution of a linear problem after the convergence of each nonlinear step to compute the response gradient with respect to each variable. This approach is not only computationally efficient but also produces gradients that are accurate and consistent with the computed responses.

Algorithms for finding the design point, such as the iHL-RF algorithm described in Section 5.2.3, typically require repeated computations of the limit-state function and its gradient with respect to the basic variables. Figure 12.1 shows the flow diagram for such an algorithm that employs a FE code to compute responses $\mathbf{s}(\mathbf{y})$ and the Jacobian $\mathbf{J}_{\mathbf{s},\mathbf{y}}$. Note that the trial points are selected in the standard normal space and transformed into the original space so that the random variables can be interpreted for the FE code. In determining the trial point, the algorithm requires selecting a search direction, $\mathbf{d}$, and a step size, λ. The search direction in (5.21) can be used. However, for the step size, the Armijo rule suggested in (5.24) may require adjustment when working with a nonlinear FE code. This is because sometimes a large value of λ puts the trial point too far in the failure region, in which case the FE problem could become unstable. In Haukaas and Der Kiureghian (2004), the

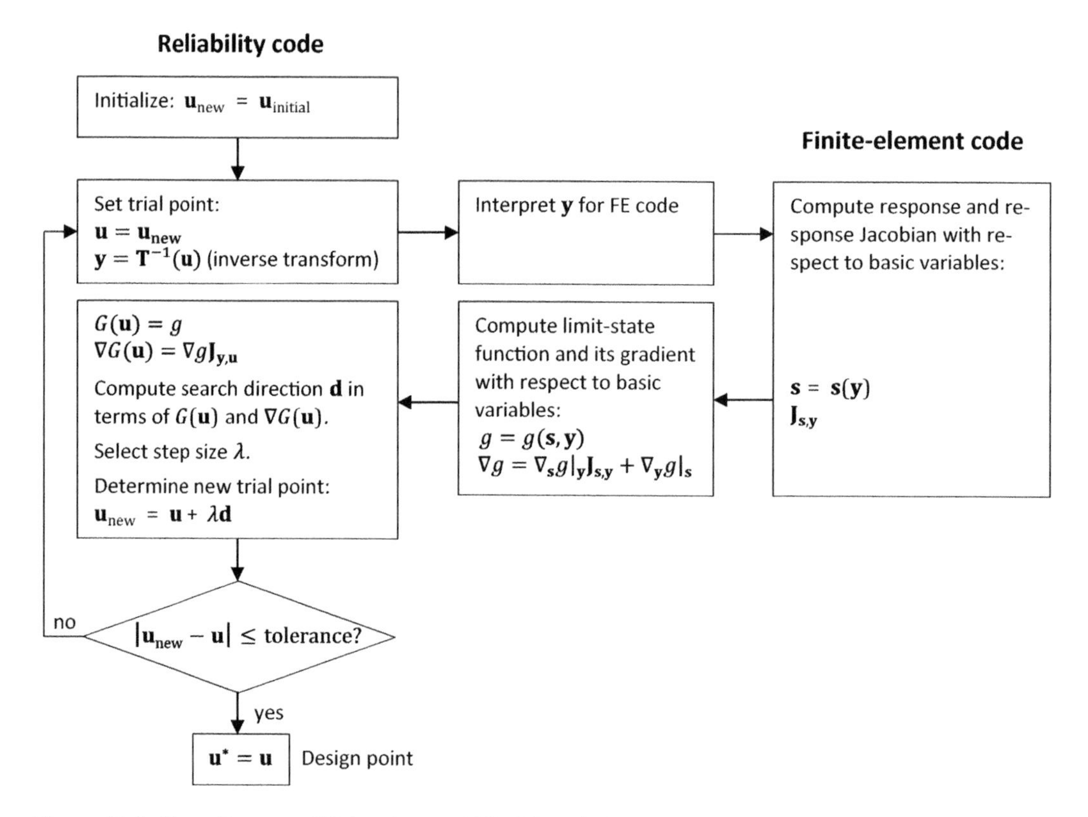

Figure 12.1 Flow diagram of finite-element FORM analysis

step size is monitored and selected such that the trial points approach the design point from within the safe domain. For example, if the failure criterion is a response quantity exceeding a threshold, the trial point along the search direction is selected such that the response quantity there does not exceed the specified threshold.

For most problems, convergence with the iHL-RF algorithm is achieved in 10–20 steps and seldom exceeds 50 or 100 steps. This is independent of the number of random variables but depends on the degree of nonlinearity of the limit-state function in the standard normal space. Thus, finding the design point requires around 10–100 repeated FE solutions, including computation of response gradients with respect to all the random variables at each solution point.

An important assumption in the FORM is that the limit-state function is continuously differentiable with respect to the basic random variables. This property is essential for the convergence of algorithms used for finding the design point. It is also necessary for defining the tangent plane at the design point. In the context of a FE formulation, this means that the derivative of a response quantity with respect to the basic variables, including load, structural property, and geometric variables, must be continuously differentiable. In the next section, we discuss the conditions under which this assumption is valid.

We saw in Chapter 7 that the classical SORM, i.e., the curvature-fitting SORM described in Section 7.2, requires computing the Hessian of the limit-state function at the design point. Computing second derivatives of response quantities in a FE formulation by finite differences is obviously problematic. Furthermore, a DDM for computing second derivatives in FE analysis has not been developed because it would require an enormous amount of effort in both formulation and code implementation. Hence, the classical SORM is not currently applicable in the context of a FE solution. The gradient-based SORM described in Section 7.3 requires repeated computations of the limit-state function and its gradients for realizations in the neighborhood of the design point. One advantage of this approach is that it finds the principal curvatures in descending order of magnitude so that the computation can be stopped when the last computed curvature is sufficiently small. This approach is viable if the FE code allows accurate and rapid computation of the response gradients. The point-fitting SORM method described in Section 7.4 requires repeated computations of the limit-state function in a secant algorithm to find the fitting points for the piecewise paraboloid surface. Assuming that each fitting point requires on average four calculations of the limit-state function, this approach requires approximately $8(n-1)$ limit-state function evaluations, where n is the number of basic random variables. This is in addition to the calculations necessary for finding the design point. Thus, the point-fitting SORM is also viable if the FE code allows accurate and rapid computation of the response gradients and the number of basic random variables is not too large.

As mentioned earlier, in FORM analysis, the number of required computations of the limit-state function is independent of the number of random variables. However, the size of $\mathbf{J}_{s,y}$ is obviously proportional to the size of $\mathbf{Y}$. Furthermore, in the gradient-based SORM or point-fitting SORM, the number of limit-state function evaluations increases with the number of basic random variables. Therefore, in order to reduce the number of repeated FE solutions, it is useful to know which uncertain quantities have a significant influence on the reliability measure. The FORM importance vector $\boldsymbol{\gamma}$ described in Section 6.5 provides the necessary information. Variables with small γ_i values can be considered deterministic (possibly set to their mean values), thus reducing the number of random variables in a subsequent SORM analysis. To reduce the number of random variables in the FORM analysis, Haukaas and Der Kiureghian (2005) suggest using importance measures obtained from a simple mean-centered, first-order, second-moment (MCFOSM) analysis (see Section 5.2.1). This measure for each basic variable is computed as the square root of the direct contribution of the variable to the variance of the limit-state function linearized at the mean point. The result is the vector $\nabla g(\mathbf{M})\mathbf{D}$, where $\nabla g(\mathbf{M})$ is the gradient vector at the mean point and $\mathbf{D}$ is the diagonal matrix of standard deviations. Thus, a single FE analysis with gradient computation at the mean point of the basic variables is sufficient to select the important variables for a subsequent FORM analysis. However, as stated by Haukaas and Der Kiureghian (2005), one should be careful with this approach when the behavior of the structure at the mean point is significantly different from that at the design point.

Among simulation methods, the ordinary Monte Carlo is easy to implement in a FE context as it requires repeated computations of the limit-state function at the sample points only. However, it can be computationally expensive for small failure probabilities. Subset simulation also requires repeated computations of the limit-state function only and is effective for small failure probabilities, particularly when the number of random variables is large. The importance-sampling method using design points can be far more efficient but requires prior finding of the design points. Directional sampling and orthogonal-plane sampling methods require root finding in sampled directions, which if a secant method is used requires repeated computations of the limit-state function. The latter method requires prior knowledge of the design point, and if directional importance sampling is used then that method also requires knowledge of the design points. Hence, these methods are viable when DDM is implemented in the FE code.

In addition to the failure probability, we are often interested in reliability sensitivities with respect to limit-state function or distribution parameters as well as in variable importance measures. As we saw in Chapter 6 in the context of FORM, the computation of these measures requires the gradient of the limit-state function with respect to parameters or variables of interest. Hence, implementation of DDM in a FE code also enables the evaluation of these sensitivity and importance measures.

Finally, an important issue in FE reliability analysis is the degree to which a FE code must be modified in order to allow reliability analysis. As shown in the flow diagram in Figure 12.1, FORM analysis requires repeated computations of the response and response gradients at selected trial points. If the FE code is DDM enabled, the needed change in the FE code is minimal. One needs to develop an "interpreter" to position the elements of the vector $\mathbf{y}$ in proper locations in the input data of the FE code. This can be done outside the FE code. In fact, the reliability and FE codes can be separate and communicate through writing/reading and interpreting input and output files. The gradient-based and point-fitting SORM and various sampling methods can also be implemented in a similar manner.

However, as we will point out later, some of the methods developed for uncertainty propagation or reliability analysis are more intrusive in the sense that they require extensive modification of the FE code. Lately, non-intrusive methods for uncertainty propagation and reliability analysis employing metamodels have been developed. These are described in Section 12.7.

12.4 The Direct-Differentiation Method

Let θ denote a parameter with respect to which we wish to compute the partial derivative $\partial \mathbf{u}/\partial \theta$ of the nodal displacements. θ may represent a material property constant, e.g., the elastic modulus, yield stress, hardening parameter, a geometry parameter such as a nodal coordinate, or a load parameter such as the magnitude of an external load at a point. To illustrate the basic idea of the DDM, we start from the simplest problem, i.e., a linear static problem defined by the matrix equilibrium equation

$$\mathbf{K}\mathbf{u} = \mathbf{P}, \tag{12.12}$$

where $\mathbf{K}$ is the stiffness matrix and $\mathbf{P}$ denotes the external load vector. Taking derivatives with respect to θ on both sides of the equation and rearranging terms, we obtain

$$\mathbf{K}\frac{\partial \mathbf{u}}{\partial \theta} = \frac{\partial \mathbf{P}}{\partial \theta} - \frac{\partial \mathbf{K}}{\partial \theta}\mathbf{u}. \tag{12.13}$$

It is clear that the solution of $\partial \mathbf{u}/\partial \theta$ in (12.13) can be obtained in the same manner as that of $\mathbf{u}$ in (12.12), with the right-hand side in (12.13) considered as the "external-load vector." Three observations are noteworthy: (a) The solution of the gradient requires assembling partial derivatives of the external-load vector and the stiffness matrix with respect to θ. Naturally, if θ is a load parameter $\partial \mathbf{K}/\partial \theta = 0$, and if θ is a material-property parameter $\partial \mathbf{P}/\partial \theta = 0$. The assembly process is similar to the assembly of the vector $\mathbf{P}$ and matrix $\mathbf{K}$, but uses partial derivative quantities at the element level. To enable DDM, these element-level partial derivatives must be implemented in the FE code. (b) The solution of the gradient involves the solution of the response, as $\mathbf{u}$ appears in the right-hand side of (12.13). Thus, the solution of the gradient should follow the solution of the response itself. (c) Usually (12.12) is solved by forward substitution of the factorized stiffness matrix. Thus, solving (12.13) only requires assembling the right-hand side and forward- and back-substituting the already-factorized stiffness matrix.

We now consider the more general problem defined in (12.1). First consider the derivative of the internal force vector $\mathbf{P}_{n+1}^{\text{int}}$ with respect to θ. For a nonlinear structure, the internal force vector is an implicit function of $\mathbf{u}_{n+1}$. Thus, using the chain rule of differentiation, we can write

$$\frac{\partial \mathbf{P}_{n+1}^{\text{int}}}{\partial \theta} = \frac{\partial \mathbf{P}_{n+1}^{\text{int}}}{\partial \mathbf{u}_{n+1}}\frac{\partial \mathbf{u}_{n+1}}{\partial \theta} + \left.\frac{\partial \mathbf{P}_{n+1}^{\text{int}}}{\partial \theta}\right|_{\mathbf{u}_{n+1}\text{ fixed}}, \tag{12.14}$$

where the derivative at fixed θ, $\partial \mathbf{P}_{n+1}^{\text{int}}/\partial \mathbf{u}_{n+1} = \mathbf{K}_{n+1}$, represents the current tangent stiffness matrix. Now differentiating both sides of (12.1) with respect to θ, using (12.14) and rearranging terms, we obtain

$$\mathbf{M}\frac{\partial \ddot{\mathbf{u}}_{n+1}}{\partial \theta} + \mathbf{C}_0\frac{\partial \dot{\mathbf{u}}_{n+1}}{\partial \theta} + \mathbf{K}_{n+1}\frac{\partial \mathbf{u}_{n+1}}{\partial \theta} = \frac{\partial \mathbf{P}_{n+1}^{\text{ext}}}{\partial \theta} - \left.\frac{\partial \mathbf{P}_{n+1}^{\text{int}}}{\partial \theta}\right|_{\mathbf{u}_{n+1}\text{ fixed}} - \frac{\partial \mathbf{M}}{\partial \theta}\ddot{\mathbf{u}}_{n+1} - \frac{\partial \mathbf{C}_0}{\partial \theta}\dot{\mathbf{u}}_{n+1}. \tag{12.15}$$

Replacing the derivatives $\partial \ddot{\mathbf{u}}_{n+1}/\partial \theta$ and $\partial \dot{\mathbf{u}}_{n+1}/\partial \theta$ with the derivatives of the right-hand sides of (12.5) and (12.6), respectively, and further rearranging terms, one arrives at

$$\begin{aligned}(a_1\mathbf{M} + a_5\mathbf{C}_0 + \mathbf{K}_{n+1})\frac{\partial \mathbf{u}_{n+1}}{\partial \theta} &= \frac{\partial \mathbf{P}_{n+1}^{\text{ext}}}{\partial \theta} - \left.\frac{\partial \mathbf{P}_{n+1}^{\text{int}}}{\partial \theta}\right|_{\mathbf{u}_{n+1}\text{ fixed}} \\ &\quad - \frac{\partial \mathbf{M}}{\partial \theta}\ddot{\mathbf{u}}_{n+1} - \frac{\partial \mathbf{C}_0}{\partial \theta}\dot{\mathbf{u}}_{n+1} \\ &\quad - \mathbf{M}\left(a_2\frac{\partial \mathbf{u}_n}{\partial \theta} + a_3\frac{\partial \dot{\mathbf{u}}_n}{\partial \theta} + a_4\frac{\partial \ddot{\mathbf{u}}_n}{\partial \theta}\right) \\ &\quad - \mathbf{C}_0\left(a_6\frac{\partial \mathbf{u}_n}{\partial \theta} + a_7\frac{\partial \dot{\mathbf{u}}_n}{\partial \theta} + a_8\frac{\partial \ddot{\mathbf{u}}_n}{\partial \theta}\right).\end{aligned} \tag{12.16}$$

The above equation expresses the desired response derivatives, $\partial \mathbf{u}_{n+1}/\partial \theta$, as a function of the solution of the response at the current time step and the response derivatives at the previous time step. It also requires formation of the partial derivatives with respect to θ of the mass and damping matrices and the external- and internal-load vectors, the latter with the nodal displacements fixed. Remarkably, the equation is linear, and it involves the dynamic stiffness matrix $a_1\mathbf{M} + a_5\mathbf{C}_0 + \mathbf{K}_{n+1}$, whose factorized form is available at the end of the iterative solution obtained from (12.9a–c). Thus, the response gradient is obtained by a single forward- and back-substitution of the factorized dynamic stiffness matrix after obtaining the solution of the nonlinear response through the iterative scheme in (12.9a–c).

We now consider the formulation of $\partial \mathbf{P}_{n+1}^{\text{int}}/\partial \theta|_{\mathbf{u}_{n+1}\text{ fixed}}$. For the sake of simplicity, we drop the subscript $n+1$, keeping in mind that all values are for time t_{n+1}. First consider the case when θ represents a material property constant and geometric nonlinearities are excluded. In that case matrix $\mathbf{B}$, defined in (12.3), is independent of θ and we have

$$\left.\frac{\partial \mathbf{P}^{\text{int}}}{\partial \theta}\right|_{\mathbf{u}\text{ fixed}} = \int_{\Omega_{\text{el}}} \mathbf{B}^{\text{T}}\left.\frac{\partial \boldsymbol{\sigma}}{\partial \theta}\right|_{\boldsymbol{\epsilon}\text{ fixed}} \mathrm{d}\mathbf{x}. \tag{12.17}$$

When geometric nonlinearities are included, $\mathbf{B}$ depends on the displacements via the element shape functions. Furthermore, if θ represents a nodal coordinate, then variations in the shape functions must also be considered. Haukaas and Der Kiureghian (2005) showed that the above equation then takes the form

$$\left.\frac{\partial \mathbf{P}^{\text{int}}}{\partial \theta}\right|_{\mathbf{u}\text{ fixed}} \cong \sum_{i=1}^{np} \omega_i \left[\left|\mathbf{J}_{\mathbf{x},\chi}\right| \left(\left.\frac{\partial \mathbf{B}^{\text{T}}}{\partial \theta}\right|_{\mathbf{u}\text{ fixed}} \boldsymbol{\sigma} + \mathbf{B}^{\text{T}} \mathbf{k} \left.\frac{\partial \boldsymbol{\epsilon}}{\partial \theta}\right|_{\mathbf{u}\text{ fixed}} + \mathbf{B}^{\text{T}} \left.\frac{\partial \boldsymbol{\sigma}}{\partial \theta}\right|_{\boldsymbol{\epsilon}\text{ fixed}} \right) + \mathbf{B}^{\text{T}} \boldsymbol{\sigma} \left|\mathbf{J}_{\mathbf{x},\chi}\right| \mathbf{J}_{\mathbf{x},\chi}^{-\text{T}} \frac{\partial \mathbf{N}_x(\chi)}{\partial \chi} \right], \tag{12.18}$$

in which $\mathbf{k} = \partial\boldsymbol{\sigma}/\partial\boldsymbol{\epsilon}$ is the current material tangent stiffness and $\mathbf{N}_x(\chi)$ is the column vector of shape functions in $\mathbf{N}(\chi)$ corresponding to the degree of freedom matching the nodal coordinate represented by θ when θ represents a nodal coordinate. The reader should consult Haukaas and Der Kiureghian (2005) for more details.

The assembly of partial derivative matrices $\partial\mathbf{M}/\partial\theta$ and $\partial\mathbf{C}_0/\partial\theta$ and vector $\partial\mathbf{P}_{n+1}^{\text{ext}}/\partial\theta$ are described in Haukaas and Der Kiureghian (2004, 2005). Furthermore, a number of important implementation issues are discussed in Liu and Der Kiureghian (1991), Zhang and Der Kiureghian (1993), Conte et al. (2003), Haukaas and Der Kiureghian (2005), and Gu et al. (2009). Formulations for specific material models and nonlinearities are presented in the above references as well as in Roth and Grigoriu (2001), Barbato and Conte (2005a, 2005b), and Barbato et al. (2006).

In the case of static analysis, (12.16) reduces to

$$\mathbf{K}_{n+1}\frac{\partial \mathbf{u}_{n+1}}{\partial \theta} = \frac{\partial \mathbf{P}_{n+1}^{\text{ext}}}{\partial \theta} - \left.\frac{\partial \mathbf{P}_{n+1}^{\text{int}}}{\partial \theta}\right|_{\mathbf{u}_{n+1}\text{ fixed}}. \tag{12.19}$$

In this case, to compute the product $\nabla_{\mathbf{s}} g|_{\mathbf{y}} \mathbf{J}_{\mathbf{s},\mathbf{y}}$ in (12.11), where $\mathbf{s}$ is the vector of response quantities and $\mathbf{y}$ is the selected outcome of the vector of random variables, it is not necessary to explicitly compute the Jacobian $\mathbf{J}_{\mathbf{s},\mathbf{y}}$. Using the so-called adjoint method (Arora and Haug, 1979), one first solves for the vector λ in

$$\mathbf{K}^{\text{T}}\lambda = \left(\nabla_{\mathbf{s}} g|_{\mathbf{y}}\right)^{\text{T}}, \tag{12.20}$$

and then computes the needed product as

$$\nabla_{\mathbf{s}} g|_{\mathbf{y}} \mathbf{J}_{\mathbf{s},\mathbf{y}} = \lambda^{\text{T}} \left(\frac{\partial \mathbf{P}^{\text{ext}}}{\partial \mathbf{y}} - \left.\frac{\partial \mathbf{P}^{\text{int}}}{\partial \mathbf{y}}\right|_{\mathbf{u}} \right), \tag{12.21}$$

where $\mathbf{K}$, $\mathbf{P}^{\text{ext}}$, and $\mathbf{P}^{\text{int}}$ are the values at the convergence of (12.9a–c). This requires a single solution of (12.20) for all random variables rather than separate solutions of (12.19) for each random variable. This method cannot be used in the case of a dynamic response because, as can be seen in (12.16), the response sensitivities in step n are needed to compute the sensitivities in step $n + 1$.

One important issue with regard to use of response sensitivities for reliability analysis is the continuity of the gradients, as that is necessary in order to find the design point in the FORM. This issue is discussed and investigated through examples by Zhang and Der Kiureghian (1993), Kleiber et al. (1997), Haukaas and Der Kiureghian (2005), and Barbato and Conte (2005b). One can show that when θ represents a material property parameter, such as yield stress or hardening modulus, then discontinuities in the response derivatives may occur when the material makes a non-smooth state transition in a loading path, such as from an elastic to a plastic state. These discontinuities are particularly evident in the sensitivities of local response quantities such as the stress at a Gauss point of an element. However, for global response quantities, such as a nodal displacement that is contributed by deformations throughout the body, the derivatives tend to be smooth. Furthermore, because of the discrete nature of time steps used in the iterative solution, discontinuities appear as a rapid rise or fall over a time step rather than a sudden jump in value. In order to overcome these sharp variations in response derivatives, Haukaas and Der Kiureghian (2004) have developed smoothened material models, some of which are implemented in the OpenSees software (McKenna et al., 2003). Haukaas and Der Kiureghian (2004) and Barbato and Conte (2005b) show comparisons of response sensitivity and reliability results for smooth and non-smooth material models.

Example 12.1 – Response Sensitivities of a Perforated Strip under Cyclic Load

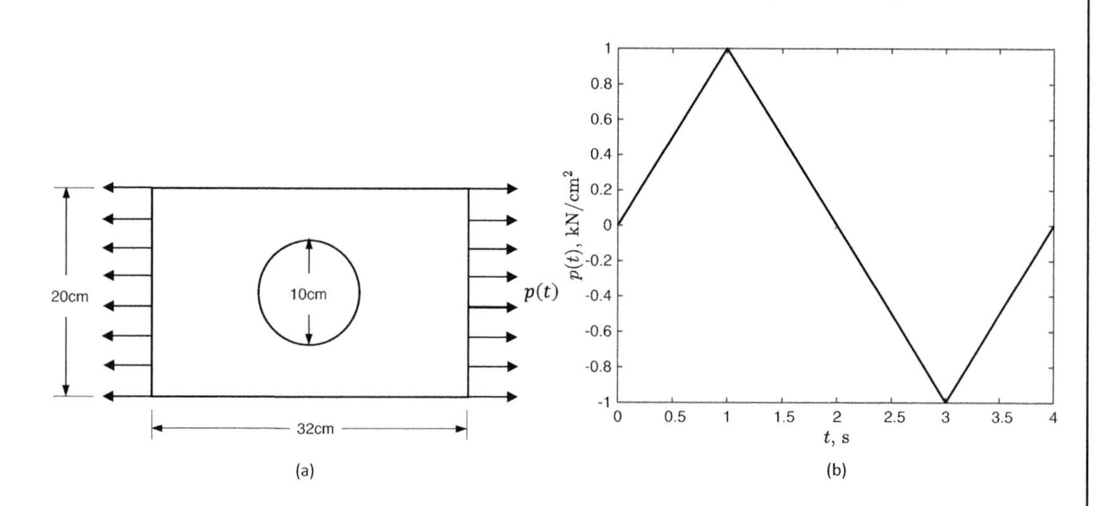

Figure 12.2 Perforated strip under quasi-static cyclic load (after Zhang and Der Kiureghian, 1993)

Consider the rectangular strip with a circular hole shown in Figure 12.2(a), which is subjected to the quasi-static cyclic load shown in Figure 12.2(b). The inertia and damping forces are considered to be negligible. The strip is assumed to be infinitely thick and in a state of plane strain. The material follows the J_2 plasticity model with the elastic modulus $E = 10^3$ kN/m^2, Poisson's ratio $\nu = 0.3$, yield stress $\sigma_y = 1$ kN/cm^2, and isotropic and kinematic hardening moduli $H_{\text{iso}} = H_{\text{kin}} = 50$ kN/cm^2. Zhang and Der Kiureghian (1993) present the details of this material model and the DDM formulation at the element level. Four-node quadrilateral finite elements with the mesh shown in Figure 12.3 for a quarter of the body are used to represent the

Example 12.1 (cont.)

strip. A time step size of $\Delta t = 0.0625$ is used in the iterative solution for the response and for the subsequent solution of the response gradients. Figure 12.4 shows the displacement response u_1 at point A and the stress response σ_{11} at point B. Observe in Figure 12.4b that yielding occurs at point B at around $t = 0.27$ s and $t = 1.96$ s.

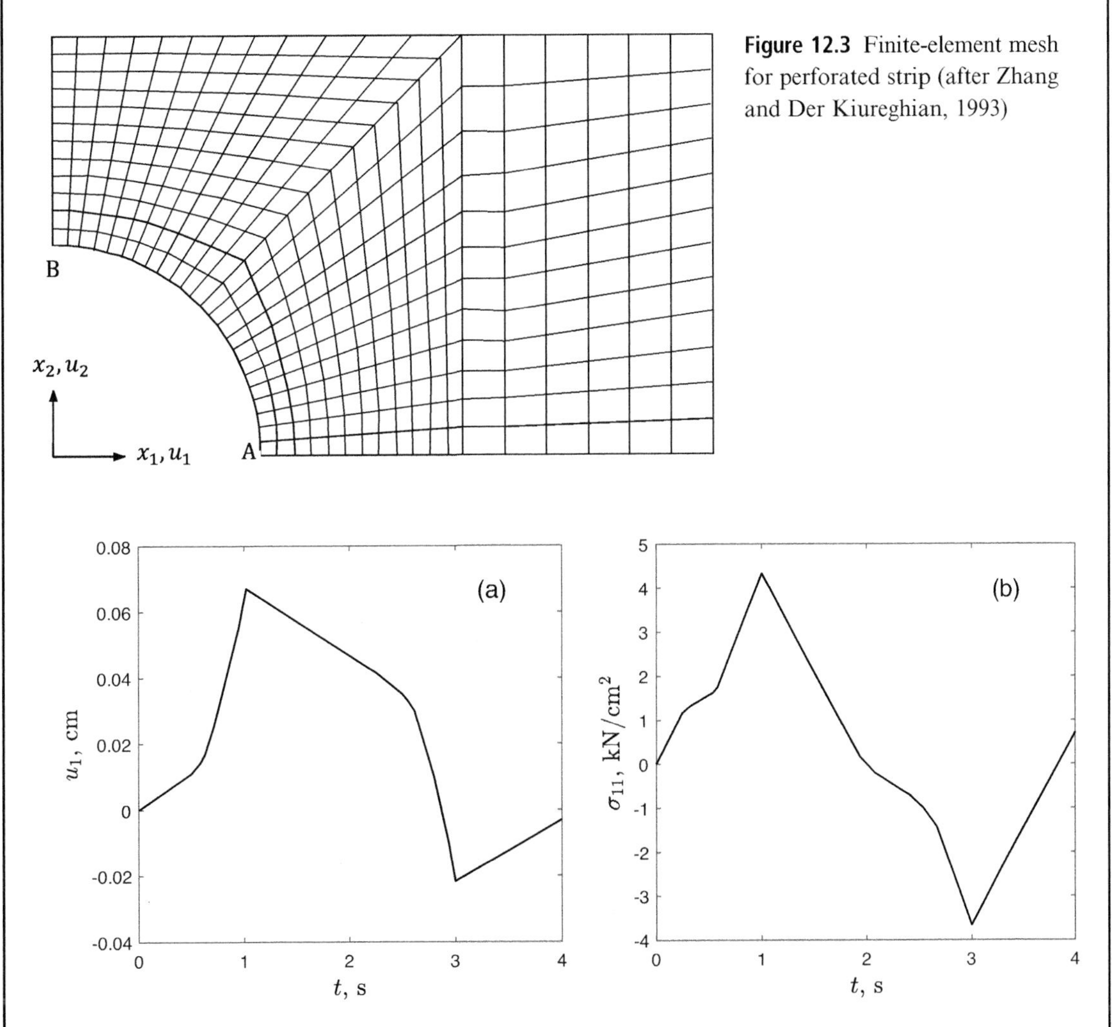

Figure 12.3 Finite-element mesh for perforated strip (after Zhang and Der Kiureghian, 1993)

Figure 12.4 Responses of perforated strip: (a) displacement u_1 at point A; (b) stress σ_{11} at point B (after Zhang and Der Kiureghian, 1993)

Figure 12.5 shows the derivatives of the displacement and stress responses with respect to the material parameters σ_y, H_{iso}, and H_{kin}, each multiplied by the parameter itself so that the plotted quantity has the same units as the corresponding response. This scaling means that

Example 12.1 (cont.)

the derivative curves are comparable if the same percentage of variation in each parameter is applied. Observe in Figure 12.5(a) that the derivatives of the displacement response with respect to all three parameters are continuous throughout the loading period. However, rapid changes in the trends are observed at the times of reversal in loading. On the other hand, the derivative of the stress σ_{11} at point B with respect to the yield stress σ_y in Figure 12.5(b) shows discontinuities at the instances $t = 0.27$ s and $t = 1.96$ s, when the material transitions from the elastic to the plastic state, though the jumps in the values appear as steep ramps because of the selected finite time step. Observe also that discontinuities in the derivative of the stress response with respect to the yield stress do not occur when the material transitions back to the elastic state upon unloading. Finally, observe that, for the present problem, the stress at point B is not sensitive to the hardening moduli, at least for the same percent variation in each parameter as in the yield stress. These derivative results, including the sizes of the discontinuities at the mentioned time instances, were verified through finite difference calculations in Zhang and Der Kiureghian (1993).

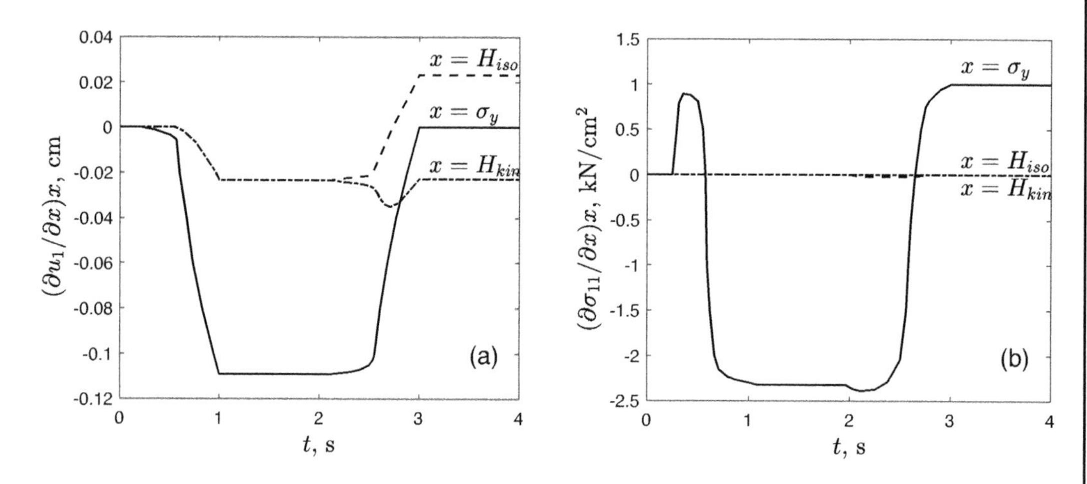

Figure 12.5 Scaled response derivatives with respect to parameters σ_y, H_{iso}, and H_{kin}: (a) derivatives of response u_1 at point A; (b) derivatives of response σ_{11} at point B (after Zhang and Der Kiureghian, 1993)

Example 12.2 – Reliability of Built-Up Column

Consider the built-up column in Figure 12.6(a), which is adopted from Liu and Der Kiureghian (1989). The column is composed of truss members with elastic material but subject to large deformations with strain energy function $E_{ss}DE_{ss}/2$, where D is the elastic modulus

Example 12.2 (cont.)

and $E_{ss} = (l^2 - L^2)/(2L^2)$ is the Green-Lagrangian strain measure, in which l and L are the current and original lengths of the member. The FE formulation accounts for geometric nonlinearity and is described in detail in the above reference. The strut members have cross-sectional areas of 1.59 in^2 and a random elastic modulus D_1. The braces and battens have cross-sectional areas of 0.938 in^2 and a random elastic modulus D_2. D_1 and D_2 are assumed to be jointly lognormally distributed with means $\mu_{D_1} = \mu_{D_2} = 30{,}000$ ksi, coefficients of variation (c.o.v.) $\delta_{D_1} = \delta_{D_2} = 0.08$, and a correlation coefficient of $\rho_{D_1D_2} = 0.3$. The x_2 coordinates of the nodes are deterministic, but the x_1 coordinates are considered random to model manufacturing imperfection. The latter coordinates are assumed to be normally distributed with means $\mu_{X_{1i}} = 60$ in or -60 in and standard deviations $\sigma_{X_{1i}} = 1.2$ in. Using the mean values of the random variables, the buckling load of the column is determined to be 2,593 kips. The column is subjected to a lateral force F_1 at midspan and an axial force F_2. These loads are assumed to be independent random variables with lognormal distributions. F_1 has mean $\mu_{F_1} = 20$ kips and standard deviation $\sigma_{F_2} = 2$ kips, while F_2 has variable mean μ_{F_2} and a c.o.v. of $\delta_{F_2} = 0.1$. Thus, 22 basic random variables $\mathbf{Y} = (D_1, D_2, X_{1,2}, \ldots, X_{1,19}, F_1, F_2)$ define the reliability problem. The limit-state function is defined by

$$g(\mathbf{Y}, u_0) = u_0 - U_{1,10}(\mathbf{Y}), \tag{E1}$$

where $U_{1,10}(Y)$ denotes the displacement at node 10 in the x_1 direction and u_0 is the specified threshold. The probability of failure is computed by the FORM and point-fitting SORM for

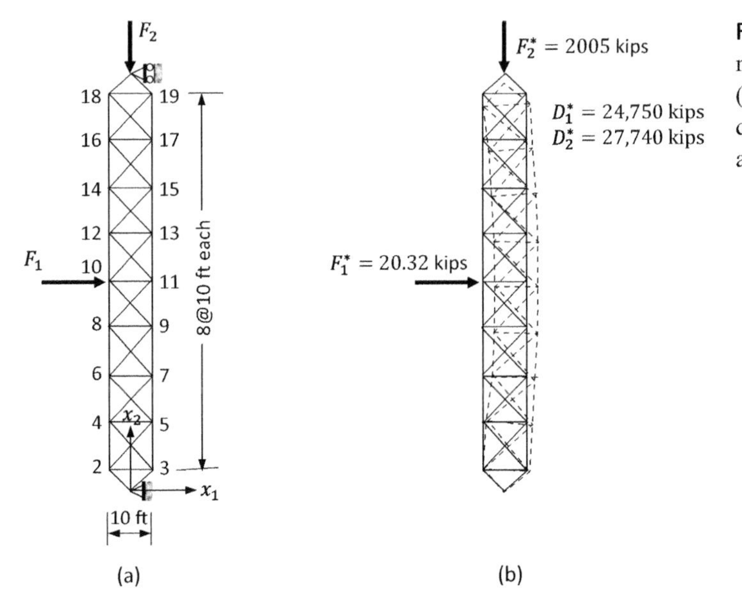

Figure 12.6 Built-up column: (a) mean undeformed configuration; (b) design-point undeformed and deformed configurations (after Liu and Der Kiureghian, 1989)

Example 12.2 (cont.)

$\mu_{F_2} = 500$ to 2,500 kips and $u_0 = 0$ to 30 in. The results from the two methods nearly coincide, suggesting that the limit-state surface in the standard normal space is nearly flat, at least in the neighborhood of the design point. Hence, we present only the results of the FORM analysis. The program CalREL in conjunction with an earlier version of the FE code FEAP (Taylor and Govindjee, 2020) with DDM functionality was used to solve this problem.

Figure 12.6(b) shows the undeformed and deformed configurations of the column and the outcomes of random variables F_1, F_2, D_1, and D_2 at the design point for $\mu_{F_2} = 1{,}500$ kips and $u_0 = 30$ in. As one would expect, the design-point values of D_1 and D_2 are below their means and those of loads F_1 and F_2 are above their respective mean values. Furthermore, although it is difficult to see in this figure, the struts on the compression side in the undeformed configuration exhibit a sine-shaped imperfection with the design-point coordinate at node 10 being $x^*_{1,10} = -59.65$ in.

Figure 12.7 shows the probability of failure as a function of u_0 for selected values of μ_{F_2}. For small values of the mean axial load, the probability of failure rapidly decreases with increasing threshold u_0. However, for large values of the mean axial load, the probability of failure is relatively insensitive to the threshold. This is an indication of instability because, under a large axial load, the lateral deformation of the column tends to become unbounded and easily exceeds any specified threshold.

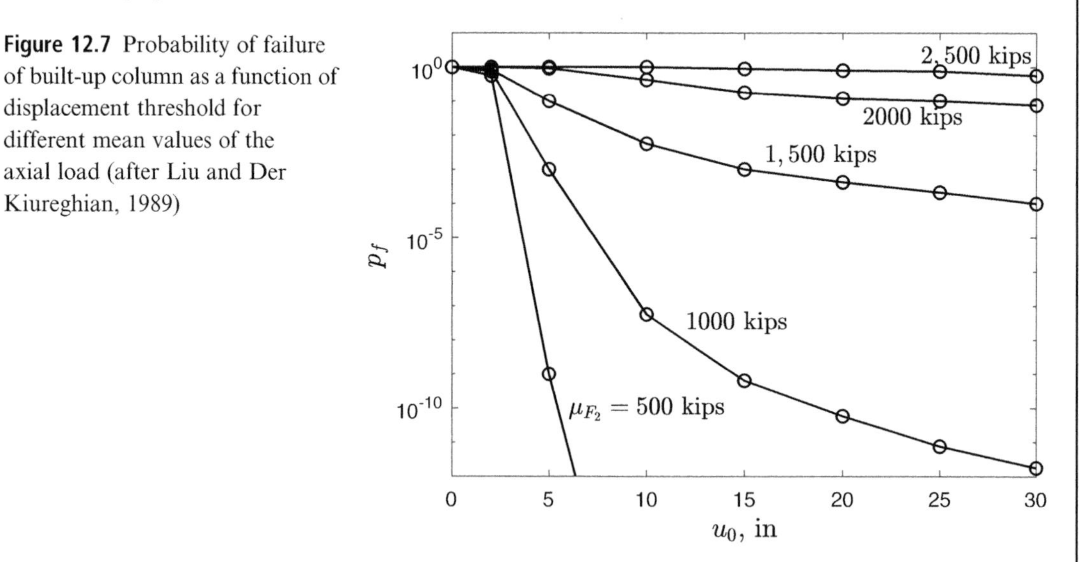

Figure 12.7 Probability of failure of built-up column as a function of displacement threshold for different mean values of the axial load (after Liu and Der Kiureghian, 1989)

Liu and Der Kiureghian (1989) also computed the scaled importance measures relative to the means of the random variables, $\boldsymbol{\delta} = (\sigma_1 \partial p_F / \partial \mu_1 \cdots \sigma_n \partial p_F / \partial \mu_n)$, for the values $(u_0, \mu_{F_2}) =$ (30 in, 500 kips), (30 in, 1,500 kips), and (5 in, 1,500 kips) of the threshold and the mean axial load. It turns out that D_1 is the most important random variable in all three

Example 12.2 (cont.)

cases, followed by F_2 in the latter two cases and followed by F_1 in the first case. The elastic modulus D_2 has relatively little importance. Among the nodal coordinates, the most important is the x_1 coordinate of node 10.

Example 12.3 – Reliability of Reinforced-Concrete Frame Building and Relative Importance of Random Variables

Consider the two-bay, two-story reinforced-concrete frame building shown in Figure 12.8, which is adopted from Haukaas and Der Kiureghian (2005). Employing the OpenSees software (McKenna et al., 2003), the frame is modeled using displacement-based beam-column elements with fiber-discretized cross sections. The uniaxial material models in Figure 12.8(b) describe the behavior of the reinforcing steel and the confined and unconfined concrete fibers. The concrete cross sections are discretized using 14 fiber elements in the in-plane direction. Interior column sections have six reinforcing bars on each side, each with an area of 500 mm^2; exterior column sections have six reinforcing bars on each side, each with an area of 250 mm^2; girder sections have four reinforcing bars on each side, each with an area of 325 mm^2. All material parameters, nodal coordinates, and lateral loads are characterized as random variables. Each member of the frame is assigned one random variable per material property. Fibers of the same material type within one element share the same random variables; similar material parameters in different members are correlated. Denoting $\mathrm{LN}(\mu, \delta, \rho)$ as a group of equicorrelated lognormal random variables of mean μ, coefficient of variation δ, and common correlation coefficient ρ, the following values are selected for the reinforcing steel: $\sigma_y = \mathrm{LN}(300\ \mathrm{N/mm^2},\ 0.05,\ 0.5)$, $E = \mathrm{LN}(200\ \mathrm{kN/mm^2},\ 0.05,\ 0.5)$, and $b = \mathrm{LN}(0.02,\ 0.10,\ 0.5)$. For the confined concrete of columns: $f'_c = \mathrm{LN}(22\ \mathrm{N/mm^2},\ 0.15,\ 0.3)$, $f'_{cu} = \mathrm{LN}(18\ \mathrm{N/mm^2},\ 0.15,\ 0.3)$, $\epsilon_c = \mathrm{LN}(0.005,\ 0.15,\ 0.3)$, and $\epsilon_{cu} = \mathrm{LN}(0.02,\ 0.15,\ 0.3)$. For the girder concrete and cover concrete of columns: $f'_c = \mathrm{LN}(16\ \mathrm{N/mm^2},\ 0.15,\ 0.3)$, $f'_{cu} = 0$, $\epsilon_c = \mathrm{LN}(0.002,\ 0.15,\ 0.3)$, and $\epsilon_{cu} = \mathrm{LN}(0.006,\ 0.15,\ 0.3)$. The lateral loads are modeled as $P_i = \mathrm{LN}(400\ \mathrm{kN},\ 0.20,\ 0.6)$, $i = 1, 2$. Deterministic gravity loads are as follows (subscripts denote node numbers; see Figure 12.8(a)): $G_2 = G_6 = G_8 = 850$ kN, $G_5 = 1,700$ kN, $G_3 = G_9 = 430$ kN. The nodal coordinates are modeled as independent normal random variables with mean values as in Figure 12.8(a) and standard deviations equal to 20 mm. In total, 104 random variables define the reliability problem.

We consider the following four limit-state functions under a static push-over analysis:

$$g_1 = 0.02 \times 8.3\ \mathrm{m} - u_{\mathrm{roof}}, \tag{E1}$$

$$g_2 = 0.02 \times 3.7\ \mathrm{m} - (u_{\mathrm{roof}} - u_{\mathrm{floor}}), \tag{E2}$$

Example 12.3 (cont.)

$$g_3 = 0.02 \times 4.6\text{ m} - u_{\text{floor}}, \tag{E3}$$

$$g_4 = 0.006 + \epsilon_{\text{compressive outermost fiber element 3}}, \tag{E4}$$

where u_{roof} refers to the horizontal displacement of the roof; u_{floor} refers to the horizontal displacement of the first floor, and $\epsilon_{\text{compressive outermost fiber element 3}}$ is the largest negative (compressive) strain in the cross section of the foundation end of element 3. The performance function g_2, for instance, defines the failure event as the exceedance of upper inter-story drift above 2% of the story height.

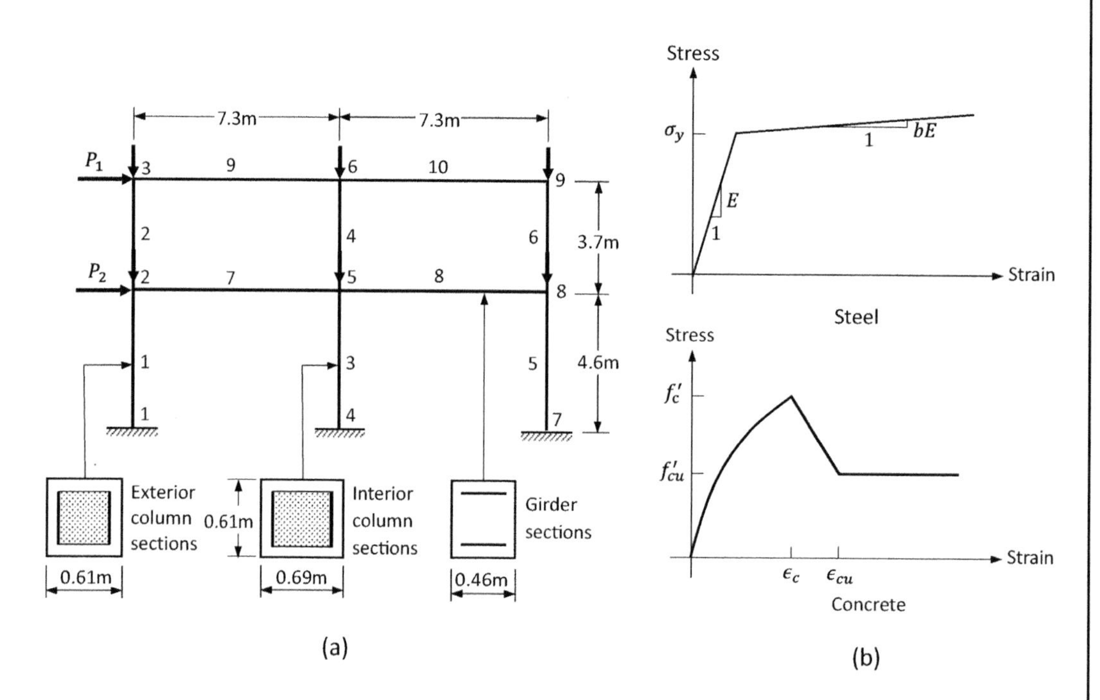

Figure 12.8 Reinforced-concrete frame: (a) geometry and node and element numbers, (b) uniaxial behavior of materials (after Haukaas and Der Kiureghian, 2005, with permission from ASCE)

FORM analysis with the above limit-state functions produces the reliability indices (first-order failure probabilities) $\beta_1 = 2.45$ ($p_{F1} = 0.00711$), $\beta_2 = 2.44$ ($p_{F1} = 0.00733$), $\beta_3 = 2.46$ ($p_{F1} = 0.00689$), and $\beta_4 = 2.49$ ($p_{F1} = 0.00639$). The results indicate fairly equal reliabilities for the four failure modes, with the upper inter-story drift mode having slightly higher probability of occurrence. The design points are found by the iHL-RF algorithm with 5–10 iterations each, where the gradients are computed by DDM. Figure 12.9 shows the nonlinear finite element response at each iteration of the algorithm for the limit-state function g_1. It is

Example 12.3 (cont.)

observed that the response converges to the design-point displacement of $0.02 \times 8.3 = 0.166$ m in the final iteration.

We examine the relative importance of the random variables for limit-state function g_1 in terms of the vectors $\widehat{\boldsymbol{\gamma}}$, $\boldsymbol{\delta}$, and $\boldsymbol{\eta}$ (see Sections 6.5 and 6.8). Recall that $\widehat{\boldsymbol{\gamma}}$ provides the relative importances of the random variables in terms of their individual contributions to the variance of the linearized limit-state function, $\boldsymbol{\delta}$ provides the relative importances in terms of scaled variations in the means, and $\boldsymbol{\eta}$ provides the relative importances in terms of scaled variations in the standard deviations. We find that the two most important random variables are the lateral loads P_1 and P_2, with γ_i values 0.86 and 0.48. Figure 12.10 shows the absolute values of the elements of $\widehat{\boldsymbol{\gamma}}$, $\boldsymbol{\delta}$, and $\boldsymbol{\eta}$ for the next 23 most important random variables. Also shown are importance values computed by the MCFOSM (marked as FOSM), which require only one gradient computation at the mean point. The purpose is to see whether the important variables can be identified by a MCFOSM analysis, thus allowing elimination of unimportant random variables in subsequent FORM analyses.

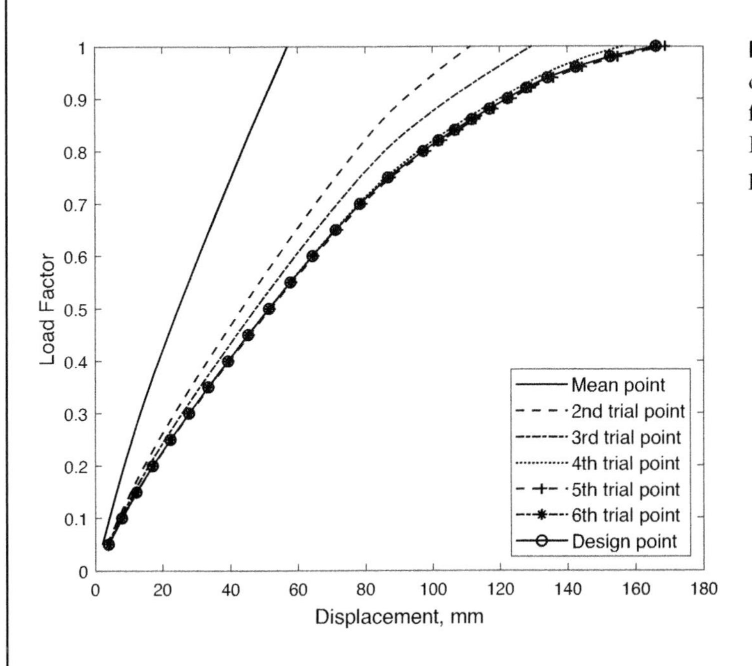

Figure 12.9 Trial-point responses of frame structure for limit-state function g_1 (after Haukaas and Der Kiureghian, 2005, with permission from ASCE)

Remarkably, several nodal coordinates are included in the top 25 most important random variables. Specifically, the horizontal coordinates of the lower columns are found to be important. The design-point realization of the geometrical imperfections, magnified by a factor of 500, is shown in Figure 12.11. This realization foresees a situation where the lower columns are all

Example 12.3 (cont.)

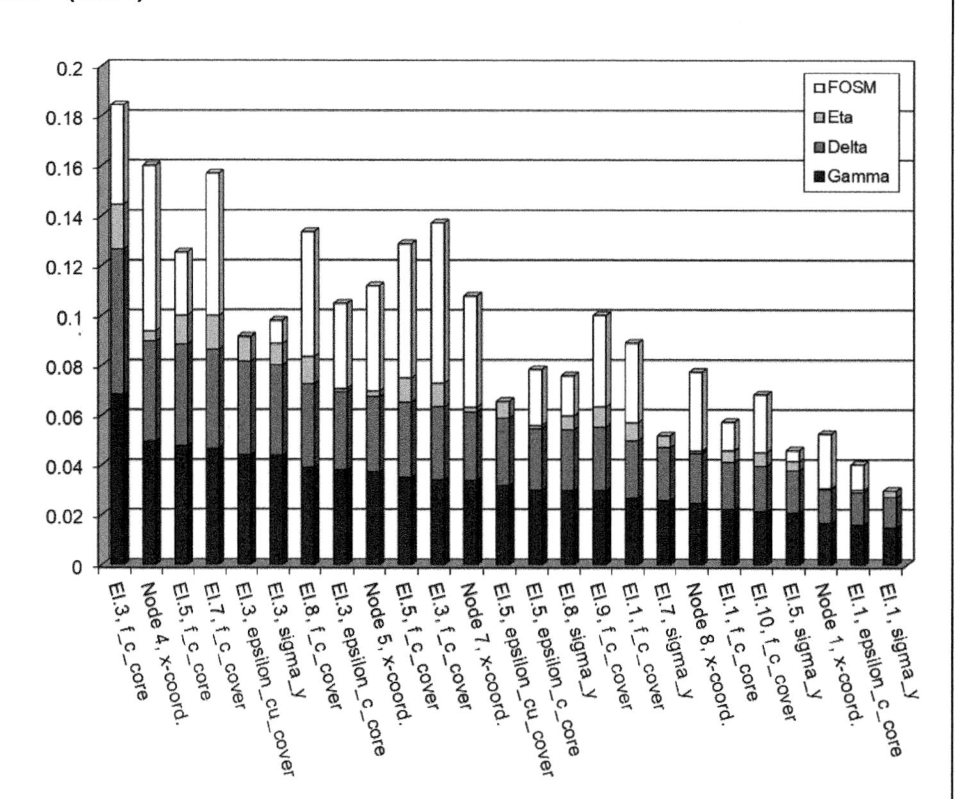

Figure 12.10 Importance vectors of frame structure for limit-state function g_1 for the 3rd–25th most important random variables. Subscripts "core" and "cover" refer to confined and unconfined concrete; "El" denotes element number (after Haukaas and Der Kiureghian, 2005, with permission from ASCE)

constructed in inclined positions in the direction of the specified displacement threshold. This suggests that the importance of the uncertainty in the nodal coordinates in a frame structure is due to the so called "P-Δ" effect: variability in the nodal coordinates of the lower columns effectively induces additional bending moments, which tend to increase the horizontal drift.

Figure 12.10 suggests that the importance rankings of random variables by the $\widehat{\boldsymbol{\gamma}}$ and $\boldsymbol{\delta}$ vectors are similar, whereas that of $\boldsymbol{\eta}$ is significantly different. Furthermore, the importance ordering obtained by MCFOSM overall is not too dissimilar from that of $\widehat{\boldsymbol{\gamma}}$, except for ϵ_{cu} and σ_y. This is because the frame is not subjected to extensive nonlinear action at the mean realization of the random variables, as can be seen in Figure 12.9. Nevertheless, the importance vector obtained by MCFOSM can be used as a guide to eliminate some of the less important random variables that are not related to the nonlinear behavior of the structure, e.g., nodal coordinates at the roof level.

Example 12.3 (cont.)

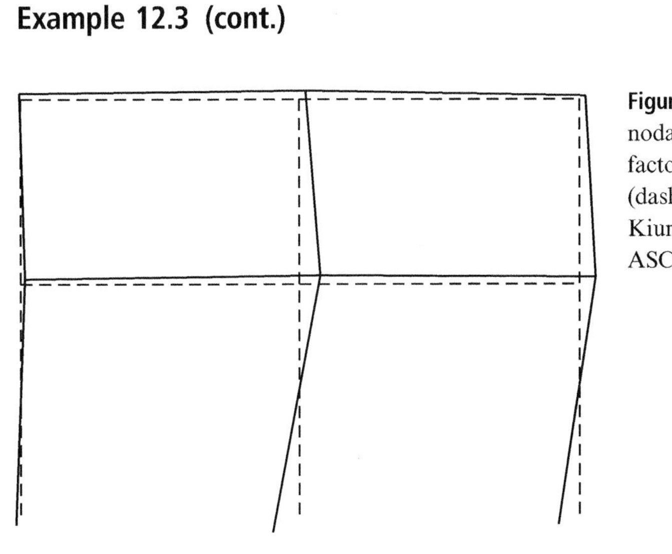

Figure 12.11 Design-point realization of nodal coordinates (solid lines; amplified by a factor of 500) compared to mean coordinates (dashed lines) (after Haukaas and Der Kiureghian, 2005, with permission from ASCE)

12.5 Discrete Representation of Random Fields

In FE analysis, it is often necessary to model material properties or distributed loads as random fields. A random field is a multi-dimensional random process characterized by its distribution and auto-correlation function. More precisely, a random field requires specification of the joint distribution of its values at any number of selected points (see Section 11.2). In particular, a Gaussian random field is one in which the joint distribution of the values at the selected points is multinormal. A homogeneous random field is one where the probabilistic characteristics are invariant of translation of the coordinate axes. In particular, the marginal distribution remains invariant of the location of the selected point and the auto-correlation function depends only on the distance between the two selected points. For probabilistic FE analysis, it is desirable to represent such random fields in terms of a finite number of random variables. This process is known as discretization of a random field. Li and Der Kiureghian (1993) and Sudret and Der Kiureghian (2000) review several discretization methods and examine their accuracy and efficiency in representing random fields.

The simplest discretization method in FE analysis is the mid-point method (Der Kiureghian and Ke, 1988). The field value over each finite element is assumed to be constant and represented by the value at the centroid of the element. The representation of the field is discontinuous at element boundaries. Of course, the values for different elements are correlated, depending on the distance between the mid-points. This formulation is applicable to non-homogenous and non-Gaussian random fields.

A more-refined representation of the field value for each element is the spatial average over the element (Vanmarcke and Grigoriu, 1983), i.e., the integral of the field over the element divided by the area or volume of the element. The representation is again discontinuous at the element boundaries. This approach is applicable to Gaussian fields only, because the distribution of the integral of a Gaussian field has the normal distribution whereas the distribution of the integral of a non-Gaussian field is unknown, though, according to the central limit theorem, it asymptotically approaches the normal distribution as the averaging domain increases in size.

A third approach is the shape-function method introduced by Liu et al. (1986b), which defines the random field over the element in terms of nodal values and shape functions similar to the description of the displacement field in the FE method. The advantage of this method relative to the previous two methods is that the field can be continuous as it passes from one element to the next. The formulation is valid only for Gaussian random fields.

Another approach for discretizing Gaussian random fields is the orthogonal series expansion (OSE) method introduced by Lawrence (1987) and further developed by Ghanem and Spanos (1991a) and Zhang and Ellingwood (1994). Ghanem and Spanos used the Karhunen–Loève (K–L) expansion to express the random field $V(\mathbf{x})$ in terms of its spectral decomposition in the form

$$V(\mathbf{x}) = \mu(\mathbf{x}) + \sigma(\mathbf{x})\sum_{i=1}^{\infty} u_i\sqrt{\lambda_i}\phi_i(\mathbf{x}), \tag{12.22}$$

where $\mu(\mathbf{x})$ and $\sigma(\mathbf{x})$ are the mean and standard deviation functions, u_i are independent standard normal random variables, and λ_i and $\phi_i(\mathbf{x})$, $i = 1, 2, \ldots$, are eigenvalues and eigenfunctions obtained from the integral equation

$$\int_{\Omega} \rho(\mathbf{x}, \mathbf{x}')\phi_i(\mathbf{x}')d\mathbf{x}' = \lambda_i\phi_i(\mathbf{x}), \tag{12.23}$$

where $\rho(\mathbf{x}, \mathbf{x}')$ is the auto-correlation coefficient function of the field values at $\mathbf{x}$ and $\mathbf{x}'$ and Ω is the domain of the random field. The eigenfunctions are normalized so that

$$\int_{\Omega} \phi_i(\mathbf{x})\phi_j(\mathbf{x})d\mathbf{x} = \delta_{ij}, \tag{12.24}$$

where δ_{ij} is the Kronecker delta ($\delta_{ii} = 1$; $\delta_{ij} = 0$ for $i \neq j$). In practice, only a few terms with the largest eigenvalues are included, to obtain an approximate representation of the field as

$$\widehat{V}(\mathbf{x}) = \mu(\mathbf{x}) + \sigma(\mathbf{x})\sum_{i=1}^{r} u_i\sqrt{\lambda_i}\phi_i(\mathbf{x}), \tag{12.25}$$

where r is the number of terms included. One can show that the variance of the error in the above representation is given by

$$\mathrm{Var}\left[V(\mathbf{x}) - \widehat{V}(\mathbf{x})\right] = \sigma^2(\mathbf{x})\left[1 - \sum_{i=1}^{r} \lambda_i\phi_i^2(\mathbf{x})\right], \tag{12.26}$$

which happens to be identical to the difference between the variances of $V(\mathbf{x})$ and $\widehat{V}(\mathbf{x})$. As the error variance cannot be negative, it follows that truncation of the series always results in underestimation of the variance of the random field.

Because of its spectral nature, the K-L expansion offers the most efficient representation of a Gaussian field, i.e., it has the smallest error for a fixed number of terms, provided the integral eigenvalue problem in (12.23) can be solved exactly. Unfortunately, the integral problem has an exact solution for very few auto-correlation coefficient functions. Therefore, one must resort to a numerical solution. Ghanem and Spanos (1991a) used a Galerkin-type approximation in which the eigenfunctions are described in terms of a set of basis functions and corresponding coefficients, converting the problem into a matrix eigenvalue problem. The approach requires discretization of the domain of the random field.

To avoid the integral eigenvalue problem in (12.23), Zhang and Ellingwood (1994) proposed using a complete set of orthonormal functions $\psi_i(\mathbf{x})$ (Legendre polynomials, in their case) to represent the Gaussian random field in the truncated form

$$\widehat{V}(\mathbf{x}) = \mu(\mathbf{x}) + \sigma(\mathbf{x})\sum\nolimits_{i=1}^{r} v_i\psi_i(\mathbf{x}), \tag{12.27}$$

where v_i are now correlated Gaussian random variables. Through an eigenvalue decomposition of their covariance matrix, these variables are linearly transformed to uncorrelated standard normal random variables, leading to an expression similar to that in (12.25) with approximate eigenvalues λ_i and eigenfunctions $\phi_i(\mathbf{x})$. Hence, their approach is equivalent to the K-L expansion when the solution to the eigenvalue problem in (12.23) must be obtained by numerical means. Hereafter, the method of Zhang and Ellingwood is referred to as the OSE method.

Finally, Li and Der Kiureghian (1993) proposed the optimal linear estimation (OLE) method. Consider a grid of nodal points $\mathbf{x}_i$, $i = 1, \ldots, n$, in the domain of the Gaussian random field and let $\mathbf{V} = [V(\mathbf{x}_1) \cdots V(\mathbf{x}_n)]^{\mathrm{T}}$ be a vector of correlated normal random variables representing the values of the field at these points. Now consider a representation of the random field as

$$\widehat{V}(\mathbf{x}) = a(\mathbf{x}) + \mathbf{b}^{\mathrm{T}}(\mathbf{x})\mathbf{V}, \tag{12.28}$$

where $a(\mathbf{x})$ is a scalar function and $\mathbf{b}(\mathbf{x})$ is an n-vector function. These functions are determined by minimizing the variance of the error $V(\mathbf{x}) - \widehat{V}(\mathbf{x})$, subject to $\widehat{V}(\mathbf{x})$ being an unbiased estimator of $V(\mathbf{x})$ in the mean, i.e., $a(\mathbf{x})$ and $\mathbf{b}(\mathbf{x})$ are obtained as the solutions to the optimization problem

$$\text{minimize Var}\left[V(\mathbf{x}) - \widehat{V}(\mathbf{x})\right] \tag{12.29a}$$

$$\text{subject to E}\left[V(\mathbf{x}) - \widehat{V}(\mathbf{x})\right] = 0. \tag{12.29b}$$

The solution is found to be

$$a(\mathbf{x}) = \mu(\mathbf{x}) - \mathbf{b}^{\mathrm{T}}(\mathbf{x})\boldsymbol{\mu}_{\mathbf{V}} \tag{12.30}$$

$$\mathbf{b}(\mathbf{x}) = \boldsymbol{\Sigma}_{\mathbf{VV}}^{-1}\boldsymbol{\Sigma}_{V(\mathbf{x})\mathbf{V}}, \tag{12.31}$$

in which $\boldsymbol{\mu}_{\mathbf{V}}$ and $\boldsymbol{\Sigma}_{\mathbf{VV}}$ are the mean vector and covariance matrix of $\mathbf{V}$, respectively, and $\boldsymbol{\Sigma}_{V(\mathbf{x})\mathbf{V}}$ is an $n \times 1$ vector containing the covariances of $V(\mathbf{x})$ with the elements of $\mathbf{V}$. Thus, the discretized random field takes the form

$$\widehat{V}(\mathbf{x}) = \mu(\mathbf{x}) + \boldsymbol{\Sigma}_{V(\mathbf{x})\mathbf{V}}^{\mathrm{T}} \boldsymbol{\Sigma}_{\mathbf{VV}}^{-1} (\mathbf{V} - \boldsymbol{\mu}_{\mathbf{V}}). \tag{12.32}$$

Because the field is expressed as a linear function of the nodal random variables, the above formulation is strictly valid only for Gaussian random fields. In that case, $\mathbf{V}$ has the multi-normal distribution. However, there is no extra effort required in applying the method to non-homogeneous Gaussian random fields, in which case the mean and variance of $\mathbf{V}$ vary with the location of the selected grid points.

The variance of the error in the above representation is given by

$$\mathrm{Var}\left[V(\mathbf{x}) - \widehat{V}(\mathbf{x})\right] = \sigma^2(\mathbf{x}) - \boldsymbol{\Sigma}_{V(\mathbf{x})\mathbf{V}}^{\mathrm{T}} \boldsymbol{\Sigma}_{\mathbf{VV}}^{-1} \boldsymbol{\Sigma}_{V(\mathbf{x})\mathbf{V}}. \tag{12.33}$$

The second term again turns out to be the variance of $\widehat{V}(\mathbf{x})$ so that the error is the difference between the exact variance and the variance of the approximated random field. It follows that, like the K-L method, the OLE representation always underestimates the variance of the random field.

With the aim of reducing the number of random variables representing the random field, Li and Der Kiureghian (1993) developed the extended OLE (EOLE) method by using the spectral decomposition of the random vector $\mathbf{V}$

$$\mathbf{V} = \boldsymbol{\mu}_{\mathbf{V}} + \sum\nolimits_{i=1}^{n} u_i \sqrt{\theta_i} \boldsymbol{\phi}_i, \tag{12.34}$$

where u_i are standard normal random variables and θ_i and $\boldsymbol{\phi}_i$ are the eigenvalues and eigenvectors of the covariance matrix obtained from

$$\boldsymbol{\Sigma}_{\mathbf{VV}} \boldsymbol{\phi}_i = \theta_i \boldsymbol{\phi}_i \tag{12.35}$$

with the eigenvectors normalized such that $\boldsymbol{\phi}_i^{\mathrm{T}} \boldsymbol{\phi}_i = \delta_{ij}$. Retaining the first $r \leq n$ terms with the largest eigenvalues in (12.34) and substituting for $\mathbf{V}$ in (12.32), one obtains

$$\widehat{V}(\mathbf{x}) = \mu(\mathbf{x}) + \boldsymbol{\Sigma}_{V(\mathbf{x})\mathbf{V}}^{\mathrm{T}} \boldsymbol{\Sigma}_{\mathbf{VV}}^{-1} \sum\nolimits_{i=1}^{r} u_i \sqrt{\theta_i} \boldsymbol{\phi}_i. \tag{12.36}$$

Owing to its spectral nature, this representation has similar efficiency to the K-L representation.

The grid points selected in the Galerkin solution of the K-L integral eigenvalue problem in (12.23) or for the OLE/EOLE methods described above need not be the same as the grid points in the FE formulation. The proper size of the grid in these discretization methods is governed by the correlation length of the random field, specifically the length over which the correlation decays to $\exp(-1)$, while the mesh size in FE analysis is usually governed by the gradient of the stress field. According to Sudret and Der Kiureghian (2000), a distance of 1/10 to 1/2 of the correlation length between the grid points of the random field is sufficient, with the smaller values needed if the field is not differentiable (see Section 11.2).

Numerical comparisons in Li and Der Kiureghian (1993) show that the OLE method is far superior to the mid-point, spatial average, and shape-function methods in terms of efficiency in representing a Gaussian random field. Here, efficiency refers to the number of random variables needed to represent the random field with a given level of accuracy. Sudret and Der Kiureghian (2000) compared the errors in representations based on the K-L, OSE, and EOLE methods. The comparison is made in terms of the relative error measure

$$\text{err}(\mathbf{x}) = \frac{\text{Var}\left[V(\mathbf{x}) - \widehat{V}(\mathbf{x})\right]}{\text{Var}[V(\mathbf{x})]}. \tag{12.37}$$

As an example, consider a one-dimensional homogeneous random field over the domain $0 \leq x \leq 10$, having the auto-correlation coefficient function $\rho(x_1 - x_2) = \exp(-|x_1 - x_2|/a)$, where $a = 5$ denotes the correlation length. For this function, a closed-form solution of the eigenfunctions $\phi_i(\mathbf{x})$ in the K-L formulation is available (Spanos and Ghanem, 1989). For the EOLE, the field is divided into 20 elements, each element having the size $a/10$. Figure 12.12 shows the error measures for K-L, OSE, and EOLE methods for $r = 2, 4, 6,$ and 10 expansion

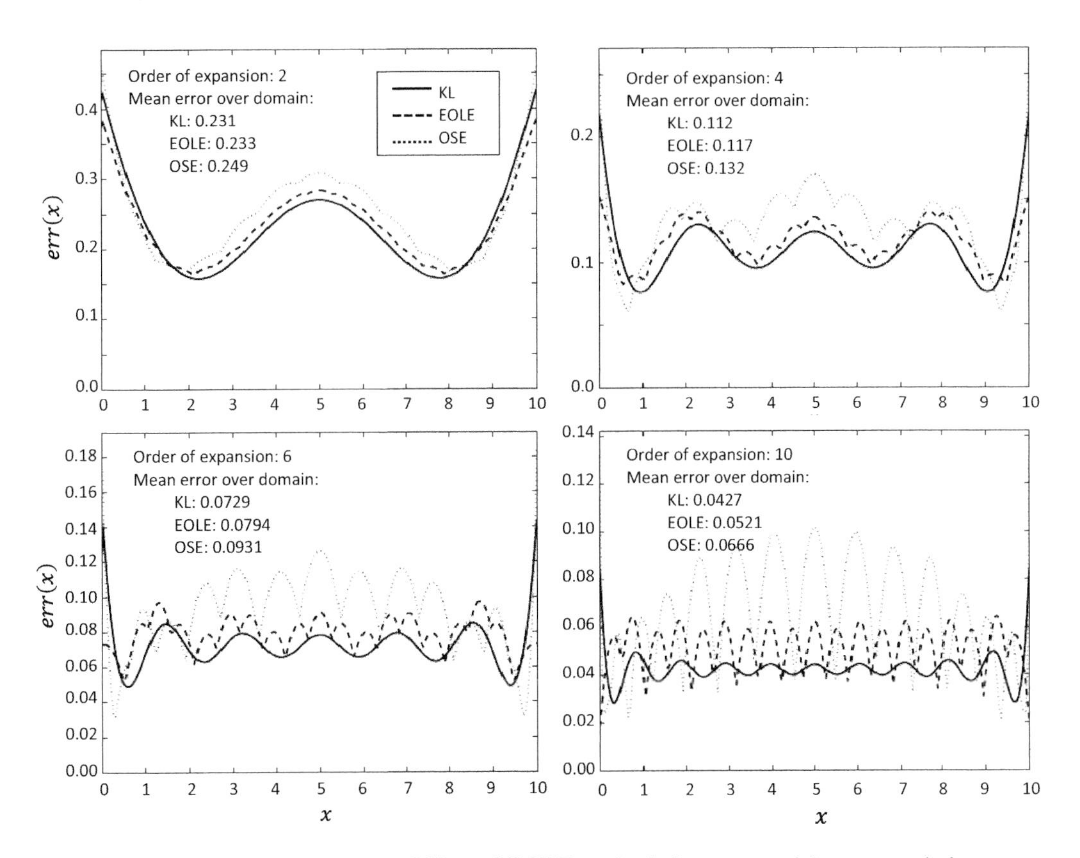

Figure 12.12 Error estimates for K-L, OSE, and EOLE methods for exponential auto-correlation coefficient function for a varying order of expansion terms (after Sudret and Der Kiureghian, 2000)

terms. Also shown in the figure are the mean errors over the domain. It can be seen that the K-L expansion has the smallest error of the three methods, except near the boundaries of the domain. It also has the smallest mean error over the domain. The error in the EOLE method is slightly higher in the interior of the domain, but smaller near the boundaries. The error in the OSE method is largest almost everywhere. It is noted that even for $r = 10$ terms, the mean error in the representation is quite large, i.e., near 5%. This is because of the assumed auto-correlation coefficient function, which corresponds to a non-differentiable process.

Figure 12.13 shows the relative errors when the auto-correlation coefficient function is of the form $\rho(x_1 - x_2) = \exp\left(-|x_1 - x_2|^2/a^2\right)$, where $a = 5$ is again the correlation length. In this case an analytical solution of the eigenfunctions in the K-L expansion is not available; hence, the OSE method represents a numerical solution of the K-L expansion. It is seen that the EOLE method has much smaller pointwise as well as mean errors relative to the OSE method. It is also observed that using $r = 4$ terms in the expansion is sufficient to represent the field with negligible error. This shows that the number of terms needed in the expansion critically depends on the form of the auto-correlation coefficient function. Investigations by both Li and Der Kiureghian (1993) and Sudret and Der Kiureghian (2000) also revealed that, for the EOLE method, it is advantageous to use a fine grid of points in defining the underlying OLE discretization for the same order of expansion r of the EOLE formulation.

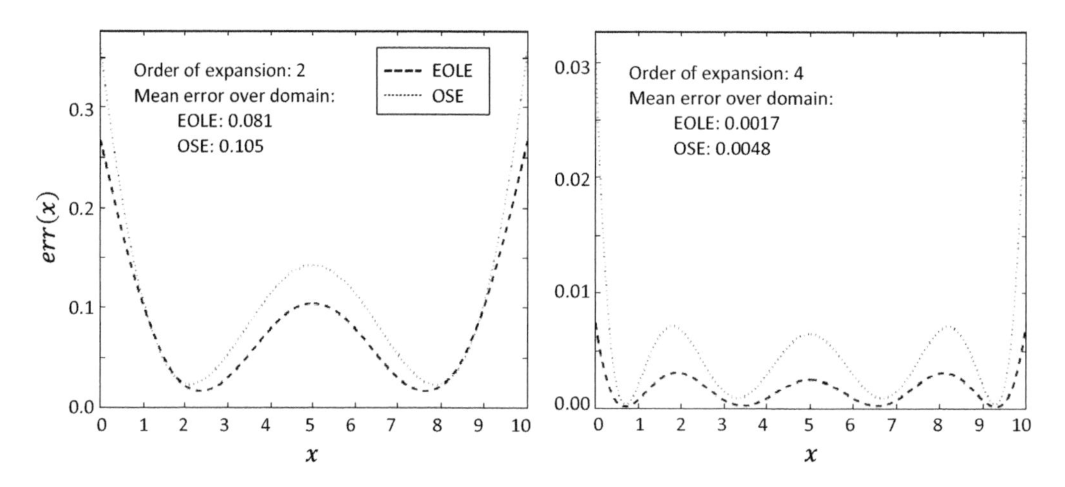

Figure 12.13 Error estimates for OSE and EOLE methods for squared exponential auto-correlation coefficient function for a varying order of expansion terms (after Sudret and Der Kiureghian, 2000)

Other than the mid-point method, all the discretization methods described above are strictly applicable to Gaussian random fields only. In particular, because the representations in (12.25), (12.27), (12.32), and (12.36) are in the form of linear functions of random variables, they preserve the type of distribution only if the random field is Gaussian. On the other hand, many field quantities modeled in FE are by their nature non-Gaussian. Examples are random fields of material properties, such as elastic modulus and yield stress, or geometric quantities, such as thickness, which cannot admit the negative values that are allowed by the normal distribution. In such cases, it is possible to define the random field as a marginal transformation of a Gaussian field (Grigoriu, 1984b; Li and Der Kiureghian, 1993). A Nataf-type transformation is particularly effective in this regard. For example, if $V(\mathbf{x})$ denotes a Gaussian random field with mean $\mu(\mathbf{x})$, standard deviation $\sigma(\mathbf{x})$, and auto-correlation coefficient function $\rho(\mathbf{x}, \mathbf{x}')$, then $\exp[V(\mathbf{x})]$ is a lognormal random field with mean $\exp[\mu(\mathbf{x}) + \sigma^2(\mathbf{x})/2]$, c.o.v. $\sqrt{\exp[\sigma^2(\mathbf{x})] - 1}$, and an auto-correlation coefficient function nearly equal to $\rho(\mathbf{x}, \mathbf{x}')$. (The exact expression is the inverse of the relation for the lognormal-lognormal case in Table A3.5 in Chapter 3.) With the marginal transformation defined, one performs the discretization in the Gaussian space and then transforms the discretized field into the non-Gaussian space. This transformation usually needs to be performed at the integration points of the FE mesh.

Example 12.4 – Composite Plate with Random Field Initial Damage

This problem is adopted from Zhang and Der Kiureghian (1997). We consider the composite plate in Figure 12.14(a), which is made of two perfectly bonded materials. The plate is subject to a uniformly distributed edge load of time-varying intensity $p(t)$, as shown in Figure 12.14(b). The plate is 1 cm thick and is under plane stress conditions. For FE analysis, the plate is represented by 128 quadrilateral four-node elements, as shown in Figure 12.14(c).

An isotropic ductile plastic damage model is employed to describe the accumulation of plastic strain and damage in the material (Lamaitre and Chaboche, 1990). The surface density of voids at a point $\mathbf{x}$ and time t is described by the index $D(\mathbf{x}, t)$, independent of the orientation of the surface. Adopting tensor notation, effective stresses are given by

$$\tilde{\sigma}_{ij} = \frac{\sigma_{ij}}{1 - D}, \tag{E1}$$

where σ_{ij} denote nominal stresses. The stress-strain relations are given by

$$\tilde{\sigma}_{ij} = D_{ijkl}\left(\epsilon_{kl} - \epsilon_{kl}^{p}\right), \tag{E2}$$

where D_{ijkl} are elastic constants and ϵ_{kl} and ϵ_{kl}^{p} denote the total and plastic strains, respectively. The evolution of the plastic strain and the damage index in time are governed by the rules

$$\dot{\epsilon}_{ij}^{p} = \dot{\lambda}\frac{\partial f}{\partial \tilde{\sigma}_{ij}}, \tag{E3}$$

$$\dot{D}(\mathbf{x}, t) = -\dot{\lambda}\frac{\partial \varphi^*}{\partial Y}, \tag{E4}$$

where $\dot{\lambda}$ is the rate factor of plastic deformation, f is the plastic potential or yield function, φ^* is the damage dissipation potential, and

$$-Y = \frac{1}{2E(1 - D)}\left[(1 + \nu)\tilde{s}_{ij}\tilde{s}_{ij} + 3(1 - 2\nu)\tilde{p}^2\right], \tag{E5}$$

Example 12.4 (cont.)

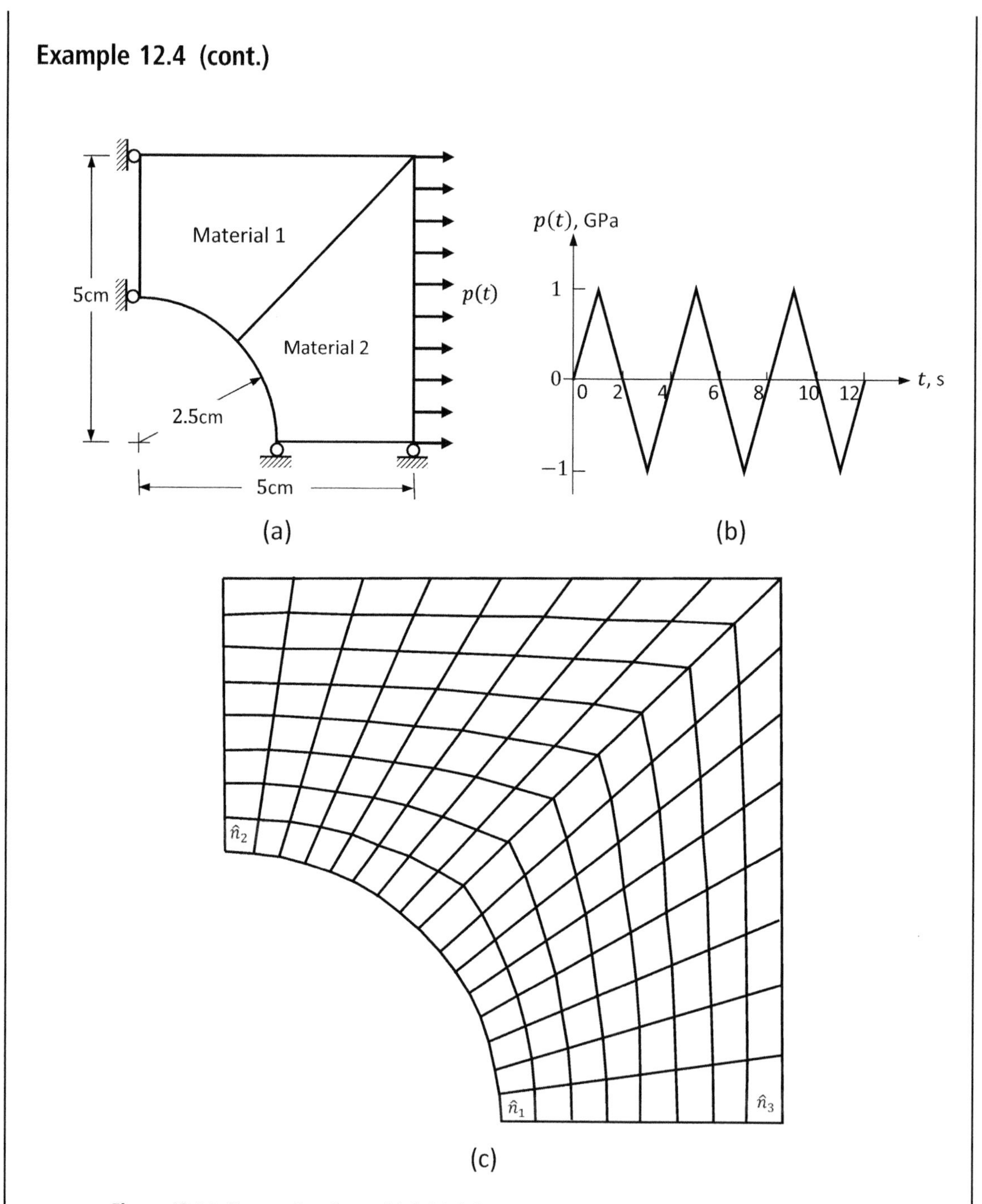

Figure 12.14 Composite plate with initial damage under cyclic loading (after Zhang and Der Kiureghian, 1997)

Example 12.4 (cont.)

is the conjugate of $D(\mathbf{x}, t)$, where E is the elastic modulus, ν is the Poisson's ratio, $\tilde{p} = \tilde{\sigma}_{kk}/3$ is the effective hydrostatic stress, and $\tilde{s}_{ij} = \tilde{\sigma}_{ij} - \delta_{ij}\tilde{p}$, where δ_{ij} denotes the Kronecker delta, are the effective deviatoric stresses. Assuming J_2 plasticity with kinematic hardening,

$$f = \frac{1}{2}\left(\tilde{s}_{ij} - \alpha_{ij}\right)\left(\tilde{s}_{ij} - \alpha_{ij}\right) - \frac{2}{3}\sigma_y^2, \tag{E6}$$

where $\alpha_{ij} = 2H_{\text{kin}}\epsilon_{kl}^p/3$ are the back stresses, H_{kin} is the kinematic hardening coefficient, and σ_y is a parameter related to the yield stress. For the damage dissipation potential, we employ the model suggested by Lemaitre and Chaboche (1990)

$$\varphi^* = \frac{S_0}{s_0 + 1}\left(\frac{-Y}{S_0}\right)^{s_0+1}\frac{1}{1-D}, \tag{E7}$$

where S_0 and s_0 are material constants.

For the example plate, the following property values are assumed: For material 1, $E = 200$ GPa, $\nu = 0.3$, $H_{\text{kin}} = 10$ GPa, $\sigma_y = 250$ MPa, $s_0 = 1$, and $S_0 = 0.06$ MPa; for material 2, all material constants are the same, except $S_0 = 0.16$ MPa. With a larger S_0 value, material 2 has a slower damage accumulation rate.

For each material, the initial damage index $D(\mathbf{x}, 0)$ is assumed to be a homogeneous lognormal random field with mean $\mu(\mathbf{x}) = 0.12$, standard deviation $\sigma(\mathbf{x}) = 0.04$, and auto-correlation coefficient function $\rho(\mathbf{x}_1 - \mathbf{x}_2) = \exp\left(-\|\mathbf{x}_1 - \mathbf{x}_2\|^2/a^2\right)$, where $a = 2$ cm is the correlation length. The random fields for the two materials are assumed to be statistically independent. Transforming the random fields to the Gaussian space and using the EOLE method, each material random field is represented by 81 nodal normal random variables, which are then reduced through eigenvalue decomposition of the covariance matrix to 15 variables per material random field. Hence, the problem is defined in terms of 30 standard normal random variables $\mathbf{u} = (u_1, \ldots, u_{30})$.

For reliability analysis, considering that damage accumulates in time and, hence, $D(\mathbf{x}, t)$ is a non-decreasing function of t, we select the limit-state function

$$g(\mathbf{u}) = D_c - \max_n \bar{D}_n(\mathbf{x}, T), \tag{E8}$$

where $T = 12$ s is the duration of loading, $\bar{D}_n(\mathbf{x}, T)$ denotes the spatially averaged damage index over the nth element, and $D_c = 0.28$ is the safe threshold. This is clearly a series system reliability problem requiring consideration of all elements. However, stress analysis at the mean point of the random variables reveals that the three elements marked with indices $\hat{n}_1$, $\hat{n}_2$, and $\hat{n}_3$ in Figure 12.14(c) are the most critical elements. Hence, the problem is reformulated in terms of the three limit-state functions

Example 12.4 (cont.)

$$g_i(\mathbf{u}) = D_c - \bar{D}_{\hat{n}_i}(\mathbf{x}, T), \quad i = 1, 2, 3. \tag{E9}$$

Considering the strong correlation between failure events in neighboring elements, this is expected to provide a close lower bound to the failure probability of the entire plate. Performing FORM analysis with the above limit-state functions, the reliability indices $\mathbf{B} = (1.99, 1.97, 2.29)$ and correlation matrix (product of $\widehat{\boldsymbol{\alpha}}_i$ vectors)

$$\mathbf{R} = \begin{bmatrix} 1 & 0.076 & 0.387 \\ 0.076 & 1 & 0.052 \\ 0.387 & 0.052 & 1 \end{bmatrix} \tag{E10}$$

are obtained. Note that the failure events within the same material (elements $\hat{n}_1$ and $\hat{n}_3$) are more closely correlated than those in different materials. The realizations of the damage index random fields at the three design points are depicted in Figure 12.15. These represent the most likely realizations of the damage index if failure in each of the three elements were to occur. Using the FORM multinormal probability estimate for series systems in (8.65), we obtain $p_{F1} = 0.056$ with a generalized reliability index of $\beta_g = 1.59$. No more than 20 FE analyses, including gradient calculations by DDM, were used to converge to each of these design points. An earlier version of program FEAP (Taylor and Govindjee, 2020) together with CalREL was used to solve this problem.

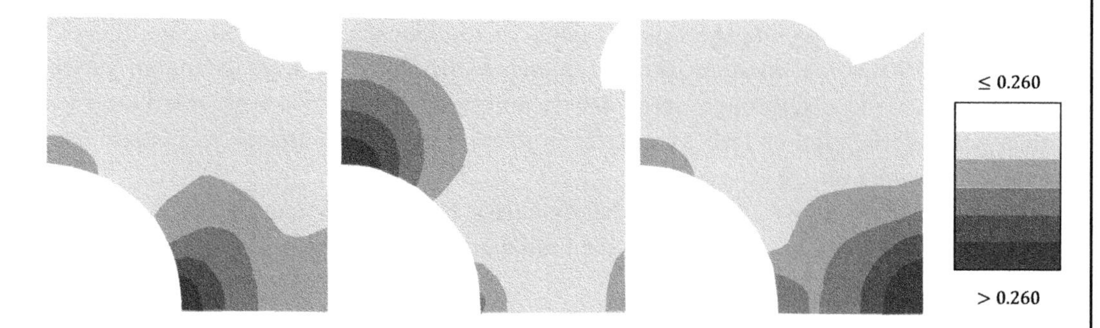

Figure 12.15 Design-point realizations of the damage index for three elements (after Zhang and Der Kiureghian, 1997)

In this problem, the potential locations of failure within the plate were evident because of stress concentrations. This allowed us to focus on three elements and formulate the reliability problem as a series system with three components. More generally, when the stress field is relatively uniformly distributed, the potential locations of failure cannot be predetermined and

Example 12.4 (cont.)

all elements must be considered. The problem then becomes one of space-variant reliability analysis. Der Kiureghian and Zhang (1999) present a solution approach for such a problem by generalizing the outcrossing approach described in Section 11.5.1 to two dimensions.

12.6 The Spectral Stochastic Finite-Element Method

The spectral stochastic finite element method (SSFEM) was originally proposed by Ghanem and Spanos (1991a) and has received widespread attention and further development by numerous researchers. The method is suitable for problems involving random fields. Although it is primarily appropriate for second-moment analysis, we present a brief account of the method in this section because, as described in Section 12.7, more recent developments allow it to be used for reliability analysis as well.

SSFEM relies on two discretizations aside from that of the FE formulation: One to represent the random field(s) of interest through the K-L expansion in terms of a finite number of standard normal random variables (although other expansion methods, such as EOLE, can also be used), and the other to represent the response vector in terms of polynomial functions of the standard normal random variables representing the input random fields. For the latter, the polynomial chaos (PC) expansion involving Hermite polynomials is used. Here, we provide a brief review of the formulation for the case of a linear elastic structure under deterministic loads with the Young's modulus represented as a random field. The governing equation of equilibrium for this case is (12.12).

For a linear elastic body, the element stiffness matrix is given by

$$\mathbf{k}_{\mathrm{el}} = \int_{\Omega_{\mathrm{el}}} \mathbf{B}^{\mathrm{T}}\mathbf{D}\mathbf{B}\mathrm{d}\Omega_{\mathrm{el}}, \tag{12.38}$$

where $\mathbf{D}$ is the elasticity matrix and $\mathbf{B}$ is the strain-displacement matrix defined earlier. Because the Young's modulus can be factored out of $\mathbf{D}$, the elasticity matrix at point $\mathbf{x}$ can be written as $\mathbf{D}(\mathbf{x}) = E(\mathbf{x})\mathbf{D}_0$, where $E(\mathbf{x})$ is the Young's modulus random field and $\mathbf{D}_0$ is a deterministic matrix. Using the K-L expansion of $E(\mathbf{x})$ as in (12.25), substituting the resulting expression of $\mathbf{D}(\mathbf{x})$ in (12.38), and assembling over the finite elements, the structure stiffness matrix is eventually obtained in the form

$$\mathbf{K}(\mathbf{u}) = \mathbf{K}_0 + \sum\nolimits_{i=1}^{r}\mathbf{K}_i u_i, \tag{12.39}$$

where

$$\mathbf{K}_0 = \bigcup\nolimits_{\mathrm{el}} \int_{\Omega_{\mathrm{el}}} \mu_E(\mathbf{x})\mathbf{B}^{\mathrm{T}}\mathbf{D}_0\mathbf{B}\mathrm{d}\Omega_{\mathrm{el}}, \tag{12.40}$$

$$\mathbf{K}_i = \bigcup_{\text{el}} \int_{\Omega_{\text{el}}} \sigma_E(\mathbf{x})\sqrt{\lambda_i}\phi_i(\mathbf{x})\mathbf{B}^{\text{T}}\mathbf{D}_0\mathbf{B}\text{d}\Omega_{\text{el}}, \tag{12.41}$$

in which $\mu_E(\mathbf{x})$ and $\sigma_E(\mathbf{x})$ are, respectively, the mean and standard deviation functions of the Young's modulus random field, λ_i and $\phi_i(\mathbf{x})$ are the eigenvalues and eigenfunctions of its K-L expansion, and u_i are independent standard normal random variables.

Because the stiffness matrix is a function of the standard normal random variables u_i, $i = 1, \ldots, r$, the nodal displacement vector must also be a function of these random variables. In the PC expansion, the vector of nodal displacement is written in the form

$$\mathbf{U}(\mathbf{u}) = \sum_{j=0}^{P-1} \mathbf{U}_j \Psi_j(\mathbf{u}), \tag{12.42}$$

where $\mathbf{U}_j$ is a vector of unknown coefficients and $\Psi_j(\mathbf{u})$ are Hermite polynomials in terms of u_i having the properties $\Psi_0(\mathbf{u}) = 1$, $\text{E}[\Psi_j(\mathbf{u})] = 0$ for $0 < j$, and $\text{E}[\Psi_j(\mathbf{u})\Psi_k(\mathbf{u})] = 0$ for $j \neq k$. Table 12.1 lists the Hermite polynomials for two standard normal random variables ($r = 2$) up to order 3. The number of terms, P, in the above expansion depends on the number r of the terms retained in the K-L expansion and the selected highest order, p, of the polynomials in $\Psi_j(\mathbf{u})$; it is given by (Sudret and Der Kiureghian, 2000)

$$P = \sum_{k=0}^{p} \binom{r+k-1}{k} \tag{12.43}$$

where $\binom{:}{:}$ denotes the binomial coefficient. For example, for $r = 4$ and $p = 3$, $P = 35$. This number rapidly grows with increasing r and/or p.

Table 12.1 Hermite polynomials for $r = 2$ standard normal random variables up to order $p = 3$

j	p	$\Psi_j(\mathbf{u})$
0	0	1
1	1	u_1
2		u_2
3	2	$u_1^2 - 1$
4		$u_1 u_2$
5		$u_2^2 - 1$
6	3	$u_1^3 - 3u_1$
7		$u_2(u_1^2 - 1)$
8		$u_1(u_2^2 - 1)$
9		$u_2^3 - 3u_2$

Using (12.39) and (12.42), the residuals in the governing equations of equilibrium are

$$\epsilon(r,P) = \left(\mathbf{K_0} + \sum_{i=1}^{r} \mathbf{K}_i u_i\right)\left(\sum_{j=0}^{P-1} \mathrm{U}_j \Psi_j(\mathbf{u})\right) - \mathbf{P}, \tag{12.44}$$

where $\mathbf{P}$ denotes the external load vector. For given r and p, the best approximation is obtained by minimizing the residuals in the mean-square sense. This is equivalent to requiring that the residual be orthogonal to $\Psi_j(\mathbf{u}), j = 0, \ldots, P-1$. The minimization eventually leads to the set of equations for the unknown coefficients $\mathbf{U}_j$ (see Sudret and Der Kiureghian, 2000 for details)

$$\begin{bmatrix} \mathbf{K}_{0,0} & \cdots & \mathbf{K}_{0,P-1} \\ \mathbf{K}_{1,0} & \cdots & \mathbf{K}_{1,P-1} \\ \vdots & & \vdots \\ \mathbf{K}_{P-1,0} & \cdots & \mathbf{K}_{P-1,P-1} \end{bmatrix} \begin{Bmatrix} \mathrm{U}_0 \\ \mathrm{U}_1 \\ \vdots \\ \mathrm{U}_{P-1} \end{Bmatrix} = \begin{Bmatrix} \mathbf{P} \\ \mathbf{0} \\ \vdots \\ \mathbf{0} \end{Bmatrix}, \tag{12.45}$$

in which $\mathbf{K}_{j,k} = \sum_{i=0}^{r} \mathrm{E}\left[\xi_i \Psi_j(\mathbf{u}) \Psi_k(\mathbf{u})\right] \mathbf{K}_i$. Observe that the size of the problem to solve now is P times the size of the deterministic FE problem. Therefore, in practice, this formulation can be used only for small values of r and p.

Once the unknown coefficients are determined by solving (12.45), the nodal displacements are available as linear functions of the Hermite polynomials as in (12.42). Thus, the properties of the Hermite polynomials can be used to compute the statistical moments of the nodal displacements. In particular, U_0 is the approximate estimate of the mean vector and $\sum_{j=1}^{P-1} \mathrm{U}_j \mathrm{U}_j^{\mathrm{T}} \mathrm{E}\left[\Psi_j^2(\mathbf{u})\right]$ is the covariance matrix. Higher moments can also be computed with decreasing accuracy. Attempts have been made to also compute the reliability (Ghanem and Spanos, 1991b). However, small failure probabilities are sensitive to the tail of distributions and, as shown in Sudret and Der Kiureghian (2002), one needs a large number of terms P to accurately capture the behavior in the tail. For this reason, the formulation presented in this section is appropriate as a method for the analysis of uncertainty propagation.

The main drawbacks of the above approach are (a) the rapidly growing size of the problem in (12.45) with an increasing number of random variables or degree of the chaos polynomials, (b) the intrusive nature of the solution (one needs to significantly modify the FE code), and (c) the inability to capture the behavior of the response in the tail region of the distribution. In recent years, modifications of this method have been proposed that, to various degrees, address all three mentioned drawbacks. Brief accounts of these methods are presented in the following section.

12.7 Response-Surface Methods

When the limit state for a reliability problem involves an implicit response function that requires time-consuming calculations, such as the use of a FE code, one approach is to construct an analytical function that approximates the implicit response function and then use it to perform reliability analysis. The approximating function is known as a *response surface*, though the terms *metamodel* or *surrogate model* have also been used. As the approximating function has an analytical form, it is easily computed and, hence, any reliability method can be used, including FORM, SORM, and various sampling techniques. The main challenge then is the construction of a response surface that accurately fits the implicit response function, at least in the critical parts of the outcome space of the random variables, while using a manageable number of FE calculations.

The first application of the response-surface method to finite-element reliability analysis was by Faravelli (1989). Let $s(\mathbf{y})$ denote a response quantity of interest expressed as a function of the outcome of an n-vector of basic random variables $\mathbf{Y}$. Faravelli used a second-order quadratic function of the form

$$\widehat{s}(\mathbf{y}) = a + \mathbf{b}^{\mathrm{T}}\mathbf{y} + \mathbf{y}^{\mathrm{T}}\mathbf{c}\mathbf{y} \tag{12.46}$$

to construct a surrogate model for $s(\mathbf{y})$, where the scalar a, the n-vector $\mathbf{b} = \{b_i\}$, and the $n \times n$ matrix $\mathbf{c} = [c_{ij}], i,j = 1, \ldots, n$, collect the $1 + (3n + n^2)/2$ unknown parameters of the model. In order to reduce the number of computations, the expression in (12.46) may contain only those basic random variables that are deemed important by the analyst. The model parameters are estimated by calculating the true response function $s(\mathbf{y})$ at a set of *experimental design* points $\widehat{\mathbf{y}}_i$, $i = 1, \ldots, N$, and minimizing a measure of the error between the values of the true response function and its surrogate model at these points. When not all of the basic variables are included in (12.46), for the purpose of computing the true function values, the missing variables are randomly sampled at the experimental design points in accordance with their probability distributions. Usually, the least-squares method is used to minimize the sum of squares of the errors between the true response values and the predictions by the surrogate model. In general, the error contains two components: a lack of fit component and a random component, the latter arising when not all of the basic random variables are represented in the surrogate model. The main computational effort in this method lies in evaluating by FE analysis the true function values at the experimental design points. This depends on the number of selected experimental design points, which in turn depends on the number of random variables and the number of terms in the selected polynomial surrogate model.

Various schemes are available for selecting the experimental design points (Box et al., 1978). These include the widely used 2^n factorial design, which consists of the mean point and all combinations of the mean plus or minus one standard deviation for each random variable. When n is large, this design may be augmented by pairs of points along each axis on the sides of the mean point. When the number of experimental points is equal to the number of uncertain parameters, the design is said to be fully saturated. In that case, the surrogate model exactly fits the true response function at the design points (provided no random variables are missing from the response-surface model); however, there is no way to measure the error in the model at points away from the experimental design points.

For reliability analysis, fitting the response surface around the mean point of the random variables is clearly not a wise choice, as the failure event is usually far from the mean point. Hence, several investigators have suggested adaptive schemes to move the experimental points toward the region of the limit-state surface that contributes most to the failure probability. Bucher and Bourgund (1990) suggested an iterative scheme whereby one finds the design point (the most likely failure point in the standard normal space) for the first approximation of the response surface and then moves the center of the experimental design toward this point and repeats the procedure until convergence is achieved. To reduce the number of analyses, they excluded the cross-terms in (12.46). Rajashekhar and Ellingwood (1993) suggested ways of including experimental design points in the lower distribution tails of capacity variables and the upper distribution tails of demand variables and demonstrated the importance of including cross-terms. Kim et al. (2005), Kang et al. (2010), Taflanidis and Cheung (2012), and Goswami et al. (2016) offered additional improvements and recommended using the moving least squares method as a way of reducing the amount of computation in the adaptive response surface method.

A main drawback of the above approaches is that the selected form of the surrogate model has no intrinsic relation to the true response function and may not provide a good fit; furthermore, no measure of the underlying error in the fit is available. To provide more accuracy, higher-order polynomial terms may be included, but this rapidly increases the number of unknown parameters to be estimated and, hence, the number of FE analyses to be performed. Leonel et al. (2011) showed that, for a crack-propagation problem, a direct FORM solution, for which the gradients were computed by finite differences, was in fact computationally more efficient than the response-surface method using a polynomial surrogate model.

To improve the response-surface method, it is desirable to select an approximating surrogate model that has a flexible form, so its fit is better guided by the data, and to have a measure of the goodness of the fit. It is also desirable to select experimental design points that maximally enhance the accuracy of the surrogate model where it matters, i.e., in the regions of the outcome space that contribute most to the failure probability. The remainder of this section focuses on recent developments in this direction.

Kriging is a statistical technique originally developed in the geosciences for constructing a random field model for a ground property value using measurements at a finite number of experimental points (Matheron, 1963). In the absence of measurement error, the mean of the estimated random field model coincides with the measured values at the experimental design points; the variance vanishes at these points. Between the experimental design points, the variance increases with distance, thus providing a measure of the uncertainty in the predicted value. This technique was later used to develop surrogates for computationally expensive computer models, where the “measurements” now were computer runs for specific instances of the model (Sacks et al., 1989). Applications have included predicting the response values at unmeasured points, optimizing a functional of responses, and tuning the computer model to physical data. Below, we describe the application of this approach to structural reliability analysis where a FE code provides the computer model.

Bichon et al. (2008) used kriging to adaptively fit a Gaussian random field to the response function $s(\mathbf{x})$ obtained from a FE code using a set of experimental design points. Working in the standard normal space, the response function is modeled as

$$\widehat{S}(\mathbf{u}) = \mathbf{f}(\mathbf{u})^{\mathrm{T}}\boldsymbol{\beta} + Z(\mathbf{u}), \tag{12.47}$$

where $\mathbf{u}$ denotes coordinates in the n-dimensional standard normal space, $\mathbf{f}(\mathbf{u}) = [f_1(\mathbf{u}) \cdots f_q(\mathbf{u})]^{\mathrm{T}}$ a q-vector of *trend functions*, $\boldsymbol{\beta} = [\beta_1 \cdots \beta_q]^{\mathrm{T}}$ a q-vector of *trend coefficients*, and $Z(\mathbf{u})$ a homogeneous Gaussian random field with a prior mean of 0 and prior auto-covariance function $\Sigma(\mathbf{u}_1 - \mathbf{u}_2) = \sigma^2\rho(\mathbf{u}_1 - \mathbf{u}_2)$, where σ^2 denotes the prior variance and $\rho(\mathbf{u}_1 - \mathbf{u}_2)$ the auto-correlation coefficient function. Let $\widehat{\mathbf{u}}_i$, $i = 1, \ldots, N$, denote a set of experimental design points at which the exact response values are computed by FE analyses and let the N-vector $\mathbf{S} = [S(\widehat{\mathbf{u}}_1) \cdots S(\widehat{\mathbf{u}}_N)]^{\mathrm{T}}$ collect these response values. For the auto-correlation function, Bichon et al. (2008) selected the parametrized form $\rho(\mathbf{u}_1 - \mathbf{u}_2) = \exp\left[-\sum_{i=1}^{n}\theta_i(u_{i1} - u_{i2})^2\right]$, where $\boldsymbol{\theta} = [\theta_1 \cdots \theta_n]^{\mathrm{T}}$ are unknown parameters controlling the correlation lengths along the coordinate axes. Note that $\mathbf{f}(\mathbf{u})^{\mathrm{T}}\boldsymbol{\beta}$ represents the prior mean of $\widehat{S}(\mathbf{u})$. In their application, Bichon et al. employed *ordinary* kriging, which uses a single, constant trend function equal to unity so that $\mathbf{f}(\mathbf{u})^{\mathrm{T}}\boldsymbol{\beta} = \beta$, where β is unknown. More generally, when the trend is a function of $\mathbf{u}$, the formulation is called *universal* kriging.

Sacks et al. (1989) showed that the best linear unbiased predictor of the updated mean of the model in (12.47) is

$$\mu_{\widehat{S}}(\mathbf{u}) = \mathbf{f}(\mathbf{u})^{\mathrm{T}}\widehat{\boldsymbol{\beta}} + \mathbf{r}(\mathbf{u})^{\mathrm{T}}\mathbf{R}^{-1}\left(\mathbf{S} - \mathbf{F}\widehat{\boldsymbol{\beta}}\right), \tag{12.48}$$

where $\mathbf{r}(\mathbf{u}) = [\rho(\mathbf{u} - \widehat{\mathbf{u}}_1) \cdots \rho(\mathbf{u} - \widehat{\mathbf{u}}_N)]^{\mathrm{T}}$ is an N-vector collecting the correlation coefficients between a point $\mathbf{u}$ and the experimental design points $\widehat{\mathbf{u}}_i$, $\mathbf{R}$ is an $N \times N$ matrix collecting the correlation coefficients $\rho(\widehat{\mathbf{u}}_i - \widehat{\mathbf{u}}_j)$ at the experimental design points, $\mathbf{F} = [\mathbf{f}(\widehat{\mathbf{u}}_1) \cdots \mathbf{f}(\widehat{\mathbf{u}}_N)]^{\mathrm{T}}$ is an $N \times q$ matrix collecting the trend functions at the experimental design points, and

$$\widehat{\boldsymbol{\beta}} = \left(\mathbf{F}^{\mathrm{T}}\mathbf{R}^{-1}\mathbf{F}\right)^{-1}\mathbf{F}^{\mathrm{T}}\mathbf{R}^{-1}\mathbf{S} \tag{12.49}$$

is the generalized least-square estimate of $\boldsymbol{\beta}$. The updated variance of $\widehat{S}(\mathbf{u})$ is given by

$$\sigma^2_{\widehat{S}}(\mathbf{u}) = \sigma^2\left(1 - \left[\mathbf{f}(\mathbf{u})^{\mathrm{T}}\ \mathbf{r}(\mathbf{u})^{\mathrm{T}}\right]\begin{bmatrix}\mathbf{0} & \mathbf{F}^{\mathrm{T}} \\ \mathbf{F} & \mathbf{R}\end{bmatrix}^{-1}\begin{Bmatrix}\mathbf{f}(\mathbf{u}) \\ \mathbf{r}(\mathbf{u})\end{Bmatrix}\right). \tag{12.50}$$

The above analysis is conditioned on σ^2 and the correlation parameters $\boldsymbol{\theta}$. The maximum likelihood estimate of σ^2 is given by

$$\widehat{\sigma}^2 = \frac{1}{N}\left(\mathbf{S} - \mathbf{F}\widehat{\boldsymbol{\beta}}\right)^{\mathrm{T}}\mathbf{R}^{-1}\left(\mathbf{S} - \mathbf{F}\widehat{\boldsymbol{\beta}}\right). \tag{12.51}$$

Furthermore, the maximum likelihood estimate of $\boldsymbol{\theta}$ is obtained as

$$\widehat{\boldsymbol{\theta}} = \max_{\boldsymbol{\theta}} \left\{ -\frac{1}{N} \ln \left(\det \mathbf{R} \right) - \ln \left(\widehat{\sigma}^2 \right) \right\}, \tag{12.52}$$

where $(\det \mathbf{R})$ denotes the determinant of $\mathbf{R}$. We note that when $\widehat{\boldsymbol{\beta}}$ from (12.49) is substituted in (12.51) and the result substituted in (12.52), the expression in (12.52) is then only a function of the data and the correlation parameters and can be maximized to determine $\widehat{\boldsymbol{\theta}}$. These values are then substituted in (12.51) to compute $\widehat{\sigma}^2$.

With the parameters $\widehat{\boldsymbol{\beta}}$, $\widehat{\sigma}^2$, and $\widehat{\boldsymbol{\theta}}$ estimated, the updated mean function in (12.48) provides a prediction of the true response function $S(\mathbf{u})$ and the variance function in (12.50) provides an estimate of the uncertainty in this prediction. As mentioned earlier, provided that all the random variables are included in $\mathbf{u}$, the mean function $\mu_{\widehat{S}}(\mathbf{u})$ coincides with $S(\mathbf{u})$ at the experimental design points and the variance $\sigma^2_{\widehat{S}}(\mathbf{u})$ is zero at these points.

Bichon et al. (2008) used kriging for reliability analysis in conjunction with an active learning scheme to select additional experimental design or "training" points. They assume a limit-state function of the form $G(\mathbf{u}) = \bar{s} - S(\mathbf{u})$, where $\bar{s}$ denotes the failure threshold for the considered response quantity $S(\mathbf{u})$, which is an implicit function requiring costly FE computations. They start the analysis for constructing a surrogate model of $S(\mathbf{u})$ with an initial set of experimental design points, recommending the same number of points as for a quadratic response function, i.e., $N = 1 + \left(3n + n^2\right)/2$. They then enhance the experimental design by adding a new training point through maximization of an *expected feasibility* function defined as

$$\mathrm{EF}\left[\widehat{S}(\mathbf{u})\right] = \int_{\bar{s}-\epsilon}^{\bar{s}+\epsilon} [\epsilon - |\bar{s} - s|] f_{\widehat{S}}(s) \mathrm{d}s, \tag{12.53}$$

where $f_{\widehat{S}}(s)$ is the normal PDF of $\widehat{S}(\mathbf{u})$ with the mean and variance as computed from (12.48) and (12.50), respectively, and ϵ is a parameter proportional to the standard deviation $\sigma_{\widehat{S}}(\mathbf{u})$. The expected feasibility function tends to have a large value at locations $\mathbf{u}$ where the mean of $\widehat{S}(\mathbf{u})$ falls near the threshold $\bar{s}$ and at locations where the variance $\sigma^2_{\widehat{S}}(\mathbf{u})$ is large. This allows a balance between "exploitation" at locations near the limit-state surface and "exploration" at locations with large uncertainty in the prediction. With the true response function $S(\mathbf{u})$ determined at the new training point $\widehat{\mathbf{u}}_{N+1}$, the fitting of the surrogate model is repeated, and a new training point is sought by maximizing the expected feasibility function with the new mean and variance functions. This procedure is continued until the maximum of the feasibility function over the outcome space is sufficiently small. (Bichon et al. recommend the threshold be 0.001.) The updated mean function $\mu_{\widehat{S}}(\mathbf{u})$ is then considered as the surrogate model of the response function $S(\mathbf{u})$ and reliability analysis is performed using the explicit limit-state function $\widehat{G}(\mathbf{u}) = \bar{s} - \mu_{\widehat{S}}(\mathbf{u})$. Because the limit-state function is easy to compute, any reliability analysis method may be used, including sampling methods. Bichon et al. (2008) applied this method to a FE problem showing good accuracy in comparison with Latin hypercube sampling, in which a relatively small number of FE calculations are performed to construct the surrogate model. However, their example contained only two random variables. Bichon et al. (2011) extended the above approach to system reliability problems involving up to 11 random variables and 10 failure modes, again showing good accuracy with a relatively small number of calls to a FE code.

One difficulty with the above approach is that the maximization of the feasibility function is carried out over the entire $\mathbf{u}$ space, which can be computationally demanding when the number of random variables is large. Echard et al. (2011) developed an active learning kriging method coupled with Monte Carlo simulation, which avoids the need to optimize over the entire $\mathbf{u}$ space. They first generate a large sample of points $\mathbf{u}_i$, $i = 1, \ldots, N$, for Monte Carlo analysis. A small subset of these points is then used as experimental design points to construct an initial surrogate kriging model. Next, from the remaining Monte Carlo points, they select the one that is most optimal as the next training point. For this purpose, in addition to the expected feasibility function (12.53), they explore the *learning function*

$$U(\mathbf{u}) = \frac{\left|\widehat{G}(\mathbf{u})\right|}{\sigma_{\hat{S}}(\mathbf{u})}, \tag{12.54}$$

which essentially measures the predicted value of the limit-state function in units of its standard deviation. Note that $U(\mathbf{u})$ is small when the sampling point is near the predicted limit-state surface ($\widehat{G}(\mathbf{u})$ is near 0), or where the variance of the prediction is large. In the first case, selecting such a point for training is informative because even a small correction of the model may change the sign of the limit-state function and, hence, affect the Monte Carlo estimate of the failure probability. (Recall that the Monte Carlo estimate of the failure probability is the fraction of sample points with negative or zero limit-state function values.) The second case indicates a large uncertainty in the value of the surrogate model and, hence, in the algebraic sign of the true limit-state function. Such a point is worthy of exploration in order to reduce the uncertainty in the sign of the limit-state function. Thus, among the remaining Monte Carlo points, the one with the smallest $U(\mathbf{u})$ value is selected for enriching the experimental design. With the new training point added, an improved surrogate model is obtained, and the process is continued until the minimum of $U(\mathbf{u})$ exceeds a predetermined value (Echard et al. [2011] recommend $\min\{U(\mathbf{u})\} \geq 2$ as the stopping criterion). With the improved surrogate model determined, all the sample points are used to estimate the failure probability by Monte Carlo analysis with the surrogate model. If necessary, additional sampling points are generated so that the c.o.v. of the probability estimate is sufficiently small. Fauriat and Gayton (2014) extended the above approach to system reliability problems showing somewhat more efficient results than Bichon et al. (2011).

An alternative formulation for adaptively developing a surrogate model has been proposed by Blatman and Sudret (2011) using a sparse PC expansion. The coefficients in the expansion are estimated by least-mean-square regression using a set of experimental design points where the exact FE model is computed. They reduce the number of terms in the PC expansion by using a *hyperbolic index set*, which drops high-order cross-terms, and employing the *least-angle regression* method to automatically identify significant coefficients in the expansion. Using these techniques, they obtain a sparse PC expansion of the response quantity that maintains only the significant terms while avoiding overfitting. They then adaptively enhance the design by adding terms to the polynomial expansion and new training points to the experimental design, while monitoring a measure of the error in the surrogate model. In numerical examples with linear FE problems, they are able to compute up to the fourth moment of the response quantity (with decreasing accuracy for the higher moments) with far fewer calculations of the exact FE model than a full PC model would require.

More recently, Schöbi et al. (2015) and Schöbi et al. (2017) have combined polynomial chaos and kriging to develop a surrogate model for reliability analysis that provides more accuracy for the same number of experimental design points than a polynomial chaos or kriging model alone. They represent the trend function $\mathbf{f}(\mathbf{u})^{\mathrm{T}}\boldsymbol{\beta}$ in (12.47) with a sparse polynomial chaos model of the form $\sum_{\tau\in\mathbf{T}}\beta_\tau\Psi_\tau(\mathbf{u})$, where $\Psi_\tau(\mathbf{u})$ are a set of orthonormal polynomials (they use Hermite polynomials), β_τ are coefficients determined through least-square regression, and $\mathbf{T}$ is an index set. They use the ideas developed by Blatman and Sudret (2011) to select a sparse set of significant polynomials. The adaptive Monte Carlo kriging method of Echard et al. (2011), together with the learning function in (12.54), is used to enrich the experimental design points and improve the surrogate model. Schöbi et al. (2015) show that, as a result of the combined formulation, the polynomial chaos trend models the global features of the true response function while the random field $Z(\mathbf{u})$ captures the local variations, so that the combined model provides more accuracy for the same number of experimental design points. Schöbi et al. (2017) apply the method to a truss problem, showing that small probabilities of failure can be estimated with good accuracy and a small number of calls to the FE code.

It is evident that response-surface methods for FE reliability analysis are rapidly evolving. The trend is toward adaptive methods, where an initial experimental design is enriched through selection of new training points guided by a learning function. The resulting surrogate models can be accurate in regions of importance and yield accurate reliability estimates with a relatively small number of FE calculations. An important feature of these approaches is the non-intrusive nature of the formulation: One need not modify an existing FE code to perform reliability analysis. Furthermore, the formulation is independent of the nature of the FE problem, i.e., linear or nonlinear, stress, flow, or heat transfer problems can be handled in the same manner. However, the existing applications are relatively trivial FE problems with small number of random variables. More experience is needed to evaluate the applicability and efficiency of these methods for complex real-world problems involving large numbers of uncertain variables. Furthermore, while these methods may provide more accurate estimates of the failure probability than FORM, they lack the richness of FORM in providing valuable additional information such as sensitivities with respect to distribution and limit-state parameters and measures of relative importance of the random variables in the context of reliability analysis.

As a final observation, it is worth mentioning that, other than sampling and direct-integration methods, all structural reliability methods can be viewed as response-surface methods. In particular, in FORM and SORM we replace the limit-state surface in the standard normal space with first- and second-order response-surface models, respectively. The power of these methods lies in the fact that these approximations are carried out at an optimal point, i.e., the point in the failure domain with the highest probability density in the standard normal space.

12.8 Concluding Remarks

The integration of FE and reliability methods is the logical "next step" in the development of methods for safety and reliability assessment of complex structural systems. Much has been accomplished during the past three decades, but the field is still in a research mode and applications are relatively simple, academic problems, mostly for the demonstration and verification of new methods. Applications to real-world problems motivated by engineering need have yet to be carried out. Ideally, FE codes used in practice would have capabilities for defining input quantities as random variables or random fields and for uncertainty propagation and reliability analysis. We are not there yet.

This chapter presented a number of methods available for FE reliability analysis. For solution by FORM and SORM, it is necessary to have the capability to compute the gradients of response quantities with respect to the random variables of the problem, including those representing discretized random fields. For this purpose, the DDM was presented as an efficient way of computing response gradients. It was shown that this method essentially requires the solution of a linear problem, even when the structure itself is nonlinear. However, this approach requires significant intrusion into the FE code, specifically the implementation of partial derivative routines at the element level with respect to material properties and nodal coordinates as well as partial derivatives of load vectors. On the other hand, such a capability in a FE code can be useful beyond reliability analysis, e.g., for optimal design, model fitting, and sensitivity analysis. Other methods for computation of gradients in a computer code are also available, which can be less intrusive. These include the *complex-step* derivative approximation (Martins et al., 2003), which employs an imaginary perturbation in the variable of interest, and *algorithmic differentiation* (Griewank and Walther, 2008), which employs the chain rule of differentiation for an iterative algorithm for evaluating a function. To the author's knowledge, neither of these methods has yet been explored in the context of FE reliability analysis.

An alternative to a direct application of FORM and SORM for finite-element reliability analysis is the response-surface method, where a surrogate analytical model of the response function(s) is developed in the outcome space of the random variables. This is done by fitting to a set of experimental design points, computing the response function for the specific outcomes of the random variables at that point. As the surrogate model is easy to compute, any reliability method, including FORM, SORM, and sampling methods, can be used to compute the failure probability. Promising recent applications of this approach employ adaptive kriging, where the surrogate model is gradually improved by selecting additional experimental points guided by a learning function. An important advantage of this approach is that it is not intrusive and requires runs of the FE code at the experimental design points only. Another advantage is that it is independent of the nature of the FE code and can be applied to linear as well as nonlinear problems in a variety of fields.

It is fair to say that the field of FE reliability analysis is in its early phase of development. There is enormous potential for innovation and further development. With increasing computing power, it is quite reasonable to predict that within 10 years we could have commercial FE codes that will have capability for reliability analysis of real-world, complex structural problems.

13 Reliability Methods for Stochastic Structural Dynamics

13.1 Introduction

Dynamic analysis of structures subjected to stochastic excitation is a classical topic initially introduced in texts by Crandall and Mark (1963), Lin (1967), and Bolotin (1984), with intense development in recent decades, see, e.g., Roberts and Spanos (1990), Soong and Grigoriu (1993), Lin and Cai (1995), and Lutes and Sarkani (2004). It is not the intention to present a full development of this topic in a broad sense in this chapter. Rather, the purpose is to present the topic as it relates to the structural reliability methods described in this book, in particular the first-order reliability method (FORM). The chapter offers a new way of looking at the stochastic structural dynamics problem, which is different from the classical approaches and provides the opportunity to see certain problems in a new light and thereby develop new solutions to old or unsolved problems.

In order to make use of time-variant reliability methods such as FORM, it is necessary to describe random processes representing input excitations or structural response in terms of a finite number of standard normal random variables. For this purpose, the chapter begins by describing discrete representations of random processes. Gaussian processes, both stationary and non-stationary, are directly represented in terms of standard normal random variables. For non-Gaussian processes, we consider those that can be formulated through translation or filtering of Gaussian processes. The discretization of the non-Gaussian process is performed on the underlying Gaussian process, which is then translated into the non-Gaussian space.

Next are described the response of a linear structure subjected to a discretized Gaussian excitation. This leads to geometric interpretations of the excitation and response in the space of the standard normal random variables representing the discretized excitation. Next, we consider the response of a linear system to a non-Gaussian excitation and obtain the statistics of the non-Gaussian response. An example compares the responses of a linear system to comparable Gaussian and non-Gaussian excitations. Then we consider the stochastic response of nonlinear systems, for which the tail-equivalent linearization method (TELM) is developed. Using the basic ideas behind FORM, the tail probability of the response of the nonlinear system is approximated in first order by the tail probability of a linear system. This leads to the definition of the tail-equivalent linear system (TELS). After describing several important properties of the TELS, attention is focused on various measures of the reliability, including mean up-crossing rate and first-passage probability for stationary and nonstationary excitations. Several example applications demonstrate the theoretical concepts developed in this chapter.

13.2 Discrete Representation of Random Processes

Let $F(t)$ denote a Gaussian random process having the mean function $\mu_F(t)$. We wish to represent the process in terms of a finite number of standard normal random variables, U_i, $i = 1, \ldots, n$. A variety of methods are available for this purpose. Virtually all of them take the form

$$\begin{aligned}\widehat{F}(t) &= \mu_F(t) + \sum\nolimits_{i=1}^{n} U_i s_i(t) \\ &= \mu_F(t) + \mathbf{s}(t)\mathbf{U}\end{aligned} \tag{13.1}$$

where $\mathbf{s}(t) = [s_1(t) \cdots s_n(t)]$ is a row vector of deterministic shape functions and $\mathbf{U} = [U_1 \cdots U_n]^{\mathrm{T}}$. (Note that $\mathbf{s}(t)$ is the gradient vector of $\widehat{F}(t)$ with respect to $\mathbf{U}$. It is our convention to show gradients as row vectors.) The notation $\widehat{F}(t)$ is used because, with a finite number of terms, the above expression can represent only an approximation of $F(t)$. In Section 12.5, several methods for discretizing random fields (i.e., multidimensional random processes) were presented that had the above form. These included the K-L expansion with $s_i(t) = \sqrt{\lambda_i}\phi_i(t)$ and the EOLE method with $s_i(t) = \boldsymbol{\Sigma}_{F(t)\mathbf{F}}^{\mathrm{T}} \boldsymbol{\Sigma}_{\mathbf{FF}}^{-1} \sqrt{\theta_i} \boldsymbol{\phi}_i$, where the terms are defined as in Section 12.5. Here, we describe other time- and frequency-domain discretization methods that are more appropriate for stochastic dynamic analysis.

First consider a zero-mean, band-limited white-noise process $W(t)$ having the power-spectral density (PSD)

$$\begin{aligned}\Phi(\omega) &= \Phi_0, \quad |\omega| \leq \Omega, \\ &= 0, \quad\;\; \Omega \leq |\omega|,\end{aligned} \tag{13.2}$$

where Ω denotes the upper frequency limit in rad/s. The process has the variance $\sigma^2 = 2\Phi_0\Omega$. A simple way to discretize this process is to select a set of equally spaced discrete time points $t_i = (i - 1)\Delta t$, $i = 1, \ldots, n$, and represent the process through a sequence of random pulses defined by

$$\widehat{W}_i = \frac{1}{\Delta t}\int_{t_{i-1}}^{t_i} W(t)\mathrm{d}t, \, t_{i-1} < t \leq t_i, i = 1, \ldots, n. \tag{13.3}$$

To properly include the frequency range, we select $\Delta t = \pi/\Omega$ s. One can show that $\widehat{W}_i$ has a mean of zero and the variance σ^2. Setting $U_i = \widehat{W}_i/\sigma$, $i = 1, \ldots, n$, as the standard normal random variables, the discretized process has the form in (13.1) with

$$s_i(t) = \sigma, \quad t_{i-1} < t \le t_i, i = 1, \ldots, n,$$
$$= 0, \qquad \text{elsewhere.} \tag{13.4}$$

A realization of this process with $\sigma = 1$ and $\Delta t = 0.02$ s appears in Figure 13.1.

The discretized white-noise process can be used to define non-white processes through filtering. Let $h_f(t)$ denote the unit impulse-response function (IRF) of a linear filter. As defined more precisely in the following section, this function represents the response of the filter to a unit impulse applied at time zero. The discretized process then has the form in (13.1) with

$$s_i(t) = \sigma \int_{t_{i-1}}^{t_i} h_f(t - \tau) d\tau, t_{i-1} < t, \quad i = 1, \ldots, n,$$
$$= 0, \qquad \text{elsewhere.} \tag{13.5}$$

The resulting process $\widehat{F}(t)$ starts from zero at $t = 0$ and asymptotically approaches a stationary state, provided the filter is damped. As an example, Figure 13.2 shows the realization of a random process with Δt and σ as in the above and the filter IRF $h_f(t) = \left[\omega_f \left(1 - \zeta_f^2\right)^{-1/2}\right] \exp(-\zeta_f \omega_f t) \sin \left[\omega_f \left(1 - \zeta_f^2\right)^{1/2} t\right]$, where $\omega_f = 5\pi$ rad/s and $\zeta_f = 0.05$ are used for the filter frequency and damping ratio, respectively. $n = 251$ standard normal random variables were used to generate this sample function.

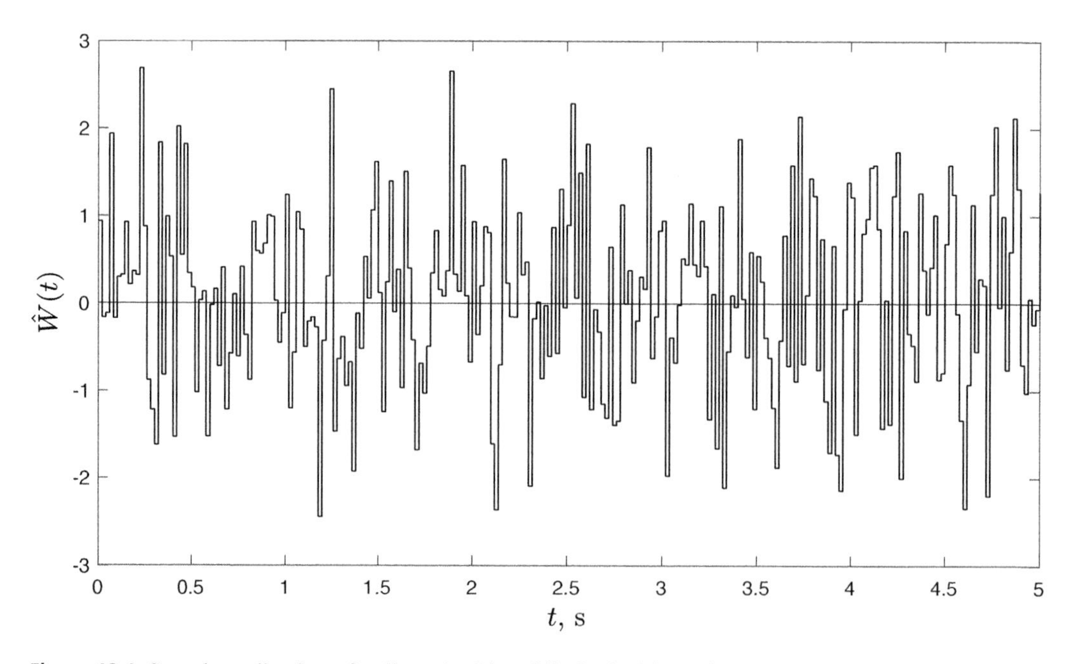

Figure 13.1 Sample realization of a discretized band-limited white-noise process

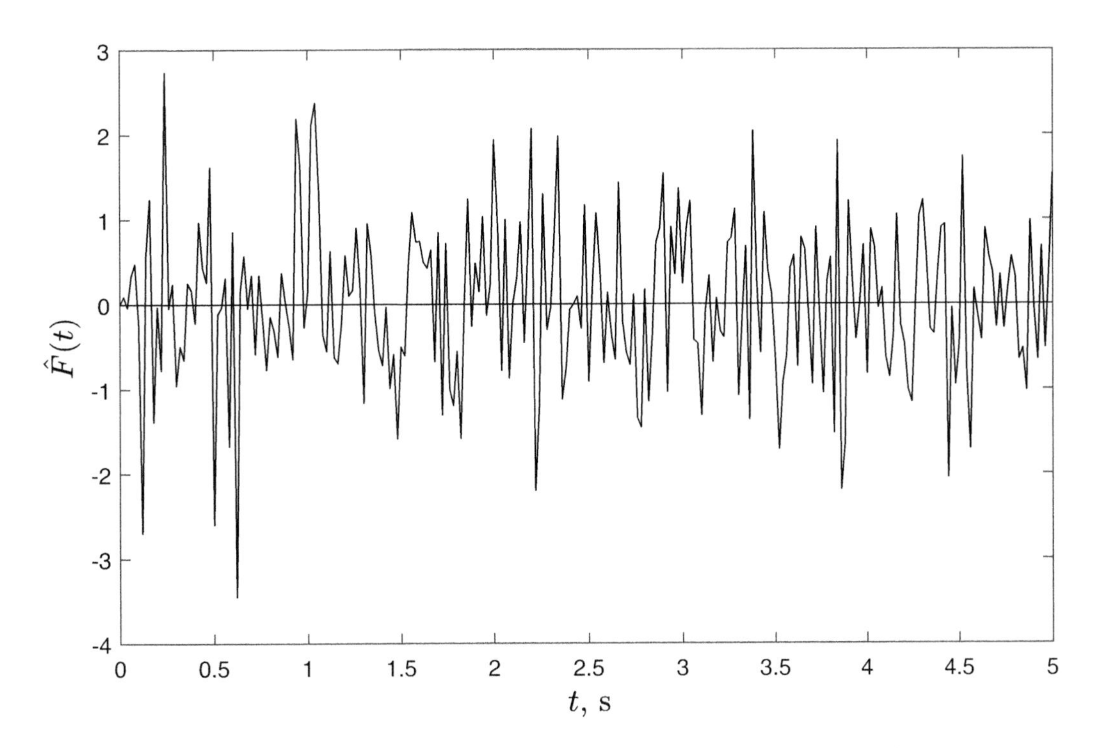

Figure 13.2 Sample realization of a process obtained by filtering a discretized white-noise process

To model a non-stationary process, one can modulate the filtered white-noise process with a time-varying function $q(t)$, in which case the shape functions take the form

$$\begin{aligned} s_i(t) &= \sigma \int_{t_{i-1}}^{t_i} q(\tau)h(t-\tau)\mathrm{d}\tau, \quad && t_{i-1} < t, \\ &= 0, && \text{elsewhere.} \end{aligned} \tag{13.6}$$

Alternatively, one can modulate the resulting process itself, in which case the shape functions take the form

$$\begin{aligned} s_i(t) &= \sigma q(t) \int_{t_{i-1}}^{t_i} h(t-\tau)\mathrm{d}\tau, \quad && t_{i-1} < t, \\ &= 0, && \text{elsewhere.} \end{aligned} \tag{13.7}$$

The above processes are non-stationary in the time domain but not in the frequency domain. To achieve non-stationarity in the frequency domain, Rezaeian and Der Kiureghian (2008) used a filter with a time-varying frequency.

Grigoriu and Balopoulou (1993) and Grigoriu (2006) suggested the use of the sine cardinal function, commonly known as the sinc function, for representing and simulating band-limited random processes. More recently, Broccardo and Der Kiureghian (2018) proposed the use of this representation in reliability analysis. The discrete representation for a band-limited, white-noise process has the form in (13.1) with

$$s_i(t) = \sigma \ \mathrm{sinc}_{\Delta t}(t - i\Delta t), \quad i = 1, \ldots n, \tag{13.8}$$

where $\Delta t = \pi/\Omega$ is a time increment with Ω denoting the band limit and

$$\mathrm{sinc}_{\Delta t}(t) = \frac{\sin\left(\frac{\pi t}{\Delta t}\right)}{\left(\frac{\pi t}{\Delta t}\right)} \tag{13.9}$$

is the sinc function. As we will see later, one advantage of this approach is that Δt and, therefore, Ω can be set independently of the time step needed for the integration of the dynamic response. This helps reduce the number of random variables needed in the reliability analysis. As with the representation in (13.5), the resulting white noise can be filtered and modulated to produce a colored and non-stationary process.

For the processes in Figures 13.1 and 13.2, we selected $\Delta t = 0.02$, which implies that frequencies up to $\Omega = \pi/0.02 = 50\pi$ rad/s are included. Suppose we have a structure that is not affected by frequencies beyond 20π rad/s. We can use discretization employing the sinc function with $\Delta t = \pi/20\pi = 0.05$ s. Figure 13.3 shows a sample realization of such a process having standard deviation $\sigma = 1$. $n = 100$ standard normal random variables were used to generate this sample function (as opposed to $n = 251$ used to generate the sample in Figure 13.2), which was then plotted using a time increment of 0.01 s. Naturally, this sample lacks the high frequency content that is included in the realization in Figure 13.2, which evidently has no influence on the response of the structure. Note that for numerical integration

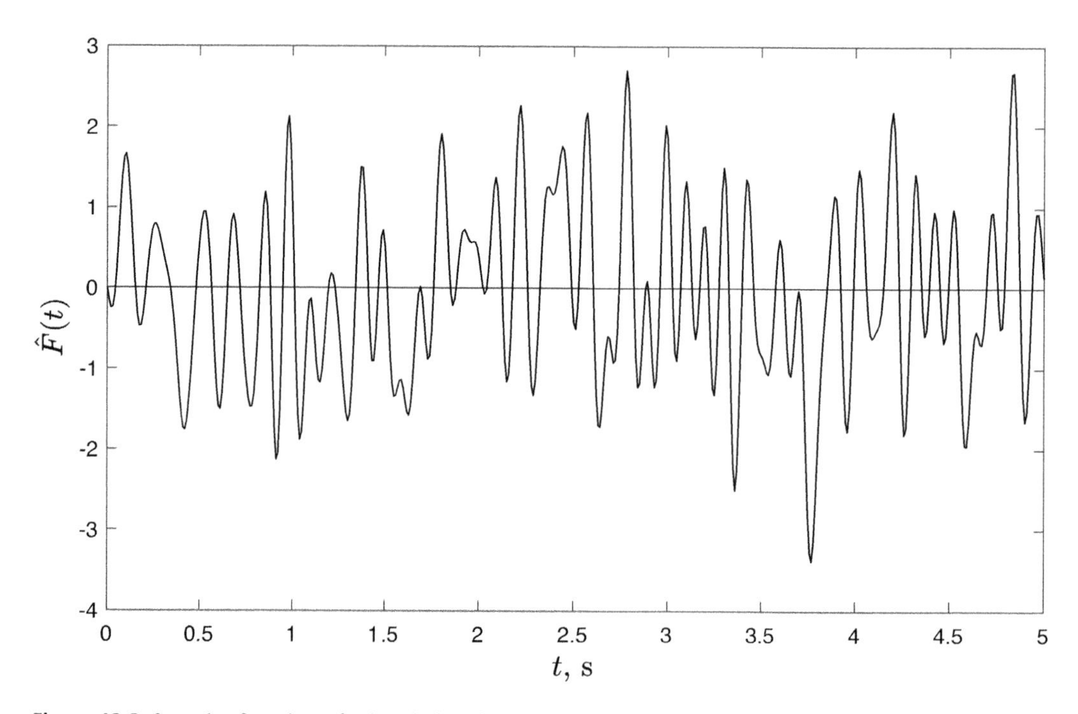

Figure 13.3 Sampler function of a banded white-noise process generated using the sinc function

of the response, one can use any time step that is necessary to achieve the desired accuracy, independent of the time step used for discretizing the input process employing the sinc function.

When the excitation process is stationary, an alternative formulation is to discretize the PSD in the frequency domain. Let $\omega_i = \pm i\Delta\omega$, $i = 1, \ldots, n$ be a set of equally spaced points along the frequency axis, where $0 < \Delta\omega$ denotes the frequency increment. As shown by Rice (1944, 1945) and Shinozuka and Jan (1972), such a discretization in the frequency domain leads to an expression similar to (13.1) with the basis functions

$$\mathbf{s}(t) = [\sigma_1 \cos(\omega_1 t) \cdots \sigma_n \cos(\omega_n t) \;\; \sigma_1 \sin(\omega_1 t) \cdots \sigma_n \sin(\omega_n t)], \tag{13.10}$$

where $\sigma_i^2 = 2\Phi(\omega_i)\Delta\omega$, $i = 1, \ldots, n$, and $\Phi(\omega)$ is the PSD. Note that the number of terms now is $2n$ to account for the symmetric nature of the PSD relative to the zero-frequency axis. When $\sigma_i^2 = \sigma^2/n$ for all i, the above representation produces a band-limited white noise with upper frequency $\Omega = n\Delta\omega$. One can show that the process represented by the above discretization is periodic with the period being $T = 2\pi/\Delta\omega$ s. That is, the values of the process at times t, $t + T$, $t + 2T$, etc., are identical. Thus, $\Delta\omega$ must be selected such that T is sufficiently long for the intended analysis. This is in contrast to the time-domain discretization, where Δt should be sufficiently small to include frequencies up to the desired upper limit $\Omega = \pi/\Delta t$.

The discrete representation (13.1) offers an opportunity to see the random process in a new light (Der Kiureghian, 2000). Assuming the mean is zero, the process is represented as the inner product of two vectors, $\widehat{F}(t) = \mathbf{s}(t) \cdot \mathbf{U}$. These vectors represent two distinct characteristics of the process: the n-vector of random variables $\mathbf{U}$ represents the randomness of the process, and the deterministic n-vector function $\mathbf{s}(t)$ represents the evolution of the process in time. Consider the n-dimensional space defined by orthonormal vectors $\hat{\mathbf{e}}_i$, $i = 1, \ldots, n$. An outcome of $\mathbf{U}$ in this space is defined by the vector $\mathbf{u} = u_1\hat{\mathbf{e}}_1 + \cdots + u_n\hat{\mathbf{e}}_n$. Likewise, the realization of $\mathbf{s}(t)$ at time t is the vector $\mathbf{s}(t) = s_1(t)\hat{\mathbf{e}}_1 + \cdots + s_n(t)\hat{\mathbf{e}}_n$. While $\mathbf{u}$ is fixed in time, $\mathbf{s}(t)$ evolves in time. As the covariance matrix of $\mathbf{U}$ is the identity matrix, the auto-covariance function of the discretized process is given by the inner product $\Sigma_{\hat{F}\hat{F}}(t, t') = \mathbf{s}(t) \cdot \mathbf{s}(t')$ and the standard deviation function is given by the Euclidean norm $\sigma_{\hat{F}}(t) = \|\mathbf{s}(t)\|$. Furthermore, let $\gamma(t, t')$ denote the angle between $\mathbf{s}(t)$ and $\mathbf{s}(t')$ at time instances t and t'. Then $\rho_{\hat{F}\hat{F}}(t, t') = \cos[\gamma(t, t')]$ is the auto-correlation coefficient function of the process. Also, let $\theta(t)$ denote the angle between $\mathbf{u}$ and $\mathbf{s}(t)$. Then $\widehat{F}(t) = \|\mathbf{s}(t)\| \cos[\theta(t)]$. When the process is stationary, the trajectory of the tip of the vector $\mathbf{s}(t)$ falls on a hypersphere of radius $\sigma_{\hat{F}}$. In that case, the variability in $\widehat{F}(t)$ is entirely due to the time evolution of the angle $\theta(t)$. Some of these geometric properties are depicted in Figure 13.4.

As mentioned earlier, the representation in (13.1) is valid only for Gaussian processes. There are several options for representing non-Gaussian processes. These include translation processes, of which the Nataf class of processes is a special case, and the use of nonlinear filters. In either case, it is possible to perform the discretization in the Gaussian space and then transform the discretized process (by translation or filtering) to obtain the corresponding realization of the non-Gaussian process. An example later in this chapter demonstrates one such application.

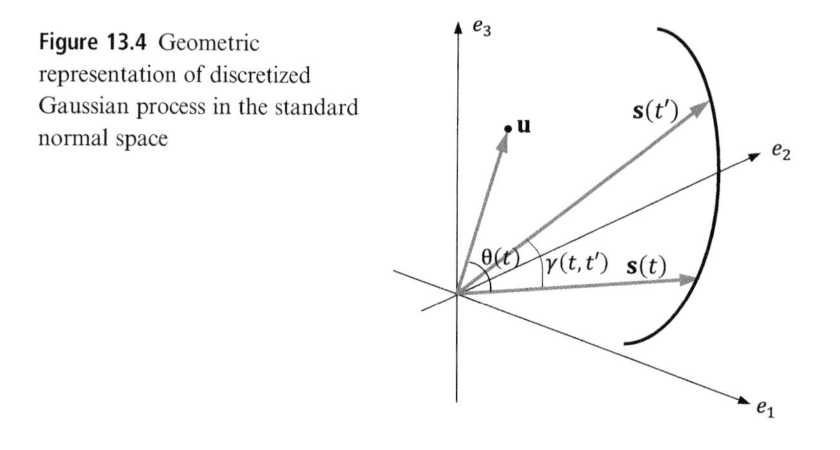

Figure 13.4 Geometric representation of discretized Gaussian process in the standard normal space

13.3 Response of Linear System to Gaussian Excitation

Consider the response $X(t)$ of a general linear system subjected to a Gaussian excitation $F(t)$. We assume the system is subjected to the excitation starting from its rest condition at time $t = 0$. Figure 13.5 shows a conceptual depiction of the system. For an input-output pair $[F(t), X(t)]$, the linear system is completely characterized by its IRF, $h(t)$. This function represents the response $X(t)$ when the input $F(t)$ is a unit-impulse function, i.e., when $F(t) = \delta(t)$, where $\delta(t)$ is the Dirac Delta function having the properties $\delta(0) = \infty$, $\delta(t) = 0$ for $t \neq 0$, and $\int_{-\epsilon}^{+\epsilon} \delta(t)\mathrm{d}t = 1$. The system is said to be causal when $h(t) = 0$ for $t < 0$, and stable when $\lim_{t\to\infty} h(t) = 0$. All systems considered in this chapter are causal and stable.

Given the IRF, the response of the system is given by the Duhamel's integral

$$X(t) = \int_0^t F(\tau)h(t-\tau)\mathrm{d}\tau. \tag{13.11}$$

Replacing $F(\tau)$ by its discretized form in (13.1), we have

$$X(t) = \mu_X(t) + \mathbf{a}(t)\mathbf{U}, \tag{13.12}$$

where $\mu_X(t) = \int_0^t \mu_F(\tau)h(t-\tau)d\tau$ is the mean response and $\mathbf{a}(t) = [a_1(t) \cdots a_n(t)]$ is the gradient row vector of $X(t)$ with respect to $\mathbf{U}$, having the elements

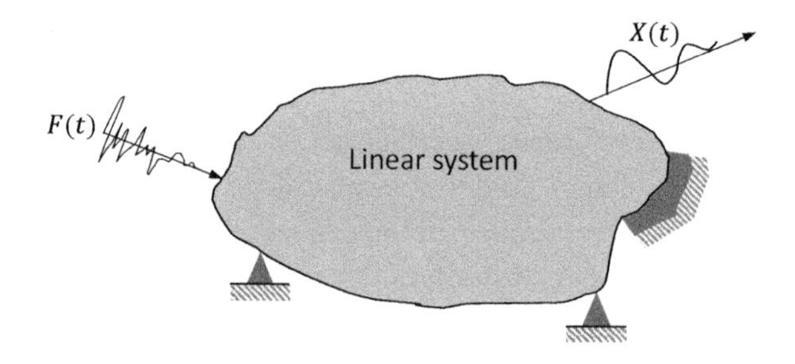

Figure 13.5 Input-output pair in a linear system

$$a_i(t) = \int_0^t s_i(\tau)h(t-\tau)\mathrm{d}\tau. \tag{13.13}$$

Observe that the response process is a linear function of the standard normal random variables and, hence, is Gaussian, as expected. Furthermore, (13.12) has the same discretized form as the excitation process in (13.1) with the shape functions $s_i(t)$ replaced by $a_i(t)$. It follows that the geometric representations described in Figure 13.4 for the excitation process also apply to the response process with $\mathbf{s}(t)$ replaced by $\mathbf{a}(t)$. In particular, $\Sigma_{XX}(t,t') = \mathbf{a}(t)\cdot\mathbf{a}(t')$ denotes the autocovariance function of the response process, $\sigma_X(t) = \|\mathbf{a}(t)\|$ is the standard deviation function, and $X(t) = \|\mathbf{a}(t)\|\cos[\theta(t)]$, where $\theta(t)$ is the angle between vectors $\mathbf{a}(t)$ and $\mathbf{u}$. Additionally, the cross-covariance function between the input excitation $\widehat{F}(t)$ and the response $X(t')$ is given by the inner product $\Sigma_{\hat{F}X}(t,t') = \mathbf{s}(t)\cdot\mathbf{a}(t')$ and the cross-correlation coefficient function is identical to the cosine of the angle between the two vectors.

Now consider the event $\{\mu_X(t_x) + x \le X(t_x)\}$, where $t_0 \le t_x \le t_n$ and x denotes a selected threshold above the mean response. Using the discretized form of the response in (13.12), this event is equivalent to $\{G(\mathbf{U}, x, t_n) \le 0\}$, where

$$G(\mathbf{U}, x, t_x) = x - \mathbf{a}(t_x)\mathbf{U} \tag{13.14}$$

is the limit-state function. Owing to the linearity of the limit-state function, the FORM solution of this problem is readily available:

$$\mathbf{u}^*(x, t_x) = \frac{x\mathbf{a}(t_x)}{\|\mathbf{a}(t_x)\|^2} \text{ design point,} \tag{13.15}$$

$$\widehat{\boldsymbol{\alpha}}(x, t_x) = \frac{\mathbf{a}(t_x)}{\|\mathbf{a}(t_x)\|} \text{ unit normal vector,} \tag{13.16}$$

$$\beta(x, t_x) = \frac{x}{\|\mathbf{a}(t_x)\|} \text{ reliability index,} \tag{13.17}$$

$$\Pr\{\mu_X(t_x) + x \le X(t_x)\} = \Phi[-\beta(x, t_x)] \text{ tail probability.} \tag{13.18}$$

Observe that the design point and the reliability index are proportional to the threshold x. Furthermore, substituting $\mathbf{u}^*(x, t_n)$ for $\mathbf{U}$ in (13.1) and (13.12), we obtain the design-point excitation and response, which are the most likely realizations of these quantities that give rise to the event $\{\mu_X(t_x) + x \le X(t_x)\}$, as

$$\hat{f}^*(t) = \mu_F(t) + x\frac{\mathbf{s}(t)\mathbf{a}(t_x)^{\mathrm{T}}}{\|\mathbf{a}(t_x)\|^2} \tag{13.19}$$

$$x^*(t) = \mu_X(t) + x\frac{\mathbf{a}(t)\mathbf{a}(t_x)^{\mathrm{T}}}{\|\mathbf{a}(t_x)\|^2}, \tag{13.20}$$

respectively.

It was shown by Koo et al. (2005) that the design-point response for an elastic single-degree-of-freedom oscillator (linear or nonlinear) subjected to a zero-mean white-noise excitation is identical to the mirror image of the free-vibration response of the oscillator when it is released from the initial conditions $X(0) = x$ and $\dot{X}(0) = 0$, provided t_x is sufficiently large so that the response reaches the stationary state. This result approximately holds for non-stationary and colored excitations, and for oscillators having a non-Gaussian response. As shown in an example below, this finding is useful for determining the design-point excitation for such cases.

Another result that is the cornerstone for developing the TELM for nonlinear systems is that, knowing the design point, we can determine the vector $\mathbf{a}(t_x)$ from

$$\mathbf{a}(t_x) = \frac{x\mathbf{u}^*(x, t_x)^{\mathrm{T}}}{\|\mathbf{u}^*(x, t_x)\|^2}. \tag{13.21}$$

This relation is verified by observing that $x = \mathbf{a}(t_x)\mathbf{u}^*(x, t_x)$ at the design point, and by post-multiplying the transpose of this relation by $\mathbf{u}^*(x, t_x)$. Although x appears in the numerator of (13.21), $\mathbf{a}(t_x)$ is actually independent of the threshold because for a linear system the design point is proportional to x, as can be seen in (13.15).

An alternative way to characterize a linear system for a given input-output pair is through the *frequency-response function* (FRF). The FRF, denoted $H(\omega)$, represents the complex-valued amplitude of the steady-state response of the linear system to the complex harmonic excitation $F(t) = \exp(\mathrm{i}\omega t)$, where $\mathrm{i} = \sqrt{-1}$ denotes the imaginary unit. The FRF exists as long as the system is stable, i.e., as long as $\lim_{t\to\infty} h(t) = 0$. One can show that the IRF and FRF are Fourier-transform pairs:

$$H(\omega) = \int_{-\infty}^{+\infty} h(t)\exp(-\mathrm{i}\omega t)\mathrm{d}t, \tag{13.22}$$

$$h(t) = \frac{1}{2\pi}\int_{-\infty}^{+\infty} H(\omega)\exp(+\mathrm{i}\omega t)\mathrm{d}\omega. \tag{13.23}$$

This characterization of the system is useful when the input and output processes are stationary. In that case, the PSD of the response, $\Phi_{XX}(\omega)$, is given in terms of the PSD of the input excitation, $\Phi_{FF}(\omega)$, by the product rule

$$\Phi_{XX}(\omega) = |H(\omega)|^2\Phi_{FF}(\omega). \tag{13.24}$$

As Vanmarcke (1972) showed, many statistical quantities of the stationary response are given in terms of the *spectral moments* defined by

$$\lambda_m = \int_{-\infty}^{+\infty} |\omega|^m\Phi_{XX}(\omega)\mathrm{d}\omega, m = 0, 1, 2, 4. \tag{13.25}$$

In particular, λ_0 is the variance of the response, λ_2 and λ_4 are the variances of the first and second derivatives of the response, and λ_1 is the covariance between the response and the derivative of its Hilbert transform. Later in this chapter, we will use the spectral moments to compute various measures of the response that are of interest in reliability analysis.

The above representation of a linear system is convenient when the input excitation is discretized in the frequency domain, as in (13.10). The results in (13.15)–(13.21) are still valid, provided t_x is sufficiently long so that $a_i(t_x)$, obtained from (13.13), achieves a steady harmonic state.

13.4 Response of Linear System to Non-Gaussian Excitation

The stochastic response of a system is non-Gaussian when the excitation is non-Gaussian or when the system is nonlinear. As mentioned earlier, a non-Gaussian excitation can be defined through translation or nonlinear filtering of a Gaussian process, with the latter discretized by one of the methods described in Section 13.2. In either case, when the excitation is non-Gaussian, the response of the linear system is an implicit function of random variables $\mathbf{U}$ used to represent the underlying Gaussian process in a discrete form. To highlight this implicit dependence, we write the response process as $X(t, \mathbf{U})$. The results (13.12)–(13.21) in the preceding section are no longer valid. However, given the limit-state function

$$G(\mathbf{U}, x, t_x) = x - X(t_x, \mathbf{U}), \tag{13.26}$$

one can determine the probability that the response at time t_x will exceed the threshold x by reliability methods, including the FORM and second-order reliability methods (SORM). For this purpose, we need to use an optimization algorithm such as the iHL-RF (see Section 5.2.3) to determine the design point $\mathbf{u}^*(x, t)$, from which the reliability index and design-point realizations of the excitation and response can be determined. Recall that the optimization algorithm requires repeated evaluations of the limit-state function and its gradient with respect to the random variables $\mathbf{U}$ at a sequence of trial outcome points $\mathbf{u}_i$, $i = 1, 2, \ldots$, until convergence is achieved. Because in (13.26) $\mathbf{U}$ only appears in the response function, the algorithm essentially requires repeated evaluations of the response $X(t_x, \mathbf{U})$ and its gradient with respect to $\mathbf{U}$ at the trial points. This requires linear dynamic analysis for the selected realizations of the discretized non-Gaussian excitation, $\widehat{F}(t, \mathbf{u}_i)$, and the gradients with respect to $\mathbf{u}_i$. The direct differentiation method described in Section 12.4 is used to compute the gradients.

Example 13.1 – Linear Oscillator Subjected to Gaussian and Non-Gaussian Excitations

Consider the dynamic response of a linear, single-degree-of-freedom oscillator subjected to base motion, which is governed by the differential equation

$$m\ddot{X}(t) + c\dot{X}(t) + kX(t) = -mF(t), \tag{E1}$$

where m is the mass, c the damping coefficient, k the stiffness and $F(t)$ the base acceleration. Let $\omega_0 = \sqrt{k/m} = 5\pi$ rad/s denote the natural frequency of the oscillator and $\xi = c/(2m\omega_0) = 0.05$ denote the damping ratio. We consider two cases for the excitation: (a) $F(t) = W(t)$

Example 13.1 (cont.)

a band-limited, zero-mean, Gaussian white noise with spectral density $S = 0.03$ m²/s³ for $|\omega| \leq 50\pi$ rad/s and 0 elsewhere; and (b) $F(t)$ is a shifted lognormal random process with the same median, variance, and bandwidth as $W(t)$, obtained by translating the latter. Of interest is the reliability of the oscillator in exceeding the response threshold $x = 0.05$ m at the end of a 10 s duration of loading, i.e., $t_x = 10$ s.

To accommodate the frequency range of the input, we set $\Delta t = \pi/50\pi = 0.02$ s and $t_i = (i-1)\Delta t$, $i = 1, \ldots, n$, with $n = 501$ for a 10 s duration of excitation and $t_x = t_n$. For case (a), the variance of the band-limited white-noise process is $\sigma^2 = 2 \times 0.03 \times 50\pi = (3.07)^2$ m²/s⁴. The discretized process is represented in the form (13.1) with zero mean and $s_i(t)$ as defined in (13.4). For case (b), the discretized process is defined in terms of a series of point values $\widehat{F}(t_i) = \exp[\lambda + \zeta U_i] - c$, $i = 1, \ldots, n$, with $\lambda = 1.63$, $\zeta = 0.5$, $c = 5.08$, and U_i denoting standard normal random variables. One can easily verify that this process has the same median and variance as $W(t)$. Figure 13.6 compares the point-in-time probability density functions (PDFs) of the two input processes and Figure 13.7 shows their sample realizations. Observe that the realization of the shifted lognormal process exhibits a lower bound around $-c$ and large peaks on the positive side due to the long right tail of the lognormal distribution.

The well-known IRF of the linear oscillator subjected to base motion is given by

$$
\begin{aligned}
h(t) &= \frac{1}{\omega_d} \exp(-\xi\omega_0 t) \sin(\omega_d t), \quad && 0 \leq t, \\
&= 0, && t < 0,
\end{aligned} \tag{E2}
$$

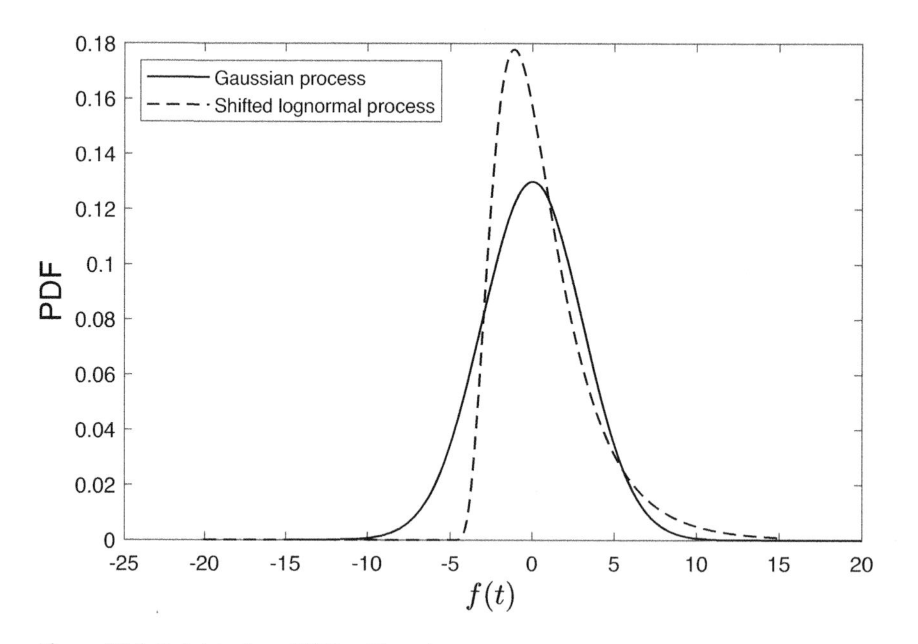

Figure 13.6 Point-in-time PDFs of input processes

Example 13.1 (cont.)

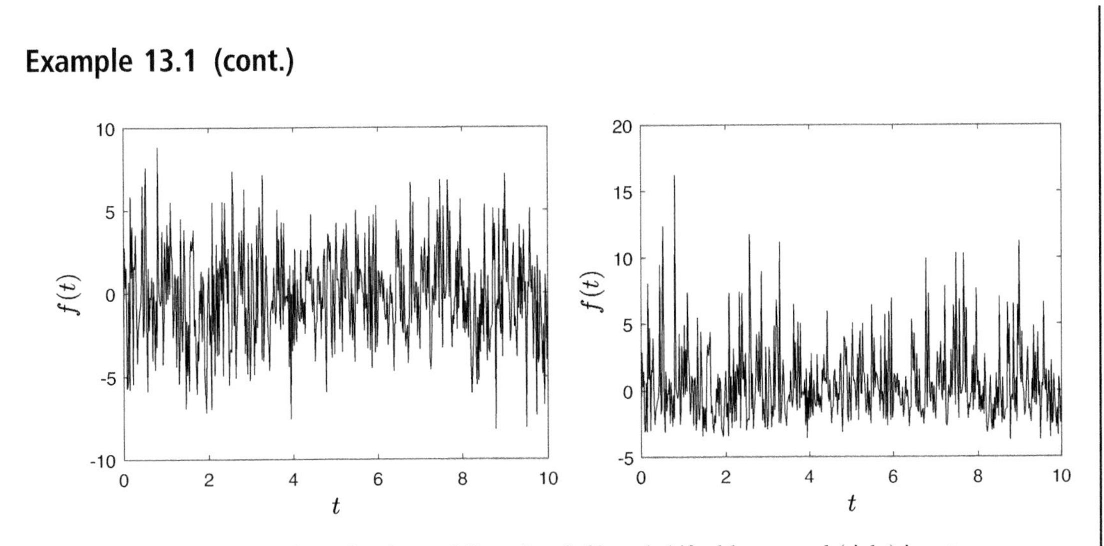

Figure 13.7 Sample realizations of Gaussian (left) and shifted lognormal (right) input process

where $\omega_d = \omega_0\sqrt{1-\zeta^2}$ is the damped frequency. Considering that the IRF is practically constant over the small time interval $(t, t+\Delta t)$, from (3.13), $a_i(t_x) = \sigma h(t_x - t_i)\Delta t$ is obtained for case (a). This result is now used in (13.15)–(13.20) to obtained quantities of interest for the response to the Gaussian excitation. In particular, for the threshold $x = 0.05$ m, we obtain $\beta(x, t_x) = 3.21$ and the design-point excitation and response as shown in Figure 13.8. Observe that the design-point excitation starts from zero (the most likely outcome of a zero-mean Gaussian distribution) and then gradually builds up to achieve the required response $x = 0.05$ m at time $t_x = 10$ s. The design point excitation value at t_x is zero. This is because the input process has no memory and a non-zero excitation value at this time instance will not affect the response at the same time. Furthermore, zero is the most likely value of a zero-mean Gaussian distribution. The trajectory of the design-point response in the right-side chart in Figure 13.8 is identical to the mirror image of the free-vibration response of the oscillator when it is released from the initial position $X(0) = 0.05$ m and $\dot{X}(0) = 0$. This has to do with the fact that, for a zero-mean Gaussian white-noise excitation, the most likely trajectory to achieve a given response threshold is one that requires minimum input energy. This property also holds for the free-vibration response of an elastic oscillator because it follows a trajectory that dissipates minimum energy.

Now consider the response to the shifted lognormal excitation process. In this case, the response is obtained from Duhamel's integral (3.11) with $F(t)$ replaced by its discrete representation $\widehat{F}(t_i, U_i) = \exp[\lambda + \zeta U_i] - c$, $i = 1, \ldots, n$. The limit-state function now is a nonlinear function of the standard normal random variables, $G(\mathbf{U}, x, t_x) = x - X(t_x, \mathbf{U})$. We obtain the design point by use of the iHL-RF algorithm starting from the design point of the case with the Gaussian excitation. Convergence is achieved in four steps. The reliability index is $\beta = 3.77$ and the design-point excitation and response are as shown in Figure 13.9.

Example 13.1 (cont.)

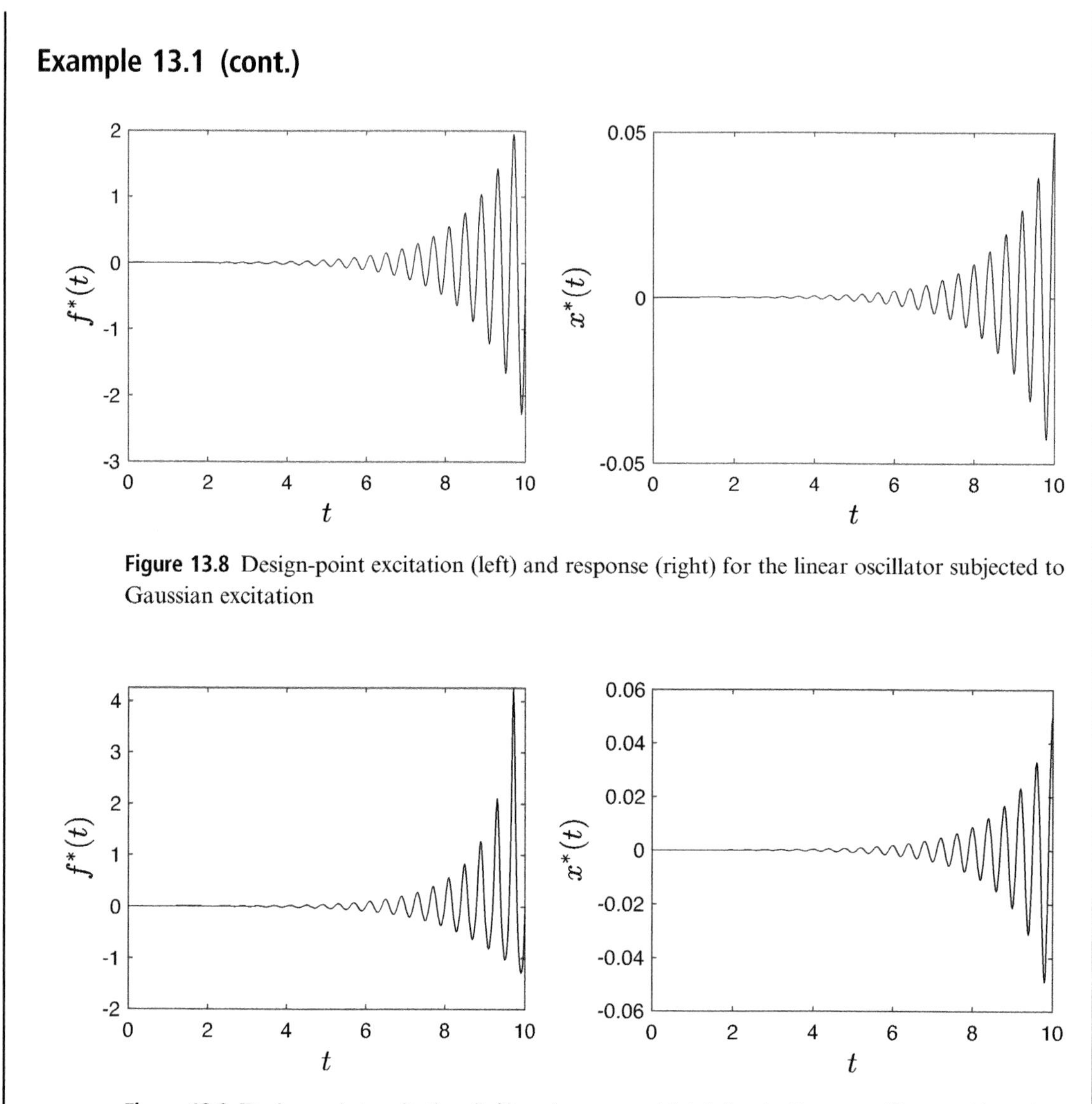

Figure 13.8 Design-point excitation (left) and response (right) for the linear oscillator subjected to Gaussian excitation

Figure 13.9 Design-point excitation (left) and response (right) for the linear oscillator subjected to the shifted lognormal excitation

Comparing the respective results in Figures 13.8 and 13.9, we see that, although the design-point excitations are considerably different, the design-point responses are quite similar. We note that the design point excitation starts from zero because the median of the shifted lognormal distribution, which is zero, maps onto the origin in the standard normal space, which is the most likely point in that space.

As a final result of interest, Figure 13.10 compares the point-in-time reliability indices as functions of the threshold for the two cases. For the Gaussian excitation, as shown in (13.17),

Example 13.1 (cont.)

the reliability index is proportional to the threshold x. However, for the shifted lognormal excitation, the line is curved, indicating the non-Gaussian nature of the response.

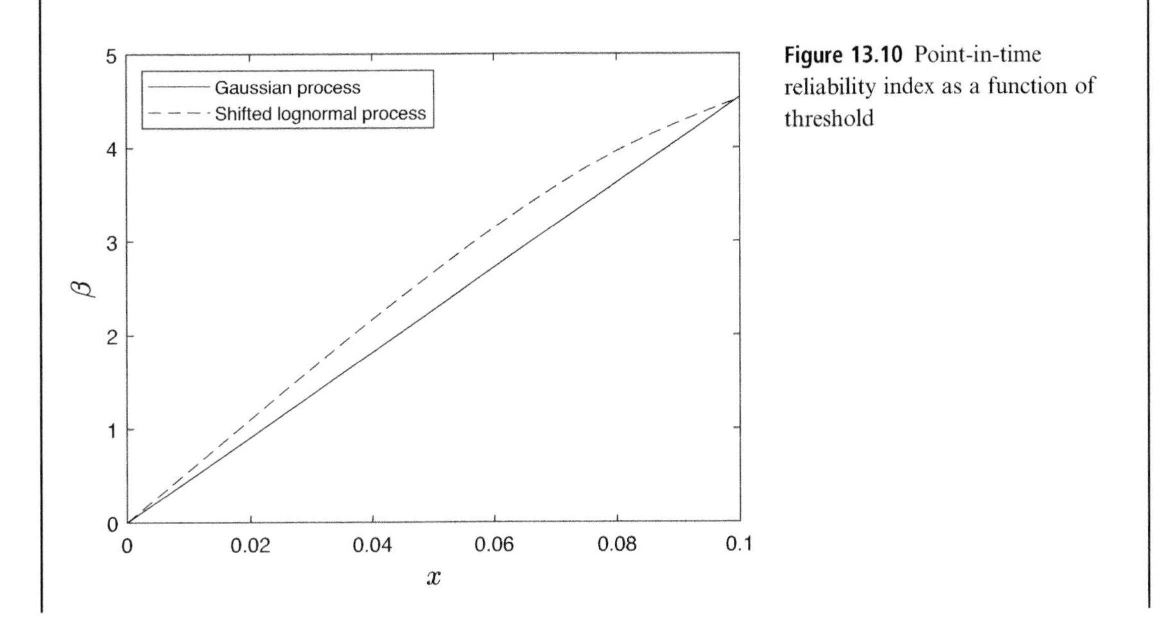

Figure 13.10 Point-in-time reliability index as a function of threshold

13.5 Tail-Equivalent Linearization for Nonlinear Stochastic Dynamic Analysis

Structural failures typically occur when the structure is well into its nonlinear regime. Yielding, fracture, and large deformation are typical nonlinear behaviors that precede structural failure. Furthermore, loads arising from man-made phenomena such as traffic, or natural hazards such as earthquakes, severe winds, and ocean waves, are stochastic in nature and induce dynamic effects. Hence, nonlinear stochastic dynamic analysis is an important topic for the reliability and safety assessment of structures.

The topic of nonlinear stochastic structural dynamic analysis has been the focus of much research and development in the past several decades. Methods developed include the Fokker-Planck equation, stochastic averaging, moment closure, perturbation, path integration, and equivalent linearization. Recent accounts of these methods can be found in the texts by Roberts and Spanos (1990), Soong and Grigoriu (1993), Lin and Cai (1995), Lutes and Sarkani (2004), and the paper by Iourtchenko et al. (2008), among others. Among these methods, the equivalent linearization method (ELM) has gained wide popularity because of its versatility in application to general multi-degree-of-freedom nonlinear systems. The other methods, though more accurate, are largely restricted to specialized systems or forms of the excitation and are difficult to apply in practice. The Monte Carlo simulation method (Shinozuka and Jan, 1972) is without restriction; however, in application it is often computationally prohibitively demanding.

In the ELM, the nonlinear system of interest is replaced by an equivalent linear system, the parameters of which are determined by minimizing a measure of the discrepancy between the responses of the nonlinear and linear systems (Caughey, 1963). The measure of discrepancy most often used is the mean-square error between the two responses (Atalik and Utku, 1976; Wen, 1976), although an energy-based measure has also been considered (Zhano et al., 1991). The solution requires an iterative scheme because the parameters of the linear system, e.g., the elements of the stiffness, damping, and mass matrices, are determined as functions of the second moment of its response, which in turn depends on these parameters. Furthermore, the method requires an assumption regarding the probability distribution of the nonlinear response and most often the Gaussian distribution is used. As a result, while the method can be quite accurate in estimating the second moment of the response, the probability distribution can be far from correct, particularly in the tail region. It follows that estimates of such response statistics as crossing rates and the tail distribution of the maximum response, which are of particular interest in reliability analysis, can be grossly inaccurate at high thresholds.

The method described in this section is also an ELM. However, rather than defining the linear system by minimizing the mean-square error in the response, it is defined by matching the tail probability of the linear system response to the first-order approximation of the tail probability of the nonlinear response – hence, the TELM. The method was first introduced in Fujimura and Der Kiureghian (2007) and further developed in many subsequent publications as described below.

Consider the response of a multi-degree-of-freedom nonlinear system to a scalar excitation $F(t)$ defined by the governing equation

$$\mathbf{M}\ddot{\mathbf{Y}}(t) + \mathbf{C}\dot{\mathbf{Y}}(t) + \mathbf{R}(t) = \mathbf{P}F(t), \tag{13.27}$$

where $\mathbf{Y}(t)$ denotes the vector of nodal displacements with a superposed dot indicating the derivative with respect to time, $\mathbf{M}$ is the mass matrix, $\mathbf{C}$ the viscous damping matrix, $\mathbf{R}(t)$ the vector of resisting forces, which for a nonlinear system in general depends on the entire displacement history, and $\mathbf{P}$ a load-distribution vector that allocates the scalar forcing function $F(t)$ to the degrees of freedom of the structure. We assume $F(t)$ is represented in a discrete form in terms of a set of standard normal random variables $\mathbf{U}$, as in (13.1) for a Gaussian excitation, or as a nonlinear function of $\mathbf{U}$ through translation of a Gaussian process, as described in Section 13.4. In either case, $\mathbf{Y}(t)$ is a nonlinear function of $\mathbf{U}$ and we denote it $\mathbf{Y}(t, \mathbf{U})$.

Let $X(t)$ denote a generic response quantity, such as an internal force, stress or deformation at a critical point of the structure. In general, $X(t)$ is a function of the vector of nodal displacements $\mathbf{Y}$, which itself is an implicit function of the random variables $\mathbf{U}$. Thus, showing the implicit dependence on $\mathbf{U}$, we write the generic response quantity as $X(t, \mathbf{U})$.

We first consider the failure event defined as the exceedance of the generic response quantity above the threshold x at time t_x. This is a reliability problem defined by the limit-state function

$$G(\mathbf{U}, x, t_x) = x - X(t_x, \mathbf{U}). \tag{13.28}$$

A FORM solution of this problem is obtained by finding the design point $\mathbf{u}^*(x, t_x) = \min\{\|\mathbf{u}\| | G(\mathbf{u}, x, t_x) \leq 0\}$ and computing the unit vector $\widehat{\boldsymbol{\alpha}}(x, t_x) = -\nabla_{\mathbf{u}} G(\mathbf{u}^*, x, t_x)/\|\nabla_{\mathbf{u}} G(\mathbf{u}^*, x, t_x)\|$, where $\nabla_{\mathbf{u}} G(\mathbf{u}^*, x, t_x) = -\nabla_{\mathbf{u}} X(t_x, \mathbf{u}^*)$, and reliability index $\beta(x, t_x) = \widehat{\boldsymbol{\alpha}}(x, t_x)\mathbf{u}^*(x, t_x)$. The first-order approximation of the tail probability then is given by $\Pr[x \leq X(t_x, \mathbf{U})] \cong p_{F1}(x, t_x) = \Phi[-\beta(x, t_x)]$. As we saw in Chapter 6, this approximation essentially replaces the limit-state surface $G(\mathbf{u}, x, t_x) = 0$ with the tangent hyperplane $\beta(x, t_x) - \widehat{\boldsymbol{\alpha}}(x, t_x)\mathbf{u} = 0$ at the design point.

We saw in Section 13.3 that the limit-state surface for the tail probability of the response of a linear system subjected to a discretized Gaussian excitation is a hyperplane in the space of the standard normal random variables $\mathbf{U}$. Furthermore, knowing the design point (the point of perpendicular projection from the origin), the gradient vector $\mathbf{a}(t_x)$ of the hyperplane is determined from (13.21). We define the *tail-equivalent linear system* (TELS) for the particular response quantity $X(t, \mathbf{U})$ of the nonlinear system, for the specified threshold x and time $t = t_x$, as the linear system whose tail probability is defined by the hyperplane $\beta(x, t_x) - \widehat{\boldsymbol{\alpha}}(x, t_x)\mathbf{u} = 0$. As shown below, this uniquely defines the TELS in terms of its IRF. Note that the TELS is specific for a selected excitation process, nonlinear response quantity, threshold x, and time t_x. The TELS in general may vary if any of these four quantities change. Figure 13.11 illustrates this definition of the TELS.

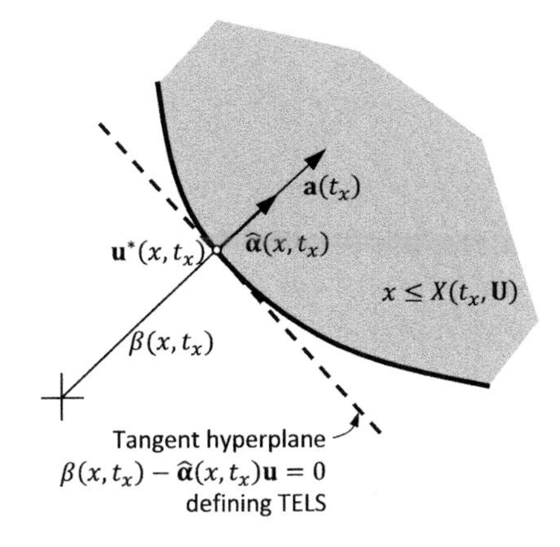

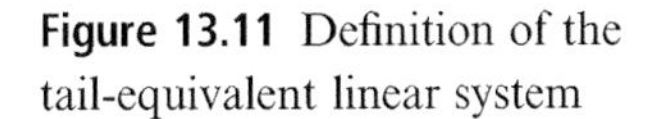
Figure 13.11 Definition of the tail-equivalent linear system

How do we identify the TELS? Let $t_i = (i-1)\Delta t$, $i = 1, \ldots, n$, be the points at which the input excitation has been discretized and set $t_x = t_n$. For now, we consider one of the discretization methods with $s_i(t)$ defined as in (13.4)–(13.7), so that the input is Gaussian. Using a simple rectangular integration rule, (13.13) is written as

$$\sum_{j=1}^{n} h(t_n - t_j) s_i(t_j) \Delta t = a_i(t_n), \quad i = 1, \ldots, n. \tag{13.29}$$

Note that Δt must be sufficiently small to accommodate not only the bandwidth of the excitation but also the accuracy of the above integration. The above expression represents n linear equations for the n unknowns $h(t_i)$, $i = 0, 1, \ldots, n-1$. Writing in a matrix form, we have

$$\mathbf{Sh} = \mathbf{a}(t_n), \tag{13.30}$$

where

$$\mathbf{S} = \Delta t \begin{bmatrix} s_1(t_1) & s_1(t_2) & \ldots & s_1(t_n) \\ 0 & s_2(t_2) & \ldots & s_2(t_n) \\ \vdots & \vdots & \vdots & \vdots \\ 0 & 0 & 0 & s_n(t_n) \end{bmatrix} \tag{13.31}$$

and

$$\mathbf{h} = \begin{bmatrix} h(t_n - t_1) \\ h(t_n - t_2) \\ \vdots \\ h(t_n - t_n) \end{bmatrix}. \tag{13.32}$$

Observe that $\mathbf{S}$ is an upper-triangular matrix because, according to (13.4)–(13.7), $s_i(t) = 0$ for $t < t_i$. Furthermore, the diagonal elements $s_i(t_i)$, $i = 1, \ldots, n$, are non-zero. Thus, the set of equations (13.36) are easily solved, providing a unique numerical solution of the IRF of the TELS at the discrete times t_i, $i = 1, \ldots, n$. Note that in the case of a band-limited white-noise excitation, $s_i(t_j) = 0$ for $i \neq j$ so that $\mathbf{S}$ is a diagonal matrix.

When the sinc function is used to represent the discretized excitation process using the basis functions in (13.8), the time increment may not be sufficiently small to obtain an accurate numerical integration. Let Δt_s denote the time increment used in discretizing the input excitation via the sinc function and Δt denote the increment used in the numerical integration. Also let $m = \Delta t_s / \Delta t$ be an integer. Because the IRF of the TELS is itself a band-limited function, it can be represented by the sinc function interpolation rule

$$h(t) \cong \sum_{k=1}^{n} h(k\Delta t_s) \, \mathrm{sinc}_{\Delta t_s}(t - k\Delta t_s). \tag{13.33}$$

Note that $n\Delta t_s = mn\Delta t = t_n$. Substituting (13.8) and (13.33) in (13.29), we have

$$\sum_{j=1}^{nm}\left[\sum_{k=1}^{n}h(k\Delta t_s)\,\mathrm{sinc}_{\Delta t_s}(t_n - t_j - k\Delta t_s)\right]\sigma\ \mathrm{sinc}_{\Delta t_s}(t - j\Delta t)\Delta t = a_i(t_n), i = 1, \ldots, n. \tag{13.34}$$

Observe that the unknowns in this equation are the interpolation values $h(k\Delta t_s)$, $k = 1, \ldots, n$. The set of n equations are still in the form of (13.30), but $\mathbf{S}$ now is a full matrix. With the interpolation values determined, the IRF of the TELS is constructed using (13.33).

When the excitation and response processes are stationary, it is more convenient to discretize the input processes in the frequency domain with the basis functions in (13.10). As Garré and Der Kiureghian (2010) have shown, in that case it is possible to directly obtain the FRF of the TELS in terms of the information available at the design point. Let $\mathbf{a}(t_x) = [a_1^c(t_x) \cdots a_n^c(t_x)\, a_1^s(t_x) \cdots a_n^s(t_x)]$ denote the gradient vector of the hyperplane tangent at the design point obtained from (13.13), where $a_i^c(t_x)$ are associated with the cosine terms and $a_i^s(t_x)$ are associated with the sine terms in (13.10). Knowing the design point $\mathbf{u}^*(x, t_x)$ of the nonlinear system, the gradient vector $\mathbf{a}(t_x)$ is determined from (13.21). Considering that the FRF is complex valued, we write it in the form $H(\omega) = |H(\omega)|[\cos(\varphi) + \mathrm{i}\sin(\varphi)]$, where $|H(\omega)|$ is the modulus and φ is the phase angle. Garré and Der Kiureghian (2010) have shown that $a_i^c(t_x) = \sigma_i|H(\omega_i)|\cos(\omega_i t_x + \varphi_i)$ and $a_i^s(t_x) = \sigma_i|H(\omega_i)|\sin(\omega_i t_x + \varphi_i)$ so that the modulus and phase angle of the FRF are obtained in terms of the gradient vector in a discretized form as

$$|H(\omega_i)| = \frac{\sqrt{a_i^c(t_x)^2 + a_i^s(t_x)^2}}{\sigma_i}, \quad i = 1, \ldots n, \tag{13.35}$$

$$\varphi_i = \omega_i t_x + \tan^{-1}\left[\frac{a_i^s(t_x)}{a_i^c(t_x)}\right], \quad i = 1, \ldots n. \tag{13.36}$$

It is noted that the modulus of the FRF obtained from (13.35) is independent of time, as it should be for a stationary response.

With the TELS determined in terms of either its IRF or its FRF, various statistical measures of the nonlinear response for threshold x and time t_x can be determined in terms of the Gaussian response of the TELS. Of course, for a stationary response, these statistics will be invariant of the selected time, t_x, provided it is sufficiently long for the stationary state to have been reached. In particular, the complementary cumulative distribution function (CDF) of the response at time t_x is determined by performing reliability analysis for a sequence of increasing thresholds $x_1 < x_2 < \cdots < x_m$. The negative of the sensitivity of the tail probability with respect to the threshold then provides the corresponding PDF. This requires finding the design points for the sequence of thresholds. As suggested by Fujimura and Der Kiureghian (2007), once the design points for the lowest two thresholds are determined, warm starting trial points for the remaining thresholds can be obtained by extrapolation from the preceding two design points, thus reducing the amount of computation needed.

The above approach is applicable when the excitation is non-Gaussian and the system is nonlinear. However, the TELS now accounts not only for the nonlinear behavior of the system but also for the non-Gaussian nature of the excitation. More specifically, the nonlinearity in the dependence of $X(t, \mathbf{U})$ on $\mathbf{U}$ is not only from the nonlinearity in the response but also from the nonlinear dependence of the non-Gaussian excitation process $F(t, \mathbf{U})$ on $\mathbf{U}$. TELM addresses the combined effect of these nonlinearities.

Other response statistics of particular interest in reliability analysis include level crossings and the first-passage probability. These are described in later sections. But first we describe the specific properties of the TELS that have an important bearing on the accuracy and effectiveness of the TELM approximation.

13.5.1 Properties of the Tail-Equivalent Linear System

For a given nonlinear system, the TELS depends on the excitation process, the selected response quantity, threshold x, and time t_x. A change in any of these quantities may change the TELS. This is different from the classical ELM, where a parametrized linear system is defined for all response quantities regardless of the threshold. More specifically, the characteristics of the TELM and TELS are as follows:

1) For a given threshold x and time t_x, the tail probability of the TELS is equal to the first-order approximation of the tail probability of the nonlinear system response.
2) Unlike the ELM, the TELM is a non-parametric linearization method: For a given input-output pair, the equivalent linear system is determined numerically in terms of its IRF or FRF. Fujimura and Der Kiureghian (2007) investigated the nature of the TELS considering first-, second-, and third-order linear systems defined by differential equations involving the corresponding order derivatives. Their findings suggest that the TELS for a nonlinear system in general is a high-order system. Because the TELS is determined in terms of its IRF or FRF, there is no need to determine its order or parameters.
3) For a given input-output pair and time t_x, the TELS depends critically on the selected threshold x. Although the response of the TELS for each threshold is Gaussian, at different thresholds we have different TELSs and, therefore, different Gaussian distributions. As a result, the predicted distribution of the nonlinear response in general is non-Gaussian. Through this dependence of the TELS on the threshold, TELM is able to capture the non-Gaussian distribution of the nonlinear system response. To highlight this property, henceforth the IRF of the TELS is denoted as $h(t, x)$ and its FRF is as $H(\omega, x)$, where x denotes the selected threshold.
4) For a stationary response, the TELS is independent of the selected time. However, a sufficiently long time t_x should be selected in the limit-state function so that the nonlinear system response reaches stationary state. For nonstationary response, the TELS mildly depends on the selected time point. In that case, the TELS determined at a critical time, e.g., shortly after the time where the variance of the input excitation peaks, can be conservatively selected for reliability analysis. An improved approximation using

evolutionary processes is investigated by Broccardo and Der Kiureghian (2013, 2021) and is described in a later section. Alternatively, time-varying TELS can be identified by performing TELM analysis at a sequence of time points.

5) For broad-band excitations, the TELS is mildly dependent on the frequency content of the input excitation. For such excitations, the TELS determined for response to a white-noise excitation can be used as an approximation. The white-noise approximation may not work well for narrow-band excitation processes.
6) Fujimura and Der Kiureghian (2007) showed that the scaling of the excitation has no influence on the TELS. That is, the TELS developed for a selected nonlinear response quantity, threshold x, and time t_x for the excitations $F(t)$ and $cF(t)$ are identical for any $0 < c$. This, of course, does not mean that the response of the nonlinear system is independent of the intensity of the excitation. That dependence is accounted for through the response of the TELS to the specified input. Furthermore, Fujimura and Der Kiureghian (2007) have shown that the reliability index and the gradient vector for the response to the scaled excitation are $\beta(x, t_x)/c$ and $c\mathbf{a}(t_x)$, respectively, where $\beta(x, t_x)$ and $\mathbf{a}(t_x)$ are the solutions for the reference excitation $F(t)$. This property is useful for generating the fragility curve for a nonlinear system, which is defined as the conditional probability of failure, e.g., exceeding a critical response threshold, for a range of values of the intensity of the excitation. Specifically, the TELS determined for a given intensity of excitation is used for a range of intensities to generate the fragility curve. Such an application is demonstrated in Der Kiureghian and Fujimura (2009). It is noted that the conventional ELM is dependent on the scaling of the excitation and, therefore, would require repeated analyses in such an application.

The question of the existence and uniqueness of the TELS for a given nonlinear system input-output pair, threshold x, and time t_x may arise. For the TELS to exist, there must be a tangent plane at the design point $\mathbf{u}^*(x, t)$. This means that the limit-state function and, therefore, the nonlinear system response $X(t, \mathbf{u})$ must be differentiable with respect to $\mathbf{u}$. As described in Section 12.4, this requirement is satisfied when the loading trajectory of the nonlinear system response is differentiable, i.e., when the transitions between material states (except for elastic unloading) are smooth. These transitions are not smooth for common inelastic material models such as bi-linear elasto-plastic and J2 plasticity. For this reason, smooth versions of these models have been developed and implemented in the OpenSees software (McKenna et al., 2003). In practice, TELM also works well when non-smooth state transitions occur gradually over many material fibers or elements, thus effectively smoothing the overall response transitions.

In order for the TELS to be unique, it is necessary that there be a single design point, i.e., a single global minimum-distance point from the origin. It is possible to encounter threshold values at which two or more points on the limit-state surface have equal minimum distances from the origin. Clearly, at such thresholds, TELS lacks uniqueness. Furthermore, the FORM approximation of the tail probability may not be accurate in case of multiple design points (see Section 6.9). A simple remedy for this situation is to skip the threshold values where multiple design points occur and construct a matching approximation by interpolation between neighboring thresholds.

The following example demonstrates some of the characteristics described above.

Example 13.2 – Hysteretic Oscillator Subjected to Band-Limited White-Noise Excitation

This example is adopted from Fujimura and Der Kiureghian (2007). Consider a hysteretic oscillator subjected to base motion with its response governed by the differential equation

$$m\ddot{X}(t) + c\dot{X}(t) + k[\alpha X(t) + (1 - \alpha)Z(t) = -mF(t), \tag{E1}$$

where $m = 300{,}000$ kg, $c = 150$ kN s/m, and $k = 21{,}000$ kN/m are the mass, damping, and stiffness values, respectively; α is a parameter that controls the degree of hysteresis and here $\alpha = 0.1$ is assumed; $F(t)$ is the ground acceleration, which is assumed to be a banded white-noise process with spectral density S; and $Z(t)$ is a variable characterizing the hysteresis and governed by the Bouc-Wen (Bouc, 1963; Wen, 1976) hysteresis model

$$\dot{Z}(t) = -\gamma|\dot{X}(t)||Z(t)|^{n-1}Z(t) - \eta|Z(t)|^n\dot{X}(t) + A\dot{X}(t), \tag{E2}$$

where the parameters are selected as $n = 3$, $A = 1$, and $\gamma = \eta = 1/\left(2\sigma_0^n\right)$, in which $\sigma_0^2 = \pi S m^2/(ck)$ is the mean-square response of the linear ($\alpha = 1$) oscillator. The linear oscillator has natural frequency $\omega_0 = \sqrt{k/m} = 8.37$ rad/s, corresponding to a natural period of $T_0 = 0.751$ s, and damping ratio $\zeta = c/\left(2\sqrt{km}\right) = 0.03$. We consider the reliability problems defined by the response thresholds $x = \sigma_0$ and $3\sigma_0$, both at time $t_x = 12$ s. The banded white-noise excitation is discretized using $\Delta t = 0.02$ s and the basis functions in (13.4). After obtaining the design point, the gradient vector $\mathbf{a}(t_x)$ is computed from (13.21). This is then used in the set of equations (13.30) to compute the IRF of the TELS, $h(t, x)$. The inverse-Fourier transform of the IRF then gives the corresponding FRF, $H(\omega, x)$. Note that $h(t, x)$ and $H(\omega, x)$ are independent of the scale of the excitation. Hence, any positive value for S produces the same result.

Figure 13.12 shows plots of the IRF and FRF of the TELS for the two selected response thresholds of the hysteretic oscillator, as well as the same functions for the linear ($\alpha = 1$) oscillator. Although the IRF are computed for a duration of 12 s, during which they decay to a zero value indicating that the stationary state has been reached, only the first 5 s are shown for clarity of the graph. It can be seen that the IRF (left graph) of the TELS of the hysteretic oscillator shows a longer period of oscillations and a faster decay of amplitudes with increasing threshold. The first is due to the softening effect of the hysteretic rule; the second is due to the dissipation of energy through inelastic action. These effects are also evident in the FRF (right graph), where the peak at the natural frequency of the oscillator decays in amplitude and slightly shifts toward lower frequencies with increasing threshold; furthermore, the spectral amplitudes at lower frequencies increase with increasing hysteretic action. It is also worth

Example 13.2 (cont.)

noting the emergence of a resonant frequency about twice the frequency of the oscillator in the case of the FRF of the TELS for the higher threshold. Such effects are known to occur in nonlinear dynamics.

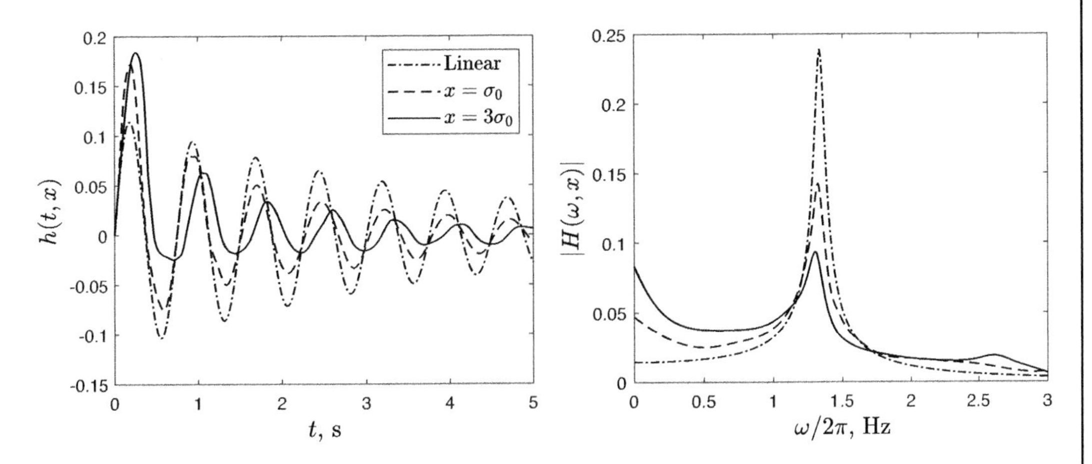

Figure 13.12 IRF (left) and FRF (right) of the TELS of the hysteretic oscillator for two thresholds and of the corresponding linear oscillator (after Fujimura and Der Kiureghian, 2007)

Now we consider the distribution of the oscillator response for the band-limited white-noise base acceleration for the spectral density $S = 0.0117$ m^2/s^3, which roughly corresponds to a mean response spectral acceleration of 0.5 g at the natural period of the linear ($\alpha = 1$) oscillator with a damping ratio of 0.05. Figure 13.13 shows the complementary CDF (left) and PDF (right) of the response at $t_x = 12$ s as functions of the normalized threshold x/σ_0. Also shown in the figure are the distributions of the responses of the linear ($\alpha = 1$) oscillator and the linear oscillator obtained from the classical ELM. Furthermore, mean and mean plus or minus one standard deviation estimates obtained from 100,000 Monte Carlo simulations are shown as confidence bars. The following observations on this figure are noteworthy: (a) The TELM approximation closely matches the Monte Carlo results, indicating the accuracy of this linearization method for the considered hysteretic oscillator and input excitation. (b) The distributions of the linear and hysteretic oscillators are vastly different, indicating the significance of the nonlinear behavior. In particular, the hysteretic behavior results in much smaller tail probabilities, of course at the expense of inelastic deformation. (c) The distribution based on the ELM is also vastly different from the TELM and Monte Carlo results, indicating that this approach is not well suited for predicting the distribution of the nonlinear response, particularly in the tail region. (d) In the semi-log chart, the Gaussian PDF plots as a parabola. This is evident for the PDFs predicted for the linear oscillator and the

Example 13.2 (cont.)

ELM. However, the PDF predicted by TELM and matched by the Monte Carlo results clearly has a non-parabolic trend, thus indicating that the distribution of the response of the hysteretic oscillator is not Gaussian. As mentioned earlier, TELM is able to predict the non-Gaussian distribution of the nonlinear response because of the dependence of the TELS on the considered threshold, as demonstrated in Figure 13.12 through the dependence of the IRF and FRF on the threshold.

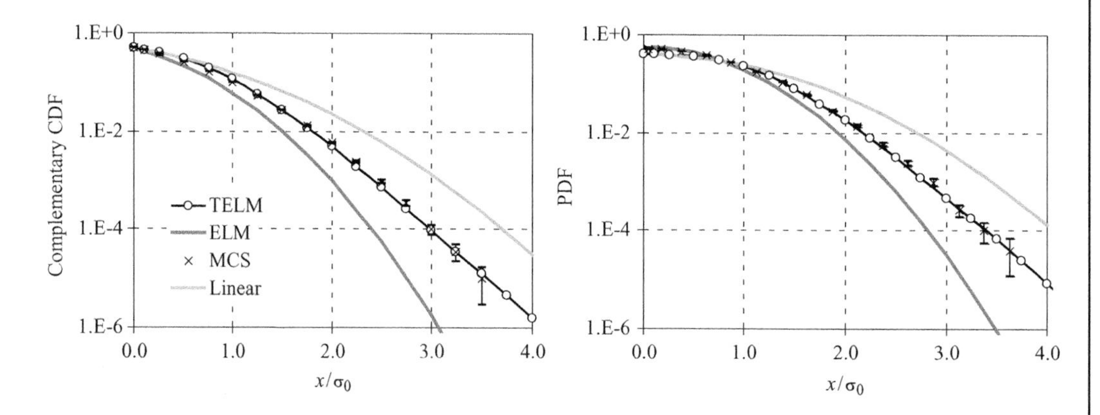

Figure 13.13 Complementary CDF (left) and PDF (right) of hysteretic oscillator response at $t_x = 12$ s to white-noise base acceleration (after Fujimura and Der Kiureghian, 2007)

As mentioned earlier, the TELS for a nonlinear system response is independent of the scaling of the excitation process. This property is useful in computing fragility curves. The fragility curve is defined as the conditional probability of failure of a system given a measure of the demand. Suppose we are interested in determining the conditional tail probability of the stationary response of the hysteretic oscillator above the threshold $x = 3\sigma_0$ at time $t_x = 12$ s for given values of the intensity of the input ground acceleration process. For the intensity measure, we select the mean spectral acceleration at the natural frequency of the linear ($\alpha = 1$) oscillator, which is related to the spectral density of the input band-limited white-noise process (see Fujimura and Der Kiureghian, 2007). We can use the TELS determined for any intensity of the input process for the entire range of intensity values. Here, we use the FRF in Figure 13.12 for the threshold $x = 3\sigma_0$, which has been generated for the PSD value $S = 1$. Figure 13.14 shows the fragility estimates obtained by TELM, ELM, and Monte Carlo. Observe that the TELM result closely matches the Monte Carlo result, and that the result based on ELM is far from accurate. Note also that the ELM requires identifying a new parametrized linear system for each input intensity level. The optimization algorithm used for ELM did not converge for higher intensity values, thus the ELM curve is truncated.

Example 13.2 (cont.)

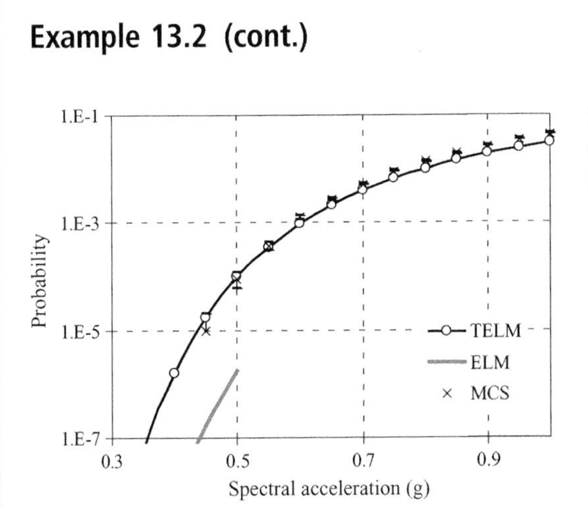

Figure 13.14 Fragility curve for the tail probability at threshold $x = 3\sigma_0$ at time $t_x = 12$ s (after Fujimura and Der Kiureghian, 2007)

13.6 Level Crossings of the Response Process

From the viewpoint of reliability assessment, of particular interest in stochastic dynamic analysis are the crossings of the response process above a given threshold. The mean rate of crossing a threshold x as a function of time, denoted $v(x, t)$, is given by Rice's well-known formula (Rice, 1944, 1945)

$$v(x, t) = \int_0^{+\infty} \dot{x} f_{X\dot{X}}(x, \dot{x}, t) \mathrm{d}\dot{x}, \tag{13.37}$$

where $f_{X\dot{X}}(x, \dot{x}, t)$ denotes the joint PDF of the response and its derivative at time t. When the response process is Gaussian with zero mean, the joint distribution is given by

$$f_{X\dot{X}}(x, \dot{x}, t) = \frac{1}{2\pi\sigma_X\sigma_{\dot{X}}\sqrt{1-\rho^2}} \exp\left[-\frac{1}{2(1-\rho^2)}\left(\frac{x^2}{\sigma_X^2} - \frac{2\rho x\dot{x}}{\sigma_X\sigma_{\dot{X}}} + \frac{\dot{x}^2}{\sigma_{\dot{X}}^2}\right)\right], \tag{13.38}$$

where $\sigma_X = \sigma_X(t)$ and $\sigma_{\dot{X}} = \sigma_{\dot{X}}(t)$, respectively, are the standard deviations of the response and its time derivative at time t and $\rho = \rho_{X\dot{X}}(t, t)$ is the cross-correlation coefficient between the two processes. Substituting (13.38) in (13.37), after a lengthy derivation, one obtains

$$\begin{aligned} v(x, t) = {} & \frac{\sqrt{1-\rho^2}}{2\pi}\frac{\sigma_{\dot{X}}}{\sigma_X} \exp\left[-\frac{1}{2(1-\rho^2)}\left(\frac{x}{\sigma_X}\right)^2\right] \times \left\{1 + \frac{\sqrt{\pi}\rho}{\sqrt{2(1-\rho^2)}}\frac{x}{\sigma_X} \times \right. \\ & \left. \exp\left[\frac{\rho^2}{2(1-\rho^2)}\left(\frac{x}{\sigma_X}\right)^2\right] \mathrm{erfc}\left[-\frac{\rho}{\sqrt{2(1-\rho^2)}}\frac{x}{\sigma_X}\right]\right\}, \end{aligned} \tag{13.39}$$

where $\text{erfc}(x) = (2/\sqrt{\pi}) \int_x^\infty \exp(-u^2) du$ is the complementary error function. When the response process is stationary, the above result reduces to

$$\nu(x) = \frac{1}{2\pi} \frac{\sigma_{\dot{X}}}{\sigma_X} \exp\left[-\frac{1}{2}\left(\frac{x}{\sigma_X}\right)^2\right], \tag{13.40}$$

which, of course, is time invariant. When the response is non-Gaussian, which occurs when the excitation is non-Gaussian or the system is nonlinear, getting hold of the joint distribution $f_{X\dot{X}}(x, \dot{x}, t)$ or an analytical solution for $\nu(x, t)$ is difficult or infeasible.

The concept of level crossings was introduced in Section 11.5.1 in connection with the down-crossings of the limit-state function below the zero level. Furthermore, in the context of time-variant reliability analysis, the limit formula (11.34) was developed. For the mean up-crossing rate of a scalar response process above a threshold x at time t, the limit formula takes the form

$$\nu(x, t) = \lim_{\Delta t \to 0} \frac{\Pr\{X(t, \mathbf{U}) - x \leq 0 \cap x - X(t + \Delta t, \mathbf{U}) \leq 0\}}{\Delta t}. \tag{13.41}$$

The event in the numerator sets the condition that the response at time t is below the threshold x and at time $t + \Delta t$ is above the same threshold. Therefore, the numerator is the probability of making one or more up-crossings during the interval $(t, t + \Delta t]$. As shown in Section 11.5.1, for a small Δt, this probability is equal to the expected number of crossings during that interval. Thus, divided by Δt and as Δt approaches zero, the ratio provides the mean rate of up-crossings of the threshold x at time t.

The numerator in (13.41) represents a parallel system reliability problem with two components. The FORM solution of this probability is the bivariate normal CDF $\Phi_2[-\beta(x, t), -\beta(x, t + \Delta t), \rho(x, t, \Delta t)]$, where $\beta(x, t)$ and $\beta(x, t + \Delta t)$ are the reliability indices associated with the limit-state function $G(\mathbf{U}, x, t) = x - X(t, \mathbf{U})$ at times t and $t + \Delta t$, respectively, and $\rho(x, t, \Delta t) = \widehat{\boldsymbol{\alpha}}(x, t) \cdot \widehat{\boldsymbol{\alpha}}(x, t + \Delta t)$ is the correlation coefficient between the linearized limit-state functions, in which $\widehat{\boldsymbol{\alpha}}(x, t)$ and $\widehat{\boldsymbol{\alpha}}(x, t + \Delta t)$ are the corresponding unit normal vectors. A simple approximation of this probability was given in (11.36), which we repeat here for the sake of convenience:

$$\nu(x, t) \cong \frac{1}{\Delta t} \exp\left[-\frac{\beta(x, t)\beta(x, t + \Delta t)}{2}\right] \left\{\frac{1}{4} + \frac{\sin^{-1}[\rho(x, t, \Delta t)]}{2\pi}\right\}. \tag{13.42}$$

When the response process is stationary, the reliability index is independent of time and the correlation coefficient depends only on the time lag Δt so that the above formula reduces to

$$\nu(x) \cong \frac{1}{\Delta \tau} \exp\left[-\frac{\beta(x)^2}{2}\right] \left\{\frac{1}{4} + \frac{\sin^{-1}[\rho(x, \Delta t)]}{2\pi}\right\}. \tag{13.43}$$

The above analysis requires finding two design points, one at t and one at $t + \Delta t$. As Koo et al. (2005) have shown, when the response process is stationary, the design point at $t + \Delta t$ is identical to the design point at t translated by Δt. For a nonstationary response, the translated design point provides a warm starting trial point for quick convergence to the exact design point.

Another alternative for computing the mean up-crossing rate is to use (13.39), or (13.40) in the case of a stationary process, for the TELS response. (Adjustments in these formulas must be made when the response process has a non-zero mean.) This is because the TELS is a linear system and it is subjected to a Gaussian excitation. This requires linear random vibration analysis to determine $\sigma_X(t)$, $\sigma_{\dot{X}} = \sigma_{\dot{X}}(t)$, and $\rho = \rho_{X\dot{X}}(t,t)$. As the IRF and FRF of the TELS are available, this analysis can use well-established methods in linear random vibration theory (see, e.g., Lutes and Sarkani, 2004). In particular, in the case of a zero-mean response process and the system starting from its rest condition,

$$\sigma_X^2(t) \cong \int_0^t \int_0^t \Sigma_{FF}(\tau_1, \tau_2) h(t - \tau_1) h(t - \tau_2) \mathrm{d}\tau_1 \mathrm{d}\tau_2, \tag{13.45}$$

$$\sigma_{\dot{X}}^2(t) \cong \int_0^t \int_0^t \Sigma_{FF}(\tau_1, \tau_2) \dot{h}(t - \tau_1) \dot{h}(t - \tau_2) \mathrm{d}\tau_1 \mathrm{d}\tau_2, \tag{13.46}$$

$$\rho_{X\dot{X}}(t,t) \cong \frac{1}{\sigma_X(t)\sigma_{\dot{X}}(t)} \int_0^t \int_0^t \Sigma_{FF}(\tau_1, \tau_2) h(t - \tau_1) \dot{h}(t - \tau_2) \mathrm{d}\tau_1 \mathrm{d}\tau_2, \tag{13.47}$$

where $\dot{h}(\cdot)$ denotes the derivative of the IRF with respect to time. In the case of a stationary response, $\sigma_X^2 = \lambda_0$, $\sigma_{\dot{X}}^2 = \lambda_2$, and $\rho_{X\dot{X}}(t,t) = 0$, where λ_0 and λ_2 are the zeroth- and second-order spectral moments obtained from (13.24)–(13.25) in terms of the input PSD and the FRF of the TELS. In this case, working in the frequency domain and using the FRF is more convenient.

The above approach based on random vibration theory is strictly a FORM approximation, because it uses the TELS. On the other hand, (13.41) is not limited to the FORM. For example, one can construct point-fitting SORM approximations (see Section 7.4) of the limit-state surfaces $X(t, \mathbf{u}) - x = 0$ and $x - X(t + \Delta t, \mathbf{u}) = 0$ and then use these surfaces in conjunction with directional or orthogonal-plane sampling to compute the probability content in the wedge domain (see Sections 9.9 and 9.10 and Figure 9.9). In so doing, it is useful to note that only those elements of vector $\mathbf{u}$ used in discretizing the input process that are close to the considered time point need to be considered in developing the second-order approximating surfaces. This is because, as evident in the left-hand charts in Figures 13.8 and 13.9, the design point excitation has practically zero values for the early elements of $\mathbf{u}$ and, therefore, these elements have little influence on the limit-state surfaces. From a physical standpoint, this has to do with the fact that the early phase of the excitation has little influence on a later response because of the effect of damping and energy dissipation, particularly in an inelastic system.

Example 13.3 – Mean Up-Crossing Rate of Linear Oscillator Subjected to Gaussian and Non-Gaussian Excitations

We compute the mean up-crossing rate of the oscillator in Example 13.1 for a range of thresholds at time $t_x = 10$ s, assuming the response at that time has reached the stationary state. Figure 13.15 compares the results for the two cases of the input excitation. For the stationary response to the Gaussian excitation (solid line), the FORM solution is exact because the limit-state function is linear in $\mathbf{U}$. For this case, a good theoretical approximation is given by the mean up-crossing rate of the oscillator subjected to a white-noise excitation with the spectral density $S = 0.03$ m^2/s^3 (essentially disregarding the band limit of the input process, which has practically no effect on the response of the oscillator with frequency $\omega_0 = 5\pi$ rad/s). This solution is given by (13.40) with $\sigma_X^2 = \pi S/(2\xi\omega_0^3)$ and $\sigma_{\dot{X}}^2 = \pi S/(2\xi\omega_0)$. Numerical comparison shows that this result is identical to that obtained from (13.43) up to three significant digits.

For the shifted lognormal excitation, no exact solution of the mean up-crossing rate exists. The FORM approximation in Figure 13.15 (dashed line) indicates that the response of the linear oscillator to this non-Gaussian excitation has fewer up-crossings than the response to the Gaussian excitation. Given the fatter right tail of the lognormal distribution, as seen in Figure 13.6, and the large peaks appearing in the sample realization in Figure 13.7, this result might be surprising. However, one should note that the shifted lognormal excitation has much smaller cycles in the negative direction (compare the realizations for the Gaussian and shifted lognormal processes in Figure 13.7). Thus, the combined effect of larger cycles in the positive direction and smaller cycles in the negative direction results in a smaller number of up-crossings of the response to the shifted lognormal excitation compared to the response to the Gaussian excitation.

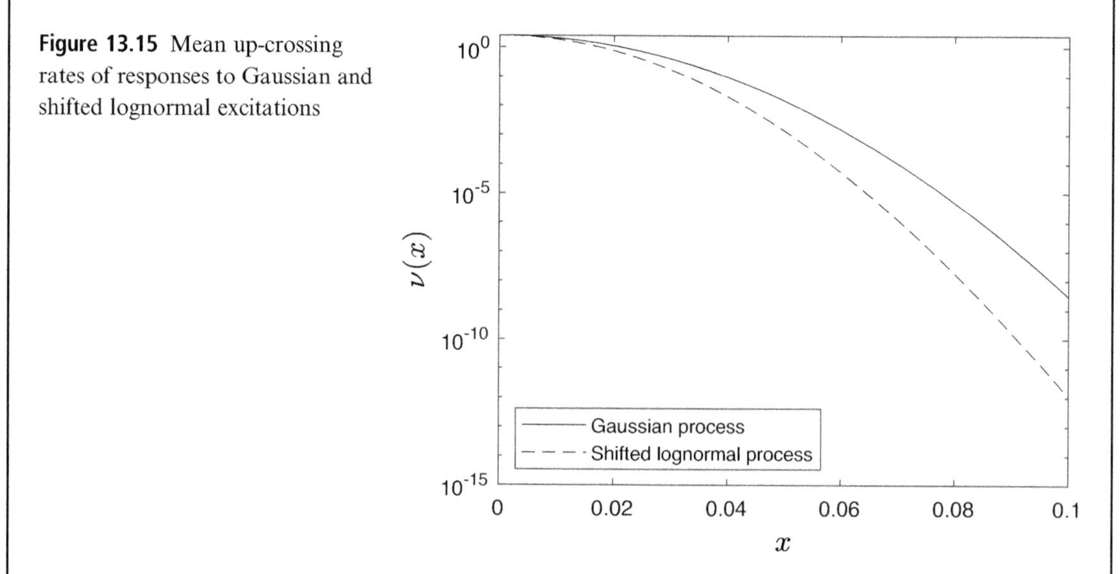

Figure 13.15 Mean up-crossing rates of responses to Gaussian and shifted lognormal excitations

Example 13.4 – Mean Up-Crossing Rate of Hysteretic Oscillator

We compute the mean up-crossing rate of the hysteretic oscillator in Example 13.2 for a range of the normalized threshold x/σ_0 at time $t_x = 12$ s, assuming the response at that time has reached the stationary state. Figure 13.16 compares the results based on TELM and the random vibration (RV) formula in (13.40), TELM and the limit formula (LF) in (13.43), ELM, and Monte Carlo simulation. Also shown for comparison purposes is the mean up-crossing rate of the linear ($\alpha = 1$) oscillator. Observe that the two TELM results practically coincide and that they are in close agreement with the Monte Carlo result. The ELM result is far from the Monte Carlo result, once again indicating the inability of this method to give good results in the tail region of the response, and the result for the linear oscillator is vastly different, indicating the significant effect of the nonlinearity in the response of the oscillator.

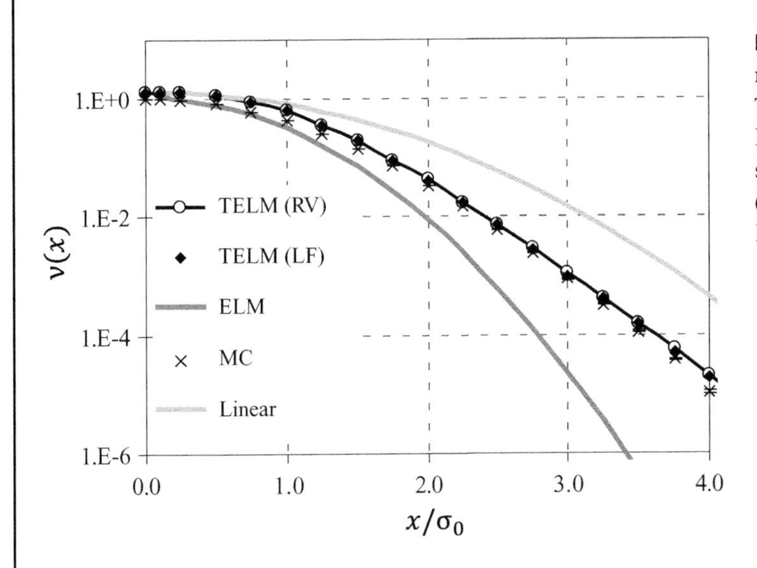

Figure 13.16 Mean up-crossing rate of hysteretic oscillator using TELM with RV formula, TELM LF, ELM, Monte Carlo (MC) simulation, and linear oscillator (after Fujimura and Der Kiureghian, 2007)

13.7 The First-Passage Probability

A response quantity of special interest in reliability analysis is the maximum of the stochastic response over a given interval of time. If this quantity is above a safe threshold, then failure is said to have occurred. This problem is closely related to the well-known first-passage probability problem in stochastic process theory.

Let T_x denote the time to the first excursion of the response process $X(t)$ above the threshold x, including at time $t = 0$. Also let $X_\tau = \max_{0 \le t < \tau} X(t)$ denote the maximum value of $X(t)$ over the interval $[0, \tau)$. The probability distributions of these two random variables are related through

$$F_{T_x}(\tau) = 1 - F_{X_\tau}(x). \tag{13.48}$$

No exact solution of this distribution exists, even for stationary Gaussian processes.

In Section 11.5.1, we derived an upper bound to the probability that a time-variant limit-state function would down-cross the zero level. A similar argument for the excursions of process $X(t)$ above the threshold x leads to the upper bound

$$\begin{aligned} F_{T_x}(\tau) &= 1 - F_{X_\tau}(x) \\ &\leq \Pr[x \leq X(0)] + \int_0^\tau \nu(x,t)\mathrm{d}t, \end{aligned} \tag{13.49}$$

where $\nu(x,t)$ is the mean rate of up-crossings of $X(t)$ above the threshold x. Often the first term on the right-hand side of the above expression is much smaller than the second, and can be neglected.

For high thresholds, the up-crossing events are rare. Following the argument in Section 11.6.3, it is then reasonable to assume that these events constitute a Poisson process. Dropping the probability of failure at $t = 0$, an approximation of the first-passage probability then is given by

$$\begin{aligned} F_{T_x}(\tau) &= 1 - F_{X_\tau}(x) \\ &\cong 1 - \exp\left[-\int_0^\tau \nu(x,t)\mathrm{d}t\right]. \end{aligned} \tag{13.50}$$

In the case of a stationary process, the mean up-crossing rate is constant and the above approximation reduces to

$$\begin{aligned} F_{T_x}(\tau) &= 1 - F_{X_\tau}(x) \\ &\cong 1 - \exp[-\nu(x)\tau]. \end{aligned} \tag{13.51}$$

For the special case of a zero-mean, stationary Gaussian process, an improved approximation of the first-passage probability was derived by Vanmarcke (1975). The approximation accounts for the bandwidth of the process and the effect of "clumping" of the up-crossing events, which gives rise to dependence between these events and makes the Poisson assumption invalid. The solution is given as

$$\begin{aligned} F_{T_x}(\tau) &= 1 - F_{X_\tau}(x) \\ &\cong 1 - \left[1 - \exp\left(-\frac{r^2}{2}\right)\right] \exp\left[-\nu(0)\tau \frac{1 - \exp\left(-\sqrt{2\pi}\delta^{1.2} r\right)}{\exp\left(r^2/2\right) - 1}\right], \end{aligned} \tag{13.52}$$

where $r = x/\sqrt{\lambda_0}$ is the normalized threshold, $\nu(0) = \sqrt{\lambda_2/\lambda_0}/(2\pi)$ is the mean rate of zero-level up-crossings, and $\delta = \sqrt{1 - \lambda_1^2/(\lambda_0\lambda_2)}$ is a measure of the bandwidth of the process, in which λ_m, $m = 0, 1, 2$, are the spectral moments defined in (13.25).

In some applications with a zero-mean response process, the failure event is defined as the up-crossing of the positive threshold x or down-crossing of the negative threshold $-x$. This is equivalent to defining the maximum response as $X_\tau = \max_{0 \le t < \tau} |X(t)|$. In that case, some adjustments in the above expressions must be made. Specifically, the term $\Pr[x \le X(0)]$ in (13.49) must be replaced by $\Pr[x \le X(0)] + \Pr[X(0) \le -x]$ and the term $\nu(x,t)$ in (13.49)–(13.51) must be replaced by $\nu(x^+,t) + \nu(-x^-,t)$, where the first term denotes the up-crossings of level x and the second term denotes the down-crossings of level $-x$. Furthermore, in (13.52), $\nu(0)$ must be replaced by $2\nu(0)$ and $\sqrt{2\pi}$ must be replaced by $\sqrt{\pi/2}$.

In TELM analysis, the upper bound in (13.49) and the Poisson-based approximation in (13.50) employ the mean up-crossing rate, $\nu(x,t)$. Considering that the response of the TELS for a given threshold is Gaussian, a FORM approximation of the mean up-crossing rate can be computed by using (13.39). Alternatively, a FORM or SORM approximation of the mean up-crossing rate can be obtained by using the limit formula (13.42). For a nonstationary process, because $\nu(x,t)$ usually varies smoothly over time, it is sufficient to compute it at a few time points within the interval $[0,\tau)$ and construct an approximation using interpolation. For a stationary process, the approximation in (13.51) with $\nu(x)$ given by (13.40) or the improved approximation in (13.52) can be used. This analysis requires use of conventional random vibration methods employing the FRF of the TELS to compute the spectral moments. It is noted that the TELS varies with the threshold and, therefore, the RV analysis will have to be repeated for each new threshold value. As mentioned before, it is through this dependence of the TELS on the threshold that TELM is able to capture the non-Gaussian distribution of the response process. Nevertheless, since the response of the TELS for each threshold is Gaussian, we are able to use formulas that were developed strictly for Gaussian processes.

An alternative way to compute the distribution of the maximum response is to use the series-system formulation $1 - F_{X_\tau}(x) = \Pr(x \le X_\tau) \ge \Pr\left[\bigcup_{i=1}^{n} \{x \le X(t_i)\}\right]$, where t_i are a set of selected time points within the interval $[0,\tau)$. Obviously, for any finite set of time points, the series-system probability provides a lower bound to the tail probability of the maximum response. As described in Section 8.10 and equation (8.65), the FORM approximation of the series-system probability requires computation of the multinormal probability involving the reliability indices and correlation coefficients, the latter obtained as scalar products of the unit normal vectors, at the design points associated with the discrete time points t_i. For a stationary process, the successive design points can be obtained by time shifting, which greatly reduces the computational effort. This approach is used in an example below. Another alternative is to conduct importance sampling using the design points, as described by Au and Beck (2001a).

Example 13.5 – Distribution of Maximum Response of Linear Oscillator Subjected to Gaussian and Non-Gaussian Excitations

Figure 13.17 shows the distributions of the maxima of stationary responses of the linear oscillator of Example 13.1 over a 10 s duration for the Gaussian and shifted-lognormal excitations. The Poisson approximation in (13.51) together with the mean up-crossing rates

Example 13.5 (cont.)

in Example 13.3 are used. The PDFs are generated through numerical differentiation. Observe that the shifted lognormal distribution is likely to produce a smaller maximum response.

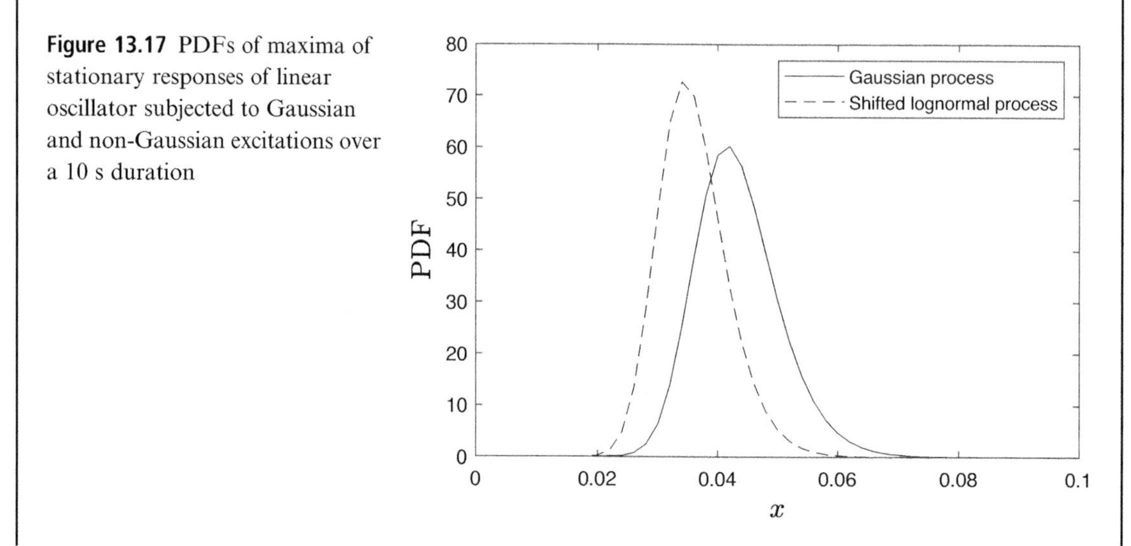

Figure 13.17 PDFs of maxima of stationary responses of linear oscillator subjected to Gaussian and non-Gaussian excitations over a 10 s duration

Example 13.6 – Distribution of Maximum Response of Hysteretic Oscillator

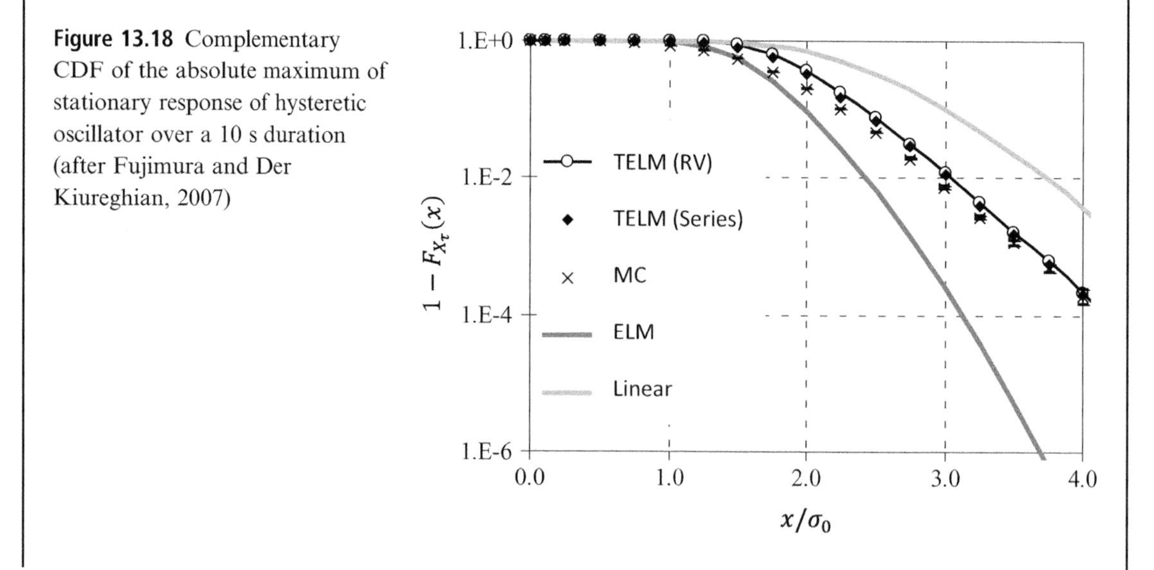

Figure 13.18 Complementary CDF of the absolute maximum of stationary response of hysteretic oscillator over a 10 s duration (after Fujimura and Der Kiureghian, 2007)

Example 13.6 (cont.)

Figure 13.18 shows the complementary CDF of the absolute maximum of the stationary response of the hysteretic oscillator over a 10 s duration as a function of the normalized threshold x/σ_0. Solutions shown are obtained by TELM (RV) using Vanmarcke's formula (13.52) and the series-system approximation (Series), mean and mean plus or minus one standard deviation bounds obtained from 100,000 Monte Carlo (MC) simulations, the result based on ELM, and the solution obtained for the linear ($\alpha = 1$) oscillator. The two TELM solutions practically coincide and closely agree with the Monte Carlo result, particularly at high thresholds. Not surprisingly, the result based on ELM is far from the Monte Carlo result at high thresholds. The result for the linear oscillator is also vastly different, indicating the significant effect of nonlinearity on the response.

Example 13.7 – Application of TELM to a Multi-Degree-of-Freedom Hysteretic System

Adopted from Fujimura and Der Kiureghian (2007), this example demonstrates the application of TELM to a multi-degree-of-freedom (MDOF) hysteretic system subjected to a stationary filtered white-noise base acceleration. Consider the six-story shear building with rigid floors modeled in Figure 13.19 (left). The floor masses and total initial story stiffnesses are shown in the figure. The hysteretic behavior of the columns in each story is idealized by a bi-linear model with the yield displacement set at 0.01 m. To assure differentiability of the response, a smoothed bi-linear model as described by Haukaas and Der Kiureghian (2006) is used. This model essentially replaces the sharp transition from the elastic to the plastic state with a circular segment with an adjustable radius that is tangent to the elastic and plastic loading trajectories. Sample hysteresis loops for the first and sixth stories are shown in Figure 13.19 (right). Rayleigh damping with 5% damping ratio in modes 1 and 2 is assumed. The input excitation is modeled with the basis functions in (13.5) using the Kanai-Tajimi filter defined by the IRF

$$h_f(t) = \exp\left(-\zeta_f\omega_f t\right)\left[\frac{\left(2\zeta_f^2 - 1\right)\omega_f}{\sqrt{1-\zeta_f^2}}\sin\left(\omega_f\sqrt{1-\zeta_f^2}t\right) - 2\zeta_f\omega_f\cos\left(\omega_f\sqrt{1-\zeta_f^2}t\right)\right] \quad \text{(E1)}$$

with $\omega_f = 5\pi$ rad/s and $\zeta_f = 0.4$. The spectral amplitude of the underlying band-limited white noise is set at $S = 0.0156$ m^2/s^3, which roughly corresponds to a mean response spectral acceleration of 1 g at the natural period of the linear structure with 5% damping. For the analysis, $\Delta t = 0.02$ s is used, which effectively cuts off frequencies beyond 25 Hz. The target

Example 13.7 (cont.)

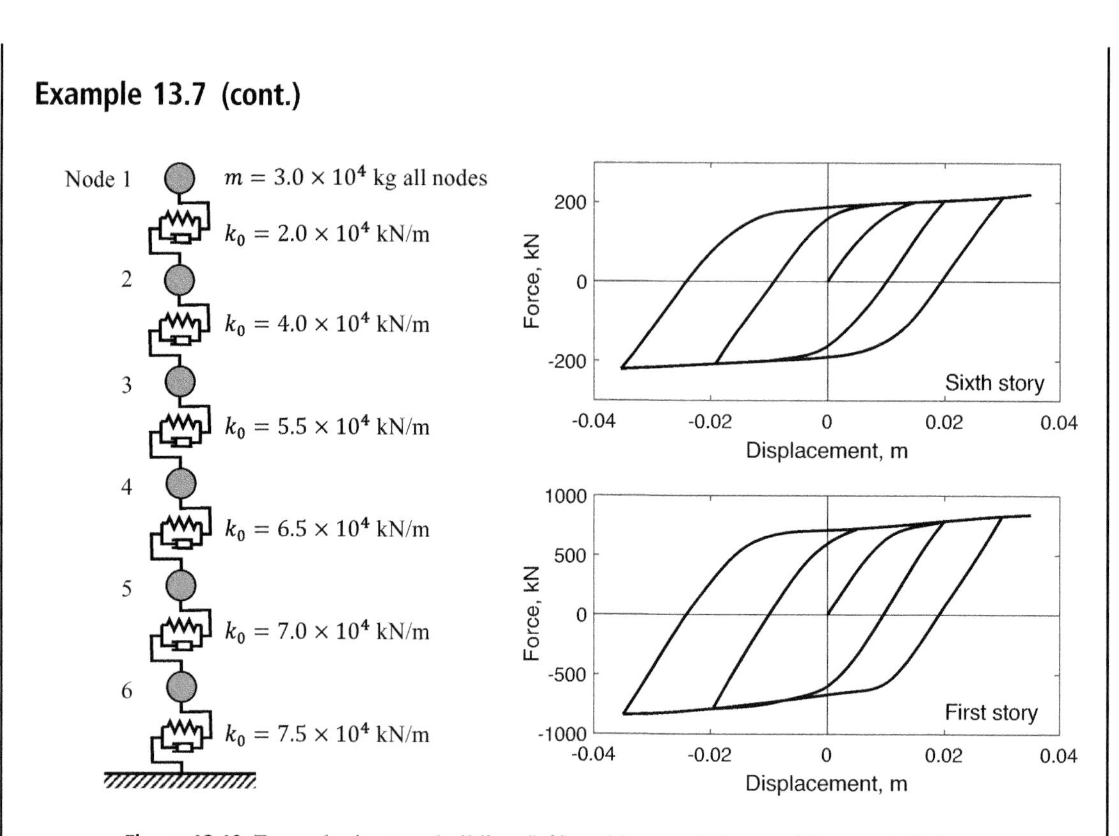

Figure 13.19 Example six-story building (left) and hysteresis loops of first- and sixth-story columns (right) (after Fujimura and Der Kiureghian, 2007)

time for determining the TELS is set at $t_x = 10$ s, by which time the response has well reached the stationary state. The objective is to compute the probability distributions of the maximum absolute values of the first and sixth inter-story drifts over a 10 s duration of stationary response.

Figure 13.20 shows the FRFs of the TELSs for thresholds $x = 0.01$ m, 0.02 m, and 0.03 m of the first and sixth inter-story drifts. Also shown are the FRFs of the linear system with initial stiffness. As the threshold increases, for the first inter-story drift, the dominant peak in the FRF of the TELS becomes smaller and broader and slightly shifts toward lower frequencies. Furthermore, the low-frequency portion of the FRF is amplified with the increasing threshold. The same trends are observed for the dominant peak in the FRF of the sixth inter-story drift; however, the second peak in the FRF becomes larger and broader with increasing threshold, indicating increasing significance of the higher modes of vibration for this response quantity.

Figure 13.21 shows the complementary CDFs of the maximum absolute inter-story drifts over a 10 s stationary response to the filtered white-noise excitation as computed using Vanmarcke's formula (13.52). Included are results based on the TELM, the TELM but using the TELS obtained for a white-noise excitation (denoted as TELM-WN), Monte Carlo (MC) with 100,000 simulations, and the linear structure with the initial stiffness. It is observed that

Example 13.7 (cont.)

the TELM and TELM-WN provide nearly identical results, indicating that the TELS developed for a white-noise input can be used as a good approximation for other wide-band excitations. The comparison with Monte Carlo results indicates fair accuracy for the first-story drift and good accuracy for the sixth-story drift. This type of order-of-magnitude accuracy in small-tail probabilities is often sufficient for engineering purposes. In any case, this level of accuracy is far superior to that obtained with the conventional ELM, as previous examples have demonstrated. It is noted that the result for the linear system is vastly different, indicating the significant effect of nonlinear behavior for this structure.

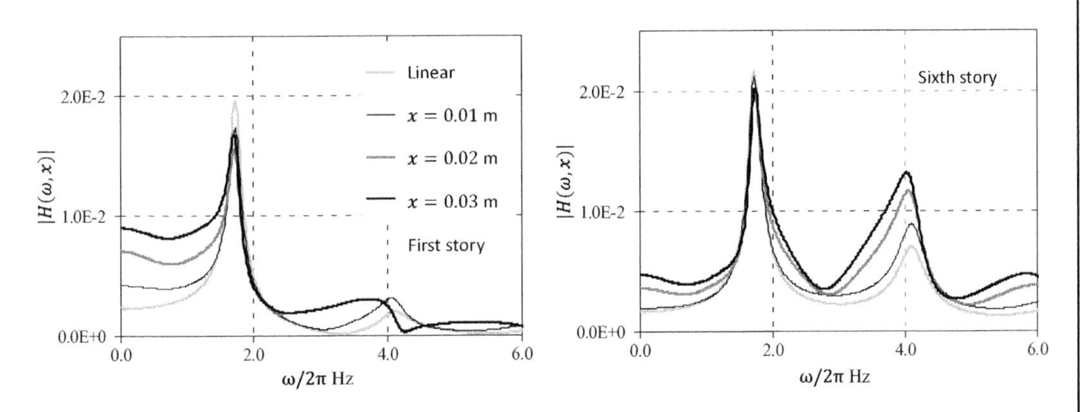

Figure 13.20 FRFs of the TELSs for selected thresholds of the first and sixth inter-story drifts (after Fujimura and Der Kiureghian, 2007)

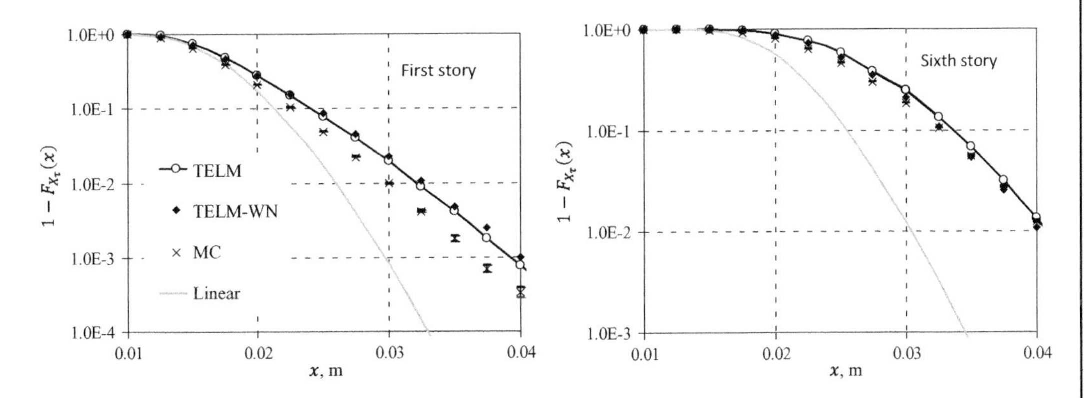

Figure 13.21 Complementary CDFs of the maximum absolute inter-story drifts of the first and sixth stories over a 10 s stationary response (after Fujimura and Der Kiureghian, 2007)

13.8 TELM with Multiple Excitations

So far, we have considered applications of TELM to structures subjected to a scalar Gaussian excitation process, $F(t)$. Suppose the structure is subjected to a vector of m statistically independent Gaussian excitation processes $\mathbf{F}(t) = [F_1(t) \cdots F_m(t)]^{\mathrm{T}}$ with the response of the nonlinear structure governed by the differential equation

$$\mathbf{M}\ddot{\mathbf{Y}}(t) + \mathbf{C}\dot{\mathbf{Y}}(t) + \mathbf{R}(t) = \mathbf{P}\mathbf{F}(t), \tag{13.53}$$

where the load-distribution matrix $\mathbf{P}$ now is of size $N \times m$ with N denoting the number of degrees of freedom of the structure. Suppose we discretize each of the processes $F_i(t)$ according to (13.1) with n_i denoting the number of random variables used for representing the process. The set of excitation processes now is represented by a vector of standard normal variables $\mathbf{U}$ having $n = n_1 + \cdots + n_m$ elements.

For a linear structure, using the principle of superposition, the generic response is given by

$$X(t, \mathbf{U}) = \sum_{i=1}^{m} \left[\mu_{Xi}(t) + \sum_{j=1}^{n_i} a_{ij}(t) U_{ij}\right], \tag{13.54}$$

where

$$\mu_{X_i}(t) = \int_0^t \mu_{F_i}(\tau) h_i(t-\tau)\mathrm{d}\tau, \quad i = 1, \ldots, m, \tag{13.55}$$

is the mean response for the ith excitation and

$$a_{ij}(t) = \int_0^t s_{ij}(\tau) h_i(t-\tau)\mathrm{d}\tau, \quad i = 1, \ldots, m, \quad j = 1, \ldots, n_i, \tag{13.56}$$

are the elements of the gradient vector for the ith excitation, in which $s_{ij}(\tau)$ denotes the jth basis function and U_{ij} denotes the corresponding standard normal random variable. Furthermore, $h_i(t)$ denotes the IRF of the linear system for the input-output pair $[F_i(t), X(t)]$. It is seen that the response is still in the form of (13.12) but with an expanded gradient vector $\mathbf{a}(t) = [a_{1,1}(t) \cdots a_{1,n_1}(t) \cdots a_{m,1}(t) \cdots a_{m,n_m}(t)]$ and vector of standard normal random variables $\mathbf{U} = [U_{1,1} \cdots U_{1,n_1} \cdots U_{m,1} \cdots U_{m,n_m}]^{\mathrm{T}}$.

For the nonlinear system, the generic response $X(t, \mathbf{U})$ for any realization $\mathbf{u}$ of $\mathbf{U}$ is computed by numerically solving the set of equations (13.53) for the combined action of the m excitations. The design point for the limit-state function (13.28) is obtained by use of an appropriate optimization algorithm such as iHL-RF. As we have seen, this requires repeated computations of the response and its gradient, possibly computed by the direct-differentiation method (DDM), with respect to selected realizations of the standard normal random variables. Once the design point $\mathbf{u}^*(x, t_x)$ for a threshold x at time t_x is obtained, the expanded gradient vector $\mathbf{a}(t)$ of the TELS is computed from (13.21). The relevant elements of this vector and the corresponding basis functions are then used to compute the IRF for each input-output pair $[F_i(t), X(t)]$, $i = 1, \ldots, m$. For example, for a time-domain discretization with the basis functions according to (13.4)–(13.7), the IRF for the ith excitation is obtained by solving for $h_i(t)$ in the set of n_i equations

$$\sum\nolimits_{k=1}^{n_i} h_i(t_{n_i} - t_k)s_{ij}(t_k)\Delta t_k = a_{ij}(t_{n_i}), \quad j = 1, \ldots, n_i. \tag{13.57}$$

Note that the time increments Δt_k can be different for different excitation processes, but they should each be sufficiently small to accommodate the bandwidth of the process as well as the accuracy of the numerical integration in (13.57). A similar formulation of the IRF based on the discretization using the sinc function can be developed by expanding (13.34). For the frequency-domain discretization according to (13.10), the FRF for every input-output pair is obtained by using the appropriate elements of the gradient vector in (13.35)–(13.36).

The above analysis applies when the input Gaussian processes $F_i(t)$, $i = 1, \ldots, m$, are statistically independent. When these processes are correlated, as characterized by their cross-correlation coefficient functions $\rho_{F_iF_j}(t_i, t_j)$, $i,j = 1, \ldots, m$, two options are available. One is to transform the processes to be uncorrelated and then discretize the uncorrelated processes. A well-known example is the components of ground motion that can be made uncorrelated through a rotation of the axes to the so-called principal directions (Penzien and Watabe, 1975). The second approach is to discretize the correlated processes into a set of standard normal random variables $\mathbf{U}$, where the subsets belonging to different inputs are correlated with the correlation matrix determined in terms of the cross-correlation coefficient functions $\rho_{F_iF_j}(t_i, t_j)$. One then applies a linear transformation to $\mathbf{U}$ to obtain a vector of uncorrelated standard normal random variables. The TELM approximation is performed in the space of the transformed random variables. One such application is presented in Wang and Der Kiureghian (2016) for multi-support structures, e.g., bridges, subjected to spatially varying support motions.

Example 13.8 – Application of TELM to a Hysteretic Structure Subjected to Bi-Component Earthquake Excitation

Adopted from Broccardo and Der Kiureghian (2016), this example demonstrates the application of TELM to a MDOF hysteretic system subjected to a bi-component stochastic ground motion. Consider the one-story, one-bay structure in Figure 13.22, which is subjected to two horizontal components of ground motion in directions x and y. The structure has a rigid roof and three degrees of freedom: translations in directions x and y of the center of the roof mass and counterclockwise rotation of the roof. The lateral resisting force system consists of two symmetric shear walls in direction x and one shear wall in direction y, the latter having eccentricity e relative to the center of the roof mass. The three shear walls have in-plane inelastic behavior governed by the Bouc-Wen model described in Example 13.2, and negligible stiffness in the out-of-plane direction. The system parameters are: roof mass $m = 3.6 \times 10^6$ kg; damping coefficient of each wall in x direction $c_x = 8.05 \times 10^4$ kNs/m and in y direction $c_y = 1.61 \times 10^4$ kNs/m; initial stiffness of each wall in x direction $k_x = 1.5 \times 10^4$ kN/m and in y direction $k_y = 3 \times 10^3$ kN/m; roof dimensions $a = 6$ m and $b = 10$ m, eccentricity

Example 13.8 (cont.)

$e = 0.9$ m, and hysteretic model parameters $\alpha = 0.1$, $n = 5$, $\gamma = \eta = \left(2u_{\text{yld}}\right)^{-n}$, $A = 1$, where $u_{\text{yld}} = 0.04$ m is the yield deformation. Figure 13.22 shows the three modes of vibration of the linear ($\alpha = 1$) structure and hysteretic loops of the walls in x and y directions.

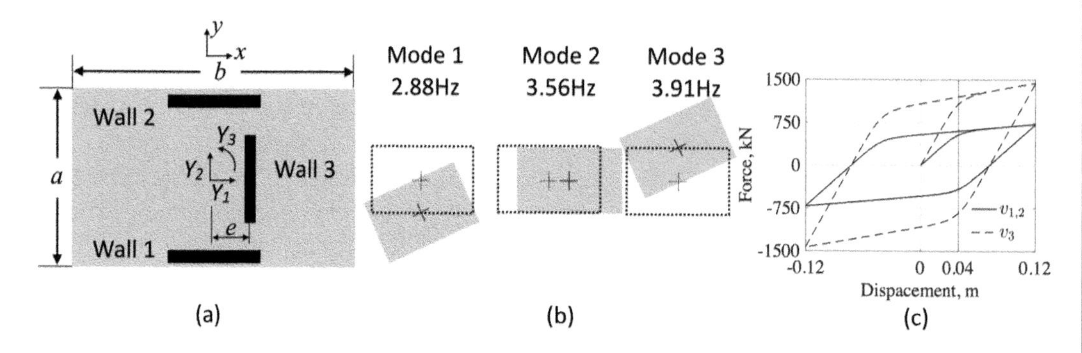

Figure 13.22 Example one-story, one-bay structure subjected to bi-component base motion: (a) plan view, (b) vibration modes of linear structure, (c) hysteretic loops (after Broccardo and Der Kiureghian, 2016, with permission from ASCE)

The input excitations in directions x and y are modeled as statistically independent, stationary Gaussian processes defined by the Kanai-Tajimi PSD

$$\Phi(\omega) = S\frac{\omega_f^4 + 4\zeta_f^2\omega_f^2\omega^2}{\left(\omega_f^2 - \omega^2\right)^2 + 4\zeta_f^2\omega_f^2\omega^2}, \tag{13.58}$$

where $\omega_f = 6$ rad/s, $\zeta_f = 0.6$, and $S = 0.25$ m^2/s^3 are assumed for both components. This is the frequency-domain equivalent of the filter IRF defined in Example 13.7 and is widely used in earthquake engineering to model the ground acceleration. The adopted frequency and damping values are characteristic of earthquake motions on firm ground. Discretization is done in the frequency domain using a cut-off frequency of $\Omega = 20\pi$ rad/s and a frequency increment of $\Delta\omega = 0.2\pi$ rad/s. Thus, $2 \times 20\pi/0.2\pi = 200$ standard normal random variables define each component for a total of 400 for the two components.

The response quantity of interest, $X(t, \mathbf{U})$, is the deformation of Wall 1 in the x direction. This is obtained as a function of the responses at the nodal degrees of freedom of the structure. Figure 13.23 (left) shows the design-point excitations for $x = 0.12$ m at $t_x = 12$ s. Shown are the trajectory of the design-point ground acceleration vector $\boldsymbol{F}^*(t) = \left[F_x^*(t), F_y^*(t)\right]^{\text{T}}$ as well as the individual components in units of gravity acceleration. Observe that much larger accelerations are achieved in the in-plane direction of Wall 1. Interestingly, the design-point

Example 13.8 (cont.)

excitations do not have zero values at the target time t_x. This has to do with the fact that the Kanai–Tajimi process has memory. That is, the values of $\mathbf{F}(t)$ at different times are correlated. Hence, although a non-zero value of the input process at time t_x does not influence the response at the same time, because of the correlation with the values of the process at the earlier times, the most likely realization of the input vector process that yields the response x at time t_x is one that has non-zero values at t_x.

Figure 13.23 (right) shows the design-point response for the considered threshold and time. It shows the trajectory of the center of the roof mass and the time history of the deformation of Wall 1. Observe that the design-point trajectory of the center of the roof ends with a large displacement in the x direction. Also observe that the design-point wall deformation ends at $x = 0.12$ m at $t_x = 12$ s, but with a positive slope, indicating the presence of kinematic energy at the end point. This again has to do with the memory property of the colored input process.

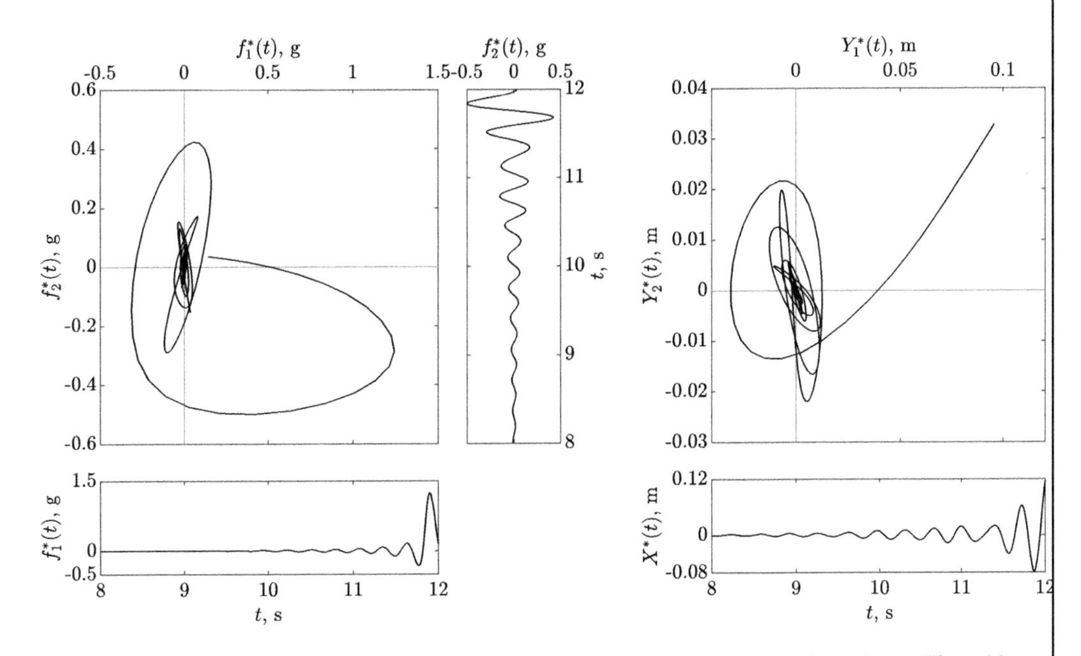

Figure 13.23 Design-point excitation (left) and response (right) (after Broccardo and Der Kiureghian, 2016, with permission from ASCE)

Example 13.8 (cont.)

Figure 13.24 shows the FRFs of the response quantity with respect to the two input components. Results are shown for three thresholds. For the low threshold $x = 0.01$ m the response is essentially linear, whereas for thresholds $x = 0.06$ m and $x = 0.12$ m significant inelastic behavior is observed: Resonant peaks are dampened due to dissipation of energy by inelastic action and FRF values at lower frequencies are amplified on account of the softening. It is observed that the FRF for the input component in direction x has much larger values than that in direction y. It is also noteworthy that the second mode of the linear structure dominates the FRF of the input in the x direction while the first and third modes dominate the FRF of the input in the y direction.

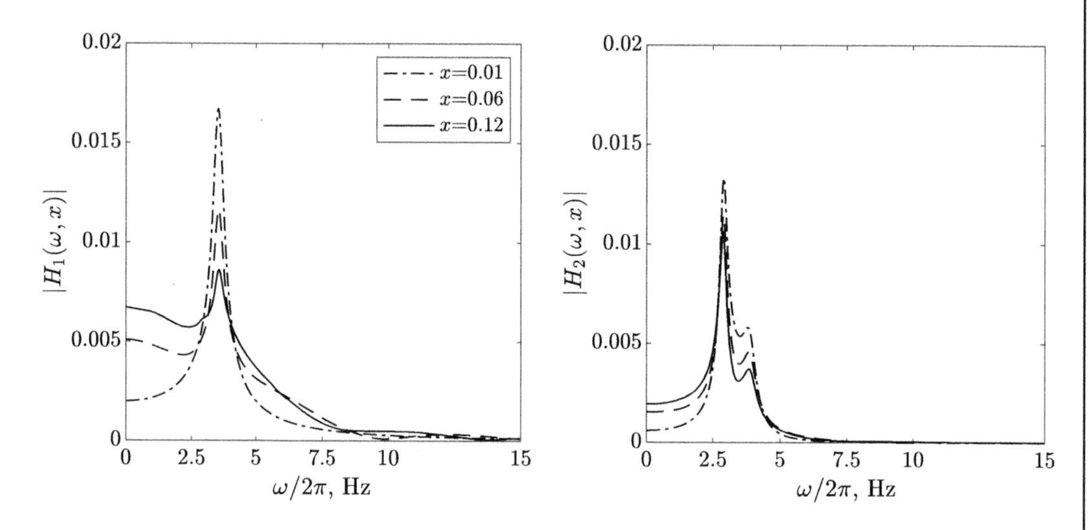

Figure 13.24 FRFs of TELS with respect to inputs in directions x (left) and y (right) for selected threshold values (after Broccardo and Der Kiureghian, 2016, with permission from ASCE)

Figure 13.25 shows the complementary CDFs of the response at time t_x and the maximum response over a 10 s stationary duration. It shows results based on TELM, one-standard-deviation confidence intervals of 100,000 Monte Carlo (MC) simulations, and the results for the linear ($\alpha = 1$) system. Although the behaviors of the inelastic and linear systems are vastly different, in this case the tail probabilities are not significantly different. Finally, Figure 13.26 shows the reliability index for the considered response quantity for threshold $x = 0.12$ m at time $t_x = 12$ s as a function of the normalized eccentricity e/b. It shows that the reliability index steadily decreases with increasing eccentricity of the third wall.

Example 13.8 (cont.)

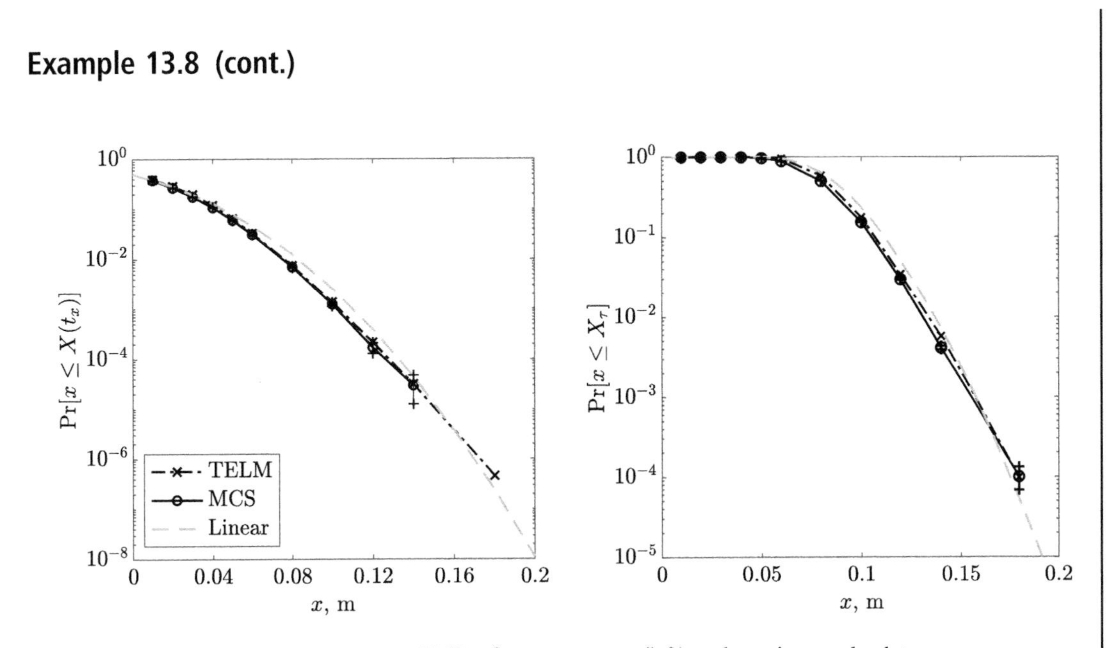

Figure 13.25 Complementary CDFs of response at t_x (left) and maximum absolute response over a 10 s stationary duration (right) (after Broccardo and Der Kiureghian, 2016, with permission from ASCE)

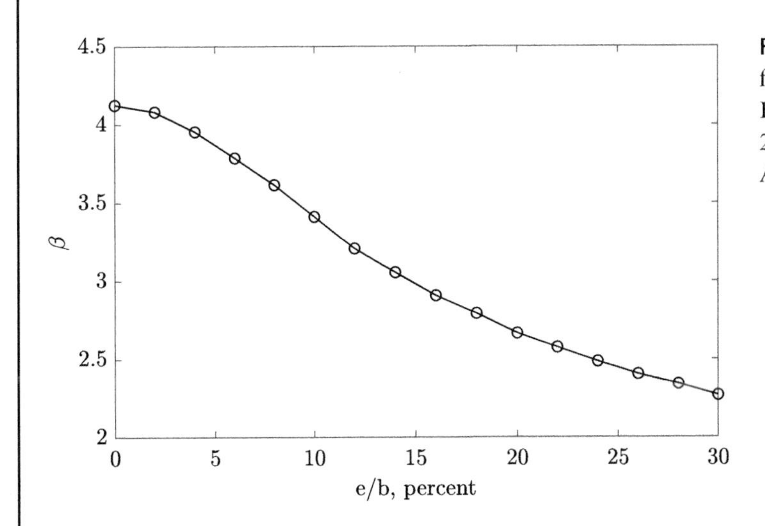

Figure 13.26 Reliability index as a function of eccentricity (after Broccardo and Der Kiureghian, 2016, with permission from ASCE)

13.9 Evolutionary TELM

A convenient way to represent a nonstationary input process $F(t)$ is through the evolutionary power spectral density (EPSD) function defined by Priestley (1967)

$$\Phi_{FF}(\omega, t) = |A(\omega, t)|^2 \Phi(\omega), \tag{13.59}$$

where $A(\omega, t)$ is a possibly complex-valued, time-frequency modulating function and $\Phi(\omega)$ the PSD of an underlying stationary process. Both $\Phi(\omega)$ and $A(\omega, t)$ must be symmetric functions over the frequency domain $(-\infty < \omega < +\infty)$. One can show that, in the case of a zero-mean process, the auto-covariance function of $F(t)$ is given by

$$\Sigma_{FF}(t_1, t_2) = \int_{-\infty}^{+\infty} A(\omega, t_1) A^*(\omega, t_2) \Phi(\omega) \exp\left[i\omega(t_1 - t_2)\right] d\omega, \tag{13.60}$$

where the superposed asterisk denotes the complex conjugate. Setting $t_1 = t_2 = t$, one obtains the variance function

$$\sigma_F^2(t) = \int_{-\infty}^{+\infty} |A(\omega, t)|^2 \Phi(\omega) d\omega. \tag{13.61}$$

It is seen that not only the variance but also the frequency content of the process varies with time. A special class of the above process arises when the modulating function is a separable function of frequency and time, $A(\omega, t) = q(t) A(\omega)$. Then (13.60) simplifies to

$$\sigma_F^2(t) = |q(t)|^2 \int_{-\infty}^{+\infty} |A(\omega)|^2 \Phi(\omega) d\omega. \tag{13.62}$$

It is clear that in this case the frequency content is uniformly modulated in time.

One can show that the response $X(t)$ of a linear system to an evolutionary input process is also evolutionary with its EPSD given by

$$\Phi_{XX}(\omega, t) = |M(\omega, t)|^2 \Phi(\omega), \tag{13.63}$$

where

$$M(\omega, t) = \int_0^t A(\omega, t - \tau) h(\tau) \exp(-i\omega\tau) d\tau \tag{13.64}$$

is the new modulating function with $h(t)$ being the IRF of the linear system for the particular input-output pair. Note that $M(\omega, t)$ in general is a complex-valued function, even when $A(\omega, t)$ is real-valued. Because $A(\omega, t - \tau) = 0$ for $t < \tau$ and $h(\tau) = 0$ for $\tau < 0$, the limits of the above integral can be changed to $-\infty$ and $+\infty$. Comparing with (13.22), we see that the function $M(\omega, t)$ represents a generalization of the FRF of the linear system for stationary response. For that reason, we name this function evolutionary FRF (EFRF). Note, however, that while the FRF is purely a characterization of the linear system for the particular input-output pair, the EFRF also incorporates information about the input excitation through the modulating function $A(\omega, t)$.

Suppose we discretize the underlying stationary process with PSD $\Phi(\omega)$ in the frequency domain as in (13.1) with the basis functions defined in (13.10). The discretized evolutionary process then has the form in (13.1) with the vector of basis functions

$$\mathbf{s}(t) = [\sigma_1|A(\omega_1,t)|\cos(\omega_1 t)\cdots\sigma_n|A(\omega_n,t)|\cos(\omega_n t) \\ \sigma_1|A(\omega_1,t)|\sin(\omega_1 t)\cdots\sigma_n|A(\omega_n,t)|\sin(\omega_n t)]. \tag{13.65}$$

Furthermore, as shown by Broccardo and Der Kiureghian (2013), the response of the linear system has the form in (13.12) with the gradient vector being

$$\mathbf{a}(t) = [\sigma_1|M(\omega_1,t)|\cos(\omega_1 t)\cdots\sigma_n|M(\omega_n,t)|\cos(\omega_n t) \\ \sigma_1|M(\omega_1,t)|\sin(\omega_1 t)\cdots\sigma_n|M(\omega_n,t)|\sin(\omega_n t)]. \tag{13.66}$$

It follows that the relations in (13.35) and (13.36) can now be applied to the EFRF; i.e., the modulus and phase angle of the EFRF are given by

$$|M(\omega_i,t)| = \frac{\sqrt{a_i^c(t)^2 + a_i^s(t)^2}}{\sigma_i}, \quad i = 1,\ldots,n, \tag{13.67}$$

$$\varphi_i = \omega_i t + \tan^{-1}\left[\frac{a_i^s(t)}{a_i^c(t)}\right], \quad i = 1,\ldots,n. \tag{13.68}$$

where the superscript c and s on the elements of $\mathbf{a}(t)$ indicate terms associated with cosine and sine terms.

In TELM analysis with a nonstationary input process, ideally one should find the design point for a sequence of time points over the duration of the excitation. However, as mentioned earlier and showed in example applications, the TELS for a given threshold is insensitive to the frequency content of the broad-band excitations as well as any scaling of the excitation. On this basis, a crude approximation is to use the TELS developed for a critical time point for the entire duration of the excitation. The evolutionary TELM approach adopts an approximation in between these two extremes. Specifically, the IRF or FRF of the TELS for the underlying stationary process (typically a band-limited white noise) is first determined. This is then used in (13.63), if the IRF is known, or in (13.66)–(13.67) if the FRF is known, to compute the corresponding EFRF of the TELS for the specific evolutionary input. Observe that, as opposed to using a single, time-invariant TELS, the TELS obtained by evolutionary TELM analysis evolves with time. Furthermore, using the gradient vector $\mathbf{a}(t)$ obtained from (13.65), the design point $\mathbf{u}^*(x,t)$, the unit normal vector $\hat{\boldsymbol{\alpha}}(x,t)$, and the reliability index $\beta(x,t)$ for different time points are determined from (13.15)–(13.17), respectively. These values can be used to compute the tail probability and the mean up-crossing rate for the considered threshold x as functions of time. For example, the mean up-crossing rate is computed from the limit formula (13.42) by using the reliability indices and unit normal vectors for two closely spaced time points. Having the mean up-crossing rate as a function of time allows the Poisson approximation of the first-passage time to be used in (13.50). Alternatively, the first-passage probability can be expressed as a series system at a large number of closely spaced time points and solved in terms of the multinormal CDF or by efficient importance sampling, as suggested by Au and Beck (2001a).

Example 13.9 – Application of TELM to a Hysteretic Oscillator Subjected to Evolutionary Excitation

Adopted from Broccardo and Der Kiureghian (2013), this example demonstrates application of the evolutionary TELM to an oscillator having the Bouc–Wen hysteresis model of Example 13.2. The linear ($\alpha = 1$) oscillator has natural frequency $\omega_0/2\pi = 1$ Hz and a 5% damping ratio. Hysteretic parameter values are $\alpha = 0.1$, $n = 3$, $A = 1$, and $\gamma = \eta = 1/(2\sigma_0)$, where σ_0 is the standard deviation of the response of the linear oscillator to a white-noise excitation of spectral density S_0, as described in Example 13.2. The oscillator is subjected to an evolutionary process defined by an underlying white noise of power spectral density S_0 and the separable modulating function $A(\omega, t) = q(t)A(\omega)$, where

$$q(t) = \frac{1}{\sqrt{2}}\left\{1 + \sin\left[\pi\left(\frac{t}{t_x} - \frac{1}{2}\right)\right]\right\}^{1/2}, \quad 0 \leq t \leq 2t_x \tag{E1}$$

$$A(\omega) = \left[\frac{\omega_f^4 + 4\zeta_f^2\omega_f^2\omega^2}{\left(\omega_f^2 - \omega^2\right)^2 + 4\zeta_f^2\omega_f^2\omega^2}\right]^{1/2}. \tag{E2}$$

Note that the time-modulating function has a peak equal to unity at $t = t_x$ and that the frequency-modulating function is proportional to the square root of the Kanai-Tajimi PSD. Two sets of parameter values are considered, as listed in Table 13.1. Excitation I is a broad-band process characterized by a filter frequency of 4 Hz and a damping ratio of 0.9, and Excitation II is a relatively narrow-band process with a filter frequency matching the frequency of the oscillator and a damping ratio of 0.6. Also listed in the table are values of S_0, t_x, the frequency-modulating function at the natural frequency of the linear oscillator, and the standard deviation of each process at t_x. For reliability analysis, the threshold $x = 3\sigma_0$ is considered.

Table 13.1 Selected parameters of three excitation models

	t_x, s	S_0, m²/s³	$\lvert A(\omega_0)\rvert^2$	$\omega_f/2\pi$, Hz	ζ_f	$\sigma_F(t_x)$, g
Excitation I	10	0.899	1	4	0.9	0.320
Excitation II	10	0.590	1	1	0.6	0.178

For each excitation, we perform two sets of analysis. One is the ordinary TELM analysis at selected time points. The other is evolutionary TELM analysis. For the latter analysis, the IRF

Example 13.9 (cont.)

is first determined for the underlying banded white-noise process for $x = 3\sigma_0$ and $t_x = 10$ s. This is used in (13.63) to determine the EFRF. Using the EFRF, the gradient vector $\mathbf{a}(t)$ is determined from (13.65), which then allows computing the reliability quantities of interest from (13.15)–(13.17) for any time point of interest. Note that, as opposed to the ordinary TELM that requires repeated analyses to determine the IRFs and design points at all selected time points, the evolutionary TELM requires only one IRF and one design point to be determined. The results for other time points are computed from knowledge of the corresponding gradient vector values obtained from (13.65).

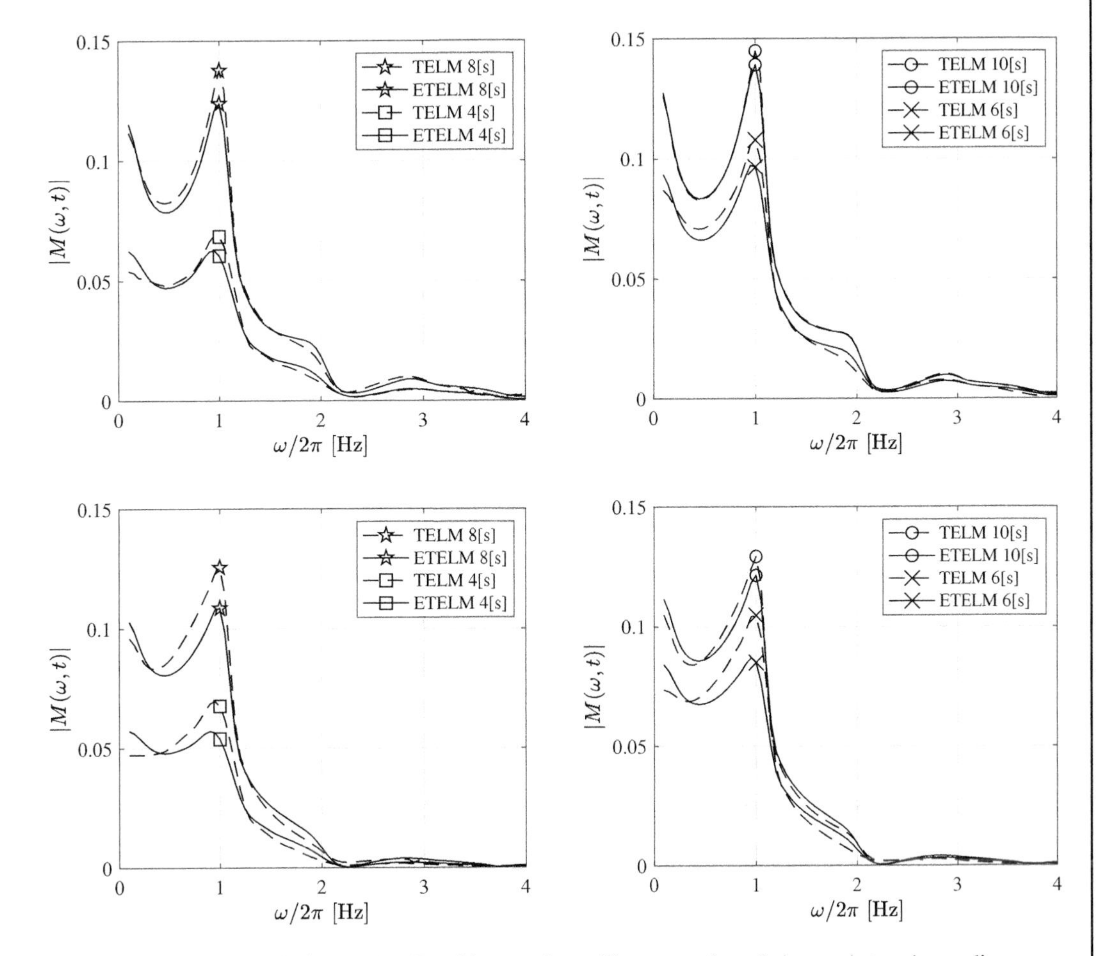

Figure 13.27 Evolutionary FRFs of hysteretic oscillator at selected time points using ordinary TELM and evolutionary TELM for the broad-band excitation (top charts) and narrow-band excitation (bottom charts) (after Broccardo and Der Kiureghian, 2013)

Example 13.9 (cont.)

Figure 13.27 shows the EFRFs of the oscillator response at selected time points. The charts on the left are for the broad-band excitation and those on the right are for the narrow-band excitation. As mentioned above, for each time point, the result from the ordinary TELM employs the IRF computed specifically for that time point, while the evolutionary TELM results all employ the IRF identified for $t_x = 10$ s. Observe that the results based on the two methods closely match, indicating the efficacy of the evolutionary TELM approximation.

Figure 13.28 compares the reliability indices for the tail probability $\Pr[x \leq X(t)]$ as a function of time. The results for ordinary TELM are computed by finding the design point at the selected time, while the results based on the evolutionary TELM are computed using the design point for $t_x = 10$ s and the identified gradient vector in (13.17). Observe that there is practically a perfect match between the two sets of results, again demonstrating the efficacy of the evolutionary TELM approximation. Also observe that the reliability index steadily decreases as the time-modulating function in (E1) builds up to its maximum value at $t = 10$ s.

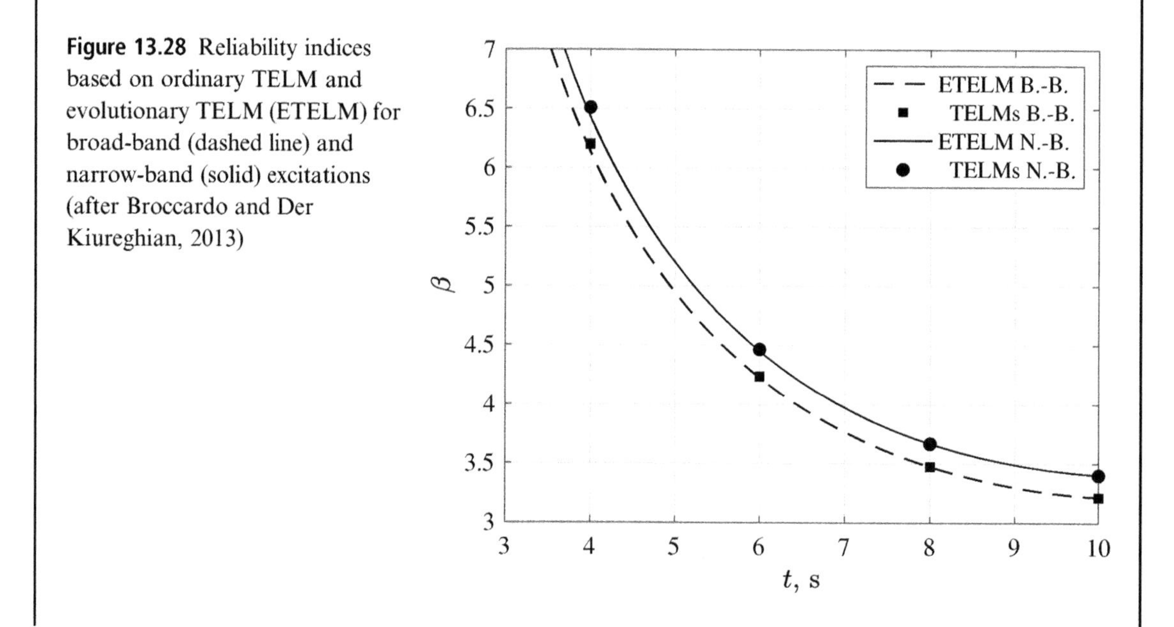

Figure 13.28 Reliability indices based on ordinary TELM and evolutionary TELM (ETELM) for broad-band (dashed line) and narrow-band (solid) excitations (after Broccardo and Der Kiureghian, 2013)

13.10 Concluding Remarks

As mentioned in the introduction to this chapter, the topic of stochastic structural dynamics is vast, and this chapter does not come close to presenting the topic in a thorough and comprehensive manner. Rather, the objective was to present some aspects of that topic, particularly for nonlinear structures, that relate to the time-invariant reliability methods developed in the earlier chapters of this book. In particular, the TELM is founded on the fundamental notions of the FORM, though it also takes advantage of important concepts from linear structural dynamics theory, namely the characterization of such systems in terms of the IRF or the FRF. The approach also employs discrete representations of random processes, which has been a topic of interest for a long time, particularly in connection with simulation of stochastic dynamics.

TELM provides an alternative method for equivalent linearization of nonlinear systems, which aims at providing improved results in the tail region of the probability distribution of the nonlinear response. Owing to the dependence of the TELS on the considered threshold, the approach is able to capture the non-Gaussian distribution of the nonlinear response. This aspect of TELM is a distinct advantage over the conventional ELM, which is able to approximate the variance of the response well, but not the tail of the probability distribution. Another advantage is in developing fragility curves conditioned on a measure of intensity of the excitation by using a single equivalent linear system. In contrast, the conventional ELM requires a new equivalent linear system to be developed for each intensity level of the excitation. TELM also provides insight into the physics of the problem, particularly in terms of the design-point excitation and response, which are the realizations of the input and output processes that have the highest likelihood of giving rise to the tail event of interest.

On the other hand, TELM has a number of computational challenges and shortcomings. It requires the solution of a constrained optimization problem that involves repeated computations of the nonlinear response and its gradient with respect to random variables defining the discretized excitation process. The linearization requires continuous differentiability of the response, which in turn requires smooth transition of material states. Hence, it cannot be applied to such common material models as the bi-linear elasto-plastic material without smoothing the transition between the elastic and plastic states. The gradient of the response with respect to input random variables can be computed by the DDM described in Section 12.4. Other efficient methods for computing gradients are available and can be developed, including complex differentiation (Martins et al., 2003) and algorithmic differentiation (Griewank and Walther, 2008). A gradient-free method for finding the design point in TELM analysis has been developed by Alibrandi and Der Kiureghian (2012).

A shortcoming that TELM shares with FORM is that its accuracy deteriorates when the number of random variables is large (see Section 6.4). However, although the number of random variables in TELM analysis is typically large (in hundreds), the number that induces nonlinearity in the limit-state surface is usually small. This is evident from the design-point excitations depicted in Figures 13.9 and 13.23, where significant amplitudes in the excitation occur only during a few seconds before the target time of interest. Another shortcoming shared with FORM is that there is no measure of the error involved in the approximation. To address these issues, Alibrandi and Mosalam (2017), Alibrandi and Koh (2017), and Broccardo et al. (2016) have developed an alternative definition of the hyperplane that characterizes the TELS. Rather than defining it as the tangent plane at the design point, they define it as a secant hyperplane that best separates a set of sampled safe and failure points in the standard normal space. This definition does not require differentiability of the response function, nor gradient computations, but it requires knowledge of a "critical direction," which is typically aligned with the direction of the design point. Furthermore, they are able to provide a measure of the error in the probability estimate through a band of uncertainty around the secant hyperplane.

TELM is one method among several that are being developed for nonlinear stochastic dynamic analysis. Its relevance in this book is its close relation to reliability methods. Further developments in this area are anticipated in the coming years.

14 Reliability-Based Design Optimization

14.1 Introduction

Optimization in the design of structural and mechanical systems aims at producing designs that optimize a measure of quality, quantity, or performance, subject to certain requirements. Design optimization is essential for effective use of resources and the maximization of net benefits. The same problem arises in selecting options for retrofitting or maintaining of existing structures. The problem is typically formulated in terms of an objective function that is to be maximized or minimized and a set of constrains, all expressed in terms of a set of design parameters.

As we have argued throughout this book, uncertainties are omnipresent in the characteristics of structures and the demands placed on them. Randomness in structural properties and dimensions and in deterioration of capacities, the stochastic nature of loads, errors in mathematical models describing structural behavior and response, and statistical uncertainties arising from sparse data and measurement errors all give rise to a lack of certainty in how a particular design of a structure will perform in real life. Therefore, optimization of structural design should properly account for these uncertainties. In the past, uncertainties have been accounted for in the design process in a crude and indirect manner; for example, by using factored but deterministic values of capacities and loads, i.e., below mean values for capacities and above mean values for loads. A mathematically rigorous approach is to formulate the optimal design problem in conjunction with a reliability formulation that properly accounts for the uncertainties underlying the design problem. This approach is known as reliability-based design optimization (RBDO).

Royset et al. (2001a) define three classes of RBDO problems:

P_1: Minimize the cost of the design, subject to reliability and structural constraints;
P_2: Maximize the reliability of the structure, subject to cost and structural constraints;
P_3: Minimize the initial cost plus the expected cost of failure, repair, and maintenance, subject to reliability and structural constraints.

In P_1, the cost of the design may be expressed in terms of the monetary value needed to build the design, the volume of materials used in the design, or any other surrogate for the cost of realizing the design. Structural constraints are typically limitations on the range of expected responses of the structure, primarily to assure functionality, as well as constraints on the design parameters to make sure they are realistic values. The constraint on the reliability can be an admissible upper bound on the probability of failure or loss of functionality, or a lower bound on the corresponding generalized reliability index. Of course, a given problem can have multiple structural and reliability constraints, the latter for different failure modes or cut sets.

P_2 aims at achieving maximum reliability for a given cost, the latter expressed in terms of a monetary value, volume of material, or other such quantity. A system with multiple failure modes may be formulated as a multi-objective optimization problem, where one seeks to simultaneously maximize the reliability measures of the different failure modes. Often it is not possible to simultaneously achieve these maxima. In that case, one approach is the so-called Pareto optimal solution, which is a solution for which no objective function can be further improved without degrading one or more of the other objective functions.

P_3 is the ultimate problem in optimal design, where the aim is to minimize the total expected cost, including the cost of the design and the cost of potential failures, as well as other costs such as maintenance and repair during a projected lifetime. The cost of failure enters as an expected cost, which is the product of the cost of failure and the probability of the event. Thus, the objective function combines costs and reliability measures. As before, the constraints can be structural as well as upper bounds on the probabilities of failure or loss of functionality to assure minimum safety and serviceability.

Algorithms for solving optimization problems typically rely on repeated computations of the objective and constraint functions for selected trial values of the design parameters. When the design parameters are continuous, gradients of the objective and constraint functions with respect to the design parameters can be used in selecting a search direction. Continuous differentiability of the objective and constraint functions is then essential for convergence to a solution point. Additionally, if the objective and constraint functions are convex, then solution to a global optimal point is guaranteed. Otherwise, the convergence can be to a local solution.

Early studies of RBDO employed the first-order reliability method (FORM) approximation of the probability of failure in the objective or constraint functions (see, e.g., Enevoldsen and Sorensen, 1994). This approach is now known as the reliability index approach (RIA) (Tu et al., 1999). As parameter sensitivities are readily available in FORM, a gradient-based optimization algorithm is used. However, the FORM approximation of the failure probability may not be continuously differentiable. As the design parameters are varied, discontinuities in the derivative of the FORM approximation may occur when the design point jumps from one location to another, while the limit-state surface in the standard normal space varies with the parameters. Because of this, the optimization algorithm may fail to converge. Note also that the calculations for each trial design point in the optimization algorithm involve computing the FORM failure probability or reliability index, which itself involves an optimization algorithm to find the linearization point. Hence, this approach requires nested optimization algorithms with no assurance of convergence.

To avoid the nested optimization approach, Madsen and Friis Hansen (1992) and Kuschel and Rackwitz (1997, 2000) formulated a single-loop design optimization problem, which includes the optimality conditions for finding the design point in the standard normal space as constraints in the formulation. Similar "single-loop" RBDO formulations have been proposed by Liang et al. (2004, 2007) and used by Nguyen et al. (2010, 2011) for reliability-based topology optimization. This approach can be more efficient in terms of the number of limit-state function evaluations. However, it requires derivatives of the design-point optimality conditions, which implies second derivatives of the limit-state function. Additionally, it suffers from a lack of continuity of the derivatives because it is based on the FORM.

The following example demonstrates the lack of continuity of the derivatives of the FORM reliability index with respect to a design parameter.

Example 14.1 – Lack of Continuity of the Derivative of FORM Approximation

Consider the limit-state function in the standard normal space

$$G(\mathbf{U}, v) = 4 + (v - 3)U_1 - 0.5U_1^2 - U_2, \tag{E1}$$

where v is a design parameter with a value confined within the range $[1, 4]$. Figure 14.1 shows plots of the limit-state surface $G(\mathbf{U}, v) = 0$ for selected values of v in the feasible range. Also shown are the design points and their trajectory as v varies from 1 to 4. For $v = 3$, two design

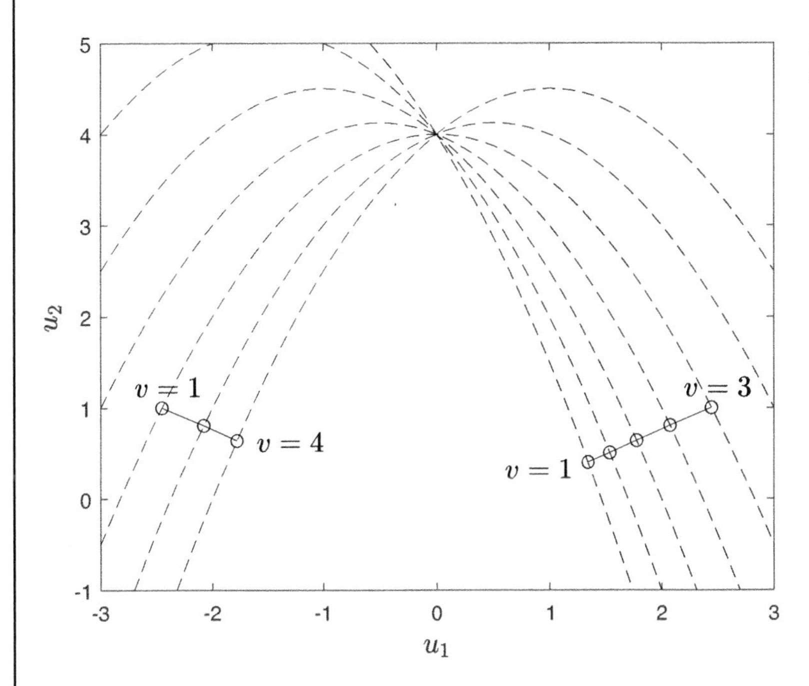

Figure 14.1 Limit-state surfaces and the trajectory of design points in the standard normal space with varying design parameter v

Example 14.1 (cont.)

points with coordinates $\mathbf{u}^* = [2.45, 1.00]^T$ and $\mathbf{u}^* = [-2.45, 1.00]^T$ are obtained with identical reliability indices $\beta = 2.65$. The corresponding gradient vectors are $\nabla G = [-2.45, -1.00]$ and $\nabla G = [2.45, -1.00]$. Furthermore, the derivative of the limit-state function with respect to v at the design point is $\partial G/\partial v = u_1^*$. Using these values in (6.41), we obtain $\mathrm{d}\beta/\mathrm{d}v = 0.926$ for the first design point and $\mathrm{d}\beta/\mathrm{d}v = -0.926$ for the second design point. Recall that these are both for $v = 3$. It is clear that the derivative of the reliability index makes a large jump at this value of v. Figure 14.2 shows a plot of the derivative as a function of v, which clearly demonstrates the lack of continuity of the derivative of the FORM reliability index with respect to the design parameter. As mentioned earlier, such a discontinuity in the gradient of the reliability index with respect to a design parameter makes gradient-based optimization algorithms inoperable for RBDO problems that include the FORM reliability index in their objective or constraint functions.

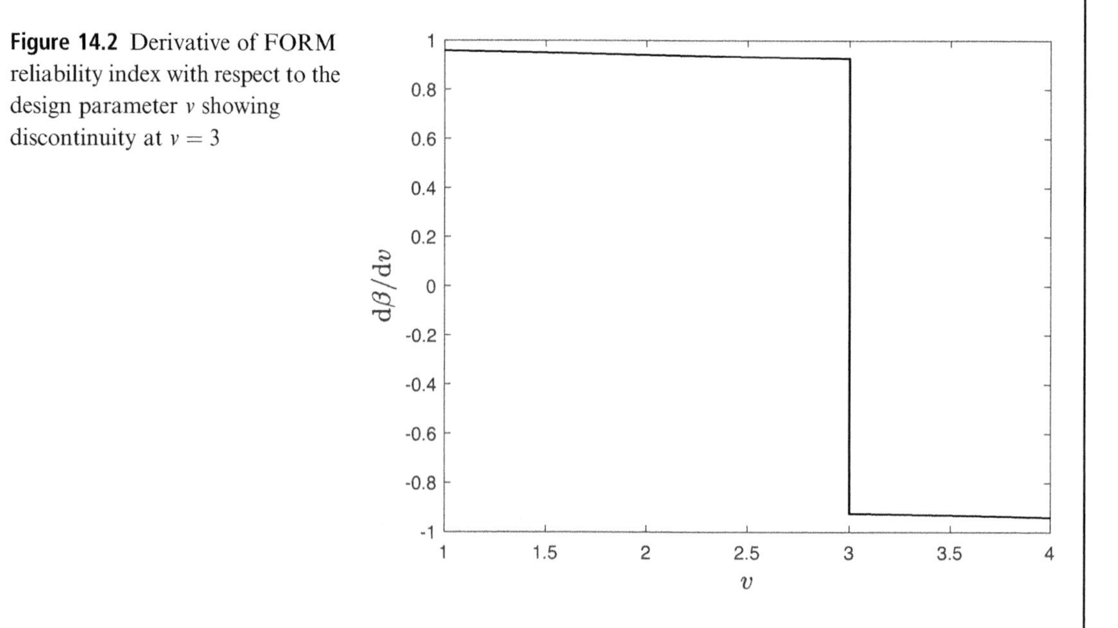

Figure 14.2 Derivative of FORM reliability index with respect to the design parameter v showing discontinuity at $v = 3$

Tu et al. (1999) proposed an alternative formulation of the RBDO problem, in which they replaced the reliability index by what they called a "performance measure," expressed as a quantile of the limit-state function. The formulation, called the performance measure approach (PMA), essentially uses a measure that is the inverse of the measure used in the RIA formulation. The PMA provides computational advantage and has been used by many investigators. However, no study has demonstrated that the performance measure in terms of the quantile of the limit-state function has continuous derivatives with respect to the design parameters. Hence, the PMA may suffer from the same convergence problem as the RIA formulation.

In order to overcome the lack of continuous differentiability of the FORM approximation, Kirjner-Neto et al. (1998) and Royset et al. (2001a, 2001b, 2006) reformulated the RBDO problem in terms of the limit-state function, avoiding explicit use of the failure probability or the reliability index in the objective or constraint functions. They only require continuous differentiability of the limit-state function(s) as well as those of the objective and structural constraint functions. Their formulations lead to single-loop deterministic optimization problems that can be solved by semi-infinite optimization algorithms. Furthermore, through a set of simple adjustments, Royset et al. (2001a, 2001b, 2006) are able to accommodate more-exact estimates of the failure probability, including second-order reliability method (SORM) approximations and estimates from simulation methods.

Another trend in RBDO is the use of surrogate or meta models in place of the true limit-state function, particularly when the latter requires expensive computations such as the use of a finite-element code. The most promising methods employ the Gaussian process kriging approach described in Section 12.7, where simulation points are gradually added based on a measure of the model improvement they provide (Dubourg et al., 2011; Bichon et al., 2013). Of course, such models must be developed in the combined space of the random variables and design parameters.

Use of sampling methods for RBDO analysis poses the problem of random fluctuations in the probability estimate, which render gradient computations by finite differences problematic. However, as described in Section 9.12, it is possible to directly estimate the gradient of the failure probability by sampling. Nevertheless, it is necessary to demonstrate that the random fluctuations in both the probability estimate and the gradient do not inhibit convergence of the optimization algorithm. Royset and Polak (2004a, 2004b, 2007) developed a rigorous formulation using Monte Carlo importance sampling. Their formulation is described later in this chapter, along with an example application.

More recently, Rockafellar and Royset (2010, 2015) advocated the use of buffered probability as a mathematically convenient surrogate for the failure probability. They argue that their selected measure is more appropriate for design decision-making because it accounts for the shape of the far tail of the probability distribution of the limit-state function. Furthermore, the measure has properties that facilitate the use of well-established optimization algorithms. A brief account of this approach is presented in Section 14.7.

This chapter focuses on aspects of the theory and methods of solution of RBDO that are closely related to reliability methods described in earlier chapters of this book. In particular, these include the use of FORM, SORM, sampling methods, and surrogate models in conjunction with optimization algorithms to solve the RBDO problems defined above. Attention is focused on formulations that have proven convergence properties.

14.2 Problem Formulation

As before, let $\mathbf{X} = [X_1, \ldots, X_n]^{\mathrm{T}}$ denote the n-vector of random variables that collects the uncertain quantities affecting the structural design problem of interest and

$\mathbf{U} = [U_1, \ldots, U_n]^T$ denote its transformation to the standard normal space. Also, let the m-vector $\mathbf{v} = [v_1, \ldots, v_m]^T$ collect a set of deterministic design parameters. The reliability of a structural system with N components is defined in terms of a set of limit-state functions $g_i(\mathbf{X}, \mathbf{v})$ or, equivalently after transformation to the standard normal space, $G_i(\mathbf{U}, \mathbf{v})$, $i = 1, \ldots, N$. Also let the function $f_0(\mathbf{v})$ denote a cost function in terms of the design parameters and $f_j(\mathbf{v}), j = 1, \ldots, J$, denote a set of J structural constraint functions, including allowable bounds on the design parameters. The system failure event is defined as $\mathcal{F} = \left\{\bigcup_k \bigcap_{i \in \mathcal{C}_k} [G_i(\mathbf{U}, \mathbf{v}) \leq 0]\right\}$, where $\mathcal{C}_k$ denotes the kth min-cut set. As we saw before, this includes the special cases of a component reliability problem that has only one limit-state function, a series system in which each min-cut set has only one component, and a parallel system that has only one cut set. Furthermore, let $p_i(\mathbf{v}) = \Pr[G_i(\mathbf{U}, \mathbf{v}) \leq 0]$, $i = 1, \ldots, N$, denote the failure probability of component i and $p_{\text{sys}}(\mathbf{v}) = \Pr(\mathcal{F})$ the failure probability of the system for design $\mathbf{v}$ with $\beta_i(\mathbf{v}) = \Phi^{-1}[1 - p_i(\mathbf{v})]$ and $\beta_{\text{sys}}(\mathbf{v}) = \Phi^{-1}\left[1 - p_{\text{sys}}(\mathbf{v})\right]$ denoting the corresponding generalized reliability indices, respectively.

Existing RBDO studies are mostly limited to component and series-system problems, though formulations for general systems have also been developed (Royset and Polak, 2007). The focus in this chapter will be on optimal design for component and series-system structural reliability problems.

Royset et al. (2001a, 2001b) formulate problem P_1 as follows:

$$\text{P}_1: \ \min_{\mathbf{v}} \left\{f_0(\mathbf{v}) \mid \beta_{0,i} \leq \beta_i(\mathbf{v}), i = 1, \ldots, N; f_j(\mathbf{v}) \leq 0, j = 1, \ldots, J\right\}. \tag{14.1}$$

This formulation minimizes the cost of the design while imposing a lower bound, denoted $\beta_{0,i}$, on the generalized reliability index of each component i, in addition to the structural constraints defined by the inequalities $f_j(\mathbf{v}) \leq 0$. The latter may include bounds on member sizes, slenderness ratios, or other constraints dictated by the functional requirements of the design. Note that the reliability constraints in this formulation are expressed in terms of component reliabilities, with possibly different lower bounds depending on the severity of the consequences if each component fails. If a lower bound on the system reliability is to be imposed, then we define problem $\text{P}_{1,\text{sys}}$

$$\begin{aligned} \text{P}_{1,\text{sys}}: \quad & \min_{\mathbf{v}} \{f_0(\mathbf{v}) \mid \beta_{0,\text{sys}} \leq \beta_{\text{sys}}(\mathbf{v}), \beta_{0,i} \leq \beta_i(\mathbf{v}), i = 1, \ldots, N; \\ & f_j(\mathbf{v}) \leq 0, j = 1, \ldots, J\}, \end{aligned} \tag{14.2}$$

where, in addition to the bounds on the component reliabilities, the lower bound $\beta_{0,\text{sys}}$ is imposed on the system generalized reliability index.

Royset et al. (2001a, 2001b) formulate problem P_2 as follows:

$$\text{P}_2: \ \max_{\mathbf{v}} \left\{ \min_i \beta_i(\mathbf{v}) \middle| f_j(\mathbf{v}) \leq 0, j = 1, \ldots, J\right\}. \tag{14.3}$$

This formulation maximizes the reliability of the least reliable component, subject to structural constraints. The constraints may also include an upper bound on the cost of the design. The system version of this problem is formulated as

$$\mathrm{P}_{2,\mathrm{sys}}:\ \max_{\mathbf{v}} \left\{\beta_{\mathrm{sys}}(\mathbf{v}) \middle| f_j(\mathbf{v}) \leq 0, j = 1, \ldots, J\right\}, \tag{14.4}$$

which aims at maximizing the system reliability index, subject to the structural and possibly cost constraints.

To formulate problem P_3, let $c_0(\mathbf{v})$ denote the cost of realizing the design $\mathbf{v}$, possibly also including the cost of monitoring and maintenance of the structure, and $c_i(\mathbf{v})$, $i = 1, \ldots, N$, denote the expected cost if component i were to fail. Royset et al. (2001b) formulate problem P_3 as follows:

$$\begin{aligned}\mathrm{P}_3:\ \min_{\mathbf{v}}\ \{&c_0(\mathbf{v})+\textstyle\sum_{i=1}^{N} c_i(\mathbf{v})p_i(\mathbf{v}) \mid p_i(\mathbf{v}) \leq p_{0,i}, i = 1, \ldots, N; \\ &f_j(\mathbf{v}) \leq 0, j = 1, \ldots, J\},\end{aligned} \tag{14.5}$$

where $p_{0,i}$ is an upper bound imposed on the failure probability of component i. The above formulation assumes that the expected costs of component failures are additive. The formulation includes lower bounds on component reliabilities as well as structural constraints. Of course, the initial cost, $c_0(\mathbf{v})$, and the failure costs, $c_i(\mathbf{v})$, could themselves be functions of other random variables, such as uncertain costs of materials and labor. In such cases, these cost functions represent expected values over the distributions of these random variables. These expectations must be computed outside the optimization problem.

A systems version of the above problem is formulated as (Royset et al. 2006)

$$\mathrm{P}_{3,\mathrm{sys}}:\ \min_{\mathbf{v}}\ \{c_0(\mathbf{v})+c_{\mathrm{sys}}(\mathbf{v})p_{\mathrm{sys}}(\mathbf{v}) \mid p_{\mathrm{sys}}(\mathbf{v}) \leq p_{0,\mathrm{sys}}; f_j(\mathbf{v}) \leq 0, j = 1, \ldots, J\}, \tag{14.6}$$

where $p_{0,\mathrm{sys}}$ is the upper bound on the system failure probability and $c_{\mathrm{sys}}(\mathbf{v})$ denotes the expected cost of failure of the system. This formulation does not impose constraints on component-level probabilities.

Obviously, other formulations can be conceived. For example, problem $\mathrm{P}_{3,\mathrm{sys}}$ may include constraints on individual component reliabilities, or P_3 may include a constraint on the system reliability. Royset et al. (2006) additionally formulate a problem at the level of a portfolio of structural systems. Such a formulation is of interest, for example, in optimizing the design or retrofit of a collection of structures. For our purposes, the above formulations are sufficient to provide a range of RBDO problems to study.

A somewhat different problem is optimizing the topology of a structure, subject to reliability constraints. Although the objective in such a problem might be to minimize cost, e.g., the volume of material used, the design involves selecting the shape of the structure rather than minimizing a cost function of design parameters. Among many other studies, this problem has been investigated by Maute and Frangopol (2003), Kharmanda et al. (2004), and Nguyen et al. (2010, 2011).

14.3 Solution by the Decoupling Approach

Virtually all RBDO studies assume that the cost functions $f_0(\mathbf{v})$, $c_0(\mathbf{v})$, $c_i(\mathbf{v})$, etc. and the constraint functions $f_j(\mathbf{v})$ have continuous first-order derivatives with respect to the design parameters $\mathbf{v}$, and that the limit-state functions $G_i(\mathbf{U},\mathbf{v})$ have continuous second-order mix partial derivatives $\partial^2 G_i(\mathbf{u},\mathbf{v})/\partial u_j \partial v_k$ with respect to the outcomes $\mathbf{u}$ of $\mathbf{U}$ and the design parameters $\mathbf{v}$. We also assume that every neighborhood of a point $\mathbf{u}$ satisfying $G_i(\mathbf{u},\mathbf{v}) = 0$ contains points $\mathbf{u}$ such that $G_i(\mathbf{u},\mathbf{v}) \neq 0$. This assumption assures that $G_i(\mathbf{u},\mathbf{v}) = 0$ is a curve for $n = 2$, a surface for $n = 3$, and a hyper-surface for $n > 3$, where n is the dimension of $\mathbf{U}$.

As mentioned before, there is no guarantee that the failure probabilities $p_i(\mathbf{v})$ and $p_{\text{sys}}(\mathbf{v})$, the corresponding generalized reliability indices $\beta_i(\mathbf{v})$ and $\beta_{sys}(\mathbf{v})$, respectively, or their various approximations are continuously differentiable with respect to the design parameters $\mathbf{v}$. In particular, we know that the FORM and SORM approximations are not differentiable at points where the trajectory of the design point as a function of $\mathbf{v}$ makes a jump. Furthermore, probability estimates based on simulation randomly fluctuate and a finite-difference method for computing derivatives is not reliable, although direct computation of derivatives by simulation as described in Section 9.12 is a possibility.

The decoupling method developed by Royset et al. (2001a, 2001b, 2006) and described in this section avoids direct use of approximate estimates of the failure probability or the reliability index. Aside from addressing the non-differentiability of the probability and reliability index approximations, an important advantage is that the calculations for optimization and reliability analyses are entirely decoupled, so that one can use any suitable algorithm to solve each subproblem regardless of the algorithm used for solving the other subproblem. Furthermore, as described below, through a set of adjustment factors, one can improve the accuracy of the solution by using more advanced reliability analysis methods, including simulation.

14.3.1 Solution of Problems P_1 and $P_{1,\text{sys}}$

First consider the case where the limit-state functions are affine in $\mathbf{U}$, i.e., $G_i(\mathbf{U},\mathbf{v}) = a_i(\mathbf{v}) + \mathbf{b}_i(\mathbf{v})^{\mathrm{T}}\mathbf{U}$, $i = 1, \ldots, N$, where $a_i(\mathbf{v})$ is a scalar function and $\mathbf{b}_i(\mathbf{v})$ is an n-vector function of $\mathbf{v}$. We assume $0 < a_i(\mathbf{v})$, which implies that, in the standard normal space, the origin is located in the safe domain. It follows (see (5.10)) that the reliability index for the ith affine limit-state function is given by

$$\beta_i(\mathbf{v}) = \frac{a_i(\mathbf{v})}{\|\mathbf{b}_i(\mathbf{v})\|}. \tag{14.7}$$

Also define the n-dimensional balls $\mathcal{B}_{r_i} = \{\mathbf{u} \mid \|\mathbf{u}\| = r_i\}$ of radii r_i, $i = 1, \ldots, N$, in the standard normal space and the functions

$$\psi_{r_i}(\mathbf{v}) = \max\{-G_i(\mathbf{u},\mathbf{v}) \mid \mathbf{u} \in \mathcal{B}_{r_i}\}, i = 1, \ldots, N. \tag{14.8}$$

Now we consider the following reformulation of problem P_1:

$$P_{1,r}\text{: } \min\left\{f_0(\mathbf{v}) \mid \psi_{r_i}(\mathbf{v}) \le 0, i = 1, \ldots, N; f_j(\mathbf{v}) \le 0, j = 1, \ldots, J\right\}. \tag{14.9}$$

Compared to the original problem P_1, problem $P_{1,r}$ replaces the reliability constraints of P_1 with the requirement that the limit-state functions remain non-negative within the respective balls $\mathcal{B}_{r_i}$. Observe that problem $P_{1,r}$ is a purely deterministic optimization problem that does not involve probabilities or reliability indices. It is a semi-infinite optimization problem because it has a finite number of design parameters and an infinite number of constraints. (Observe that $\psi_{r_i}(\mathbf{v}) \leq 0$ must be satisfied for all points $\mathbf{u}$ within each ball, thus an infinite number of constraints.) Well-tested algorithms are available for solution of this class of optimization problems, see, e.g., Polak (1997). As shown by Royset et al. (2001a), when $r_i = \beta_{0,i}$ for all i, then the solution of $P_{1,r}$ is identical to the solution of P_1 for affine limit-state functions. To see this, observe that for an affine limit-state function $\psi_{r_i}(\mathbf{v}) = -a_i(\mathbf{v}) + r_i\|\mathbf{b}_i(\mathbf{v})\|$. By requiring $\psi_{r_i}(\mathbf{v}) \leq 0$ in $P_{1,r}$, we have $r_i \leq a_i(\mathbf{v})/\|\mathbf{b}_i(\mathbf{v})\| = \beta_i(\mathbf{v})$ for any feasible design. Thus, by setting $r_i = \beta_{0,i}$, we satisfy the conditions $\beta_{0,i} \leq \beta_i(\mathbf{v})$ for the affine limit-state functions for all i.

A consequence of the above is that problems $P_{1,r}$ and P_1 have identical solutions if the reliability constraints in the P_1 are expressed in terms of the FORM approximation. In other words, for nonaffine limit-state functions, solving $P_{1,r}$ with $r_i = \beta_{0,i}$ for all i satisfies the reliability constraints in P_1 to first-order approximation. Furthermore, to obtain an approximately optimal solution that satisfies the constraints for the generalized reliability indices for nonaffine limit-state functions, we solve $P_{1,r}$ for different radii r_i until those constraints are satisfied for the generalized reliability indices. If the design vector obtained as the solution of $P_{1,r}$ yields a generalized reliability index $\beta_i(\mathbf{v})$ less than the specified threshold $\beta_{0,i}$, r_i is proportionally increased. Conversely, if $\beta_i(\mathbf{v})$ is greater than $\beta_{0,i}$, r_i is proportionally reduced. Owing to the normally close relation between the first order and the generalized reliability indices, the number of trial values of r_i to achieve convergence to an approximate design solution is usually small. Of course, it is not guaranteed that the design found is truly optimal with respect to the generalized reliability indices. However, because the design is optimal with respect to the FORM reliability indices, it will probably be sufficiently close to the true optimal solution for the generalized reliability indices. The generalized reliability indices may be computed by any higher-order reliability analysis method, including SORM and simulation.

Problem $P_{1,\text{sys}}$ is solved approximately by $P_{1,r}$ in the following manner:

1) Solve $P_{1,r}$ with $r_i = \beta_{0,i}$ neglecting the system-level constraint. Note that $\beta_{0,\text{sys}} \leq \beta_{0,i}$ must be specified.
2) Compute $\beta_{\text{sys}}(\mathbf{v})$ for the resulting design. If $\beta_{0,\text{sys}} \leq \beta_{\text{sys}}(\mathbf{v})$, the solution is found. If not, increase $\beta_{0,i}$ for the active component with the smallest reliability index and repeat the analysis until the system-level constraint is satisfied.

For the case with a system reliability constraint but no component reliability constraints, proceed as follows:

1) Select a $\beta_{0,i}$ value somewhat larger than $\beta_{0,\text{sys}}$ for all components and solve $P_{1,r}$.
2) Compute $\beta_{\text{sys}}(\mathbf{v})$ for the resulting design. If $\beta_{0,\text{sys}} = \beta_{\text{sys}}(\mathbf{v})$, the solution is found. If $\beta_{0,\text{sys}} < \beta_{\text{sys}}(\mathbf{v})$, reduce $\beta_{0,i}$ proportionally and re-solve. If $\beta_{\text{sys}}(\mathbf{v}) < \beta_{0,\text{sys}}$, increase $\beta_{0,i}$ proportionally and re-solve.

The above procedures adjust the constraint on the critical failure mode (the component with the largest failure probability) until the system constraint is satisfied. Because the probability of failure of a series system is dominated by the probability of failure of its least reliable component, usually only a few iterations are necessary to converge. The final result is a design that satisfies all the constraints but is only approximately optimal. Owing to the dominance of the critical failure mode, it is expected that, for all practical purposes, the design will be sufficiently close to the true optimal solution with respect to the series-system reliability constraint.

Example 14.2 – Design of Column for Minimum Cost

Adopted from Royset et al. (2001a), this example solves problem $\mathbf{P}_1$ for a single component. Consider a short column with a rectangular cross section of dimensions b and h, which is subjected to biaxial bending moments M_1 and M_2 and axial force P. Assuming an elastic-perfectly plastic material, the reliability of the column is defined by the limit-state function

$$g(\mathbf{X}, \mathbf{v}) = 1 - \frac{4M_1}{bh^2 Y} - \frac{4M_2}{b^2 h Y} - \left(\frac{P}{bhY}\right)^2, \tag{E1}$$

where Y denotes the yield strength of the material, $\mathbf{X} = [M_1, M_2, P, Y]^{\mathrm{T}}$ is the vector of random variables, and $\mathbf{v} = [b, h]^{\mathrm{T}}$ is the vector of design parameters. We assume M_1, M_2, P, and Y are statistically independent lognormal random variables with the means and coefficients of variation (c.o.v.) listed in Table 14.1.

Suppose the column is to be designed for minimum cross-sectional area $A = bh$, subject to $3 \leq \beta(b, h)$, $0 \leq b, h$, and $0.5 \leq b/h \leq 2$. The last constraint is imposed to bound the aspect ratio of the cross section. To design the column, we solve problem $\mathbf{P}_{1,r}$ with $f_0(b, h) = bh$, $f_1(b, h) = -b$, $f_2(b, h) = -h$, $f_3(b, h) = b/h - 2$, $f_4(b, h) = 0.5 - b/h$, $r = 3$, and $G(\mathbf{U}(\mathbf{X}), \mathbf{v}) = g(\mathbf{X}, \mathbf{v})$, where $\mathbf{U} = \mathbf{U}(\mathbf{X})$ is the probability transformation. The result is $\widehat{b} = 0.346$ m and $\widehat{h} = 0.553$ m. The corresponding FORM reliability index is $\beta_1\left(\widehat{b}, \widehat{h}\right) = 3.001$, which, within an error tolerance of 0.001, is equal to the assumed value of r, as expected. However, the generalized reliability index, obtained by "exact" reliability analysis using Monte Carlo simulation (with 5% c.o.v.), is $\beta_{\mathrm{MC}}\left(\widehat{b}, \widehat{h}\right) = 2.82$, which violates the reliability constraint.

Table 14.1 Means and c.o.v. of random variables

Variable	Mean	c.o.v.
M_1	250 kNm	0.30
M_2	125 kNm	0.30
P	2,500 kN	0.20
Y	40 MPa	0.10

Example 14.2 (cont.)

The optimization solution is now repeated using $r = 3^2/2.82 = 3.19$. The solution of $P_{1,r}$ now is $\hat{b} = 0.334$ m and $\hat{h} = 0.586$ m with $\beta_1(\hat{b}, \hat{h}) = 3.19$ and $\beta_{MC}(\hat{b}, \hat{h}) = 3.00$, the latter satisfying the reliability constraint. The cross-sectional area for this design is $\widehat{b}\widehat{h} = 0.195$ m^2.

For information about the specific semi-infinite optimization algorithms used for solving this and the next example, the reader is referred to Royset et al. (2001a).

Example 14.3 – Design of a Frame for Minimum Cost

Adopted from Royset et al. (2001a), this example considers the minimum-weight design of a one-bay frame under random vertical load V and horizontal load H, similar to that shown in Figure 8.11 of Example 8.7. The columns of the frame have rectangular cross sections of width b and depth h_1 and the beam has a rectangular cross section of width b and depth h_2. The material of the frame is elasto-plastic with yield stress Y, which is also a random variable. Plastic hinges may form in the lower ends of the columns and in the beam. Under the applied loads, the frame may collapse in any of the "sway," "beam," or "combined" failure mechanisms shown at the bottom of Figure 8.11. Thus, the frame constitutes a series structural system. The objective is to determine the optimal dimensions (b, h_1, h_2) for minimum material volume, subject to a constraint on the reliability of the system.

Using the virtual work method, limit-state functions defining the three failure modes of the structural system are derived as

$$g_1(\mathbf{X}, \mathbf{v}) = 0.5bh_1^2Y + 0.5bh_2^2Y - 5H, \tag{E1}$$

$$g_2(\mathbf{X}, \mathbf{v}) = bh_2^2Y - 5V, \tag{E2}$$

$$g(\mathbf{X}, \mathbf{v}) = 0.5bh_1^2Y + bh_2^2Y - 5H - 5V, \tag{E3}$$

where $\mathbf{X} = [H, V, Y]^T$ is the vector of random variables and $\mathbf{v} = [b, h_1, h_2]^T$ is the vector of deterministic design parameters. The random variables are assumed to be statistically independent and to have the distributions listed in Table 14.2.

The objective function (material volume of the frame) is $f_0(\mathbf{v}) = 10(bh_1 + bh_2)$. We impose a minimum of 0.2 m on each member dimension and a maximum aspect ratio of 2 on each cross section. Furthermore, we require a "strong-column-weak-beam" requirement so that hinges form in the beam, not in the columns. These requirements are expressed as the constraint functions $f_1(\mathbf{v}) = 0.2 - h_1$, $f_2(\mathbf{v}) = 0.2 - h_2$, $f_3(\mathbf{v}) = 0.2 - b$, $f_4(\mathbf{v}) = h_1 - 2b$, and $f_5(\mathbf{v}) = h_2 - h_1$. We also require the minimum system generalized reliability index $\beta_{0,sys} = 2.5$.

Example 14.3 (cont.)

Table 14.2 Distributions of random variables

Variable	Distribution	Mean	c.o.v.
H	Gumbel	50 kN	0.30
V	Gumbel	60 kN	0.30
Y	Lognormal	25 MPa	0.10

This problem is of type $P_{1,\text{sys}}$, which is solved by repeatedly solving $P_{1,r}$ for trial values of r. It is sufficient to constrain all the failure modes of the series system by a single r value, i.e., $r_i = r$, $i = 1, 2, 3$. This effectively imposes a reliability constraint on the critical failure mode. Aiming at a value a little greater than 2.5, $r = 2.68$ is selected. Solving $P_{1,r}$ for this value of r, the design $\widehat{\mathbf{v}}$ in the first row of Table 14.3 is obtained. Performing FORM analysis for this design produces $\beta_{1,\text{sys}}(\widehat{\mathbf{v}}) = 2.43$. The design does not satisfy the system reliability constraint in the first-order sense. Repeating the analysis using $r = (2.68 \times 2.50)/2.43 = 2.76$ produces the results in the second row of Table 14.3. Again, the system reliability constraint is not satisfied in the first-order sense. The analysis is repeated with $r = (2.76 \times 2.50)/2.47 = 2.80$, producing the results in the third row of Table 14.3. This design satisfies the system reliability constraint in the first-order sense.

Now suppose we wish to satisfy the system reliability constraint not in the first-order sense, but "exactly," based on an estimate of the system failure probability obtained by Monte Carlo simulation (with 2% c.o.v.). For the design in the third row of Table 14.3, Monte Carlo simulation yields $\beta_{\text{MC}}(\widehat{\mathbf{v}}) = 2.45$, which does not satisfy the reliability constraint. Repeating the analysis using $r = (2.80 \times 2.50)/2.45 = 2.86$ produces the results in the fourth row of Table 14.3. Thus, the design $\widehat{\mathbf{v}} = (0.200, 0.352, 0.352)$ m satisfies the system reliability constraint "exactly" in the Monte Carlo simulation sense.

Table 14.3 Design of ductile frame for minimum cost (after Royset et al., 2001a, with permission from ASCE)

r	$\widehat{b}$, m	$\widehat{h}_1$, m	$\widehat{h}_2$, m	$\beta_{1,\text{sys}}(\widehat{\mathbf{v}})$	$\beta_{\text{MC,sys}}(\widehat{\mathbf{v}})$
2.68	0.202	0.347	0.344	2.43	–
2.76	0.201	0.348	0.348	2.47	–
2.80	0.200	0.350	0.350	2.50	2.45
2.86	0.200	0.352	0.352	2.55	2.50

14.3.2 Solution of Problems P_2 and $P_{2,sys}$

Consider the reformulation of problem P_2 defined in (14.3) as

$$P_{2,r}:\ \min\left\{\psi_r(\mathbf{v})\middle|f_j(\mathbf{v})\leq 0, j=1,\ldots,J\right\}, \tag{14.10}$$

where

$$\psi_r(\mathbf{v})=\max\{-G_i(\mathbf{u},\mathbf{v})|i=1,\ldots,N;\mathbf{u}\in\mathcal{B}_r\}, \tag{14.11}$$

in which $\mathcal{B}_r$ denotes a ball of radius r. Royset et al. (2001a) have shown that problem $P_{2,r}$ is identical to problem P_2 when the limit-state functions are affine, i.e., when $G_i(\mathbf{U},\mathbf{v})=a_i(\mathbf{v})+\mathbf{b}_i(\mathbf{v})^{\mathrm{T}}\mathbf{U}$, $i=1,\ldots,N$. To see this, first recall that for an affine limit-state function the reliability index is given by (14.7). Thus, P_2 aims to maximize the minimum of $a_i(\mathbf{v})/\|\mathbf{b}_i(\mathbf{v})\|$ or minimize the maximum of $\|\mathbf{b}_i(\mathbf{v})\|/a_i(\mathbf{v})$ over all i. Let $\hat{\imath}$ be the index for the critical component so that $\|\mathbf{b}_{\hat{\imath}}(\mathbf{v})\|/a_{\hat{\imath}}(\mathbf{v})=\max_{1\leq i\leq N}\|\mathbf{b}_i(\mathbf{v})\|/a_i(\mathbf{v})$. Because it is the algebraic sign of the limit-state function and not its value that matters, the affine limit-state function with $0<a_i(\mathbf{v})$ can be reformulated as $G_i(\mathbf{U},\mathbf{v})=1+\mathbf{b}_i(\mathbf{v})^{\mathrm{T}}\mathbf{U}/a_i(\mathbf{v})$. The maximum of $-G_i(\mathbf{u},\mathbf{v})$ within a ball of radius r is then given by $-1+r\|\mathbf{b}_i(\mathbf{v})\|/a_i(\mathbf{v})$ so that by maximizing over i we have $\psi_r(\mathbf{v})=\{-1+r\|\mathbf{b}_{\hat{\imath}}(\mathbf{v})\|/a_{\hat{\imath}}(\mathbf{v})\}$. For any $0<r$, the minimum of the latter function occurs at the same point as the minimum of $\|\mathbf{b}_{\hat{\imath}}(\mathbf{v})\|/a_{\hat{\imath}}(\mathbf{v})$. It follows that, for affine limit-state functions, problems P_2 and $P_{2,r}$ have identical solutions for any $0<r$. Observe that, like $P_{1,r}$, problem $P_{2,r}$ is a purely deterministic semi-infinite optimization problem.

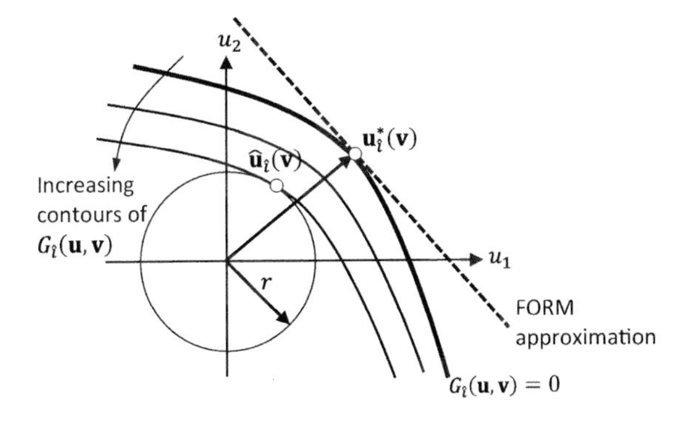

Figure 14.3 Geometric interpretation of problems P_2 and $P_{2,r}$ for nonaffine limit-state functions (after Royset et al., 2001a, with permission from ASCE)

A geometric interpretation of problems P_2 and $P_{2,r}$ with the FORM approximation helps us understand the situation for nonaffine limit-state functions. With this approximation, P_2 finds the optimal design by maximizing the distance from the origin in the $\mathbf{u}$ space to the design point (the nearest point on the limit-state surface) of the critical failure mode. As shown in Figure 14.3 for a case with two random variables, this distance is $\beta_{1,\hat{\imath}}(\mathbf{v})=\|\mathbf{u}_{\hat{\imath}}^*(\mathbf{v})\|$, where $\hat{\imath}$ is the index for the critical failure mode and $\mathbf{u}_{\hat{\imath}}^*(\mathbf{v})$ the corresponding design point.

In contrast, $P_{2,r}$ finds the optimal design by minimizing $\psi_r(\mathbf{v})$ or, equivalently, maximizing the minimum value of the limit-state function for the critical failure mode for $\mathbf{u}$ values within a ball of radius r. As shown in Figure 14.3, this minimum occurs at a point $\hat{\mathbf{u}}_{\hat{i}}(\mathbf{v})$, which is different from $\mathbf{u}_{\hat{i}}^*(\mathbf{v})$. It is clear from Figure 14.3 that the two approaches would produce designs identical to those from the FORM approximation if $r = \beta_{1,\hat{i}}(\mathbf{v})$ is selected. In that case, points $\mathbf{u}_{\hat{i}}^*(\mathbf{v})$ and $\hat{\mathbf{u}}_{\hat{i}}(\mathbf{v})$ coincide, and the design that maximizes $\beta_{1,\hat{i}}(\mathbf{v})$ also minimizes $\psi_r(\mathbf{v})$. Furthermore, based on the geometric interpretation in Figure 14.3, it is expected that the solution of $P_{2,r}$ for a nonaffine limit-state surface would be insensitive to the values of r in the neighborhood of $\beta_{1,\hat{i}}(\mathbf{v})$. For practical implementation, a set of r values, r_s, $s = 1, \ldots, S$, is selected with magnitudes in the neighborhood of the anticipated FORM reliability index of the critical failure mode, and the corresponding designs $\hat{\mathbf{v}}_s$ are determined by solving $P_{2,r}$. The approximate optimal design solution is then obtained as

$$\hat{\mathbf{v}} = \arg\ \max\{\beta_{\hat{i}}(\hat{\mathbf{v}}_s)|s = 1, \ldots, S\} \tag{14.12}$$

where $\beta_{\hat{i}}(\hat{\mathbf{v}}_s)$ is the generalized reliability index of the critical failure mode associated with design $\hat{\mathbf{v}}_s$ obtained by an appropriate reliability method.

Problem $P_{2,\text{sys}}$ is approximately solved by $P_{2,r}$ in a manner similar to the above. The only change is that $\beta_{\hat{i}}(\hat{\mathbf{v}}_s)$ in (14.12) is replaced by $\beta_{\text{sys}}(\hat{\mathbf{v}}_s)$, the generalized reliability index of the series system, which is computed by an appropriate reliability method. Because the failure probability of the series system is dominated by the failure probability of its critical mode, designs based on P_2 and $P_{2,\text{sys}}$ are often nearly identical.

Example 14.4 – Design of Column for Maximum Reliability

Table 14.4 Column design for maximum reliability (after Royset et al., 2001a, with permission from ASCE)

r	$\hat{b}$	$\hat{h}$	$\beta_1(\hat{\mathbf{v}})$	$\beta_2(\hat{\mathbf{v}})$	$\beta_{MC}(\hat{\mathbf{v}})$
2.0	0.309	0.606	2.90	2.69	2.70
2.5	0.310	0.605	2.90	2.69	2.70
3.0	0.310	0.605	2.90	2.69	2.70
3.5	0.310	0.605	2.90	2.69	2.70

Suppose the column in Example 14.2 is to be designed for maximum reliability, subject to its cross-sectional area not exceeding 0.1875 m^2. We solve problem $P_{2,r}$ with the constraints f_1, f_2, f_3, and f_4 of Example 14.1 and the additional constraint on the cross-sectional area $f_5(\mathbf{v}) = bh - 0.1875$ m^2. Table 14.4 summarizes the results of the optimal design solutions for a range of r values. Listed are the design points $\hat{\mathbf{v}} = \left(\hat{b}, \hat{h}\right)$ and the corresponding reliability indices computed by FORM, $\beta_1(\hat{\mathbf{v}})$, SORM, $\beta_2(\hat{\mathbf{v}})$, and Monte Carlo simulation (with 5% c.o.v.), $\beta_{MC}(\hat{\mathbf{v}})$. Note that the range of r values encompasses the first-order reliability index.

Example 14.4 (cont.)

It is observed that the optimal solution and the reliability indices are virtually invariant of the value of r. Furthermore, the reliability indices by FORM are significantly different from those by SORM and Monte Carlo simulation, indicating that the limit-state surface is strongly nonaffine. We know from the analysis in the preceding section that for an affine limit-state function, the assumed value of r is immaterial. The results in Table 14.4 indicate that this property also approximately holds for nonaffine limit-state functions. Based on the results in Table 14.4, the optimal dimensions of the column in this case are $\widehat{b} = 0.310$ m and $\widehat{h} = 0.605$ m with the optimal generalized reliability index being $\beta_{MC}(\widehat{\mathbf{v}}) = 2.70$.

14.3.3 Solution of Problems P_3 and $P_{3,sys}$

Problems P_3 and $P_{3,sys}$ in (14.5) and (14.6), respectively, contain probability terms in both the objective function and as constraints. Ordinary optimization algorithms cannot directly solve this problem with assurance of convergence. Following Royset et al. (2006), we reformulate this problem in two steps. First, we expand the design vector by adding the parameters $\mathbf{a} = [a_1, \ldots, a_N]^T$ that represent component failure probabilities. The reformulated problem is defined as

$$\bar{P}_3: \quad \min_{(\mathbf{v},\,\mathbf{a})}\{c_0(\mathbf{v})+\sum\nolimits_{i=1}^{N} c_i(\mathbf{v})a_i \mid p_i(\mathbf{v}) \le a_i, 0 \le a_i \le p_{0,i}, \\ i = 1, \ldots, N;\ f_j(\mathbf{v}) \le 0, j = 1, \ldots, J\}. \tag{14.13}$$

Problems P_3 and $\bar{P}_3$ are equivalent based on the following argument: Assuming $0 < c_i(\mathbf{v})$ for all i, suppose that a constraint $p_i(\mathbf{v}) \le a_i$ is inactive, i.e., $p_{i'}(\mathbf{v}) < a_{i'}$ for $1 \le i' \le N$. Then, because $0 < c_{i'}(\mathbf{v})$, the objective function can be reduced in value by decreasing $a_{i'}$ without violating the corresponding constraint. It follows that at the solution of $\bar{P}_3$, all constraints $p_i(\mathbf{v}) \le a_i$ must be active, i.e., $p_i(\mathbf{v}) = a_i$ for all i. Therefore, the objective functions of $\bar{P}_3$ and P_3 are identical. Furthermore, the constraints in $\bar{P}_3$ impose the condition $p_i(\mathbf{v}) = a_i \le p_{0,i}$ for all i. Hence, P_3 and $\bar{P}_3$ have identical solutions. In case $c_i(\mathbf{v}) = 0$ for some i for all designs, those terms vanish in the objective functions of both P_3 and $\bar{P}_3$ and their equivalence remains intact.

We have removed the probability terms in the objective function of $\bar{P}_3$ but they still exist in the constraints. To remove them, we reformulate problem $\bar{P}_3$ as follows: Let $\mathbf{t} = (t_1, \ldots, t_N)$ be a set of positive parameters. We define problem $\bar{P}_{3,\mathbf{t}}$ as

$$\bar{P}_{3,\mathbf{t}}: \quad \min_{(\mathbf{v},\,\mathbf{a})}\{c_0(\mathbf{v})+\sum\nolimits_{i=1}^{N} c_i(\mathbf{v})a_i \mid \psi_{i,t_i}(\mathbf{v}, a_i) \le 0, 0 \le a_i \le p_{0,k}, \\ i = 1, \ldots, N;\ f_j(\mathbf{v}) \le 0, j = 1, \ldots, J\}, \tag{14.14}$$

where

$$\psi_{i,t_i}(\mathbf{v}, a_i) = \max\{-G_i(\mathbf{u}, \mathbf{v}) \mid \mathbf{u} \in \mathcal{B}_r\}, \tag{14.15}$$

in which $\mathcal{B}_r$ is the ball of radius $r = \Phi^{-1}(1 - a_i)t_i$ in the $\mathbf{u}$ space and $\Phi^{-1}(\cdot)$ denotes the inverse of the standard normal cumulative distribution function (CDF). Suppose for now that $t_i = 1$. If $\psi_{i,1}(\mathbf{v}, a_i) \leq 0$, then according to (14.15) the limit-state function must be non-negative for all realizations of $\mathbf{u}$ in a ball of radius $\Phi^{-1}(1 - a_i)$. This implies that the FORM reliability index for component i is greater than or equal to $\Phi^{-1}(1 - a_i)$, i.e., $\beta_{1,i}(\mathbf{v}) \geq \Phi^{-1}(1 - a_i)$, or that $p_{1,i}(\mathbf{v}) = \Phi[-\beta_{1,i}(\mathbf{v})] \leq a_i$, where $p_{1,i}(\mathbf{v})$ is the FORM approximation of the failure probability. It follows that: (a) if the limit-state functions are affine in $\mathbf{U}$, then the solution of $\bar{\mathrm{P}}_{3,\mathbf{t}}$ with $t_i = 1$ for all i is identical to the solution of $\bar{\mathrm{P}}_3$ and, therefore, of P_3; and (b) if the limit-state functions are nonaffine, then the solution of $\bar{\mathrm{P}}_{3,\mathbf{t}}$ is identical to the solution of $\bar{\mathrm{P}}_3$ or P_3 with the probability terms replaced with their FORM approximations. Higher-order approximations, e.g., SORM or Monte Carlo simulation, can be accounted for by adjusting the parameters t_i. Specifically, if for a particular solution of $\bar{\mathrm{P}}_{3,\mathbf{t}}$ the FORM approximation overestimates the failure probability $p_i(\mathbf{v})$, the parameter t_i is adjusted downward, whereas if it underestimates the failure probability then t_i is adjusted upward. The solution of $\bar{\mathrm{P}}_{3,\mathbf{t}}$ with the adjusted parameters must then be checked with the selected higher-order reliability method to make sure that the probability constraints are all satisfied. It should be clear that this method can produce only an approximate optimal design.

Observe that $\bar{\mathrm{P}}_{3,\mathbf{t}}$ is a purely deterministic semi-infinite optimization problem. Hence, well-tested semi-infinite optimization algorithms can be used to solve this problem. The reader is referred to Royset et al. (2006) and references therein for details of an algorithm used for solving this problem.

Problem $\mathrm{P}_{3,\mathrm{sys}}$ in (14.6) is solved in a similar manner with one exception, as described below. First, we reformulate the problem as

$$\begin{aligned}\bar{\mathrm{P}}_{3,\mathrm{sys}}\colon \min_{(\mathbf{v},a)} \{&c_0(\mathbf{v})+c_{\mathrm{sys}}(\mathbf{v})a \mid p_{\mathrm{sys}}(\mathbf{v}) \leq a, 0 \leq a \leq p_{0,\mathrm{sys}};\\ &f_j(\mathbf{v}) \leq 0, j = 1, \ldots, J\},\end{aligned} \tag{14.16}$$

where a is an additional design parameter. Problems $\bar{\mathrm{P}}_{3,\mathrm{sys}}$ and $\mathrm{P}_{3,\mathrm{sys}}$ have identical solutions based on the same argument as that made for the equivalence of $\bar{\mathrm{P}}_3$ and P_3. Next, for a positive parameter t, we define the reformulated problem

$$\begin{aligned}\bar{\mathrm{P}}_{3,\mathrm{sys},t}\colon \min_{(\mathbf{v},a)} \{&c_0(\mathbf{v})+c_{\mathrm{sys}}(\mathbf{v})a \mid \psi_t(\mathbf{v}, a) \leq 0, 0 \leq a \leq p_{0,\mathrm{sys}};\\ &f_j(\mathbf{v}) \leq 0, j = 1, \ldots, J\},\end{aligned} \tag{14.17}$$

where

$$\psi_t(\mathbf{v}, a) = \max_{1 \leq i \leq N} \max\{-G_i(\mathbf{u}, \mathbf{v}) | \mathbf{u} \in \mathcal{B}_r\}, \tag{14.18}$$

in which $\mathcal{B}_r$ is the ball of radius $r = \Phi^{-1}(1 - a)t$ in the $\mathbf{u}$ space. Observe that (14.18) includes a maximization over the components of the series system. This is where this formulation for the system departs from that of $\bar{\mathrm{P}}_{3,\mathbf{t}}$. To provide an approximate solution of $\mathrm{P}_{3,\mathrm{sys}}$, we make the following argument: Suppose we solve $\bar{\mathrm{P}}_{3,\mathrm{sys},t}$ with $t = 1$. Then $\psi_t(\mathbf{v}, a) \leq 0$ implies that the limit-state function of the most critical component is non-negative for all realizations of

$\mathbf{u}$ in a ball of radius $\Phi^{-1}(1-a)$, or that $p_{1,i'}(\mathbf{v}) = \Phi[-\beta_{1,i'}(\mathbf{v})] \le a$, where i' is the index for the critical component and $p_{1,i'}(\mathbf{v})$ and $\beta_{1,i'}(\mathbf{v})$ are the FORM approximations of its failure probability and reliability index, respectively. Of course, this does not mean that the system probability constraint $p_{\text{sys}}(\mathbf{v}) \le p_{0,\text{sys}}$ is satisfied, even in the first-order sense. However, owing to the dominant contribution of the critical component to the series system probability, design changes are expected to result in similar variations in $p_{i'}(\mathbf{v})$ and $p_{\text{sys}}(\mathbf{v})$. Hence, problem $\bar{\text{P}}_{3,\text{sys},t}$ is solved for a range of values of t around 1, and the approximate optimal design is obtained as the one with the smallest objective function value that satisfies the system probability constraint for any desired reliability method. The details of a semi-infinite algorithm that solves the deterministic problem $\bar{\text{P}}_{3,\text{sys},t}$ are given in Royset et al. (2006).

Example 14.5 – Design of Reinforced-Concrete Girder

This problem was originally studied by Lin and Frangopol (1996) and Frangopol et al. (1997). The version presented here is based on Royset et al. (2006).

We consider the design of the reinforced-concrete girder shown in Figure 14.4 that is typically used in highway bridges. The design variables are $\mathbf{v} = (A_s, b, h_f, b_w, h_w, A_v, s_1, s_2, s_3)$, where A_s = area of the tension steel reinforcement, b = width of the flange, h_f = thickness of the flange, b_w = width of the web, h_w = height of the web, A_v = area of the shear reinforcement (= twice the cross-section area of a stirrup), and $s_1, s_2,$ and s_3 = spacings of the shear reinforcement in intervals 1, 2, and 3 as shown in Figure 14.4, respectively.

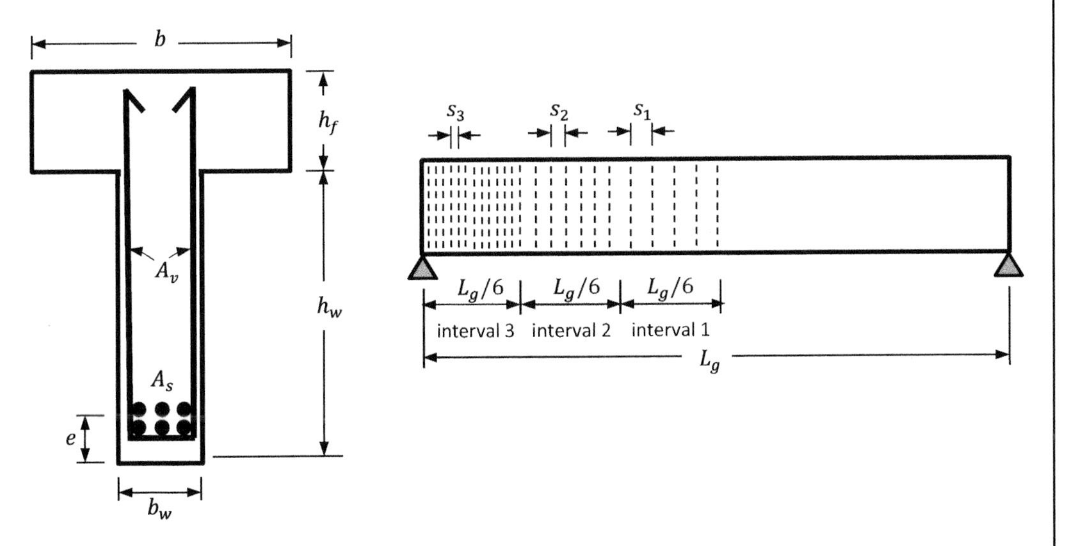

Figure 14.4 Reinforced-concrete bridge girder cross section (left) and side view (right) (after Royset et al., 2006, with permission from ASCE)

Example 14.5 (cont.)

The random variables describing the loading and material properties are collected in the vector $\mathbf{X} = \left[f_y, f'_c, P_D, M_L, P_{S1}, P_{S2}, P_{S3}, W\right]^T$, where f_y = yield strength of the reinforcement, f'_c = compressive strength of concrete, P_D =dead load excluding the weight of the girder, M_L = live load bending moment, P_{S1}, P_{S2}, and P_{S3} = live load shear forces in intervals 1, 2, and 3, respectively (see Figure 14.4), and W = unit weight of concrete. Following Lin and Frangopol (1996), all the random variables are considered to be statistically independent and normally distributed with the means and c.o.v. listed in Table 14.5. The length of the girder is $L_g = 18.30$ m, and the distance from the bottom fiber to the centroid of the tension reinforcement is $e = 0.1$ m, see Figure 14.4.

The girder is assumed to have four failure modes corresponding to failures due to the bending moment in the mid span and the shear forces in intervals 1, 2, and 3. Each failure mode is described by a limit-state function. The limit-state function for the bending failure mode is given by

Table 14.5 Description of normal random variables.

Variable	Description	Mean	c.o.v.
f_y	Yield strength of reinforcement, Pa	413.4×10^6	0.150
f'_c	Compressive strength of concrete, Pa	27.56×10^6	0.150
P_D	Dead load excluding girder, N/m	13.57×10^3	0.200
M_L	Live load bending moment, Nm	929×10^3	0.243
P_{S1}	Live load shear in interval 1, N	138.31×10^3	0.243
P_{S2}	Live load shear in interval 2, N	183.39×10^3	0.243
P_{S3}	Live load shear in interval 3, N	228.51×10^3	0.243
W	Unit weight of concrete, N/m^3	22.74×10^3	0.100

$$g_1(\mathbf{X}, \mathbf{v}) = 1 - \frac{M_L}{\omega(\mathbf{X}, \mathbf{v})} - \frac{P_D L_g^2}{8\omega(\mathbf{X}, \mathbf{v})} - \frac{(bh_f + b_w h_w) W L_g^2}{8\omega(\mathbf{X}, \mathbf{v})}, \tag{E1}$$

where $\omega(\mathbf{X}, \mathbf{v}) = A_s f_y [h_f + h_w - e - \eta(\mathbf{X}, \mathbf{v})/2]$ and $\eta(\mathbf{X}, \mathbf{v}) = A_s f_y / (0.85 f'_c b)$. The limit-state functions for the shear failure modes in intervals 1, 2, and 3 are respectively described by

$$g_{j+1}(\mathbf{X}, \mathbf{v}) = 1 - \frac{P_{Sj}}{\kappa_j(\mathbf{X}, \mathbf{v})} - \frac{P_D L_g}{6\kappa_j(\mathbf{X}, \mathbf{v})/j} - \frac{(bh_f + b_w h_w) W L_g}{6\kappa_j(\mathbf{X}, \mathbf{v})/j}, \quad j = 1, 2, 3, \tag{E2}$$

where $\kappa_j(\mathbf{X}, \mathbf{v}) = 8.45 b_w (h_f + h_w - e)(f'_c/\gamma)^{0.5}/0.0254^2 + A_v f_y (h_f + h_w - e)/S_j$. All variables are expressed in SI units. The girder constitutes a series system with four failure modes.

Example 14.5 (cont.)

The objective is to design the girder according to the specifications in American Association of State Highway and Transportation Officials (AASHTO, 1992). However, some AASHTO specifications are of the form $f(\mathbf{v}) \leq 1$ if $h(\mathbf{v}) \leq 0$, otherwise $f(\mathbf{v}) \leq 2$, where $f(\mathbf{v})$ and $h(\mathbf{v})$ are continuous functions. To address this problem, Royset et al. (2006) considered several cases. For example, case 1 has the constraints $f(\mathbf{v}) \leq 1$ and $h(\mathbf{v}) \leq 0$, and case 2 has the constraints $f(\mathbf{v}) \leq 2$ and $-h(\mathbf{v}) \leq 0$. The optimal design for each case is found independently, and the design with the smallest value of the objective function is the solution. Furthermore, some AASHTO constraints include random variables. In such cases, the mean values of the random variables are used in the constraint functions. In total, the problem involves four cases and 28 constraint functions. These are described in Appendix 2 of Royset et al. (2006).

We first design the girder for minimum initial cost, subject to the system failure probability constraint $p_{\text{sys}}(\mathbf{v}) \leq 0.00135$ (corresponding to a constraint $3 \leq \beta_{\text{sys}}(\mathbf{v})$ on the generalized reliability index) and the deterministic constraints according to the AASHTO requirements mentioned above. This is a $\text{P}_{1,\text{sys}}$ problem, as defined earlier. Let $C_s = 50$ and $C_c = 1$ respectively be the unit costs of the steel reinforcement and concrete per cubic meter in an appropriate monetary unit. We employ the cost function defined by Li and Frangopol (1996)

$$c_0(\mathbf{v}) = 0.75 C_s L_g A_s + C_s n_s A_v \left(h_f + h_w - e + 0.5 b_w\right) + C_c L_g \left(b h_f + b_w h_w\right), \qquad \text{(E3)}$$

where $n_s = L_g(1/s_1 + 1/s_2 + 1/s_3)/3$ is the total number of stirrups. The first term represents the cost of the bending reinforcement. The factor 0.75 appears because of the assumption that the total amount of bending reinforcement is placed only within a length $L_g/2$ centered at the middle of the girder, and the remaining part is reinforced with $0.5A_s$. The second and third terms represent the costs of shear reinforcement and concrete, respectively.

The problem is solved using the algorithm based on problem $\text{P}_{1,r}$ as described in Section 14.3.1, where the system probability of failure is computed by Monte Carlo simulation with 1% c.o.v. The design solution is summarized in the second column of Table 14.6 along with the optimized cost value and the resulting failure probability.

Now suppose we wish to minimize the initial cost plus the expected cost of system failure, subject to the same constraints as described above. Let $c_{\text{sys}}(\mathbf{v}) = 500 c_0(\mathbf{v})$ be the cost of system failure. This is problem $\text{P}_{3,\text{sys}}$ and we solve it by using the algorithm described in the preceding section for problem $\bar{\text{P}}_{3,\text{sys},t}$. The design solution is summarized in the third column of Table 14.6 along with the initial, failure, and total expected costs and the probability of system failure computed by Monte Carlo simulation with 1% c.o.v. It is seen that this design requires a wider flange and substantially more steel reinforcement, resulting in an almost 14% increase in initial cost. On the other hand, the probability of failure is almost one order of magnitude smaller than that of the first design. Obviously, this is due to the assumed high cost of failure.

Example 14.5 (cont.)

Next, suppose the girder is subject to corrosion of its longitudinal reinforcement. We adopt the corrosion model used by Frangopol et al. (1997), where the diameter $D_b(t)$ of the longitudinal reinforcing bar at time t is given by

$$\begin{aligned} D_b(t) &= D_{b0} - 2v(t - T_i), \quad & T_i < t \\ &= D_{b0}, & \text{otherwise,} \end{aligned} \tag{E4}$$

where $D_{b0} = 0.025$ m denotes the initial diameter, v is the corrosion rate, and T_i is the corrosion initiation time. The factor 2 in the above expression considers the fact that the reinforcing bar is subject to corrosion from all sides. We assume $T_i = A + Bc_a$, where A is a lognormal random variable with a mean of 5 years and 20% c.o.v. representing the time it takes to initiate corrosion with the existing concrete cover, B is a lognormal random variable with mean 300 years/m and 20% c.o.v., representing the additional time it takes to initiate corrosion per meter additional concrete cover, and c_a denotes the concrete cover in addition to the existing cover. c_a is considered a design variable and is included in $\mathbf{v}$. The corrosion rate v is lognormally distributed with mean 4×10^{-5} m/year and 30% c.o.v. The random variables A, B, and v are assumed to be statistically independent and are included in vector $\mathbf{X}$.

To account for the reinforcing bar corrosion, A_s in the limit-state function (E1) is replaced by $A_s'(t) = A_s[1 - 2v(t - T_i)/D_b]^2$. The problem now is a time-varying reliability problem; however, as the deterioration is monotonic, the girder will have failed if it is in the failed state at the end of its lifetime, which we assume to be 60 years. Thus, the problem is solved with t set to 60 years. Furthermore, the term $C_c L_g b_w c_a$ is added to the cost function in (E3) to account for the cost of the added concrete cover and the constraints $-c_a \le 0$ and $c_a - 0.05 \le 0$ are added to limit the additional concrete cover in the range from 0 to 0.05 m. The problem now is $\text{P}_{3,\text{sys}}$ with the probability constraint $p_{\text{sys}}(\mathbf{v}) \le 0.00135$ applied at the end of the 60-year lifetime.

The fourth column in Table 14.6 summarizes the design solution along with the initial, failure, and total expected costs and the probability of system failure computed by Monte Carlo simulation with 1% c.o.v. Observe that, compared to the previous two cases, this design requires a substantially wider flange and more steel reinforcement. The optimal additional concrete cover is at the maximum value allowed. The resulting increase in the initial cost is almost 50%, indicating the needed investment to guard against the more likely failure due to deterioration. Furthermore, the expected cost of failure has increased by 72%, while the probability of system failure has increased by 31% but is still well below the allowable upper bound.

Finally, suppose that it is decided to maintain the structure in intervals of 20 years, i.e., at 20 and 40 years after its construction. The times of maintenance can be incorporated as design parameters but in this example we fix them. Let $m_i \in [0, 1]$, $i = 1, 2$, be two design parameters characterizing the maintenance effort at 20 years and 40 years, respectively, with $m_i = 0$ denoting no maintenance and $m_i = 1$ denoting full maintenance, i.e., restoration to the initial

Example 14.5 (cont.)

Table 14.6 Optimal design of reinforced-concrete girder (after Royset et al., 2006, with permission from ASCE)

Parameter	Case 1	Case 2	Case 3	Case 4
A_s, m^2	0.00983	0.0116	0.0161	0.0144
b, m	0.418	0.492	0.686	0.612
h_f, m	0.415	0.415	0.415	0.415
b_w, m	0.196	0.196	0.197	0.196
h_w, m	0.785	0.785	0.785	0.785
A_v, m^2	0.000186	0.000227	0.000255	0.000255
s_1, m	0.508	0.502	0.549	0.550
s_2, m	0.224	0.226	0.246	0.247
s_3, m	0.140	0.142	0.154	0.155
c_a, m	N/A	N/A	0.050	0.050
m_1	N/A	N/A	N/A	0.105
m_2	N/A	N/A	N/A	0.243
Initial cost	13.664	15.558	20.434	18.678
Expected failure cost	N/A	1.459	2.514	1.824
Maintenance cost	N/A	N/A	N/A	1.699
Total expected cost	13.664	17.017	22.948	22.201
$p_{\text{sys}}(\mathbf{v})$	0.00131	0.000188	0.000246	0.000195

state of the structure. Furthermore, we consider m_1 as the fraction of the aging of the structure that is restored from its initial state at $t = 0$ to the first maintenance action. Thus, $40 - 20m_1$ years is the effective age of the structure before the second maintenance action at $t = 40$ years. Similarly, m_2 is the fraction of the aging of the structure from initial construction to the second maintenance action that is mitigated by the second maintenance effort, i.e., $20 + (40 - 20m_1)(1 - m_2)$ years is the effective age of the structure at $t = 60$ years. We add m_1 and m_2 to the vector of design parameters $\mathbf{v}$.

We aim to design the structure and the maintenance plan by imposing the constraint $p_{\text{sys}}(\mathbf{v}) \leq 0.00135$ on the system failure probability over the 60-year lifetime. This probability is obtained as the probability of the union of the failure events during the intervals 0–20 years, 20–40 years, and 40–60 years. As mentioned earlier, because of the monotonic effect of deterioration, the event of failure within each interval is identical to the failure event at the end of the interval. Thus, the problem is defined as a series system with $3 \times 4 = 12$ limit-state functions. The design is subject to the same deterministic constraints as in the third case, with the additional constraints $-m_i \leq 0$ and $m_i - 1 \leq 0$, $i = 1, 2$. Furthermore, the cost of maintenance is $c_m(\mathbf{v}) = c_y[20m_1 + (40 - 20m_1)m_2]$, in which $c_y = 0.15$ represents the cost of complete restoration of the girder after one year of corrosion. Because this cost does not depend on the

Example 14.5 (cont.)

failure probability, we incorporate it into the initial cost. This is a $P_{3,sys}$ problem and we solve it with the algorithm described in the preceding section.

The design solution for the case with the maintenance option is summarized in the fifth column of Table 14.6 along with the initial, failure, maintenance, and total expected costs and the probability of system failure computed by Monte Carlo simulation with 1% c.o.v. Observe that the optimal solution calls for modest maintenance at 20 years ($m_1 = 0.105$) and more substantial maintenance at 40 years ($m_2 = 0.243$). Compared to the previous case, the initial cost and the cost of failure are smaller but there is the added cost of maintenance, resulting in a 3% reduction in the total expected cost compared to the case without maintenance. Furthermore, the probability of failure is 21% smaller than in the case without maintenance.

14.4 Sampling-Based RBDO

When the computation of the limit-state function is not expensive, a viable alternative for reliability analysis is Monte Carlo simulation or other sampling methods. We have seen in Chapter 9 that the probability estimate in these methods randomly fluctuates and asymptotically approaches the true failure probability, provided the generated sample of random variables is truly random and aperiodic. Because of these fluctuations, it is not possible to compute the gradient of the failure probability by finite differences. However, as described in Section 9.12, under certain conditions it is possible to directly compute the gradient of the failure probability by sampling. Sampling estimates of the failure probability and its gradient can be used in gradient-based algorithms to solve RBDO problems. In this section, we describe two such methods based on the work of Royset and Polak (2004a, 2004b, 2007).

Royset and Polak (2004a, 2004b) developed an RBDO formulation for problem P_3 employing Monte Carlo simulation with importance sampling. Their method is based on the following assumptions:

1) The limit-state functions $G_i(\mathbf{U}, \mathbf{v})$, $i = 1, \dots, N$, are continuous and continuously differentiable with respect to $\mathbf{U}$ and $\mathbf{v}$.
2) The value of each limit-state function in every neighborhood of the limit-state surface is non-zero. This assures that $G_i(\mathbf{u}, \mathbf{v}) = 0$ is a line for $n = 2$, a surface for $n = 3$, and a hypersurface for $n > 3$, where n is the dimension of $\mathbf{U}$.
3) Given $G_i(\mathbf{u}, \mathbf{v}) = 0$, it is possible to uniquely solve for one of the elements of $\mathbf{u}$, say u_1, in terms of $\mathbf{v}$ and the remaining elements $\bar{\mathbf{u}} = [u_2, \dots, u_n]^{\mathrm{T}}$. We let $u_1 = h_i(\bar{\mathbf{u}}, \mathbf{v})$ denote this solution for the ith limit-state function.
4) For all $\bar{\mathbf{u}}$ and $\mathbf{v}$, $\partial G_i(\mathbf{u}, \mathbf{v})/\partial u_1 \neq 0$.

Assume $G_i(\mathbf{U},\mathbf{v}) \le 0$ for $U_1 \le h_i(\bar{\mathbf{u}},\mathbf{v})$. If this is not the case, then U_1 can be replaced by $-U_1$ to satisfy this condition with no effect on the reliability problem. The probability of failure of the ith component is now written as

$$p_i(\mathbf{v}) = \int_{-\infty}^{+\infty} \cdots \int_{-\infty}^{+\infty} \Phi(u_1)\phi(u_2)\cdots\phi(u_n)\mathrm{d}u_2\cdots\mathrm{d}u_n, \tag{14.19}$$

where $u_1 = h_i(\bar{\mathbf{u}},\mathbf{v})$ and $\phi(\cdot)$ is the standard normal probability density function (PDF). Furthermore, the gradient of the above probability with respect to $\mathbf{v}$ is

$$\nabla_{\mathbf{v}} p_k(\mathbf{v}) = -\int_{-\infty}^{+\infty} \cdots \int_{-\infty}^{+\infty} \phi(u_1)\frac{\nabla_{\mathbf{v}} G_i(u_1,\bar{\mathbf{u}},\mathbf{v})}{\partial G_i(u_1,\bar{\mathbf{u}},\mathbf{v})/\partial u_1}\phi(u_2)\cdots\phi(u_n)\mathrm{d}u_2\cdots\mathrm{d}u_n, \tag{14.20}$$

where the quotient with the negative sign in the front of the integrals represents the gradient of $h_i(\bar{\mathbf{u}},\mathbf{v})$ obtained by differentiating $G_i[h_i(\bar{\mathbf{u}},\mathbf{v}),\bar{\mathbf{u}},\mathbf{v}] = 0$ with respect to $\mathbf{v}$.

Monte Carlo simulation can be used to estimate the above quantities. We generate a sample $\bar{\mathbf{u}}_j$, $j = 1, \ldots, N_s$, of the standard normal random variables $\bar{\mathbf{U}}$. The corresponding estimators then are

$$\widehat{p}_i(\mathbf{v}) = \frac{1}{N_s}\sum_{j=1}^{N_s}\Phi(u_{1,j}) \tag{14.21}$$

and

$$\widehat{\nabla p}_i(\mathbf{v}) = -\frac{1}{N_s}\sum_{j=1}^{N_s}\phi(u_{1,j})\frac{\nabla_{\mathbf{v}} G_i(u_{1,j},\bar{\mathbf{u}}_j,\mathbf{v})}{\partial G_i(u_{1,j},\bar{\mathbf{u}}_j,\mathbf{v})/\partial u_1}, \tag{14.22}$$

where $u_{1,j} = h_i(\bar{\mathbf{u}}_j,\mathbf{v})$. Alternatively, importance sampling can be used to speed up the convergence of the estimators. One option for the importance sampling density is a non-standard normal distribution in the $\bar{\mathbf{U}}$ space centered at a point $\mathbf{M} = [\mu_2, \ldots, \mu_n]^{\mathrm{T}}$ and a diagonal covariance matrix with equal standard deviations σ. After a transformation of variables, the above estimators then take the form

$$\widehat{p}_i(\mathbf{v}) = \frac{1}{N_s}\sum_{j=1}^{N_s}\Phi(u_{1,j})\frac{\phi(\sigma u_{2,j}+\mu_2)\cdots\phi(\sigma u_{n,j}+\mu_n)\sigma^{n-1}}{\phi(u_{2,j})\cdots\phi(u_{n,j})} \tag{14.23}$$

and

$$\widehat{\nabla p}_i(\mathbf{v}) = -\frac{1}{N_s}\sum_{j=1}^{N_s}\phi(u_{1,j})\frac{\nabla_{\mathbf{v}} G_i(u_{1,j},\sigma\bar{\mathbf{u}}_j+\mathbf{M},\mathbf{v})}{\partial G_i(u_{1,j},\sigma\bar{\mathbf{u}}_j+\mathbf{M},\mathbf{v})/\partial u_1}\frac{\phi(\sigma u_{2,j}+\mu_2)\cdots\phi(\sigma u_{n,j}+\mu_n)\sigma^{n-1}}{\phi(u_{2,j})\cdots\phi(u_{n,j})}, \tag{14.24}$$

where $u_{1,j} = h_i(\sigma\bar{\mathbf{u}}_j + \mathbf{M},\mathbf{v})$ and $u_{i,j}$, $i = 2, \ldots, n$, is the ith element of $\bar{\mathbf{u}}_j$. Note that with this formulation the sampling is still carried out in terms of the standard normal random variables $\bar{\mathbf{U}}$, but the transformation accounts for the translated sampling density. The parameters μ_i may be selected to coincide with above-mean values of demand variables and below-mean values of capacity variables. In most cases $\sigma = 1$ is a good choice, but a larger value may be selected to assure coverage of important points in the $\bar{\mathbf{U}}$ space.

Royset and Polak (2004a, 2004b) showed that the probability estimator in (14.21) is unbiased. They also derived a measure of the error in the probability estimate. Furthermore, they developed an algorithm for solving problem P_3, with the above sampling estimates of the probability of failure and its gradient, that employs a sequence of approximating solutions with increasing accuracy. Hence, instead of setting a high precision for the optimization solution in advance, they find a solution with a moderate tolerance and continue sampling while gradually tightening the tolerance. The solution for each step in the sequence is used as a warm starting point for the next sequence until the desired tolerance is achieved. They also show that the optimization problem can be solved by standard quadratic programming algorithms. The following example demonstrates this approach.

Example 14.6 – Design of Column for Maximum Reliability

Consider the rectangular column problem in Example 14.2 aimed at designing for minimum cross-sectional area, subject to a minimum reliability index of 3, which corresponds to the constraint $p(\mathbf{v}) \leq 0.00135$ on the failure probability. Royset and Polak (2004a) solve the problem by using the importance sampling method described above. Because the random variables are statistically independent, there is a one-to-one correspondence between the variables in the original and standard normal spaces. For U_1, they select the bending moment M_1 and solve for it in terms of the other random variables in $g(M_1, M_2, P, Y) = 0$. For the sampling density, they use the mean vector $\mathbf{M} = [2, 2, -1]^T$ and standard deviation $\sigma = 1.01$. Note that the center of the sampling density corresponds to two logarithmic standard deviations above the median values of M_2 and P and one logarithmic standard deviation below the median of the capacity variable Y.

Table 14.7 summarizes the results of the analysis. Listed are the cumulative number of samples, the cumulative number of iterations after each sampling sequence of the algorithm, and the resulting designs and estimated failure probabilities. The final probability estimate is computed with a 0.5% c.o.v. Compared to the solution obtained in Example 14.2, it is

Table 14.7 Results for design of rectangular column (after Royset and Polak, 2004a)

N_i	Cumulative number of iterations	$(\widehat{b}, \widehat{h})$	$c_0(\mathbf{v})$	p_{N_i}
1000	1	(1.000, 1.000)	0.252	0.00000
1000	68	(0.357, 0.551)	0.197	0.00135
5000	95	(0.325, 0.603)	0.196	0.00130
2.5×10^4	99	(0.319, 0.613)	0.196	0.00130
1.25×10^5	103	(0.315, 0.619)	0.195	0.00135
6.25×10^5	106	(0.313, 0.624)	0.195	0.00134
3.125×10^6	108	(0.313, 0.624)	0.195	0.00135

Example 14.6 (cont.)

interesting to note that, while the values of the optimal cost function and the failure probability estimate are identical, the optimal cross-sectional dimensions are somewhat different. It is evident the two approaches have converged to two different but equally viable solutions.

Royset and Polak (2007) formulated a quadratic program to solve problem $P_{1,sys}$ by use of directional sampling. Their formulation is based on the following assumptions:

1) The limit-state functions $G_i(\mathbf{U}, \mathbf{v})$, $i = 1, \ldots, N$, are continuous and continuously differentiable with respect to $\mathbf{U}$ and $\mathbf{v}$.
2) For all relevant designs, the complement of the system failure domain in the standard normal space is "star-shaped." This means that the origin lies within the safe domain and that any radial line emanating from the origin in the standard normal space has at most one intersection with the limit-state surface of the system.
3) For all relevant designs, there is a finite distance from the origin to the limit-state surface in all radial directions. That is, the complement of the failure domain is bounded. If this is not the case, an artificial limit-state function $G_{N+1}(\mathbf{u}, \mathbf{v}) = \rho - \|\mathbf{u}\|$ with a sufficiently large ρ is added to satisfy this condition. Note that this essentially enlarges the failure domain, but if ρ is sufficiently large the effect on the failure probability is negligible.
4) Only one limit-state function is active at each point on the boundary of the system failure domain. That is, limit-state surfaces for different active components do not coincide.

Like the approach using Monte Carlo importance sampling, the tolerance for convergence of the optimization algorithm is gradually tightened as the number of samples is increased. The following example demonstrates one such application.

Example 14.7 – Design of Reinforced-Concrete Girder for Minimum Cost

Royset and Polak (2007) solved Case 1 of Example 14.5 by use of directional sampling. Recall that the problem involves a series system with four limit-state functions, eight random variables, nine design parameters and 28 constraint functions, including the probability constrain $p_{sys}(\mathbf{v}) \leq 0.00135$. To make the failure domain bounded, $\rho = 8$ is used to define an artificial limit-state function of the form described in item (3) above. Furthermore, a sample of 25,000 is used, with the convergence criteria tightened after 200, 1,600, 5,400, and 12,800 simulations. For each directional sample, the distance to the limit-state surface is solved by using MATLAB's root-finding command. Furthermore, MATLAB's `Quadprog` command for quadratic programming is used to solve the optimization problem.

The solution of the problem yields a design with an optimal cost of 13.288 and an estimated failure probability of $p_{sys}(\mathbf{v}) = 0.00135$ with 2% c.o.v. In contrast, the solution in Example 14.5 with the decoupling approach yielded a design with the cost 13.664 and an estimated failure

Example 14.7 (cont.)

probability $p_{sys}(\mathbf{v}) = 0.00131$ with 1% c.o.v. Thus, the decoupling solution yields a slightly more reliable but 2.8% more costly design. We attribute this to the fact that the decoupling approach is exact for affine limit-state functions whereas the application with nonaffine limit-state functions, which is the case here, relies on a heuristic. As a result, the solution by the decoupling approach is only approximately optimal. On the other hand, the decoupling approach is computationally far more efficient, particularly if FORM or SORM are used.

14.5 RBDO Employing Surrogate Models

As described in Section 12.7, when the limit-state function is expensive to compute, one possible approach to reliability analysis is to construct a surrogate mathematical model of the limit-state function (also called response surface or metamodel) and use it in conjunction with sampling methods to compute the failure probability. Because the sampling involves the surrogate model that is easy to compute, the main computational effort lies in constructing a surrogate model that accurately represents the limit-state surface. Section 12.7 described several methods for constructing surrogate models for finite-element reliability analysis by use of kriging and polynomial chaos expansion. In recent years, this approach has also been explored for RBDO analysis.

The surrogate model for RBDO must consider the augmented space of the random variables and design parameters, $(\mathbf{X}, \mathbf{v})$. Indeed, Bichon et al. (2013) propose three approaches for developing and utilizing the surrogate model in conjunction with their kriging method described in Section 12.7 (Bichon et al. 2008). The first approach employs nested reliability and optimization analyses with separate surrogate models for the objective function in the space of $\mathbf{v}$ and for the limit-state function in the space of $\mathbf{X}$ for each given design $\mathbf{v}$. This approach requires repeated constructions of the surrogate model of the limit-state function at each iteration of the design parameters. The second and third methods employ a single surrogate model in the augmented space $(\mathbf{X}, \mathbf{v})$, the difference between the two methods being how frequently the reliability model is updated in the iterative steps of the optimization algorithm. They have applied and compared the three methods for several example problems. Although all three methods appear to be more efficient (in terms of the number of computations of the true limit-state function) than nested RBDO methods employing FORM, the authors report on difficulties having to do with scaling to larger numbers of random variables and/or design parameters, accuracy of the reliability measures, and convergence issues.

Dubourg et al. (2011) considered the RBDO problem P_1 in (14.1) with the design parameters $\mathbf{v}$ being parameters in the distribution of random variables $\mathbf{X}$. That is, the

PDF of $\mathbf{X}$ is defined as $f_{\mathbf{X}}(\mathbf{x}|\mathbf{v})$, where $\mathbf{v}$ are distribution parameters such as means and variances. With this definition, the limit-states in the original space are functions only of $\mathbf{X}$, i.e., $g_i = g_i(\mathbf{X})$, $i = 1, \ldots, N$; however, the limit-state functions in the standard normal space, $G_i(\mathbf{U}, \mathbf{v}) = g_i[\mathbf{X}(\mathbf{U})]$, are functions of $\mathbf{U}$ as well as $\mathbf{v}$. This is because the transformation $\mathbf{U} = \mathbf{U}(\mathbf{X})$ to the standard normal space and its inverse, $\mathbf{X} = \mathbf{X}(\mathbf{U})$, depend on the distribution of $\mathbf{X}$ and, therefore, on the design parameters $\mathbf{v}$. It should also be clear that the failure probabilities, $p_i(\mathbf{v}) = \Pr[g_i(\mathbf{X}) \leq 0] = \Pr[G_i(\mathbf{U}, \mathbf{v}) \leq 0]$, and the corresponding generalized reliability indices, $\beta_i(\mathbf{v}) = \Phi^{-1}[1 - p_i(\mathbf{v})]$, are functions of $\mathbf{v}$. Dubourg et al. (2011) suggest that any design parameter that is not a parameter of the distribution of $\mathbf{X}$ can be included by introducing a fictitious random variable with a small variance and a mean value representing the design parameter.

Because the limit-state functions in the original space are functions only of $\mathbf{X}$, it is possible to construct surrogate models for RBDO analysis in the space of $\mathbf{X}$ alone. In fact, each limit-state surface $g_i(\mathbf{x}) = 0$ remains invariant of the design parameters $\mathbf{v}$. However, the distribution of $\mathbf{X}$ depends on $\mathbf{v}$, so the region of importance of each limit-state surface, e.g., the most likely failure point, varies with $\mathbf{v}$.

Dubourg et al. (2011) constructed a surrogate model of the limit-state surface in the $\mathbf{X}$ space using the kriging Gaussian process interpolation method described in Section 12.7. They started with an initial sample of experimental design points and then added new points by sampling within a confidence interval of the predicted mean limit-state surface. The RBDO analysis proceeds while the surrogate model is improved, until appropriate convergence criteria are satisfied. For each design, failure probabilities and their gradients are computed by an importance sampling method similar to that described in Section 9.12 but applied to the surrogate model. Thus, the main computational effort in this approach is the construction of the surrogate model, which requires computation of the true limit-state function values at the experimental points. A subtle but important point in this regard is that, due to the dependence of the distribution of $\mathbf{X}$ on the design parameters, the accuracy of the surrogate model of the limit-state surface should be assured for all realistic design alternatives. Thus, the surrogate model must be accurate over a broader domain of the space of $\mathbf{X}$ than would be necessary for reliability analysis alone. To achieve this objective, Dubourg et al. (2011) assign distributions to the design parameters and work in the space of the "updated" random variables $\widetilde{\mathbf{X}}$ that incorporate the parameter uncertainties, i.e., similar to the predictive analysis described in Section 10.3 (see (10.9)). This effectively broadens the distribution of $\mathbf{X}$ so that the experimental design points are distributed over the entire domain of viable designs.

14.6 Buffered Failure Probability Approach

Rockafellar and Royset (2010, 2015) used the concept of value at risk from the field of financial engineering to formulate an alternative measure of structural safety, named *buffered failure probability*, that offers certain advantages over the conventional failure probability for RBDO analysis. This definition is described below together with the arguments made by the authors for its preference over the failure probability. In order to remain consistent with our earlier definition of the limit-state function, the formulation presented here is somewhat different from that in Rockafellar and Royset (2010).

We describe the case with a single limit-state function $G(\mathbf{U}, \mathbf{v})$. The generalization for a system problem is a relatively simple extension of the component case (see Rockafellar and Royset, 2010). Considering $G(\mathbf{U}, \mathbf{v})$ as a random variable and assuming its CDF is continuous and strictly increasing for all $\mathbf{v}$, we define $q_\alpha(\mathbf{v})$ as its α-quantile so that $\Pr[G(\mathbf{U}, \mathbf{v}) \leq q_\alpha(\mathbf{v})] = \alpha$. With this definition, the probability of failure is given as $p(\mathbf{v}) = \alpha_0$, where α_0 corresponds to a zero value of the quantile, i.e., $q_{\alpha_0}(\mathbf{v}) = 0$. Next, the α-superquantile, denoted $\bar{q}_\alpha(\mathbf{v})$, is defined as the conditional expectation of the limit-state function, given $G(\mathbf{U}, \mathbf{v}) \leq q_\alpha(\mathbf{v})$, i.e.,

$$\bar{q}_\alpha(\mathbf{v}) = \mathrm{E}[G(\mathbf{U}, \mathbf{v}) \,|G(\mathbf{U}, \mathbf{v}) \leq q_\alpha(\mathbf{v})]. \tag{14.25}$$

The buffered failure probability, denoted $\bar{p}(\mathbf{v})$, is now defined as the value of α for which $\bar{q}_\alpha(\mathbf{v}) = 0$. That is,

$$\bar{p}(\mathbf{v}) = \Pr[G(\mathbf{U}, \mathbf{v}) \leq q_\alpha(\mathbf{v})], \tag{14.26}$$

$$\bar{q}_\alpha(\mathbf{v}) = 0. \tag{14.27}$$

Note that this implies $\bar{q}_{\bar{p}(\mathbf{v})}(\mathbf{v}) = 0$. Figure 14.5 illustrates the idea of the buffered failure probability, showing the PDF of the random variable $G = G(\mathbf{U}, \mathbf{v})$ for a given design together with the α-quantile, the α-superquantile that has a zero value, and the areas underneath the PDF corresponding to the failure probability and the buffered failure probability. It should be clear that for any design $\mathbf{v}$

$$p(\mathbf{v}) \leq \bar{p}(\mathbf{v}). \tag{14.28}$$

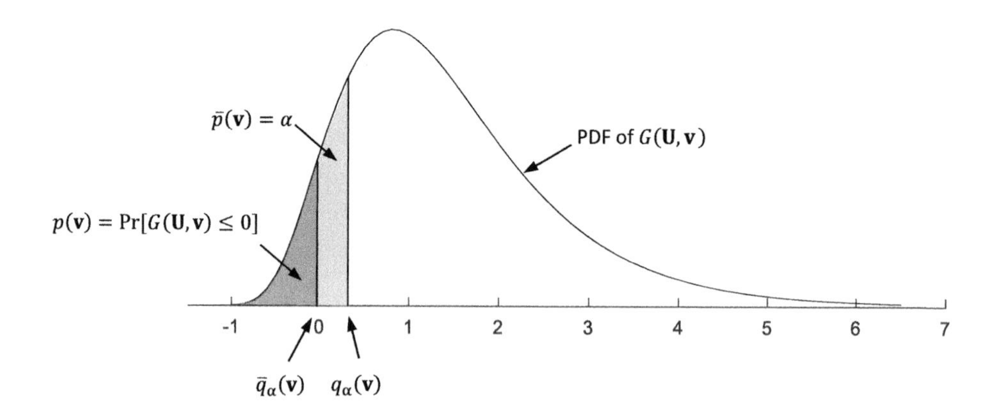

Figure 14.5 Definition of buffered failure probability

Thus, the buffered failure probability provides a conservative measure of the failure probability.

Rockafellar and Royset (2010) make the following arguments in favor of the buffered failure probability:

- Whereas the failure probability only considers the sign of the limit-state function, the buffered failure probability considers the shape of the lower tail of the distribution of the limit-state function. A longer tail in the negative range will yield a larger buffered failure probability than a shorter tail of equal area. Rockafellar and Royset argue that, through this property, the buffered failure probability accounts for the "degree" or extent of failure. They argue that an optimal design should consider the likelihoods for the extents of failure that are characterized by the shape of the lower tail of the PDF of the limit-state function. The buffered failure probability does this in an indirect way.
- There are difficulties in computing the failure probability. Simulation methods can be costly for small failure probabilities, while approximations by FORM and SORM have unknown accuracies and may require nested optimization analysis. In contrast, the buffered failure probability is relatively easy to compute by sampling methods.
- Although under rather general conditions the failure probability is continuously differentiable, computing its gradient by simulation is costly. Furthermore, approximations such as FORM and SORM may not have continuous derivatives. In contrast, the buffered failure probability leads to a RBDO problem with continuously differentiable reliability constraints as long as the limit-state function is continuously differentiable.
- Convexity of the objective and constraint functions in an optimization problem is desirable because one then has assurance of convergence to a global solution. In general, it is not possible to determine whether the failure probability $p(\mathbf{v})$ is convex, even if the limit-state function $G(\mathbf{u}, \mathbf{v})$ is convex. Thus, when used in a constraint function, the failure probability does not "preserve" the convexity of the RBDO problem – a convex limit-state function may result in a nonconvex failure probability. The same situation holds when the generalized reliability index or its FORM or SORM approximations are used to define the constraint functions. In contrast, the buffered failure probability preserves the convexity of the RBDO problem, if the limit-state function as well as the objective and cost functions are convex.

On the other hand, Rockafellar and Royset admit that the buffered failure probability cannot be used in problems involving the expected cost of failure because there is no basis for computing the expected cost of failure in terms of that probability value. Furthermore, a shortcoming that is not mentioned by the authors is that the buffered failure probability lacks invariance relative to the formulation of the limit-state function. That is, equivalent but different formulations of the limit-state function (e.g., $g = X_1 - X_2$ versus $g = 1 - X_2/X_1$) may lead to different shapes of the lower tail of its PDF and, hence, different estimates of the buffered failure probability. This could lead to different optimal designs for equivalent but different formulations of the limit-state function. This problem of lack of invariance relative to the formulation of the limit-state function is similar to that observed for the mean-centered, first-order, second-moment (MCFOSM) reliability measure, as demonstrated in Example 5.1.

Without providing details, below we present the reformulation of problem $\mathbf{P}_1$ for a single component in terms of the buffered failure probability. Rockafellar and Royset (2010) have shown that the solution of $\mathbf{P}_1$ is identical to the solution of the problem

$$\begin{aligned}\mathbf{BP}_{1,N_s}\!: \min_{\mathbf{v},\bar{\mathbf{z}}}\Big\{f_0(\mathbf{v})\ \Big|\ & z_0+\frac{1}{N_s\alpha_0}\sum\nolimits_{i=1}^{N_s} z_i\le 0;\ -G(\mathbf{u}_i,\mathbf{v})-z_0\le \\ & z_i, z_i\ge 0, i=1,\ldots,N_s; f_j(\mathbf{v})\le 0, j=1\ldots,J\Big\}.\end{aligned} \tag{14.29}$$

In the above, α_0 is the upper bound on the buffered failure probability, $\bar{\mathbf{z}}=(z_0,\ldots,z_{N_s})$ is an auxiliary set of design parameters, and $\mathbf{u}_i$, $i=1,\ldots,N_s$, is a set of N_s sample values of $\mathbf{U}$ obtained by Monte Carlo sampling. Observe that the above formulation involves only the limit-state function in the reliability constraints. Importantly, if the objective $f_0(\mathbf{v})$ and structural constraints $f_j(\mathbf{v})\le 0, j=1,\ldots,J$, are convex functions of $\mathbf{v}$ and the limit-state function $G(\mathbf{u},\mathbf{v})$ is a convex function of $\mathbf{v}$ for all $\mathbf{u}$, then the above is a convex optimization problem and its solution to the global minimum is guaranteed. Also note that, because the buffered failure probability is greater than the failure probability, any solution of the above problem is a feasible solution of $\mathbf{P}_1$ when the latter is defined in terms of the failure probability. Furthermore, it is possible to adjust the bound on the buffered failure probability so that the reliability constraint is satisfied in terms of the true failure probability. Of course, in that case the design solution is only approximately optimal. An example in Rockafellar and Royset (2010) demonstrates such an application.

14.7 Concluding Remarks

RBDO is a field that is rapidly evolving. In this chapter, we presented three classes of RBDO problems involving reliability measures in the constraints, the objection function, or both. Of the vast collection of works in this field, we focused on those that offer rigorous treatment of both reliability analysis and the applicable optimization algorithms. This is because it is important to know the convergence properties of the optimization algorithm if it is to be used for real-world problems in practice. Unfortunately, the vast majority of publications in this field focus only on the reliability aspect and disregard the conditions under which the optimization problem is solvable. Demonstration of convergence in a few academic example problems is not sufficient proof that an RBDO algorithm will converge for other, more elaborate, problems that are encountered in practice. For this reason, we have focused on methods that reformulate the RBDO problem so that it can be solved by well-established algorithms, or on methods that provide rigorous treatment of the convergence properties of the RBDO algorithm. In recent years, scholars with deep expertise in optimization theory have become interested in the RBDO problem. This bodes well for further development of the field.

15 Bayesian Network for Reliability Assessment and Updating

15.1 Introduction

A *Bayesian network* (BN), also known as a *Bayes net* or *belief network*, is a graphical representation of a set of random variables and an efficient means for codifying their joint distribution. It provides a visual description of the random variables and the interdependencies among them. A BN consists of nodes representing random variables and directed links between pairs of nodes representing dependence between the corresponding random variables. A BN may be used for inference purposes, e.g., determining the distribution of a subset of the random variables upon receiving information (evidence) on other random variables, or for learning purposes, e.g., estimating model parameters or determining the topology of a graph based on observed data on a set of random variables. When augmented with decision and utility nodes, a BN is called an *influence diagram* and is used for decision-making in accordance with the von Neumann–Morgenstern utility theory (von Neumann and Morgenstern, 1947).

BNs are used in a number of fields, including statistics, artificial intelligence, health monitoring, early warning systems, seismology, systems modeling and analysis, and risk assessment. Their use in structural and system reliability analysis is currently limited, but there is a vast potential for further development.

This chapter presents a brief introduction to BNs. The focus is on aspects of the graphical approach that are relevant to structural and system reliability methods. We will refrain from extending the presentation to influence diagrams because that falls beyond the scope of this book, even though that is a natural extension of BNs. For further reading on BNs and influence diagrams, the interested reader is referred to well-known texts, including Pearl (1988), Cowell et al. (1999), and Jensen and Nielsen (2007). In the field of civil engineering, the earliest applications were reported by Friis-Hansen (2001) and Mahadevan et al. (2001).

15.2 Elements of a Bayesian Network

A BN consists of nodes representing random variables (or vectors of random variables) and directed links describing dependencies among the random variables. A BN is a directed acyclic graph in the sense that the links are directed but cannot form a cycle in the graph. Normally, the dependence structure follows causal reasoning. That is, if random variable X_1 causes random variable X_2, then the link is directed from X_1 to X_2. However, this is not a strict requirement. Sometimes random variables are dependent without one causing the other. Even when X_1 causes X_2 it is permissible to have a link from X_2 to X_1. However, in that case, additional links may be necessary to correctly describe the dependence structure among the random variables. Additional links complicate the network and increase the computational effort. The causal BN also has the advantage of being consistent under intervention. That is, when adding a parent node that affects a child node in a causal manner, the rest of the BN does not need to be modified if the BN is causal. Hence, whenever possible, it is best to follow the causal relationships.

A BN can represent discrete, continuous, or mixed random variables. When continuous random variables are jointly normal and the functional relations between them are liner, exact inference can be carried out using the special properties of the multinormal distribution (see Section 3.2). However, more generally, inference in BNs with continuous random variables is carried out approximately using sampling methods. For discrete random variables with finite numbers of states, algorithms for exact inference are available. This chapter will focus on BNs with discrete random variables or continuous random variables that are discretized. However, Section 15.11 introduces BNs with mixed random variables, where the continuous random variables are handled by structural reliability methods.

Figure 15.1 illustrates a simple BN with five discrete random variables X_i, $i = 1, \ldots, 5$, shown as circular nodes. Considering the directed links and assuming that this is a causal network, random variable X_1 causes X_4, which in turn causes X_5. Furthermore, random variables X_1 and X_2 jointly cause random variable X_3. It is common to use a familial terminology in describing the nodes of a BN. Thus, we say X_1 is a *parent* of X_4 and X_4 is a *child* of X_1. Furthermore, X_5 is a *descendant* of X_1. Similarly, X_1 and X_2 are the parents of X_3, while the latter is their child. Parents of a common child are called *spouses*. Nodes

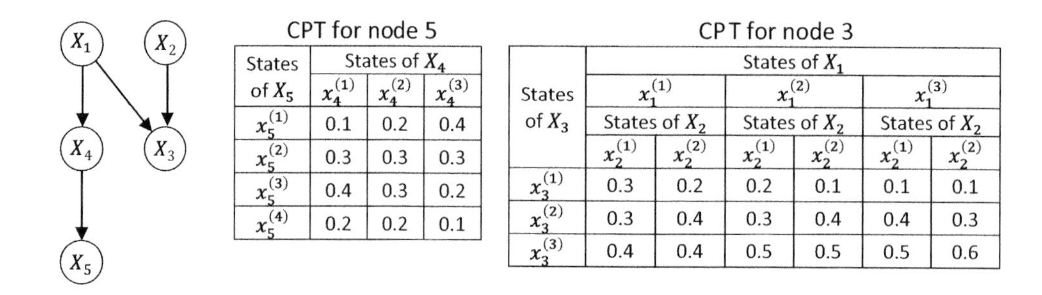

CPT for node 5

States of X_5	States of X_4		
	$x_4^{(1)}$	$x_4^{(2)}$	$x_4^{(3)}$
$x_5^{(1)}$	0.1	0.2	0.4
$x_5^{(2)}$	0.3	0.3	0.3
$x_5^{(3)}$	0.4	0.3	0.2
$x_5^{(4)}$	0.2	0.2	0.1

CPT for node 3

States of X_3	States of X_1					
	$x_1^{(1)}$		$x_1^{(2)}$		$x_1^{(3)}$	
	States of X_2		States of X_2		States of X_2	
	$x_2^{(1)}$	$x_2^{(2)}$	$x_2^{(1)}$	$x_2^{(2)}$	$x_2^{(1)}$	$x_2^{(2)}$
$x_3^{(1)}$	0.3	0.2	0.2	0.1	0.1	0.1
$x_3^{(2)}$	0.3	0.4	0.3	0.4	0.4	0.3
$x_3^{(3)}$	0.4	0.4	0.5	0.5	0.5	0.6

Figure 15.1 Simple BN with CPTs for nodes 3 and 5

without parents are *root* nodes. In the example BN, X_1 and X_2 are spouses as well as root nodes.

To each node of the BN we attach a *conditional probability table* (CPT) listing the conditional probability mass function (PMF) of the random variable given the states of its parents. Let $x_i^{(j)}$, $i = 1, \ldots, n, j = 1, \ldots, m_i$, denote the jth state of random variable X_i. The CPT lists the conditional probability values $\Pr\left[X_i = x_i^{(j)}|\text{pa}(X_i)\right]$, where $\text{pa}(X_i)$ denotes the states of the parents of X_i. For example, the CPT attached to node X_5 lists the conditional probability values $p\left(x_5^{(j)}|x_4^{(k)}\right) = \Pr\left[X_5 = x_5^{(j)}|X_4 = x_4^{(k)}\right]$ for every j, $j = 1, \ldots, m_5$, and every k, $k = 1, \ldots, m_4$. Likewise, the CPT attached to node X_3 lists the conditional probabilities $p\left(x_3^{(j)}|x_1^{(k)}, x_2^{(l)}\right) = \Pr\left[X_3 = x_3^{(j)}|X_1 = x_1^{(k)}, X_2 = x_2^{(l)}\right]$ for every j, k, and l. Example tables are shown in Figure 15.1 for $m_1 = 3$, $m_2 = 2$, $m_3 = 3, m_4 = 3$, and $m_5 = 4$. Observe that the probability values in each column must add up to 1. It should be clear that, if every node in the BN has m states, the number of entries in the CPT for a node with M parents is m^{M+1}. This can be a large number when a node has many parents. Hence, in developing BNs, it is desirable to have as few parents to each node as possible. Methods to achieve this goal are described later in this chapter. For root nodes that have no parents, the CPT is constructed using the marginal PMF.

We saw in Section 2.5 that the joint PMF of a set of n discrete random variables can be written in a factorized form in terms of n conditional PMFs

$$p(x_1, x_2, \ldots, x_n) = p(x_1)p(x_2|x_1)p(x_3|x_1, x_2)\ldots p(x_n|x_1, \ldots, x_{n-1}), \tag{15.1}$$

where, for the sake of simplicity of notation, we have removed the superscripts indicating the specific state of each random variable. We will use the superscripts only when it is necessary for clarity.

By definition, a node in a BN is conditionally independent of its non-descendants, given its parents. As an example, for the BN in Figure 15.1, given X_1 and X_2, X_3 is independent of X_4 and X_5. It follows that if the random variables in a BN are topologically numbered, i.e., parents before children, then the joint PMF of the random variables can be written in the form

$$p(x_1, x_2, \ldots, x_n) = \prod\nolimits_{i=1}^{n} \Pr[X_i = x_i|\text{pa}(X_i)]. \tag{15.2}$$

The difference between (15.1) and (15.2) is that the conditioning in the latter is only on the parents of the node. This is allowed because, given the parents, the node is independent of its non-descendants, and also because the dependence with the descendants is accounted for in the conditional PMFs of the children nodes. As an example, for the BN in Figure 15.1, we have

$$p(x_1, x_2, \ldots, x_5) = p(x_1)p(x_2)p(x_3|x_1, x_2)p(x_4|x_1)p(x_5|x_4). \tag{15.3}$$

Observe that the dependence between X_1 and its descendent X_5 is accounted for through the last two terms of the above expression. Depending on the topology of the BN, the form in

(15.2) can be significantly more efficient (in terms of the number of states to consider) than that in (15.1). This is further elaborated below.

Suppose that in our example BN in Figure 15.1, each node has m states. In that case, to define the joint PMF, we need to specify $m^5 - 1$ probability values. (The -1 is there because the probability values of a PMF have to add up to 1 and, hence, the very last value is determined by subtracting the sum of previous values from 1.) On the other hand, to specify the conditional PMFs on the right-hand side of (15.3), we need to specify $(m-1)+(m-1)+(m^3-m^2)+(m^2-m)+(m^2-m) = m^3+m^2-2$ probability values. For example, for $m = 10$, the joint PMF requires specification of 99,999 values whereas the BN requires specification of 1,098 values.

The above argument naturally raises the question of whether the conditional distributions are available in practice. It turns out that, in most cases in structural and system reliability analysis, the available information indeed is in terms of the conditional distributions rather than the joint distribution. For example, structural capacities are often provided in terms of fragility functions, which are conditional probabilities of being in various states of damage given the magnitude of the demand. Likewise, the outcome of a test to determine the state of a structural component is usually expressed in terms of the test likelihood, which is the conditional probability of the test outcome given the true state of the component. Also, when the dependence between random variables is causal, it is natural to express the cause-and-effect relationship in terms of conditional probabilities. Finally, when statistical uncertainty is present and we consider distribution parameters as random variables, the distribution of the random variable can be considered conditional on the parameter values.

An important property of the BN is that it can be constructed according to an object-oriented formulation (Koller and Pfeffer, 1997). A group of nodes are defined as a BN object with input and output variables (attributes of the object) and internal variables. Given the input and output variables, the internal nodes are independent of the rest of the graph. BN objects are embedded in a higher-level BN, with which they communicate through the interface input and output variables. This facilitates a clearer graphical representation of the BN and more efficient inference. We will show BN objects as rounded rectangular nodes in examples later in this chapter.

Example 15.1 – Causal Versus Non-Causal BNs

Consider the state of a structure under an external demand. Let C denote the uncertain structural capacity, D denote the random demand placed on the structure, and S denote the state of the structure. Let the latter be a discrete random variable with two outcomes, $S = \{F, \bar{F}\}$, where F denotes the failure of the structure and $\bar{F}$ denotes its complement, the survival of the structure. We assume the capacity and demand are statistically independent. It is clear that the state of the structure is caused by the combination of the capacity and demand values. Hence, a causal formulation of the BN for the three random variables is as shown in Figure 15.2(a). Note that the directed links are from C and D to S, as dictated by the causal

Example 15.1 (cont.)

relationships. Assuming C and D are discrete or discretized, the joint PMF of the three random variables for this BN configuration is $p(c, d, s) = p(s|c, d)p(c)p(d)$.

Now suppose we reverse the directions of the links and make them from S to C and D, as shown in Figure 15.2(b). This is admissible as long as we draw a link between C and D. The need for this additional link is evident from the following argument: If the structure is in the failed state, i.e., $S = F$, then we must have $\{C \leq D\}$, and if the structural is in the survival state, i.e., $S = \bar{F}$, then we must have $\{C > D\}$. Thus, given S, C and D are dependent. This necessitates the link between the nodes C and D. The joint PMF of the three random variables for this BN configuration is $p(c, d, s) = p(d|c, s)p(c|s)p(s)$.

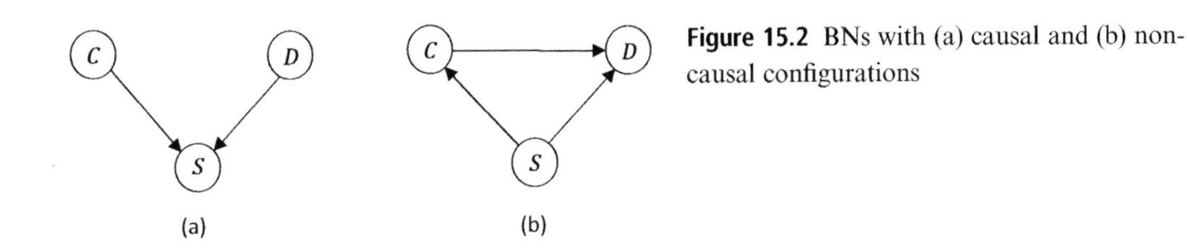

Figure 15.2 BNs with (a) causal and (b) non-causal configurations

15.3 D-Separation Rules

As we have seen, the topology of a BN determines the dependence structure between the nodes. This includes the conditions under which nodes are statistically independent. When nodes are statistically independent, entering evidence on one node does not alter the probability distribution of the other node. Knowing this information greatly facilitates efficient updating of CPTs upon evidence entered at a node. In this section, we introduce *d-separation* (*d* for *directed*) rules that describe the conditional independence between the nodes of a BN (Pearl, 1988, Russell and Norvig, 2003).

For a node in a BN, the set of parents, children, and spouse nodes constitute the *Markov blanket* of the node. As an example, for the BN in Figure 15.1, nodes X_2, X_3, and X_4 constitute the Markov blanket of node X_1, while the Markov blanket of node X_4 consists of nodes X_1 and X_5. Given its Markov blanket, a node in a BN is conditionally independent (d-separated) of all the other nodes in the graph. Hence, for the example BN, given the states of X_2, X_3, and X_4, random variable X_1 is independent of X_5. Furthermore, d-separation rules apply to three types of nodal connections, as follows:

- *Serial connection*: A set of nodes $\mathbf{X}$ is connected to another set of nodes $\mathbf{Z}$ through an intermediary node Y by a serial (head-to-tail) connection, as shown in Figure 15.3(a). The node sets $\mathbf{X}$ and $\mathbf{Z}$ are d-separated if the state of node Y is known with certainty. In that case, transfer of information between the two sets of nodes is blocked, i.e., observing the states of one or more of the nodes in **X** does not alter the joint distribution of nodes **Z**.

- *Diverging connection*: A set of nodes **X** is connected to another set of nodes **Z** through an intermediary node Y by a diverging (tail-to-tail) connection, as shown in Figure 15.3(b). The node sets **X** and **Z** are d-separated if the state of node Y is known with certainty. In that case, transfer of information between the two sets of nodes **X** and **Z** is blocked.
- *Converging connection*: A set of nodes **X** is connected to another set of nodes **Z** through an intermediary node Y by a converging (head-to-head) connection, as shown in Figure 15.3(c). The node sets **X** and **Z** are d-separated if neither node Y nor any of its descendants has received evidence. In that case, transfer of information between the two sets of nodes **X** and **Z** is blocked.

The d-separation property plays an important role in the development of efficient inference algorithms for BNs. It is worth mentioning that, given the input and output variables into a BN object, the internal nodes of the object are d-separated from the rest of the BN.

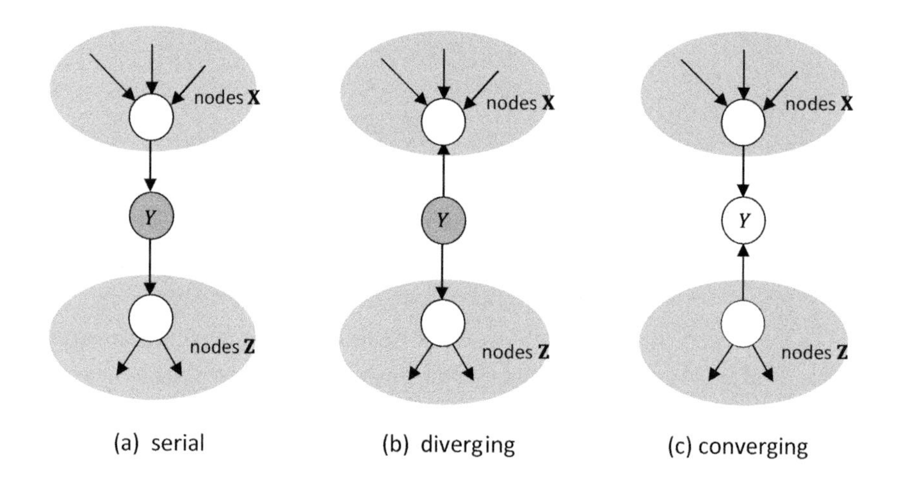

Figure 15.3 Rules for d-separation of nodes **X** and **Z** through intermediate node Y. Gray-shaded Y nodes are known with certainty; unshaded Y node has received no evidence

15.4 Discretization of Continuous Random Variables

As mentioned earlier, exact inference algorithms exist for discrete random variables with a finite number of states, whereas for continuous random variables, with the exception of the multinormal case with linear relations between the random variables and a special case with a mixture of exponential distributions, one must resort to approximate, typically sampling-based algorithms. Hence, wherever possible, it is desirable to work with discrete random variables. When the BN includes continuous random variables, one option is to replace them with "equivalent" discrete random variables through a process of discretization. These kinds of random variables are called *interval* random variables because the states indicate outcomes within specified intervals. Methods for this purpose are described by Neil et al. (2007), Straub (2009), Bensi et al. (2011a), and Zwirglmaier and Straub (2016).

As suggested in Straub (2009), nodes should be discretized from the top down, i.e., parent nodes before children nodes. First consider a root node, i.e., a node having no parents, representing a continuous random variable X with the marginal cumulative distribution function (CDF) $F_X(x)$. Let $\left[x^{(L)}, x^{(U)}\right]$ denote the range of outcomes of the random variable. We divide this range into m mutually exclusive and collectively exhaustive intervals defined by their boundaries $\left(x^{(j)}, x^{(j+1)}\right]$, $j = 1, \ldots, m$, with $x^{(1)} = x^{(L)}$ and $x^{(m+1)} = x^{(U)}$. Let $\widehat{X}$ denote the discretized interval random variable and $\widehat{x}^{(j)}$ represent its jth interval. The probability mass function (PMF) of $\widehat{X}$ is given by

$$p_{\widehat{X}}\left(\widehat{x}^{(j)}\right) = F_X\left(x^{(j+1)}\right) - F_X\left(x^{(j)}\right), j = 1, \ldots, m, \tag{15.4}$$

where $p_{\widehat{X}}\left(\widehat{x}^{(j)}\right)$ is interpreted as the probability $\Pr\left[x^{(j)} < X \leq x^{(j+1)}\right]$, i.e., the probability that the random variable falls within the jth interval. The number and widths of the intervals should be selected judiciously. The number of intervals, m, directly affects the size of the CPT of the discretized node and those of its children, hence directly affecting the computational demand imposed on the BN. The intervals need not be of equal size. They can be narrower where there is a larger probability mass, or where the outcomes of the random variable are important for the specific problem at hand. For example, if X denotes a load value in a BN aimed at assessing the risk of structural failure, then it would be advisable to select a finer discretization in the upper tail of the distribution.

The situation is more complicated when the node has parents because the marginal distribution of the node is not available. Let $\text{pa}(X) = (\mathbf{Y}, \mathbf{Z})$ denote the parents of X, where $\mathbf{Y}$ are discrete and $\mathbf{Z}$ are continuous random variables. Also let $F_{X|\mathbf{Y},\mathbf{Z}}(x|\mathbf{y}, \mathbf{z})$ denote the conditional CDF of X given $\mathbf{Y} = \mathbf{y}$ and $\mathbf{Z} = \mathbf{z}$. We assume the continuous parents have been discretized and are represented by interval random variables $\widehat{\mathbf{Z}}$. Let $\widehat{\mathbf{z}}^{(k)}$ represent the kth hyper-interval in the outcome space of $\mathbf{Z}$ for the selected intervals of the individual variables. Following Straub (2009), we approximate the conditional CDF of X within the hyper-interval by

$$F_{X|\mathbf{Y},\widehat{\mathbf{Z}}}\left(x|\,\mathbf{y}, \widehat{\mathbf{z}}^{(k)}\right) = \int_{\widehat{\mathbf{z}}^{(k)}} F_{X|\mathbf{Y},\mathbf{Z}}(x|\mathbf{y}, \mathbf{z}) f_{\mathbf{Z}}\left(\mathbf{z}|\widehat{\mathbf{z}}^{(k)}\right) \mathrm{d}\mathbf{z}, \tag{15.5}$$

where $f_{\mathbf{Z}}\left(\mathbf{z}|\widehat{\mathbf{z}}^{(k)}\right)$ is the joint PDF of $\mathbf{Z}$ given that its outcome falls within the kth hyper-interval. The above expression is an approximation because it disregards possible dependence between the parent sets $\mathbf{Y}$ and $\mathbf{Z}$. Furthermore, we normally do not have the joint PDF $f_{\mathbf{Z}}(\mathbf{z}|\widehat{\mathbf{z}})$. Generalizing the formulation by Straub to the case of multiple parents, it is suggested to use the uniform distribution over the hyper-interval $\widehat{\mathbf{z}}^{(k)}$. That is, we approximate the joint PDF by

$$f_{\mathbf{Z}}(\mathbf{z}|\widehat{\mathbf{z}}) = \frac{1}{\int_{\widehat{\mathbf{z}}^{(k)}} \mathrm{d}\mathbf{z}}. \tag{15.6}$$

This formulation disregards any statistical dependence that may exist among the elements of $\mathbf{Z}$ within the hyper-interval. If the hyper-interval falls in the extreme tails of one or more random variables, then an appropriate exponential distribution may be used for that variable to better represent the rapidly decaying probability density in the tail. Naturally, these approximations improve when finer intervals are selected.

In order to discretize X, it is necessary to select appropriate intervals. We first determine the range $\left[x^{(L)}, x^{(U)}\right]$ of outcomes of X for all outcomes of its parents. We then divide this range into intervals with boundaries $\left(x^{(j)}, x^{(j+1)}\right], j = 1, \ldots, m$, with $x^{(1)} = x^{(L)}$ and $x^{(m+1)} = x^{(U)}$, as described earlier. The conditional PMF of X given its parents is then obtained as

$$p_{\hat{X}|\mathbf{Y},\hat{\mathbf{Z}}}\left(\widehat{x}^{(j)}|\,\mathbf{y}, \widehat{\mathbf{z}}^{(k)}\right) = F_{X|\mathbf{Y},\hat{\mathbf{Z}}}\left(x^{(j+1)}|\,\mathbf{y}, \widehat{\mathbf{z}}^{(k)}\right) - F_{X|\mathbf{Y},\hat{\mathbf{Z}}}\left(x^{(j)}|\,\mathbf{y}, \widehat{\mathbf{z}}^{(k)}\right), j = 1, \ldots, m. \tag{15.7}$$

This PMF is then used to construct the CPT for node X. Bensi et al. (2011a) perform the necessary calculations in (15.5)–(15.7) by sampling. Note that this calculation is performed once during the construction of the CPTs of the discretized nodes and does not affect subsequent inference analyses by the BN.

15.5 Inference in Bayesian Network

The most common use of BNs is for the purpose of inference. This takes two forms: *Forward* or *predictive inference*, aimed at determining the probability distribution of a node for given marginal distributions of root nodes and conditional distributions of children nodes, and *backward* or *diagnostic inference*, aimed at determining the posterior distribution of a node when evidence is entered at other nodes. The latter is the classical Bayesian updating problem but done on a large scale and for a complex model.

Suppose in the BN of Figure 15.1, we wish to determine the posterior PMF of node X_2, given the observations $X_3 = x_3$ and $X_4 = x_4$. Using Bayes' rule, we have

$$p(x_2|x_3, x_4) = \frac{p(x_2, x_3, x_4)}{p(x_3, x_4)}. \tag{15.8}$$

The PMFs in the numerator and denominator of the above expression are obtained by *marginalization*, i.e., by summing up the joint PMF of the random variables over all the states of the unwanted random variables, i.e.,

$$p(x_2, x_3, x_4) = \sum_{x_1} \sum_{x_5} p(x_1, x_2, x_3, x_4, x_5), \tag{15.9}$$

$$p(x_3, x_4) = \sum_{x_1} \sum_{x_2} \sum_{x_5} p(x_1, x_2, x_3, x_4, x_5). \tag{15.10}$$

For a large BN with many states to each node, the above sums can be daunting. For example, if each node of the BN has m states, the number of terms to add to compute each of the above joint PMFs, including all states of the retained variables, is m^5. For $m = 10$, this is 100,000 terms, which is a very large number for such a small BN. It can be seen that the computational effort increases exponentially as a function of the number of nodes in the BN. The BN structure provides a means for performing this calculation in a more efficient manner. Below, we briefly describe two algorithms for this purpose, namely, the *node elimination* algorithm and the *junction tree* algorithm.

We demonstrate the node elimination algorithm by applying it to the marginalization problem in (15.10). Using the factorization of the joint PMF in (15.3), we first eliminate node X_1 to obtain the joint distribution of $X_2, \ldots, X_5$:

$$\begin{aligned} p(x_2, \ldots, x_5) &= \sum\nolimits_{x_1} p(x_1)p(x_2)p(x_3|x_1, x_2)p(x_4|x_1)p(x_5|x_4) \\ &= p(x_2)p(x_5|x_4)\sum\nolimits_{x_1} p(x_1)p(x_3|x_1, x_2)p(x_4|x_1) \\ &= p(x_2)p(x_5|x_4)\phi(x_2, x_3, x_4). \end{aligned} \tag{15.11}$$

In the second line we have moved the summation operator as far to the right as possible. This is because the summation need only be performed over the CPTs that include variable X_1. The result of the sum, $\phi(x_2, x_3, x_4) = \sum_{x_1} p(x_1)p(x_3|x_1, x_2)p(x_4|x_1) = p(x_3, x_4|x_2)$, is called a *potential* and is stored as a table over the random variables remaining after the summation. Next, we eliminate random variable X_2:

$$\begin{aligned} p(x_3, x_4, x_5) &= \sum\nolimits_{x_2} p(x_2)p(x_5|x_4)\phi(x_2, x_3, x_4) \\ &= p(x_5|x_4)\sum\nolimits_{x_2} p(x_2)\phi(x_2, x_3, x_4) \\ &= p(x_5|x_4)\phi(x_3, x_4), \end{aligned} \tag{15.12}$$

where $\phi(x_3, x_4) = p(x_3, x_4)$ is the new potential. Finally, we eliminate node X_5:

$$\begin{aligned} p(x_3, x_4) &= \phi(x_3, x_4)\sum\nolimits_{x_5} p(x_5|x_4) \\ &= \phi(x_3, x_4). \end{aligned} \tag{15.13}$$

Note that the summation over the conditional PMF of X_5 equals 1. The number of additions to perform assuming m states for each node is m^4 for eliminating X_1, m^3 for eliminating X_2, and zero for eliminating X_5. The total number of additions, $m^4 + m^3$, is smaller than that using the joint PMF as in (15.9) for $2 \le m$. For $m = 10$, it is nearly an order of magnitude smaller. On the other hand, observe that we need to store each potential table for the computation in the next step. The largest potential determines the needed additional memory size. In this case, the largest potential table is of size m^3.

The order in which the variables are eliminated affects the size of the CPTs and potentials that must be multiplied at each stage of the node elimination algorithm. In the above example, the largest memory demand comes from the product of three CPTs involving four nodes when eliminating X_1. If, instead, we first eliminate X_5, we have

$$\begin{aligned} p(x_1, x_2, x_3, x_4) &= p(x_1)p(x_2)p(x_3|x_1, x_2)p(x_4|x_1)\sum\nolimits_{x_5} p(x_5|x_4) \\ &= p(x_1)p(x_2)p(x_3|x_1, x_2)p(x_4|x_1). \end{aligned} \tag{15.14}$$

Next, we eliminate X_2:

$$\begin{aligned} p(x_1, x_3, x_4) &= p(x_1)p(x_4|x_1)\sum\nolimits_{x_2} p(x_2)p(x_3|x_1, x_2) \\ &= p(x_1)p(x_4|x_1)\phi(x_1, x_3). \end{aligned} \tag{15.15}$$

And finally, we eliminate X_1:

$$\begin{aligned} p(x_3, x_4) &= \sum\nolimits_{x_1} p(x_1)p(x_4|x_1)\phi(x_1, x_3) \\ &= \phi(x_3, x_4). \end{aligned} \tag{15.16}$$

The number of additions now is zero for eliminating X_5, m^3 for eliminating X_2, and m^3 for eliminating X_1, for a total of $2m^3$, which for $m = 10$ is 50 times smaller than that required for the marginalization using the joint PMF and 5.5 times smaller than that required for the previous elimination order. The trick is to select nodes for elimination by pushing the summation operator as far to the right as possible. As x_1 appears in three of the CPTs, x_2 in two CPTs, and x_5 in only one CPT, advantage is gained by summing in the order X_5, X_2, and then X_1.

Marginalization by the node elimination algorithm is query sensitive. That is, the calculations must be repeated for each query of interest. This is because the nodes corresponding to the query must not be eliminated and, therefore, must be the last nodes in the elimination order. In the above example, if we were next interested in the joint PMF of X_1 and X_5, the algorithm would need to be repeated because the calculations performed to determine the joint distribution of X_3 and X_4 are of no use for determining the joint distribution of X_1 and X_5. As a result, the node elimination algorithm is efficient from the viewpoint of computer memory but inefficient in terms of computation time.

An alternative method that facilitates reuse of node elimination results is the junction tree algorithm. This algorithm can be viewed as a generalization of the node elimination algorithm. Like in that algorithm, nodes are eliminated in a particular order. However, summations are not performed. Instead, a data structure is created known as a *junction tree*, which contains subsets of the random variables in the BN known as *cliques*. Each clique consists of the eliminated node and its neighbors and is associated with a potential table. The cliques are stored in the computer memory and are reused to facilitate efficient inference for a broad set of queries without reconstructing the potential tables. Naturally, the size of the largest clique and the total sum of all clique tables determines the memory demand. This approach is efficient from a computational standpoint for repeated queries but is more demanding on memory. Details of both algorithms can be found in Jensen and Nielsen (2007).

Example 15.2 – Forward and Backward Inference

This is a simple example to demonstrate inference by marginalization and the node elimination algorithm. Consider the state of a structure in the case of an impending earthquake, which depends on the condition of its foundation. Existing information suggests that the foundation could be in good (G) or bad (B) condition with probabilities 0.8 and 0.2, respectively. If the foundation is in the good state, the probability of failure event (F) is estimated as 10^{-3}, whereas if the foundation is in the bad state the probability of failure is 10^{-2}. Information about the state of the foundation can be obtained by performing non-destructive tests of the surrounding soil and the foundation. Test 1 has two possible outcomes: an indication that the state of the foundation is good (IG) and an indication that the state of the foundation is bad (IB). Test 2 also provides indications of good and bad states (IG and IB) but may also produce an inconclusive (IN) result.

Because the state of the foundation causes the test outcomes as well as the state of the structure, we consider node C representing the condition of the foundation as a parent node to the other three nodes, denoted respectively as T_1, T_2, and S. Figure 15.4 shows the causal formulation of the BN together with the CPT of each node. Note that the CPTs for the two tests represent the *test likelihood* matrices, i.e., the conditional probabilities of the test results, given the true state of the foundation. Normally, the test likelihoods are obtained by calibration of the test equipment and measurement devices. Given the configuration of the BN, the joint PMF is formulated as

$$p(c, t_1, t_2, s) = p(t_1|c)p(t_2|c)p(s|c)p(c), \tag{E1}$$

where the marginal and conditional PMFs are the CPTs given in Figure 15.4.

As a first item of interest, we perform forward inference to determine the probability of failure of the structure before any test is performed. For this purpose, we need to marginalize the PMF of node S by eliminating nodes C, T_1, and T_2. Using the expression in (E1), we write

$$\begin{aligned} p(s = \mathrm{F}) &= \sum_{t_1}\sum_{t_2}\sum_{c} p(t_1|c)p(t_2|c)p(s = \mathrm{F}|c)p(c) \\ &= \sum_{c} p(s = \mathrm{F}|c)p(c) \\ &= 10^{-3} \times 0.8 + 10^{-2} \times 0.2 \\ &= 0.0028. \end{aligned} \tag{E2}$$

Observe that the summations over t_1 and t_2 yield 1, leading to the result in the second line, which is nothing but a statement of total probability.

Now suppose Test 1 is performed and it indicates that the condition of the foundation is bad. We need to perform backward inference to determine the posterior probability of failure, given the evidence on node T_1. For this purpose, we need to marginalize the PMFs $p(t_1, s)$ and $p(t_1)$ for $t_1 = \mathrm{IB}$ and $s = \mathrm{F}$. The steps of the node elimination algorithm are

Example 15.2 (cont.)

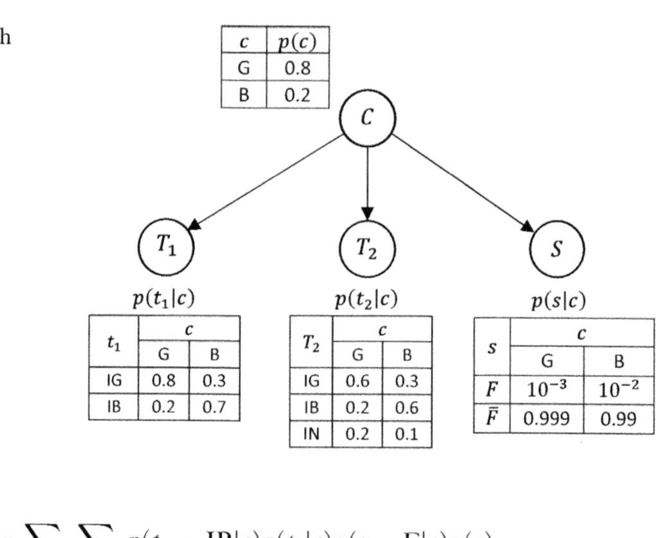

Figure 15.4 BN model for structure with uncertain foundation condition

$$\begin{aligned} p(t_1 = \text{IB}, s = \text{F}) &= \sum\nolimits_{t_2} \sum\nolimits_{c} p(t_1 = \text{IB}|c)p(t_2|c)p(s = \text{F}|c)p(c) \\ &= \sum\nolimits_{c} p(t_1 = \text{IB}|c)p(s = \text{F}|c)p(c) \\ &= 0.2 \times 10^{-3} \times 0.8 + 0.7 \times 10^{-2} \times 0.2 \\ &= 0.00156 \end{aligned} \tag{E3}$$

and

$$\begin{aligned} p(t_1 = \text{IB}) &= \sum\nolimits_{t_2} \sum\nolimits_{s} \sum\nolimits_{c} p(t_1 = \text{IB}|c)p(t_2|c)p(s|c)p(c) \\ &= \sum\nolimits_{c} p(t_1 = \text{IB}|c)p(c) \\ &= 0.2 \times 0.8 + 0.7 \times 0.2 \\ &= 0.3. \end{aligned} \tag{E4}$$

Thus, the posterior probability of failure is

$$\begin{aligned} p(s = \text{F}|t_1 = \text{IB}) &= \frac{p(s = \text{F}, t_1 = \text{IB})}{p(t_1 = \text{IB})} \\ &= \frac{0.00156}{0.3} \\ &= 0.00520. \end{aligned} \tag{E5}$$

As expected, the updated probability of failure is larger than the unconditional probability of failure in (E2) in the light of the unfavorable test result.

Example 15.2 (cont.)

Now suppose Test 2 is also conducted and provides an inconclusive result. We now need to marginalize the PMFs $p(t_1, t_2, s)$ and $p(t_1, t_2)$ for $t_1 = \text{IB}$, $t_2 = \text{IN}$, and $s = \text{F}$. The steps of the node elimination algorithm are

$$\begin{aligned} p(t_1 = \text{IB}, t_2 = \text{IN}, s = \text{F}) &= \sum\nolimits_c p(t_1 = \text{IB}|c)p(t_2 = \text{IN}|c)p(s = \text{F}|c)p(c) \\ &= 0.2 \times 0.2 \times 10^{-3} \times 0.8 + 0.7 \times 0.1 \times 10^{-2} \times 0.2 \\ &= 0.000172 \end{aligned} \tag{E6}$$

and

$$\begin{aligned} p(t_1 = \text{IB}, t_2 = \text{IN}) &= \sum\nolimits_s \sum\nolimits_c p(t_1 = \text{IB}|c)p(t_2 = \text{IN}|c)p(s|c)p(c) \\ &= \sum\nolimits_c p(t_1 = \text{IB}|c)p(t_2 = \text{IN}|c)p(c) \\ &= 0.2 \times 0.2 \times 0.8 + 0.7 \times 0.1 \times 0.2 \\ &= 0.0460. \end{aligned} \tag{E7}$$

Thus, the posterior probability of failure is

$$\begin{aligned} p(s = \text{F}|t_1 = \text{IB}, t_2 = \text{IN}) &= \frac{p(s = \text{F}, t_1 = \text{IB}, t_2 = \text{IN})}{p(t_1 = \text{IB}, t_2 = \text{IN})} \\ &= \frac{0.000172}{0.0460} \\ &= 0.00374. \end{aligned} \tag{E8}$$

It is noteworthy that the updated probability of failure after obtaining an inconclusive result from Test 2 is smaller than the updated estimate after Test 1 alone. This has to do with the fact that the probability of Test 2 producing an inconclusive result is twice as large when the true condition of the foundation is good as when it is bad.

Example 15.3 – BN Model of Seismic Hazard at a Site

This example, adopted from Bensi et al. (2011a), exemplifies the application of BN to modeling a natural hazard. Consider the site of a building in the neighborhood of several active faults, which are idealized as straight lines in Figure 15.5. An earthquake may occur on any of the faults, having a random magnitude and location. We idealize the earthquake source as a rupture of random length, depending on the earthquake magnitude, and a random location within the fault. The intensity of the ground motion at the site of interest mainly depends on the earthquake magnitude, the distance from the site to the rupture, various source characteristics, e.g., source mechanism, direction of propagation of the rupture, and local soil properties.

Example 15.3 (cont.)

In the field of earthquake engineering, ground-motion-prediction equations (GMPEs) have been developed by regression analysis of large amounts of recorded data from multiple earthquakes (Abrahamson et al., 2008). These predictive equations have the general form

$$\ln S = f(M, R, \mathbf{V}, \boldsymbol{\theta}) + \epsilon_m + \epsilon_r, \tag{E1}$$

where S is a measure of intensity (peak ground acceleration, velocity, or displacement; or a response spectral ordinate at a selected frequency), $f(\cdot)$ is the regression formula and gives the logarithmic mean of the intensity measure, M is the magnitude, R is the distance to the source (typically the distance from the site to the nearest point on the rupture), $\mathbf{V}$ is a set of variables characterizing the source and the local soil properties, $\boldsymbol{\theta}$ are the regression parameters, and ϵ_m and ϵ_r are zero-mean normal random variables representing the regression errors, the former characterizing the inter-event variability, i.e., the variability from earthquake to earthquake, and the latter characterizing the intra-event variability, i.e., the variability from site to site for the same earthquake. The form of the regression formula and the variances of the error terms may depend on the type of the fault mechanism, e.g., strike-slip, dip-slip, normal, or reverse faulting.

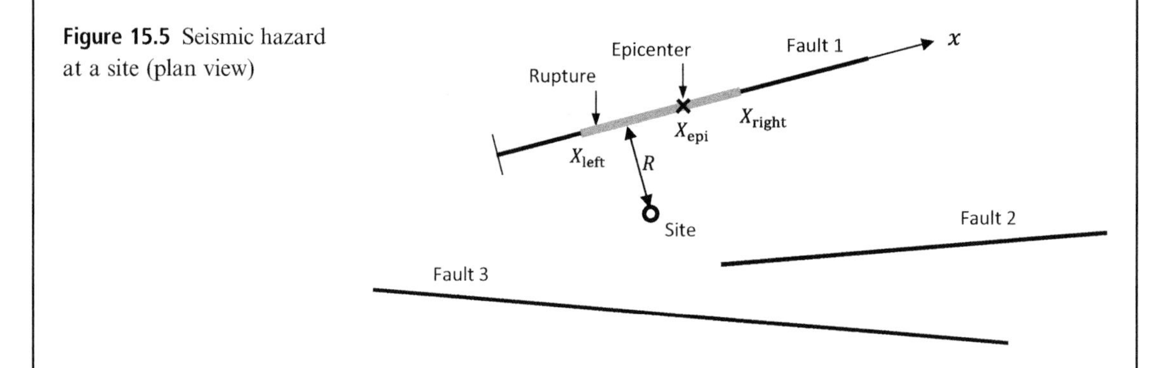

Figure 15.5 Seismic hazard at a site (plan view)

The rupture on the fault initiates at a point called the epicenter and may propagate in either or both directions along the fault, but usually does not extend beyond the ends of the fault. Based on data from past earthquakes, empirical models of the form

$$\ln L = a + bM + \epsilon_L \tag{E2}$$

have been developed (Wells and Coppersmith, 1994), where L is the rupture length, (a, b) are regression coefficients, and ϵ_L is a zero-mean normal random variable representing the regression error. The regression coefficients and the variance of the error depend on the type of faulting.

We wish to develop a BN model for predicting the intensity of the ground motion at the site of the building for an impending earthquake of engineering interest, say an earthquake with a

Example 15.3 (cont.)

magnitude equal to or greater than 4. Figure 15.6(a) shows the BN model, which we now describe in detail. We start with the root node, *Source*. This is a discrete node indicating the fault on which the earthquake will occur. One may assume that the likelihood of occurrence of the earthquake on each fault is proportional to the mean rate of occurrence of earthquakes on the fault. Thus, if v_i denotes the mean rate of earthquakes of magnitude 4 or greater on fault i, then the PMF for the *Source* node is $p(Source = i) = v_i/\sum_{j=1}^{N_s} v_j$, where N_s is the number of faults. The *Source* node is a parent to node M representing the magnitude of the impending earthquake. M has a continuous distribution, which usually depends on the length and type of the fault. The Gutenberg and Richter (1944) magnitude recurrence law implies an exponential distribution. Here, we may shift the distribution to start from magnitude 4 and truncate it to

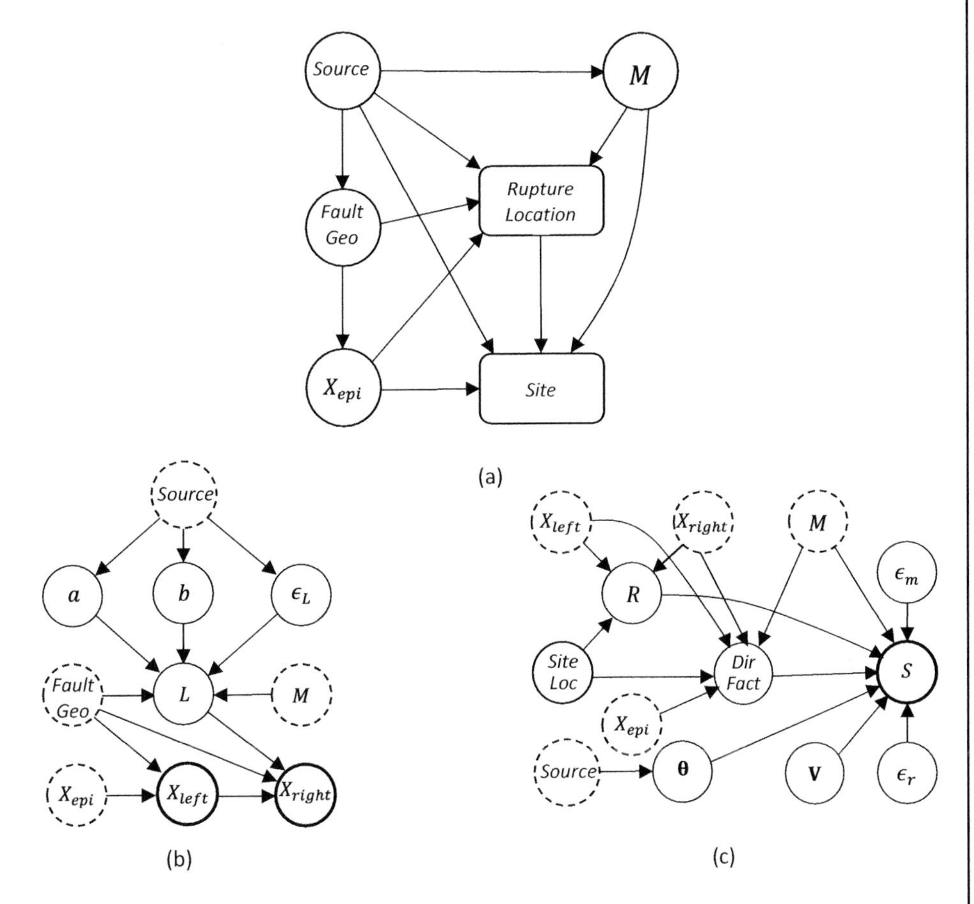

Figure 15.6 BN for seismic hazard at a site: (a) higher-level BN, (b) BN object *Rupture Location*, and (c) BN object *Site*

Example 15.3 (cont.)

account for the maximum magnitude that each fault may produce depending on its length. For reasons described in Section 15.4, we will need to discretize this node. For that purpose, we need to determine the range of M for all faults and select intervals within that range. If the range of magnitudes for a particular fault is narrower than the selected total range, then the extra intervals have zero probability. For more details on the discretization of the distribution of magnitudes see Bensi et al. (2011a).

The *Source* node is also a parent to the node *Fault Geo* (fault geometry). Given the source, this is a deterministic node, unless we wish to account for uncertainties in the location of the fault. The node describes the geometry of each fault relative to the building site. Next is node X_{epi}, which is a child of *Fault Geo* and describes the coordinates of the epicenter (the point where the rupture initiates) within the fault (see Figure 15.5). The epicenter may occur anywhere along the fault. Typically, a uniform distribution is assumed, though a non-uniform distribution may be used if it is believed that certain portions of the fault are more or less likely to rupture. In any case, this distribution will also need to be discretized in the manner described in Section 15.4. Note that the selected range of X_{epi} must cover the entire range of possible values of the coordinate for all the faults, but shorter faults will have zero probability values for intervals beyond their lengths.

Next, we need to describe the location of the rupture within the fault. As mentioned before, the rupture may propagate on either or both sides of the epicenter (see Figure 15.5), but it cannot extend beyond the known ends of the fault. In the BN in Figure 15.6(a), the node for the location of the rupture is shown as a rounded rectangle named *Rupture Location*, representing a BN object. Figure 15.6(b) shows the BN behind this object. The input variables to this object, shown as nodes with dashed boundaries, are *Source*, M, *Fault Geo*, and X_{epi}. The internal variables, shown as nodes with thin solid boundaries, are the regression coefficients (a, b), the regression error ϵ_L, and the length of the rupture L. The latter is determined from (E2) for given values of the *Source*, M, and *Fault Geo* nodes. Note that L cannot be greater than the fault length. Finally, given *Fault Geo*, X_{epi}, and L, the coordinates of the rupture along the fault are determined. This is done by assuming a uniform distribution of the location of the epicenter along the rupture, subject to the rupture not extending beyond the known ends of the fault. The coordinates X_{left} and X_{right} of the rupture are the output variables, shown as nodes with thick solid boundaries in the BN object.

Figure 15.6(c) shows the BN behind the object *Site* in Figure 15.6(a) for the intensity measure at the site. The input variables are *Source* (to account for the dependence of the regression formula in E1 on the type of faulting), M, the coordinates X_{left} and X_{right} of the rupture, and X_{epi}. Having the *Site Loc* (site location) as an internal node, using the coordinates X_{left} and X_{right} of the rupture, the distance R from the site to the nearest point on the rupture is calculated. This node, as well as regression parameters $\boldsymbol{\theta}$, error terms ϵ_m and ϵ_r, and variables $\mathbf{V}$, which are all internal nodes, together with M, are all parents of the only output node S,

Example 15.3 (cont.)

whose value is computed according to (E1) given its parents' states. An additional node, *Dir Fact* (directivity factor), is included in this object. As is well known in earthquake engineering, under certain conditions when the fault rupture propagates toward a site, constructive interference of seismic waves arriving from intermittent segments of the rupture may cause a large-amplitude, long-period velocity pulse at the site. This is known as the forward directivity effect (Abrahamson, 2000). Conversely, when the rupture propagates away from the site, the ground motion at the site is likely to have a small amplitude but long duration. In the current practice, this effect is accounted for by applying a correction factor to the GMPE (Somerville et al., 1997). The factor depends on the geometry of the site relative to the fault rupture, principally the angle between the fault line and the line connecting the site to the epicenter, and the length of the rupture propagating toward the site. The BN object in Figure 15.6(c) includes this factor as a child of M, X_{left}, X_{right}, X_{epi}, and *Site Loc*. It serves as an additional internal parent to node S. The *Site* object includes several internal nodes that are continuous and need to be discretized. These include R, ϵ_m, ϵ_r, $\mathbf{V}$ (e.g., if soil properties are uncertain), and the *Dir Fact*. The output node, S, must also be discretized.

More detailed BN modeling of the seismic hazard, including the effects of liquefaction, fault rupture and spatial variability of ground motion, is described in Bensi et al. (2011a).

15.6 BN Modeling of Components

The simplest way to model a component in a BN is by use of fragility functions. For a two-state component, the fragility function defines the conditional probability of failure given a measure of the demand placed on the component. Let S denote the demand placed on the component, $C \in (0, 1)$ denote the state of the component with $C = 0$ indicating the fail state and $C = 1$ indicating the safe state, and $F(s) = \Pr(C = 0|S = s)$ denote the fragility function. The BN in Figure 15.7(a) then describes this component problem, where the conditional probabilities entering the CPT of the component node for given $S = s$ are $F(s)$ for $C = 0$ and $1 - F(s)$ for $C = 1$.

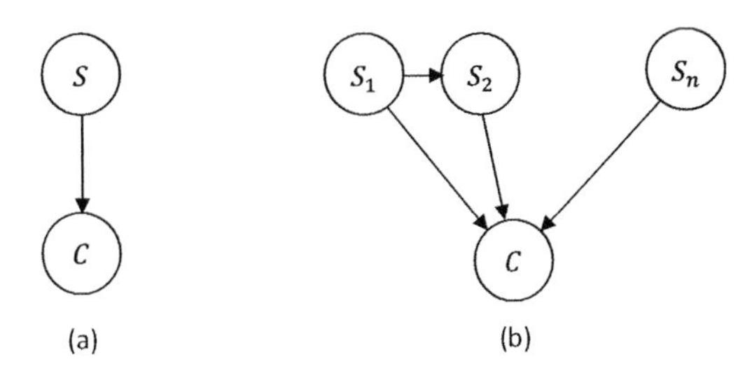

Figure 15.7 BNs for (a) single-demand and (b) multiple-demand components

More generally, the component may have multiple states with fragility functions describing the transitions between the states. Let $C \in (0, 1, \ldots, K)$ denote the states of the component with increasing values of the index indicating decreasing states of damage, with $C = 0$ indicating complete failure and $C = K$ indicating the intact state. Also let

$$F_k(s) = \Pr(C \leq k | S = s), \quad k = 0, 1, \ldots, K \tag{15.17}$$

denote the fragility function representing the transition between states k and $k + 1$. Note that $F_K(s) = 1$. The CPT for the component node is then constructed by using the probability values

$$\Pr(C = k) = F_k(s) - F_{k-1}(s), \quad k = 0, 1, \ldots, K \tag{15.18}$$

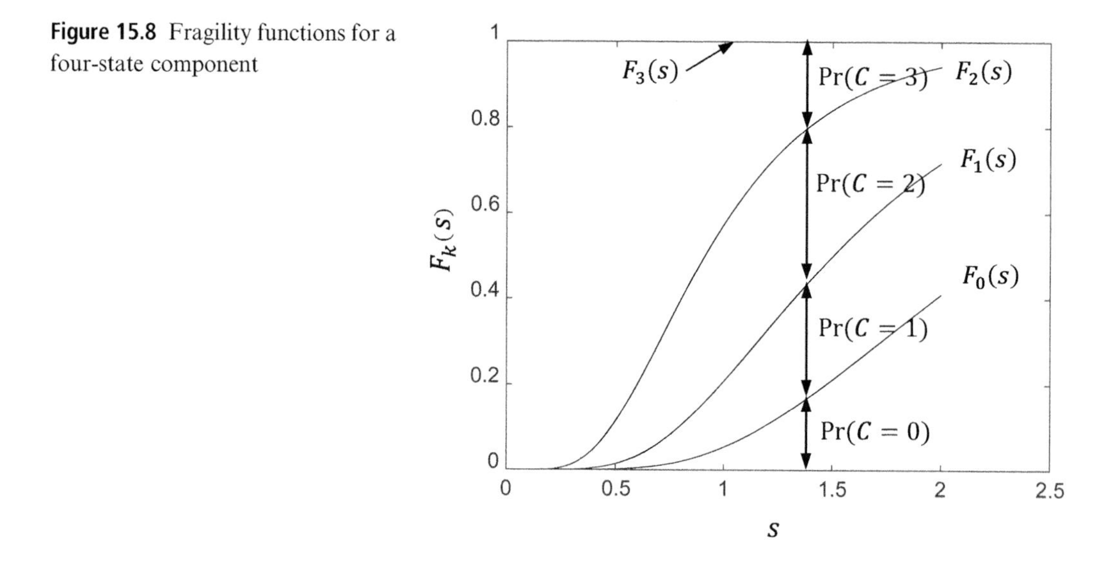

Figure 15.8 Fragility functions for a four-state component

with $F_{-1}(s) = 0$ and $F_K(s) = 1$. Figure 15.8 shows fragility curves and probability assignments for a component with $K = 3$, i.e., four states.

A component might be subject to more than one demand value. It is possible then to define fragility surfaces $F_k(\mathbf{s}) = \Pr(C \leq k | \mathbf{S} = \mathbf{s})$, $k = 0, \ldots, K$, in the outcome space of the demand vector $\mathbf{S}$ that delineate the transitions between various states of the component. As an example, Gardoni et al. (2002) have developed a bivariate fragility surface for two-state (fail/safe) reinforced-concrete bridge columns under the combined effects of deformation and shear force demands. In the BN model, all we need to do is include additional parent demand nodes to the component node, as shown in Figure 15.7(b). Note, however, that the demands might be statistically dependent, in which case links between the demand nodes must be included, as shown for demands S_1 and S_2 in Figure 15.7(b). For example, in the case of a bridge column, the shear and flexural loads acting on the column are generally correlated because they arise from common loads.

More generally, it may not be possible to distinguish uncertain variables affecting the state of a component as demand and capacity values. (See the discussion in Section 4.4.) In that case, we model the transitions between the states of the component with limit-state surfaces $g_k(\mathbf{x}) = 0,\ k = 0, \ldots, K$, where $\mathbf{x}$ denotes the outcome of the set of random variables $\mathbf{X}$ that affect the state of the component. The component node in the BN now has random variables $\mathbf{X}$ as its parents. Following the convention that we have used throughout this book for the failure event, the CPT of the component node is then constructed using the probability values

$$\Pr(C = k) = \Pr[g_k(\mathbf{x}) \leq 0] - \Pr[g_{k-1}(\mathbf{x}) \leq 0], \quad k = 0, 1, \ldots, K, \tag{15.19}$$

with $\Pr[g_{-1}(\mathbf{x}) \leq 0] = 0$ and $\Pr[g_K(\mathbf{x}) \leq 0] = 1$. These probabilities can be computed by the methods of structural reliability. Later in this chapter we will further develop this approach by enhancing the BN with structural reliability analysis methods.

15.7 BN Modeling of Systems

As defined earlier, a system is a collection of possibly interdependent components, each performing a function, such that the functioning state of the system depends on the functioning states of its constituent components. As an example, we consider the system of eight components in Figure 15.9(a), where the objective is to reach the "sink" from the "source." Each component has two states, open or closed. One simple way to model the

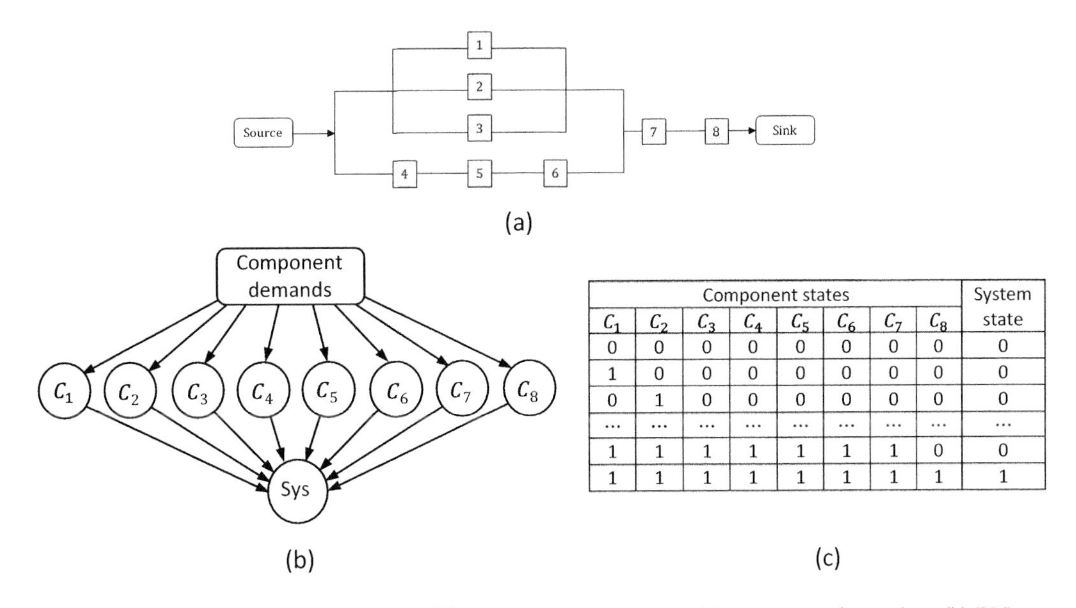

Component states								System state
C_1	C_2	C_3	C_4	C_5	C_6	C_7	C_8	
0	0	0	0	0	0	0	0	0
1	0	0	0	0	0	0	0	0
0	1	0	0	0	0	0	0	0
...	...	...	...	...	...	...	...	...
1	1	1	1	1	1	1	0	0
1	1	1	1	1	1	1	1	1

Figure 15.9 Example two-state system with two-state components: (a) system configuration, (b) BN model, (c) CPT of system node

system BN is to make the node representing the system state a child of the nodes representing the component states, as shown in Figure 15.9(b), where the demands on the components are shown as a BN object. The system node here is a deterministic node because, given the states of the components, we know the state of the system with certainty. Shown in Figure 15.9(c) is the CPT of the system node, where 0 indicates the fail state and 1 indicates the working state for both the components and the system. For a system with N components, this table has 2^N rows. More generally, for a system with m-state components, the number of rows is m^N. We see that this table can be extremely large for a system with a large number of components. For the example system with eight two-state components, the number of rows is $2^8 = 256$. Note that we only need to store the column listing the system states, because the component states can be ordered in a systematic manner and reproduced for a given row number.

The large size of the CPT for the system node is because of the large number of parents. One way to address this is to introduce intermediate nodes representing the min-link sets (MLSs) or min-cut sets (MCSs) of the system. In the BN terminology, introducing intermediate nodes is known as *divorcing* the parents. Figure 15.10 shows the BN model for the example system with the MLS formulation on the left and the MCS formulation on the right. The system has the MLSs $\{(1,7,8),(2,7,8),(3,7,8),(4,5,6,7,8)\}$ and MCSs $\{(1,2,3,4),(1,2,3,5),(1,2,3,6),(7),(8)\}$. The number of parents to the system node now is smaller in both formulations. However, a large system is likely to have many MLSs and MCSs, which may themselves contain large numbers of components, in which case we will have the same problem of large CPTs.

Two approaches have been developed to address the large size of the CPT for the system node in the BN. In Tien and Der Kiureghian (2016), data-compression techniques are used to store the system CPT in an efficient manner. Recall that we need only store the column of the CPT that lists the states of the system. For a two-state system, this is a column of 0s and 1s. In the data-compression method, advantage is taken of run-length encoding (consecutive repetitions of 0s or 1s) and phrase encoding (repeated patterns of 0s and 1s) to store the column in a highly efficient manner. Furthermore, algorithms are developed to perform inference analysis and compute intermediate potential tables, which are also stored in a compressed form, without decompressing the CPT or the intermediate tables. Example investigations have shown that this approach requires vastly reduced memory and is only mildly dependent on the number of components. On the other hand, calculations with compressed tables take longer. Hence, there is a trade-off between the memory demand and computation time. The inference algorithm is further refined in Tien and Der Kiureghian (2017) in order to reduce the computation time. Recently, Tong and Tien (2017) applied this method to a flow system with multi-state components.

Bensi et al. (2013) developed a method to construct chain-like BN topologies for two-state systems with two-state components. Define a *survival path sequence* (SPS) as a chain of events, corresponding to an MLS, in which the terminal event in the sequence indicates whether or not all the components in the MLS are in the survival state. An SPS is comprised of a chain of *survival path events* (SPEs), each associated with a component and describing

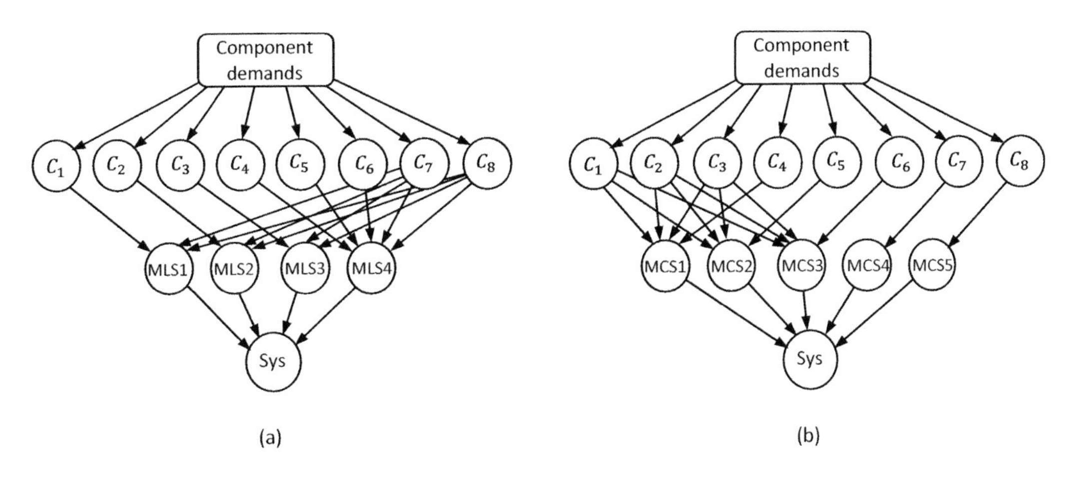

Figure 15.10 Alternative BN models of the example system: (a) min-link set formulation, (b) min-cut set formulation

the state of the sequence up to that event. SPEs are represented in the BN by nodes labeled $E_{s,i}$, the subscript i indicating the association with component i. The state of $E_{s,i}$ is defined as

$$\begin{aligned} E_{s,i} &= 1 \quad \text{if } \{C_i = 1 \cap E_{s,\text{pa}(i)} = 1\} \\ &= 0 \quad \text{otherwise,} \end{aligned} \tag{15.20}$$

where $E_{s,\text{pa}(i)}$ is the parent node to $E_{s,i}$. As before, 0 indicates the fail state and 1 the working state.

Now consider a series system. As such a system has only one MLS (all components together form the sole MLS of the system), it has only one SPS, as shown in the BN in Figure 15.11(a) for a system with N components. The system state is identical to the state of its sole parent, $E_{s,N}$. Observe that in this BN, nodes have at most two parents, regardless of the number of components. In contrast, a parallel system has as many MLSs and, therefore, SPSs as components. As a result, the formulation in terms of SPSs is not helpful for a parallel system.

Next, define a *failure path sequence* (FPS) as a chain of events, corresponding to an MCS, in which the terminal event in the sequence indicates whether or not all the components in the MCS are in the fail state. An FPS is comprised of a chain of *failure path events* (FPEs), each associated with a component and describing the state of the sequence up to that event. FPEs are represented in the BN by nodes labeled $E_{f,i}$, the subscript i indicating the association with component i. The state of $E_{f,i}$ is defined as

$$\begin{aligned} E_{f,i} &= 0 \quad \text{if } \{C_i = 0 \cap E_{f,\text{pa}(i)} = 0\} \\ &= 1 \quad \text{otherwise,} \end{aligned} \tag{15.21}$$

where $E_{f,\text{pa}(i)}$ is the parent node to $E_{f,i}$. Now consider a series system. Because such a system has only one MCS (all components together form the sole MCS of the system), it has only one FPS, as shown in the BN in Figure 15.11(b) for a system with N components. The system state is identical to the state of its sole parent, $E_{f,N}$. Observe again that in this BN, nodes have at most two parents, regardless of the number of components. In contrast, a series system has as many MCSs and, therefore, FPSs as components. Hence, the formulation in terms of FPSs is not helpful for a series system.

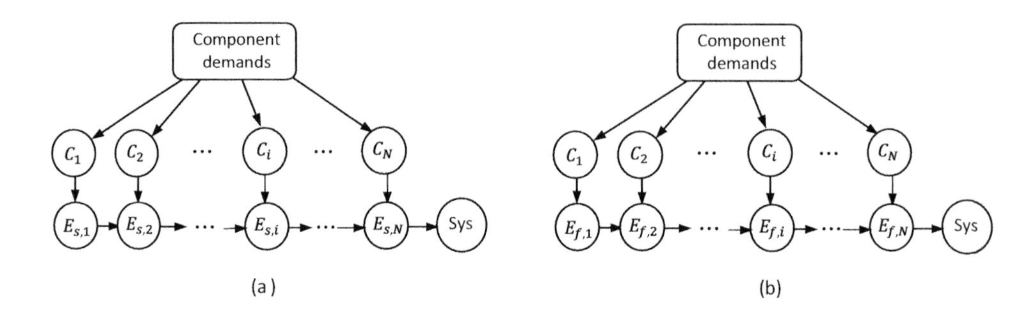

Figure 15.11 Chain-like BN topologies for (a) series system, (b) parallel systems (after Bensi et al., 2013)

As shown in Chapter 8, a general two-state system can be represented as a parallel system of its MLSs or a series system of its MCSs. Hence, a combination of the SPSs and FPSs can be used to represent a general system in a chain-like BN topology. In an MLS formulation, each MLS is represented by an SPS. The terminal nodes of the SPSs are considered as "components" of a parallel system and a chain-like FPS similar to that shown in Figure 15.11b is used to represent this parallel system, leading to the system node. In the MCS formulation, the terminal nodes of the FPSs are considered as "components" of a series system and a chain-like SPS similar to that in Figure 15.11(a) is used to represent the series system, leading to the system node. The following example demonstrates this approach.

Example 15.4 – Efficient BN Topology for a Two-State System

Consider the eight-component system depicted in Figure 15.9(a). As described before, this system has four MLSs: $\{(1,7,8),(2,7,8),(3,7,8),(4,5,6,7,8)\}$. Figure 15.12(a) shows the chain-like BN formulation using these MLSs, where for the sake of simplicity of the graph the demand object is not shown. The SPEs are denoted as $E^j_{s,i}$, where i is the index for the associated component and j is the index for the MLS. Note that components 7 and 8 are common to all the SPSs. The four terminal nodes of the SPSs, $E^j_{s,8}$, $j = 1,2,3,4$, are considered the "components" of a parallel system. Using these nodes, an FPS is constructed with $E_{f,j}$ as the child of the terminal node for the jth SPS. The states of this FPE are defined as in (15.21) with $E^j_{s,8}$ representing the component. Observe that nodes in this BN have no more than two parents.

Example 15.4 (cont.)

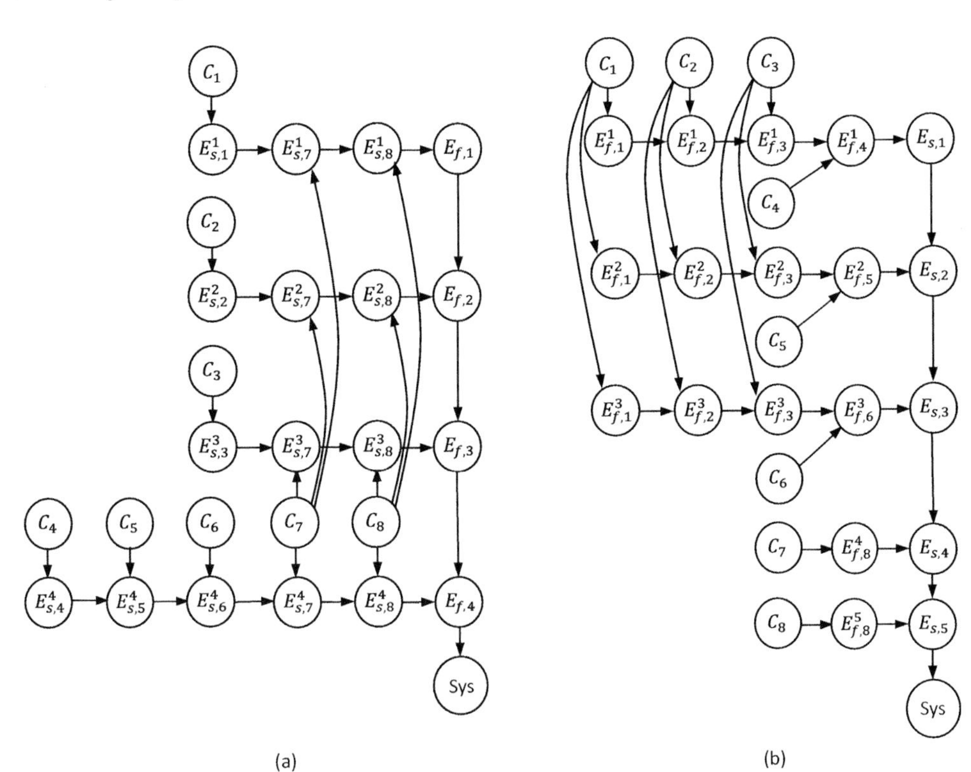

Figure 15.12 Efficient BN models of example system: (a) MLS formulation, (b) MCS formulation (after Bensi et al., 2013)

Figure 15.12(b) shows the chain-like BN formulation for the example system using the MCS formulation. The system has five MCSs: $\{(1,2,3,4),(1,2,3,5),(1,2,3,6),(7),(8)\}$. As shown in the figure, the terminal nodes of the five FPSs are considered as the "components" of a series system, and an SPS is constructed to represent this series system, leading to the system node. Observe again that the BN nodes have no more than two parents each.

Bensi et al. (2013) further develop the above approach, including its application to multistate systems and optimization of the order of components in the SPS or FPS. They also develop a heuristic for the use of super components to further enhance the efficiency of the BN model for systems. The reader is referred to the original paper for the details.

15.8 BN Modeling of Random Fields

As seen in Chapters 11, 12, and 13, some structural and system reliability problems involve random processes or fields. ("Random field" is the name we give to a multidimensional random process.) As we have seen, a random field is a collection of correlated random variables indexed by time or space coordinates with the correlation structure specified by an autocovariance function. Typically, the correlation (and dependence) between the values of the random field at two points decays with increasing separation between the points. This implies that, if we observe the field value at a point, we can learn about the values of the field at nearby points, but distant points of the field will not be affected by that information. The BN can be a powerful tool for this type of probability updating when values of a random field are observed.

To provide a motivation for the problem, consider the seismic hazard model described in Example 15.3. Suppose, instead of the seismic hazard at a single site, we are interested in predicting the ground-motion intensity values at multiple sites, say the locations of critical components of a spatially distributed infrastructure system such as a transportation network with roadways, bridges, tunnels, embankments, etc. In such a case, the intra-event error term ϵ_r in (E1) of Example 15.3, which describes the variability in the ground motion prediction equation for different locations during an earthquake, should be considered as a random field. Observe that if the intensity of ground motion at a site is measured to be higher (lower) than the median value, i.e., if the outcome of ϵ_r is positive (negative), one would expect that the intensities at nearby sites also be higher (lower) than the median value. On the other hand, for sites distant from the observation site, the information provided by the measurement would have little or no influence on our prediction of the variability around the median intensity value. A BN model of the random field error term ϵ_r can provide a framework for updating the demands placed on the components of a spatially distributed infrastructure system when the ground motions at selected sites are measured, provided an efficient BN model can be developed.

Let $X(\mathbf{s})$ denote a Gaussian random field with mean function $\mu_X(\mathbf{s})$ and autocovariance function $\kappa_{XX}(\mathbf{s}, \mathbf{s}')$, where $\mathbf{s}$ and $\mathbf{s}'$ are spatial coordinates. (If the field is non-Gaussian defined by translation of a Gaussian field, the BN is developed for the underlying Gaussian field.) Also let $\mathbf{s}_i$, $i = 1, \ldots, n$, denote a set of points of interest in the random field, where measurements are made or predictions are desired. Without loss of generality, it is convenient to work with the normalized random field

$$Y(\mathbf{s}) = \frac{X(\mathbf{s}) - \mu(\mathbf{s})}{\sigma(\mathbf{s})}, \tag{15.22}$$

where $\sigma^2(\mathbf{s}) = \kappa_{XX}(\mathbf{s}, \mathbf{s})$ is the variance function. The Gaussian random field $Y(\mathbf{s})$ has zero mean and unit variance and is completely defined by its autocorrelation coefficient function $\rho(\mathbf{s}, \mathbf{s}') = \kappa_{XX}(\mathbf{s}, \mathbf{s}')/[\sigma(\mathbf{s})\sigma(\mathbf{s}')]$. Let $\mathbf{R} = [\rho_{ij}]$, $\rho_{ij} = \rho(\mathbf{s}_i, \mathbf{s}_j)$, $i, j = 1, \ldots, n$, denote the correlation matrix of the vector of random variables $\mathbf{Y} = [Y(\mathbf{s}_1), Y(\mathbf{s}_2), \ldots, Y(\mathbf{s}_n)]^{\mathrm{T}}$ at the selected points. In general, the correlation matrix $\mathbf{R}$ is full. If the random variables $\mathbf{Y}$ are selected as the nodes of a BN, then, because of the pairwise dependence of the random variables, there must be directed links between all pairs of the nodes, as shown in Figure 15.13. Note that the nodes have increasing number of parents with $Y(\mathbf{s}_n)$ having

n parents. For large n, this makes the BN model of the random field extremely complicated and computationally demanding. Of course, it is possible to delete the links between pairs of nodes corresponding to distant points in the random field. However, we need a logical framework for this purpose. In this section, we describe a method developed by Bensi et al. (2011b) for efficient but approximate BN representation of random variables drawn from a Gaussian random field.

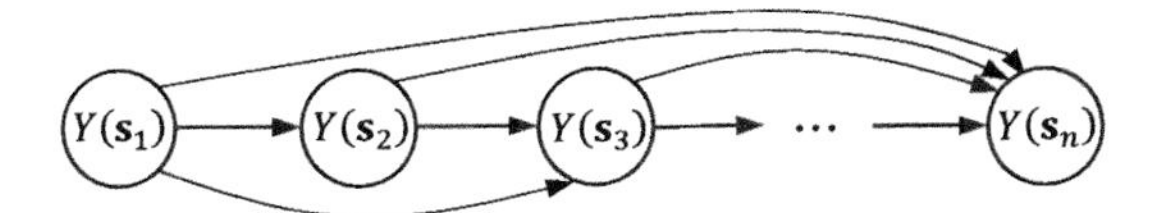

Figure 15.13 BN model of random variables drawn from a Gaussian random field at selected points (after Bensi et al., 2011b)

It is well known that an n-vector of Gaussian random variables $\mathbf{Y}$ can be expressed as a linear function of an n-vector of statistically independent standard normal random variables $\mathbf{U}$, provided the correlation matrix $\mathbf{R}$ is non-singular (see Section 3.6.3). Here, because the means of $\mathbf{Y}$ are zero and its variances are unity, the transformation takes the form

$$\mathbf{Y} = \mathbf{TU}, \tag{15.23}$$

where $\mathbf{T} = [t_{ij}]$ is an $n \times n$ transformation matrix having elements t_{ij}, $i, j = 1, \ldots, n$. Matrix $\mathbf{T}$ can be determined as the eigenmatrix of $\mathbf{R}$ or its Cholesky decomposition (see Appendix 3B). In the latter case, $\mathbf{T}$ is a lower-triangular matrix so that $Y(\mathbf{s}_i)$ is a function of only the first i elements of $\mathbf{U}$. For this case, a BN model of random variables $\mathbf{Y}$ and $\mathbf{U}$ appears as in Figure 15.14. Observe that the number of parents of the $Y(\mathbf{s}_i)$ nodes is increasing with the index i so that node $Y(\mathbf{s}_n)$ still has n parents. So, the transformed formulation does not provide an advantage over the BN model in Figure 15.13. If the eigen-decomposition formulation is used, in general $\mathbf{T}$ is a full matrix and, hence, links are directed from every U_j node to every $Y(\mathbf{s}_i)$ node. In both cases, the dependence between the elements of $\mathbf{Y}$ is modeled through their common parents.

As $Y(\mathbf{s}_i) = t_{i1}U_1 + \cdots + t_{in}U_n$, there is a one-to-one correspondence between the elements t_{ij} of the transformation matrix and the directed links between nodes U_j and $Y(\mathbf{s}_i)$. Specifically, if $t_{ij} = 0$ then there is no link from U_j to $Y(\mathbf{s}_i)$. Furthermore, if all elements t_{ij}, $i = 1, \ldots, n$, of the jth column of $\mathbf{T}$ are zero, then no links originate from node U_j and that node can be eliminated. This suggests that an approximate but more efficient BN model of the random field can be developed by replacing by zero some of the elements of the

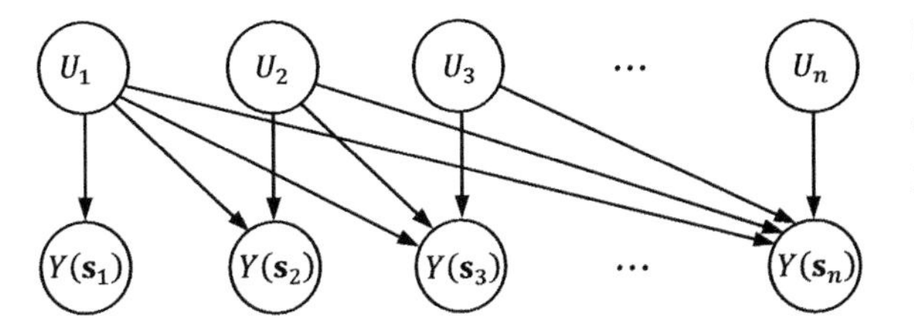

Figure 15.14 BN model of random variables using Cholesky decomposition of the correlation matrix (after Bensi et al., 2011b)

transformation matrix or entire columns that have small values. A problem with this approach is that errors occur in the variances of $Y(\mathbf{s}_i)$, which must equal unity. To compensate for this error, we write the approximate transformation in the form

$$\widehat{\mathbf{Y}} = \mathbf{DV} + \widehat{\mathbf{T}}\widehat{\mathbf{U}}, \tag{15.24}$$

where $\widehat{\mathbf{Y}}$ is the approximated vector of random field variables, $\widehat{\mathbf{T}}$ is the approximated transformation matrix having size $n \times m$, $m \leq n$, after elimination of zero columns of $\mathbf{T}$, $\widehat{\mathbf{U}}$ is the m-vector of standard normal random variables obtained by dropping the elements of $\mathbf{U}$ corresponding to the zero columns of $\mathbf{T}$, $\mathbf{V}$ is an n-vector of statistically independent standard normal random variables, and $\mathbf{D}$ is a diagonal matrix with element d_{ii}, $i = 1, \ldots, n$, selected such that

$$d_{ii}^2 + \sum_{j=1}^{m} \hat{t}_{ij}^2 = 1, \tag{15.25}$$

where $\hat{t}_{ij}$ is the (i,j) element of $\widehat{\mathbf{T}}$. The BN model of the approximated random field variables is now constructed as shown in Figure 15.15. Depending on how many links and nodes are removed, this BN can potentially be far more efficient than those in Figures 15.13 and 15.14. Observe that the dependence between nodes $Y(\mathbf{s}_i)$ is achieved through their common parents $\widehat{U}_j$. The role of nodes V_i is to restore the unit variance of the nodes $Y(\mathbf{s}_i)$.

Figure 15.15 Approximate, efficient BN model of random field variables (after Bensi et al., 2011b)

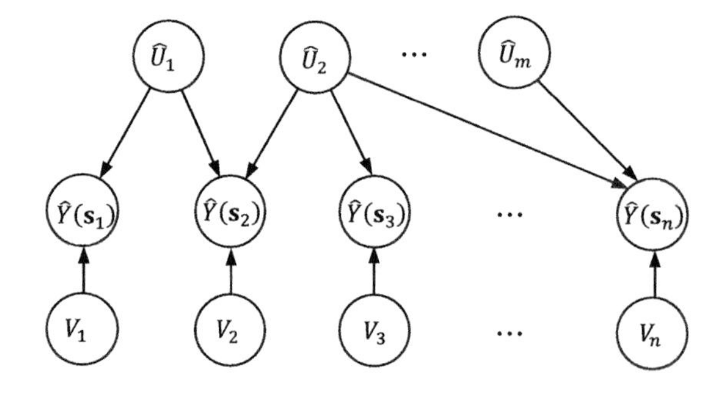

To guide the selection of elements and columns of the transformation matrix, to set them equal to zero, Bensi et al. (2011b) introduce node and link importance measures. The importance measure for node j is defined as

$$N_j = \sum_{k=1}^{n} \sum_{i=1}^{n} |t_{ij} t_{kj}|. \tag{15.26}$$

Without the absolute-value sign, the double sum represents the sum of contributions of node U_j to the correlation matrix $\mathbf{R}$. However, because the products $t_{ij} t_{kj}$ can have positive or negative signs and both are important, the absolute value is used to define the measure. Thus, N_j provides a measure of the information contributed by node U_j to the correlation matrix that will be lost if node U_j is removed. The importance measure for the link directed from node U_j to node $Y(\mathbf{s}_i)$ is defined as

$$L_{ij} = \sum_{k=1}^{n} \left| t_{ij} t_{kj} \right| - t_{ij}^2. \tag{15.27}$$

Without the absolute-value sign, the above sum is the contribution of the term t_{ij} and, hence, the link from U_j to $Y(\mathbf{s}_i)$, to the correlation matrix. Again, because the product terms can have positive or negative signs and both are important, the absolute-value sign is used. L_{ij} can be seen as a measure of the information contributed by the link from node U_j to node $Y(\mathbf{s}_i)$ that will be lost if this link is removed. In the method proposed by Bensi et al. (2011b), essentially, nodes and links are removed in the order of smallest values of N_j and L_{ij} first, with the variances corrected according to (15.24)–(15.25).

Bensi et al. (2011b) explore and compare seven different methods for eliminating nodes and links. These include methods employing the transformation matrix defined by the eigenvalue or Cholesky decomposition, and methods with the transformation matrix constructed through an optimization algorithm. For the latter, first the desired number m of the parent nodes is selected. Then, the elements of the approximate transformation matrix are determined by solving the constrained nonlinear optimization problem

$$\text{Minimize: } \sum_{i=1}^{n-1} \sum_{j=i+1}^{n} \left[\rho_{ij} - \sum_{k=1}^{m} \hat{t}_{ik} \hat{t}_{jk} \right]^2 \tag{15.28a}$$

$$\text{Subject to: } \sum_{k=1}^{m} \hat{t}_{ik}^2 \leq 1, i = 1, \ldots, n. \tag{15.28b}$$

This optimization problem minimizes the squared error in the representation of the correlation coefficients for the given number of parents m. The resulting approximate transformation matrix is then used in (15.24)–(15.25) to correct the variances. Extensive investigations by Bensi et al. (2011b) show that the optimization approach provides the best results with gradual link elimination being slightly more accurate than the node elimination approach for the same number of parents. The following example demonstrates an application of this method to post-earthquake risk assessment for an infrastructure system.

Example 15.5 – Post-Earthquake Risk Assessment of an Infrastructure System

This example, adopted from Bensi et al. (2011a, 2015), develops and applies a BN for post-earthquake risk assessment of a hypothetical model of a segment of the proposed California high-speed rail (HSR) system. At the time of the study, the system was in its early design phase. Hence, the adopted model is hypothetical and highly idealized. For this reason, the application and the results derived from it should be considered as a conceptual illustration of what can be achieved with the BN framework rather than viewed as results relevant to the actual HSR system, if and when it becomes a reality. For the sake of brevity, only novel aspects of the model are described. Further details and results can be found in the original publications.

Example 15.5 (cont.)

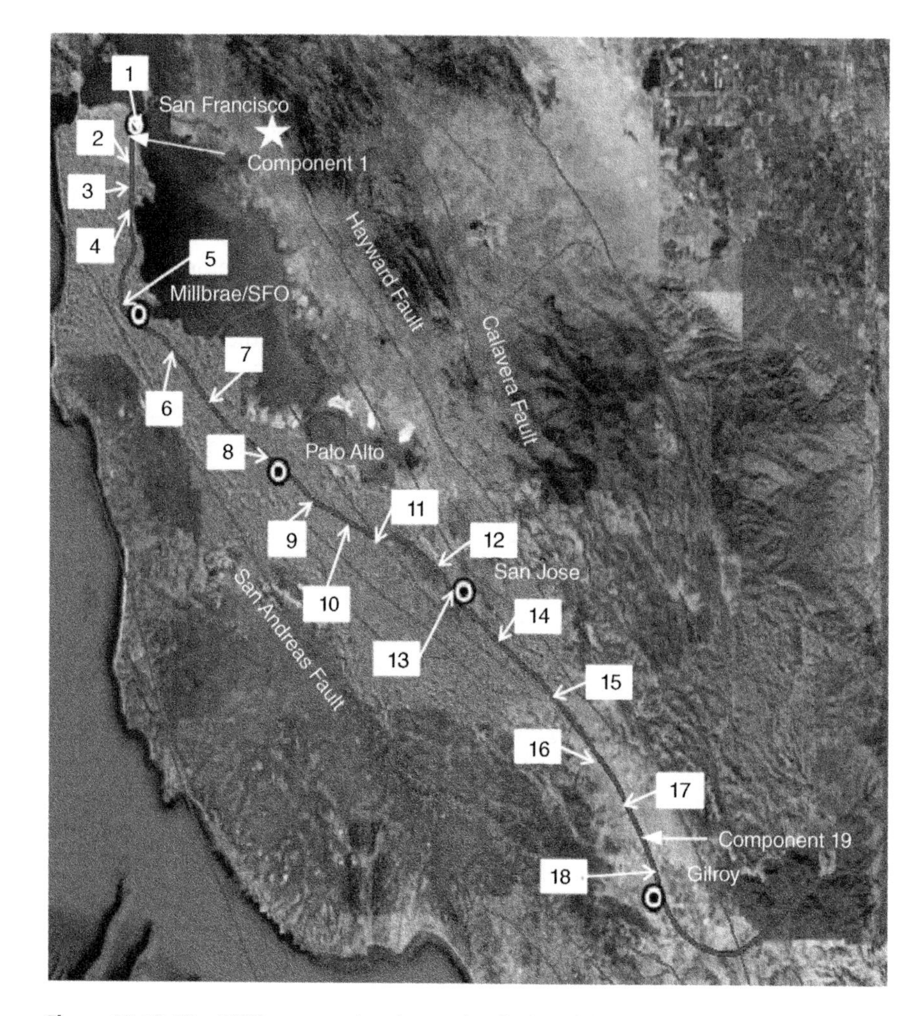

Figure 15.16 The HSR system showing active faults, cities with stations, system components and ground-motion-prediction points (after Bensi et al., 2015, with permission from ASCE)

We consider the northern segment of the proposed HSR system, from San Francisco to Gilroy. This segment is situated in a highly seismic region with several active faults, including the San Andreas, Hayward, and Calaveras Faults, in the vicinity. Figure 15.16 shows the configuration of the system and the seismic environment. The system consists of different types of point- or distributed-site components along the route, including tunnels, trenches, embankments, and aerial structures (bridges). These are idealized into 17 distributed-site (numbers 1–11, 13, and 15–19) and 2 point-site (numbers 12 and 14) components. Eighteen

Example 15.5 (cont.)

ground-motion-prediction points (GMPPs) define the end points of these components, as shown in Figure 15.16. For example, component 1, located in San Francisco, is a tunnel of 4 km length and stretching from GMPP 1 to GMPP 2, and component 19 is a 9.5 km embankment north of Gilroy, stretching from GMPP 17 to GMPP 18. Three states for each component are considered: undamaged, slightly damaged, and moderately/severely damaged. Undamaged components are assumed to be fully operational. Slight damage is associated with reduced performance, i.e., trains must slow down when traversing the component. Moderate or severe damage implies complete loss of service. Fragility models for these component states are adapted from the literature, as described in Bensi et al. (2011a).

We define system performance as the ability to travel from San Francisco to Gilroy and from there on to Southern California. However, if a northern segment of the system is not passable, passengers can use alternative transportation means to bypass the damaged segment and board the train further down the line to complete their journey to Gilroy and on to Southern California. With this in mind, four system-performance criteria are considered: ability to travel with normal/reduced speed from (1) San Francisco to Gilroy, (2) Millbrae to Gilroy, (3) Palo Alto to Gilroy, and (4) San Jose to Gilroy. Travel at normal speed between San Jose and Gilroy is possible if none of components 15–19 is damaged; travel must be at reduced speed if any of these components is slightly damaged; and ability to travel is lost if any of these components is moderately or severely damaged. Travel from Palo Alto to Gilroy similarly depends on the states of components 8–14, but also on the state of travel from San Jose to Gilroy. Similarly, travel from Millbrae to Gilroy depends on the states of components 5–7 as well as the state of travel from Palo Alto to Gilroy, and that from San Francisco to Gilroy depends on the states of components 1–4 and the state of travel from Millbrae to Gilroy.

Figure 15.17 shows the BN objects for the intra-event random field term $\epsilon_r(\mathbf{s})$ (left) and the components and system model (right). Observe that for the former, only six U_i nodes with 18 links are used to approximately describe the random field at the 18 GMPPs with at most three parents per node. For the system, the survival-path-sequence approach described in Section 15.8 is used to formulate the BN with no more than two parents per node. The objects for the seismic hazard and component states are similar to those described in Example 15.3 and Section 15.7, respectively, and are not shown here. We recall that the inter-event error term ϵ_m is a parent to ground motion intensity nodes at all sites (see Figure 15.6c). For details, see Bensi et al. (2011b, 2015).

A BN framework such as the one for the HSR system is particularly useful when used in near-real-time application for risk assessment and decision-making, as information is gradually received right after a disaster. To illustrate such an application, suppose the following sequence of evidence cases (ECs) are realized for the above system:

EC1: Shortly after the earthquake event, we learn from the Berkeley Seismological Laboratory that the earthquake had a magnitude $M = 6.8$ and an epicenter located on Hayward fault, 30 km from its north end, shown as a star in Figure 15.16.

Example 15.5 (cont.)

EC2: A little later we learn that a recording at GMPP 2 indicated a spectral acceleration of $S_2 = 0.45 - 0.50$ g at 1 Hz frequency.

EC3: A little later we learn that the ground motion at GMPP 9 was "pulse-like." This observation suggests that the rupture on the fault propagated southward from the epicenter.

EC4: A little later we learn that component 17 has experienced severe damage.

As we will see below, the above evolving ECs all reveal pessimistic information. In reality, ECs can describe both optimistic and pessimistic observations. However, the uniformly negative observations here are easier to illustrate. Below, we show posterior results for selected nodes in the BN for the above ECs.

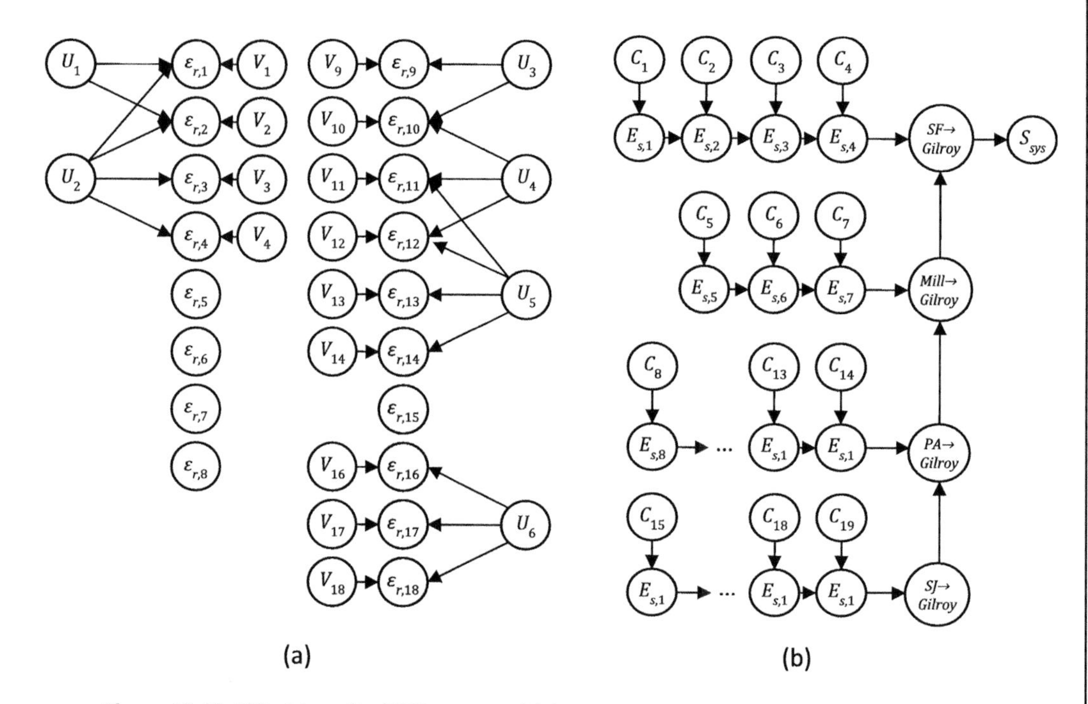

Figure 15.17 BN objects for HSR system: (a) intra-event error random field term $\epsilon_r(\mathbf{s})$, (b) components and system (after Bensi et al., 2015, with permission from ASCE)

Figure 15.18 shows the prior (EC1) and posterior (EC2–EC4) distributions of the discretized ground-motion intensities (spectral acceleration at 1 Hz frequency in units of gravity acceleration) at GMPPs 1, 6, 11, and 16 under the above evolving set of information. Note that the last interval in each distribution collects the probability masses for the entire tail. First examine

Example 15.5 (cont.)

the posterior distributions for EC1 and EC2. For GMPP 1, we see that the prior distribution under EC1 has a mode around 0.1–0.2 g. Upon updating for EC2, the distribution significantly shifts toward higher intensity values with the mode now being around 0.2–0.3 g. This is partly because of the positive correlation between the ground motion intensities at all nodes, due to the common parent ϵ_m. However, for GMPP 1, the shift is more due to the proximity of GMPP 1 to GMPP 2: an unexpectedly high intensity at GMPP 2 suggests that the intensity at the nearby GMPP 1 must also be high. This is the effect of the spatial autocorrelation of the random field $\epsilon_r(\mathbf{s})$. Comparing the prior and posterior distributions for EC2 at GMPPs 6, 11, and 16, we see that there is a much smaller shift toward higher intensities. Because these GMPPs are far from GMPP 2, the shift is solely due to the common parent ϵ_m.

Now consider the posterior distributions after EC3. The observation that the motion at GMPP 9 was "pulse-like" suggests that a good portion of the earthquake rupture on Hayward fault propagated southward from the epicenter. This direction of the rupture has no effect on the GMPPs that are located on the northern segment of the system, while the GMPPs in the

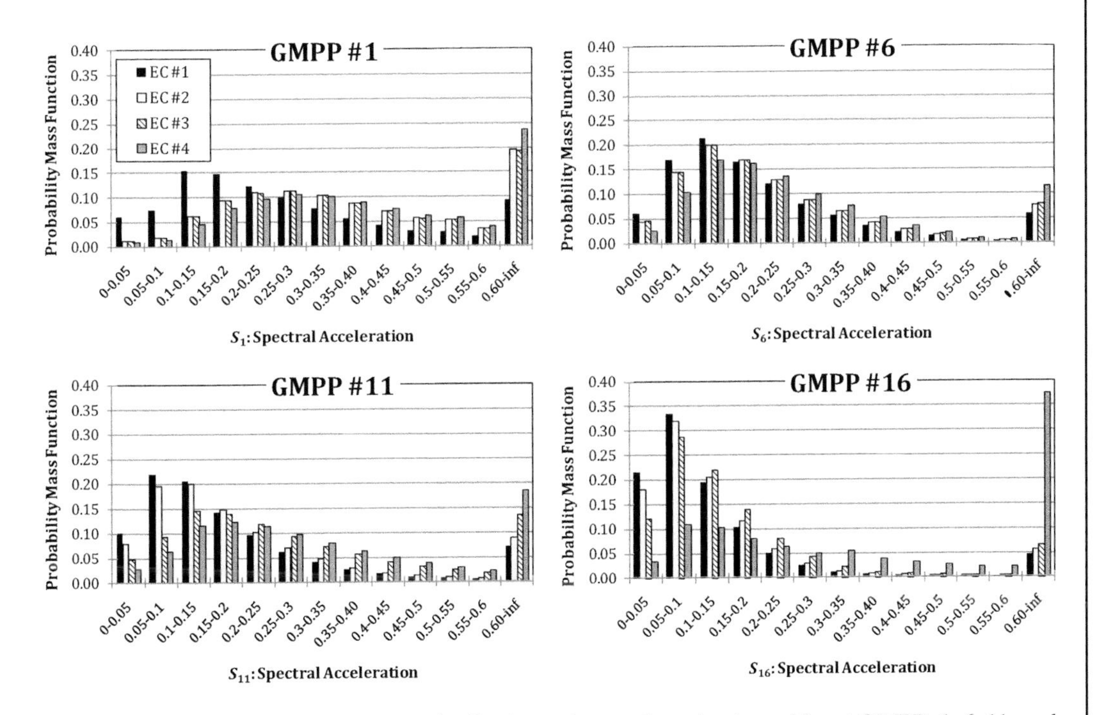

Figure 15.18 Prior and posterior distributions of ground-motion intensities at GMPPs 1, 6, 11, and 16 under EC1–EC4 (after Bensi et al., 2015, with permission from ASCE)

Example 15.5 (cont.)

southern segment will be influenced by the directivity effect. Indeed, examining the posterior distributions at the selected GMPPs in Figure 15.18, we see that the intensity distributions at GMPPs 1 and 6 are not affected by the observation, whereas the posterior distributions at GMPPs 11 and 16 have significantly shifted upwards on account of the probable directivity effect.

Finally, consider EC4. The failure of component 17 could be due to inadequate capacity, or high intensity of the ground motion. The latter likelihood sharply shifts the distribution of the intensity at GMPP 16, which is located at one end of component 17, upward. In contrast, the posterior distributions at GMPPs 1, 6, and 11 shift by much less, all because of having ϵ_m as a common parent.

Next, we consider the probabilities of moderate or severe damage of the individual components of the system under the evolving information, shown in Figure 15.19. Under EC1, the prior probabilities are more or less uniform, with somewhat smaller probabilities for components at larger distances from the epicenter. After EC2, the posterior probabilities show sharp increases for components in the vicinity of GMPP 2, due to the random field correlation effect, and smaller but uniform increases for the other components, due to the common ϵ_m parent.

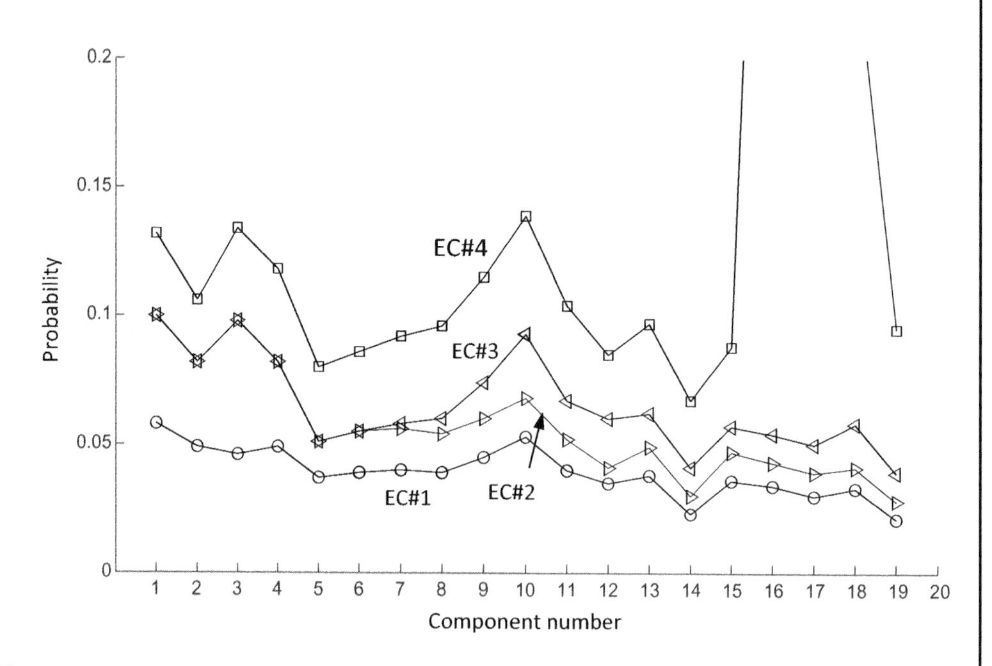

Figure 15.19 Prior and posterior probabilities of moderate to severe damage of components under evolving state of information (after Bensi et al., 2015, with permission from ASCE)

Example 15.5 (cont.)

Under EC3, increases are observed in the probabilities only for components positioned in the southern segment of the system, due to the likely directivity effect. Finally, under EC4, increases in probabilities for all components are observed, due to the common ϵ_m parent, but increases are larger for components near the severely damaged component 17, due to the random field effect of $\epsilon_r(\mathbf{s})$.

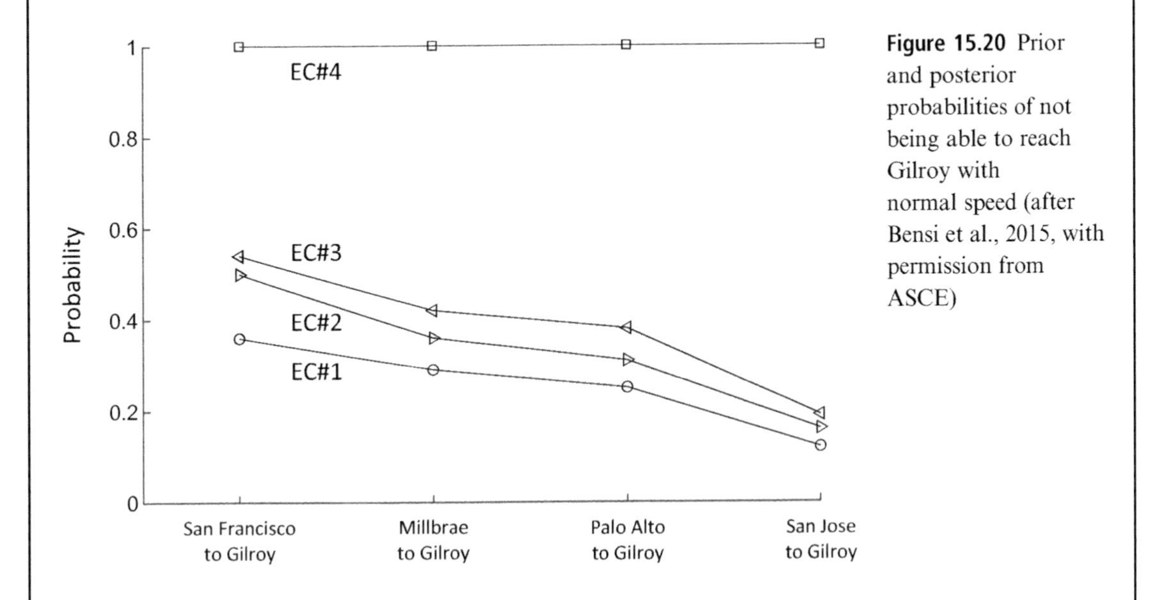

Figure 15.20 Prior and posterior probabilities of not being able to reach Gilroy with normal speed (after Bensi et al., 2015, with permission from ASCE)

Finally, Figure 15.20 shows the prior and posterior probabilities of one or more damaged components along the travel path, which would force the HSR system to slow down or stop operation. Observe that the probability of not being able to reach Gilroy with normal speed is larger the farther away from Gilroy we are. Obviously, this is due to the larger number of components that can potentially fail along the path. With increasing unfavorable evidence, all posterior probabilities increase. Under EC4, the posterior probabilities of not reaching Gilroy are, of course 1, because of the severe damage experienced by component 17.

This example demonstrates the power of BN as a tool for information processing in the context of risk assessment. Thanks to exact inference algorithms that are available for BNs with discrete or discretized nodes, this kind of analysis can be performed in near-real-time. In this context, the BN framework can be highly useful in pre-, during-, and post-hazard risk assessment and decision-making. Bensi et al. (2011a, 2015) provide more examples, including a decision framework for determining the optimal order of inspecting components of a system after a seismic hazard. Other applications of BNs can be found in Faber et al. (2002) for the decommissioning of offshore facilities, Straub and Grêt-Regamey (2006) for avalanche risk

Example 15.5 (cont.)

assessment, Bayraktarli and Faber (2011) for managing seismic hazard in cities, Langseth and Portinale (2007) for system reliability analysis, Nishijima and Faber (2007) for typhoon risk management, Castelletti and Soncini-Sessa (2007) for water resources management, and Sättele et al. (2015) for an early-warning system for debris flow hazard.

15.9 Dynamic Bayesian Network

As described in Chapters 11–13, many problems in structural reliability involve random processes, wherein variables influencing the state of the system randomly vary along a time or space coordinate. A special class of these problems arises when the processes involved are Markovian in nature. A process is said to be Markovian when its state at a time or space coordinate t_2 given its state at an earlier time or space coordinate t_1 is independent of the states at all instances prior to t_1. As an example, the response of a linear oscillator to a stochastic excitation is Markovian because, given the position and velocity of the oscillator at time t_1, the response at a later time t_2 is independent of the response history prior to time t_1. This is not the case for a hysteretic oscillator, for which the system characteristics and, hence, the response at a given time depend on the entire prior history of the response. Given a set of time points t_i, $i = 0, 1, \ldots, n$, with t_0 denoting the initial time, a Markov process is completely defined by describing the distribution of the initial state and the conditional probability distribution at time t_i given the state at time t_{i-1}, $i = 1, \ldots, n$. The latter is known as the transition probability distribution. The Markov process is said to be homogeneous if, for equally spaced time points, the transition probability distribution is identical for all i.

A dynamic BN (DBN) is a generalization of a Markov process. It consists of a sequence of BNs associated with time or space instances called *slices*, where the nodes in the $(i-1)$th slice are parents to the corresponding nodes in the ith slice. Thus, the transition probability distributions define the CPTs of the nodes in each slice. For a homogeneous case, the CPTs remain independent of the slice index i. Figure 15.21 shows a simple DBN for a Markov process $X(t)$ at time instances t_i, $i = 0, 1, \ldots, n$, with $X_0 = X(t_0)$ indicating the initial state and $X_i = X(t_i)$ indicating the state at time instance t_i. The pair of nodes (X_i, Y_i) represents the ith slice of the DBN. The variables Y_i, $i = 0, 1, \ldots, n$, are internal, and children of the respective X_i. They may represent, for example, measurements (exact or inexact) of the process values at the respective time instances. Inference can be made for the states of the Markov process at any time step for any evidence on Y_i. Although the algorithms mentioned in Section 15.5 are applicable to DBNs, specialized algorithms are available that take advantage of the specific structure of DBNs to more efficiently perform inference analysis (Murphy, 2002; Russell and Norvig, 2003).

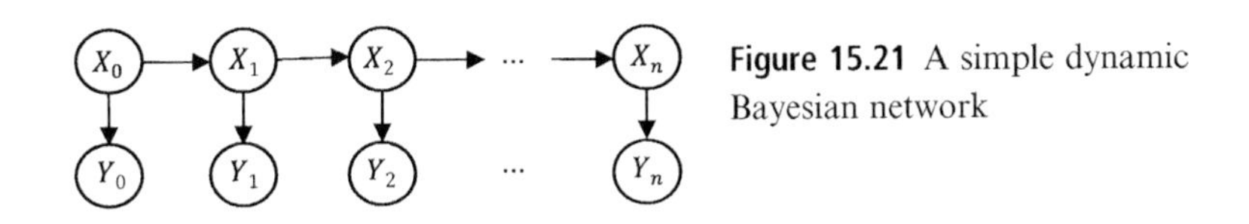

Figure 15.21 A simple dynamic Bayesian network

A stochastic model often involves time-invariant uncertain variables or parameters, such as those representing system properties, model errors, or uncertainties in distribution parameters. This kind of uncertainty, which is common to all time instances and causes dependence among the states of the process, invalidates the Markov assumption. However, conditioned on such time-invariant variables or parameters, the process can be Markovian if the assumptions mentioned above hold conditionally. This kind of dependence caused by time-invariant uncertainties poses no problem for modeling and analysis by DBN because the nodal CPTs are already based on conditional distributions. The following example demonstrates an application of the DBN.

Example 15.6 – DBN Modeling and Inference of Structural Deterioration

Taken from Straub (2009), this example shows the modeling of structural deterioration by a DBN and inference as various quantities are measured or observed. Adopted from Ditlevsen and Madsen (1996), the example considers the fracture mechanics model of crack growth in an infinite plate in accordance with Paris' law

$$\frac{\mathrm{d}a(n)}{\mathrm{d}n} = C\left[\Delta S\sqrt{\pi a(n)}\right]^m, \tag{E1}$$

where n is the number of stress cycles, $a(n)$ is the crack length after n cycles, ΔS is the stress range (here assumed to be the same for all cycles), and C and m are model parameters. With the initial condition $a(0) = a_0$, the solution of this differential equation provides the crack size as a function of the number of cycles (Ditlevsen and Madsen, 1996)

$$a(n) = \left[\left(1 - \frac{m}{2}\right)C\Delta S^m \pi^{m/2} n + a_0^{(1-m/2)}\right]^{(1-m/2)^{-1}}. \tag{E2}$$

The event of failure after n cycles is described by the limit-state function

$$g(a_0, \Delta S, C, m) = a_c - a(n), \tag{E3}$$

where a_c is the critical crack length. Distributions of the four random variables are described in Table 15.1. Note that $\ln C$ and m are jointly normal with a correlation coefficient of -0.9.

Because the crack length after n cycles depends only on the initial crack length (not the history of its growth), conditioned on the time-invariant random variables ΔS, C, and m, the crack growth process $a(n)$ is Markovian. Let $t = 1, \ldots, T$ denote the set of selected time instances and a_t the crack length at time t. Then, using (E2) and the Markov property, the crack length at time step t can be written as

Example 15.6 (cont.)

Table 15.1 Description of random variables for example problem (after Straub, 2009, with permission from ASCE)

Variable	Distribution	Mean	Standard deviation	Correlation coefficient
a_0, mm	Exponential	1	1	–
ΔS, N/mm^2	Normal	60	10	
$\ln C, m$[a]	Bi-normal	$(-33, 3.5)$	$(0.47, 0.3)$	$\rho_{\ln C, m} = -0.9$

[a] Parameter values are for dimensions specified in Newtons and millimeters.

$$a_t = \left[\left(1 - \frac{m}{2}\right) C \Delta S^m \pi^{m/2} \Delta n + a_{t-1}^{(1-m/2)}\right]^{(1-m/2)^{-1}}, \quad t = 1, \ldots, T, \tag{E4}$$

where Δn is the number of cycles between time steps $t-1$ and t. In the following analysis $\Delta n = 10^5$ cycles is used. To simplify the DBN model, Straub (2009) takes advantage of the fact that random variables C and ΔS^m appear as a product. Thus, he defines the intermediate random variable $q = (1 - m/2) C \Delta S^m \pi^{m/2} \Delta n$, which clearly also depends on m. The time-invariant random variables now are m and q. All random variables in this problem are continuous. To facilitate inference by use of an exact algorithm, Straub (2009) discretizes the random variables following the procedure described in Section 15.4. Details of the number and sizes of intervals are given in the paper and will not be repeated here. Suffice to say that the smallest number of selected intervals is 30 for m and the largest is 80 for a_t. These variables are treated as interval random variables, as described in Section 15.4.

Figure 15.22 shows the DBN for the example problem. On the far upper left are the nodes for the time-invariant random variables C, ΔS, and m. These are parents to random variable q, which is a function of the three random variables as defined in the preceding paragraph. The CPT of node q is determined by simulating realizations of C, ΔS, and m in their respective intervals. Note that node m is also a parent of node C because of their joint normal distribution. The CPT of node C is determined from the conditional normal distribution given m. Also on the left is node a_0 of the initial crack size. This is a root node and its CPT is obtained by discretizing the exponential distribution described in Table 15.1.

A typical slice of the DBN in Figure 15.22 consists of nodes m_t, q_t, a_t, Z_t, and E_t. As m_t and q_t are time invariant, their distributions do not change from time slice to time slice and, therefore, their CPTs are identity matrices. Nodes a_t are deterministic and their values are determined from (E4) for the given values of the parents m_t, q_t, and a_{t-1}. Nodes Z_t represent the results of inspections at selected time instances. Specifically, Z_t is a two-state $(0, 1)$ variable indicating whether at time instance t a defect is detected $(Z_t = 1)$ or not $(Z_t = 0)$. The probability of detecting a defect depends on the size of the crack and is given by

$$\Pr(Z_t = 1 | a_t) = 1 - \exp\left(-\frac{a_t}{10}\right). \tag{E5}$$

Example 15.6 (cont.)

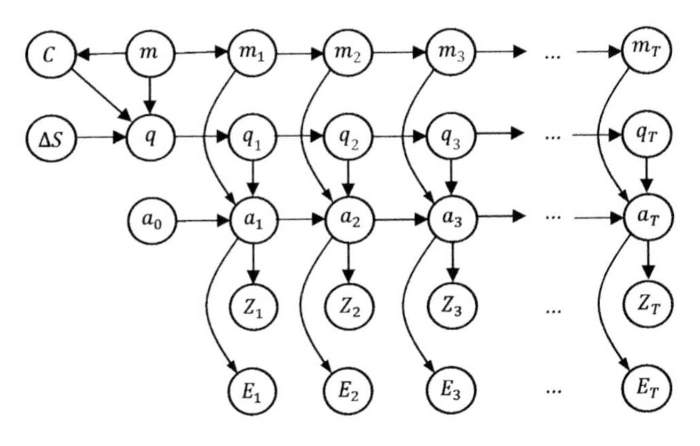

Figure 15.22 DBN of the structural deterioration problem (after Straub, 2009, with permission from ASCE)

Hence, for a given a_t, the CPT of Z_t consists of the probabilities $\exp(-a_t/10)$ and $1 - \exp(-a_t/10)$. Finally, node E_t denotes the state of the structure at time instance t. E_t is a two-state variable with $\{E_t = 0\}$ indicating the failure state defined as $\{a_c \leq a_t\}$ and $\{E_t = 1\}$ indicating its complement, the survival event.

Figure 15.23 shows the generalized reliability index for the failure event as a function of the number of cycles. Shown are the estimates based on the DBN, the 95% confidence interval based on Monte Carlo simulations (labeled as MCS in the figure), and the second-order

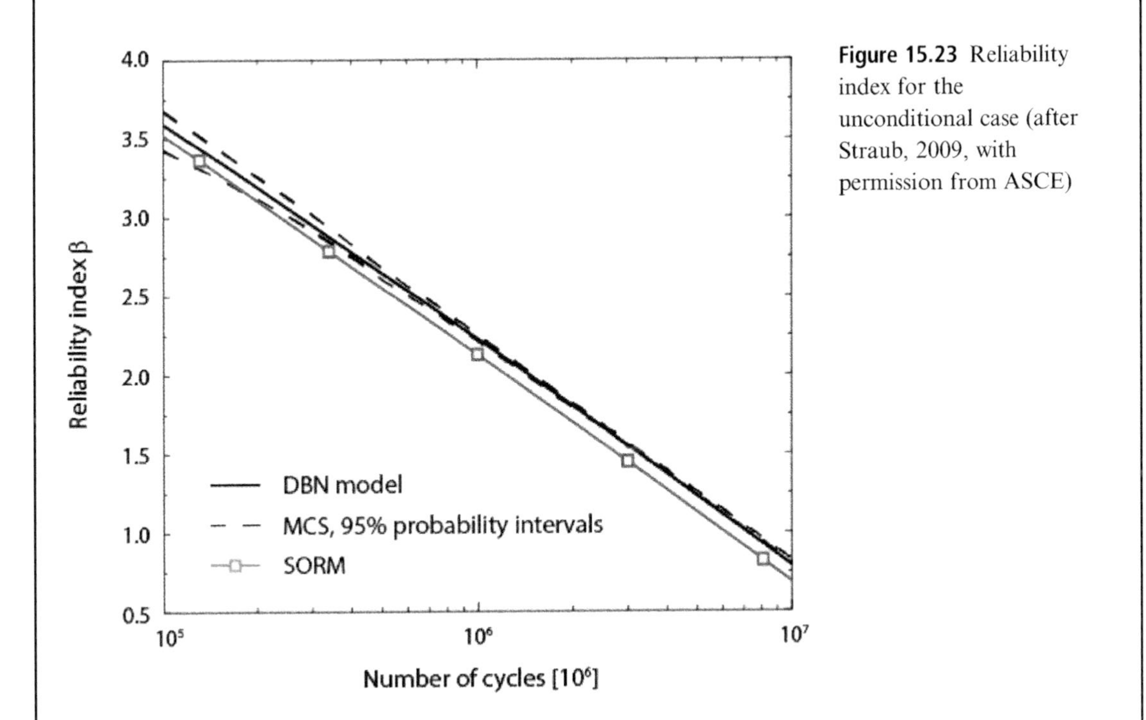

Figure 15.23 Reliability index for the unconditional case (after Straub, 2009, with permission from ASCE)

Example 15.6 (cont.)

reliability method (SORM). The DBN result shows the probability of the event $\{E_t = 0\}$ for the sequence of time slices. The close agreement between the DBN and Monte Carlo simulation results indicates that the discretization of random variables is adequate. Note that other than the discretization, no other approximation is involved in the DBN analysis. In contrast, the SORM result uses the continuous distributions but employs a second-order approximation of the limit-state surface at the design point. The results reported in Figure 15.23 are unconditional, i.e., do not involve any observations.

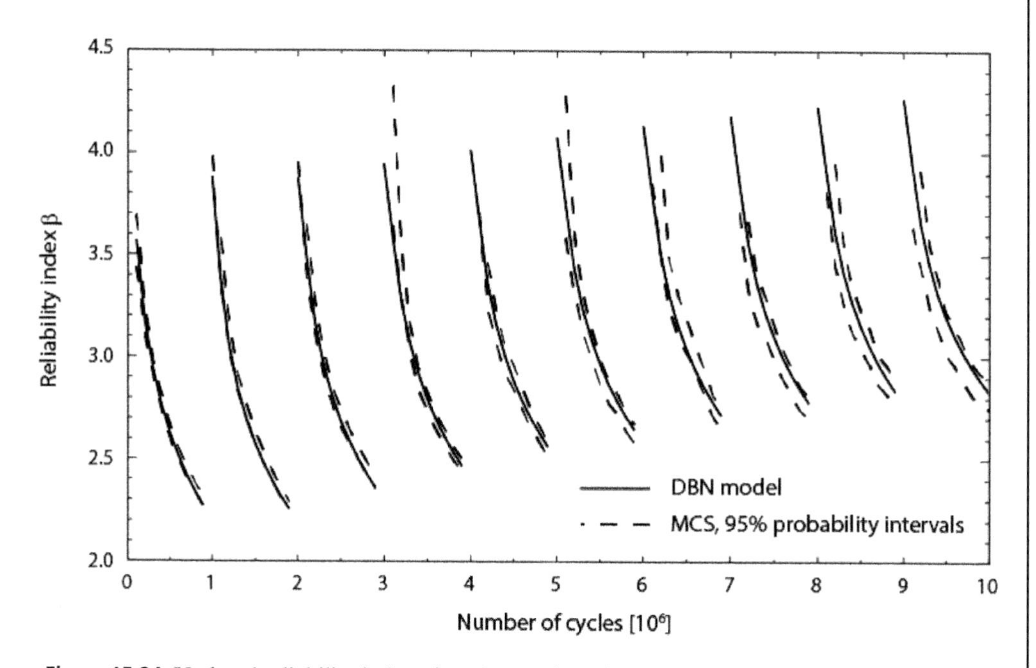

Figure 15.24 Updated reliability index after observation of no defect upon inspection after each set of 10^6 cycles of loading (after Straub, 2009, with permission from ASCE)

Figure 15.24 shows updated results for the generalized reliability index, where no defect is detected upon inspection after each set of 10^6 cycles of loading, i.e., $Z_{10} = 0$, $Z_{20} = 0$, etc., are observed. We see that the reliability index decreases with increasing load cycles until a no-defect detection is made, at which point the reliability index makes a sharp jump. The results based on DBN and Monte Carlo simulations are in close agreement.

As we have seen earlier, it is possible to update the distribution of each node in the BN upon receiving evidence on other nodes. Figure 15.25 shows the complementary CDF of node ΔS for different evidence cases. The solid curve shows the result with no evidence (the prior

Example 15.6 (cont.)

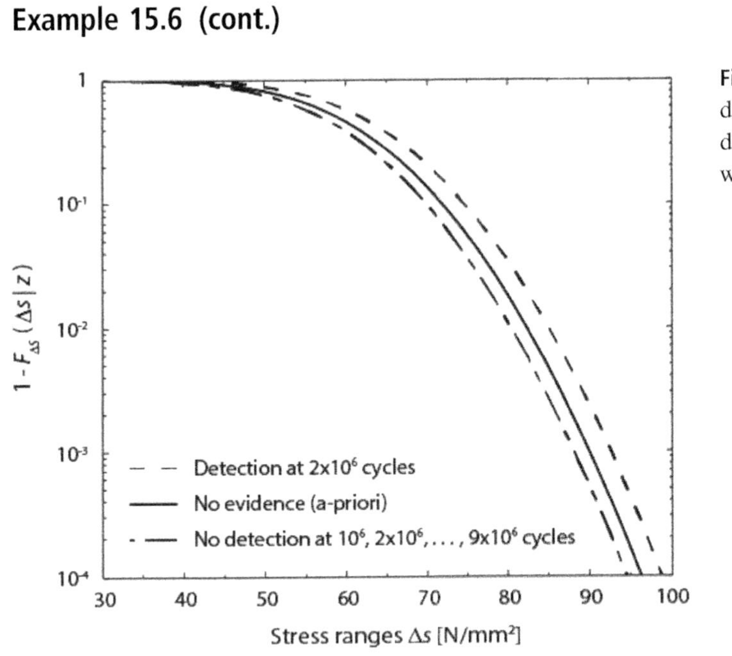

Figure 15.25 Prior and posterior distributions of stress range for different ECs (after Straub, 2009, with permission from ASCE)

distribution). The dashed line is based on the observation of a defect at 2×10^6 cycles of loading. We can see that the likelihood for larger values of the stress range has increased. The dash-dotted line shows the posterior distribution of the stress range after repeated observations of no defect after 10^6, 2×10^6, ..., 9×10^6 cycles of loading. Here, the posterior distribution shows smaller likelihoods for large values of the stress range.

Further examples of the approach described in this example for deteriorating structures can be found in Straub (2009) and Luque and Straub (2016).

15.10 Bayesian Network Enhanced by Structural Reliability Methods

As we have seen, BNs provide a powerful tool for probabilistic analysis with discrete random variables, including exact inference upon receiving information on selected nodes. Modeling of the BN is intuitive, and analysis can be performed in near-real-time, even by individuals who lack deep expertise in probabilistic analysis. In this context, the BN framework is an ideal tool for risk management of infrastructure systems, particularly for risk assessment and decision-making in preparation for, during, or post-hazard events. On the other hand, structural reliability methods (SRMs), such as the first-order reliability method (FORM), SORM, and simulation methods, can handle continuous random variables with arbitrary dependence. However, there are certain limitations to these methods, including the approximate nature of FORM and SORM and the difficulty in estimating small probabilities by simulation techniques. The problem of updating upon observation of rare events, including equality events that have zero probability, is cumbersome or infeasible to solve with these methods. Furthermore, SRMs require expertise and specialized software that may not be suitable for use by non-experts or in real-time applications. The advantages and disadvantages of BN and SRMs appear to be complementary. The question then arises whether it is possible to combine these two approaches into a unified framework that takes advantage of the strengths of each method and avoids their weaknesses.

Based on the work in Straub and Der Kiureghian (2010a, 2010b), this section describes a BN approach enhanced by SRMs. The nodes of the enhanced BN (eBN) represent discrete or continuous random variables. A method is developed to eliminate the continuous nodes, arriving at a reduced BN (rBN) that has only discrete nodes and can be solved by exact inference algorithms. This requires computing the CPTs of the discrete children of continuous nodes by use of SRMs. The rBN can be used in near-real-time applications for inference and decision-making.

Consider a BN having nodes that represent discrete random variables Y_i, $i = 1, \ldots, n_Y$, with finite numbers of states, and nodes that represent vectors of continuous random variables $\mathbf{X}_j$, $j = 1, \ldots, n_X$. The conditional distributions at the discrete nodes are defined by the conditional PMFs $p[y_i|\mathrm{pa}(Y_i)]$ and at the continuous nodes they are defined by the conditional joint probability density functions (PDFs) $f[\mathbf{x}_j|\mathrm{pa}(\mathbf{X}_j)]$. The variables in each vector $\mathbf{X}_j$ can be dependent, as defined by their joint PDF. Furthermore, the discrete random variables and vectors of continuous random variables that are represented by different nodes in the BN are generally dependent in accordance with the topology of the BN. We define the eBN as a special class of BNs with discrete and continuous nodes, with the following restriction: If a discrete node Y_i is a child of continuous nodes, then its states are defined either as subdomains within the outcome space of its continuous parents or by its conditional PMF that is parametrized by the continuous parent variables. Furthermore, we take the following operational strategy: We discretize any continuous node for which evidence is to be entered or for which inference is to be sought. This is because these nodes must be present in the rBN after elimination of the continuous nodes in order to perform inference analysis. The remaining continuous nodes are not necessary for inference analysis and can be eliminated, as described next.

Following Shachter (1986), we define a *barren node* in the BN as one that has no children and that neither receives evidence nor is the subject of inference. If $\mathbf{X}_j$ is a barren node in an eBN, we can remove it, together with all the links directing to it, without affecting the joint distribution of the remaining nodes. It follows that a continuous node $\mathbf{X}_j$ can be removed from the eBN if the eBN can be altered to make $\mathbf{X}_j$ a barren node.

Consider the eBN in Figure 15.26(a) with seven discrete nodes $Y_1, \ldots, Y_7$ and one continuous node $\mathbf{X}_1$. For clarity, hereafter we show continuous nodes in gray. Observe that the latter has one discrete parent, Y_3, and two discrete children, Y_5 and Y_6. Shachter (1986) showed that the direction of a link between two nodes in a BN can be reversed if no directed cycle is created, provided links are added from the parents of each node to the other node. This rule can be used to make $\mathbf{X}_1$ a barren node. For that purpose, we should reverse the directions of the links from $\mathbf{X}_1$ to Y_5 and Y_6. Observe that if we start by reversing the direction of the link from $\mathbf{X}_1$ to Y_6, the directed cycle $Y_5 \rightarrow Y_6 \rightarrow \mathbf{X}_1 \rightarrow Y_5$ will be created. Hence, we choose to first reverse the direction of the link from $\mathbf{X}_1$ to Y_5, an action that does not create a cycle. This requires a link to be added from the parent of Y_5, node Y_4, to node $\mathbf{X}_1$, and a link from the parent of $\mathbf{X}_1$, node Y_3, to node Y_5. The result is the eBN in Figure 15.26(b). Next, we reverse the direction of the link from $\mathbf{X}_1$ to Y_6. Note that this action does not create a directed cycle in the revised eBN in Figure 15.26(b). The result is shown in Figure 15.25(c), where links are added from the parents of $\mathbf{X}_1$, nodes Y_3 and Y_4, to node Y_6. Node $\mathbf{X}_1$ now is a barren node and can be removed together with all the links directed to it to arrive at the rBN with only discrete nodes shown in Figure 15.26(d).

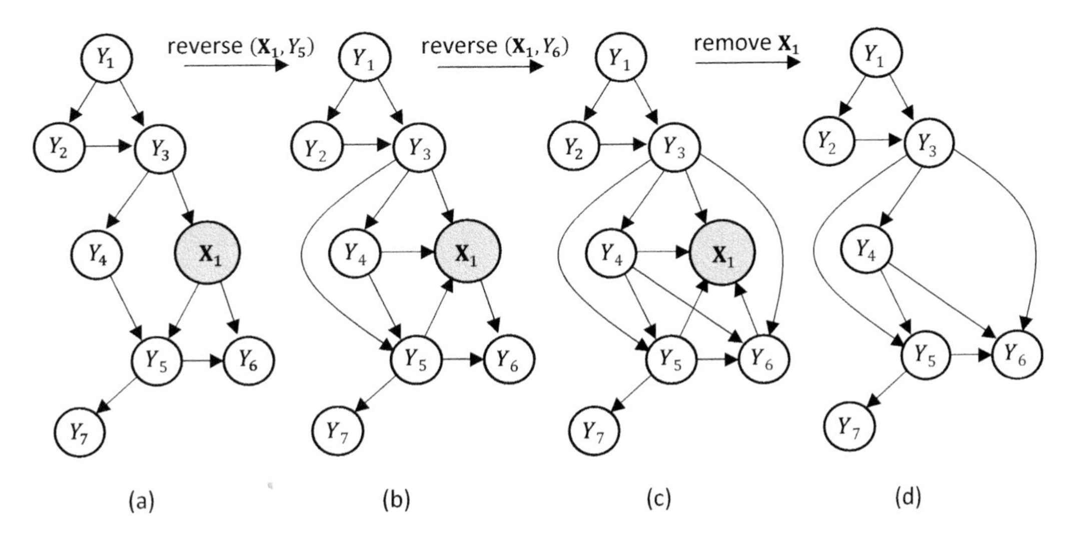

Figure 15.26 Removal of continuous node $\mathbf{X}_1$ in eBN to arrive at rBN (after Straub and Der Kiureghian, 2010a, with permission from ASCE)

Observe that in going from the eBN in Figure 15.26(a) to the rBN in Figure 15.26(d), only the children of $\mathbf{X}_1$ have experienced changes: namely, they have acquired additional discrete parents. Specifically, they have acquired the discrete parents and spouses of $\mathbf{X}_1$ as additional parents. Naturally, the CPTs of these children nodes in the rBN are different from those in the eBN and need to be determined. Recall from Section 15.3 that the parents, children, and spouses of $\mathbf{X}_1$ constitute its Markov blanket. Hence, the Markov blanket of a continuous node has a direct bearing on the formation of its discrete children in the rBN and the corresponding CPTs. On the other hand, the parents and, therefore, the CPTs of the discrete nodes that are not children of continuous nodes remain unchanged in going from the eBN to the rBN.

Let $\mathbf{X} = (\mathbf{X}_1, \ldots, \mathbf{X}_{n_X})$ denote the set of all continuous random variables, $\mathbf{Y} = (Y_1, \ldots, Y_{n_Y})$ the set of all discrete random variables, $\mathbf{Y}_C$ the subset of $\mathbf{Y}$ that are children of one or more continuous nodes, and $\mathbf{Y}_{NC}$ the remaining discrete nodes that have no continuous parents in the eBN. Also let $\text{pa}(Y_i)$ denote the set of parents of the discrete node Y_i in the eBN and $\text{pa}'(Y_i)$ denote its parents in the rBN. Obviously, the joint PMFs of $\mathbf{Y}$ in the eBN and rBN should be the same. Thus, using the factored form of the PMF in (15.1), we can write

$$\begin{aligned} p(\mathbf{y}) &= \prod\nolimits_{Y_i \in \mathbf{Y}} p[y_i|\text{pa}(Y_i)] \\ &= \prod\nolimits_{Y_i \in \mathbf{Y}_C} p[y_i|\text{pa}'(Y_i)] \prod\nolimits_{Y_i \in \mathbf{Y}_{NC}} p[y_i|\text{pa}(Y_i)], \end{aligned} \tag{15.29}$$

where the first line is for the eBN and the second line for the rBN. In the second line, we have made use of the fact that the CPTs of nodes in $\mathbf{Y}_{NC}$ remain unchanged in going from the eBN to the rBN. We need to determine the CPTs $p[y_i|\text{pa}'(Y_i)]$ for all $Y_i \in \mathbf{Y}_C$.

For the eBN, we can write

$$p(\mathbf{y}) = \int_{\mathbf{x}_1} \cdots \int_{\mathbf{x}_{n_X}} \prod\nolimits_{Y_i \in \mathbf{Y}_C} p[y_i|\text{pa}(Y_i)] \prod\nolimits_{Y_i \in \mathbf{Y}_{NC}} p[y_i|\text{pa}(Y_i)] \prod\nolimits_{\mathbf{X}_j \in \mathbf{X}} f\left[\mathbf{x}_j|\text{pa}\left(\mathbf{X}_j\right)\right] \text{d}\mathbf{x}_1 \cdots \text{d}\mathbf{x}_{n_X}. \tag{15.30}$$

Because $\prod_{Y_i \in \mathbf{Y}_{NC}} p[y_i|\text{pa}(Y_i)]$ is independent of $\mathbf{X}$, it can be taken out of the integral. Then, substituting the second line of (15.29) for the left-hand side of the above equation and canceling the identical terms, we arrive at

$$\prod\nolimits_{Y_i \in \mathbf{Y}_C} p[y_i|\text{pa}'(Y_i)] = \int_{\mathbf{x}_1} \cdots \int_{\mathbf{x}_{n_X}} \prod\nolimits_{Y_i \in \mathbf{Y}_C} p[y_i|\text{pa}(Y_i)] \prod\nolimits_{\mathbf{X}_j \in \mathbf{X}} f\left[\mathbf{x}_j|\text{pa}\left(\mathbf{X}_j\right)\right] \text{d}\mathbf{x}_1 \cdots \text{d}\mathbf{x}_{n_X}. \tag{15.31}$$

The above equality holds for any set of states of the variables $Y_i \in \mathbf{Y}_C$ and the states of their parents in the rBN, which constitute the Markov blankets of the continuous random variables $\mathbf{X}$. For these given states, the above integral can be written in the form

$$\prod\nolimits_{Y_i \in \mathbf{Y}_C} p[y_i|\text{pa}'(Y_i)] = \int_{\mathbf{x}} \Pr(E|\mathbf{x}) f\left(\mathbf{x}|\mathbf{y}_p\right) \text{d}\mathbf{x}, \tag{15.32}$$

where $\mathbf{y}_p$ is the set of states of the discrete parents of $\mathbf{X}$ and E is the event describing the set of states of the discrete random variables in $\mathbf{Y}_C$ and their parents in the eBN. The integral on the right-hand side is of the form solvable by SRMs. Specifically, if the states of the discrete random variables in $\mathbf{Y}_C$ are defined as subdomains in the outcome space of $\mathbf{X}$, then the integral takes the form

$$\prod\nolimits_{Y_i \in \mathbf{Y}_C} p[y_i|\text{pa}'(Y_i)] = \int_{\mathbf{x} \in \Omega} f\left(\mathbf{x}|\mathbf{y}_p\right) \text{d}\mathbf{x}, \tag{15.33}$$

where Ω is the intersection of the domains for all the states in $\mathbf{y}_p$. The right-hand side is a parallel-system reliability problem and can be solved by standard SRMs. If $\mathbf{X}$ represents parameters in the PMF of $\mathbf{Y}_C$, then (15.32) can be solved by the nested reliability analysis approach described in Section 10.5 (see (10.45) and related discussion) with the limit-state function $g(\mathbf{X}) = U + \beta(\mathbf{X})$, with U being a standard normal random variable and $\beta(\mathbf{X}) = \Phi^{-1}[1 - \Pr(E|\mathbf{X})]$ the conditional reliability index.

The number of SRM solutions to compute all the needed CPTs in the rBN equals the product of the number of states of all the discrete nodes in all the Markov blankets of $\mathbf{X}$. This number can be large. However, as demonstrated in Straub and Der Kiureghian (2010a), separate calculations can be done for the discrete nodes in *Markov envelopes* of continuous nodes. A Markov envelope is constructed as follows: Determine the Markov blanket of a continuous node. If the blanket includes another continuous node (as parent, child, or spouse), add the Markov blanket of the second node to that of the first node. Continue this process until the last Markov blanket does not contain additional continuous nodes. The union of the Markov blankets then constitutes the Markov envelope. The number of needed SRM solutions then is the sum of the products of all the states of the discrete nodes in each Markov envelope. This concept is illustrated in Figure 15.27 for an eBN with six discrete and four continuous nodes. Observe that the Markov blanket of $\mathbf{X}_1$ includes $\mathbf{X}_2$, whose Markov blanket in turn includes $\mathbf{X}_3$. However, X_4 is not a member of the Markov blankets of any of the first three continuous nodes. Hence, this eBN has two Markov envelopes, which are shown in the figure with dashed boundaries. If m_i denotes the number of states of Y_i, then the number of SRM solutions needed to construct the CPTs of the rBN resulting from this eBN is $m_1m_2m_3m_4m_5 + m_5m_6$. This is substantially less than $m_1m_2m_3m_4m_5m_6$, which would be the case if the idea of Markov envelopes is not used. A number of other strategies to optimize the rBN and reduce the number of SRM calculations are described in Straub and Der Kiureghian (2010a). These include strategies for selecting the orders of link reversals and node eliminations, selected discretization of continuous variables, causal modeling of the eBN and maintaining causality in the rBN, and divorcing of variables, i.e., introducing intermediate variables to reduce the number of parents to a node.

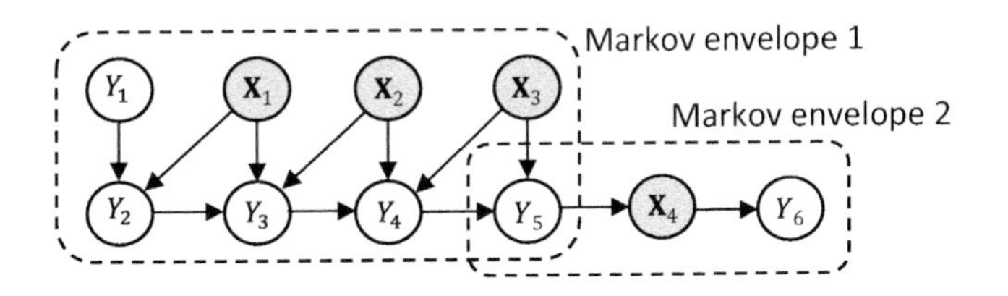

Figure 15.27 Concept of Markov envelopes of continuous variables in eBN (after Straub and Der Kiureghian, 2010a, with permission from ASCE)

Example 15.7 – Reliability Updating of Ductile Frame under Time-Varying Load

Adopted from Straub and Der Kiureghian (2010b), this example demonstrates the use of the eBN and rBN for reliability assessment and updating of the one-bay ductile frame considered in several examples in this book, starting from Example 8.7. As we saw in that example, the

Example 15.7 (cont.)

frame is a series system with three failure modes characterized by sway, beam, and combined plastic mechanisms as shown in Figure 8.11. The limit-state functions are as defined in (E1)–(E3) of Example 8.7 and involve seven random variables: Plastic moment capacities R_i, $i = 1, \ldots, 5$, at potential locations of formation of plastic hinges, the horizontal load H, and the vertical load V. Table 8.4 describes the probability distributions of these random variables. Note that R_i are equicorrelated, jointly lognormal random variables. H and V are statistically independent of each other and of R_i.

We first consider the basic problem of calculating the reliability of the frame. Figure 15.28 shows the eBN and the corresponding rBN for the problem. Because R_i are equicorrelated, it is possible to represent their dependence through a common parent node U_R, representing a standard normal random variable. The conditional distribution of R_i given U_R is the lognormal distribution with the CDF

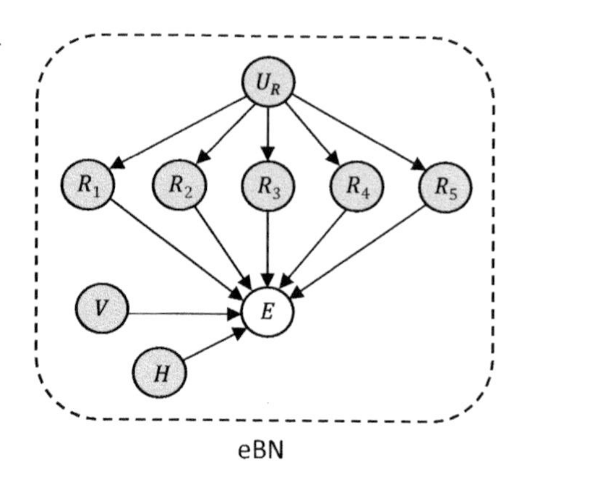

Figure 15.28 Models of eBN and rBN for unconditional reliability analysis of ductile frame (after Straub and Der Kiureghian, 2010b, with permission from ASCE)

$$F_{R_i}(r_i|u_r) = \Phi\left(\frac{\ln r_i - u_r\sqrt{\rho} - \lambda_R}{\sqrt{\zeta_R^2 - \rho}}\right), \quad i = 1, \ldots, 5, \tag{E1}$$

where λ_R and ζ_R are the parameters of the common marginal lognormal distribution of R_i and ρ is the common correlation coefficient between the log values $\ln R_i$ and $\ln R_j$. Node E in the eBN in Figure 15.28 is a two-state discrete random variable with $E = 0$ indicating the failure state and $E = 1$ indicating the survival state of the frame. All the nodes are in one Markov envelope. We see that the rBN in this case is the sole node E. To determine its CPT, we must perform reliability analysis of the frame. Obviously, the eBN–rBN framework does not provide an advantage in this case.

Now suppose we are able to measure the bending moment capacities R_4 and R_5. The measured values are given by

$$M_i = R_i + \epsilon_i, \quad i = 4, 5, \tag{E2}$$

Example 15.7 (cont.)

where ϵ_i is a N(0,15) kNm random variable representing the measurement error. Because evidence is to be entered for random variables R_4 and R_5, the corresponding nodes should be discretized. As described in Straub and Der Kiureghian (2010a), when discretizing a continuous node in the eBN, it is convenient to introduce replicas of the node as continuous children of the discretized node. Doing so, the original children of the continuous node become children of the replicated nodes and, therefore, no changes in their conditional distributions occur. This strategy is used in the eBN shown in Figure 15.29, where the white nodes R_4 and R_5 are the discretized interval nodes and R'_{4a}, R'_{4b}, R'_{5a}, and R'_{5b} are replicas of the continuous nodes. Observe that the conditional distributions of the child nodes E, M_4, and M_5 are not affected by the discretization. Details on how to construct the CPTs of the discretized interval nodes are described in Section 15.4; more detailed information is given in Straub (2009) and Straub and Der Kiureghian (2010a).

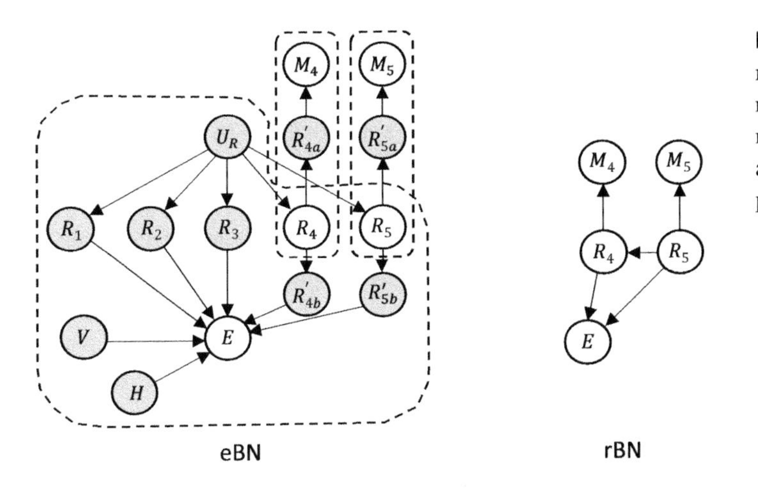

Figure 15.29 Models of eBN and rBN for reliability updating with measurement of bending moment capacities (after Straub and Der Kiureghian, 2010b, with permission from ASCE)

The eBN in Figure 15.29 has three Markov envelopes, indicated by dashed boundaries. The corresponding rBN is shown on the right-hand side. In discretizing R_4 and R_5, the 21 intervals (50:10:250) kNm are selected for each. As a result, $21 \times 21 = 441$ SRMs must be performed to construct the CPT of node E. Each of these is an SRM with R'_{4b} and R'_{5b} specified within the selected intervals of the discrete nodes R_4 and R_5. The CPTs for nodes M_4 and M_5 for given values of their parents are calculated by use of (E2) and discretizing the distribution of ϵ_i. Once the rBN is constructed, reliability updating of the frame for any measured values $M_4 = m_4$ and $M_5 = m_5$ is a trivial calculation. Table 15.2 shows the results for the updated generalized reliability index and failure probability obtained by the rBN and compares them with Monte Carlo simulation results for several cases of measured values of the plastic bending moment capacities. Also shown are the results for the case of no measurements described earlier. It can

Example 15.7 (cont.)

be seen that the results obtained by the rBN are in close agreement with the Monte Carlo simulation results.

Table 15.2 Updated reliability index (failure probability) for measured plastic bending moment capacities (after Straub and Der Kiureghian, 2010b, with permission from ASCE)

	Measurements (kNm)			
Method of analysis	No measurement	$m_4 = 50$, $m_5 = 100$	$m_4 = 150$, $m_5 = 100$	$m_4 = 150$, $m_5 = 200$
eBN–rBN	1.94 (0.026)	0.70 (0.24)	1.80 (0.036)	2.45 (0.0071)
Monte Carlo simulations 90% confidence interval	1.93–1.94	0.63–0.77	1.78–1.81	2.44–2.48

Next, we consider the case where the horizontal load varies with time. Assuming it is an environmental load, we consider the annual maximum values over a 20-year service life denoted as H_t, $t = 1, 2, \ldots, 20$. The annual maxima are assumed to be statistically independent and to have identical Gumbel distributions with mean $U_H + 9$ kN and standard deviation 20 kN, where U_H is the mode of the distribution (parameter v in Table A2 of Chapter 2). To account for statistical uncertainty, U_H is considered to be a random variable having the lognormal distribution with mean 35 kN and coefficient of variation of 0.286. Note that this common uncertain parameter induces dependence among the annual maximum horizontal load values. The discrete nodes E_t, $t = 1, \ldots, 20$, now represent the states of the structural system at the end of each year during its 20-year lifetime. All other random variables have the same distributions as before. Our objective is to assess the life-cycle reliability of the system for selected ECs.

In modeling the eBN, Straub and Der Kiureghian (2010b) use the strategy of divorcing in order to simplify the eBN model. Specifically, an intermediate node Q is introduced that separates nodes H_t from other parent nodes of E_t. Node Q can be interpreted as the residual capacity of the frame for the horizontal load after application of the vertical load. Furthermore, node Q is discretized in order to split the Markov envelopes of the eBN. In addition, nodes H_t and U_H are discretized with the objective of entering evidence at these nodes or making inference on them. The eBN model and corresponding rBN are shown in Figure 15.30. Note that continuous replicas of all discretized nodes are included. For details on the discretization of the nodes and computation of the CPTs, the reader is referred to the original reference.

Figure 15.31(a) shows the results for the system reliability as a function of time for selected ECs. Included are the case of no evidence; the observation that the structure has survived for

Example 15.7 (cont.)

up to five years; the observations that the structure has survived for up to five years and the horizontal load on the fifth year is measured to be $h_5 = 80$ kN; and the same evidence plus the observation that the horizontal load was measured in each of the first four years to be 30 kN.

Observe that the evidence that the frame has survived up to the fifth year significantly increases the reliability index relative to the unconditional case. The increase is smaller when the high value $h_5 = 80$ kN is measured on the fifth year, and is larger when additionally $h_i = 30$ kN, $i = 1, 2, 3, 4$, are measured. The measurements of the horizontal load are informative for reliability estimates in future years because of the statistical dependence among the annual maximum load values caused by the statistical uncertainty in the common uncertain parameter U_H.

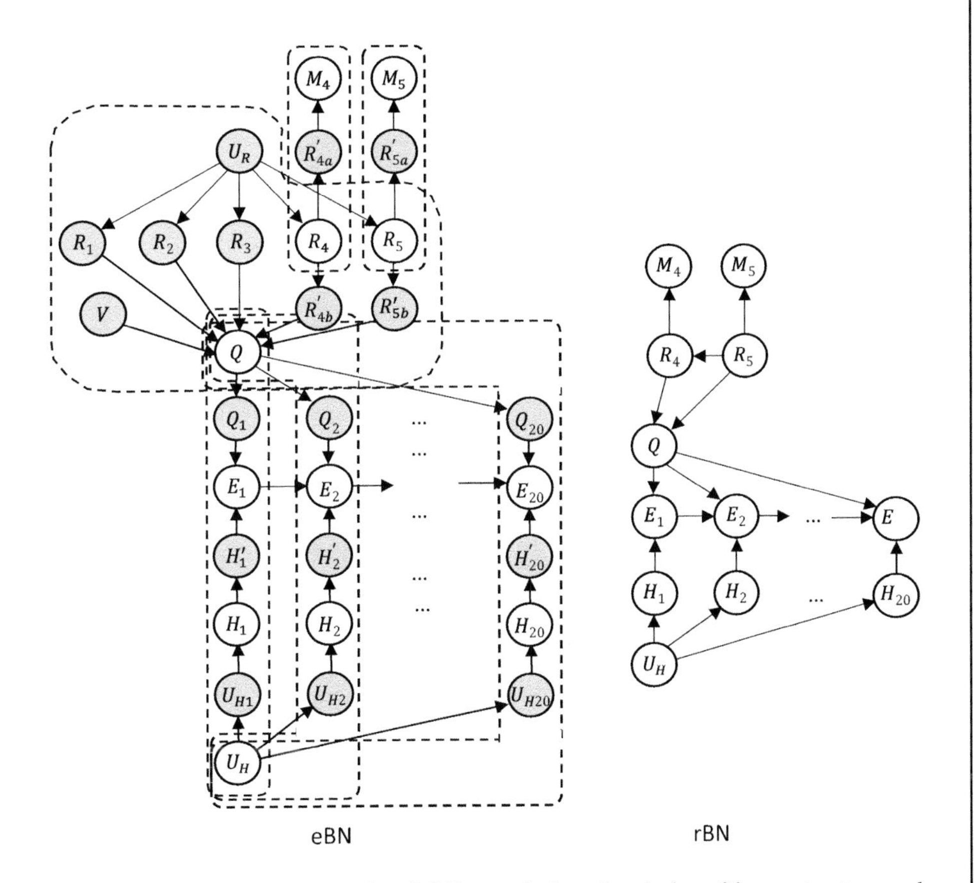

Figure 15.30 eBN and rBN for life-cycle reliability analysis and updating of frame structure under time-varying load (after Straub and Der Kiureghian, 2010b, with permission from ASCE)

Example 15.7 (cont.)

Figure 15.31(b) shows the poster PDFs of the uncertain parameter U_H for different evidence cases. Included are the case of no evidence (the prior lognormal distribution) and the last two ECs described above. Observe that the measurement $h_5 = 80$ kN moves the distribution of U_H toward higher values, while additional observations of low values of the horizontal load during the first four years move the distribution of U_H toward lower values.

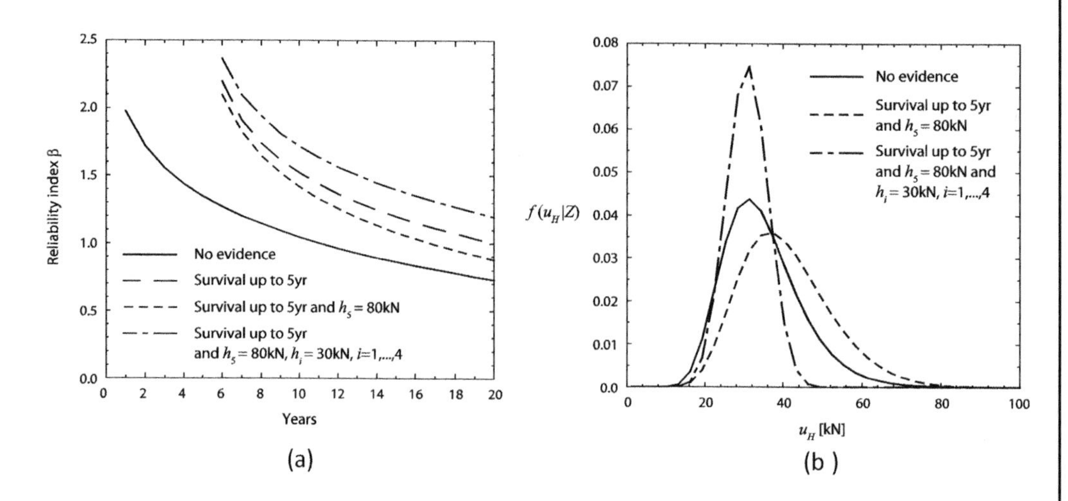

Figure 15.31 Updated results for selected evidence cases: (a) generalized reliability index as a function of time, (b) posterior distribution of U_H (after Straub and Der Kiureghian, 2010b, with permission from ASCE)

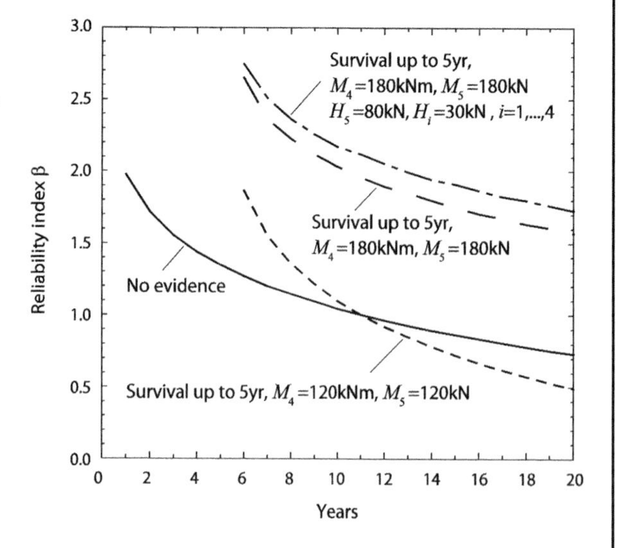

Figure 15.32 Updated generalized reliability index for selected ECs, including measurement of capacities (after Straub and Der Kiureghian, 2010b, with permission from ASCE)

Example 15.7 (cont.)

Finally, Figure 15.32 shows estimates of the generalized reliability index for different evidence cases, including measurements of plastic moment capacities R_4 and R_5. Observe that the evidence that the frame survived up to the fifth year increases the reliability estimate relative to the no-evidence case for the next five years, even when the measured capacities are below the mean value (small dashed line). However, this estimate shows a rapid decrease in the reliability index for the later years. With higher-than-mean measurements of the capacities, the reliability index is much higher and remains high in later years.

Straub and Der Kiureghian (2010b) perform further analysis of this example, including value-of-information analysis for the measurement of capacities. Furthermore, an eBN framework for analysis of deteriorating infrastructure systems consisting of a collection of structural systems is presented. The interested reader is referred to the original paper.

15.11 Concluding Remarks

Bayesian networks have been in existence for more than 30 years. The method originated with Pearl's 1988 book in the field of artificial intelligence, but soon found application in a variety of fields including data science, medicine, biology, semantics, image processing, risk assessment, and health monitoring. In civil engineering, applications have been developed during the past 20 years. However, it has been a slow process. The topic is seldom taught in civil engineering curricula, even in courses dealing with the applications of probability and statistics in civil engineering. However, there is enormous potential for the use of this technique in a variety of civil engineering applications that deal with information processing and decision-making. With the advent of sensor technology and the possibility of measuring all kinds of attributes and easy transmission of the information for processing by computers, BNs can provide the needed framework for near-real-time updating and decision support systems in the operation of large constructed facilities. These may include the handling of uncertainties in the design, construction, and operation of facilities; reliability and risk assessment for natural and man-made hazards and decision-making for mitigation of losses in preparation, during and after such hazards; health monitoring of constructed facilities; and the management of infrastructure systems under uncertain and evolving information conditions. This chapter included a number of examples that demonstrated the potential for such applications in civil engineering.

On the other hand, there are difficulties that need to be addressed through research and further development. In particular, BN models for complex engineering systems are difficult to develop and the computational demands for large systems can be prohibitive, particularly for near-real-time applications. There is enormous room to develop new ideas and techniques for facilitating the use of BNs in the civil engineering domain. For doctoral students, this topic remains a fertile ground for research and innovation.

References

Abrahamson, N. (2000). Effects of rupture directivity on probabilistic seismic hazard analysis. *Proceedings of the 6th International Conference on Seismic Zonation*, Earthquake Engineering Research Institute, Palm Springs, CA.

Abrahamson, N., G. Atkinson, D. Boore, Y. Bozorgnia, K. Campbell, B. Chiou, I.M. Idriss, W. Silva, and R. Youngs (2008). Comparisons of the NGA ground-motion relations. *Earthquake Spectra*, **24**(1):45–66.

Alibrandi, U., and A. Der Kiureghian (2012). A gradient-free method for determining the design point in nonlinear stochastic dynamic analysis. *Probabilistic Engineering Mechanics*, **28**:2–10.

Alibrandi, U., and K. Mosalam (2017). Equivalent linearization methods for stochastic dynamic analysis using linear response surfaces. *Journal of Engineering Mechanics*, ASCE. https://doi.org/10.1061/(ASCE)EM.1943-7889.0001264.

Alibrandi, U., and C.G. Koh (2017). Stochastic dynamic analysis of floating production systems using the First Order Reliability Method and the Secant Hyperplane Method. *Ocean Engineering*, **137**:68–77.

American Association of State Highway and Transportation Officials (AASHTO) (1992). *Standard specifications for highway bridges* (15th ed.). Washington DC.

Anders, G.J. (1990). *Probability concepts in electric power systems*. John Wiley & Sons, New York, NY.

Ambartzumian, R., A. Der Kiureghian, V. Ohanian, and H. Sukiasian (1998). Multinormal probability by sequential conditioned importance sampling: theory and application. *Probabilistic Engineering Mechanics*, **13**:299–308.

Ang, A.H.-S., and M. Amin (1968). Reliability of structures and structural systems. *Journal of the Engineering Mechanics Division*, ASCE, **94**:671–694.

Ang, A.H.-S., and C.A. Cornell (1974). Reliability bases of structural safety and design. *Journal of Structural Engineering*, ASCE, **100**:1755–1769.

Ang, A.H.-S., and W.-H. Tang (1975). *Probability concepts in engineering planning and design, vol. I – basic principles*. John Wiley & Sons, New York, NY.

Ang, A.H.-S., and W.-H. Tang (1984). *Probability concepts in engineering planning and design, vol. II – decision, risk and reliability*. John Wiley & Sons, New York, NY.

Atalik, T.S., and S. Utku (1976). Stochastic linearization of multi-degree of freedom nonlinear systems. *Earthquake Engineering & Structural Dynamics*, **4**:411–420.

Au, S.K., and J.L. Beck (1999). A new adaptive importance sampling scheme for reliability calculations. *Structural Safety*, **21**:135–158.

Au, S.K., and J.L. Beck (2001a). First excursion probabilities for linear systems by very efficient importance sampling. *Probabilistic Engineering Mechanics*, **16**:193–207.

Au, S.K., and J.L. Beck (2001b). Estimation of small failure probabilities in high dimensions by subset simulation. *Probabilistic Engineering Mechanics*, **16**:263–277.

Au, S.-K., and Y. Wang (2014). *Engineering risk assessment with subset simulation*. John Wiley & Sons, New York, NY.

Augusti, G., A. Baratta, and F. Casciati (1984). *Probabilistic methods in structural mechanics*. Chapman and Hall, London, UK.

Arora, J.S., and E.J. Haug (1979). Methods of design sensitivity analysis in structural optimization. *AIAA Journal*, **17**:970–974.

Beacher, G.B., and T.S. Ingra (1981). Stochastic FEM in settlement predictions, *Journal of the Geotechnical Engineering Division*, ASCE, **107**:449–463.

Blatman, G., and B. Sudret (2011). Adaptive sparse polynomial chaos expansion based on least angle regression. *Journal of Computational Physics*, **230**:2345–2367.

Barbato, M., and J.P. Conte (2005a). Finite element response sensitivity analysis: a comparison between force-based and displacement-based frame element models. *Computer Methods in Applied Mechanics and Engineering*, **194**:1479–1512.

Barbato, M., and J.P. Conte (2005b). Smooth versus non-smooth material constitutive models in gradient-based optimization for structural reliability analysis. *Proceedings of the 9th International Conference on Structural Safety and Reliability*, Rome, Italy.

Barbato, M., Q. Gu, and J.P. Conte (2006). Response sensitivity and probabilistic response analysis of reinforced concrete frame structures. *Proceedings of the 8th National Conference on Earthquake Engineering*, San Francisco, CA.

Barlow, R., and F. Proschan (1975). *Statistical theory of reliability and life testing*. Holt, Rinehart and Winston, Inc., New York, NY.

Bayraktarli, Y.Y., and M. Faber (2011). Bayesian probabilistic network approach for managing earthquake risks of cities. *Georisk*, **5**:2–24.

Beck, J.L., and L.S. Katafygiotis (1998). Updating models and their uncertainties. I: Bayesian statistical framework. *Journal of Engineering Mechanics*, ASCE, **124**:455–461.

Belyaev, Y.K. (1968). On the number of exits across the boundary of a region by a vector stochastic process. *Theory of Probability Applications*, **13**:320–324.

Benjamin, J. (1964). *Statistics of civil engineers*. McGraw-Hill, New York, NY.

Benjamin, J. (1968a). Probabilistic models for seismic force design. *Journal of Structural Division*, ASCE, **94**:1175–1196.

Benjamin, J. (1968b). Probabilistic structural and design. *Journal of Structural Division*, ASCE, **94**:1665–1679.

Benjamin, J.R., and C.A. Cornell (1970). *Probability, statistics and decision for civil engineers*, McGraw-Hill, New York, NY.

Bensi, M., A. Der Kiureghian, and D. Straub (2011a). A Bayesian network methodology for infrastructure seismic risk assessment and decision support. *Report No. 2011/02*, Pacific Earthquake Engineering Research Center, University of California, Berkeley, CA.

Bensi, M., A. Der Kiureghian, and D. Straub (2011b). Bayesian network modeling of correlated random variables drawn from a Gaussian random field. *Structural Safety*, **33**:317–332.

Bensi, M., A. Der Kiureghian, and D. Straub (2013). Efficient Bayesian network modeling of systems. *Reliability Engineering & System Safety*, **112**:200–213.

Bensi, M., A. Der Kiureghian, and D. Straub (2015). Framework for post-earthquake risk assessment and decision making for infrastructure systems. *ASCE-ASME Journal of Risk and Uncertainty in Engineering Systems, Part A: Civil Engineering*, **1** (online).

Bertsimas, D., and J.N. Tsitsiklis (1997). *Introduction to linear optimization*. Athena Scientific, Belmont, MA.

Bichon, B.J., M.S. Eldred, S. Mahadevan, and J.M. McFarland (2013). Efficient global surrogate modeling for reliability-based design optimization. *Journal of Mechanical Design*, **135**:011009–1.

Bichon, B.J., M.S. Eldred, L.P. Swiler, S. Mahadevan, and J.M. McFarland (2008). Efficient global reliability analysis for nonlinear implicit performance function. *AIAA Journal*, **46**:2459–2468.

Bichon, B.J., J.M. McFarland, and S. Mahadevan (2011). Efficient surrogate models for reliability analysis of systems with multiple failure modes. *Reliability Engineering & System Safety*, **96**: 1386–1395.

Bjerager, P. (1988). Probability integration by directional simulation. *Journal of Engineering Mechanics*, ASCE, **114**:1285–1302.

Bjerager, P., and S. Krenk (1989). Parametric sensitivity in first-order reliability theory. *Journal of Engineering Mechanics*, ASCE, **115**:1577–1582.

Bolotin, V.V. (1967). Statistical aspects in the theory of structural stability. *Proceedings of the International Conference on Dynamic Stability of Structures*, Northwestern University, Evanston, IL, pp. 67–81.

Bolotin, V.V. (1971). *Application of theory of probability for reliability of structures.* Stroiizdat Publishers, Moscow, Russia. (In Russian.)

Bolotin, V.V. (1984). *Random vibration of elastic systems.* Springer Science & Business Media, Berlin, Germany. (Translated from Russian.)

Bouc, R. (1963). Forced vibration of mechanical systems with hysteresis. *Proceeding of the 4th Conference on Nonlinear Oscillation*, Academia Publishing, Prague, Czechoslovakia, p. 315. [Abstract.]

Bourinet, J.-M. (2017). FORM sensitivities to distribution parameters with the Nataf transformation. In *Risk and reliability analysis: theory and applications*, P. Gardoni (Ed.), Springer, New York, NY.

Bourinet, J.-M., and M. Lemaire (2008). FORM sensitivities to correlation: application to fatigue crack propagation based on Virkler data. In *Proceedings of the 4th International ASRANet Colloquium*, P.K. Das (Ed.), Athens, Greece.

Box, G.E.P., W.G. Hunter, and J.S. Hunter (1978). *Statistics for experimenters.* John Wiley and Sons, New York, NY.

Box, G.E.P., and G.C. Tiao (1973). *Bayesian inference in statistical analysis.* Addison-Wesley, Reading, MA.

Breitung, K. (1984). Asymptotic approximations for multinormal integrals. *Journal of Engineering Mechanics*, ASCE, **110**:357–366.

Breitung, K., and M. Hohenbichler (1989). Asymptotic approximations for multivariate integrals with an application to multinormal probabilities. *Journal of Multivariate Analysis*, **30**:80–97.

Broccardo, M., U. Alibrandi, Z. Wang, and L. Garré (2016). The tail-equivalent linearization method for nonlinear stochastic processes, genesis and developments. In *Risk and reliability analysis: Theory and applications*, P. Gardoni (Ed.), Springer, New York, NY.

Broccardo, M., and A. Der Kiureghian (2013). Non-stationary stochastic dynamic analysis by tail-equivalent linearization. *Proceedings of the 11th International Conference on Structural Safety and Reliability*, New York, NY.

Broccardo, M., and A. Der Kiureghian (2016). Multi-component nonlinear stochastic dynamic analysis by tail-equivalent linearization. *Journal of Engineering Mechanics*, ASCE, **142**:04015100.

Broccardo, M., and A. Der Kiureghian (2018). Simulation of stochastic processes by sinc basis functions and application in TELM analysis. *Journal of Engineering Mechanics*, ASCE, **144**:04017154.

Broccardo, M., and A. Der Kiureghian (2021). Nonlinear stochastic dynamic analysis by evolutionary tail-equivalent linearization method. *Structural Safety*, to appear.

Bucher, C.G. (1988). Adaptive sampling – an iterative fast Monte Carlo procedure. *Structural Safety*, **5**:119–126.

Bucher, C.G., and U. Bourgund (1990). A fast and efficient response surface approach for structural reliability problems. *Structural Safety*, **7**:57–66.

Byun, J.-E., and J. Song (2020). Bounds on reliability of larger systems by linear programming with delayed column generation. *Journal of Engineering Mechanics*, ASCE, **146**:04020008.

Castelletti, A., and R. Soncini-Sessa (2007). Bayesian networks in water resource modelling and management. *Environmental Modelling & Software*, **22**:1073–1074.

Caughey, T.K. (1963). Equivalent linearization techniques. *Journal of the Acoustical Society of America*, **35**:1706–1711.

Chang, Y., and Y. Mori (2013). A study on the relaxed linear programming bounds method for system reliability. *Structural Safety*, **41**:64–72.

Chehade, F.E.H., and R. Younes (2020). Structural reliability software and calculation tools: a review. *Innovative Infrastructure Solutions*, **5**:29.

Cheney, W., and D. Kincaid (2009). *Linear algebra: Theory and applications*. Jones and Bartlett Publishers, Sudbury, MA.

Conte, J.P., P.K. Vijalapura, and M. Meghella (2003). Consistent finite-element response sensitivity analysis. *Journal of Engineering Mechanics*, ASCE, **129**:1380–1393.

Cornell, C.A. (1967). Bounds on the reliability of structural systems. *Journal of Structural Division*, ASCE, **93**:171–200.

Cornell, C.A. (1969). A probability-based structural code. *Journal of American Concrete Institute*, **66**:974–985.

Corder, G.W., and D.I. Foreman (2014). *Nonparametric statistics: a step-by-step approach*, John Wiley & Sons, New York, NY.

Cowell R.G., A.P. Dawid, S.L. Lauritzen, and D.J. Spiegelhalter (1999). *Probabilistic Networks and Expert Systems*. Springer, New York, NY.

Crandall, S., and W.D. Mark (1963). *Random vibration in mechanical systems*. Academic Press, Cambridge, MA.

De Finetti, B. (1975). *Theory of probability*. John Wiley & Sons, New York, NY.

Deodatis, G. (1990). Bounds on response variability of stochastic finite element systems: effect of statistical dependence. *Probabilistic Engineering Mechanics*, **5**:88–98.

Deodatis, G. (1991). Weighted integral method. I: stochastic stiffness matrix. *Journal of Engineering Mechanics*, ASCE, **117**:1851–1864.

Deodatis, G., and M. Shinozuka (1991). Weighted integral method. I: response variability and reliability. *Journal of Engineering Mechanics*, ASCE, **117**:1865–1877.

Der Kiureghian, A. (1981). A response spectrum method for random vibration analysis of MDF systems. *Earthquake Engineering Structural Dynamics*, **9**:419–435.

Der Kiureghian, A. (1989). Measures of structural safety under imperfect states of knowledge. *Journal of Structural Engineering*, ASCE, **115**:1119–1140.

Der Kiureghian, A. (2000). The geometry of random vibrations and solutions by FORM and SORM. *Probabilistic Engineering Mechanics*, **15**:81–90.

Der Kiureghian, A. (2005). First- and second-order reliability methods. Chapter 14 in *Engineering design reliability handbook*, E. Nikolaidis, D.M. Ghiocel, and S. Singhal (Eds.), CRC Press, Boca Raton, FL.

Der Kiureghian, A. (2006). System reliability revisited. *Proceedings of the 3rd ASRANet International Colloquium*, Glasgow, UK.

Der Kiureghian, A. (2008). Analysis of structural reliability under parameter uncertainties. *Probabilistic Engineering Mechanics*, **23**:351–358.

Der Kiureghian, A., and T. Dakessian (1998). Multiple design points in first and second-order reliability. *Structural Safety*, **20**:37–49.

Der Kiureghian, A., and M. De Stefano (1991). Efficient algorithm for second-order reliability analysis. *Journal of Engineering Mechanics*, ASCE, **117**:2904–2923.

Der Kiureghian, A., and O. Ditlevsen (2009). Aleatory or epistemic? Does it matter? *Structural Safety*, **31**:105–112.

Der Kiureghian, A., and K. Fujimura (2009). Nonlinear stochastic dynamic analysis for performance-based earthquake engineering. *Earthquake Engineering & Structural Dynamics*, **38**:719–738.

Der Kiureghian, A., T. Haukaas, and K. Fujimura (2006). Structural reliability software at the University of California, Berkeley. *Structural Safety*, **28**:44–67.

Der Kiureghian, A., and J.-B. Ke (1988). The stochastic finite element method in structural reliability. *Probabilistic Engineering Mechanics*, **3**:83–91.

Der Kiureghian, A., H.-Z. Lin, and S.-J. Hwang (1987). Second-order reliability approximations. *Journal of Engineering Mechanics*, ASCE, **113**:1208–1225.

Der Kiureghian, A., and P.-L. Liu (1986). Structural reliability under incomplete probability information. *Journal of Engineering Mechanics*, ASCE, **112**:85–104.

Der Kiureghian, A., and J. Song (2008). Multi-scale reliability analysis and updating of complex systems by use of linear programming. *Reliability Engineering and System Safety*, **93**:288–297.

Der Kiureghian, A., and Y. Zhang (1999). Space-variant finite element reliability analysis. *Computer Methods in Applied Mechanics and Engineering*, **168**:173–183.

Der Kiureghian, A., Y. Zhang, and C.-C. Li (1994). Inverse reliability problem. *Journal of Engineering Mechanics*, ASCE, **120**:1154–1159.

Dinkelbach, W. (1967). On nonlinear fractional programming. *Management Science*, **13**:492–498.

Ditlevsen, O. (1973). Structural reliability and the invariance problem. *Research Report No. 22*, Solid Mechanics Division, University of Waterloo, Waterloo, Canada.

Ditlevsen, O. (1979a). Generalized second moment reliability index. *Journal of Structural Mechanics*, **7**:435–451.

Ditlevsen, O. (1979b). Narrow reliability bounds for structural systems. *Journal of Structural Mechanics*, **7**:453–472.

Ditlevsen, O. (1981). *Uncertainty modeling*. McGraw-Hill, New York, NY.

Ditlevsen, O. (1982). Model uncertainty in structural reliability. *Structural Safety*, **1**:73–86.

Ditlevsen, O. (1983). Gaussian out-crossings from safe convex polyhedrons. *Journal of Engineering Mechanics*, ASCE, **109**:127–148.

Ditlevsen, O. (1987). On the choice of expansion point in FORM or SORM. *Structural Safety*, **4**:243–245.

Ditlevsen, O., and H.O. Madsen (1996). *Structural reliability methods*. John Wiley & Sons, New York, NY. Available for download at www.mek.dtu.dk/staff/od/books.htm.

Dubourg, V., B. Sudret, and J.M. Bourinet (2011). Reliability-based design optimization using kriging surrogates and subset simulation. *Structural & Multidisciplinary Optimization*, **44**:673–690.

Dunnett, C.W., and M. Sobel (1955). Approximations to the probability integral and certain percentage points of a multivariate analogue of Student's t-distribution. *Biometrika*, **42**:258–260.

Echard, B., N. Gayton, and M. Lemaire (2011). AK-MCS: An active learning reliability method combining Kriging and Monte Carlo simulation. *Structural Safety*, **33**:145–154.

Engelund, S., and R. Rackwitz (1993). A benchmark study on importance sampling techniques in structural reliability. *Structural Safety*, **12**:255–276.

Enevoldsen, I., and J.D. Sorensen (1994). Reliability-based optimization in structural engineering. *Structural Safety*, **15**:169–196.

Faber, M.H., I.B. Kroon, E. Kragh, D. Bayly, and P. Decosemaeker (2002). Risk assessment of decommissioning options using Bayesian networks. *Journal of Offshore Mechanics and Arctic Engineering*, **124**:231–238.

Faravelli, L. (1989). Response surface approach for reliability analysis. *Journal of Engineering Mechanics*, ASCE, **115**:2763–2781.

Fauriat, W., and N. Gayton (2014). AK-SYS: An adaptation of the AK-MCS method for system reliability. *Reliability Engineering & System Safety*, **123**:137–144.

Ferry-Borges, J., and M. Castanheta (1971). *Structural safety*. Laboratoria Nacional de Engenhera Civil, Lisbon, Portugal.

Fiessler, B., H.J. Neumann, and R. Rackwitz (1979). Quadratic limit states in structural reliability. *Journal of Engineering Mechanics*, ASCE, **105**:661–676.

Frangopol, D.M., Y.K. Lin, and A.C. Estes (1997). Life-cycle cost design of deteriorating structures. *Journal of Structural Engineering*, ASCE, **123**:1390–1401.

Freudenthal, A.M. (1947). The safety of structures. *Transactions of the American Society of Civil Engineers*, **112**:125–159.

Freudenthal, A.M. (1956). Safety and the probability of structural failure. *Transactions of the American Society of Civil Engineers*, **121**:1337–1397.

Friis-Hansen, P. (2001). *Bayesian networks as a decision support tool in marine applications*. Ph.D. thesis, Dept. of Naval Architecture and Offshore Engineering, Technical University of Denmark, Lyngby, Denmark.

Fujimura, K., and A. Der Kiureghian (2007). Tail-equivalent linearization method for nonlinear random vibration. *Probabilistic Engineering Mechanics*, **22**:63–76.

Fussell, B.J. (1973). How to hand-calculate system reliability characteristics. *IEEE Transactions on Reliability*, **R-24**:169–174.

Gardoni, P., A. Der Kiureghian, and K. Mosalam (2002). Probabilistic capacity models and fragility estimates for RC columns based on experimental observations. *Journal of Engineering Mechanics*, ASCE, **128**:1024–1038.

Garré, L., and A. Der Kiureghian (2010). Tail-equivalent linearization method in frequency domain and application to marine structures. *Marine Structures*, **23**:322–338.

Genz, A. (1992). Numerical computation of multivariate normal probabilities. *Journal of Computational and Graphical Statistics*, **1**:141–149.

Genz, A., and F. Bretz (2002). Comparison of methods for the computation of multivariate t probabilities. *Journal of Computational and Graphical Statistics*. **11**:950–971.

Ghanem, R., and P. Spanos (1991a). *Stochastic finite elements – A spectral approach*. Springer Verlag, New York, NY.

Ghanem, R., and P. Spanos (1991b). Spectral stochastic finite-element formulation for reliability analysis. *Journal of Engineering Mechanics*, ASCE, **117**:2351–2372.

Gollwitzer, S., and R. Rackwitz (1988). An efficient numerical solution to the multinormal integral. *Probabilistic Engineering Mechanics*, **3**:98–101.

Golub, G.H., and C.F. Van Loan (1996). *Matrix Computations* (3rd ed.). Johns Hopkins University Press, Baltimore, MD.

Goswami, S., S. Ghosh, and S. Chakraborty (2016). Reliability analysis of structures by iterative improved response surface method. *Structural Safety*, **60**:56–66.

Goudreau, G.L., and R.L. Taylor (1972). Evaluation of numerical integration methods in elastodynamics. *Computer Methods in Applied Mechanics and Engineering*, **2**:69–97.

Griewank, A., and A. Walther (2008). *Evaluating derivatives: Principles and techniques of algorithmic differentiation* (2nd ed.). Society for Industrial and Applied Mathematics, Philadelphia, PA.

Grigoriu, M. (Ed.) (1984a). *Risk, structural engineering and human error*. University of Waterloo Press, Waterloo, Canada.

Grigoriu, M. (1984b). Crossings of non-Gaussian translation processes. *Journal of Engineering Mechanics*, ASCE, **110**:610–620.

Grigoriu, M., (2006). Evaluation of Karhunen-Loeve, Spectral and Sampling Representations for Stochastic Processes. *Journal of Engineering Mechanics*, ASCE, **132**:179–189.

Grigoriu, M., and S. Balopoulou (1993). A simulation method for stationary Gaussian random functions based on the sampling theorem. *Probabilistic Engineering Mechanics*, **8**:239–254.

Gu, Q., J.P. Conte, A. Elgamal, and Z. Yang (2009). Finite element response sensitivity analysis of multi-yield-surface J2 plasticity model by direct differentiation method. *Computer Methods in Applied Mechanics and Engineering*, **198**:2272–2285.

Gumbel, E.J. (1958). *Statistics of extremes*. Columbia University Press, New York, NY.

Gutenberg, B., and C.F. Richter (1944). Frequency of earthquakes in California. *Bulletin of the Seismological Society of America*, **34**:185–188.

Hagen, O., and L. Tvedt (1991). Vector process out-crossing as parallel system sensitivity measure. *Journal of Engineering Mechanics*, ASCE, **117**:2201–2220.

Hailperin, T. (1965). Best possible inequalities for the probability of a logical function of events. *American Mathematical Monthly*, **72**:343–359.

Haldar, A., and S. Mahadevan (2000a). *Probability, reliability, and statistical methods in engineering design*. John Wiley & Sons, New York, NY.

Haldar, A., and S. Mahadevan (2000b). *Reliability assessment using stochastic finite element analysis*. John Wiley & Sons, New York, NY.

Handa, K., and K. Anderson (1981). Application of finite element methods in the statistical analysis of structures. *Proceedings of the 3rd International Conference on Structural Safety and Reliability*, Trondheim, Norway, pp. 409–417.

Harbitz, A. (1986). An efficient sampling method for probability of failure calculation. *Structural Safety*, **3**:109–115.

Hasofer, A.M., and N.C. Lind (1974). Exact and invariant second-moment code format. *Journal of Engineering Mechanics Division*, ASCE, **100**:111–121.

Haukaas, T., and A. Der Kiureghian (2004). Finite element reliability and sensitivity methods for performance-based earthquake engineering. *PEER Report 2003/14*, Pacific Earthquake Engineering Research Center, University of California, Berkeley, CA, April.

Haukaas, T., and A. Der Kiureghian (2005). Parameter sensitivity and importance measures in nonlinear finite element reliability analysis. *Journal of Engineering Mechanics*, ASCE, **131**:1013–1026.

Haukaas, T., and A. Der Kiureghian (2006). Strategies for finding the design point in nonlinear finite element reliability analysis. *Probabilistic Engineering Mechanics*, **21**:133–147.

Henley, E.J., and H. Kumamoto (1981). *Reliability engineering and risk assessment*. Prentice-Hall, Englewood Cliffs, NJ.

Hilber, H., T.J.R. Hughes, and R.L. Taylor (1977). Improved numerical dissipation for the time integration algorithms in structural dynamics. *Earthquake Engineering & Structural Dynamics*, **5**:283–292.

Hisada, T., and S. Nakagiri (1981). Stochastic finite element method developed for structural safety and reliability. *Proceedings of the 3rd International Conference on Structural Safety and Reliability*, Trondheim, Norway, pp. 395–408.

Hohenbichler, M., S. Gollwitzer, W. Kruse, and R. Rackwitz (1987). New light on first- and second-order reliability methods. *Structural Safety*, **4**:267–284.

Hohenbichler, M., and R. Rackwitz (1981). Non-normal dependent vectors in structural safety. *Journal of Engineering Mechanics*, ASCE, **107**:1227–1238.

Hohenbichler, M., and R. Rackwitz (1983). First-order concepts in system reliability. *Structural Safety*, **1**:177–188.

Hohenbichler, M., and R. Rackwitz (1986a). Sensitivity and importance measures in structural reliability. *Civil Engineering Systems*, **3**:203–210.

Hohenbichler, M., and R. Rackwitz (1986b). Asymptotic crossing rate of Gaussian vector processes into intersections of failure domains. *Probabilistic Engineering Mechanics*, **1**:177–179.

Hohenbichler, M., and R. Rackwitz (1988). Improvement of second-order reliability estimates by importance sampling. *Journal of Engineering Mechanics*, ASCE, **114**:2195–2199.

Hunter, D. (1976). An upper bound for the probability of a union. *Journal of Applied Probability*, **13**:597–603.

Igusa, T., and A. Der Kiureghian (1985). Dynamic characterization of two degree-of-freedom equipment-structure systems. *Journal of Engineering Mechanics*, ASCE, **111**:1–19.

Iourtchenko, D., E. Mo, and A. Naess (2008). Reliability of strongly nonlinear single degree of freedom dynamic systems by the path integration method. *Journal of Applied Mechanics*, **75**: 061016.

Jang, Y.-S., N. Sitar, and A. Der Kiureghian (1994). Reliability analysis of contaminant transport in saturated porous media. *Water Resources Research*, **30**:2435–2448.

Jensen, F.V., and T.D. Nielsen (2007). *Bayesian networks and decision graphs*. Springer-Verlag, New York, NY.

Jiao, G., and T. Moan (1990). Methods of reliability model updating through additional events. *Structural Safety*, **9**:139–153.

Kang, S.-C., H.-M. Koh, and J.F. Choo (2010). An efficient response surface method using moving least squares approximation for structural reliability analysis. *Probabilistic Engineering Mechanics*, **25**:365–371.

Kang, W.-H., and J. Song (2010). Evaluation of multivariate normal integrals for general systems by sequential compounding. *Structural Safety*, **32**:35–41.

Karamchandani, A., P. Bjerager, and C.A. Cornell (1989). Methods to estimate parametric sensitivity in structural reliability analysis. *Proceedings of the 5th ASCE-EMD Specialty Conference*, Blacksburg, VA, pp. 86–89.

Katafygiotis, L.S., and K.M. Zuev (2008). Geometric insight into the challenges of solving high-dimensional reliability problems. *Probabilistic Engineering Mechanics*, **23**:208–218.

Kawashima, K., and K. Aizawa (1986). Modification of earthquake response spectra with respect to damping ratio. *Proceedings of the 3rd US National Conference on Earthquake Engineering*, Charleston, SC, **II**:1107–1116.

Kim, T., and J. Song (2018). Generalized reliability importance measure (GRIM) using Gaussian mixture. *Reliability Engineering & System Safety*, **173**:105–115.

Kim, C., S. Wang, and K.K. Choi (2005). Efficient response surface modeling by using moving least-squares method and sensitivity. *AIAA Journal*, **43**:2404–2411.

Kirjner-Neto, C., E. Polak, and A. Der Kiureghian (1998). An outer approximations approach to reliability-based optimal design of structures. *Journal of Optimization Theory & Application*, **98**:1–17.

Kleiber, M., H. Antunez, T.D. Hien, and P. Kowalczyk (1997). *Parameter sensitivity in nonlinear mechanics*. John Wiley & Sons, West Sussex, UK.

Kleiber, M., and T.D. Hien (1992). *The stochastic finite element method: Basic perturbation technique and computer implementation*, John Wiley & Sons, New York, NY.

Knuth, D. (1998). *Seminumerical algorithms: The art of computer programming*, Vol. 2 (3rd ed.). Addison-Wesley, Reading, MA.

Konishi, I., and M. Shinozuka (1956). A consideration on safety of structures by plastic and stochastic theory. *Transactions of the Japan Society of Civil Engineers*, **33**:7–11.

Kharmanda, G., N. Olhoff, A. Mohamed, and M. Lemaire (2004). Reliability-based topology optimization. *Structural & Multidisciplinary Optimization*, **26**:295–307.

Koller, D., and A. Pfeffer (1997). Object-oriented Bayesian networks. *Proceedings of the 13th Conference on Uncertainty in Artificial Intelligence*, D. Geiger and P.P. Shenoy (Eds.), Morgan Kaufmann Publishers, San Francisco, CA, pp. 302–313.

Kotz, S. (1975). Multivariate distribution at a cross road. In *Statistical Distributions in Scientific Work*, G.P. Patil, S. Kotz, and J.K. Ord (Eds.), D. Reidel Publishing Co., Dordrecht, Holland, **1**:247–270.

Kotz, S., N. Balakrishnan, and N.L. Johnson (2004). *Continuous multivariate distributions, volume 1, models and applications* (2nd ed.). John Wiley & Sons, New York, NY.

Koo, H., and A. Der Kiureghian (2003). FORM, SORM and simulation techniques for nonlinear random vibrations. *Report No. UCB/SEMM-2003/01*, Department of Civil & Environmental Engineering, University of California, Berkeley, CA.

Koo, H., A. Der Kiureghian, and K. Fujimura (2005). Design-point excitation for nonlinear random vibrations. *Probabilistic Engineering Mechanics*, **20**:136–147.

Kounias, E.G. (1968). Bounds for the probability of a union, with applications. *Annals of Mathematical Statistics*, **39**:2154–2158.

Kuschel, N., and R. Rackwitz (1997). Two basic problems in reliability-based structural optimization. *Mathematical Methods of Operations Research*, **46**:309–333.

Kuschel, N., and R. Rackwitz (2000). Optimal design under time-variant reliability constraints. *Structural Safety*, **22**:113–127.

Lamaire, M. (2005). *Structural reliability*. John Wiley & Sons, New York, NY.

Langseth, H., and L. Portinale (2007). Bayesian networks in reliability. *Reliability Engineering & System Safety*, **92**:92–108.

Lawrence, M. (1987). Basis random variables in finite element analysis. *International Journal of Numerical Methods in Engineering*, **24**:1849–1863.

Lee, Y.-J., and J. Song (2011). Risk analysis of fatigue-induced sequential failures by branch-and-bound method employing system reliability bounds. *Journal of Engineering Mechanics*, ASCE, **137**:807–821.

Lemaitre, J., and J.L. Chaboche (1990). *Mechanics of solid materials*. Cambridge University Press, Cambridge, UK.

Leonel, E.D., A.T. Beck, and W.S. Venturini (2011). On the performance of response surface and direct coupling approaches in solution of random crack propagation problems. *Structural Safety*, **33**:261–274.

Li, C.-C., and A. Der Kiureghian (1993). Optimal discretization of random fields. *Journal of Engineering Mechanics*, ASCE, **119**:1136–1154.

Liang, J., Z.P. Mourelatos, and E. Nikolaidis (2007). A single-loop approach for system reliability-based design optimization. *Journal of Mechanical Design*, **129**:1215–1224.

Liang, J., Z.P. Mourelatos, and J. Tu (2004). A single-loop method for reliability-based design optimization. *Proceedings of ASME Design Engineering Technical Conference*, Salt Lake City, UT, pp. 1–12.

Lin, H.-Z., and A. Der Kiureghian (1987). Second-order system reliability using directional simulation. *Proceedings of the 5th International Conference on Applications of Statistics and Probability in Soil and Structural Engineering*, Vancouver, Canada, **2**:930–938.

Lin, Y.K. (1967). *Probabilistic theory of structural dynamics*, McGraw-Hill, New York, NY.

Lin, Y.K., and G.Q. Cai (1995). *Probabilistic structural dynamics: advanced theory and applications*, McGraw-Hill, New York, NY.

Lin, K.Y., and D.M. Frangopol (1996). Reliability-based optimum design of reinforced concrete girders. *Structural Safety*, **18**:239–258.

Lind, N. (1971). Consistent partial safety factors. *Journal of Structural Division*, ASCE, **97**:1651–1669.

Lindley, D.V. (2014). *Understanding uncertainty*. John Wiley & Sons, Hoboken, NJ.

Liu, P.-L., and A. Der Kiureghian (1986). Multivariate distribution models with prescribed marginals and covariances. *Probabilistic Engineering Mechanics*, **1**:105–112.

Liu, P.-L., and A. Der Kiureghian (1989). Finite-element reliability methods for geometrically nonlinear stochastic structures. *Report No. UCB/SEMM-89/05*, Department of Civil Engineering, University of California, Berkeley, CA.

Liu, P.-L., and A. Der Kiureghian (1990). Optimization algorithms for structural reliability. *Structural Safety*, **9**:161–177.

Liu, P.-L., and A. Der Kiureghian (1991). Finite-element reliability of geometrically nonlinear uncertain structures. *Journal of Engineering Mechanics*, ASCE, **117**:1806–1825.

Liu, W.K., T. Belytschko, and A. Mani (1986a). Probabilistic finite elements for nonlinear structural dynamics. *Computer Methods in Applied Mechanics and Engineering*, **56**:61–86.

Liu, W.K., T. Belytschko, and A. Mani (1986b). Random field finite elements. *International Journal for Numerical Methods in Engineering*, **23**:1831–1845.

Luenberger, D.G. (1986). *Introduction to linear and nonlinear programming*. Addison-Wesley, Reading, MA.

Luque, J., and D. Straub (2016). Reliability analysis and updating of deteriorating systems with dynamic Bayesian networks. *Structural Safety*, **62**:34–46.

Lutes, L.D., and S. Sarkani (2004). *Random vibrations: analysis of structural and mechanical systems*, Elsevier Butterworth-Heinemann, Burlington, MA.

Madsen, O. (1987). Model updating in reliability theory. *Proceedings of the 5th International Conference on Applications of Statistics and Probability in Soil and Structural Engineering*, Vancouver, Canada, **1**:564–577.

Madsen, H.O. (1988). Omission sensitivity factors. *Structural Safety*, **5**:35–45. Also see comment by P. Bjerager, *Structural Safety*, **7**:77–79.

Madsen, H., and P. Friis Hansen (1992). A comparison of some algorithms for reliability based structural optimization and sensitivity analysis. In *Reliability and Optimization of Structural Systems*, R. Rackwitz and P. Thoft-Christensen (Eds.), Proceedings of the 4th IFIP WG 7.5 Conference, Munich, Germany, pp. 443–451.

Madsen, H. O., S. Krenk, and N. C. Lind (1986). *Methods of structural safety*. Prentice-Hall, Englewood Cliffs, NJ.

Madsen, H.O., and L. Tvedt (1990). Methods for time-dependent reliability and sensitivity analysis. *Journal of Engineering Mechanics*, ASCE, **116**:2118–2135.

Maes, M.A., K. Breitung, and D. Dupuis (1993). Asymptotic importance sampling. *Structural Safety*, **12**: 167–186.

Mahadevan, S., R. Zhang, and N. Smith (2001). Bayesian networks for system reliability reassessment. *Structural Safety*, **23**:231–251.

Martins, J.R.R.A., P. Sturdza, and J.J. Alonso (2003). The complex-step derivative approximation. *ACM Transactions on Mathematical Software*, **28**:245–262.

Matheron, G. (1963). Principles of geostatistics. *Economic Geology*, **58**:1246–1266.

Maute, K., and D.M. Frangopol (2003). Reliability-based design of MEMS mechanisms by topology optimization. *Computers & Structures*, **81**:813–824.

Mayer M. (1926). *Safety of structures*. Springer, Germany (in German).

McKenna, F., G.L. Fenves, and M.H. Scott (2003). Open system for earthquake engineering simulation. http://opensees.berkeley.edu, Pacific Earthquake Engineering Research Center, Univ. of California, Berkeley, CA.

Melchers, R.E. (1987). *Structural reliability: analysis and prediction*. John Wiley & Sons, New York, NY.

Melchers, R.E. (1989). Importance sampling in structural systems. *Structural Safety*, **6**:3–10.

Melchers, R.E. (1990). Search-based importance sampling. *Structural Safety*, **9**:117–128.

Melchers, R.E. (1992). Load-space formulation for time-dependent structural reliability. *Journal of Engineering Mechanics*, ASCE, **118**:853–870.

Melchers, R.E. (1999). *Structural reliability: analysis and prediction* (2nd ed.). John Wiley & Sons, New York, NY.

Melchers, R.E., and A.T. Beck (2018). *Structural reliability analysis and prediction* (3rd ed.). John Wiley & Sons, New York, NY.

Moehle, J. P., K. Elwood, and H. Sezen (2000). Shear failure and axial load collapse of existing reinforced concrete columns. *Proceedings of the 2nd US-Japan Workshop on Performance-Based Design Methodology for Reinforced Concrete Building Structures*, Sapporo, Japan, pp. 241–255.

Morgenstern, D. (1956). Einfache beispiele zweidimensionaler verteilungen. *Miteilingsblatt fur Mathematische Statistik*, **8**:234–235 (in German).

Murphy, K.P. (2001). The Bayes net toolbox for Matlab. *Computing Science and Statistics*, **33**:56167286.

Murphy, K.P. (2002). *Dynamic Bayesian networks: Representation, inference and learning*. Ph.D. thesis, University of California, Berkeley, CA.

Murphy, K.P. (2007). Software for graphical models: a review. *Bulletin of the International Society for Bayesian Analysis*, **14**:13–15.

Murphy, K. (2014). Software packages for graphical models. www.cs.ubc.ca/~murphyk/Software/bnsoft.html.

Nataf, A. (1962). Determination des distribution dont les marges sont donnees. *Comptes*

Rendus de l'Academie des Sciences, **225**:42–43.

Neal, B.G. (1985). *The plastic methods of structural analysis* (3rd ed.). Chapman and Hall, London, UK.

Neal, R.M. (1998). Regression and classification using Gaussian process priors. *Bayesian Statistics*, **6**:475–501.

Neil, M., M. Tailor, and D. Marquez (2007). Inference in hybrid Bayesian networks using dynamic discretization. *Statistics and Computing*, **17**:219–233.

Nelson, R.B. (2006). *An introduction to copulas* (2nd ed.). Springer, New York, NY.

Newmark, N.M. (1959). A method of computation for structural dynamics. *Journal of Engineering Mechanics Division*, ASCE, **85**:67–94.

Nguyen, T.H., G.H. Paulino, and J. Song (2010). A computational paradigm for multiresolution topology optimization (MTOP). *Structural & Multidisciplinary Optimization*, **41**:525–539.

Nguyen, T.H., J. Song, and G.H. Paulino (2011). Single-loop system reliability-based topology optimization considering statistical dependence between limit-states. *Structural & Multidisciplinary Optimization*, **44**:593–611.

Nikolaidis, E., D.M. Ghiocel, and S. Singhal (Eds.) (2005). *Engineering design reliability handbook*, CRC Press, Boca Raton, FL.

Nishijima, K., and M. Faber (2007). A Bayesian framework for typhoon risk management. *Proceedings of the 12th International Conference on Wind Engineering*, Cairns, Australia.

Nowak, A.S. (Ed.) (1986). *Modeling human error.* ASCE, New York, NY.

Nowak, A.S., and K.R. Collins (2012). *Reliability of structures.* CRC Press, Boca Raton, FL.

Pandey, M.D. (1998). An effective approximation to evaluate multinormal integrals. *Structural Safety*, **20**:51–67.

Papaioannou, I., W. Betz, K. Zwirglmaier, and D. Straub (2015). MCMC algorithms for subset simulation. *Probabilistic Engineering Mechanics.* **41**:89–103.

Pearce, H.T., and Y.-K. Wen (1985). On linearization points for nonlinear combination of stochastic load processes. *Structural Safety*, **2**:169–176.

Pearl, J. (1988). *Probabilistic reasoning in intelligent systems: networks of plausible inference.* Morgan Kaufmann Publishers, Inc., San Francisco, CA.

Penzien, J., and M. Watabe (1975). Characteristics of 3-dimensional earthquake ground motion. *Earthquake Engineering & Structural Dynamics*, **3**:365–374.

Phoon, K.K., S.T. Quek, Y.K. Chow, and S.L. Lee (1990). Reliability analysis of pile settlement. *Journal of Geotechnical Engineering*, ASCE, **116**:1717–1734.

Polak, E. (1997). *Optimization: Algorithms and consistent approximations.* Springer, New York, NY.

Press, W.H., S.A. Teukolsky, W.T. Vetterling, and B.P. Flannery (1986). *Numerical recipes: the art of scientific computing.* Cambridge University Press, New York, NY.

Priestley, M.B. (1967). Power spectral analysis of non-stationary random processes. *Journal of Sound & Vibration*, **6**:86–97.

Pugsley A.G. (1951). Concepts of safety in structural engineering. *Journal of the Institution of Civil Engineers*, **36**:5–31.

Rackwitz, R., and B. Fiessler (1978). Structural reliability under combined load sequences. *Computers & Structures*, **9**:489–494.

Rajashekhar, M.R., and B.R. Ellingwood (1993). A new look at the response surface approach for reliability analysis. *Structural Safety*, **12**:205–220.

Rausand, M., and A. Høyland (2004). *System reliability theory: models, statistical methods, and applications* (2nd ed.). John Wiley & Sons, Hoboken, NJ.

Rezaeian, S., and A. Der Kiureghian (2008). A stochastic ground motion model with separable temporal and spectral nonstationarities. *Earthquake Engineering & Structural Dynamics*, **37**:1565–1584.

Rice, O.C. (1944). Mathematical analysis of random noise. *Bell System Technical Journal*, **23**:282–332.

Rice, O.C. (1945). Mathematical analysis of random noise. *Bell System Technical Journal*, **24**:46–156.

Roberts, J.B., and P.D. Spanos (1990). *Random vibration and statistical linearization*. John Wiley & Sons, New York, NY.

Rockafellar, R.T., and J.O. Royset (2010). On buffered failure probability in design and optimization of structures. *Reliability Engineering & System Safety*, **95**:499–510.

Rockafellar, R.T., and J.O. Royset (2015). Engineering decisions under risk averseness. *ASCE-ASME Journal of Risk & Uncertainty in Engineering Systems, Part A: Civil Engineering*, **1** (online).

Rosenblatt, M. (1952). Remarks on a multivariate transformation. *Annals of Mathematical Statistics*, **23**:470–472.

Rosenblueth, E. (1975). Point estimates for probability moments. *Proceedings of the National Academy of Sciences*, **72**:3812–3814.

Rosenblueth, E. (1976). Towards optimum design through building codes. *Journal of Structural Engineering Division*, ASCE, **102**:591–607.

Rosenblueth, E., and L. Esteva (1972). Reliability basis for some Mexico codes. *Special Publication SP-31*, American Concrete Institute, pp. 141–155.

Ross, S.M. (2021). *Introduction to probability and statistics for engineers and scientists* (6th ed.). Academic Press, London, UK.

Roth, C., and M. Grigoriu (2001). Sensitivity analysis of dynamic systems subjected to seismic loads. *Report No. MCEER-01-0003*, Multidisciplinary Center for Earthquake Engineering Research, State University of New York, Buffalo, NY.

Royset, J.O., A. Der Kiureghian, and E. Polak (2001a). Reliability-based optimal design of series structural systems. *Journal of Engineering Mechanics*, ASCE, **127**:607–614.

Royset, J.O., A. Der Kiureghian, and E. Polak (2001b). Reliability-based optimal structural design by the decoupling approach. *Reliability Engineering & System Safety*, **73**:213–221.

Royset, J.O., A. Der Kiureghian, and E. Polak (2006). Optimal design with probabilistic objective and constraints. *Journal of Engineering Mechanics*, ASCE, **132**:107–118.

Royset, J.O., and E. Polak (2004a). Implementable algorithm for stochastic optimization using sample average approximations. *Journal of Optimization Theory & Applications*, **122**:157–184.

Royset, J.O., and E. Polak (2004b). Reliability-based optimal design using sample average approximations. *Probabilistic Engineering Mechanics*, **19**:331–343.

Royset, J.O., and E. Polak (2007). Extensions of stochastic optimization results to problems with system failure probability functions. *Journal of Optimization Theory & Applications*, **133**:1–18.

Rubinstein, R.Y. (1981). *Simulation and the Monte Carlo method*. John Wiley & Sons, New York, NY.

Russell, S.J., and P. Norvig (2003). *Artificial intelligence: A modern approach* (2nd ed.). Prentice-Hall, Englewood Cliffs, NJ.

Sacks, J., W.J. Welch, T.J. Mitchell, and H.P. Wynn (1989). Design and analysis of computer experiments. *Statistical Science*, **4**:409–423.

Sättele, M., M. Bründl, and D. Straub (2015). Reliability and effectiveness of early warning for natural hazards: Concepts and application to debris flow warning. *Reliability Engineering & System Safety*, **142**:192–202.

Segal, I.E. (1938). Fiducial distribution of several parameters with applications to a normal system. *Proceedings of Cambridge Philosophical Society*, **34**:41–47.

Schöbi, R., B. Sudret, and S. Marelli (2017). Rare event estimation using polynomial-chaos kriging. *ASCE-ASME Journal of Risk & Uncertainty in Engineering Systems, Part A: Civil Engineering*, **3** (online).

Schöbi, R., B. Sudret, and J. Wiart (2015). Polynomial-chaos-based kriging. *International Journal of Uncertainty Quantification*, **5**:171–193.

Schuëller, G.I., H.J. Pradlwarter, and P.S. Koutsourelakis (2004). A critical appraisal of reliability estimation procedures for high dimensions. *Probabilistic Engineering Mechanics*, **19**:463–474.

Schuëller, G.I., and R. Stix (1987). A critical appraisal of methods to determine failure probabilities. *Structural Safety*, **4**:293–309.

Shachter, R.D. (1986). Evaluating influence diagrams. *Operations Research*, **34**:871–882.

Shinozuka, M. (1964). Probability of structural failure under random loading. *Journal of Engineering Mechanics Division*, ASCE, **90**:147–171.

Shinozuka, M. (1972). Methods of safety and reliability analysis. *Proceedings of the International Conference on Structural Safety and Reliability*, Washington DC, pp. 11–45.

Shinozuka, M. (1983). Basic analysis of structural safety. *Journal of Structural Engineering*, ASCE, **109**:721–740.

Shinozuka, M., and C.M. Jan (1972). Digital simulation of random processes and its applications. *Journal of Sound & Vibrations*, **25**:111–128.

Sitar, N., J. D. Cawlfield, and A. Der Kiureghian (1987). First-order reliability approach to stochastic analysis of subsurface flow and contaminant transport. *Water Resources Research*, **23**:794–804.

Sobol, I.M. (2001). Global sensitivity indices for nonlinear mathematical models and their Monte Carlo estimates. *Mathematics and Computers in Simulation*, **55**:271–280.

Somerville, P.G., N.F. Smith, R.W. Graves, and N. Abrahamson (1997). Modification of empirical strong ground motion attenuation relations to include the amplitude and duration effects of rupture directivity. *Seismological Research Letters*, **68**:199–222.

Song, J., and A. Der Kiureghian (2003a). Bounds on system reliability by linear programming. *Journal of Engineering Mechanics*, ASCE, **129**:627–636.

Song, J., and A. Der Kiureghian (2003b). Bounds on system reliability by linear programming and applications to electrical substations. *Proceedings of the 9th International Conference on Applications of Probability and Statistics in Civil Engineering*, San Francisco, CA, A. Der Kiureghian, S. Madanat and J.M. Pestana (Eds.), IOS Press, **1**:111–118.

Song, J., and A. Der Kiureghian (2005). Importance measures for systems with dependent components. *Proceedings of the 9th International Conference on Structural Safety and Reliability*, IOS Press, Rome, Italy, pp. 1431–1438.

Song, J., and W-H. Kang (2009). System reliability and sensitivity under statistical dependence by matrix-based system reliability method. *Structural Safety*, **31**:148–156.

Soong, T.T., and M. Grigoriu (1993). *Random vibration of mechanical and structural systems*, Prentice Hall, Englewood Cliffs, NJ.

Spanos, P.D., and R.G. Ghanem (1989). Stochastic finite element expansion for random media. *Journal of Engineering Mechanics*, ASCE, **115**:1035–1053.

Straub, D. (2009). Stochastic modeling of deterioration processes through dynamic Bayesian networks. *Journal of Engineering Mechanics*, ASCE, **135**:1089–1099.

Straub, D. (2011). Reliability updating with equality information. *Probabilistic Engineering Mechanics*, **26**:254–258.

Straub, D., and A. Der Kiureghian (2010a). Bayesian network enhanced with structural reliability methods: Methodology. *Journal of Engineering Mechanics*, ASCE, **136**:1248–1258.

Straub, D., and A. Der Kiureghian (2010b). Bayesian network enhanced with structural reliability methods: Application. *Journal of Engineering Mechanics*, ASCE, **136**:1259–1270.

Straub, D., and A. Grêt-Regamey (2006). A Bayesian probabilistic framework for avalanche modelling based on observations. *Cold Regions Science and Technology*, **46**:192–203.

Straub, D., and I. Papaioannou (2015). Bayesian updating with structural reliability methods.

Journal of Engineering Mechanics, ASCE, **141**:0401434.

Sudret, B. (2008). Analytical derivation of the outcrossing rate in time-variant reliability problems. *Structure and Infrastructure Engineering*, **4**:353–362.

Sudret, B., and A. Der Kiureghian (2000). Stochastic finite element methods and reliability: A state-of-the-art report. *Report No. UCB/SEMM-2000/08*, Department of Civil & Environmental Engineering, University of California, Berkeley, CA.

Sudret, B., and A. Der Kiureghian (2002). Comparison of finite element reliability methods. *Probabilistic Engineering Mechanics*, **17**:337–348.

Takada, T. (1990). Weighted integral method in stochastic finite element analysis. *Probabilistic Engineering Mechanics*, **5**:146–156.

Taniguchi, T., A. Der Kiureghian, and M. Melkumyan (2008). Effect of tuned mass damper on displacement demand of base-isolated structures. *Engineering Structures*, **30**:3478–3488.

Taylor, R.L., and S. Govindjee (2020). *FEAP – Finite element analysis program*. http://projects.ce.berkeley.edu/feap/manual_86.pdf, University of California, Berkeley, CA.

Taflanidis, A.A., and S.-H. Cheung (2012). Stochastic sampling using moving least squares response surface approximations. *Probabilistic Engineering Mechanics*, **28**:216–224.

Thoft-Christensen, P., and M. Baker (1982). *Structural reliability theory and its applications*. Springer-Verlag, Berlin, Germany.

Thoft-Christensen, P., and Y. Murotsu (1986). *Applications of structural system reliability theory*. Springer-Verlag, Berlin, Germany.

Tien, I., and A. Der Kiureghian (2016). Algorithms for Bayesian network modeling and reliability assessment of infrastructure systems. *Reliability Engineering & System Safety*, **156**:134–147.

Tien, I., and A. Der Kiureghian (2017). Reliability assessment of critical infrastructure using Bayesian networks. *Journal of Infrastructure Systems*, ASCE, **23**:04017025.

Tong, Y., and I. Tien (2017). Algorithm for Bayesian network modeling, inference, and reliability assessment for multistate flow network. *Journal of Computing in Civil Engineering*, ASCE, **31**:04017051.

Tu, J., K.K. Choi, and Y.H. Park (1999). A new study on reliability-based design optimization. *Journal of Mechanical Design*, **121**:557–564.

Tvedt, L. (1990). Distribution of quadratic forms in normal space: application to structural reliability. *Journal of Engineering Mechanics*, ASCE, **116**:1183–1197.

Valdebenito, M.A., H.J. Pradlwarter, and G.I. Schuëller (2010). The role of the design point for calculating failure probabilities in view of dimensionality and structural nonlinearities. *Structural Safety*, **32**:101–111.

Vanmarcke, E.H. (1972). Properties of spectral moments with applications to random vibration. *Journal of Engineering Mechanics Division*, ASCE, **98**:425–226.

Vanmarcke, E.H. (1975). On the distribution of the first-passage time for normal stationary processes. *Journal of Applied Mechanics*, **42**:215–220.

Vanmarcke, E.H., and M. Grigoriu (1983). Stochastic finite element analysis of simple beams. *Journal of Engineering Mechanics*, ASCE, **109**:1203–1214.

Veneziano, D. (1979). New index of reliability. *Journal of Engineering Mechanics Division*, ASCE, **105**: 277–296.

Veneziano, D., M. Grigoriu, and C.A. Cornell (1977). Vector process models for system reliability. *Journal of Engineering Mechanics Division*, ASCE, **103**:441–460.

Vesely, W.E., and D.M. Rasmuson (1984). Uncertainties in nuclear probabilistic risk analysis. *Risk Analysis*, **4**:313–322.

von Neumann, J., and O. Morgenstern (1947). *Theory of games and economic behavior* (2nd rev. ed.). Princeton University Press, Princeton, NJ.

Wang, Z., M. Broccardo, and A. Der Kiureghian (2015). An algorithm for finding a sequence of design points in reliability analysis. *Structural Safety*, **58**:52–59.

Wang, Z., and A. Der Kiureghian (2016). Tail-equivalent linearization of inelastic multi-support structures subjected to spatially varying stochastic ground motion. *Journal of Engineering Mechanics*, ASCE, **142**.

Wang, P., C. Hu, and B.D. Youn (2011). A generalized complementary intersection method (GCIM) for system reliability analysis. *Journal of Mechanical Design*, **133**:071003.

Wei, P., Z. Lu, and B. Ren (2013). Reliability analysis of structural system with multiple failure modes and mixed uncertain input variables. *Proceedings of the Institute of Mechanical Engineering Part C, Journal of Mechanical Engineering Science*, **227**:1441–1453.

Wells, D., and K. Coppersmith (1994). New empirical relationships among magnitude, rupture length, rupture width, rupture area, and surface displacement. *Bulletin of the Seismological Society of America*, **84**:974–1002.

Wen, Y.-K. (1976). Method for random vibration of hysteretic systems. *Journal of Engineering Mechanics Division*, ASCE, **102**:249–263.

Wen, Y.-K. (1977). Statistical combination of extreme loads. *Journal of Structural Division*, ASCE, **103**:1079–1093.

Wen, Y.-K. (1990). *Structural load modeling and combination for performance and safety evaluation*. Elsevier, Amsterdam, the Netherlands.

Wen, Y.K. (1993). Reliability-based design under multiple loads. *Structural Safety*, **13**:3–19.

Wen, Y.K., and H.-C. Chen (1987). On fast integration for time variant structural reliability. *Probabilistic Engineering Mechanics*, **2**:156–162.

Winterstein, S., and P. Bjerager (1987). The use of higher moments in reliability estimation. *Proceedings of the 5th International Conference on Applications of Statistics and Probility in Soil and Structural Engineering*, Vancouver, Canada, 2:1027–1036.

Wu, Y.-T., H.R. Millwater, and T.A. Cruse (1990). Advanced probabilistic analysis method for implicit performance functions. *AIAA Journal*, **28**:1663–1669.

Yamazaki, F., M. Shinozuka, and G. Dasgupta (1988). Neumann expansion for stochastic finite element analysis. *Journal of Engineering Mechanics*, ASCE, **114**:1335–1354.

Zhang, Y.C. (1993). High-order reliability bounds for series systems and application to structural systems. *Computers & Structures*, **46**:381–386.

Zhang, Y., and A. Der Kiureghian (1993). Dynamic response sensitivity of inelastic structures. *Computer Methods in Applied Mechanics and Engineering*, **108**:23–36.

Zhang, Y., and A. Der Kiureghian (1995). Two improved algorithms for reliability analysis. In *Reliability and Optimization of Structural Systems*, R. Rackwitz, G. Augusti, and A. Borri (Eds.), Proceedings of the 6th IFIP WG 7.5 working conference on reliability and optimization of structural systems, pp. 297–304.

Zhang, Y., and A. Der Kiureghian (1997). Finite element reliability methods for inelastic structures. *Report No. UCB/SEMM-97/05*, Department of Civil & Environmental Engineering, University of California, Berkeley, CA.

Zhang, J., and B. Ellingwood (1994). Orthogonal series expansion of random fields in reliability analysis. *Journal of Engineering Mechanics*, ASCE, **120**:2660–2677.

Zhano, X.T., I. Elishakoff, and R.C. Zhano (1991). A new stochastic linearization technique based on minimum mean square deviation of potential energies. In *Stochastic structural dynamics: new theoretical developments*, Y.K. Lin and I. Elishakoff (Eds.), Springer Verlag, Berlin, Germany, pp. 327–338.

Zhao, Y.-G., and T. Ono (2001). Moment methods for structural reliability. *Structural Safety*, **23**:47–75.

Zienkiewicz, O.C., R.L. Taylor, and D. Fox (2013a). *The finite element method for solid and structural mechanics* (7th ed.). Elsevier, Oxford, UK.

Zienkiewicz, O.C., R.L. Taylor, and J.Z. Zhu (2013b). *The finite element method: Its basis and fundamentals* (7th ed.). Elsevier, Oxford, UK.

Index